TANGENTS AND SECANTS

30. $\int \tan u \, du = \ln |\sec u| + C$

31. $\int \sec u \, du = \ln |\sec u + \tan u| + C$

32. $\int \tan^2 u \, du = \tan u - u + C$

33. $\int \sec^2 u \, du = \tan u + C$

34. $\int \sec u \tan u \, du = \sec u + C$

35. $\int \tan^3 u \, du = \frac{1}{2} \tan^2 u + \ln |\cos u| + C$

36. $\int \sec^3 u \, du = \frac{1}{2} \sec u \tan u + \frac{1}{2} \ln |\sec u + \tan u| + C$

37. $\int \tan^n u \, du = \frac{\tan^{n-1} u}{n-1} - \int \tan^{n-2} u \, du$

38. $\int \sec^n u \, du = \frac{\sec^{n-2} u \tan u}{n-1} + \frac{n-2}{n-1} \int \sec^{n-2} u \, du$

COTANGENTS AND COSECANTS

39. $\int \cot u \, du = \ln |\sin u| + C$

40. $\int \csc u \, du = \ln |\csc u - \cot u| + C$

41. $\int \cot^2 u \, du = -\cot u - u + C$

42. $\int \csc^2 u \, du = -\cot u + C$

43. $\int \csc u \cot u \, du = -\csc u + C$

44. $\int \cot^3 u \, du = -\frac{1}{2} \cot^2 u - \ln |\sin u| + C$

45. $\int \csc^3 u \, du = -\frac{1}{2} \csc u \cot u + \frac{1}{2} \ln |\csc u - \cot u| + C$

46. $\int \cot^n u \, du = -\frac{\cot^{n-1} u}{n-1} - \int \cot^{n-2} u \, du$

47. $\int \csc^n u \, du = -\frac{\csc^{n-2} u \cot u}{n-1} + \frac{n-2}{n-1} \int \csc^{n-2} u \, du + C$

HYPERBOLIC FUNCTIONS

48. $\int \sinh u \, du = \cosh u + C$

49. $\int \cosh u \, du = \sinh u + C$

50. $\int \tanh u \, du = \ln (\cosh u) + C$

51. $\int \coth u \, du = \ln |\sinh u| + C$

52. $\int \operatorname{sech} u \, du = \tan^{-1}(\sinh u) + C$

53. $\int \operatorname{csch} u \, du = \ln \left|\tanh \frac{1}{2} u\right| + C$

54. $\int \operatorname{sech}^2 u \, du = \tanh u + C$

55. $\int \operatorname{csch}^2 u \, du = -\coth u + C$

56. $\int \operatorname{sech} u \tanh u \, du = -\operatorname{sech} u + C$

57. $\int \operatorname{csch} u \coth u \, du = -\operatorname{csch} u + C$

58. $\int \sinh^2 u \, du = \frac{1}{4} \sinh 2u - \frac{1}{2} u + C$

59. $\int \coth^2 u \, du = \frac{1}{4} \sinh 2u + \frac{1}{2} u + C$

60. $\int \tanh^2 u \, du = u - \tanh u + C$

61. $\int \coth^2 u \, du = u - \coth u - C$

62. $\int u \sinh u \, du = u \cosh u - \sinh u + C$

63. $\int u \cosh u \, du = u \sinh u - \cosh u + C$

(*table continued at the back*)

THE GREEK ALPHABET

A	α	alpha
B	β	beta
Γ	γ	gamma
Δ	δ	delta
E	ϵ	epsilon
Z	ζ	zeta
H	η	eta
Θ	θ	theta
I	ι	iota
K	κ	kappa
Λ	λ	lambda
M	μ	mu
N	ν	nu
Ξ	ξ	xi
O	o	omicron
Π	π	pi
P	ρ	rho
Σ	σ	sigma
T	τ	tau
Υ	υ	upsilon
Φ	ϕ	phi
X	χ	chi
Ψ	ψ	psi
Ω	ω	omega

SALAS

HILLE

ETGEN

CALCULUS

SEVERAL VARIABLES

CALCULUS

SEVERAL VARIABLES

JOHN WILEY & SONS, INC

ACQUISITIONS EDITOR	Michael Boezi
ASSOCIATE PUBLISHER	Laurie Rosatone
ASSISTANT EDITOR	Jennifer Battista
MARKETING MANAGER	Julie Z. Lindstrom
SENIOR PRODUCTION EDITOR	Norine M. Pigliucci
COVER DESIGNER	Madelyn Lesure
COVER AND CHAPTER OPENING PHOTO	© Antonio M. Rosario/The Image Bank
PRODUCTION MANAGEMENT SERVICES	Hermitage Publishing Services

This book was set in New Times Roman by Hermitage Publishing Services and printed and bound by Von Hoffmann Corporation. The cover was printed by Brady Palmer.

This book is printed on acid-free paper. ♾

ISBN 0-471-44970-9
Printed in the United States of America
10 9 8 7 6 5 4 3 2 1

In fond remembrance of
EINAR HILLE

PREFACE

This text is designed for a course on multivariate calculus. While applications from the sciences, engineering, business and economics are often used to motivate or illustrate mathematical ideas, the underlying emphasis throughout is on the three basic concepts of calculus: limit, derivative, integral.

This edition is the result of a substantial joint effort with S.L. Salas. He scrutinized every sentence in every chapter, seeking improved precision and readability. His gift for writing, together with his uncompromising standards for mathematical accuracy and clarity, illuminates the beauty of the subject while increasing its accessibility to students of all levels.

FEATURES OF THE NINTH EDITION

Precision and Clarity

The emphasis is on mathematical exposition, and topics are treated in a clear and understandable manner. Mathematical statements are careful and precise; the basic concepts and important points are not obscured by excess verbiage.

Balance of Theory and Applications

Problems drawn from the physical sciences are often used to introduce basic concepts in calculus. In turn, the concepts and methods of calculus are applied to a variety of problems in the sciences, engineering, business and the social sciences. Because the presentation is flexible, instructors can vary the balance of theory and applications according to the needs of their students.

Accessibility

This text is designed to be completely accessible to the beginning calculus student without sacrificing appropriate mathematical rigor. The important theorems are explained and proved, and the mathematical techniques are justified. These may be covered or omitted according to the theoretical level desired in the course.

Visualization

The importance of visualization cannot be over-emphasized in developing students' understanding of mathematical concepts. For this reason, nearly 400 illustrations accompany the text examples and exercise sets.

Technology

The technology component of the text has been expanded and strengthened through numerous new exercises involving the use of a graphing utility or computer algebra system (CAS). Well over half of the sections have new technology exercises designed to illustrate or expand upon the material developed within the sections.

Projects

Projects with an emphasis on problem solving offer students the opportunity to investigate a variety of special topics that supplement the text material. The projects typically require an approach that involves both theory and applications, including the use of technology. Many of the projects are suitable for group-learning activities.

CONTENT AND ORGANIZATION CHANGES IN THE NINTH EDITION

In our effort to produce an even more effective text, we consulted users of the Eighth Edition and other calculus instructors. Our primary goals in preparing the Ninth Edition were the following:

1. ***Improve the exposition.*** As noted above, every topic was examined for possible improvement in the clarity and accuracy of presentation. Essentially every section in the text underwent some revision and rewriting; a number of sections and subsections were completely rewritten.
2. ***Improve the illustrative examples.*** Many of the existing examples were modified to enhance students' understanding of the material. New examples were added to sections that were rewritten or substantially revised.
3. ***Revise the exercise sets.*** Every exercise set was examined for balance between drill problems, mid-level problems, and more challenging applications and conceptual problems. In many instances, the number of routine problems was reduced and new mid-level to challenging problems were added. Technology-based problems were added to more than half of the sections.

Specific changes in content and organization made to achieve these goals and meet the needs of today's students and instructors include:

Comprehensive Review Exercise Sets

Comprehensive review exercise sets, called Skill Mastery Reviews, have been designed to test and to reinforce students' understanding of basic concepts and methods. These new exercise sets are placed at strategic points in the text and average between 70 and 80 problems per set. The Skill Mastery Reviews follow Chapters 15 and 17.

Vectors and Vector Calculus (Chapters 12 and 13)

The treatment of vectors in the plane that previously paralleled the discussion of three-dimensional vectors has been rewritten. The organization of the material in Chapter 13 has been changed: curvilinear motion and curvature are treated together in a new section

(13.5), and vector calculus in mechanics and the optional section on planetary motion are now the final two sections of the chapter.

Functions of Several Variables, Gradients, Extreme Values (Chapters 14 and 15)

Except for improvements in the exposition, Chapter 14 is unchanged. There are some changes in the organization and content of Chapter 15. The former section on local and absolute extrema and the second partials test has been separated into two sections: one on local extrema and the second partials test, the other on absolute extrema. The optional section on exact differential equations has been moved to Chapter 18. The Skill Mastery Review at the end of Chapter 15 includes exercises from Chapters 12 through Chapter 15.

Multiple Integrals; Line and Surface Integrals (Chapters 16 and 17)

The basic content and organization of these two chapters are largely unchanged, but there have been improvements in the exposition, illustrative examples and exercise sets. The Skill Mastery Review at the end of Chapter 17 covers integration concepts for functions of several variables and for vector-valued functions.

Differential Equations (Chapter 18)

Each of the sections in this chapter has been completely rewritten, and the text examples and exercise sets have been modified accordingly. The material on exact differential equations from Chapter 15 has been added.

SUPPLEMENTS

Student Aids

Answers to Odd-Numbered Exercises Answers to all the odd-numbered exercises are included at the back of the text.

Student Solutions Manual This manual contains detailed solutions to all the odd-numbered exercises in the text. ISBN: 0-471-27521-2

eGrade An online assessment system that contains a large bank of skill-building problems and solutions. Instructors can now automate the process of assigning, delivering, grading, and routing all kinds of homework, quizzes, and tests while providing students with immediate scoring and feedback on their work. Wiley *eGrade* "does the math" ... and much more. For more information, visit www.wiley.com/college/egrade

Calculus Machina A web-based, intelligent software package that solves and documents calculus problems in real time. For students, Calculus Machina is a step-by-step electronic tutor. As the student works through a particular problem online, Calculus Machina provides customized feedback from the text, allowing students to identify and learn from their mistakes more efficiently. For more information, visit www.wiley.com/college/machina

Instructor Aids

Instructor's Solutions Manual This manual contains complete solutions to all the problems in the text. ISBN: 0-471-27522-0

Test Bank A wide range of problems and their solutions are keyed to the exercise sets in the text. ISBN: 0-471-27523-9

Computerized Test Bank The Computerized Test Bank allows instructors to create, customize, and print a test containing any combination of questions from the test bank. Instructors can edit existing questions from the test bank or create new ones as needed.

PowerPoint slides Key figures and examples from each chapter are supplied in PowerPoint format for use in classroom presentation and discussion. These may easily be printed onto transparencies for use with an overhead projector.

eGrade An online assessment system that contains a large bank of skill-building problems and solutions. Instructors can now automate the process of assigning, delivering, grading, and routing all kinds of homework, quizzes, and tests while providing students with immediate scoring and feedback on their work. Wiley *eGrade* "does the math" ... and much more. For more information, visit www.wiley.com/college/egrade

Calculus Machina Instructors can use Calculus Machina to preview problems and explore functions graphically and analytically. For more information, visit www.wiley.com/college/machina

ACKNOWLEDGEMENTS

The revision of a text of this magnitude and stature requires a lot of encouragement and help. I was fortunate to have an ample supply of both from many sources.

Each edition of this text was developed from those that preceded it. The present book owes much to the people who contributed to the first eight editions, most recently: Mihaly Bakonyi, Georgia State University; Edward B. Curtis, University of Washington; Kathy Davis, University of Texas-Austin; Dennis DeTurck, University of Pennsylvania; John R. Durbin, University of Texas-Austin; Charles H. Giffen, University of Virginia-Charlottesville; Michael Kinyon, Indiana University-South Bend; Nicholas Macri, Temple University; James R. McKinney, California State Polytechnic University-Pomona; Jeff Morgan, Texas A & M University; Clifford S. Queen, Lehigh University; and Yang Wang, Georgia Institute of Technology. I am deeply indebted to all of them.

The reviewers of the Ninth Edition supplied detailed criticisms and valuable suggestions. I offer my sincere appreciation to the following individuals:

Omar Adawi	Parkland College
Boris A. Datskovsky	Temple University
Ronald Gentle	Eastern Washington University
Robert W. Ghrist	Georgia Institute of Technology
Susan J. Lamon	Marquette University
Peter A. Lappan	Michigan State University
Dean Larson	Gonzaga University
James Martino	Johns Hopkins University
Peter J. Mucha	Georgia Institute of Technology
Elvira Munoz-Garcia	University of California, Los Angeles
Ralph W. Oberste-Vorth	University of South Florida
Charles Odion	Houston Community College
Charles Peters	University of Houston
J. Terry Wilson	San Jacinto College Central

I am especially grateful to Paul Lorczak, Neil Wigley, and J. Terry Wilson, who carefully read the revised material. They provided many corrections and helpful comments.

I would also like to thank Terry Wilson for his advice and guidance in the creation of the new technology exercises in the text.

I am deeply indebted to the editorial staff at John Wiley & Sons. Everyone involved in this project has been encouraging, helpful, and thoroughly professional at every stage. In particular, Laurie Rosatone, Associate Publisher, Michael Boezi, Editor, and Jennifer Battista, Assistant Editor, provided organization and support when I needed it, and prodding when prodding was required. Special thanks go to Norine Pigliucci, Production Editor, who was patient and understanding as she guided the project through the production stages; Sigmund Malinowski, Illustration Editor, who skillfully directed the art program; and Madelyn Lesure, Design Director, whose creativity produced the attractive interior design as well as the cover.

Finally, I want to acknowledge the contributions of my wife, Charlotte; without her continued support, I could not have completed this work.

Garret J. Etgen

CONTENTS

* Denote optional section.

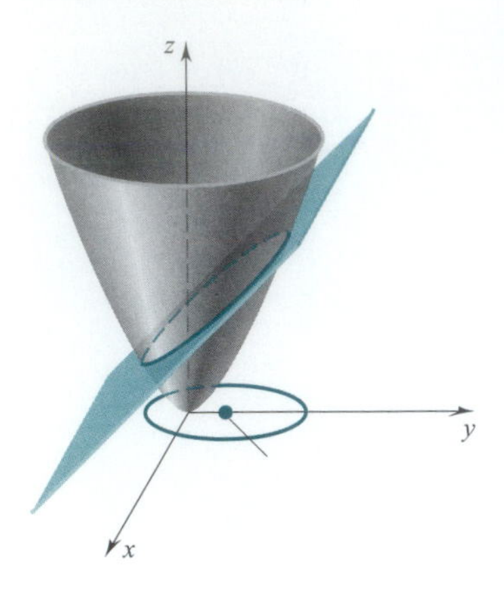

CHAPTER 14 FUNCTIONS OF SEVERAL VARIABLES 820

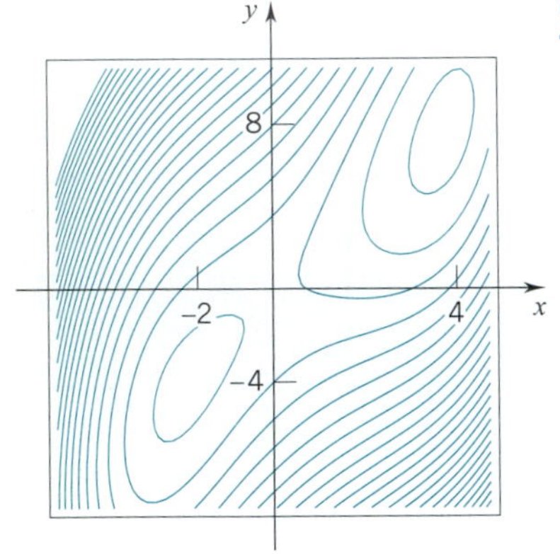

CHAPTER 15 GRADIENTS; EXTREME VALUES; DIFFERENTIALS 861

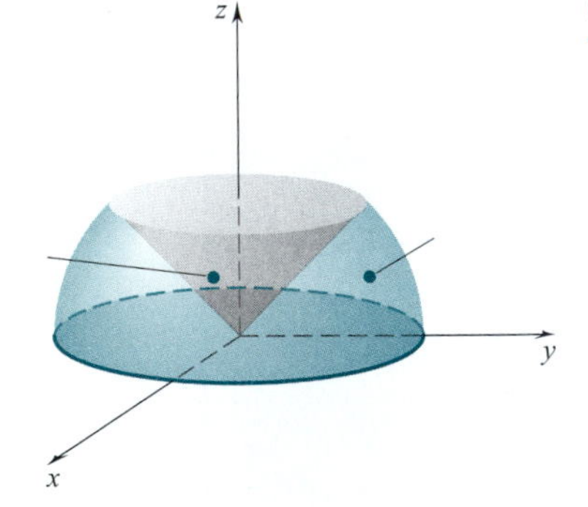

CHAPTER 16 DOUBLE AND TRIPLE INTEGRALS 942

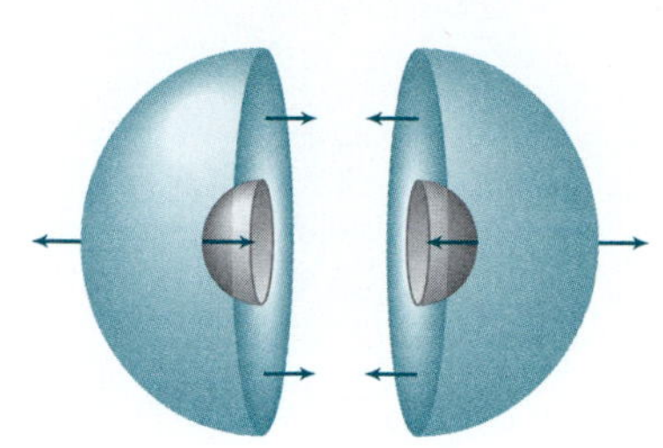

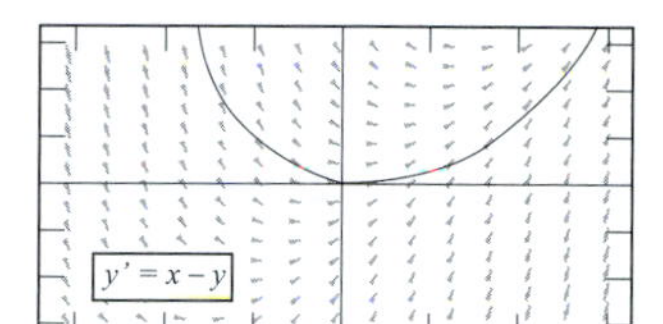
y' = x − y

SALAS

HILLE

ETGEN

CALCULUS

SEVERAL VARIABLES

CHAPTER 12

VECTORS

■ 12.1 CARTESIAN SPACE COORDINATES

To introduce a Cartesian coordinate system in three-dimensional space, we begin with a plane Cartesian coordinate O-xy. Through the point O, which we continue to call the origin, we pass a third line, perpendicular to the other two. This third line we call the z-axis. We assign coordinates to the z-axis using the same scale, assigning the z-coordinate 0 to the origin O.

For later convenience we orient the z-axis so that O-xyz forms a "right-handed" system. That is, if the index finger of the right hand points along the positive x-axis and the middle finger along the positive y-axis, then the thumb will point along the positive z-axis (see Figure 12.1.1).

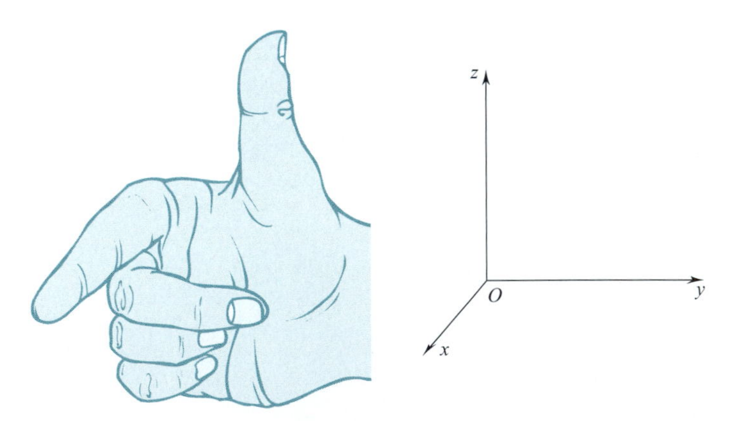

Figure 12.1.1

There are now three coordinate planes: the *xy-plane*, the *xz-plane*, and the *yz-plane*.

The point on the x-axis with x-coordinate x_0 is given space coordinates $(x_0, 0, 0)$; the point on the y-axis with y-coordinate y_0 is given space coordinates $(0, y_0, 0)$; the point on the z-axis with z-coordinate z_0 is given space coordinates $(0, 0, z_0)$.

An arbitrary point P in three-dimensional space (see Figure 12.1.2) is assigned coordinates (x_0, y_0, z_0) provided that

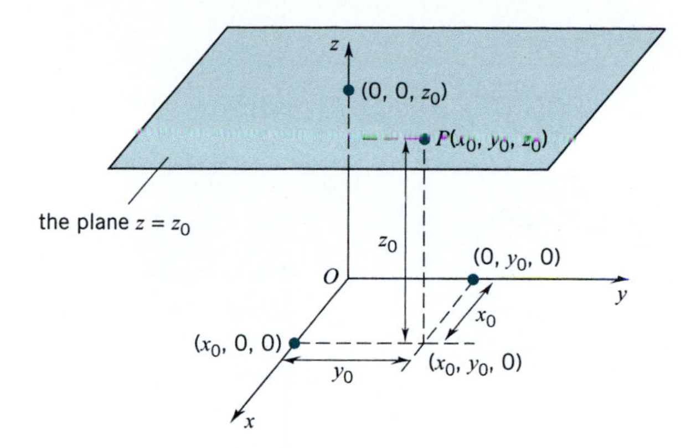

Figure 12.1.2

(1) the plane through P parallel to the yz-plane intersects the x-axis at $(x_0, 0, 0)$;

(2) the plane through P parallel to the xz-plane intersects the y-axis at $(0, y_0, 0)$;

(3) the plane through P parallel to the xy-plane intersects the z-axis at $(0, 0, z_0)$.

The space coordinates (x_0, y_0, z_0) are called the *Cartesian coordinates of P* or simply the *rectangular coordinates of P*.

A point is in the xy-plane iff it is of the form $(x, y, 0)$. Thus the equation $z = 0$ represents the xy-plane. The equation $z = z_0$ represents the set of all points (x, y, z_0), that is, the set of all points with z-coordinate z_0. This is a plane parallel to the xy-plane. Similarly, the equation $x = x_0$ represents a plane parallel to the yz-plane and the equation $y = y_0$ represents a plane parallel to the xz-plane. (See, for example, Figure 12.1.3. There we have drawn the planes $x = 1$ and $y = 3$.)

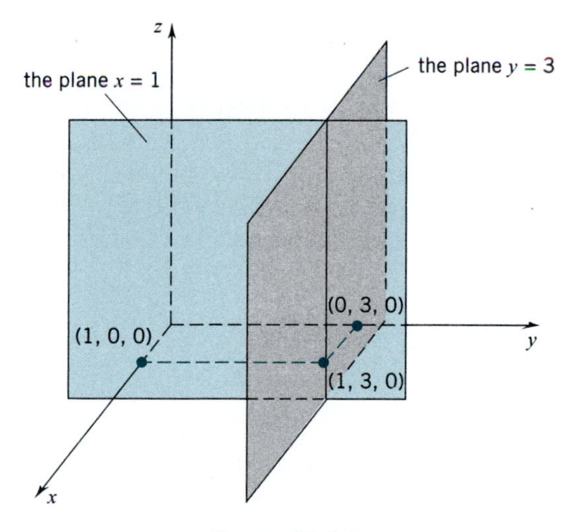

Figure 12.1.3

The Distance Formula

The distance $d(P_1, P_2)$ between two points $P_1(x_1, y_1, z_1)$ and $P_2(x_2, y_2, z_2)$ can be found by applying the Pythagorean theorem twice. With Q and R as in Figure 12.1.4, P_1P_2R

and P_1RQ are both right triangles. From the first triangle

$$[d(P_1, P_2)]^2 = [d(P_1, R)]^2 + [d(R, P_2)]^2,$$

and from the second triangle

$$[d(P_1, R)]^2 = [d(Q, R)]^2 + [d(P_1, Q)]^2.$$

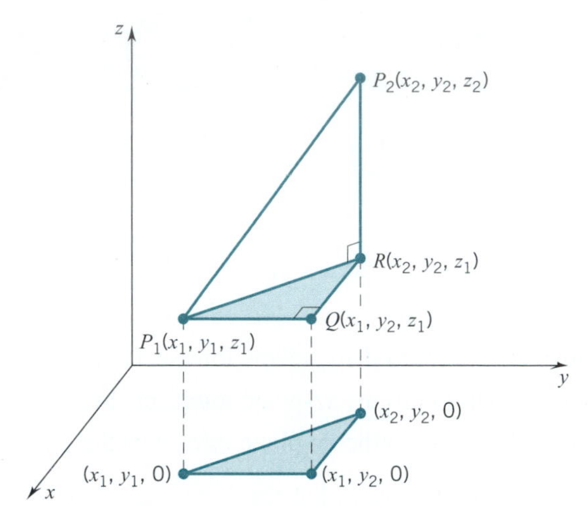

Figure 12.1.4

Combining equations,

$$\begin{aligned}[d(P_1, P_2)]^2 &= [d(Q, R)]^2 + [d(P_1, Q)]^2 + [d(R, P_2)]^2 \\ &= (x_2 - x_1)^2 + (y_2 - y_1)^2 + (z_2 - z_1)^2.\end{aligned}$$

Taking square roots, we have the distance formula:

(12.1.1)
$$d(P_1, P_2) = \sqrt{(x_2 - x_1)^2 + (y_2 - y_1)^2 + (z_2 - z_1)^2}.$$

The *sphere* of radius r centered at $P_0(a, b, c)$ is the set of all points $P(x, y, z)$ with $d(P, P_0) = r$. We can obtain an equation for this sphere by using (12.1.1):

(12.1.2)
Equation for a Sphere
$$(x - a)^2 + (y - b)^2 + (z - c)^2 = r^2.$$

See Figure 12.1.5. The equation for the sphere of radius r centered at the origin is

(12.1.3)
$$x^2 + y^2 + z^2 = r^2.$$

See Figure 12.1.6.

Example 1 The equation $(x - 5)^2 + (y + 2)^2 + z^2 = 9$ represents the sphere of radius 3 centered at the point $(5, -2, 0)$. □

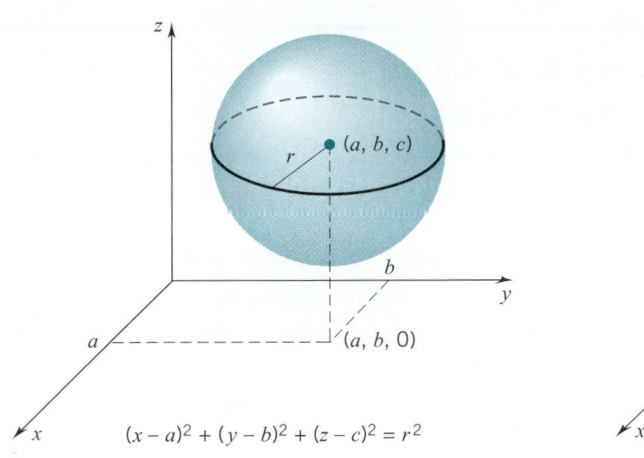

Figure 12.1.5

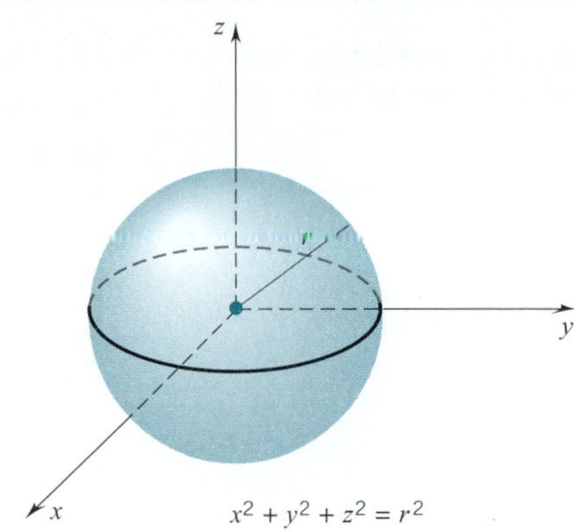

Figure 12.1.6

Example 2 Show that the equation $x^2 + y^2 + z^2 + 6x + 2y - 4z = 11$ represents a sphere. Find the center of the sphere and the radius.

SOLUTION We write the equation as $(x^2 + 6x) + (y^2 + 2y) + (z^2 - 4z) = 11$ and complete the squares. The result,

$$(x^2 + 6x + 9) + (y^2 + 2y + 1) + (z^2 - 4z + 4) = 11 + 9 + 1 + 4 = 25,$$

can be written

$$(x + 3)^2 + (y + 1)^2 + (z - 2)^2 = 25.$$

This equation represents the sphere of radius 5 centered at $(-3, -1, 2)$. ☐

Symmetry

You are already familiar with two kinds of symmetry: symmetry about a point and symmetry about a line. In space we can also speak of symmetry about a plane. These ideas are illustrated in Figure 12.1.7

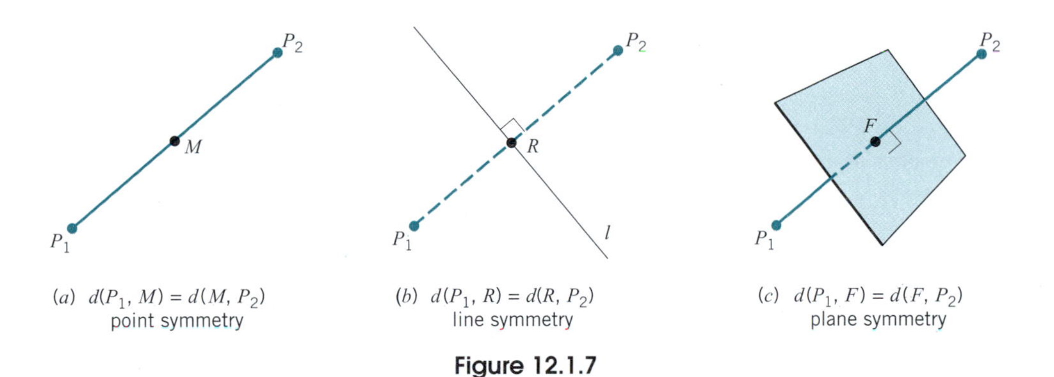

(a) $d(P_1, M) = d(M, P_2)$ point symmetry

(b) $d(P_1, R) = d(R, P_2)$ line symmetry

(c) $d(P_1, F) = d(F, P_2)$ plane symmetry

Figure 12.1.7

The endpoints $P_1(x_1, y_1, z_1)$ and $P_2(x_2, y_2, z_2)$ of the line segment $\overline{P_1P_2}$ are symmetric about the midpoint of the segment. This leads to the *midpoint formula*

(12.1.4)
$$\left(\frac{x_1 + x_2}{2}, \frac{y_1 + y_2}{2}, \frac{z_1 + z_2}{2}\right).$$

EXERCISES 12.1

Plot points A and B on a right-handed coordinate system. Then calculate the length of the line segment $\overline{AB}$ and find the midpoint.

1. $A(2,0,0), B(0,0,-4)$.

2. $A(0,-2,0), B(0,0,6)$.

3. $A(0,-2,5), B(4,1,0)$.

4. $A(4,3,0), B(-2,0,6)$.

Find an equation for the plane through $(3,1,-2)$ that satisfies the given condition.

5. Parallel to the xy-plane.

6. Parallel to the xz-plane.

7. Perpendicular to the y-axis.

8. Perpendicular to the z-axis.

9. Parallel to the yz-plane.

10. Perpendicular to the x-axis.

Find an equation for the sphere that satisfies the given conditions.

11. Centered at $(0,2,-1)$ with radius 3.

12. Centered at $(1,0,-2)$ with radius 4.

13. Centered at $(2,4,-4)$ and passes through the origin.

14. Centered at the origin and passes through $(1,-2,2)$.

15. The line segment joining $(0,4,2)$ and $(6,0,2)$ is a diameter.

16. Centered at $(2,3,-4)$ and tangent to the xy-plane.

17. Centered at $(2,3,-4)$ and tangent to the plane $x=7$.

18. Centered at $(2,3,-4)$ and tangent to the plane $y=1$.

Show that the equation represents a sphere; find the center and radius.

19. $x^2+y^2+z^2+4x-8y-2z+5=0$.

20. $3x^2+3y^2+3z^2-12x-6z+3=0$.

21. $x^2+y^2+z^2-6x+10y-2z-1=0$.

22. $4x^2+4y^2+4z^2-4x-8y-11=0$.

The points $P(a,b,c)$ and $Q(2,3,5)$ are symmetric in the sense given in Exercises 23–34. Find a,b,c.

23. About the xy-plane.

24. About the xz-plane.

25. About the yz-plane.

26. About the x-axis.

27. About the y-axis.

28. About the z-axis.

29. About the origin.

30. About the plane $x=1$.

31. About the plane $y=-1$.

32. About the plane $z=4$.

33. About the point $(0,2,1)$.

34. About the point $(4,0,1)$.

35. Find an equation for each sphere that passes through the point $(5,1,4)$ and is tangent to all three coordinate planes.

36. Find an equation for the largest sphere that is centered at $(2,1,-2)$ and intersects the sphere $x^2+y^2+z^2=1$.

37. Is the equation $x^2+y^2+z^2-4x+4y+6z+20=0$ an equation for a sphere? If so, find the center and radius. If not, why isn't it?

38. Find conditions on A,B,C,D such that the equation

$$x^2+y^2+z^2+Ax+By+Cz+D=0$$

represents a sphere.

39. Show that the points $P(1,2,3), Q(4,-5,2), R(0,0,0)$ are the vertices of a right triangle.

40. The points $(5,-1,3), (4,2,1), (2,1,0)$ are the midpoints of the sides of a triangle PQR. Find the vertices P,Q,R of the triangle.

Describe the region Ω

41. $\Omega=\{(x,y,z): x^2+y^2+z^2\le 4\}$.

42. $\Omega=\{(x,y,z): x^2+y^2+z^2>9\}$.

43. $\Omega=\{(x,y,z): 0\le x\le 1,\ 0\le y\le 2,\ 0\le z\le 3\}$.

44. $\Omega=\{(x,y,z): |x|\le 2,\ |y|\le 2,\ |z|\le 2\}$.

45. $\Omega=\{(x,y,z): x^2+y^2\le 4,\ 0\le z\le 4\}$.

46. $\Omega=\{(x,y,z): 4<x^2+y^2+z^2<9\}$.

In Exercises 47 and 48, the point R lies on the line segment that joins $P(a_1,a_2,a_3)$ and $Q(b_1,b_2,b_3)$.

47. (a) Find the coordinates of R given that

$$d(P,R)=t\,d(P,Q) \quad \text{where } 0\le t\le 1.$$

(b) Determine the value of t for which R is the midpoint of $\overline{PQ}$.

48. (a) Find the coordinates of R given that

$$d(P,R)=r\,d(R,Q) \quad \text{where } r>0.$$

(b) Determine the value of r for which R is the midpoint of $\overline{PQ}$.

C 49. Use a CAS to find the equation of the sphere that has the line segment joining the points $P(3,-2,-2)$ and $Q(-1,4,-3)$ as a diameter. Sketch the sphere.

C 50. Use a CAS to find the perimeter and the area of the triangle with vertices $P(4,-3,2), Q(-6,-2,7), R(5,-1,-2)$. Sketch the triangle.

■ 12.2 DISPLACEMENTS AND FORCES

Our purpose here is to motivate the notion of vector. A quick reading of this section will do.

Displacements

A displacement along a coordinate line can be specified by a real number and depicted by an arrow. For a displacement of a_1 units we can use the number a_1 and an arrow that begins at any number x and ends at $x + a_1$. By convention, if $a_1 > 0$, the arrow will point to the right, and if $a_1 < 0$, the arrow will point to the left (Figure 12.2.1). The *magnitude* of the displacement a_1 is defined to be the length of the arrow. Thus, the magnitude of the displacement a_1 is $|a_1|$.

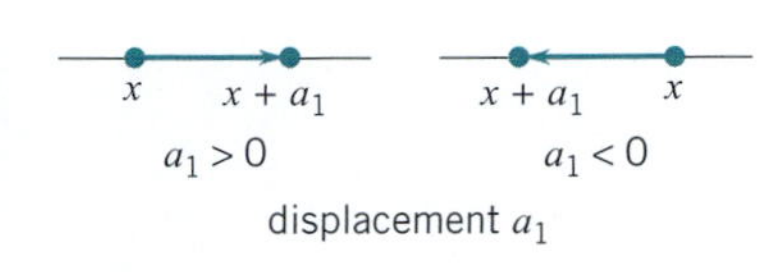

Figure 12.2.1

Displacements in the plane are more interesting. Instead of having two possible directions, there are an infinite number of possible directions. Displacements in the plane are specified by ordered pairs of real numbers. Figure 12.2.2 shows a displacement (a_1, a_2) beginning at the point (x, y) and ending at the point $(x + a_1, y + a_2)$. The magnitude of the displacement (a_1, a_2) is given by $\sqrt{a_1^2 + a_2^2}$.

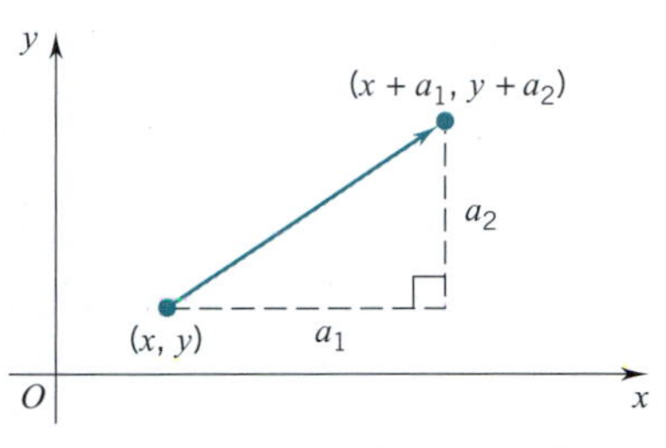

Figure 12.2.2

The most general displacements take place in space. Here ordered triples of numbers come into play. A displacement of a_1 units in x-coordinate, a_2 units in y-coordinate, and a_3 units in z-coordinate can be indicated by an arrow that begins at any point (x, y, z) and ends at the point $(x + a_1, y + a_2, z + a_3)$. This displacement is represented by the ordered triple (a_1, a_2, a_3). (Figure 12.2.3). The magnitude of the displacement (a_1, a_2, a_3) is $\sqrt{a_1^2 + a_2^2 + a_3^2}$.

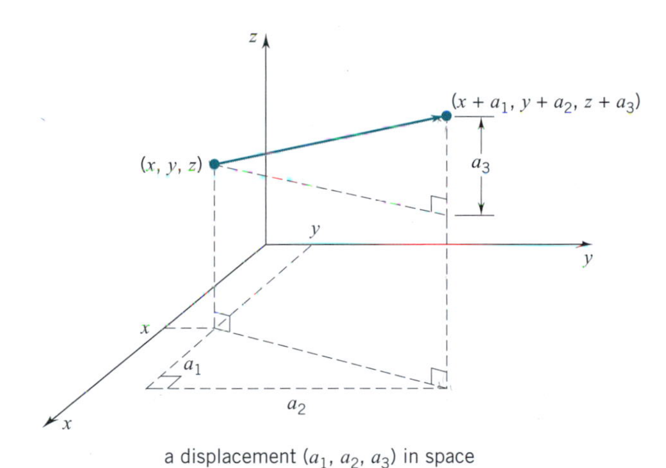

Figure 12.2.3

We can follow one displacement by another. A displacement (a_1, a_2, a_3) followed by (b_1, b_2, b_3) results in a total displacement $(a_1 + b_1, a_2 + b_2, a_3 + b_3)$. We can express this by writing

$$(a_1, a_2, a_3) + (b_1, b_2, b_3) = (a_1 + b_1, a_2 + b_2, a_3 + b_3).$$

We can picture the first displacement by an arrow from some point $P(x, y, z)$ to

$$Q(x + a_1, y + a_2, z + a_3),$$

the second displacement by an arrow from $Q(x + a_1, y + a_2, z + a_3)$ to

$$R(x + a_1 + b_1, y + a_2 + b_2, z + a_3 + b_3),$$

and then the resultant displacement by an arrow from $P(x, y, z)$ to

$$R(x + a_1 + b_1, y + a_2 + b_2, z + a_3 + b_3).$$

The three arrows then form a triangular pattern that is easy to remember (Figure 12.2.4).

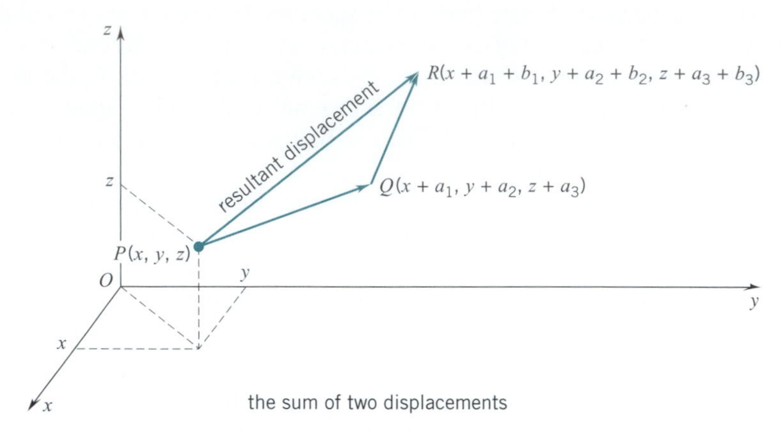

Figure 12.2.4

From a displacement (a_1, a_2, a_3) and a real number α, we can form a new displacement $(\alpha a_1, \alpha a_2, \alpha a_3)$. We view this new displacement as α times the initial displacement and write

$$\alpha(a_1, a_2, a_3) = (\alpha a_1, \alpha a_2, \alpha a_3).$$

The effect of multiplying a displacement by a number α is to change the magnitude of the displacement by a factor of $|\alpha|$:

$$\sqrt{(\alpha a_1)^2 + (\alpha a_2)^2 + (\alpha a_3)^2} = \sqrt{\alpha^2(a_1^2 + a_2^2 + a_3^2)} = |\alpha|\sqrt{a_1^2 + a_2^2 + a_3^2},$$

keeping the same direction if $\alpha > 0$, but reversing the direction if $\alpha < 0$. [If $\alpha = 0$, then $(a_1, a_2, a_3) = (0, 0, 0)$, which you can view as a displacement of length 0.] The displacement

$$2(a_1, a_2, a_3) = (2a_1, 2a_2, 2a_3)$$

is twice as long as (a_1, a_2, a_3) and has the same direction; the displacement

$$-(a_1, a_2, a_3) = (-a_1, -a_2, -a_3)$$

has the same length as (a_1, a_2, a_3) but the opposite direction; the displacement

$$-\tfrac{3}{2}(a_1, a_2, a_3) = \left(-\tfrac{3}{2}a_1, -\tfrac{3}{2}a_2, -\tfrac{3}{2}a_3\right)$$

is one and one-half times as long as (a_1, a_2, a_3) and has the opposite direction (Figure 12.2.5).

Forces

The algebraic patterns

$$(a_1, a_2, a_3) + (b_1, b_2, b_3) = (a_1 + b_1, a_2 + b_2, a_3 + b_3)$$
$$\alpha(a_1, a_2, a_3) = (\alpha a_1, \alpha a_2, \alpha a_3)$$

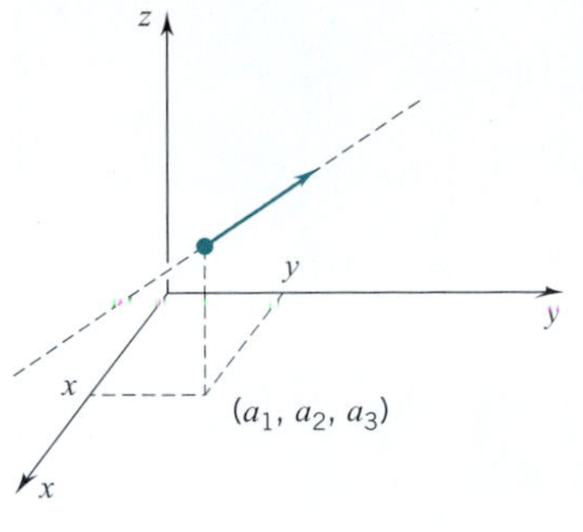

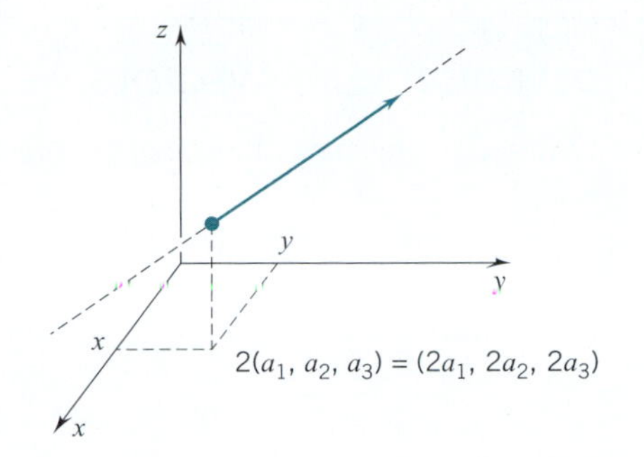

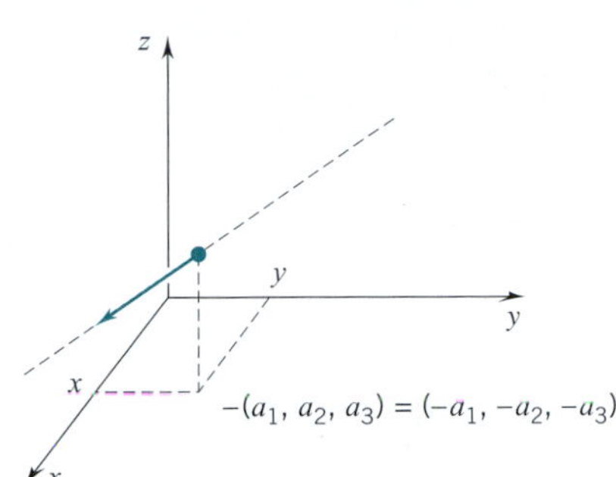

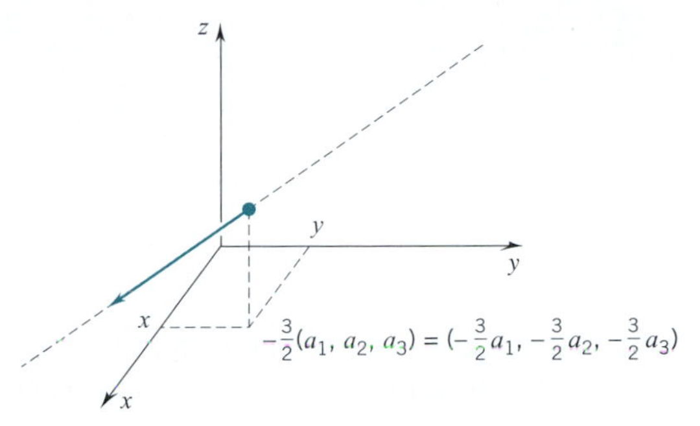

Figure 12.2.5

arise naturally in other settings; for example in the analysis of forces. A force **F** acting in three-dimensional space is completely determined by its components along the x, y, and z axes. If these components are a_1, a_2, a_3, respectively, then the force can be represented by the ordered triple (a_1, a_2, a_3). See Figure 12.2.6.

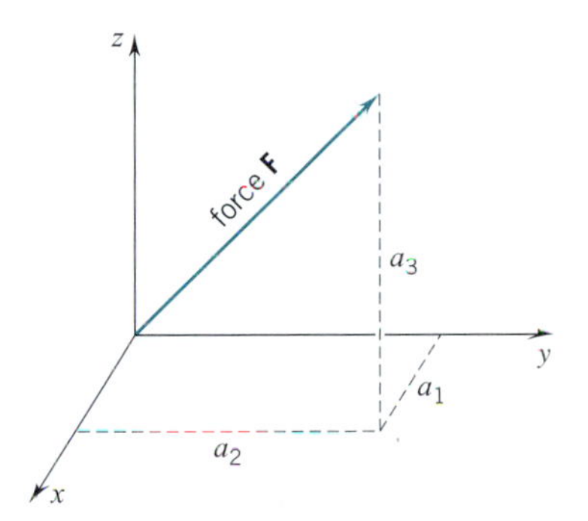

Figure 12.2.6

If two forces $\mathbf{F}_1 = (a_1, a_2, a_3)$ and $\mathbf{F}_2 = (b_1, b_2, b_3)$ are applied simultaneously at the same point, the effect is the same as that produced by the single force

$$\mathbf{F}_3 = (a_1 + b_1, a_2 + b_2, a_3 + b_3).$$

We call $\mathbf{F}_3$ the *resultant* or *total force* and write $\mathbf{F}_1 + \mathbf{F}_2 = \mathbf{F}_3$. For the ordered triples,

$$(a_1, a_2, a_3) + (b_1, b_2, b_3) = (a_1 + b_1, a_2 + b_2, a_3 + b_3).$$

Pictorially we have the usual force diagram, Figure 12.2.7. It is a parallelogram with the sides representing $\mathbf{F}_1 = (a_1, a_2, a_3)$ and $\mathbf{F}_2 = (b_1, b_2, b_3)$ and the diagonal representing

$$\mathbf{F}_3 = \mathbf{F}_1 + \mathbf{F}_2 = (a_1 + b_1, a_2 + b_2, a_3 + b_3).$$

For any force $\mathbf{F} = (a_1, a_2, a_3)$ and any real number α, the force $\alpha\mathbf{F}$ is defined by the equation

$$\alpha\mathbf{F} = (\alpha a_1, \alpha a_2, \alpha a_3).$$

Thus, once again we have

$$\alpha(a_1, a_2, a_3) = (\alpha a_1, \alpha a_2, \alpha a_3).$$

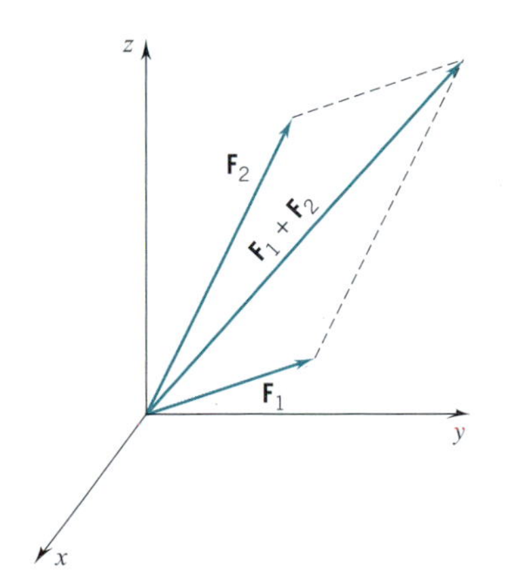

Figure 12.2.7

■ 12.3 VECTORS

The algebra of number triples that we introduced in the last section to discuss displacements and forces is so prodigiously rich in applications that it has found a firm place in the world of science and engineering, and has generated much of mathematics. It is to this mathematics that we now turn.

DEFINITION 12.3.1 VECTORS

Ordered triples of real numbers subject to addition:

$$(a_1, a_2, a_3) + (b_1, b_2, b_3) = (a_1 + b_1,\ a_2 + b_2,\ a_3 + b_3),$$

and multiplication by *scalars* (real numbers):

$$\alpha(a_1, a_2, a_3) = (\alpha a_1, \alpha a_2, \alpha a_3)$$

are called *vectors*. †

Two vectors are *equal* iff they have exactly the same *components:*

$$(a_1, a_2, a_3) = (b_1, b_2, b_3) \quad \text{iff} \quad a_1 = b_1,\ a_2 = b_2,\ a_3 = b_3.$$

Geometrically, we can depict vectors by arrows. To depict the vector (a_1, a_2, a_3), we can choose any initial point $Q = Q(x, y, z)$ and use the arrow

$$\overrightarrow{QR} \quad \text{with} \quad R = R(x + a_1, y + a_2, z + a_3). \qquad \text{(Figure 12.3.1)}$$

Usually we choose the origin as the initial point and use the arrow

$$\overrightarrow{OP} \quad \text{with} \quad P = P(a_1, a_2, a_3).\ ††$$

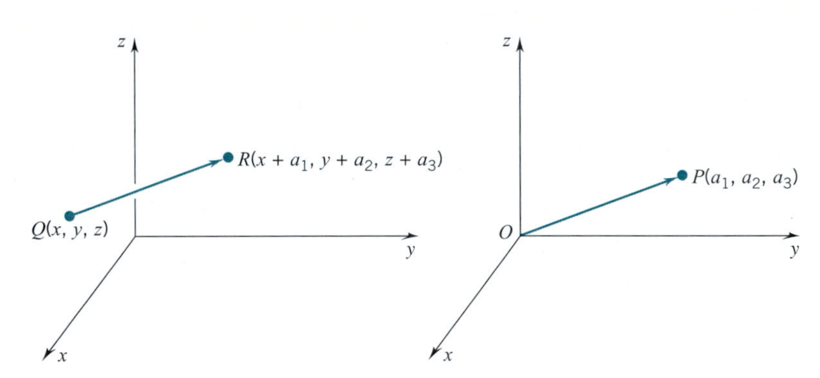

Figure 12.3.1

We use boldface letters to denote vectors. Thus for

$$\mathbf{a} = (a_1, a_2, a_3) \qquad \text{and} \qquad \mathbf{b} = (b_1, b_2, b_3)$$

we have

$$\mathbf{a} + \mathbf{b} = (a_1 + b_1, a_2 + b_2, a_3 + b_3),$$

† Strictly speaking, these are vectors in three-dimensional space. Two-dimensional vectors are represented by ordered pairs of real numbers. There are also four-dimensional vectors which are represented by ordered quadruples, five-dimensional vectors represented by ordered quintuples, and so on. We will be working in three dimensions.

†† The ordered triple $(0, 0, 0)$ will not be represented by an arrow but simply by the origin.

and, if α is a *scalar* (a real number),

$$\alpha\mathbf{a} = (\alpha a_1, \alpha a_2, \alpha a_3).$$

Addition of vectors satisfies the commutative and associative laws:

$$\mathbf{a} + \mathbf{b} = \mathbf{b} + \mathbf{a}$$
$$\mathbf{a} + (\mathbf{b} + \mathbf{c}) = (\mathbf{a} + \mathbf{b}) + \mathbf{c}.$$

These laws follow immediately from the definition of vector addition and the corresponding properties of real numbers. For the *zero vector* $(0, 0, 0)$ we will use the symbol $\mathbf{0}$. Obviously

$$0\mathbf{a} = \mathbf{0} \quad \text{for all vectors } \mathbf{a}.$$

By the vector $-\mathbf{b}$ we mean $(-1)\,\mathbf{b}$; that is,

$$-(b_1, b_2, b_3) = (-b_1, -b_2, -b_3).$$

By $\mathbf{a} - \mathbf{b}$ we mean $\mathbf{a} + (-\mathbf{b})$; that is,

$$\begin{aligned}(a_1, a_2, a_3) - (b_1, b_2, b_3) &= (a_1, a_2, a_3) + (-b_1, -b_2, -b_3)\\ &= (a_1 - b_1, a_2 - b_2, a_3 - b_3).\end{aligned}$$

Example 1 Given that $\mathbf{a} = (1, -1, 2)$, $\mathbf{b} = (2, 3, -1)$, $\mathbf{c} = (8, 7, 1)$, find **(a)** $\mathbf{a} - \mathbf{b}$. **(b)** $2\mathbf{a} + \mathbf{b}$. **(c)** $3\mathbf{a} - 7\mathbf{b}$. **(d)** $2\mathbf{a} + 3\mathbf{b} - \mathbf{c}$.

SOLUTION

(a) $\mathbf{a} - \mathbf{b} = (1, -1, 2) - (2, 3, -1) = (1 - 2, -1 - 3, 2 + 1) = (-1, -4, 3).$

(b) $2\mathbf{a} + \mathbf{b} = 2(1, -1, 2) + (2, 3, -1) = (2, -2, 4) + (2, 3, -1) = (4, 1, 3).$

(c) $3\mathbf{a} - 7\mathbf{b} = 3(1, -1, 2) - 7(2, 3, -1)$
$= (3, -3, 6) - (14, 21, -7) = (-11, -24, 13).$

(d) $2\mathbf{a} + 3\mathbf{b} - \mathbf{c} = 2(1, -1, 2) + 3(2, 3, -1) - (8, 7, 1)$
$= (2, -2, 4) + (6, 9, -3) - (8, 7, 1) = (0, 0, 0) = \mathbf{0}.$ ☐

The addition of vectors can be visualized as the "addition" of arrows by the parallelogram law (as in the case of forces). If you picture **a**, **b**, and **a** + **b** all as arrows emanating from the same point, say the origin, then **a** + **b** acts as the diagonal of the parallelogram generated by **a** and **b**. (See Figure 12.3.2.)

The addition of vectors can also be visualized as the "tail-to-head" addition of arrows (as in the case of displacements). If, instead of starting **b** at the origin, you start **b** at the tip of **a**, then **a** + **b** goes from the tail of **a** to the tip of **b** (Figure 12.3.3). These two pictorial representations of vector addition lead to the same result.

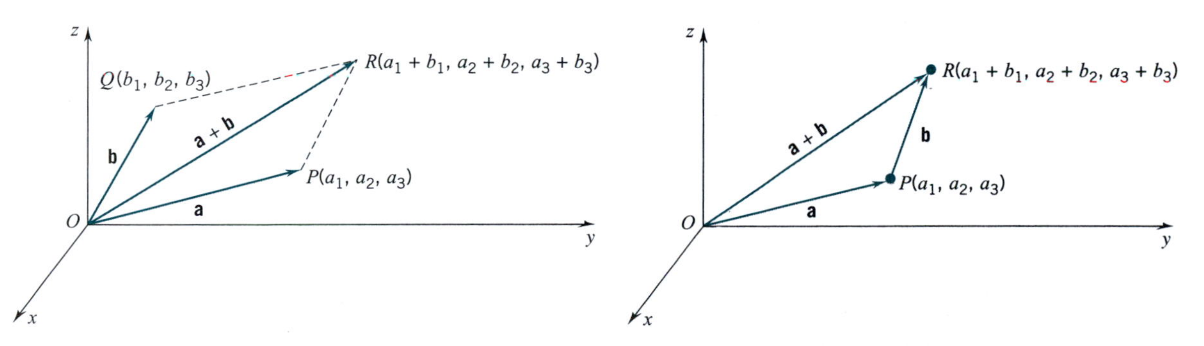

Figure 12.3.2 Figure 12.3.3

DEFINITION 12.3.2 PARALLEL VECTORS

Two nonzero vectors **a** and **b** are said to be *parallel* provided that

$$\mathbf{a} = \alpha\mathbf{b} \quad \text{for some real number } \alpha .$$

If $\alpha > 0$, **a** and **b** are said to have the *same direction;* if $\alpha < 0$, they are said to have *opposite directions*.

In the case of $\mathbf{a} = (2, -2, 6), \mathbf{b} = (1, -1, 3), \mathbf{c} = (-1, 1, -3)$ we have

$$\mathbf{a} = 2\mathbf{b} \quad \text{and} \quad \mathbf{a} = -2\mathbf{c}.$$

This tells us that **a** and **b** are parallel and have the same direction, whereas **a** and **c**, though parallel, have opposite directions. (See Figure 12.3.4.)

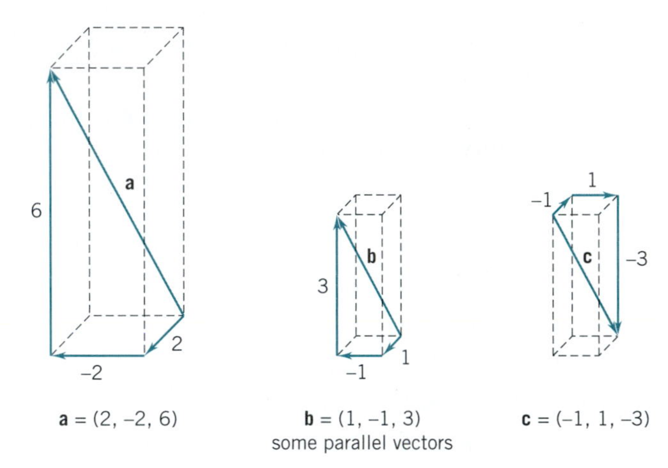

Figure 12.3.4

Definition 12.3.2 did not include the zero vector **0**. By special convention, **0** is said to be *parallel to every vector*.

(Since **0** is represented geometrically by a point, there is no geometric meaning to saying that **0** is parallel to another vector. However, it simplifies the statement of certain results to maintain that **0** is parallel to every vector. Algebraically this is warranted by the fact that **0** is a scalar multiple of every vector **b**: $\mathbf{0} = 0\mathbf{b}$.)

Example 2 Show that, if **a** and **b** are parallel to **c**, then every *linear combination*, $\alpha\mathbf{a} + \beta\mathbf{b}$, of **a** and **b** is also parallel to **c**.

SOLUTION Suppose that **a** and **b** are parallel to **c**. If $\mathbf{c} = \mathbf{0}$, then every vector is parallel to **c** and there is nothing to prove. If $\mathbf{c} \neq \mathbf{0}$, then **a** and **b** are scalar multiples of **c**:

$$\mathbf{a} = \alpha_1\mathbf{c}, \quad \mathbf{b} = \beta_1\,\mathbf{c}.$$

Then
$$\alpha\mathbf{a} + \beta\mathbf{b} = \alpha(\alpha_1\mathbf{c}) + \beta(\beta_1\mathbf{c}) = (\alpha\alpha_1 + \beta\beta_1)\mathbf{c}$$

is also parallel to **c**. ❑

DEFINITION 12.3.3 NORM

The *norm* of a vector $\mathbf{a} = (a_1, a_2, a_3)$, denoted by $||\mathbf{a}||$, is the number

$$||\mathbf{a}|| = \sqrt{a_1^2 + a_2^2 + a_3^2}.$$

The norm of **a** is also called the *length* or *magnitude* of **a**: if we represent the vector **a** by an arrow, $\overrightarrow{QR}$, with $Q = Q(x, y, z)$ and $R(x + a_1, y + a_2, z + a_3)$, then $||\mathbf{a}||$ gives the length of $\overrightarrow{QR}$. (Figure 12.3.5)

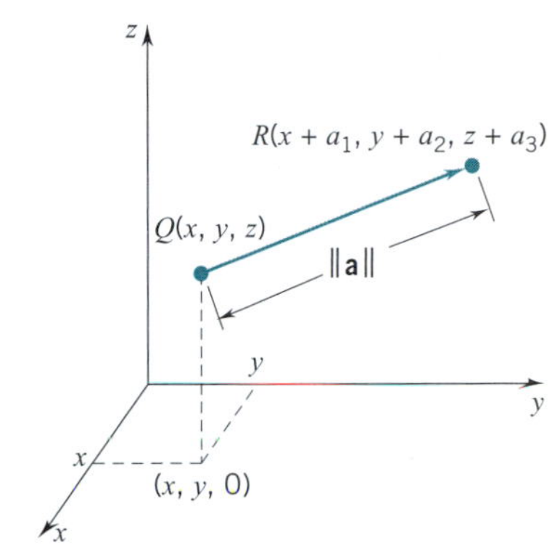

Figure 12.3.5

The norm properties of vectors are very similar to the absolute value properties of real numbers. In particular

(12.3.4)

(1) $||\mathbf{a}|| \geq 0$ and $||\mathbf{a}|| = 0$ iff $\mathbf{a} = \mathbf{0}$.

(2) $||\alpha\mathbf{a}|| = |\alpha|\ ||\mathbf{a}||$.

(3) $||\mathbf{a} + \mathbf{b}|| \leq ||\mathbf{a}|| + ||\mathbf{b}||$. (the triangle inequality)

Property (1) is obvious. Property (2) is easy to verify:

$$||\alpha\mathbf{a}|| = \sqrt{(\alpha a_1)^2 + (\alpha a_2)^2 + (\alpha a_3)^2} = |\alpha|\sqrt{a_1^2 + a_2^2 + a_3^2} = |\alpha|\ ||\mathbf{a}||.$$

We prove Property (3), the triangle inequality, in the next section, where we have "dot products" at our disposal. A proof at this time would be laborious.

You can interpret the triangle inequality as saying that the length of a side of a triangle cannot exceed the sum of the lengths of the other two sides. (See Figure 12.3.6.)

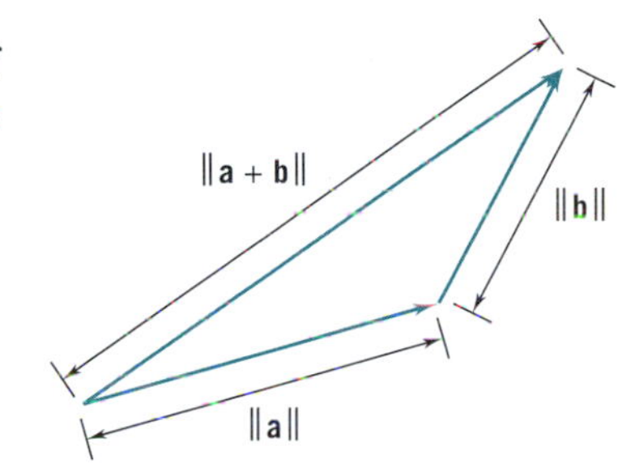

Figure 12.3.6

Example 3 Given that $\mathbf{a} = (1, -2, 3)$ and $\mathbf{b} = (-4, 1, 0)$, calculate

(a) $||\mathbf{a}||$. **(b)** $||\mathbf{b}||$. **(c)** $||\mathbf{a} + \mathbf{b}||$. **(d)** $||\mathbf{a} - \mathbf{b}||$. **(e)** $||-7\mathbf{a}||$.
(f) $||2\mathbf{a} - 3\mathbf{b}||$.

SOLUTION

(a) $||\mathbf{a}|| = \sqrt{1^2 + (-2)^2 + 3^2} = \sqrt{1 + 4 + 9} = \sqrt{14}$.

(b) $||\mathbf{b}|| = \sqrt{(-4)^2 + 1^2 + 0^2} = \sqrt{16 + 1 + 0} = \sqrt{17}$.

(c) $||\mathbf{a} + \mathbf{b}|| = ||(-3, -1, 3)|| = \sqrt{(-3)^2 + (-1)^2 + 3^2} = \sqrt{9 + 1 + 9} = \sqrt{19}$.

(d) $||\mathbf{a} - \mathbf{b}|| = ||(5, -3, 3)|| = \sqrt{5^2 + (-3)^2 + 3^2} = \sqrt{25 + 9 + 9} = \sqrt{43}$.

(e) $||-7\mathbf{a}|| = |-7|\ ||\mathbf{a}|| = 7\sqrt{14}$.

(f) $||2\mathbf{a} - 3\mathbf{b}|| = ||2(1, -2, 3) - 3(-4, 1, 0)||$
$= ||(14, -7, 6)|| = \sqrt{14^2 + (-7)^2 + 6^2}$
$= \sqrt{196 + 49 + 36} = \sqrt{281}$. □

To multiply a nonzero vector by a nonzero scalar α is to change its length by a factor of $|\alpha|$,

$$||\alpha\mathbf{a}|| = |\alpha|\ ||\mathbf{a}||,$$

keeping the same direction if $\alpha > 0$ and reversing the direction if $\alpha < 0$. The vector obtained from $\mathbf{a}$ simply by reversing its direction is the vector $(-1)\mathbf{a} = -\mathbf{a}$. (See Figure 12.3.7.)

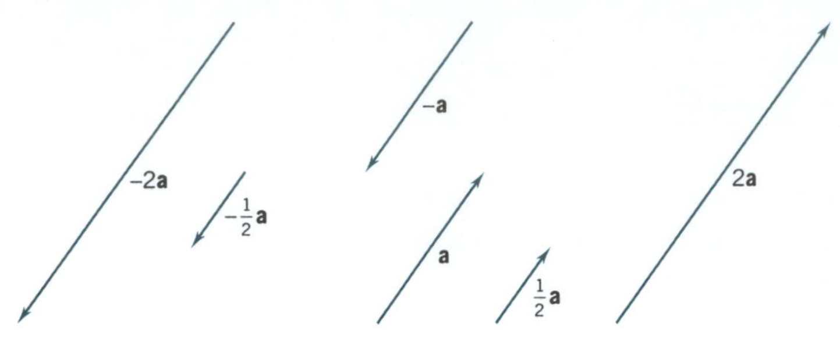

Figure 12.3.7

Since $\mathbf{a} - \mathbf{b} = \mathbf{a} + (-\mathbf{b})$, we can draw the vector $\mathbf{a} - \mathbf{b}$ by drawing $-\mathbf{b}$ and adding it to the vector $\mathbf{a}$ (Figure 12.3.8).

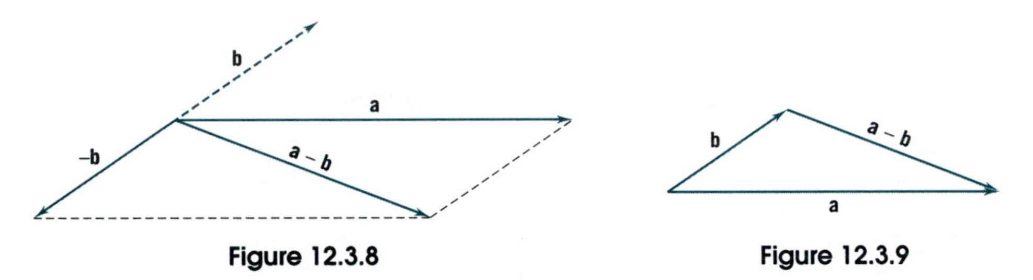

Figure 12.3.8 **Figure 12.3.9**

We can obtain the same result more easily by noting that $\mathbf{a} - \mathbf{b}$ is the vector that we must add to $\mathbf{b}$ to obtain $\mathbf{a}$ (Figure 12.3.9).

Vectors of norm 1 are called *unit vectors*. If $\mathbf{b}$ is a nonzero vector, then there is a unit vector $\mathbf{u_b}$ that has the direction of $\mathbf{b}$. To find $\mathbf{u_b}$, note that

$$||\mathbf{u_b}|| = 1 \quad \text{and} \quad \mathbf{u_b} = \alpha\mathbf{b} \quad \text{for some } \alpha > 0.$$

It follows that

$$1 = ||\mathbf{u_b}|| = ||\alpha\mathbf{b}|| = |\alpha|\,||\mathbf{b}|| = \alpha||\mathbf{b}||.$$

Thus

$$\alpha = \frac{1}{||\mathbf{b}||} \quad \text{and consequently} \quad \mathbf{u_b} = \frac{1}{||\mathbf{b}||}\mathbf{b} = \frac{\mathbf{b}}{||\mathbf{b}||}.$$

While

$$\mathbf{u_b} = \frac{\mathbf{b}}{||\mathbf{b}||}$$

is the unit vector in the direction of $\mathbf{b}$,

$$-\mathbf{u_b} = -\frac{\mathbf{b}}{||\mathbf{b}||}$$

is the unit vector in the opposite direction.

We single out for special attention the vectors

$$\mathbf{i} = (1, 0, 0), \quad \mathbf{j} = (0, 1, 0), \quad \mathbf{k} = (0, 0, 1).$$

These vectors all have norm 1 and, if pictured as emanating from the origin, lie along the positive coordinate axes. They are called *the unit coordinate vectors*. (See Figure 12.3.10)

Every vector can be expressed as a linear combination of the unit coordinate vectors:

(12.3.5) $\quad$ for $\mathbf{a} = (a_1, a_2, a_3)$ $\quad$ we have $\quad \mathbf{a} = a_1\,\mathbf{i} + a_2\,\mathbf{j} + a_3\,\mathbf{k}.$

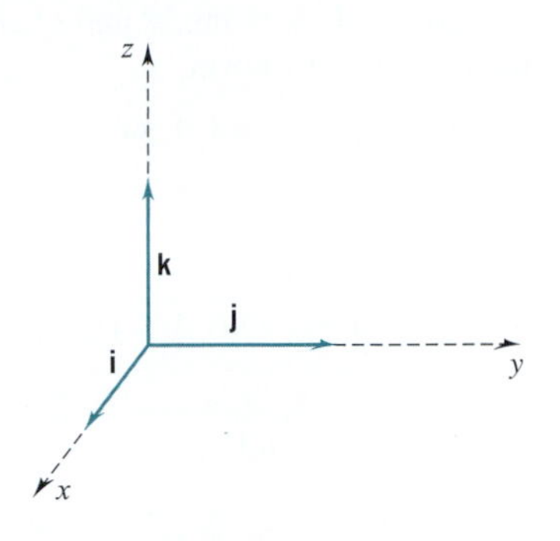

Figure 12.3.10

PROOF

$$\begin{aligned}(a_1, a_2, a_3) &= (a_1, 0, 0) + (0, a_2, 0) + (0, 0, a_3)\\ &= a_1(1,0,0) + a_2(0,1,0) + a_3(0,0,1)\\ &= a_1\,\mathbf{i} + a_2\,\mathbf{j} + a_3\,\mathbf{k}. \quad \square\end{aligned}$$

The numbers a_1, a_2, a_3 are called the **i**, **j**, **k** components of the vector **a**.

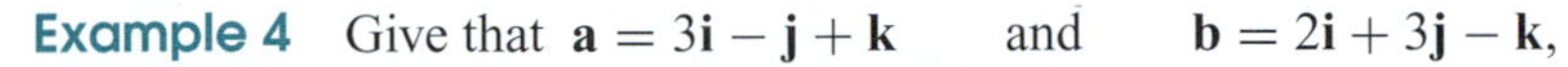

Example 4 Give that $\mathbf{a} = 3\mathbf{i} - \mathbf{j} + \mathbf{k}$ $\quad$ and $\quad \mathbf{b} = 2\mathbf{i} + 3\mathbf{j} - \mathbf{k}$,

(1) Express $2\mathbf{a} - \mathbf{b}$ as a linear combination of **i**, **j**, **k**.

(2) Calculate $||2\mathbf{a} - \mathbf{b}||$.

(3) Find the unit vector $\mathbf{u_c}$ in the direction of $\mathbf{c} = 2\mathbf{a} - \mathbf{b}$.

SOLUTION

(1) $2\mathbf{a} - \mathbf{b} = 2(3\mathbf{i} - \mathbf{j} + \mathbf{k}) - (2\mathbf{i} + 3\mathbf{j} - \mathbf{k})$
$\qquad = 6\mathbf{i} - 2\mathbf{j} + 2\mathbf{k} - 2\mathbf{i} - 3\mathbf{j} + \mathbf{k} = 4\mathbf{i} - 5\mathbf{j} + 3\mathbf{k}.$

(2) $||2\mathbf{a} - \mathbf{b}|| = ||4\mathbf{i} - 5\mathbf{j} + 3\mathbf{k}|| = \sqrt{16 + 25 + 9} = \sqrt{50} = 5\sqrt{2}.$

(3) $\mathbf{u_c} = \dfrac{2\mathbf{a} - \mathbf{b}}{||2\mathbf{a} - \mathbf{b}||} = \dfrac{1}{5\sqrt{2}}(4\mathbf{i} - 5\mathbf{j} + 3\mathbf{k}). \quad \square$

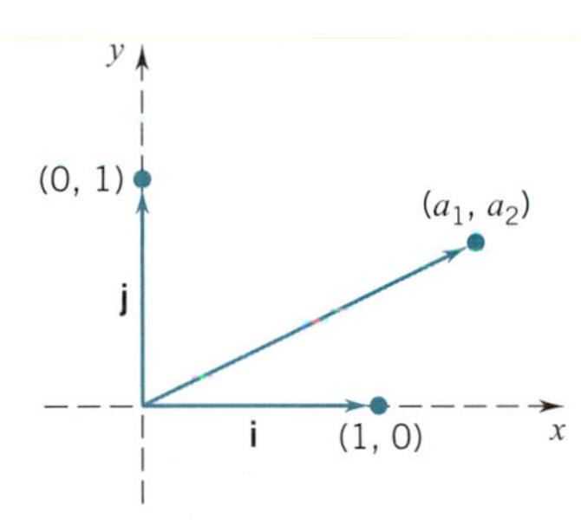

Figure 12.3.11

Remark: Vectors in the Plane A vector $\mathbf{a} = a_1\,\mathbf{i} + a_2\,\mathbf{j} + a_3\,\mathbf{k}$ for which $a_3 = 0$ is a vector in the xy-plane. Such a vector can be written more simply as $\mathbf{a} = a_1\,\mathbf{i} + a_2\,\mathbf{j}$ and identified with the ordered pair (a_1, a_2).

The unit coordinate vectors $\mathbf{i} = (1, 0, 0)$ and $\mathbf{j} = (0, 1, 0)$ are vectors in the xy-plane and are identified with the ordered pairs $(1, 0)$ and $(0, 1)$, respectively. These are the unit coordinate vectors and we will continue to denote them by **i** and **j**. As in three-dimensional space, every vector $\mathbf{a} = (a_1, a_2)$ in the plane can be expressed as a linear combination of the unit coordinate vectors:

(12.3.6) $\qquad \mathbf{a} = a_1(1, 0) + a_2(0, 1) = a_1\,\mathbf{i} + a_2\,\mathbf{j}.$ $\qquad$ (Figure 12.3.11)

The definitions and results stated for vectors in space hold for vectors in the plane as well. In particular, if $\mathbf{a} = (a_1, a_2)$ and $\mathbf{b} = (b_1, b_2)$ are vectors in the plane and α is a scalar, then

(1) $\mathbf{a} = \mathbf{b}$ $\quad$ iff $\quad a_1 = b_1 \quad a_2 = b_2.$

(2) $\mathbf{a} + \mathbf{b} = (a_1 + b_1, a_2 + b_2) = (a_1 + b_1)\,\mathbf{i} + (a_2 + b_2)\,\mathbf{j}.$

(3) $\alpha\mathbf{a} = (\alpha a_1, \alpha a_2) = \alpha a_1\,\mathbf{i} + \alpha a_2\,\mathbf{j}.$

(4) $\mathbf{0} = (0, 0)$ is the zero vector.

(5) $||\mathbf{a}|| = \sqrt{a_1^2 + a_2^2}$ is the norm, or magnitude of **a**.

We will use ordered pairs (a_1, a_2) or the form (12.3.6) to treat two dimensional problems.

EXERCISES 12.3

In Exercises 1–4, points P and Q are given. Find the vector $\overrightarrow{PQ}$ and determine its norm.

1. $P(1,-2,5)$, $Q(4,2,3)$.

2. $P(4,-2,0)$, $Q(2,4,0)$.

3. $P(0,3,1)$, $Q(0,1,0)$.

4. $P(-4,0,7)$, $Q(0,3,-1)$.

In Exercises 5–8, set $\mathbf{a}=(1,-2,3)$, $\mathbf{b}=(3,0,-1)$, $\mathbf{c}=(-4,2,1)$. Find:

5. $2\mathbf{a}-\mathbf{b}$.

6. $2\mathbf{b}+3\mathbf{c}$.

7. $-2\mathbf{a}+\mathbf{b}-\mathbf{c}$.

8. $\mathbf{a}+3\mathbf{b}-2\mathbf{c}$. Changed

Simplify the linear combinations.

9. $(2\mathbf{i}-\mathbf{j}+\mathbf{k})+(\mathbf{i}-3\mathbf{j}+5\mathbf{k})$.

10. $(6\mathbf{j}-\mathbf{k})+(3\mathbf{i}-\mathbf{j}+2\mathbf{k})$.

11. $2(\mathbf{j}+\mathbf{k})-3(\mathbf{i}+\mathbf{j}-2\mathbf{k})$.

12. $2(\mathbf{i}-\mathbf{j})+6(2\mathbf{i}+\mathbf{j}-2\mathbf{k})$.

Calculate the norm of the vector.

13. $3\mathbf{i}+4\mathbf{j}$.

14. $\mathbf{i}-\mathbf{j}$.

15. $2\mathbf{i}+\mathbf{j}-2\mathbf{k}$.

16. $6\mathbf{i}+2\mathbf{j}-\mathbf{k}$.

17. $\frac{1}{2}(\mathbf{i}+4\mathbf{j})-(\frac{3}{2}\mathbf{i}+\mathbf{k})$.

18. $(\mathbf{i}-\mathbf{j})+2(\mathbf{j}-\mathbf{i})+(\mathbf{k}-\mathbf{j})$.

19. Let

$$\mathbf{a}=\mathbf{i}-\mathbf{j}+2\mathbf{k},\quad \mathbf{b}=2\mathbf{i}-\mathbf{j}+2\mathbf{k},$$
$$\mathbf{c}=3\mathbf{i}-3\mathbf{j}+6\mathbf{k},\quad \mathbf{d}=-2\mathbf{i}+2\mathbf{j}-4\mathbf{k}.$$

(a) Which vectors are parallel?
(b) Which vectors have the same direction?
(c) Which vectors have opposite directions?

20. (*Important*) Prove the following version of the triangle inequality:

$$\big|\,\|\mathbf{a}\|-\|\mathbf{b}\|\,\big|\le\|\mathbf{a}-\mathbf{b}\|.$$

HINT: $\mathbf{a}=(\mathbf{a}-\mathbf{b})+\mathbf{b}$

Find the unit vector in the direction of $\mathbf{a}$.

21. $\mathbf{a}=(3,-4,0)$.

22. $\mathbf{a}=-2\mathbf{i}+3\mathbf{j}$.

23. $\mathbf{a}=\mathbf{i}-2\mathbf{j}+2\mathbf{k}$.

24. $\mathbf{a}=(2,1,2)$.

In Exercises 25 and 26, find the unit vector in the direction opposite to the direction of $\mathbf{a}$.

25. $\mathbf{a}=-\mathbf{i}+3\mathbf{j}+2\mathbf{k}$.

26. $\mathbf{a}=2\mathbf{i}-\mathbf{k}$.

27. Label the vectors.

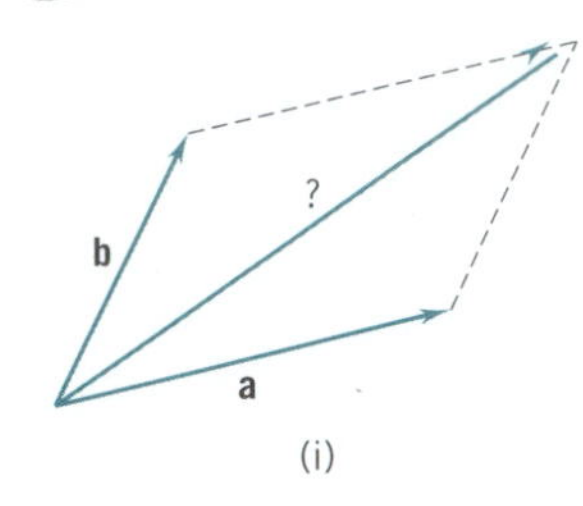

(i)

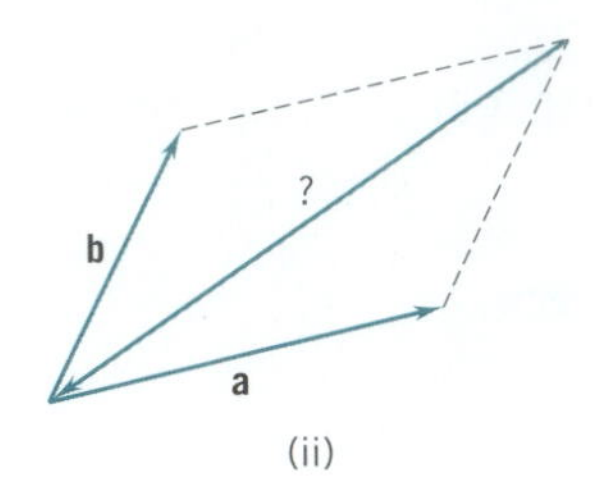

(ii)

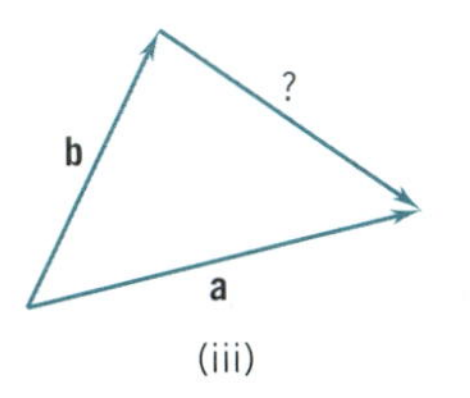

(iii)

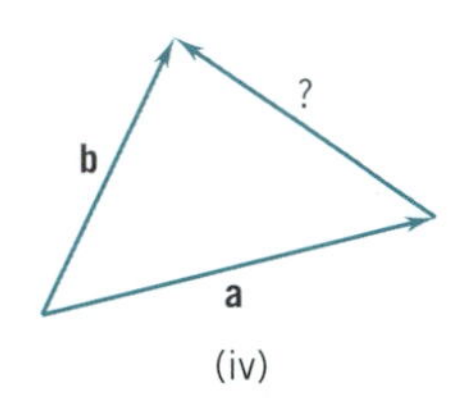

(iv)

28. Let $\mathbf{a}=(1,1,1)$, $\mathbf{b}=(-1,3,2,)$, $\mathbf{c}=(-3,0,1)$, $\mathbf{d}=(4,-1,1)$.

(a) Express $\mathbf{a}+2\mathbf{b}+3\mathbf{c}+4\mathbf{d}$ as a linear combination of $\mathbf{i},\mathbf{j},\mathbf{k}$.
(b) Find scalars A,B,C such that $\mathbf{d}=A\mathbf{a}+B\mathbf{b}+C\mathbf{c}$.

29. Let $\mathbf{a}=(2,0,-1)$, $\mathbf{b}=(1,3,5)$, $\mathbf{c}=(-1,1,1)$, $\mathbf{d}=(1,1,6)$.

(a) Express $\mathbf{a}-3\mathbf{b}+2\mathbf{c}+4\mathbf{d}$ as a linear combination of $\mathbf{i},\mathbf{j},\mathbf{k}$.
(b) Find scalars A,B,C such that $\mathbf{d}=A\mathbf{a}+B\mathbf{b}+C\mathbf{c}$.

30. Find α given that $3\mathbf{i}+\mathbf{j}-\mathbf{k}$ and $\alpha\mathbf{i}-4\mathbf{j}+4\mathbf{k}$ are parallel.

31. Find α given that $3\mathbf{i}+\mathbf{j}$ and $\alpha\mathbf{j}-\mathbf{k}$ have the same length.

32. Find the unit vector in the direction of $\mathbf{i}-2\mathbf{j}+2\mathbf{k}$.

33. Find α given that $\|\alpha\mathbf{i}+(\alpha-1)\mathbf{j}+(\alpha+1)\mathbf{k}\|=2$.

34. Find the vector of norm 2 in the direction of $\mathbf{i}+2\mathbf{j}-\mathbf{k}$.

35. Find the vectors of norm 2 parallel to $3\mathbf{j}+2\mathbf{k}$.

36. Express $\mathbf{c}$ in terms of $\mathbf{a}$ and $\mathbf{b}$, given that the tip of $\mathbf{c}$ bisects the line segment.

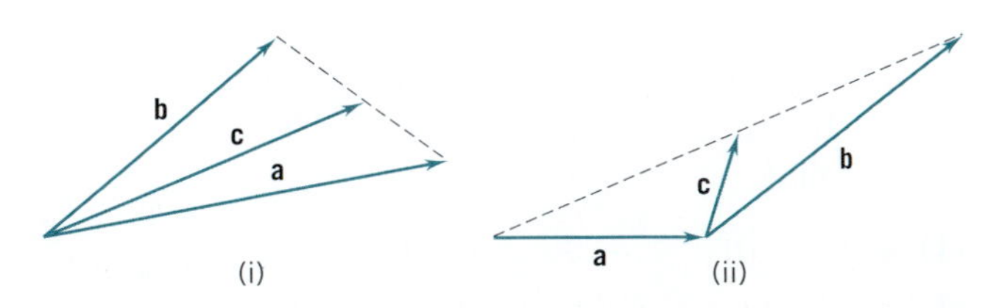

(i) (ii)

37. Let $\mathbf{a}$ and $\mathbf{b}$ be nonzero vectors such that

$$\|\mathbf{a}-\mathbf{b}\|=\|\mathbf{a}+\mathbf{b}\|.$$

(a) What can you conclude about the parallelogram generated by $\mathbf{a}$ and $\mathbf{b}$?

(b) Show that, if $\mathbf{a} = a_1\mathbf{i}+a_2\mathbf{j}+a_3\mathbf{k}$ and $\mathbf{b} = b_1\mathbf{i}+b_2\mathbf{j}+b_3\mathbf{k}$, then

$$a_1b_1 + a_2b_2 + a_3b_3 = 0.$$

38. (a) Show that, if **a** and **b** have the same direction, then

$$\|\mathbf{a}+\mathbf{b}\| = \|\mathbf{a}\| + \|\mathbf{b}\|.$$

(b) Does this equation necessarily hold if **a** and **b** are only parallel?

39. Let P and Q be two points in space and let M be the midpoint of the line segment $\overline{PQ}$. Let $\mathbf{p} = \overrightarrow{OP}$, $\mathbf{q} = \overrightarrow{OQ}$, and $\mathbf{m} = \overrightarrow{OM}$.

(a) Show that $\mathbf{m} = \mathbf{p} + \frac{1}{2}(\mathbf{q} - \mathbf{p})$.

(b) Derive the midpoint formula (12.1.4).

40. Let P and Q be two points in space and let R be the point on $\overline{PQ}$ which is twice as far from P as it is from Q. Let $\mathbf{p} = \overrightarrow{OP}$, $\mathbf{q} = \overrightarrow{OQ}$, and $\mathbf{r} = \overrightarrow{OR}$. Prove that $\mathbf{r} = \frac{1}{3}\mathbf{p} + \frac{2}{3}\mathbf{q}$.

A vector **r** emanating from the origin is called a *radius* vector. Each radius vector determines a unique point of space: the point at the tip of the vector. Conversely, each point of space determines a unique radius vector: the radius vector whose tip falls on that point. This one-to-one correspondence between the set of all radius vectors and the set of all points in three-dimensional space enables us to use radius vectors to specify sets in space. Thus, for example, the radius-vector equation $\|\mathbf{r}\| = 3$ can be used to represent the sphere of radius 3 centered at the origin: the sphere consists of the tips of all the radius vectors **r** that satisfy that equation.

41. Write a radius-vector equation or inequality for each of the following sets.

(a) The sphere of radius 3 centered at $P(a_1, a_2, a_3)$.

(b) The set of all points on or inside the sphere of radius 2 centered at the origin. (This set is called the *ball* of radius 2 about the origin.)

(c) The ball of radius 1 about the point $P(a_1, a_2, a_3)$.

42. Write a radius-vector equation for each of the following sets.

(a) The set of all points equidistant from $P(a_1, a_2, a_3)$ and $Q(b_1, b_2, b_3)$. (This set forms a plane.)

(b) The set of all points the sum of whose distances from $P(a_1, a_2, a_3)$ and $Q(b_1, b_2, b_3)$ is a constant $k > d(P, Q)$. [Such a set is an example of an *ellipsoid*. An ellipsoid is a three-dimensional analogue of the ellipse.] What happens if $k = d(P, Q)$? If $k < d(P, Q)$?

■ 12.4 THE DOT PRODUCT

In this section we introduce the first of two products that we define for vectors.

Introduction

We begin with two nonzero vectors

$$\mathbf{a} = a_1\mathbf{i} + a_2\mathbf{j} + a_3\mathbf{k}, \quad \mathbf{b} = b_1\mathbf{i} + b_2\mathbf{j} + b_3\mathbf{k}.$$

How can we tell from the components of these vectors whether these vectors meet at right angles? To explore this question we draw Figure 12.4.1. By the Pythagorean theorem, **a** and **b** meet at right angles iff

$$\|\mathbf{a}\|^2 + \|\mathbf{b}\|^2 = \|\mathbf{b} - \mathbf{a}\|^2.$$

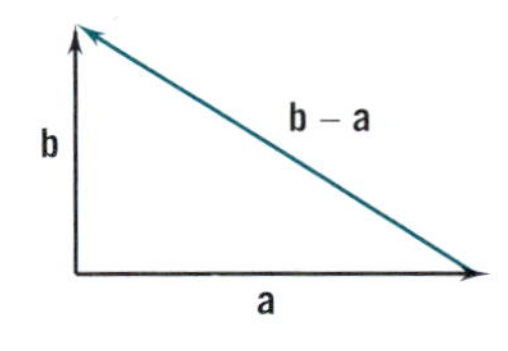

Figure 12.4.1

In terms of components this equation reads

$$(a_1^2 + a_2^2 + a_3^2) + (b_1^2 + b_2^2 + b_3^2) = (b_1 - a_1)^2 + (b_2 - a_2)^2 + (b_3 - a_3)^2,$$

which, as you can readily check, simplifies to

$$a_1b_1 + a_2b_2 + a_3b_3 = 0.$$

The expression $a_1b_1 + a_2b_2 + a_3b_3$ is widely used in geometry and in physics. It has a name, the *dot product* of **a** and **b**, and there is a special notation for it, $\mathbf{a} \cdot \mathbf{b}$. The notion is so important that it deserves a formal definition.

Definition of the Dot Product

DEFINITION 12.4.1

For any two vectors

$$\mathbf{a} = a_1\,\mathbf{i} + a_2\,\mathbf{j} + a_3\,\mathbf{k} \qquad \text{and} \qquad \mathbf{b} = b_1\,\mathbf{i} + b_2\,\mathbf{j} + b_3\,\mathbf{k},$$

we define the *dot product* $\mathbf{a} \cdot \mathbf{b}$ by setting

$$\mathbf{a} \cdot \mathbf{b} = a_1b_1 + a_2b_2 + a_3b_3.$$

Remark The dot product of two vectors in n-dimensional space, $n = 2, 4, 5, \ldots,$ is defined in the same way. In particular, for $\mathbf{a} = a_1\mathbf{i} + a_2\mathbf{j}$ and $\mathbf{b} = b_1\mathbf{i} + b_2\mathbf{j}$ in the xy-plane,

$$\mathbf{a} \cdot \mathbf{b} = a_1b_1 + a_2b_2.$$

The properties of the dot product presented in this section for three-dimensional vectors also hold for vectors in n-dimensional space. ❑

Example 1 For $\mathbf{a} = 2\mathbf{i} - \mathbf{j} + 3\mathbf{k}, \qquad \mathbf{b} = -3\mathbf{i} + \mathbf{j} + 4\mathbf{k}, \qquad \mathbf{c} = \mathbf{i} + 3\mathbf{j},$

we have

$$\begin{aligned}
\mathbf{a} \cdot \mathbf{b} &= (2)(-3) + (-1)(1) + (3)(4) = -6 - 1 + 12 = 5,\\
\mathbf{a} \cdot \mathbf{c} &= (2)(1) + (-1)(3) + (3)(0) = 2 - 3 = -1,\\
\mathbf{b} \cdot \mathbf{c} &= (-3)(1) + (1)(3) + (4)(0) = -3 + 3 = 0.
\end{aligned}$$

The last equation tells us that **b** and **c** meet at right angles. (Verify this by drawing a figure.) ❑

Because $\mathbf{a} \cdot \mathbf{b}$ is not a vector, but a scalar, it is sometimes called the *scalar product* of **a** and **b**. We will continue to call it the dot product and speak of "dotting **a** with **b**."

Properties of the Dot Product

If we dot a vector with itself, we obtain the square of its norm:

(12.4.2)
$$\mathbf{a} \cdot \mathbf{a} = \|\mathbf{a}\|^2.$$

PROOF

$$\mathbf{a} \cdot \mathbf{a} = a_1a_1 + a_2a_2 + a_3a_3 = a_1^2 + a_2^2 + a_3^2 = \|\mathbf{a}\|^2. \quad ❑$$

The dot product of any vector with the zero vector is zero:

(12.4.3)
$$\mathbf{a} \cdot \mathbf{0} = 0, \qquad \mathbf{0} \cdot \mathbf{a} = 0.$$

PROOF

$$(a_1)(0) + (a_2)(0) + (a_3)(0) = 0, \qquad (0)(a_1) + (0)(a_2) + (0)(a_3) = 0. \quad ❑$$

The dot product is commutative:

(12.4.4) $$\mathbf{a}\cdot\mathbf{b}=\mathbf{b}\cdot\mathbf{a},$$

and scalars can be factored:

(12.4.5) $$\alpha\mathbf{a}\cdot\beta\mathbf{b}=\alpha\beta(\mathbf{a}\cdot\mathbf{b}).$$

PROOF

$$\mathbf{a}\cdot\mathbf{b}=a_1b_1+a_2b_2+a_3b_3=b_1a_1+b_2a_2+b_3a_3=\mathbf{b}\cdot\mathbf{a},$$

and
$$\begin{aligned}\alpha\mathbf{a}\cdot\beta\mathbf{b}&=(\alpha a_1)(\beta b_1)+(\alpha a_2)(\beta b_2)+(\alpha a_3)(\beta b_3)\\&=\alpha\beta(a_1b_1+a_2b_2+a_3b_3)=\alpha\beta(\mathbf{a}\cdot\mathbf{b}).\quad\square\end{aligned}$$

The dot product satisfies the following distributive laws:

(12.4.6) $$\mathbf{a}\cdot(\mathbf{b}+\mathbf{c})=\mathbf{a}\cdot\mathbf{b}+\mathbf{a}\cdot\mathbf{c},\qquad(\mathbf{a}+\mathbf{b})\cdot\mathbf{c}=\mathbf{a}\cdot\mathbf{c}+\mathbf{b}\cdot\mathbf{c}.$$

PROOF

$$\begin{aligned}\mathbf{a}\cdot(\mathbf{b}+\mathbf{c})&=a_1(b_1+c_1)+a_2(b_2+c_2)+a_3(b_3+c_3)\\&=a_1b_1+a_1c_1+a_2b_2+a_2c_2+a_3b_3+a_3c_3\\&=(a_1b_1+a_2b_2+a_3b_3)+(a_1c_1+a_2c_2+a_3c_3)\\&=\mathbf{a}\cdot\mathbf{b}+\mathbf{a}\cdot\mathbf{c}.\end{aligned}$$

The second equation can be verified in a similar manner. $\square$

Example 2 Given that

$||\mathbf{a}||=1,\quad ||\mathbf{b}||=3,\quad ||\mathbf{c}||=4,\quad \mathbf{a}\cdot\mathbf{b}=0,\quad \mathbf{a}\cdot\mathbf{c}=1,\quad \mathbf{b}\cdot\mathbf{c}=-2,$

find (a) $3\mathbf{a}\cdot(\mathbf{b}+4\mathbf{c})$. (b) $(\mathbf{a}-\mathbf{b})\cdot(2\mathbf{a}+\mathbf{b})$. (c) $[(\mathbf{b}\cdot\mathbf{c})\,\mathbf{a}-(\mathbf{a}\cdot\mathbf{c})\,\mathbf{b}]\cdot\mathbf{c}$.

SOLUTION

(a) $3\mathbf{a}\cdot(\mathbf{b}+4\mathbf{c})=(3\mathbf{a}\cdot\mathbf{b})+(3\mathbf{a}\cdot4\mathbf{c})=3(\mathbf{a}\cdot\mathbf{b})+12(\mathbf{a}\cdot\mathbf{c})=12.$

(b)
$$\begin{aligned}(\mathbf{a}-\mathbf{b})\cdot(2\mathbf{a}+\mathbf{b})&=(\mathbf{a}\cdot2\mathbf{a})+(\mathbf{a}\cdot\mathbf{b})+(-\mathbf{b}\cdot2\mathbf{a})+(-\mathbf{b}\cdot\mathbf{b})\\&=2(\mathbf{a}\cdot\mathbf{a})+(\mathbf{a}\cdot\mathbf{b})-2(\mathbf{b}\cdot\mathbf{a})-(\mathbf{b}\cdot\mathbf{b})\\&=2||\mathbf{a}||^2+(\mathbf{a}\cdot\mathbf{b})-2(\mathbf{a}\cdot\mathbf{b})-||\mathbf{b}||^2\\&=2+0-2(0)-9=-7.\end{aligned}$$

(c)
$$\begin{aligned}[(\mathbf{b}\cdot\mathbf{c})\,\mathbf{a}-(\mathbf{a}\cdot\mathbf{c})\,\mathbf{b}]\cdot\mathbf{c}&=[(\mathbf{b}\cdot\mathbf{c})\,\mathbf{a}\cdot\mathbf{c}]-[(\mathbf{a}\cdot\mathbf{c})\,\mathbf{b}\cdot\mathbf{c}]\\&=(\mathbf{b}\cdot\mathbf{c})(\mathbf{a}\cdot\mathbf{c})-(\mathbf{a}\cdot\mathbf{c})(\mathbf{b}\cdot\mathbf{c})=0.\quad\square\end{aligned}$$

Geometric Interpretation of the Dot Product

We begin with a triangle with sides a, b, c (Figure 12.4.2). If θ were $\frac{1}{2}\pi$, the pythagorean theorem would tell us that $c^2=a^2+b^2$. The law of cosines,

$$c^2=a^2+b^2-2ab\cos\theta,$$

is a generalization of the Pythagorean theorem.

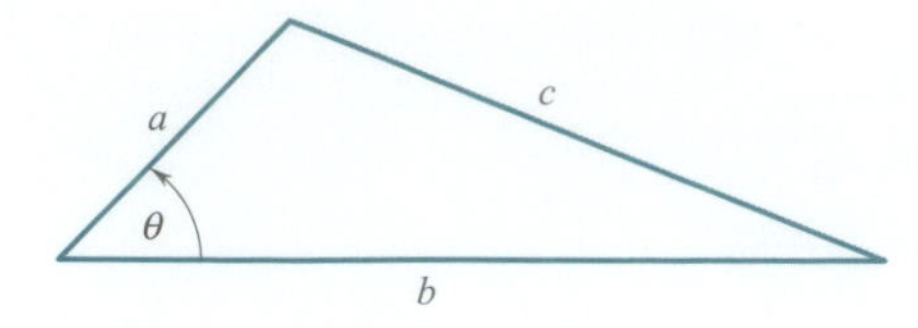

Figure 12.4.2

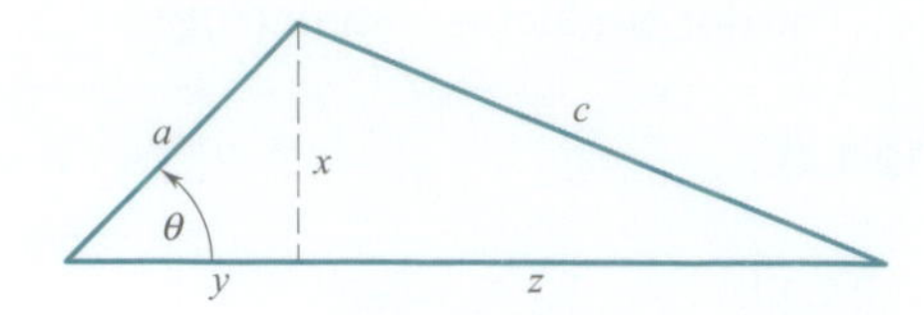

Figure 12.4.3

To derive the law of cosines, we drop a perpendicular to side b (Figure 12.4.3). We then have

$$c^2 = z^2 + x^2 = (b - y)^2 + x^2 = b^2 - 2by + y^2 + x^2.$$

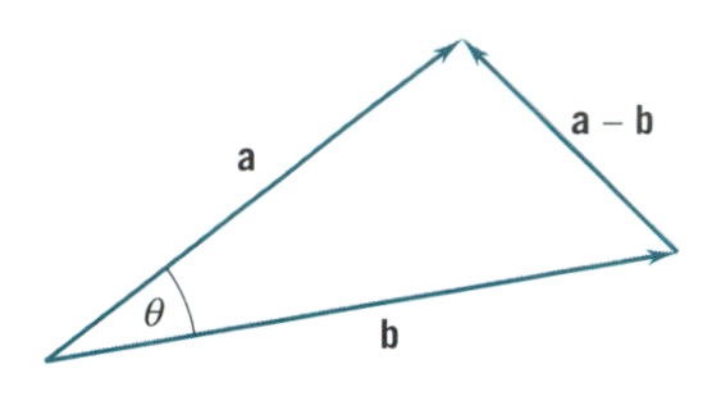

Figure 12.4.4

From the figure, $y^2 + x^2 = a^2$ and $y = a\cos\theta$. Therefore

$$c^2 = a^2 + b^2 - 2ab\cos\theta,$$

as asserted. (What if the angle θ is obtuse? We leave that case to you.)

Now back to dot products. If neither $\mathbf{a}$ nor $\mathbf{b}$ is zero, we can interpret $\mathbf{a} \cdot \mathbf{b}$ from the triangle of Figure 12.4.4. The lengths of the sides are $||\mathbf{a}||$, $||\mathbf{b}||$, $||\mathbf{a} - \mathbf{b}||$. By the law of cosines

$$||\mathbf{a} - \mathbf{b}||^2 = ||\mathbf{a}||^2 + ||\mathbf{b}||^2 - 2||\mathbf{a}||\,||\mathbf{b}||\cos\theta.$$

This gives

$$2||\mathbf{a}||\,||\mathbf{b}||\cos\theta = ||\mathbf{a}||^2 + ||\mathbf{b}||^2 - ||\mathbf{a} - \mathbf{b}||^2 = a_1^2 + a_2^2 + a_3^2 + b_1^2 + b_2^2 + b_3^2 - (a_1 - b_1)^2 - (a_2 - b_2)^2 - (a_3 - b_3)^2 = 2(a_1b_1 + a_2b_2 + a_3b_3) = 2(\mathbf{a} \cdot \mathbf{b}),$$

and thus

(12.4.7)
$$\mathbf{a} \cdot \mathbf{b} = ||\mathbf{a}||\,||\mathbf{b}||\cos\theta.$$

By convention, θ, the angle between $\mathbf{a}$ and $\mathbf{b}$, is measured in radians and taken from 0 to π; no negative angles.

From (12.4.7) you can see that the dot product of two vectors depends on the norms of the vectors and on the angle between them. For vectors of a given norm, the dot product measures the extent to which the vectors agree in direction. As the difference in direction increases, the dot product decreases:

If $\mathbf{a}$ and $\mathbf{b}$ have the same direction, then $\theta = 0$ and

$$\mathbf{a} \cdot \mathbf{b} = ||\mathbf{a}||\,||\mathbf{b}||; \qquad (\cos 0 = 1)$$

this is the largest possible value for $\mathbf{a} \cdot \mathbf{b}$.

If $\mathbf{a}$ and $\mathbf{b}$ meet at right angles, then $\theta = \frac{1}{2}\pi$ and

$$\mathbf{a} \cdot \mathbf{b} = 0. \qquad (\cos \tfrac{1}{2}\pi = 0)$$

If $\mathbf{a}$ and $\mathbf{b}$ have opposite directions, then $\theta = \pi$ and

$$\mathbf{a} \cdot \mathbf{b} = -||\mathbf{a}||\,||\mathbf{b}||; \qquad (\cos\pi = -1)$$

this is the least possible value for $\mathbf{a} \cdot \mathbf{b}$.

Two vectors are said to be *perpendicular* if they lie at right angles or one of the vectors is the zero vector; in other words, two vectors are said to be perpendicular iff their dot product is zero.† In symbols

(12.4.8)
$$\mathbf{a} \perp \mathbf{b} \quad \text{iff} \quad \mathbf{a} \cdot \mathbf{b} = 0.$$
(Figure 12.4.5)

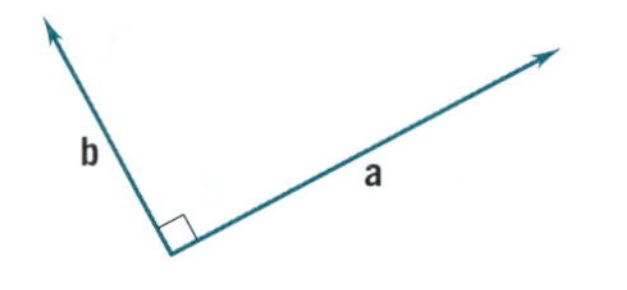

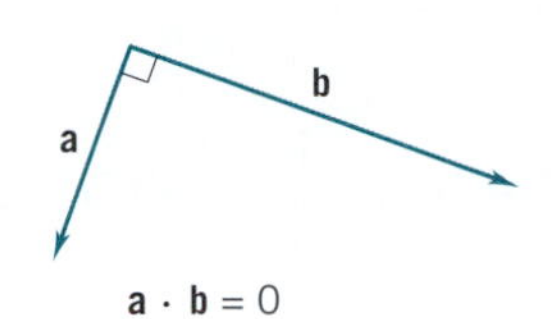

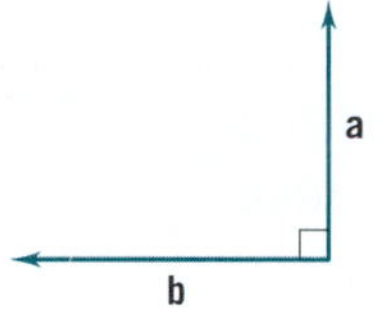

a · b = 0

Figure 12.4.5

The unit coordinate vectors are obviously mutually perpendicular:

$$\mathbf{i} \cdot \mathbf{j} = 0, \qquad \mathbf{i} \cdot \mathbf{k} = 0, \qquad \mathbf{j} \cdot \mathbf{k} = 0.$$

Example 3 Verify that the vectors $\mathbf{a} = 2\mathbf{i} + \mathbf{j} + \mathbf{k}$ and $\mathbf{b} = \mathbf{i} + \mathbf{j} - 3\mathbf{k}$ are perpendicular.

SOLUTION

$$\mathbf{a} \cdot \mathbf{b} = (2)(1) + (1)(1) + (1)(-3) = 2 + 1 - 3 = 0. \quad \square$$

Example 4 Find the value of α for which $(3\mathbf{i} - \alpha\mathbf{j} + \mathbf{k}) \perp (\mathbf{i} + 2\mathbf{j})$.

SOLUTION For the two vectors to be perpendicular, their dot product must be zero.

Since $\qquad (3\mathbf{i} - \alpha\mathbf{j} + \mathbf{k}) \cdot (\mathbf{i} + 2\mathbf{j}) = (3)(1) + (-\alpha)(2) + (1)(0) = 3 - 2\alpha,$

α must be $\frac{3}{2}$. $\square$

With **a** and **b** both different from zero, we can divide the equation

$$\mathbf{a} \cdot \mathbf{b} = \|\mathbf{a}\|\,\|\mathbf{b}\| \cos\theta$$

by $\|\mathbf{a}\|\,\|\mathbf{b}\|$ to obtain

$$\cos\theta = \frac{\mathbf{a} \cdot \mathbf{b}}{\|\mathbf{a}\|\,\|\mathbf{b}\|} = \frac{\mathbf{a}}{\|\mathbf{a}\|} \cdot \frac{\mathbf{b}}{\|\mathbf{b}\|}.$$

Writing
$$\mathbf{u_a} = \frac{\mathbf{a}}{\|\mathbf{a}\|} \quad \text{and} \quad \mathbf{u_b} = \frac{\mathbf{b}}{\|\mathbf{b}\|},$$

we have

(12.4.9)
$$\cos\theta = \mathbf{u_a} \cdot \mathbf{u_b}.$$

The cosine of the angle between the vectors is the dot product of the corresponding unit vectors.

---

† This makes the zero vector both parallel and perpendicular to every vector. There is, however, no contradiction since we do not apply the notion of "direction" to the zero vector.

Example 5 Calculate the angle between $\mathbf{a} = 2\mathbf{i} + 3\mathbf{j} + 2\mathbf{k}$ and $\mathbf{b} = \mathbf{i} + 2\mathbf{j} - \mathbf{k}$.

SOLUTION

$$\mathbf{u_a} = \frac{2\mathbf{i} + 3\mathbf{j} + 2\mathbf{k}}{||2\mathbf{i} + 3\mathbf{j} + 2\mathbf{k}||} = \frac{1}{\sqrt{17}}(2\mathbf{i} + 3\mathbf{j} + 2\mathbf{k}),$$

$$\mathbf{u_b} = \frac{\mathbf{i} + 2\mathbf{j} - \mathbf{k}}{||\mathbf{i} + 2\mathbf{j} - \mathbf{k}||} = \frac{1}{\sqrt{6}}(\mathbf{i} + 2\mathbf{j} - \mathbf{k}).$$

Therefore
$$\cos\theta = \mathbf{u_a} \cdot \mathbf{u_b} \frac{1}{\sqrt{17}} \frac{1}{\sqrt{6}}[(2\mathbf{i} + 3\mathbf{j} + 2\mathbf{k}) \cdot (\mathbf{i} + 2\mathbf{j} - \mathbf{k})].$$

Since
$$(2\mathbf{i} + 3\mathbf{j} + 2\mathbf{k}) \cdot (\mathbf{i} + 2\mathbf{j} - \mathbf{k}) = (2)(1) + (3)(2) + (2)(-1) = 6,$$

we have
$$\cos\theta = \frac{6}{\sqrt{17}\sqrt{6}} = \frac{1}{17}\sqrt{102} \cong \frac{10.1}{17} \cong 0.594.$$

Thus, $\theta \cong 0.935$ radians, which is about 54 degrees. ❑

Example 6 Show that, if **a** and **b** are both perpendicular to **c**, then every linear combination $\alpha\mathbf{a} + \beta\mathbf{b}$ is also perpendicular to **c**.

SOLUTION Suppose that **a** and **b** are both perpendicular to **c**. Then

$$\mathbf{a} \cdot \mathbf{c} = 0 \qquad \text{and} \qquad \mathbf{b} \cdot \mathbf{c} = 0.$$

It follows that

$$(\alpha\mathbf{a} + \beta\mathbf{b}) \cdot \mathbf{c} = \alpha\underbrace{(\mathbf{a} \cdot \mathbf{c})}_{0} + \beta\underbrace{(\mathbf{b} \cdot \mathbf{c})}_{0} = 0$$

and therefore $\alpha\mathbf{a} + \beta\mathbf{b}$ is perpendicular to **c**. ❑

Projections and Components

If $\mathbf{b} \neq \mathbf{0}$, then every vector **a** can be written in a unique manner as the sum of a vector $\mathbf{a}_{||}$ parallel to **b** and a vector $\mathbf{a}_{\perp}$ perpendicular to **b**:

$$\mathbf{a} = \mathbf{a}_{||} + \mathbf{a}_{\perp}. \qquad \text{(Exercise 53)}$$

The idea is illustrated in Figure 12.4.6. If **a** is parallel to **b**, then $\mathbf{a}_{||} = \mathbf{a}$ and $\mathbf{a}_{\perp} = \mathbf{0}$. If **a** is perpendicular to **b**, then $\mathbf{a}_{||} = \mathbf{0}$ and $\mathbf{a}_{\perp} = \mathbf{a}$.

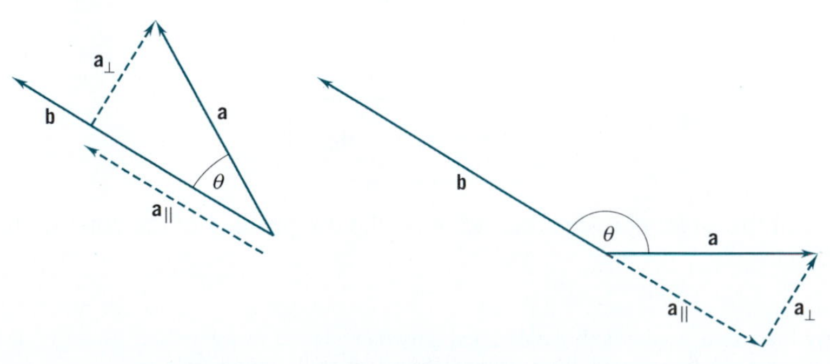

Figure 12.4.6

The vector $\mathbf{a}_{\parallel}$ is called *the projection of* **a** *on* **b** and is denoted by $\text{proj}_{\mathbf{b}}\mathbf{a}$. Since $\text{proj}_{\mathbf{b}}\mathbf{a}$ is parallel to **b**, it is a scalar multiple of the unit vector in the direction of **b**:

$$\text{proj}_{\mathbf{b}}\mathbf{a} = \lambda \mathbf{u}_{\mathbf{b}}.$$

The scalar λ is called *the component of* **a** *in the direction of* **b** (or, more briefly, the **b**-*component* of **a**) and is denoted by $\text{comp}_{\mathbf{b}}\mathbf{a}$. In symbols,

(12.4.10)
$$\text{proj}_{\mathbf{b}}\mathbf{a} = (\text{comp}_{\mathbf{b}}\mathbf{a})\,\mathbf{u}_{\mathbf{b}}.$$

The component of **a** in the direction of **b** measures the "advance" of **a** in the direction of **b**. In Figure 12.4.6 we used θ to indicate the angle between **a** and **b**. If $0 \le \theta < \frac{1}{2}\pi$, the projection and **b** have the same direction and the component is positive. If $\theta = \frac{1}{2}\pi$, the projection is **0** and the component is 0. If $\frac{1}{2}\pi < \theta \le \pi$, the projection and **b** have opposite directions and, consequently, the component is negative.

Projections and components are closely related to dot products: if $\mathbf{b} \neq \mathbf{0}$, then

(12.4.11)
$$\text{proj}_{\mathbf{b}}\mathbf{a} = (\mathbf{a} \cdot \mathbf{u}_{\mathbf{b}})\,\mathbf{u}_{\mathbf{b}} \quad \text{and} \quad \text{comp}_{\mathbf{b}}\mathbf{a} = \mathbf{a} \cdot \mathbf{u}_{\mathbf{b}}.$$

PROOF The second assertion follows immediately from the first. We will prove the first. We begin with the identity

$$\mathbf{a} = (\mathbf{a} \cdot \mathbf{u}_{\mathbf{b}})\,\mathbf{u}_{\mathbf{b}} + [\mathbf{a} - (\mathbf{a} \cdot \mathbf{u}_{\mathbf{b}})\,\mathbf{u}_{\mathbf{b}}].$$

Since the first vector $(\mathbf{a} \cdot \mathbf{u}_{\mathbf{b}})\,\mathbf{u}_{\mathbf{b}}$ is a scalar multiple of **b**, it is parallel to **b**. All we have to show now is that the second vector is perpendicular to **b**. We do this by showing that its dot product with $\mathbf{u}_{\mathbf{b}}$ is zero:

$$[\mathbf{a} - (\mathbf{a} \cdot \mathbf{u}_{\mathbf{b}})\,\mathbf{u}_{\mathbf{b}}] \cdot \mathbf{u}_{\mathbf{b}} = (\mathbf{a} \cdot \mathbf{u}_{\mathbf{b}}) - (\mathbf{a} \cdot \mathbf{u}_{\mathbf{b}})(\mathbf{u}_{\mathbf{b}} \cdot \mathbf{u}_{\mathbf{b}}) = 0. \quad \square$$

↑ $\mathbf{u}_{\mathbf{b}} \cdot \mathbf{u}_{\mathbf{b}} = \|\mathbf{u}_{\mathbf{b}}\|^2 = 1$

Example 7 Find $\text{comp}_{\mathbf{b}}\mathbf{a}$ and $\text{proj}_{\mathbf{b}}\mathbf{a}$ given that

$$\mathbf{a} = -2\mathbf{i} + \mathbf{j} + \mathbf{k} \quad \text{and} \quad \mathbf{b} = 4\mathbf{i} - 3\mathbf{j} + \mathbf{k}.$$

SOLUTION Since $\|\mathbf{b}\| = \sqrt{4^2 + (-3)^2 + 1^2} = \sqrt{26}$,

we have
$$\mathbf{u}_{\mathbf{b}} = \frac{\mathbf{b}}{\|\mathbf{b}\|} = \frac{1}{\sqrt{26}}(4\mathbf{i} - 3\mathbf{j} + \mathbf{k}).$$

Thus

$$\begin{aligned}\text{comp}_{\mathbf{b}}\mathbf{a} = \mathbf{a} \cdot \mathbf{u}_{\mathbf{b}} &= (-2\mathbf{i} + \mathbf{j} + \mathbf{k}) \cdot \frac{1}{\sqrt{26}}(4\mathbf{i} - 3\mathbf{j} + \mathbf{k}) \\ &= \frac{1}{\sqrt{26}}[(-2)(4) + (1)(-3) + (1)(1)] = -\frac{10}{\sqrt{26}} = -\frac{5}{13}\sqrt{26}\end{aligned}$$

and
$$\text{proj}_{\mathbf{b}}\mathbf{a} = (\text{comp}_{\mathbf{b}}\mathbf{a})\,\mathbf{u}_{\mathbf{b}} = -\tfrac{5}{13}(4\mathbf{i} - 3\mathbf{j} + \mathbf{k}). \quad \square$$

The following characterization of $\mathbf{a} \cdot \mathbf{b}$ is frequently used in physical applications:

(12.4.12)
$$\text{if } \mathbf{b} \neq \mathbf{0}, \quad \mathbf{a} \cdot \mathbf{b} = (\text{comp}_{\mathbf{b}}\mathbf{a})\|\mathbf{b}\|.$$

PROOF

$$\mathbf{a}\cdot\mathbf{b} = \left(\mathbf{a}\cdot\frac{\mathbf{b}}{\|\mathbf{b}\|}\right)\|\mathbf{b}\| = (\mathbf{a}\cdot\mathbf{u_b})\|\mathbf{b}\| = (\text{comp}_\mathbf{b}\,\mathbf{a})\|\mathbf{b}\|. \quad \square$$

For an arbitrary vector $\mathbf{a} = a_1\,\mathbf{i} + a_2\,\mathbf{j} + a_3\,\mathbf{k}$ we have

$$\text{comp}_\mathbf{i}\,\mathbf{a} = \mathbf{a}\cdot\mathbf{i} = a_1, \quad \text{comp}_\mathbf{j}\,\mathbf{a} = \mathbf{a}\cdot\mathbf{j} = a_2, \quad \text{comp}_\mathbf{k}\,\mathbf{a} = \mathbf{a}\cdot\mathbf{k} = a_3.$$

This agrees with our previous use of the term "component" (Section 12.3) and gives the identity

(12.4.13)
$$\boxed{\mathbf{a} = (\mathbf{a}\cdot\mathbf{i})\,\mathbf{i} + (\mathbf{a}\cdot\mathbf{j})\,\mathbf{j} + (\mathbf{a}\cdot\mathbf{k})\,\mathbf{k}.}$$

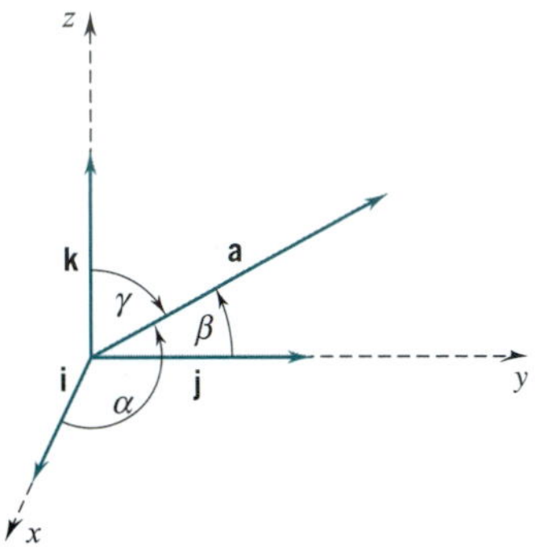

Figure 12.4.7

Direction Angles, Direction Cosines

In Figure 12.4.7 we show a nonzero vector $\mathbf{a}$. The angles α, β, γ that the vector makes with the unit coordinate vectors are called the *direction angles* of $\mathbf{a}$, and $\cos\alpha, \cos\beta, \cos\gamma$ are called the *direction cosines*. By (12.4.7)

$$\mathbf{a}\cdot\mathbf{i} = \|\mathbf{a}\|\cos\alpha, \quad \mathbf{a}\cdot\mathbf{j} = \|\mathbf{a}\|\cos\beta, \quad \mathbf{a}\cdot\mathbf{k} = \|\mathbf{a}\|\cos\gamma.$$

Thus by (12.4.13)

(12.4.14)
$$\boxed{\mathbf{a} = \|\mathbf{a}\|(\cos\alpha\,\mathbf{i} + \cos\beta\,\mathbf{j} + \cos\gamma\,\mathbf{k}).}$$

Taking the norm of both sides, we have

$$\|\mathbf{a}\| = \|\mathbf{a}\|\sqrt{\cos^2\alpha + \cos^2\beta + \cos^2\gamma}$$

and therefore

(12.4.15)
$$\boxed{\cos^2\alpha + \cos^2\beta + \cos^2\gamma = 1.}$$

The sum of the squares of the direction cosines is always 1.

For a unit vector $\mathbf{u}$, Equation 12.4.14 takes the form

(12.4.16)
$$\boxed{\mathbf{u} = \cos\alpha\,\mathbf{i} + \cos\beta\,\mathbf{j} + \cos\gamma\,\mathbf{k}.}$$

Thus, for a unit vector, the $\mathbf{i}$, $\mathbf{j}$, $\mathbf{k}$ components are simply the direction cosines.

Example 8 Find the unit vector with direction angles

$$\alpha = \tfrac{1}{4}\pi, \quad \beta = \tfrac{2}{3}\pi, \quad \gamma = \tfrac{1}{3}\pi.$$

What is the vector of norm 4 with these same direction angles?

SOLUTION The unit vector with these direction angles is

$$\cos\tfrac{1}{4}\pi\,\mathbf{i} + \cos\tfrac{2}{3}\pi\,\mathbf{j} + \cos\tfrac{1}{3}\pi\,\mathbf{k} = \tfrac{1}{2}\sqrt{2}\,\mathbf{i} - \tfrac{1}{2}\,\mathbf{j} + \tfrac{1}{2}\,\mathbf{k}.$$

The vector of norm 4 with these direction angles is

$$4(\tfrac{1}{2}\sqrt{2}\,\mathbf{i} - \tfrac{1}{2}\,\mathbf{j} + \tfrac{1}{2}\,\mathbf{k}) = 2\sqrt{2}\,\mathbf{i} - 2\,\mathbf{j} + 2\,\mathbf{k}. \quad \square$$

Example 9 Find the direction cosines of $\mathbf{a} = 2\mathbf{i} + 3\mathbf{j} - 6\mathbf{k}$. What are the direction angles?

SOLUTION Here $||\mathbf{a}|| = \sqrt{2^2 + 3^2 + (-6)^2} = 7$

so that $$2 = 7\cos\alpha, \qquad 3 = 7\cos\beta, \qquad -6 = 7\cos\gamma$$

and $$\cos\alpha = \tfrac{2}{7}, \qquad \cos\beta = \tfrac{3}{7}, \qquad \cos\gamma = -\tfrac{6}{7}.$$

Since angles between vectors are measured in radians and taken from 0 to π,

$$\alpha = \cos^{-1}\tfrac{2}{7} \cong 1.28 \text{ radians}, \quad \beta = \cos^{-1}\tfrac{3}{7} \cong 1.13 \text{ radians},$$
$$\gamma = \cos^{-1}\left(-\tfrac{6}{7}\right) \cong 2.60 \text{ radians}. \quad \square$$

Proving the Triangle Inequality

By taking absolute values, the relation

$$\mathbf{a} \cdot \mathbf{b} = ||\mathbf{a}||\,||\mathbf{b}|| \cos\theta$$

gives

(12.4.17) $$|\mathbf{a} \cdot \mathbf{b}| \leq ||\mathbf{a}||\,||\mathbf{b}||. \qquad (\cos\theta \leq 1)$$

This inequality, called *Schwarz's inequality*, enables us to give a simple proof of the triangle inequality

$$||\mathbf{a} + \mathbf{b}|| \leq ||\mathbf{a}|| + ||\mathbf{b}||.$$

PROOF

$$\begin{aligned}
||\mathbf{a} + \mathbf{b}||^2 &= (\mathbf{a} + \mathbf{b}) \cdot (\mathbf{a} + \mathbf{b}) \\
&= (\mathbf{a} \cdot \mathbf{a}) + (\mathbf{b} \cdot \mathbf{a}) + (\mathbf{a} \cdot \mathbf{b}) + (\mathbf{b} \cdot \mathbf{b}) \\
&= ||\mathbf{a}||^2 + 2(\mathbf{a} \cdot \mathbf{b}) + ||\mathbf{b}||^2 \\
&\leq ||\mathbf{a}||^2 + 2|\mathbf{a} \cdot \mathbf{b}| + ||\mathbf{b}||^2 \qquad (\mathbf{a} \cdot \mathbf{b} \leq |\mathbf{a} \cdot \mathbf{b}|) \\
&\leq ||\mathbf{a}||^2 + 2||\mathbf{a}||\,||\mathbf{b}|| + ||\mathbf{b}||^2 = (||\mathbf{a}|| + ||\mathbf{b}||)^2.
\end{aligned}$$

by Schwarz's inequality ⟶ (applies to the last step)

Taking square roots, we have

$$||\mathbf{a} + \mathbf{b}|| \leq ||\mathbf{a}|| + ||\mathbf{b}||. \quad \square$$

It is worth remarking that Schwarz's inequality, and hence the triangle inequality, can be proved by purely algebraic methods. (See Exercise 56.)

EXERCISES 12.4

Find $\mathbf{a} \cdot \mathbf{b}$.

1. $\mathbf{a} = (2, -3, 1)$, $\mathbf{b} = (-2, 0, 3)$.

2. $\mathbf{a} = (4, 2, -1)$, $\mathbf{b} = (-2, 2, 1)$.

3. $\mathbf{a} = (2, -4, 0)$, $\mathbf{b} = (1, \frac{1}{2}, 0)$.

4. $\mathbf{a} = (-2, 0, 5)$, $\mathbf{b} = (3, 0, 1)$.

5. $\mathbf{a} = 2\mathbf{i} + \mathbf{j} - 2\mathbf{k}$, $\mathbf{b} = \mathbf{i} + \mathbf{j} + 2\mathbf{k}$.

6. $\mathbf{a} = 2\mathbf{i} + 3\mathbf{j} + \mathbf{k}$, $\mathbf{b} = \mathbf{i} + 4\mathbf{j}$.

Simplify.

7. $(3\mathbf{a} \cdot \mathbf{b}) - (\mathbf{a} \cdot 2\mathbf{b})$.

8. $\mathbf{a} \cdot (\mathbf{a} - \mathbf{b}) + \mathbf{b} \cdot (\mathbf{b} + \mathbf{a})$.

9. $(\mathbf{a} - \mathbf{b}) \cdot \mathbf{c} + \mathbf{b} \cdot (\mathbf{c} + \mathbf{a})$.

10. $\mathbf{a} \cdot (\mathbf{a} + 2\mathbf{c}) + (2\mathbf{b} - \mathbf{a}) \cdot (\mathbf{a} + 2\mathbf{c}) - 2\mathbf{b} \cdot (\mathbf{a} + 2\mathbf{c})$.

11. Taking

$$\mathbf{a} = 2\mathbf{i} + \mathbf{j}, \qquad \mathbf{b} = 3\mathbf{i} - \mathbf{j} + 2\mathbf{k}, \qquad \mathbf{c} = 4\mathbf{i} + 3\mathbf{k},$$

calculate:

(a) the three dot products $\mathbf{a} \cdot \mathbf{b}$, $\mathbf{a} \cdot \mathbf{c}$, $\mathbf{b} \cdot \mathbf{c}$;

(b) the cosines of the angles between these vectors;

(c) the component of $\mathbf{a}$ (i) in the $\mathbf{b}$ direction, (ii) in the $\mathbf{c}$ direction;

(d) the projection of $\mathbf{a}$ (i) in the $\mathbf{b}$ direction, (ii) in the $\mathbf{c}$ direction.

12. Repeat Exercise 11 with $\mathbf{a} = \mathbf{j} + 3\mathbf{k}$, $\mathbf{b} = 2\mathbf{i} - \mathbf{j} + 2\mathbf{k}$, $\mathbf{c} = 3\mathbf{i} - \mathbf{k}$.

13. Find the unit vector with direction angles $\frac{1}{3}\pi, \frac{1}{4}\pi, \frac{2}{3}\pi$.

14. Find the vector of norm 2 with direction angles $\frac{1}{4}\pi, \frac{1}{4}\pi, \frac{1}{2}\pi$.

15. Find the angle between the vectors $3\mathbf{i}-\mathbf{j}-2\mathbf{k}$ and $\mathbf{i}+2\mathbf{j}-3\mathbf{k}$.

16. Find the angle between the vectors $2\mathbf{i} - 3\mathbf{j} + \mathbf{k}$ and $-3\mathbf{i} + \mathbf{j} + 9\mathbf{k}$.

17. Find the direction angles of the vector $\mathbf{i} - \mathbf{j} + \sqrt{2}\,\mathbf{k}$.

18. Find the direction angles of the vector $\mathbf{i} - \sqrt{3}\,\mathbf{k}$.

Estimate the angle between the vectors. Express your answers in radians rounded to the nearest hundredth of a radian, and in degrees to the nearest tenth of a degree.

19. $\mathbf{a} = (3, 1, -1)$, $\mathbf{b} = (-2, 1, 4)$.

20. $\mathbf{a} = (-2, -3, 0)$, $\mathbf{b} = (-6, 0, 4)$.

21. $\mathbf{a} = -\mathbf{i} + 2\mathbf{k}$, $\mathbf{b} = 3\mathbf{i} + 4\mathbf{j} - 5\mathbf{k}$.

22. $\mathbf{a} = -3\mathbf{i} + \mathbf{j} - \mathbf{k}$, $\mathbf{b} = \mathbf{i} - \mathbf{j}$.

23. Use a CAS to determine the angles and the perimeter of the triangle with vertices $P(1, 3, -2)$, $Q(3, 1, 2)$, $R(2, -3, 1)$.

24. Use a CAS to find the direction cosines and the direction angles of the vector from $P(5, 7, -2)$ to $Q(-3, 4, 1)$.

Find the direction cosines and direction angles of the vector. Express the angles in degrees rounded to the nearest tenth of a degree.

25. $\mathbf{a} = (1, 2, 2)$.

26. $\mathbf{a} = (2, 6, -1)$.

27. $\mathbf{a} = 3\mathbf{i} + 12\mathbf{j} + 4\mathbf{k}$.

28. $\mathbf{a} = 3\mathbf{i} + 5\mathbf{j} - 4\mathbf{k}$.

29. Find all the numbers x for which

$$2\,\mathbf{i} + 5\,\mathbf{j} + 2x\,\mathbf{k} \perp 6\,\mathbf{i} + 4\,\mathbf{j} - x\,\mathbf{k}$$

30. Find all the numbers x for which

$$(x\mathbf{i} + 11\mathbf{j} - 3\mathbf{k}) \perp (2x\mathbf{i} - x\mathbf{j} - 5\mathbf{k}).$$

31. Find all the numbers x for which the angle between $\mathbf{c} = x\mathbf{i} + \mathbf{j} + \mathbf{k}$ and $\mathbf{d} = \mathbf{i} + x\mathbf{j} + \mathbf{k}$ is $\frac{1}{3}\pi$.

32. Set $\mathbf{a} = \mathbf{i} + x\mathbf{j} + \mathbf{k}$ and $\mathbf{b} = 2\mathbf{i} - \mathbf{j} + y\mathbf{k}$. Compute all values of x and y for which $\mathbf{a} \perp \mathbf{b}$ and $||\mathbf{a}|| = ||\mathbf{b}||$.

33. (a) Show that $\frac{1}{4}\pi, \frac{1}{6}\pi, \frac{2}{3}\pi$ cannot be the direction angles of a vector.

(b) Show that, if $\mathbf{a} = a_1\mathbf{i} + a_2\mathbf{j} + a_3\mathbf{k}$ has direction angles $\alpha, \frac{1}{4}\pi, \frac{1}{4}\pi$, then $a_1 = 0$.

34. If a vector has direction angles $\alpha = \pi/3$, $\beta = \pi/4$, find the third direction angle γ.

35. What are the direction angles of $-\mathbf{a}$ if the direction angles of $\mathbf{a}$ are α, β, γ?

36. Suppose that the direction angles of a vector are equal. What are the angles?

37. Find the unit vectors $\mathbf{u}$ that are perpendicular to both $\mathbf{i} + 2\mathbf{j} + \mathbf{k}$ and $3\mathbf{i} - 4\mathbf{j} + 2\mathbf{k}$.

38. Find two mutually perpendicular unit vectors that are perpendicular to $2\mathbf{i} + 3\mathbf{j}$.

39. Find the angle between the diagonal of a cube and one of the edges.

40. Find the angle between the diagonal of a cube and the diagonal of one of the faces.

41. Show that

(a) $\text{proj}_{\mathbf{b}}\alpha\mathbf{a} = \alpha\,\text{proj}_{\mathbf{b}}\mathbf{a}$ for all real α, and

(b) $\text{proj}_{\mathbf{b}}(\mathbf{a} + \mathbf{c}) = \text{proj}_{\mathbf{b}}\mathbf{a} + \text{proj}_{\mathbf{b}}\mathbf{c}$.

42. Show that

(a) $\text{proj}_{\beta\mathbf{b}}\mathbf{a} = \text{proj}_{\mathbf{b}}\mathbf{a}$ for all real $\beta \neq 0$, but

(b) $\text{comp}_{\beta\mathbf{b}}\mathbf{a} = \begin{cases} \text{comp}_{\mathbf{b}}\mathbf{a}, & \text{for } \beta > 0 \\ -\text{comp}_{\mathbf{b}}\mathbf{a}, & \text{for } \beta < 0. \end{cases}$

43. (a) (*Important*) Let $\mathbf{a} \neq \mathbf{0}$. Show that $\mathbf{a} \cdot \mathbf{b} = \mathbf{a} \cdot \mathbf{c}$ does not necessarily imply that $\mathbf{b} = \mathbf{c}$, but only that $\mathbf{b}$ and $\mathbf{c}$ have the same projection on $\mathbf{a}$. Draw a figure illustrating this for $\mathbf{b}$ and $\mathbf{c}$ different from $\mathbf{0}$.

(b) Show that if $\mathbf{u} \cdot \mathbf{b} = \mathbf{u} \cdot \mathbf{c}$ for all unit vectors $\mathbf{u}$, then $\mathbf{b} = \mathbf{c}$. HINT: Consider the unit coordinate vectors.

44. What can you conclude about $\mathbf{a}$ and $\mathbf{b}$ given that

(a) $||\mathbf{a}||^2 + ||\mathbf{b}||^2 = ||\mathbf{a} + \mathbf{b}||^2$?

(b) $||\mathbf{a}||^2 + ||\mathbf{b}||^2 = ||\mathbf{a} - \mathbf{b}||^2$?

HINT: Draw figures.

45. (a) Show that

$$4(\mathbf{a} \cdot \mathbf{b}) = ||\mathbf{a} + \mathbf{b}||^2 - ||\mathbf{a} - \mathbf{b}||^2.$$

(b) Use part (a) to verify that

$$\mathbf{a} \perp \mathbf{b} \quad \text{iff} \quad ||\mathbf{a}+\mathbf{b}|| = ||\mathbf{a}-\mathbf{b}||.$$

(c) Show that, if **a** and **b** are nonzero vectors such that

$$(\mathbf{a}+\mathbf{b}) \perp (\mathbf{a}-\mathbf{b}) \quad \text{and} \quad ||\mathbf{a}+\mathbf{b}|| = ||\mathbf{a}-\mathbf{b}||,$$

then the parallelogram generated by **a** and **b** is a square.

46. Under what conditions does $|\mathbf{a} \cdot \mathbf{b}| = ||\mathbf{a}||\,||\mathbf{b}||$?

47. Given two vectors **a** and **b**, prove the *parallelogram law*:

$$||\mathbf{a}+\mathbf{b}||^2 + ||\mathbf{a}-\mathbf{b}||^2 = 2||\mathbf{a}||^2 + 2||\mathbf{b}||^2.$$

Geometric interpretation: The sum of the squares of the lengths of the diagonals of a parallelogram equals the sum of the squares of the lengths of the four sides. See the figure.

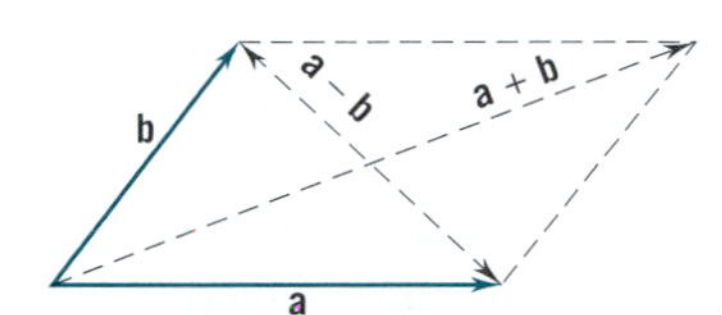

48. A *rhombus* is a parallelogram with sides of equal length. Show that the diagonals of a rhombus are perpendicular.

49. Let **a** and **b** be nonzero vectors. Show that the vector $\mathbf{c} = ||\mathbf{b}||\,\mathbf{a} + ||\mathbf{a}||\,\mathbf{b}$ bisects the angle between **a** and **b**.

50. Let θ be the angle between **a** and **b**, and let β be a negative number. Use the dot product to compute the angle between **a** and $\beta\mathbf{b}$ in terms of θ. Draw a figure to verify your answer geometrically.

51. Show that if $\mathbf{a} \perp \mathbf{b}$ and $\mathbf{a} \perp \mathbf{c}$, then $\mathbf{a} \perp (\alpha\mathbf{b} + \beta\mathbf{c})$ for all real α, β.

52. Show that, if $\mathbf{a} \,||\, \mathbf{b}$ and $\mathbf{a} \,||\, \mathbf{c}$, then $\mathbf{a} \,||\, (\alpha\mathbf{b} + \beta\mathbf{c})$ for all real α, β. ($\mathbf{a} \,||\, \mathbf{b}$ is used to indicate that **a** is parallel to **b**)

53. (*Important*) Show that, if **b** is a nonzero vector, then every vector **a** can be written in a unique manner as the sum of a vector $\mathbf{a}_{||}$ parallel to **b** and a vector $\mathbf{a}_{\perp}$ perpendicular to **b**:

$$\mathbf{a} = \mathbf{a}_{||} + \mathbf{a}_{\perp}.$$

54. let $r = f(\theta)$ be the polar equation of a curve in the plane and let

$$\mathbf{u}_r = (\cos\theta)\mathbf{i} + (\sin\theta)\mathbf{j} \qquad \mathbf{u}_\theta = (-\sin\theta)\mathbf{i} + (\cos\theta)\mathbf{j}.$$

(a) Show that $\mathbf{u}_r$ and $\mathbf{u}_\theta$ are unit vectors and that they are perpendicular.

(b) Let $P[r, \theta]$ be a point on the curve. Show that $\mathbf{u}_r$ has the same direction as the vector $\overrightarrow{OP}$ and that $\mathbf{u}_\theta$ is 90° counterclockwise from $\mathbf{u}_r$.

55. Two points on a sphere are called *antipodal* if they are opposite endpoints of a diameter. Show that, if P_1 and P_2 are antipodal points on a sphere and Q is any other point on the sphere, then $\overrightarrow{P_1Q} \perp \overrightarrow{P_2Q}$.

56. (*Important*) Give an algebraic proof of Schwarz's inequality $|\mathbf{a} \cdot \mathbf{b}| \le ||\mathbf{a}||\,||\mathbf{b}||$. HINT: If $\mathbf{b} = 0$, the inequality is trivial, so assume $\mathbf{b} \neq \mathbf{0}$. Note that for any number λ we have $||\mathbf{a} - \lambda\mathbf{b}||^2 \ge 0$. First expand this inequality using the fact that $||\mathbf{a} - \lambda\mathbf{b}||^2 = (\mathbf{a} - \lambda\mathbf{b}) \cdot (\mathbf{a} - \lambda\mathbf{b})$. After collecting terms, make the special choice $\lambda = (\mathbf{a} \cdot \mathbf{b})/||\mathbf{b}||^2$ and see what happens.

■ PROJECT 12.4 Work

If a constant force **F** is applied to an object moving in a straight line throughout a displacement **r** (see the figure), then the work done by **F** is defined by the equation

$$W = (\text{comp}_\mathbf{r}\,\mathbf{F})||\mathbf{r}||$$

This generalizes (6.5.1).

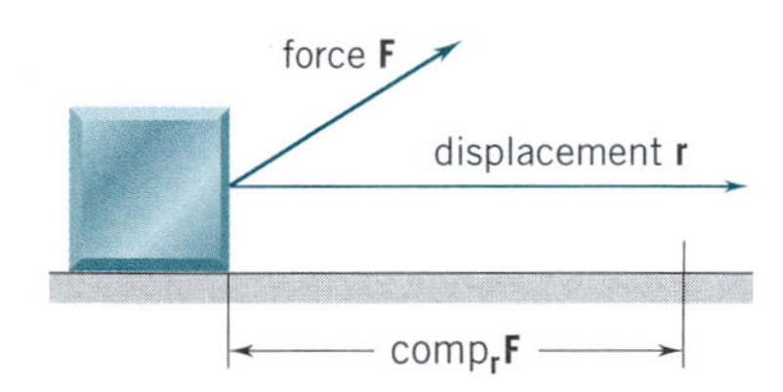

If **F** is measured in pounds and distance in feet, then work is measured in *foot-pounds*; if **F** is measured in newtons and distance in meters, then work is measured in *newton-meters* or *joules*.

Problem 1. Let the force **F** be applied throughout a displacement **r**.

a. Express the work done by **F** as a dot product.

b. What is the work done by **F** if $\mathbf{F} \perp \mathbf{r}$?

c. Show that the work done by $\mathbf{F} = ||\mathbf{F}||\,\mathbf{i}$ applied throughout the displacement $\mathbf{r} = (b - a)\mathbf{i}$ reduces to (6.5.1).

Problem 2.

(a) A sled is pulled along level ground by a force of 15 Newtons along a rope that makes an angle of 35° with the ground. Find the work done by the force in pulling the sled 50 meters. See Figure A.

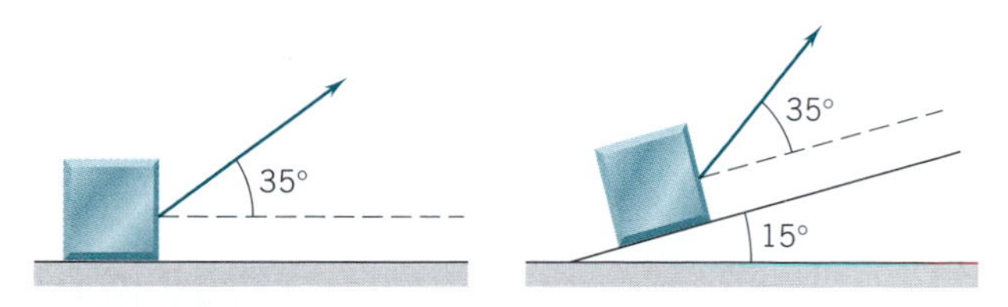

Figure A Figure B

(b) Suppose the same sled is pulled 50 meters up a hill that makes an angle of 15° with level ground. Find the work done by the force in this case. See Figure B.

Problem 3. A wooden crate is pulled along a level floor by a rope that makes an angle of 40° with the floor. If the force of friction (which acts in a direction opposite to motion) between the carton and the floor is 50 pounds, what is the minimum force that must be applied to the rope to move the crate?

Problem 4. Two forces of the same magnitude, $\mathbf{F}_1$ and $\mathbf{F}_2$, are applied throughout a displacement $\mathbf{r}$ at angles θ_1 and θ_2, respectively. Compare the work done by $\mathbf{F}_1$ to that done by $\mathbf{F}_2$ if

(a) $\theta_1 = -\theta_2$.

(b) $\theta_1 = \pi/3$ and $\theta_2 = \pi/6$.

Problem 5. What is the total work done by a constant force $\mathbf{F}$ if the object to which it is applied moves around a triangle? Justify your answer.

■ 12.5 THE CROSS PRODUCT

Everything that we have done with vectors so far (other than draw pictures) can be generalized to higher dimensions. We come now to a notion that is particular to three-dimensional space and cannot be generalized to higher dimensions.

Definition of the Cross Product

While the dot product $\mathbf{a} \cdot \mathbf{b}$ is a scalar (and as such is sometimes called the scalar product of $\mathbf{a}$ and $\mathbf{b}$), the cross product $\mathbf{a} \times \mathbf{b}$ is a vector (sometimes called the *vector product* of $\mathbf{a}$ and $\mathbf{b}$). What is the vector $\mathbf{a} \times \mathbf{b}$? We could directly write down a formula that gives the components of $\mathbf{a} \times \mathbf{b}$ in terms of the components of $\mathbf{a}$ and $\mathbf{b}$, but at this stage that would reveal little. Instead we will begin geometrically. We will define $\mathbf{a} \times \mathbf{b}$ by giving its direction and its magnitude.

The Direction of $\mathbf{a} \times \mathbf{b}$ If the vectors $\mathbf{a}$ and $\mathbf{b}$ are not parallel, they determine a plane. The vector $\mathbf{a} \times \mathbf{b}$ is perpendicular to this plane and is directed in such a way that (like $\mathbf{i}, \mathbf{j}, \mathbf{k}$) the vectors $\mathbf{a}, \mathbf{b}, \mathbf{a} \times \mathbf{b}$ form a right-handed triple. (See Figure 12.5.1.) If the index finger of the right hand points along $\mathbf{a}$ and the middle finger points along $\mathbf{b}$, then the thumb will point in the direction of $\mathbf{a} \times \mathbf{b}$. NOTE: In saying that $\mathbf{a}, \mathbf{b}, \mathbf{a} \times \mathbf{b}$ form a right-handed triple, we do not require the vectors $\mathbf{a}$ and $\mathbf{b}$ to be perpendicular (like $\mathbf{i}$ and $\mathbf{j}$); the general notion of a "right-handed triple" does not require the vectors involved to be perpendicular to each other.

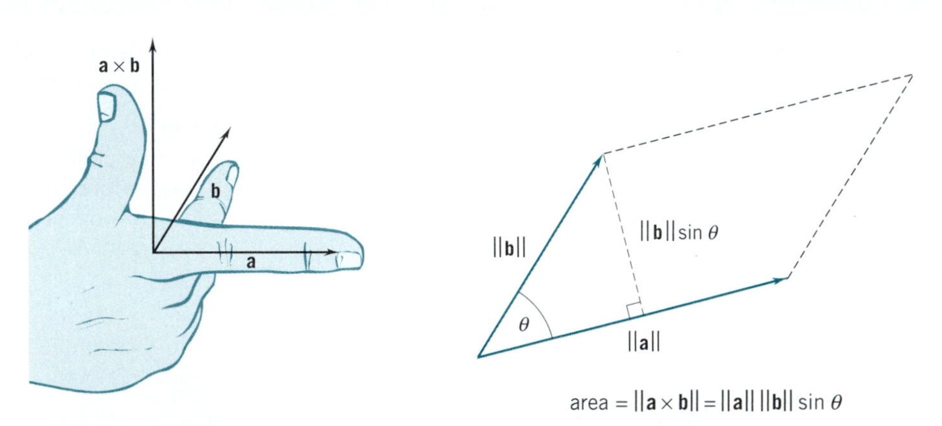

Figure 12.5.1

Figure 12.5.2

The Magnitude of $\mathbf{a} \times \mathbf{b}$ If $\mathbf{a}$ and $\mathbf{b}$ are not parallel, they form the sides of a parallelogram. (See Figure 12.5.2.) The magnitude of $\mathbf{a} \times \mathbf{b}$ is the area of this parallelogram: $\|\mathbf{a}\|\,\|\mathbf{b}\| \sin\theta$. (Recall that the area of a parallelogram with base b and height h is given by $A = bh$.)

One more point. What if $\mathbf{a}$ and $\mathbf{b}$ are parallel? Then there is no parallelogram and we define $\mathbf{a} \times \mathbf{b}$ to be $\mathbf{0}$.

We summarize all this below.

DEFINITION 12.5.1

If **a** and **b** are not parallel, then $\mathbf{a} \times \mathbf{b}$ is the vector with the following properties:

1. $\mathbf{a} \times \mathbf{b}$ is perpendicular to the plane of **a** and **b**.
2. $\mathbf{a}, \mathbf{b}, \mathbf{a} \times \mathbf{b}$ form a right-handed triple.
3. $\|\mathbf{a} \times \mathbf{b}\| = \|\mathbf{a}\|\,\|\mathbf{b}\| \sin\theta$, where θ is the angle between **a** and **b**.

If **a** and **b** are parallel, then $\mathbf{a} \times \mathbf{b} = 0$.

Properties of Right-Handed Triples

I. Note first of all that if $(\mathbf{a}, \mathbf{b}, \mathbf{c})$ forms a right-handed triple, then $(\mathbf{b}, \mathbf{c}, \mathbf{a})$ and $(\mathbf{c}, \mathbf{a}, \mathbf{b})$ also form right-handed triples (Figure 12.5.3). To maintain right-handedness we don't have to keep the vectors in the same order, but we do have to keep them in the *same cyclic order*:

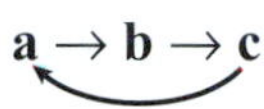

Alter the cyclic order and you reverse the orientation.

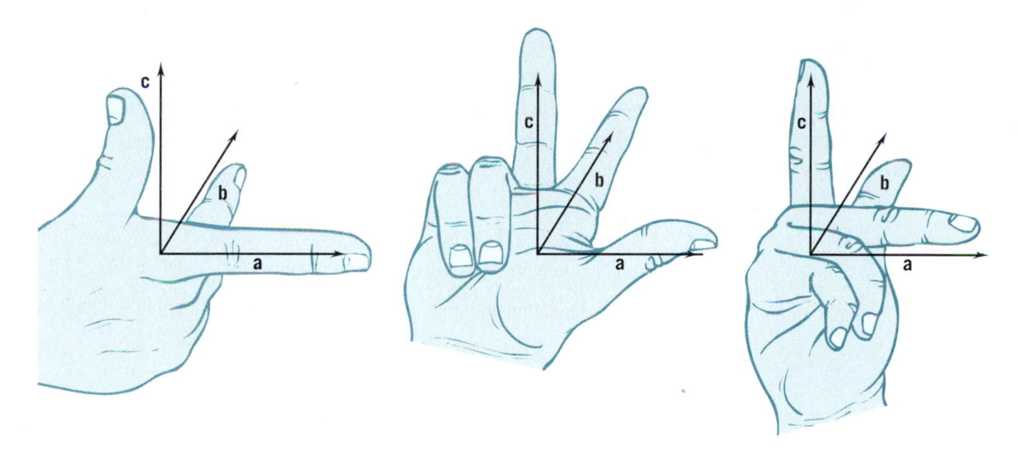

Figure 12.5.3

II. A triple $(\mathbf{a}, \mathbf{b}, \mathbf{c})$ is right-handed iff **c** and $\mathbf{a} \times \mathbf{b}$ lie on the same side of the plane determined by **a** and **b** (Figure 12.5.4). This means that $(\mathbf{a}, \mathbf{b}, \mathbf{c})$ is right-handed iff $(\mathbf{a} \times \mathbf{b}) \cdot \mathbf{c} > 0$. (Explain)

III. It follows from II that, if $(\mathbf{a}, \mathbf{b}, \mathbf{c})$ is right-handed, then $(\mathbf{a}, \mathbf{b}, -\mathbf{c})$ is not right-handed. Similarly, $(-\mathbf{a}, \mathbf{b}, \mathbf{c})$ and $(\mathbf{a}, -\mathbf{b}, \mathbf{c})$ are not right-handed. However, multiplication by positive scalars does maintain right-handedness: if $(\mathbf{a}, \mathbf{b}, \mathbf{c})$ is right-handed and α, β, γ are positive, then $(\alpha\mathbf{a}, \beta\mathbf{b}, \gamma\mathbf{c})$ is also right-handed.

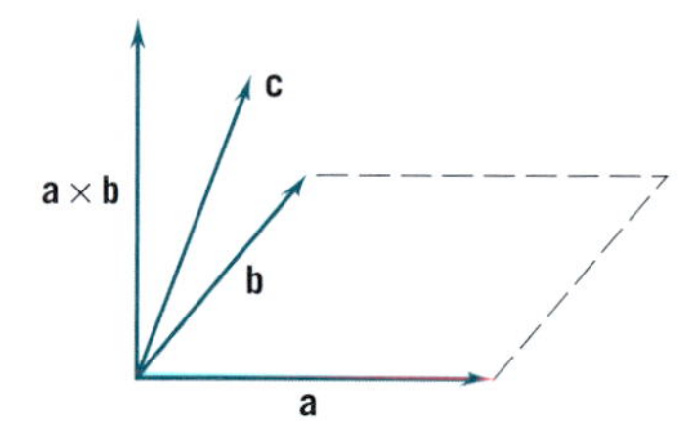

Figure 12.5.4

Properties of the Cross Product

The cross product is *anticommutative*:

(12.5.2) $$\mathbf{b} \times \mathbf{a} = -(\mathbf{a} \times \mathbf{b}).$$

To see this, note that both vectors are perpendicular to the plane determined by $\mathbf{a}$ and $\mathbf{b}$ and both have the same norm. Thus $\mathbf{b} \times \mathbf{a} = \pm(\mathbf{a} \times \mathbf{b})$. That the minus sign holds, not the plus sign, follows from observing that $\mathbf{b}, \mathbf{a}, \mathbf{b} \times \mathbf{a}$ is a right-handed triple and that $\mathbf{b}, \mathbf{a}, \mathbf{a} \times \mathbf{b}$ is not right-handed (since $\mathbf{a}, \mathbf{b}, \mathbf{a} \times \mathbf{b}$ is right-handed).

Scalars can be factored:

(12.5.3)
$$\alpha\mathbf{a} \times \beta\mathbf{b} = \alpha\beta(\mathbf{a} \times \mathbf{b}).$$

If α or β is zero, the result is obvious. We will assume that α and β are both nonzero. In this case the two vectors are perpendicular to $\mathbf{a}$, perpendicular to $\mathbf{b}$, and have the same norm. Thus $\alpha\mathbf{a} \times \beta\mathbf{b} = \pm\alpha\beta(\mathbf{a} \times \mathbf{b})$. That the positive sign holds comes from noting that $\alpha\mathbf{a}, \beta\mathbf{b}, \alpha\beta(\mathbf{a} \times \mathbf{b})$ is a right-handed triple. This is obvious if α and β are both positive. If not, two of the three coefficients $\alpha, \beta, \alpha\beta$ are negative and the other is positive. In this case, the first minus sign reverses the orientation (that is, changes right-handed to left-handed) but the second one restores it.

Finally, there are two distributive laws, the verification of which we postpone for a moment.

(12.5.4)
$$\mathbf{a} \times (\mathbf{b} + \mathbf{c}) = (\mathbf{a} \times \mathbf{b}) + (\mathbf{a} \times \mathbf{c}),$$
$$(\mathbf{a} + \mathbf{b}) \times \mathbf{c} = (\mathbf{a} \times \mathbf{c}) + (\mathbf{b} \times \mathbf{c}).$$

The Scalar Triple Product

Earlier we saw that $\mathbf{a}, \mathbf{b}, \mathbf{c}$ is a right-handed triple iff $(\mathbf{a} \times \mathbf{b}) \cdot \mathbf{c} > 0$. The expression $(\mathbf{a} \times \mathbf{b}) \cdot \mathbf{c}$ is called a *scalar triple product*. The absolute value of this number (it is a number and not a vector) has geometric significance. To describe it we refer to Figure 12.5.5. There you see a parallelepiped with edges $\mathbf{a}, \mathbf{b}, \mathbf{c}$. The absolute value of the scalar triple product gives the volume of that parallelepiped:

(12.5.5)
$$V = |(\mathbf{a} \times \mathbf{b}) \cdot \mathbf{c}|.$$

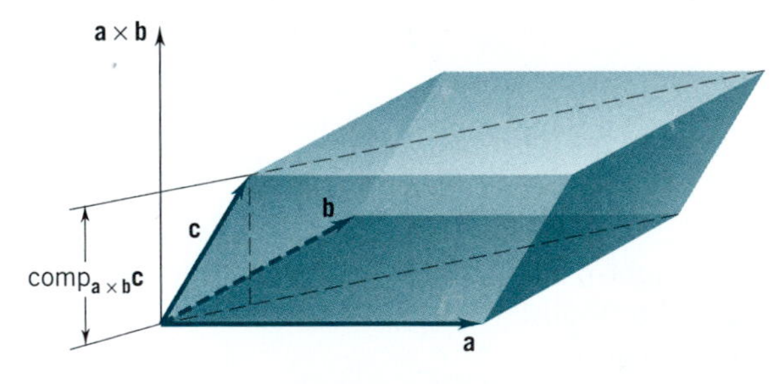

Figure 12.5.5

PROOF The area of the base is $\|\mathbf{a} \times \mathbf{b}\|$. The height is $|\text{comp}_{\mathbf{a}\times\mathbf{b}}\mathbf{c}|$. Therefore

$$V = |\text{comp}_{\mathbf{a}\times\mathbf{b}}\,\mathbf{c}|\,\|\mathbf{a} \times \mathbf{b}\| = |(\mathbf{a} \times \mathbf{b}) \cdot \mathbf{c}|. \quad \square$$

↑ (12.4.12)

Of course, we could have formed the same parallelogram using a different base (for example, using the vectors $\mathbf{c}$ and $\mathbf{a}$) with a correspondingly different height

($\text{comp}_{\mathbf{c}\times\mathbf{a}}\, \mathbf{b}$). Therefore

$$|(\mathbf{a} \times \mathbf{b}) \cdot \mathbf{c}| = |(\mathbf{c} \times \mathbf{a}) \cdot \mathbf{b}| = |(\mathbf{b} \times \mathbf{c}) \cdot \mathbf{a}|.$$

Since the $\mathbf{a}, \mathbf{b}, \mathbf{c}$ appear in the same cyclic order, the expressions inside the absolute value signs all have the same sign (Property II of right-handed triples). Therefore

(12.5.6) $$(\mathbf{a} \times \mathbf{b}) \cdot \mathbf{c} = (\mathbf{c} \times \mathbf{a}) \cdot \mathbf{b} = (\mathbf{b} \times \mathbf{c}) \cdot \mathbf{a}.$$

Verification of the Distributive Laws

We will verify the first distributive law,

$$\mathbf{a} \times (\mathbf{b} + \mathbf{c}) = (\mathbf{a} \times \mathbf{b}) + (\mathbf{a} \times \mathbf{c}).$$

The second follows readily from this one. The argument is left to you as an exercise.

Take an arbitrary vector $\mathbf{r}$ and form the dot product $[\mathbf{a} \times (\mathbf{b} + \mathbf{c})] \cdot \mathbf{r}$. We can then write

$$\begin{aligned}
[\mathbf{a} \times (\mathbf{b} + \mathbf{c})] \cdot \mathbf{r} &= (\mathbf{r} \times \mathbf{a}) \cdot (\mathbf{b} + \mathbf{c}) && (12.5.6) \\
&= [(\mathbf{r} \times \mathbf{a}) \cdot \mathbf{b}] + [(\mathbf{r} \times \mathbf{a}) \cdot \mathbf{c}] && (12.4.6) \\
&= [(\mathbf{a} \times \mathbf{b}) \cdot \mathbf{r}] + [(\mathbf{a} \times \mathbf{c}) \cdot \mathbf{r}] && (12.5.6) \\
&= [(\mathbf{a} \times \mathbf{b})] + (\mathbf{a} \times \mathbf{c})] \cdot \mathbf{r}. && (12.4.6)
\end{aligned}$$

Since this holds true for all vectors $\mathbf{r}$, it holds true for $\mathbf{i}, \mathbf{j}, \mathbf{k}$ and proves that

$$\mathbf{a} \times (\mathbf{b} + \mathbf{c}) = (\mathbf{a} \times \mathbf{b}) + (\mathbf{a} \times \mathbf{c}).$$

The Components of $\mathbf{a} \times \mathbf{b}$

You have learned a lot about cross products, but you still have not seen $\mathbf{a} \times \mathbf{b}$ expressed in terms of the components of $\mathbf{a}$ and $\mathbf{b}$. To derive the formula that does this, we need to observe one more fact, one which follows from the definition of cross product:

(12.5.7) $$\mathbf{i} \times \mathbf{j} = \mathbf{k}, \quad \mathbf{j} \times \mathbf{k} = \mathbf{i}, \quad \mathbf{k} \times \mathbf{i} = \mathbf{j}. \qquad \text{(Figure 12.5.6)}$$

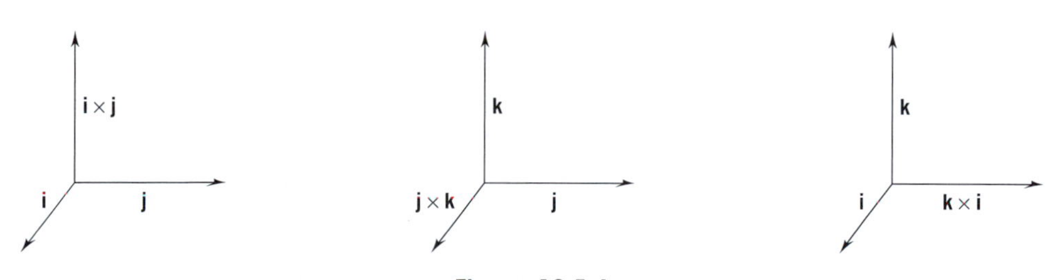

Figure 12.5.6

One way to remember these products is to arrange $\mathbf{i}, \mathbf{j}, \mathbf{k}$ in cyclic order, $\mathbf{i} \rightarrow \mathbf{j} \rightarrow \mathbf{k}$, and note that

$$[\text{each coordinate unit vector}] \times [\text{the next one}] = [\text{the third one}].$$

THEOREM 12.5.8

For vectors $\mathbf{a} = a_1\mathbf{i} + a_2\mathbf{j} + a_3\mathbf{k}$ and $\mathbf{b} = b_1\mathbf{i} + b_2\mathbf{j} + b_3\mathbf{k}$,

$$\mathbf{a} \times \mathbf{b} = (a_2b_3 - a_3b_2)\mathbf{i} - (a_1b_3 - a_3b_1)\mathbf{j} + (a_1b_2 - a_2b_1)\mathbf{k}.$$

Those of you who have studied some linear algebra will recognize that the jumble of symbols we have just written down for $\mathbf{a} \times \mathbf{b}$ can be elegantly summarized by the use of determinants. Here is Theorem 12.5.8 stated in terms of determinants. (If you are not familiar with determinants, see Appendix A-2.)

THEOREM 12.5.8′

For vectors $\mathbf{a} = a_1\mathbf{i} + a_2\mathbf{j} + a_3\mathbf{k}$ and $\mathbf{b} = b_1\mathbf{i} + b_2\mathbf{j} + b_3\mathbf{k}$,

$$\mathbf{a} \times \mathbf{b} = \begin{vmatrix} \mathbf{i} & \mathbf{j} & \mathbf{k} \\ a_1 & a_2 & a_3 \\ b_1 & b_2 & b_3 \end{vmatrix} = \begin{vmatrix} a_2 & a_3 \\ b_2 & b_3 \end{vmatrix} \mathbf{i} - \begin{vmatrix} a_1 & a_3 \\ b_1 & b_3 \end{vmatrix} \mathbf{j} + \begin{vmatrix} a_1 & a_2 \\ b_1 & b_2 \end{vmatrix} \mathbf{k}.$$

(The 3×3 determinant with $\mathbf{i}, \mathbf{j}, \mathbf{k}$ in the top row is there only as a mnemonic device.)

PROOF The hard work has all been done. With what you know about cross products now, the proof is just a matter of algebraic manipulation:

$$\begin{aligned}
\mathbf{a} \times \mathbf{b} &= (a_1\mathbf{i} + a_2\mathbf{j} + a_3\mathbf{k}) \times (b_1\mathbf{i} + b_2\mathbf{j} + b_3\mathbf{k}) \\
&= a_1b_2(\mathbf{i} \times \mathbf{j}) + a_1b_3(\mathbf{i} \times \mathbf{k}) + a_2b_1(\mathbf{j} \times \mathbf{i}) + a_2b_3(\mathbf{j} \times \mathbf{k}) + a_3b_1(\mathbf{k} \times \mathbf{i}) + a_3b_2(\mathbf{k} \times \mathbf{j}) \\
&\quad \uparrow\!\text{—} \ \mathbf{i} \times \mathbf{i} = \mathbf{j} \times \mathbf{j} = \mathbf{k} \times \mathbf{k} = \mathbf{0} \\
&= a_1b_2\mathbf{k} - a_1b_3\mathbf{j} - a_2b_1\mathbf{k} + a_2b_3\mathbf{i} + a_3b_1\mathbf{j} - a_3b_2\mathbf{i} \\
&= (a_2b_3 - a_3b_2)\mathbf{i} - (a_1b_3 - a_3b_1)\mathbf{j} + (a_1b_2 - a_2b_1)\mathbf{k} \qquad \text{(this proves Theorem 12.5.8)} \\
&= \begin{vmatrix} a_2 & a_3 \\ b_2 & b_3 \end{vmatrix} \mathbf{i} - \begin{vmatrix} a_1 & a_3 \\ b_1 & b_3 \end{vmatrix} \mathbf{j} + \begin{vmatrix} a_1 & a_2 \\ b_1 & b_2 \end{vmatrix} \mathbf{k}. \qquad \square
\end{aligned}$$

Example 1 Calculate $\mathbf{a} \times \mathbf{b}$ given that $\mathbf{a} = \mathbf{i} - 2\mathbf{j} + 3\mathbf{k}$ and $\mathbf{b} = 2\mathbf{i} + \mathbf{j} - \mathbf{k}$.

SOLUTION

$$\begin{aligned}
\mathbf{a} \times \mathbf{b} &= \begin{vmatrix} \mathbf{i} & \mathbf{j} & \mathbf{k} \\ 1 & -2 & 3 \\ 2 & 1 & -1 \end{vmatrix} = \begin{vmatrix} -2 & 3 \\ 1 & -1 \end{vmatrix} \mathbf{i} - \begin{vmatrix} 1 & 3 \\ 2 & -1 \end{vmatrix} \mathbf{j} + \begin{vmatrix} 1 & -2 \\ 2 & 1 \end{vmatrix} \mathbf{k} \\
&= -\mathbf{i} + 7\mathbf{j} + 5\mathbf{k}. \quad \square
\end{aligned}$$

Example 2 Calculate $\mathbf{a} \times \mathbf{b}$ given that $\mathbf{a} = \mathbf{i} - \mathbf{j}$ and $\mathbf{b} = \mathbf{i} + \mathbf{k}$.

SOLUTION

$$\begin{aligned}
\mathbf{a} \times \mathbf{b} &= \begin{vmatrix} \mathbf{i} & \mathbf{j} & \mathbf{k} \\ 1 & -1 & 0 \\ 1 & 0 & 1 \end{vmatrix} = \begin{vmatrix} -1 & 0 \\ 0 & 1 \end{vmatrix} \mathbf{i} - \begin{vmatrix} 1 & 0 \\ 1 & 1 \end{vmatrix} \mathbf{j} + \begin{vmatrix} 1 & -1 \\ 1 & 0 \end{vmatrix} \mathbf{k} \\
&= -\mathbf{i} - \mathbf{j} + \mathbf{k}. \quad \square
\end{aligned}$$

In Examples 1 and 2 we calculated some cross products using Theorem 12.5.8′. Of course, we can obtain the same results just by applying the distributive laws. For example, for $\mathbf{a} = \mathbf{i} - 2\mathbf{j} + 3\mathbf{k}$ and $\mathbf{b} = 2\mathbf{i} + \mathbf{j} - \mathbf{k}$, we have

$$\begin{aligned}\mathbf{a} \times \mathbf{b} &= (\mathbf{i} - 2\mathbf{j} + 3\mathbf{k}) \times (2\mathbf{i} + \mathbf{j} - \mathbf{k}) \\ &= (\mathbf{i} \times \mathbf{j}) - (\mathbf{i} \times \mathbf{k}) - 4(\mathbf{j} \times \mathbf{i}) + 2(\mathbf{j} \times \mathbf{k}) + 6(\mathbf{k} \times \mathbf{i}) + 3(\mathbf{k} \times \mathbf{j}) \\ &= \mathbf{k} + \mathbf{j} + 4\mathbf{k} + 2\mathbf{i} + 6\mathbf{j} - 3\mathbf{i} = -\mathbf{i} + 7\mathbf{j} + 5\mathbf{k}.\end{aligned}$$

Example 3 Show that the scalar triple product can be written

(12.5.9)
$$(\mathbf{a} \times \mathbf{b}) \cdot \mathbf{c} = \begin{vmatrix} a_1 & a_2 & a_3 \\ b_1 & b_2 & b_3 \\ c_1 & c_2 & c_3 \end{vmatrix}.$$

SOLUTION

$$\begin{aligned}(\mathbf{a} \times \mathbf{b}) \cdot \mathbf{c} &= \mathbf{c} \cdot (\mathbf{a} \times \mathbf{b}) \\ &= (c_1\mathbf{i} + c_2\mathbf{j} + c_3\mathbf{k}) \cdot \left(\begin{vmatrix} a_2 & a_3 \\ b_2 & b_3 \end{vmatrix} \mathbf{i} - \begin{vmatrix} a_1 & a_3 \\ b_1 & b_3 \end{vmatrix} \mathbf{j} + \begin{vmatrix} a_1 & a_2 \\ b_1 & b_2 \end{vmatrix} \mathbf{k} \right) \\ &= c_1 \begin{vmatrix} a_2 & a_3 \\ b_2 & b_3 \end{vmatrix} - c_2 \begin{vmatrix} a_1 & a_3 \\ b_1 & b_3 \end{vmatrix} + c_3 \begin{vmatrix} a_1 & a_2 \\ b_1 & b_2 \end{vmatrix}.\end{aligned}$$

This is the expansion of

$$\begin{vmatrix} a_1 & a_2 & a_3 \\ b_1 & b_2 & b_3 \\ c_1 & c_2 & c_3 \end{vmatrix}$$

by the elements of the bottom row. □

A SUGGESTION: Vectors were defined as ordered triples, and many of the early proofs were done by "breaking up" vectors into their components. This may give you the impression that the method of "breakup" and working with the components is the first thing to try when confronted with a problem that involves vectors. If it is a *computational* problem, this method may give good results. But if you have to *analyze* a situation involving vectors, particularly one in which geometry plays a role, then the "breakup" strategy is seldom the best. Think instead of using the *operations* we have defined on vectors: addition, subtraction, scalar multiplication, dot product, cross product. Being geometrically motivated, these operations are likely to provide greater understanding than breaking up everything in sight into components. □

Example 4 Let $\mathbf{a}, \mathbf{b}, \mathbf{c}$ be nonzero vectors that do not lie in the same plane. Find all the vectors $\mathbf{d}$ for which

$$(*) \qquad \mathbf{d} \cdot \mathbf{a} = \mathbf{d} \cdot \mathbf{b} = \mathbf{d} \cdot \mathbf{c}.$$

SOLUTION We could begin by writing

$$\mathbf{d} = d_1\mathbf{i} + d_2\mathbf{j} + d_3\mathbf{k}, \quad \mathbf{a} = a_1\mathbf{i} + a_2\mathbf{j} + a_3\mathbf{k}, \quad \text{and so on.}$$

Equation $(*)$ would then take the form

$$d_1a_1 + d_2a_2 + d_3a_3 = d_1b_1 + d_2b_2 + d_3b_3 = d_1c_1 + d_2c_2 + d_3c_3,$$

and we would be faced with finding all d_1, d_2, d_3 that satisfy these equations. This is a messy task.

Here is a better approach. The vectors $\mathbf{d}$ that satisfy (∗) are the vectors $\mathbf{d}$ for which

$$\mathbf{d} \cdot (\mathbf{a} - \mathbf{b}) = 0 \qquad \text{and} \qquad \mathbf{d} \cdot (\mathbf{b} - \mathbf{c}) = 0.$$

These are the vectors $\mathbf{d}$ that are perpendicular to both $\mathbf{a} - \mathbf{b}$ and $\mathbf{b} - \mathbf{c}$. One such vector is $(\mathbf{a} - \mathbf{b}) \times (\mathbf{b} - \mathbf{c})$. The vectors $\mathbf{d}$ that satisfy (∗) are the scalar multiples of that cross product. ❑

Example 5 Verify *Lagrange's identity*: $||\mathbf{a} \times \mathbf{b}||^2 + (\mathbf{a} \cdot \mathbf{b})^2 = ||\mathbf{a}||^2\,||\mathbf{b}||^2$.

SOLUTION We could begin by writing

$$\begin{aligned}
||\mathbf{a} \times \mathbf{b}||^2 &= (a_2b_3 - a_3b_2)^2 + (a_1b_3 - a_3b_1)^2 + (a_1b_2 - a_2b_1)^2 \\
(\mathbf{a} \cdot \mathbf{b})^2 &= (a_1b_1 + a_2b_2 + a_3b_3)^2 \\
||\mathbf{a}||^2\,||\mathbf{b}||^2 &= (a_1^2 + a_2^2 + a_3^2)(b_1^2 + b_2^2 + b_3^2),
\end{aligned}$$

but this would take us into a morass of arithmetic. It is much more fruitful to proceed as follows:

$$||\mathbf{a} \times \mathbf{b}|| = ||\mathbf{a}||\,||\mathbf{b}||\sin\theta \qquad \text{and} \qquad \mathbf{a} \cdot \mathbf{b} = ||\mathbf{a}||\,||\mathbf{b}||\cos\theta.$$

Therefore

$$\begin{aligned}
||\mathbf{a} \times \mathbf{b}||^2 + (\mathbf{a} \cdot \mathbf{b})^2 &= ||\mathbf{a}||^2||\mathbf{b}||^2\sin^2\theta + ||\mathbf{a}||^2||\mathbf{b}||^2\cos^2\theta \\
&= ||\mathbf{a}||^2\,||\mathbf{b}||^2(\sin^2\theta + \cos^2\theta) = ||\mathbf{a}||^2||\mathbf{b}||^2.
\end{aligned}$$ ❑

Three Important Identities

It may be tempting to think that $\mathbf{a} \times (\mathbf{b} \times \mathbf{c})$ and $(\mathbf{a} \times \mathbf{b}) \times \mathbf{c}$ are equal, that is, that the cross product satisfies the associative law. In general, this is false. For example,

$$\mathbf{i} \times (\mathbf{i} \times \mathbf{j}) = \mathbf{i} \times \mathbf{k} = -\mathbf{j} \qquad \text{but} \qquad (\mathbf{i} \times \mathbf{i}) \times \mathbf{j} = \mathbf{0} \times \mathbf{j} = \mathbf{0}.$$

What is true instead is that

(12.5.10)

$$\begin{aligned}
\mathbf{a} \times (\mathbf{b} \times \mathbf{c}) &= (\mathbf{a} \cdot \mathbf{c})\mathbf{b} - (\mathbf{a} \cdot \mathbf{b})\mathbf{c}, \\
(\mathbf{a} \times \mathbf{b}) \times \mathbf{c} &= (\mathbf{c} \cdot \mathbf{a})\mathbf{b} - (\mathbf{c} \cdot \mathbf{b})\mathbf{a}.
\end{aligned}$$

There is one more identity that we want to mention:

(12.5.11)

$$(\mathbf{a} \times \mathbf{b}) \cdot (\mathbf{c} \times \mathbf{d}) = (\mathbf{a} \cdot \mathbf{c})(\mathbf{b} \cdot \mathbf{d}) - (\mathbf{a} \cdot \mathbf{d})(\mathbf{b} \cdot \mathbf{c}).$$

The proof of this, as well as the proof of (12.5.10), is left to you in the Exercises.

Remark Dot products and cross products appear frequently in physics and in engineering. Work is a dot product. So is the power expended by a force. Torque and angular momentum are cross products. Turn on a television set and watch the dots on the screen. How they move is determined by the laws of electromagnetism. It is all based on Maxwell's four equations. Two of them specify dot products; two of them specify cross products. ❑

EXERCISES 12.5

Calculate.

1. $(\mathbf{i}+\mathbf{j})\times(\mathbf{i}-\mathbf{j})$.
2. $(\mathbf{i}-\mathbf{j})\times(\mathbf{j}-\mathbf{i})$.
3. $(\mathbf{i}-\mathbf{j})\times\mathbf{j}-\mathbf{k})$.
4. $\mathbf{j}\times(2\mathbf{i}-\mathbf{k})$.
5. $(2\mathbf{j}-\mathbf{k})\times(\mathbf{i}-3\mathbf{j})$.
6. $\mathbf{i}\cdot(\mathbf{j}\times\mathbf{k})$.
7. $\mathbf{j}\cdot(\mathbf{i}\times\mathbf{k})$.
8. $(\mathbf{j}\times\mathbf{i})\cdot(\mathbf{i}\times\mathbf{k})$.
9. $(\mathbf{i}\times\mathbf{j})\times\mathbf{k}$.
10. $\mathbf{k}\cdot(\mathbf{j}\times\mathbf{i})$.
11. $\mathbf{j}\cdot(\mathbf{k}\times\mathbf{i})$.
12. $\mathbf{j}\times(\mathbf{k}\times\mathbf{i})$.

Calculate

13. $(\mathbf{i}+3\mathbf{j}-\mathbf{k})\times(\mathbf{i}+\mathbf{k})$.
14. $(3\mathbf{i}-2\mathbf{j}+\mathbf{k})\times(\mathbf{i}-\mathbf{j}+\mathbf{k})$.
15. $(\mathbf{i}+\mathbf{j}+\mathbf{k})\times(2\mathbf{i}+\mathbf{k})$.
16. $(2\mathbf{i}-\mathbf{k})\times(\mathbf{i}-2\mathbf{j}+2\mathbf{k})$.
17. $[2\mathbf{i}+\mathbf{j}]\cdot[(\mathbf{i}-3\mathbf{j}+\mathbf{k})\times(4\mathbf{i}+\mathbf{k})]$.
18. $[(-2\mathbf{i}+\mathbf{j}-3\mathbf{k})\times\mathbf{i}]\times[\mathbf{i}+\mathbf{j}]$.
19. $[(\mathbf{i}-\mathbf{j})\times(\mathbf{j}-\mathbf{k})]\times[\mathbf{i}+5\mathbf{k}]$.
20. $[\mathbf{i}-\mathbf{j}]\times[(\mathbf{j}-\mathbf{k})\times(\mathbf{j}+5\mathbf{k})]$.
21. Find two unit vectors which are perpendicular to the vectors $\mathbf{a}=(1,3,-1)$ and $\mathbf{b}=(2,0,1)$.
22. Repeat Exercise 21 for the vectors $\mathbf{a}=(1,2,3)$ and $\mathbf{b}=(2,1,1)$.

In Exercises 23–26, find a vector $\mathbf{N}$ that is perpendicular to the plane determined by the points P, Q, R, and find the area of triangle PQR.

23. $P(0,1,0),\quad Q(-1,1,2),\quad R(2,1,-1)$.
24. $P(1,2,3),\quad Q(-1,3,2),\quad R(3,-1,2)$.
25. $P(1,-1,4),\quad Q(2,0,1),\quad R(0,2,3)$.
26. $P(2,-1,3),\quad Q(4,1,-1),\quad R(-3,0,5)$.

In Exercises 27 and 28, find the volume of the parallelepiped with the given edges.

27. $\mathbf{i}+\mathbf{j},\quad 2\mathbf{i}-\mathbf{k},\quad 3\mathbf{j}+\mathbf{k}$.
28. $\mathbf{i}-3\mathbf{j}+\mathbf{k},\quad 2\mathbf{j}-\mathbf{k},\quad \mathbf{i}+\mathbf{j}-2\mathbf{k}$.
29. Given the points $O(0,0,0), P(1,2,3), Q(1,1,2), R(2,1,1)$, find the volume of the parallelepiped with edges $\overrightarrow{OP}, \overrightarrow{OQ}$, and $\overrightarrow{OR}$.
30. Given the points $P(1,-1,4), Q(2,0,1), R(0,2,3), S(3,5,7)$, find the volume of the parallelepiped with edges $\overrightarrow{PQ}, \overrightarrow{PR}, \overrightarrow{PS}$.
31. Express $(\mathbf{a}+\mathbf{b})\times(\mathbf{a}-\mathbf{b})$ as a scalar multiple of $\mathbf{a}\times\mathbf{b}$.
32. Earlier we verified that $\mathbf{a}\times(\mathbf{b}+\mathbf{c})=(\mathbf{a}\times\mathbf{b})+(\mathbf{a}\times\mathbf{c})$. Show now that
$$(\mathbf{a}+\mathbf{b})\times\mathbf{c}=(\mathbf{a}\times\mathbf{c})+(\mathbf{b}\times\mathbf{c}).$$
33. Suppose that $\mathbf{a}\times\mathbf{i}=\mathbf{0}$ and $\mathbf{a}\times\mathbf{j}=\mathbf{0}$. What can you conclude about $\mathbf{a}$?
34. Let $\mathbf{a}=a_1\mathbf{i}+a_2\mathbf{j}$ and $\mathbf{b}=b_1\mathbf{i}+b_2\mathbf{j}$ be nonzero vectors in the xy-plane. Show that $\mathbf{a}\times\mathbf{b}$ is parallel to $\mathbf{k}$.
35. Express $(\alpha\mathbf{a}+\beta\mathbf{b})\times(\gamma\mathbf{a}+\delta\mathbf{b})$ as a scalar multiple of $\mathbf{a}\times\mathbf{b}$.
36. (a) Let $\mathbf{a}, \mathbf{b}, \mathbf{c}$ be distinct nonzero vectors. Show that
$$\mathbf{a}\times\mathbf{b}=\mathbf{a}\times\mathbf{c}\quad\text{iff}\quad \mathbf{a}\quad\text{and}\quad \mathbf{b}-\mathbf{c}\ \text{are parallel}.$$
(b) Sketch a figure depicting all the vectors $\mathbf{c}$ that satisfy the relation $\mathbf{a}\times\mathbf{b}=\mathbf{a}\times\mathbf{c}$.
37. Which of the following dot products are equal?
$$\mathbf{a}\cdot(\mathbf{b}\times\mathbf{c}),\quad \mathbf{a}\cdot(\mathbf{c}\times\mathbf{b}),\quad (\mathbf{a}\times\mathbf{b})\cdot\mathbf{c},\quad (\mathbf{c}\times\mathbf{a})\cdot\mathbf{b},$$
$$(\mathbf{b}\times\mathbf{c})\cdot\mathbf{a},\quad \mathbf{c}\cdot(\mathbf{b}\times\mathbf{a}),\quad (-\mathbf{a}\times\mathbf{b})\cdot\mathbf{c},\quad (\mathbf{a}\times-\mathbf{c})\cdot\mathbf{b}.$$
38. Show that $(\mathbf{a}\times\mathbf{b})\cdot\mathbf{b}=0$ for all vectors $\mathbf{a}$ and $\mathbf{b}$.
39. Show that the vectors $\mathbf{a}, \mathbf{b}, \mathbf{c}$ are coplanar iff
$$(\mathbf{a}\times\mathbf{b})\cdot\mathbf{c}=0.$$
40. Given that $\mathbf{a}, \mathbf{b}, \mathbf{c}$ are mutually perpendicular, show that
$$\mathbf{a}\times(\mathbf{b}\times\mathbf{c})=\mathbf{0}.$$
41. Let $\mathbf{a}\neq\mathbf{0}$. Show that if
$$\mathbf{a}\times\mathbf{b}=\mathbf{a}\times\mathbf{c}\quad\text{and}\quad \mathbf{a}\cdot\mathbf{b}=\mathbf{a}\cdot\mathbf{c},\quad\text{then}\quad \mathbf{b}=\mathbf{c}.$$
42. (a) Show that
$$\mathbf{a}\times(\mathbf{b}\times\mathbf{c})=(\mathbf{a}\cdot\mathbf{c})\mathbf{b}-(\mathbf{a}\cdot\mathbf{b})\mathbf{c}.$$
HINT: Verify that the $\mathbf{i}$ components of the two sides agree. A similar verification can be carried out for the $\mathbf{j}$ and $\mathbf{k}$ components
(b) Show that
$$(\mathbf{a}\times\mathbf{b})\times\mathbf{c}=(\mathbf{c}\cdot\mathbf{a})\mathbf{b}-(\mathbf{c}\cdot\mathbf{b})\mathbf{a}.$$
HINT: $(\mathbf{a}\times\mathbf{b})\times\mathbf{c}=-[\mathbf{c}\times(\mathbf{a}\times\mathbf{b})]$.
(c) Finally, show that
$$(\mathbf{a}\times\mathbf{b})\cdot(\mathbf{c}\times\mathbf{d})=(\mathbf{a}\cdot\mathbf{c})(\mathbf{b}\cdot\mathbf{d})-(\mathbf{a}\cdot\mathbf{d})(\mathbf{b}\cdot\mathbf{c}).$$
HINT: Set $\mathbf{c}\times\mathbf{d}=\mathbf{r}$ and use (12.5.6).
43. Let $\mathbf{a}$ and $\mathbf{b}$ be nonzero vectors with $\mathbf{a}\perp\mathbf{b}$, and set $\mathbf{c}=\mathbf{a}\times\mathbf{b}$. Express $\mathbf{c}\times\mathbf{a}$ as a multiple of $\mathbf{b}$.
44. Given that $\mathbf{u}$ is a unit vector, show that each vector $\mathbf{a}$ can be decomposed as follows into a part parallel to $\mathbf{u}$ and a part perpendicular to $\mathbf{u}$:
$$\mathbf{a}=\underbrace{(\mathbf{a}\cdot\mathbf{u})\mathbf{u}}_{\text{parallel to }\mathbf{u}}+\underbrace{(\mathbf{u}\times\mathbf{a})\times\mathbf{u}}_{\text{perpendicular to }\mathbf{u}}.$$
HINT: Use Exercise 42(b).
45. Suppose $\mathbf{a}$ and $\mathbf{b}$ are vectors such that
$$\mathbf{a}\times\mathbf{b}=\mathbf{0}\quad\text{and}\quad \mathbf{a}\cdot\mathbf{b}=0.$$
What can you conclude about $\mathbf{a}$ or $\mathbf{b}$?

Exercises 46 and 47 refer to the tetrahedron with vertices $O(0,0,0), P(a,0,0), Q(0,b,0), R(0,0,c)$ shown in the figure.

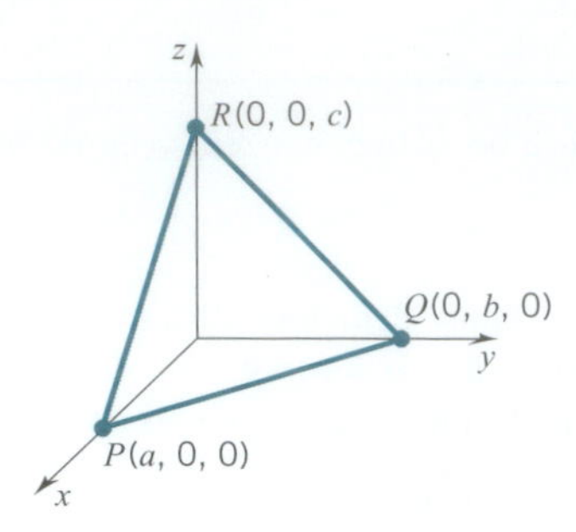

46. Use part 3 of the definition of cross product (Definition 12.5.1) to derive a formula for the area D of the face of the tetrahedron with vertices P, Q, R.

47. Let A be the area of the face opposite vertex P, let B be the area of the face opposite vertex Q, and let C be the area of the face opposite vertex R. Show that

$$A^2 + B^2 + C^2 = D^2.$$

This result is a three-dimensional version of the Pythagorean theorem.

48. Let **a**, **b** and **c** be vectors. Which of the following expressions make sense and which do not? Explain your answer in each case.

(a) $\mathbf{a} \cdot (\mathbf{b} \times \mathbf{c})$. (b) $\mathbf{a} \times (\mathbf{b} \cdot \mathbf{c})$.

(c) $\mathbf{a} \cdot (\mathbf{b} \cdot \mathbf{c})$. (d) $\mathbf{a} \times (\mathbf{b} \times \mathbf{c})$.

■ PROJECT 12.5 Torque

Suppose that a rigid body is free to rotate about a fixed point O. If a force **F** acts on the body at a point P, then the body tends to rotate about an axis through O. This effect is measured by the *torque* vector $\boldsymbol{\tau}$ which is given by

$$\boldsymbol{\tau} = \mathbf{r} \times \mathbf{F},$$

where **r** is the position vector $\overrightarrow{OP}$. The straight line through O determined by $\boldsymbol{\tau}$ is the axis of rotation. The vectors **r**, **F** and $\boldsymbol{\tau}$ form a right-handed system, and the magnitude of $\boldsymbol{\tau}$ is

$$||\boldsymbol{\tau}|| = ||\mathbf{r}||\ ||\mathbf{F}|| \sin\theta,$$

where θ is the angle between the position and force vectors (see the figure).

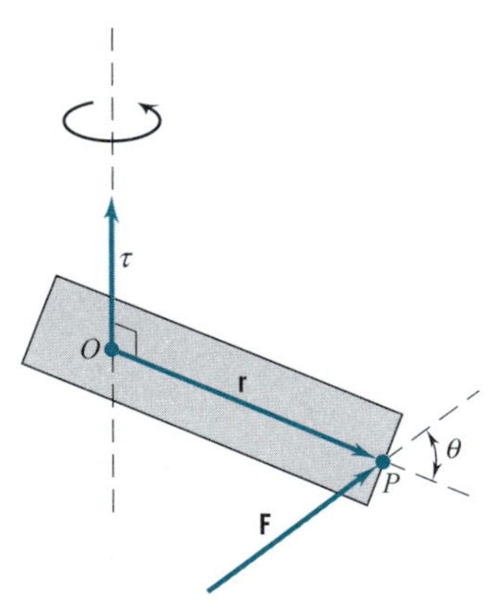

Problem 1. Find the magnitude of the torque exerted at the origin by the force $\mathbf{F} = \mathbf{i} + 2\mathbf{j} + \mathbf{k}$ applied at the point $P(1, 1, 1)$.

Problem 2. A bolt is being tightened by a 20-pound force applied to a 10-inch wrench as shown in the figure. Find the magnitude of the torque. Assuming that the wrench and the force are in the plane of the paper, in what direction will the bolt move?

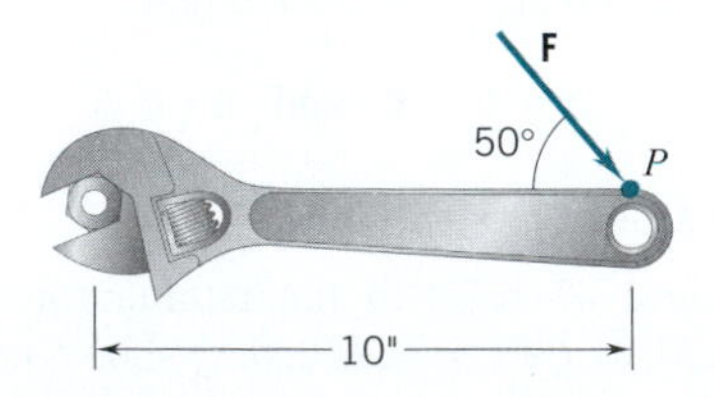

Problem 3. Repeat problem 2 if the 20-pound force is applied as shown in the figure.

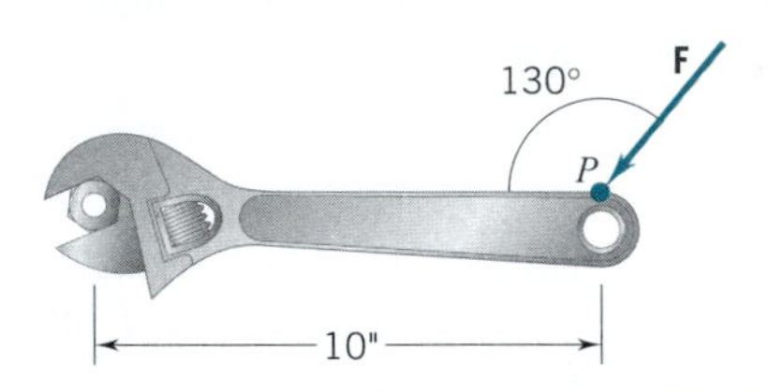

Problem 4. A bicycle with a front-wheel brake comes to a sudden stop. The horizontal braking force exerted by the brake on the front wheel is 650 newtons. The center of mass of the bicycle and its rider is 90 centimeters above the ground and 70 centimeters behind the point at which the front wheel touches the ground (see the figure).

a. What is the torque of this force about the center of mass?
b. What is the direction of the torque?
c. What rotation does it produce?

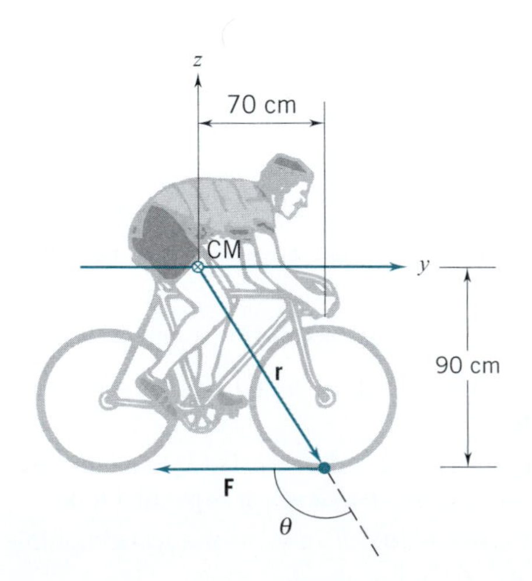

■ 12.6 LINES

Vectors used to specify position are called *position vectors*. Position vectors that emanate from the origin are known as *radius vectors*. In this section we use radius vectors to characterize lines.

Vector Parametrizations

We begin with the idea that two distinct points determine a line. In Figure 12.6.1 we have marked two points, P and Q, and the line l that they determine. To obtain a vector characterization of l, we choose the vectors $\mathbf{r}_0$ and $\mathbf{d} = \overrightarrow{PQ}$ as in Figure 12.6.2. Since we began with two distinct points, the vector $\mathbf{d}$ is nonzero.

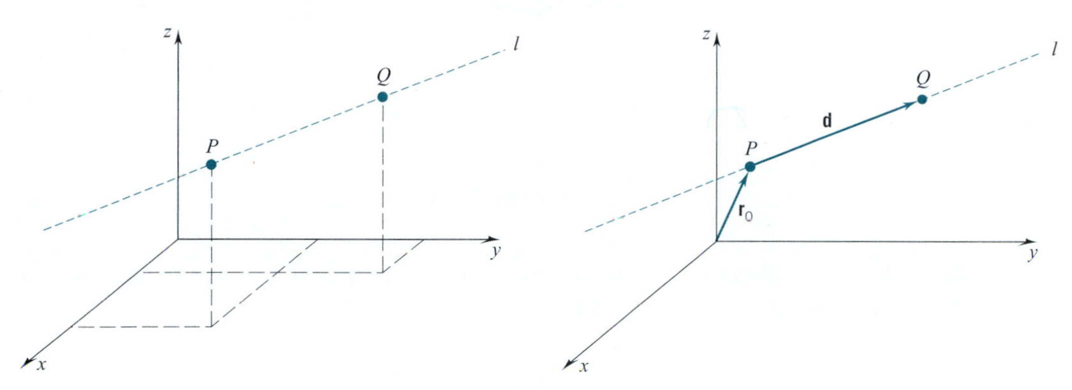

Figure 12.6.1 Figure 12.6.2

In Figure 12.6.3 we have drawn an additional vector $\mathbf{r}$. The vector that begins at the tip of $\mathbf{r}_0$ and ends at the tip of $\mathbf{r}$ is $\mathbf{r} - \mathbf{r}_0$. Therefore, the tip of $\mathbf{r}$ will fall on l iff

$$\mathbf{r} - \mathbf{r}_0 \quad \text{and} \quad \mathbf{d} \quad \text{are parallel;}$$

this in turn will happen iff

$$\mathbf{r} - \mathbf{r}_0 = t\mathbf{d} \qquad \text{for some real number } t,$$

or, equivalently, iff

$$\mathbf{r} = \mathbf{r}_0 + t\mathbf{d} \qquad \text{for some real number } t.$$

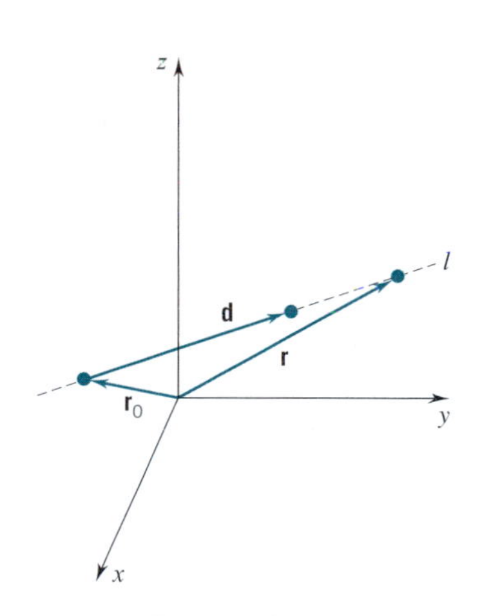

Figure 12.6.3

The vector equation

(12.6.1)
$$\boxed{\mathbf{r}(t) = \mathbf{r}_0 + t\mathbf{d}, \quad t \text{ real}}$$

parametrizes the line l: by varying t, we vary the vector $\mathbf{r}(t)$, but its tip remains on l; as t ranges over the set of real numbers, the tip of $\mathbf{r}(t)$ traces out the line l.

Now set

$$\mathbf{r}_0 = x_0\mathbf{i} + y_0\mathbf{j} + z_0\mathbf{k}, \quad \mathbf{d} = d_1\mathbf{i} + d_2\mathbf{j} + d_3\mathbf{k} \neq \mathbf{0}.$$

The tip of $\mathbf{r}_0$ is the point $P(x_0, y_0, z_0)$. The line l given by

(12.6.2)
$$\boxed{\mathbf{r}(t) = \mathbf{r}_0 + t\mathbf{d} = (x_0 + td_1)\mathbf{i} + (y_0 + td_2)\mathbf{j} + (z_0 + td_3)\mathbf{k}}$$

passes through the point $P(x_0, y_0, z_0)$ and is parallel to **d** (Figure 12.6.4). The vector **d**, which by assumption is not zero, is called a *direction vector* for l, and the components d_1, d_2, d_3 are called *direction numbers*.

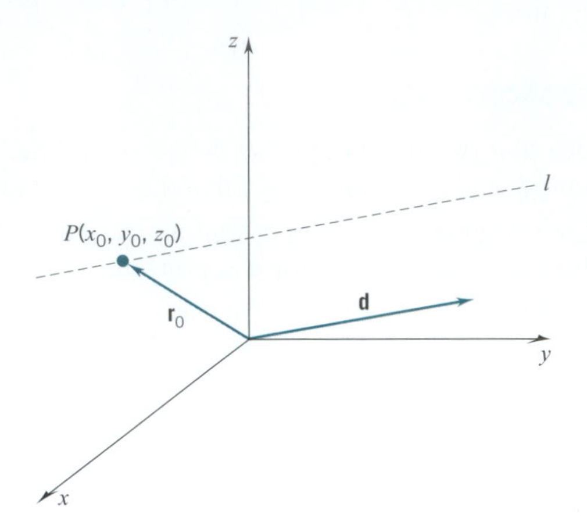

Figure 12.6.4

Remark As a direction vector for the line l determined by two distinct points $P(x_0, y_0, z_0)$ and $Q(x_1, y_2, z_3)$, we can use

$$\mathbf{d} = \overrightarrow{PQ} = (x_1 - x_0)\mathbf{i} + (y_1 - y_0)\mathbf{j} + (z_1 - z_0)\mathbf{k}.$$

The numbers $(x_1 - x_0)$, $(y_1 - y_0)$, $(z_1 - z_0)$ constitute a set of direction numbers for the line. □

Example 1 Find a vector equation that parametrizes the line that passes through the point $P(1, -1, 2)$ and is parallel to the vector $2\mathbf{i} - 3\mathbf{j} + \mathbf{k}$.

SOLUTION Here we can set

$$\mathbf{r}_0 = \mathbf{i} - \mathbf{j} + 2\mathbf{k} \qquad \text{and} \qquad \mathbf{d} = 2\mathbf{i} - 3\mathbf{j} + \mathbf{k}.$$

As a vector parametrization for the line we have

$$\mathbf{r}(t) = (\mathbf{i} - \mathbf{j} + 2\mathbf{k}) + t(2\mathbf{i} - 3\mathbf{j} + \mathbf{k}),$$

which we can rewrite as

$$\mathbf{r}(t) = (1 + 2t)\mathbf{i} - (1 + 3t)\mathbf{j} + (2 + t)\mathbf{k}. \quad □$$

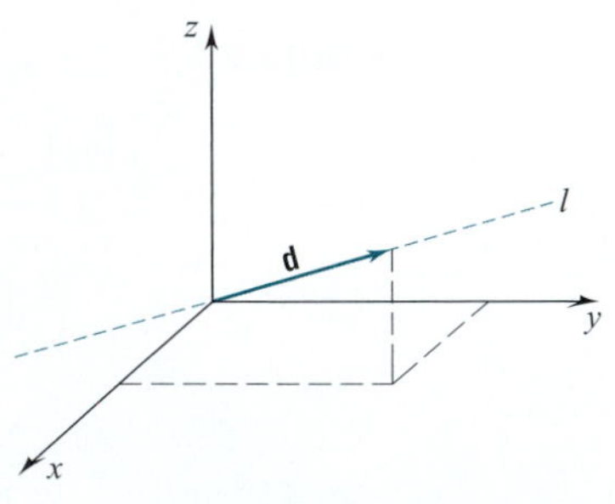

Figure 12.6.5

As a direction vector for a given line we can take any nonzero vector that is parallel to the line. Thus, if **d** is a direction vector for l, so is $\alpha\mathbf{d}$, provided that $\alpha \neq 0$. If d_1, d_2, d_3 are direction numbers for l, so are $\alpha d_1, \alpha d_2, \alpha d_3$, provided again that $\alpha \neq 0$.

The line that passes through the origin with direction vector **d** can be parametrized by the vector equation

(12.6.3) $$\mathbf{r}(t) = t\mathbf{d} = td_1\mathbf{i} + td_2\mathbf{j} + td_3\mathbf{k}.$$ (Figure 12.6.5)

There are, however, other ways to parametrize this line.

Example 2 Find all vector parametrizations (12.6.1) for the line through the origin with direction vector $\mathbf{d}$.

SOLUTION Since $\mathbf{d}$ is a direction vector, so is every vector $\alpha\mathbf{d}$ with $\alpha \neq 0$. We therefore write

$$(*) \qquad \mathbf{r}(t) = \mathbf{r}_0 + t\alpha\mathbf{d}.$$

Since the line passes through the origin, it may appear at first glance that $\mathbf{r}_0$ has to be $\mathbf{0}$, but that is not true. From the fact that the line passes through the origin we can conclude only that $\mathbf{r}_0 + t\alpha\mathbf{d} = \mathbf{0}$ is for some value of t. Call this value t_0. Then

$$\mathbf{r}_0 + t_0\alpha\mathbf{d} = \mathbf{0} \qquad \text{and thus} \qquad \mathbf{r}_0 = -t_0\alpha\mathbf{d}.$$

Substitution in $(*)$ gives

$$\mathbf{r}(t) = -t_0\alpha\mathbf{d} + t\alpha\mathbf{d} = (t - t_0)\alpha\mathbf{d}.$$

All the desired parametrizations can be written

$$\mathbf{r}(t) = (t - t_0)\alpha\mathbf{d} \quad \text{with } \alpha \text{ and } t_0 \text{ real, } \alpha \neq 0,$$

and all equations of this form parametrize that same line. ❑

Scalar Parametric Equations

It follows from (12.6.2) that the line that passes through the point $P(x_0, y_0, z_0)$ with direction numbers d_1, d_2, d_3 can be parametrized by three scalar equations:

(12.6.4)
$$x(t) = x_0 + d_1 t, \quad y(t) = y_0 + d_2 t, \quad z(t) = z_0 + d_3 t.$$

These quantities are the **i**, **j**, **k** components of the vector $\mathbf{r}(t) = \mathbf{r}_0 + t\mathbf{d}$.

Example 3 Write scalar parametric equations for the line that passes through the point $P(-1, 4, 2)$ with direction numbers $1, 2, 3$.

SOLUTION In this case the scalar equations

$$x(t) = x_0 + d_1 t, \quad y(t) = y_0 + d_2 t, \quad z(t) = z_0 + d_3 t$$

take the form

$$x(t) = -1 + t, \quad y(t) = 4 + 2t, \quad z(t) = 2 + 3t. \quad ❑$$

Example 4 What direction numbers are displayed by the parametric equations

$$x(t) = 3 - t, \quad y(t) = 2 + 4t, \quad z(t) = 1 - 5t?$$

What other direction numbers could be used for the same line?

SOLUTION The direction numbers displayed are $-1, 4, -5$. Any triple of the form

$$-\alpha, \; 4\alpha, \; -5\alpha \quad \text{with } \alpha \neq 0$$

could be used as a set of direction numbers for that same line. ❑

Symmetric Form

If the direction numbers are all nonzero, then each of the scalar parametric equations can be solved for t:

$$t = \frac{x(t) - x_0}{d_1}, \quad t = \frac{y(t) - y_0}{d_2}, \quad t = \frac{z(t) - z_0}{d_3}.$$

Eliminating the parameter t, we obtain three equations:

$$\frac{x - x_0}{d_1} = \frac{y - y_0}{d_2}, \quad \frac{y - y_0}{d_2} = \frac{z - z_0}{d_3}, \quad \frac{x - x_0}{d_1} = \frac{z - z_0}{d_3}.$$

Any two of these equations suffice; the third is redundant and can be discarded. Rather than decide which equation to discard, we simply write

(12.6.5)
$$\frac{x - x_0}{d_1} = \frac{y - y_0}{d_2} = \frac{z - z_0}{d_3}.$$

These are the equations of a line written in *symmetric form*. They can be used only if d_1, d_2, d_3 are all different from zero.

Example 5 Write equations in symmetric form for the line that passes through the point $P(x_0, y_0, z_0)$ and $Q(x_1, y_1, z_1)$. Under what conditions are the equations valid?

SOLUTION As direction numbers we can take the triple

$$x_1 - x_0, \quad y_1 - y_0, \quad z_1 - z_0.$$

We can base our calculations on $P(x_0, y_0, z_0)$ and write

$$\frac{x - x_0}{x_1 - x_0} = \frac{y - y_0}{y_1 - y_0} = \frac{z - z_0}{z_1 - z_0},$$

or we can base our calculations on $Q(x_1, y_1, z_1)$ and write

$$\frac{x - x_1}{x_1 - x_0} = \frac{y - y_1}{y_1 - y_0} = \frac{z - z_1}{z_1 - z_0}.$$

Both sets of equations are valid provided that $x_1 \neq x_0, y_1 \neq y_0, z_1 \neq z_0$. □

Equations (12.6.5) can be used only if the direction numbers are all different from zero. If one of the direction numbers is zero, then one of the coordinates is constant. As you will see, this simplifies the algebra. Geometrically, it means that the line lies on a plane that is parallel to one of the coordinate planes.

Suppose, for example, that $d_3 = 0$. Then the scalar parametric equations take the form

$$x(t) = x_0 + d_1 t, \quad y(t) = y_0 + d_2 t, \quad z(t) = z_0.$$

Eliminating t, we are left with two equations:

$$\frac{x - x_0}{d_1} = \frac{y - y_0}{d_2}, \quad z = z_0.$$

The line lies on the horizontal plane $z = z_0$ and its projection onto the xy-plane (see Figure 12.6.6) is the line l' with equation

$$\frac{x - x_0}{d_1} = \frac{y - y_0}{d_2}.$$

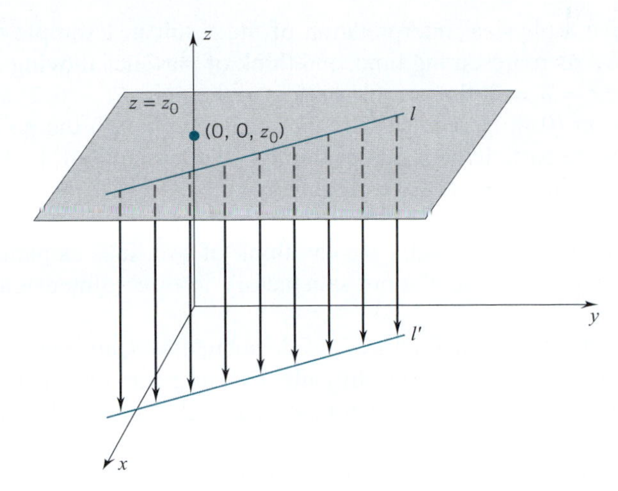

Figure 12.6.6

Intersecting Lines, Parallel Lines

Two distinct lines

$$l_1 : \mathbf{r}(t) = \mathbf{r}_0 + t\mathbf{d}, \quad l_2 : \mathbf{R}(u) = \mathbf{R}_0 + u\mathbf{D}$$

intersect iff there are numbers t and u at which

$$\mathbf{r}(t) = \mathbf{R}(u).$$

Example 6 Find the point at which the lines

$$l_1 : \mathbf{r}(t) = (\mathbf{i} - 6\mathbf{j} + 2\mathbf{k}) + t(\mathbf{i} + 2\mathbf{j} + \mathbf{k}), \quad l_2 : \mathbf{R}(u) = (4\mathbf{j} + \mathbf{k}) + u(2\mathbf{i} + \mathbf{j} + 2\mathbf{k})$$

intersect.

SOLUTION We set $\mathbf{r}(t) = \mathbf{R}(u)$

and solve for t and u:

$$(\mathbf{i} - 6\mathbf{j} + 2\mathbf{k}) + t(\mathbf{i} + 2\mathbf{j} + \mathbf{k}) = (4\mathbf{j} + \mathbf{k}) + u(2\mathbf{i} + \mathbf{j} + 2\mathbf{k}),$$
$$(1 + t)\mathbf{i} + (-6 + 2t)\mathbf{j} + (2 + t)\mathbf{k} = 2u\,\mathbf{i} + (4 + u)\mathbf{j} + (1 + 2u)\mathbf{k}$$

and therefore

$$(1 + t - 2u)\mathbf{i} + (-10 + 2t - u)\mathbf{j} + (1 + t - 2u)\mathbf{k} = \mathbf{0}.$$

This tells us that

$$1 + t - 2u = 0,$$
$$-10 + 2t - u = 0,$$
$$1 + t - 2u = 0.$$

Note that the first and third equations are the same. Solving the first two equations simultaneously, we obtain $t = 7, u = 4$. As you can verify,

$$\mathbf{r}(7) = 8\mathbf{i} + 8\mathbf{j} + 9\mathbf{k} = \mathbf{R}(4).$$

The two lines intersect at the tip of this vector, which is the point $P(8, 8, 9)$. □

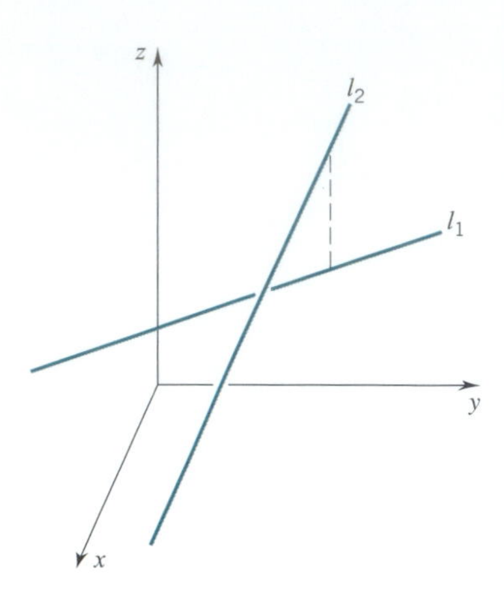

Figure 12.6.7

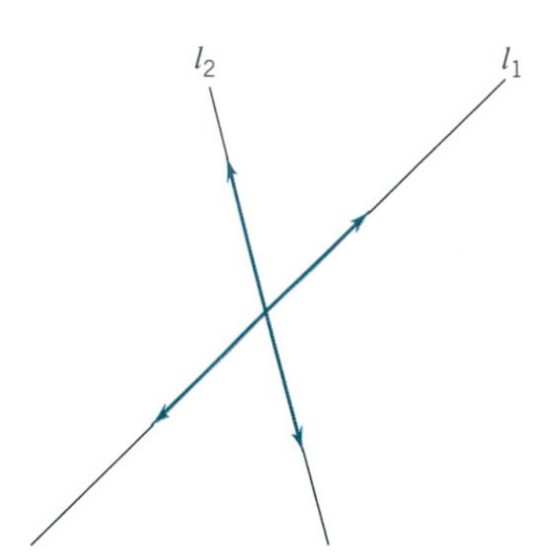

Figure 12.6.8

Remark To give a physical interpretation of the result in Example 6, think of the parameters t and u as representing time, and think of particles moving along the lines l_2 and l_2. At time $t = u = 0$, the particle on l_1 is at the point $(1, -6, 2)$ and the particle on l_2 is at the point $(0, 4, 1)$. The particle on l_1 passes through the point $P(8, 8, 9)$ at time $t = 7$, while the particle on l_2 passes through P at time $u = 4$; both particles pass through the same point P, but at different times. □

In the setting of plane geometry we can think of two lines as parallel iff they do not intersect. This point of view is not satisfactory in three-dimensional space. (See Figure 12.6.7.)

The lines l_1 and l_2 marked in Figure 12.6.7 do not intersect, and yet we would hesitate to call them parallel. We can avoid this difficulty by using direction vectors: two distinct lines are *parallel* iff their direction vectors are parallel. Nonparallel, nonintersecting lines are said to be *skew*.

If two lines l_1, l_2 intersect, we can find the angle between them by finding the angle between their direction vectors, $\mathbf{d}$ and $\mathbf{D}$. Depending on our choice of direction vectors, there are two such angles, each the supplement of the other (Figure 12.6.8). We choose the smaller of the two angles, the one with nonnegative cosine:

(12.6.6)
$$\cos\theta = |\mathbf{u_d} \cdot \mathbf{u_D}|.$$

Example 7 Earlier we verified that the lines

$$l_1 : \mathbf{r}(t) = (\mathbf{i} - 6\mathbf{j} + 2\mathbf{k}) + t(\mathbf{i} + 2\mathbf{j} + \mathbf{k}), \quad l_2 : \mathbf{R}(u) = (4\mathbf{j} + \mathbf{k}) + u(2\mathbf{i} + \mathbf{j} + 2\mathbf{k})$$

intersect at $P(8, 8, 9)$. What is the angle between these lines?

SOLUTION As direction vectors we can take

$$\mathbf{d} = \mathbf{i} + 2\mathbf{j} + \mathbf{k} \qquad \text{and} \qquad \mathbf{D} = 2\mathbf{i} + \mathbf{j} + 2\mathbf{k}.$$

Then, as you can check,

$$\mathbf{u_d} = \tfrac{1}{6}\sqrt{6}(\mathbf{i} + 2\mathbf{j} + \mathbf{k}) \qquad \text{and} \qquad \mathbf{u_D} = \tfrac{1}{3}(2\mathbf{i} + \mathbf{j} + 2\mathbf{k}).$$

It follows that

$$\cos\theta = |\mathbf{u_d} \cdot \mathbf{u_D}| = \tfrac{1}{3}\sqrt{6} \qquad \text{and} \qquad \theta \cong 0.615 \text{ radians, about } 35.26^\circ \quad \square$$

Two intersecting lines are said to be *perpendicular* if their direction vectors are perpendicular.

Example 8 Let l_1 and l_2 be the lines of the last example. These lines intersect at $P(8, 8, 9)$. Find a vector parametrization for the line l_3 that passes through $P(8, 8, 9)$ and is perpendicular to both l_1 and l_2.

SOLUTION We are given that l_3 passes through $P(8, 8, 9)$. All we need to parametrize that line is a direction vector $\mathbf{c}$. We require that $\mathbf{c}$ be perpendicular to the direction vectors of l_1 and l_2; namely, we require that

$$\mathbf{c} \perp \mathbf{d} \quad \text{and} \quad \mathbf{c} \perp \mathbf{D}, \quad \text{where} \quad \mathbf{d} = \mathbf{i}+2\mathbf{j}+\mathbf{k} \quad \text{and} \quad \mathbf{D} = 2\mathbf{i}+\mathbf{j}+2\mathbf{k}.$$

Since $\mathbf{d} \times \mathbf{D}$ is perpendicular to both $\mathbf{d}$ and $\mathbf{D}$, we can set

$$\mathbf{c} = \mathbf{d} \times \mathbf{D} = \begin{vmatrix} \mathbf{i} & \mathbf{j} & \mathbf{k} \\ 1 & 2 & 1 \\ 2 & 1 & 2 \end{vmatrix} = \begin{vmatrix} 2 & 1 \\ 1 & 2 \end{vmatrix} \mathbf{i} - \begin{vmatrix} 1 & 1 \\ 2 & 2 \end{vmatrix} \mathbf{j} + \begin{vmatrix} 1 & 2 \\ 2 & 1 \end{vmatrix} \mathbf{k} = 3\mathbf{i} - 3\mathbf{k}.$$

As a parametrization for l_3 we can write

$$\mathbf{s}(t) = (8\mathbf{i} + 8\mathbf{j} + 9\mathbf{k}) + t(3\mathbf{i} - 3\mathbf{k}). \quad \square$$

Example 9

(a) Find a vector parametrization for the line

$$l : y = mx + b \quad \text{in the } xy\text{-plane.}$$

(b) Show by vector methods that

$$l_1 : y = m_1 x + b_1 \perp l_2 : y = m_2 x + b_2 \quad \text{iff} \quad m_1 m_2 = -1.$$

SOLUTION

(a) We seek a parametrization of the form

$$\mathbf{r}(t) = \mathbf{r}_0 + t\mathbf{d}.$$

Since $P(0, b)$ lies on l, we can set

$$\mathbf{r}_0 = 0\mathbf{i} + b\mathbf{j} = b\mathbf{j}.$$

To find a direction vector, we take $x_1 \neq 0$ and note that the point $Q(x_1, mx_1 + b)$ also lies on l. (See Figure 12.6.9.) As direction numbers we can take

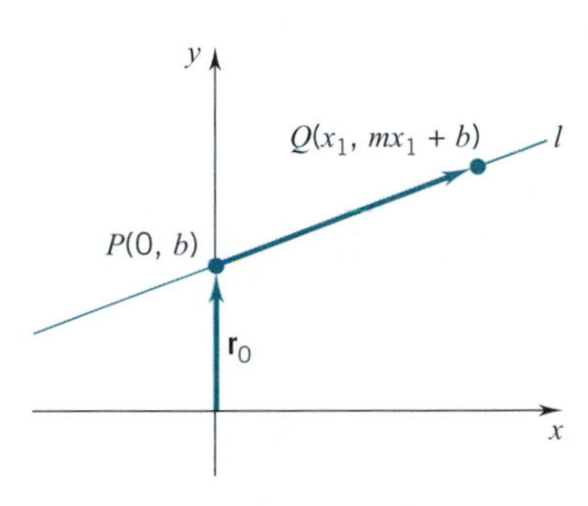

Figure 12.6.9

$$x_1 - 0 = x_1 \qquad \text{and} \qquad (mx_1 + b) - b = mx_1$$

or, more simply, 1 and m. This choice of direction numbers gives us the direction vector $\mathbf{d} = \mathbf{i} + m\mathbf{j}$. The vector equation

$$\mathbf{r}(t) = b\mathbf{j} + t(\mathbf{i} + m\mathbf{j})$$

parametrizes the line l.

(b) As direction vectors for l_1 and l_2 we have

$$\mathbf{d}_1 = \mathbf{i} + m_1\mathbf{j} \qquad \text{and} \qquad \mathbf{d}_2 = \mathbf{i} + m_2\mathbf{j}.$$

Since

$$\mathbf{d}_1 \cdot \mathbf{d}_2 = (\mathbf{i} + m_1\mathbf{j}) \cdot (\mathbf{i} + m_2\mathbf{j}) = 1 + m_1 m_2,$$

you can see that

$$\mathbf{d}_1 \cdot \mathbf{d}_2 = 0 \qquad \text{iff} \qquad m_1 m_2 = -1. \quad \square$$

Distance from a Point to a Line

In Figure 12.6.10 we have drawn a line l and a point P_1 not on l. We are interested in finding the distance $d(P_1, l)$ between P_1 and l.

Let P_0 be a point on l and let $\mathbf{d}$ be a direction vector for l. With P_0 and Q as shown in the figure, you can see that

$$d(P_1, l) = d(P_1, Q) = ||\overrightarrow{P_0P_1}|| \sin\theta$$

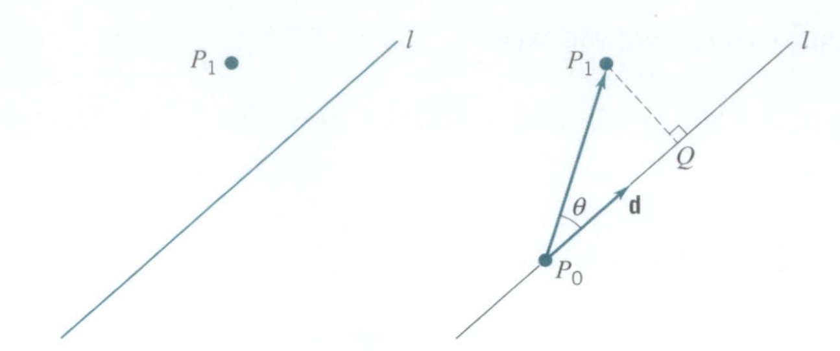

Figure 12.6.10

Since $||\overrightarrow{P_0P_1} \times \mathbf{d}|| = ||\overrightarrow{P_0P_1}||\,||\mathbf{d}||\sin\theta$, we have

(12.6.7)
$$d(P_1, l) = \frac{||\overrightarrow{P_0P_1} \times \mathbf{d}||}{||\mathbf{d}||}.$$

This elegant little formula gives the distance from a point P_1 to any line l in terms of any point P_0 on l and any direction vector $\mathbf{d}$ for l.

Computations based on this formula are left to the Exercises.

EXERCISES 12.6

1. Which of the points $P(1, 2, 0), Q(-5, 1, 5), R(-4, 2, 5)$ lie on the line

$$l : \mathbf{r}(t) = (\mathbf{i} + 2\mathbf{j}) + t(6\mathbf{i} + \mathbf{j} - 5\mathbf{k})?$$

2. Determine which of the lines are parallel:

$$l_1 : \mathbf{r}_1(t) = (\mathbf{i} + 2\mathbf{k}) + t(\mathbf{i} - 2\mathbf{j} + 3\mathbf{k}),$$
$$l_2 : \mathbf{r}_2(u) = (\mathbf{i} + 2\mathbf{k}) + u(\mathbf{i} + 2\mathbf{j} - 3\mathbf{k}),$$
$$l_3 : \mathbf{r}_3(v) = (6\mathbf{i} - \mathbf{j}) - v(2\mathbf{i} - 4\mathbf{j} + 6\mathbf{k}),$$
$$l_4 : \mathbf{r}_4(w) = (\tfrac{1}{2} + \tfrac{1}{2}w)\mathbf{i} - w\mathbf{j} + (1 + \tfrac{3}{2}w)\mathbf{k}.$$

Find a vector parametrization for the line that satisfies the given conditions.

3. Passes through $P(3, 1, 0)$ and is parallel to the line $\mathbf{r}(t) = (\mathbf{i} - \mathbf{j}) + t\mathbf{k}$.

4. Passes through $P(1, -1, 2)$ and is parallel to the line $\mathbf{r}(t) = t(3\mathbf{i} - \mathbf{j} + \mathbf{k})$.

5. Passes through the origin and $Q(x_1, y_1, z_1)$.

6. Passes through $P(x_0, y_0, z_0)$ and $Q(x_1, y_1, z_1)$.

Find a set of scalar parametric equations for the line that satisfies the given conditions.

7. Passes through $P(1, 0, 3)$ and $Q(2, -1, 4)$.

8. Passes through $P(x_0, y_0, z_0)$ and $Q(x_1, y_1, z_1)$.

9. Passes through $P(2, -2, 3)$ and is perpendicular to the xz-plane.

10. Passes through $P(1, 4, -3)$ and is perpendicular to the yz-plane.

11. Give a vector parametrization for the line that passes through $P(-1, 2, -3)$ and is parallel to the line $2(x + 1) = 4(y - 3) = z$.

12. Write equations in symmetric form for the line that passes through the origin and the point $P(x_0, y_0, z_0)$, $x_0, y_0, z_0 \neq 0$.

Determine whether the lines l_1 and l_2 are parallel, skew, or intersecting. If they intersect, find the point of intersection.

13. $l_1 : \mathbf{r}(t) = (3\mathbf{i} + \mathbf{j} + 5\mathbf{k}) + t(\mathbf{i} - \mathbf{j} + 2\mathbf{k})$,
$l_2 : \mathbf{R}(u) = (\mathbf{i} + 4\mathbf{j} + 2\mathbf{k}) + u(\mathbf{j} + \mathbf{k})$.

14. $l_1 : \mathbf{r}(t) = (-\mathbf{i} + 2\mathbf{j} + \mathbf{k}) + t(\mathbf{i} - 3\mathbf{j} + 2\mathbf{k})$,
$l_2 : \mathbf{R}(u) = (2\mathbf{i} - \mathbf{j}) + u(-2\mathbf{i} + 6\mathbf{j} - 4\mathbf{k})$.

15. $l_1 : x_1(t) = 3 + 2t,\ y_1(t) = -1 + 4t,\ z_1(t) = 2 - t$,
$l_2 : x_2(u) = 3 + 2u,\ y_2(u) = 2 + u,\ z_2(u) = -2 + 2u$.

16. $l_1 : x_1(t) = 1 + t,\ y_1(y) = -1 - t,\ z_1(t) = -4 + 2t$,
$l_2 : x_2(u) = 1 - u,\ y_2(u) = 1 + 3u,\ z_2(u) = 2u$.

17. $l_1 : x_1(t) = 1 - 6t, y_1(t) = 2 + 9t, z_1(t) = -3t$,
$l_2 : x_2(u) = 2 + 2u,\ y_2(u) = 3 - 3u,\ z_2(u) = u$.

18. $l_1 : x - 2 = \dfrac{y+1}{2} = \dfrac{z-1}{3}, \quad l_2 : \dfrac{x-5}{3} = \dfrac{y-1}{2} = z - 4.$

19. $l_1 : \dfrac{x-4}{2} = \dfrac{y+5}{4} = \dfrac{z-1}{3}, \quad l_2 : x - 2 = \dfrac{y+1}{3} = \dfrac{z}{2}.$

20. $l_1 : x_1(t) = 1 + t,\ y_1(t) = 2t,\ z_1(t) = 1 + 3t$,
$l_2 : x_2(u) = 3u,\ y_2(u) = 2u,\ z_2(u) = 2 + u$.

In Exercises 21 and 22, find the point where l_1 and l_2 intersect and find the angle between l_1 and l_2.

21. $l_1 : \mathbf{r}_1(t) = \mathbf{i} + t\mathbf{j}, \quad l_2 : \mathbf{r}_2(u) = \mathbf{j} + u(\mathbf{i} + \mathbf{j}).$

22. $l_1 : \mathbf{r}_1(t) = (\mathbf{i} - 4\sqrt{3}\,\mathbf{j}) + t(\mathbf{i} + \sqrt{3}\,\mathbf{j}),$
$l_2 : \mathbf{r}_2(u) = (4\,\mathbf{i} + 3\sqrt{3}\,\mathbf{j}) + u(\mathbf{i} - \sqrt{3}\,\mathbf{j}).$

23. Where does the line

$$\frac{x - x_0}{d_1} = \frac{y - y_0}{d_2} = \frac{z - z_0}{d_3}$$

intersect the xy-plane?

24. What can you conclude about the lines

$$\frac{x - x_0}{d_1} = \frac{y - y_0}{d_2} = \frac{z - z_0}{d_3}, \quad \frac{x - x_0}{D_1} = \frac{y - y_0}{D_2} = \frac{z - z_0}{D_3}$$

given that $d_1 D_1 + d_2 D_2 + d_3 D_3 = 0$?

25. What can you conclude about the lines

$$\frac{x - x_0}{d_1} = \frac{y - y_0}{d_2} = \frac{z - z_0}{d_3}, \quad \frac{x - x_1}{D_1} = \frac{y - y_1}{D_2} = \frac{z - z_1}{D_3}$$

given that $d_1/D_1 = d_2/D_2 = d_3/D_3$?

26. (*Important*) Let P_0, P_1 be two distinct points and let $\mathbf{r}_0, \mathbf{r}_1$ be the radius vectors that they determine:

$$\mathbf{r}_0 = \overrightarrow{OP_0}, \quad \mathbf{r}_1 = \overrightarrow{OP_1}.$$

As t ranges over the set of real numbers, $\mathbf{r}(t) = \mathbf{r}_0 + t(\mathbf{r}_1 - \mathbf{r}_0)$ traces out the line determined by P_0 and P_1. Restrict t so that $\mathbf{r}(t)$ traces out only the line segment $\overline{P_0 P_1}$.

27. Find a vector parametrization for the line segment that begins at $(2, 7, -1)$ and ends at $(4, 2, 3)$.

28. Restrict t so that the equations

$$x(t) = 7 - 5t, \quad y(t) = -3 + 2t, \quad z(t) = 4 - t$$

parametrize the line segment that begins at $(12, -5, 5)$ and ends at $(-3, 1, 2)$.

29. Determine a unit vector $\mathbf{u}$ and the values of t for which the equation

$$\mathbf{r}(t) = (6\mathbf{i} - 5\mathbf{j} + \mathbf{k}) + t\mathbf{u}$$

parametrizes the line segment that begins at $P(0, -2, 7)$ and ends at $Q(-4, 0, 11)$.

30. Suppose that the lines

$$l_1 : \mathbf{r}(t) = \mathbf{r}_0 + t\mathbf{d}, \qquad l_2 : \mathbf{R}(u) = \mathbf{R}_0 + u\mathbf{D}$$

intersect at right angles. Show that the point of intersection is the origin iff $\mathbf{r}(t) \perp \mathbf{R}(u)$ for all real numbers t and u.

31. Find scalar parametric equations for all the lines that are perpendicular to the line

$$x(t) = 1 + 2(t), \quad y(t) = 3 - 4t, \quad z(t) = 2 + 6t$$

and intersect that line at the point $P(3, -1, 8)$.

32. Suppose that $\mathbf{r}(t) = \mathbf{r}_0 + t\mathbf{d}$ and $\mathbf{R}(u) = \mathbf{R}_0 + u\mathbf{D}$ both parametrize the same line. (a) Show that $\mathbf{R}_0 = \mathbf{r}_0 + t_0\mathbf{d}$ for some real number t_0. (b) Then show that, for some real number α, $\mathbf{R}(u) = \mathbf{r}_0 + (t_0 + \alpha u)\,\mathbf{d}$ for all real u.

Find the distance from $P(1, 0, 2)$ to the indicated line.

33. The line through the origin parallel to $2\mathbf{i} - \mathbf{j} + 2\mathbf{k}$.

34. The line through $P_0(1, -1, 1)$ parallel to $\mathbf{i} - 2\mathbf{j} - 2\mathbf{k}$.

Find the distance from the point to the line.

35. $P(1, 2, 3), \quad l : \mathbf{r}(t) = \mathbf{i} + 2\mathbf{k} + t(\mathbf{i} - 2\mathbf{j} + 3\mathbf{k}).$

36. $P(0, 0, 0), \quad l : \mathbf{r}(t) = \mathbf{i} + t\mathbf{j}.$

37. $P(1, 0, 1), \quad l : \mathbf{r}(t) = 2\mathbf{i} - \mathbf{j} + t(\mathbf{i} + \mathbf{j}).$

38. Find the distance from the point $P(x_0, y_0, z_0)$ to the line $y = mx + b$ in the xy-plane.

39. What is the distance from the origin: (a) to the line that joins $P(1, 1, 1)$ and $Q(2, 2, 1)$? (b) to the line segment that joins these same points? [For part (b) find the point of the line segment $\overline{PQ}$ closest to the origin and calculate its distance from the origin.]

40. Let l be the line

$$\mathbf{r}(t) = \mathbf{r}_0 + t\mathbf{d}.$$

(a) Find the scalar t_0 for which $\mathbf{r}(t_0) \perp l$.

(b) Find the parametrizations $\mathbf{R}(t) = \mathbf{R}_0 + t\mathbf{D}$ for l in which $\mathbf{R}_0 \perp l$ and $\|\mathbf{D}\| = 1$. These are the *standard vector parametrizations* for l.

In Exercises 41 and 42, find the standard vector parametrizations (Exercise 40) for the specified line.

41. The line through $P(0, 1, -2)$ parallel to $\mathbf{i} - \mathbf{j} + 3\mathbf{k}$.

42. The line through $P(\sqrt{3}, 0, 0)$ parallel to $\mathbf{i} + \mathbf{j} + \mathbf{k}$.

43. Let A, B, C be the vertices of a triangle in the xy-plane. Given that $0 < s < 1$, determine the values of t for which the tip of the radius vector

$$\overrightarrow{OA} + s\overrightarrow{AB} + t\overrightarrow{BC}$$

lies inside the triangle. HINT: Draw a diagram.

44. (*Distance between skew lines*). The distance between two skew lines l_1 and l_2 is defined to be the minimum of the distances $d(P, Q)$ where P is a point on l_1 and Q is a point on l_2. This distance is the length of the line segment joining l_1 and l_2 and perpendicular to both lines. See the figure. Show that the distance between two skew lines is given by

$$d(l_1, l_2) = \frac{|\overrightarrow{PQ} \cdot (\mathbf{d}_1 \times \mathbf{d}_2)|}{\|\mathbf{d}_1 \times \mathbf{d}_2\|}.$$

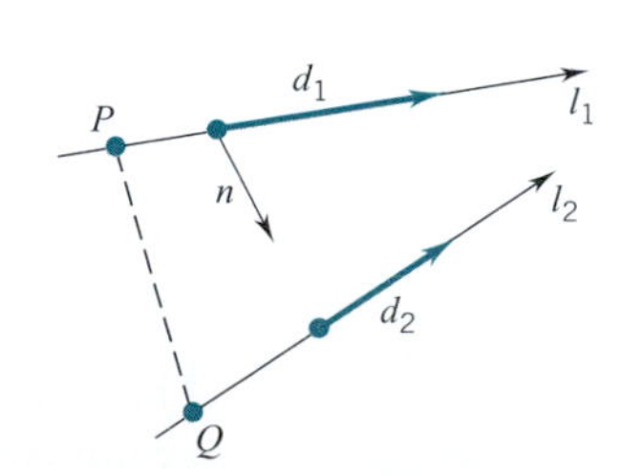

45. Show that the lines

$$l_1 : x_1(t) = 2 + t,\ y_1(t) = -1 + 3t,\ z_1(t) = 1 - 2t,$$
$$l_2 : x_2(u) = -1 + 4u,\ y_2(u) = 2 - u,\ z_2(u) = -3 + 2u.$$

are skew and find the distance between them.

46. Repeat Exercise 45 with

$$l_1 : x_1(t) = 1 + t,\ y_1(t) = -2 + 3t,\ z_1(t) = 4 - 2t,$$
$$l_2 : x_2(u) = 2u,\ y_2(u) = 3 + u,\ z_2(u) = -3 + 4u.$$

■ 12.7 PLANES

Ways of Specifying a Plane

How can we specify a plane? There are a number of ways of doing so. For example by giving three distinct points on it, so long as they are not all on the same line; by giving two distinct lines on it; or by giving a line on it and a point on it, so long as the point does not lie on the line. There is still another way to specify a plane, and that is to give a point on the plane and a nonzero vector perpendicular to the plane.

Scalar Equation of a Plane

Figure 12.7.1 shows a plane. On it we have marked a point $P(x_0, y_0, z_0)$ and, starting at that point, a nonzero vector $\mathbf{N} = A\mathbf{i} + B\mathbf{j} + C\mathbf{k}$ perpendicular to the plane. We call $\mathbf{N}$ a *normal vector*. We can obtain an equation for the plane in terms of the coordinates of P and the components of $\mathbf{N}$.

To find such an equation we take an arbitrary point $Q(x, y, z)$ in space and form the vector

$$\overrightarrow{PQ} = (x - x_0)\mathbf{i} + (y - y_0)\mathbf{j} + (z - z_0)\mathbf{k}.$$

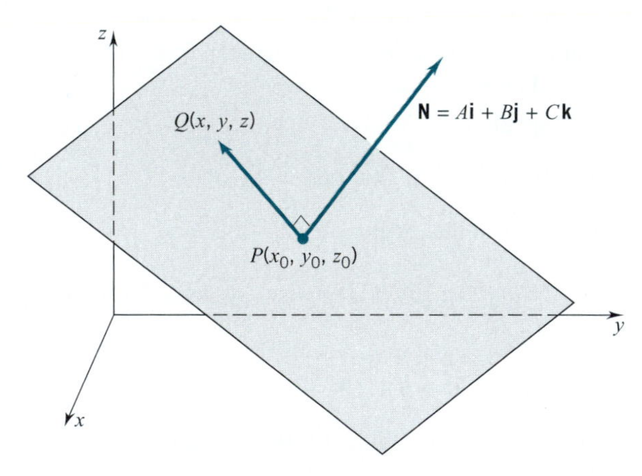

Figure 12.7.1

The point Q will lie on the given plane iff

$$\mathbf{N} \cdot \overrightarrow{PQ} = 0,$$

which is to say, iff

(12.7.1) $$A(x - x_0) + B(y - y_0) + C(z - z_0) = 0.$$

This is an equation in x, y, z for the plane that passes through $P(x_0, y_0, z_0)$ and has normal vector $\mathbf{N} = A\mathbf{i} + B\mathbf{j} + C\mathbf{k}$.

Remark If **N** is normal to a given plane, then so is every nonzero scalar multiple of **N**. Suppose we had chosen $-2\mathbf{N}$ as our normal. Then (12.7.1) would have read

$$-2A(x - x_0) - 2B(y - y_0) - 2C(z - z_0) = 0.$$

Canceling the -2, we would have the same equation we had before. It does not matter which normal we choose. All normals give equivalent equations. □

We can write (12.7.1) in the form

$$Ax + By + Cz + D = 0$$

simply by setting $D = -Ax_0 - By_0 - Cz_0$.

Example 1 Write an equation for the plane that passes through the point $P(1, 0, 2)$ and has normal vector $\mathbf{N} = 3\mathbf{i} - 2\mathbf{j} + \mathbf{k}$.

SOLUTION The general equation

$$A(x - x_0) + B(y - y_0) + C(z - z_0) = 0$$

becomes

$$3(x - 1) + (-2)(y - 0) + (z - 2) = 0,$$

which simplifies to

$$3x - 2y + z - 5 = 0. \quad \square$$

Example 2 Find an equation for the plane p that passes through $P(-2, 3, 5)$ and is perpendicular to the line l with scalar parametric equations: $x = -2 + t$, $y = 1 + 2t$, $z = 4$.

SOLUTION We can take $\mathbf{N} = \mathbf{i} + 2\mathbf{j}$ as a direction vector for l. Since p and l are perpendicular, **N** is a normal vector for p. Thus, as an equation for p, we can write

$$(x + 2) + 2(y - 3) + 0(z - 5) = 0,$$

which simplifies to

$$x + 2y - 4 = 0.$$

The last equation looks very much like the equation of a line in the xy-plane. If the context of our discussion were the xy-plane, then the equation $x + 2y - 4 = 0$ would represent a line. In this case, however, our context is three-dimensional space. Hence, the equation $x + 2y - 4 = 0$ represents the set of all points $Q(x, y, z)$, where $x + 2y - 4 = 0$ and z is unrestricted. This set forms a vertical plane that intersects the xy-plane in the indicated line (Figure 12.7.2). □

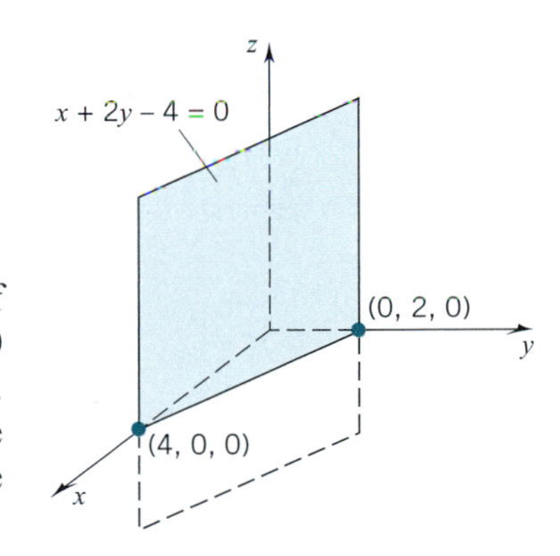

Figure 12.7.2

Example 3 Show that every equation

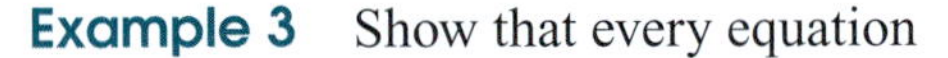

$$ax + by + cz + d = 0 \quad \text{with} \quad \sqrt{a^2 + b^2 + c^2} \neq 0$$

represents a plane in space.

SOLUTION Since $\sqrt{a^2 + b^2 + c^2} \neq 0$, the numbers a, b, c are not all zero, and therefore there exist numbers x_0, y_0, z_0 such that

$$ax_0 + by_0 + cz_0 + d = 0.\dagger$$

† Would such numbers necessarily exist if $\sqrt{a^2 + b^2 + c^2}$ were zero?

The equation

$$ax + by + cz + d = 0$$

can now be written

$$(ax + by + cz + d) - (ax_0 + by_0 + cz_0 + d) = 0,$$

and so, after factoring, we have

$$a(x - x_0) + b(y - y_0) + c(z - z_0) = 0.$$

This equation (and hence the initial equation) represents the plane through the point $P(x_0, y_0, z_0)$ with normal $\mathbf{N} = a\mathbf{i} + b\mathbf{j} + c\mathbf{k}$. The initial assumption that $\sqrt{a^2 + b^2 + c^2} \neq 0$ guarantees that $\mathbf{N} \neq \mathbf{0}$. □

Vector Equation of a Plane

We can write the equation of a plane entirely in vector notation. With

$$\mathbf{N} = A\mathbf{i} + B\mathbf{j} + C\mathbf{k}$$

and

$$\mathbf{r}_0 = x_0\mathbf{i} + y_0\mathbf{j} + z_0\mathbf{k}, \quad \mathbf{r} = x\mathbf{i} + y\mathbf{j} + z\mathbf{k},$$

Equation (12.7.1) reads

(12.7.2) $$\boxed{\mathbf{N} \cdot (\mathbf{r} - \mathbf{r}_0) = 0.}$$

This vector equation represents the plane that passes through the tip of $\mathbf{r}_0$ and has normal $\mathbf{N}$. (See Figure 12.7.3.) If the plane passes through the origin, we can take $\mathbf{r}_0 = \mathbf{0}$. Equation (12.7.2) then takes the form

(12.7.3) $$\boxed{\mathbf{N} \cdot \mathbf{r} = 0.}$$

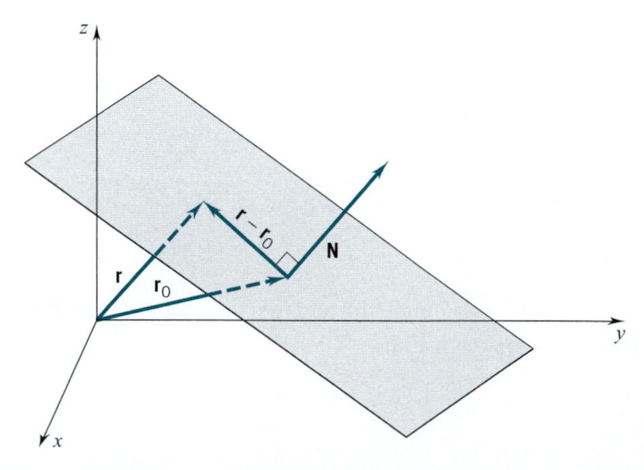

Figure 12.7.3

Collinear Vectors, Coplanar Vectors

Collinear points are points that lie on the same line; *coplanar points* are points that lie on the same plane. The terms "collinear" and "coplanar" are also applied to vectors: by definition, two vectors **a** and **b** are said to be *collinear* if there exist scalars s and t not both 0 such that

$$s\mathbf{a} + t\mathbf{b} = \mathbf{0}.$$

If $s \neq 0$, then $\mathbf{a} = -(t/s)\mathbf{b}$; if $t \neq \mathbf{0}$, then $\mathbf{b} = -(s/t)\mathbf{a}$. Collinear vectors are thus parallel. If we set

$$\mathbf{a} = \overrightarrow{PA} \qquad \text{and} \qquad \mathbf{b} = \overrightarrow{PB},$$

then the points P, A, B all fall on the same line, hence the term "collinear vectors."

Three vectors **a**, **b**, **c** are said to be *coplanar* if there exist scalars s, t, u not all zero such that

$$s\mathbf{a} + t\mathbf{b} + u\mathbf{c} = \mathbf{0}.$$

This term, too, is justified:

(12.7.4) $\mathbf{a} = \overrightarrow{PA}, \mathbf{b} = \overrightarrow{PB}, \mathbf{c} = \overrightarrow{PC}$ are coplanar vectors iff the points P, A, B, C all lie on the same plane .

PROOF Here we show that if the three vectors are coplanar, then the four points all lie on the same plane. We leave the converse to you (Exercise 45).

Suppose that the three vectors are coplanar. Then we can write

$$s\,\overrightarrow{PA} + t\,\overrightarrow{PB} + u\,\overrightarrow{PC} = \mathbf{0} \quad \text{with } s, t, u \text{ not all zero}.$$

Without loss of generality, we assume that $s \neq 0$. Then

$$\overrightarrow{PA} = -\frac{t}{s}\overrightarrow{PB} - \frac{u}{s}\overrightarrow{PC}.$$

Since $\overrightarrow{PB} \times \overrightarrow{PC}$ is perpendicular to both $\overrightarrow{PB}$ and $\overrightarrow{PC}$, we see that

$$\begin{aligned}(\overrightarrow{PB} \times \overrightarrow{PC}) \cdot \overrightarrow{PA} &= (\overrightarrow{PB} \times \overrightarrow{PC}) \cdot \left(-\frac{t}{s}\overrightarrow{PB} - \frac{u}{s}\overrightarrow{PC}\right) \\ &= -\frac{t}{s}(\overrightarrow{PB} \times \overrightarrow{PC}) \cdot \overrightarrow{PB} - \frac{u}{s}(\overrightarrow{PB} \times \overrightarrow{PC}) \cdot \overrightarrow{PC} \\ &= 0.\end{aligned}$$

But $|(\overrightarrow{PB} \times \overrightarrow{PC}) \cdot \overrightarrow{PA}|$ gives the volume of the parallelepiped with edges $\overrightarrow{PA}, \overrightarrow{PB}, \overrightarrow{PC}$. This volume can be zero only if P, A, B, C all lie on the same plane. ❑

Unit Normals

If **N** is normal to a given plane, then all other normals to that plane are parallel to **N** and hence scalar multiples of **N**. In particular there are only two normals of length 1:

$$\mathbf{u_N} = \frac{\mathbf{N}}{\|\mathbf{N}\|} \qquad \text{and} \qquad -\mathbf{u_N} = \frac{\mathbf{N}}{\|\mathbf{N}\|}.$$

These are called the *unit normals*.

Example 4 Find the unit normals for the plane $3x - 4y + 12z + 8 = 0$.

SOLUTION We can take $\mathbf{N} = 3\mathbf{i} - 4\mathbf{j} + 12\mathbf{k}$.

Since
$$\|\mathbf{N}\| = \sqrt{3^2 + (-4)^2 + 12^2} = \sqrt{169} = 13,$$

we have $\mathbf{u_N} = \frac{1}{13}(3\mathbf{i} - 4\mathbf{j} + 12\mathbf{k})$ and $-\mathbf{u_N} = -\frac{1}{13}(3\mathbf{i} - 4\mathbf{j} + 12\mathbf{k})$. □

Parallel Planes, Intersecting Planes

Two planes are called *parallel* iff their normals are parallel. If two planes, p_1 and p_2, are not parallel, we can find the angle between them by finding the angle between their normals, $\mathbf{N}_1, \mathbf{N}_2$. (See Figure 12.7.4.) Depending on our choice of normals, there are two such angles, each the supplement of the other. We will choose the smaller angle, the one with the nonnegative cosine:

(12.7.5)
$$\cos\theta = |\mathbf{u}_{\mathbf{N}_1} \cdot \mathbf{u}_{\mathbf{N}_2}|.$$

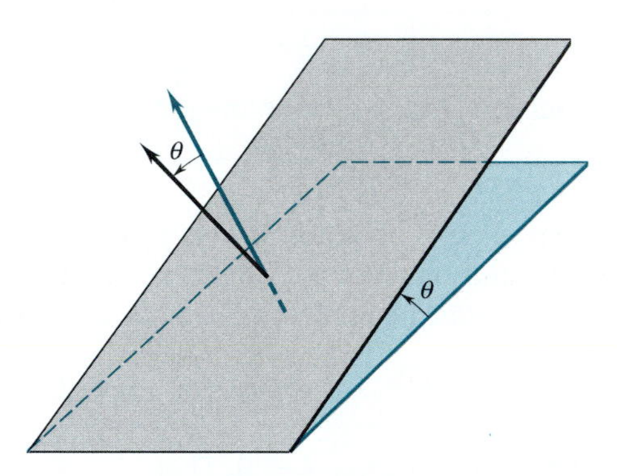

Figure 12.7.4

Example 5 Here are some planes:

$$p_1 : 2(x-1) - 3y + 5(z-2) = 0, \qquad p_2 : -4x + 6y + 10z + 24 = 0,$$
$$p_3 : 4x - 6y - 10z + 1 = 0, \qquad p_4 : 2x - 3y + 5z - 12 = 0.$$

(a) Indicate which planes are identical.

(b) Indicate which planes are distinct but parallel.

(c) Find the angle between p_1 and p_2.

SOLUTION

(a) p_1 and p_4 are identical, as you can verify by simplifying the equation of p_1.

(b) p_2 and p_3 are distinct but parallel. The planes are distinct since $P(0, 0, \frac{1}{10})$ lies on p_3 but not on p_2; they are parallel since the normals

$$-4\mathbf{i} + 6\mathbf{j} + 10\mathbf{k} \quad \text{and} \quad 4\mathbf{i} - 6\mathbf{j} - 10\mathbf{k}$$

are parallel.

(c) Taking

$$\mathbf{N}_1 = 2\mathbf{i} - 3\mathbf{j} + 5\mathbf{k} \quad \text{and} \quad \mathbf{N}_2 = -4\mathbf{i} + 6\mathbf{j} + 10\mathbf{k},$$

we have

$$\mathbf{u}_{\mathbf{N}_1} = \frac{1}{\sqrt{38}}(2\mathbf{i} - 3\mathbf{j} + 5\mathbf{k}) \qquad \text{and} \qquad \mathbf{u}_{\mathbf{N}_2} = \frac{1}{\sqrt{152}}(-4\mathbf{i} + 6\mathbf{j} + 10\mathbf{k}).$$

As you can check,

$$\cos\theta = |\mathbf{u}_{\mathbf{N}_1} \cdot \mathbf{u}_{\mathbf{N}_2}| = \tfrac{6}{19} \quad \text{and thus} \quad \theta \cong 1.25 \text{ radians, about } 71.59°. \quad \square$$

Example 6 The planes

$$p_1 : A_1x + B_1y + C_1z + D_1 = 0, \qquad p_2 : A_2x + B_2y + C_2z + D_2 = 0$$

intersect to form a line l. Find a vector parametrization for l.

SOLUTION We need to find a point P_0 on l and a direction vector for l. Finding P_0 is a matter of finding numbers x, y, z that simultaneously satisfy the equations given for p_1 and p_2. In concrete cases this is not hard, and we will not try to give a general formula for such a P_0. We will just assume that P_0 has been found and focus on finding a direction vector for l.

Since p_1 and p_2 intersect in a line, the normals

$$\mathbf{N}_1 = A_1\mathbf{i} + B_1\mathbf{j} + C_1\mathbf{k}, \quad \mathbf{N}_2 = A_2\mathbf{i} + B_2\mathbf{j} + C_2\mathbf{k}$$

are not parallel. This guarantees that $\mathbf{N}_1 \times \mathbf{N}_2$ is not $\mathbf{0}$. Since l lies on both p_1 and p_2, the line l, like the vector $\mathbf{N}_1 \times \mathbf{N}_2$, is perpendicular to both $\mathbf{N}_1$ and $\mathbf{N}_2$. This makes l parallel to $\mathbf{N}_1 \times \mathbf{N}_2$. (See Figure 12.7.5.) We can therefore take $\mathbf{N}_1 \times \mathbf{N}_2$ as a

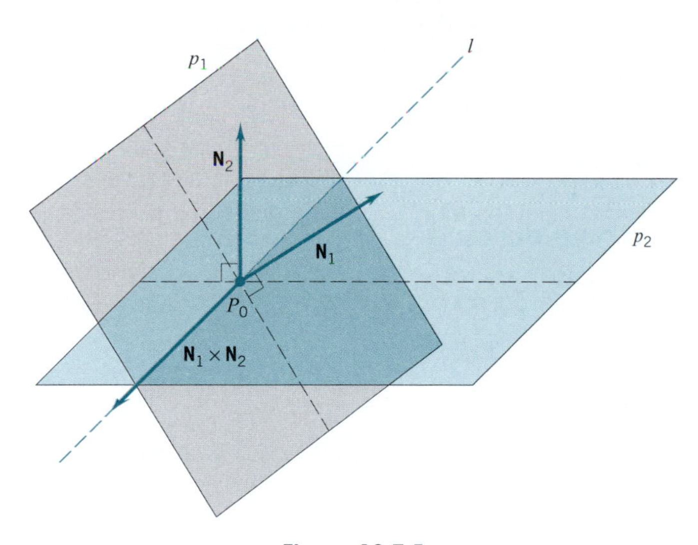

Figure 12.7.5

direction vector for l and write

$$l : \mathbf{r}(t) = \overrightarrow{OP_0} + t(\mathbf{N}_1 \times \mathbf{N}_2). \quad \square$$

Example 7 Given the planes $p_1 : 2x - 3y + 2z = 9$ and $p_2 : x + 2y - z = -4$, show that p_1 and p_2 are not parallel. Then find scalar parametric equations of the line l which is formed by the intersection of the two planes.

SOLUTION As normal vectors for p_1 and p_2, we can take $\mathbf{N}_1 = 2\mathbf{i} - 3\mathbf{j} + 2\mathbf{k}$ and $\mathbf{N}_2 = \mathbf{i} + 2\mathbf{j} - \mathbf{k}$. Since neither vector is a scalar multiple of the other, the vectors are not parallel. Therefore, p_1 and p_2 are not parallel.

As shown in Example 6, we can use

$$\mathbf{N}_1 \times \mathbf{N}_2 = \begin{vmatrix} \mathbf{i} & \mathbf{j} & \mathbf{k} \\ 2 & -3 & 2 \\ 1 & 2 & -1 \end{vmatrix} = -\mathbf{i} + 4\mathbf{j} + 7\mathbf{k}$$

as a direction vector for l. Now we need a point that lies on l. To find one, we solve the equations for p_1 and p_2 simultaneously. If, for example, we set $x = 0$ in the two equations, we get

$$-3y + 2z = 9$$
$$2y - z = -4.$$

Solving this pair of equations for y and z, we find that $y = 1$ and $z = 6$. Thus, the point $(0, 1, 6)$ is on l. As scalar parametric equations for l, we can write

$$x = -t, \quad y = 1 + 4t, \quad z = 6 + 7t. \quad \square$$

The Plane Determined by Three Noncollinear Points

Suppose now that we are given three noncollinear points P_1, P_2, P_3. These points determine a plane. How can we find an equation for this plane?

First we form the vectors $\overrightarrow{P_1P_2}, \overrightarrow{P_1P_3}$. Since P_1, P_2, P_3 are noncollinear,the vectors are not parallel. Therefore their cross product $\overrightarrow{P_1P_2} \times \overrightarrow{P_1P_3}$ can be used as a normal for the plane. We are back in a familiar situation. We have a point of the plane, say P_1, and we have a normal vector, $\overrightarrow{P_1P_2} \times \overrightarrow{P_1P_3}$. A point P will lie on the plane iff

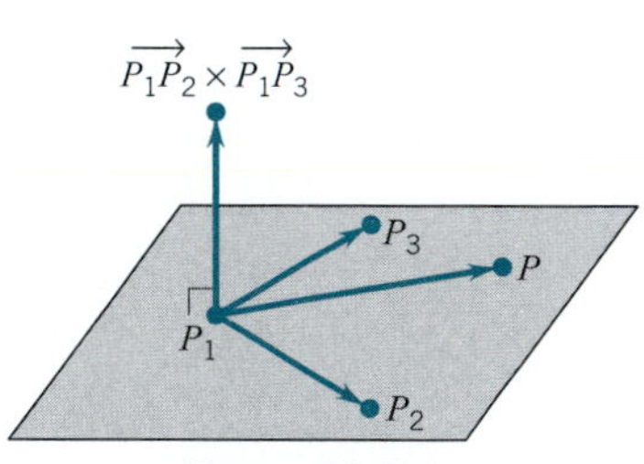

Figure 12.7.6

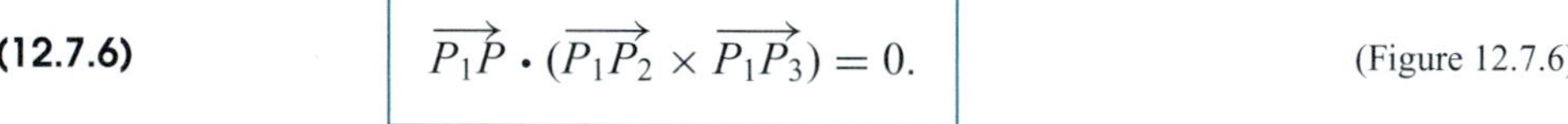

(12.7.6) $$\overrightarrow{P_1P} \cdot (\overrightarrow{P_1P_2} \times \overrightarrow{P_1P_3}) = 0.$$ (Figure 12.7.6)

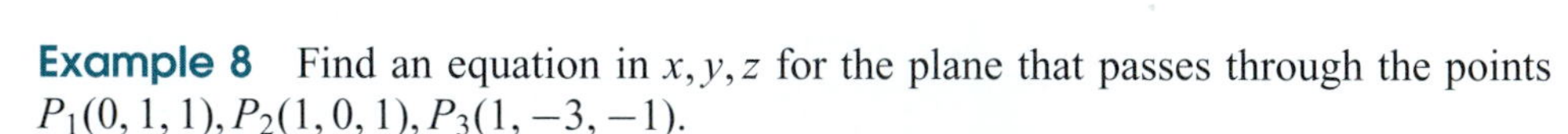

Example 8 Find an equation in x, y, z for the plane that passes through the points $P_1(0, 1, 1), P_2(1, 0, 1), P_3(1, -3, -1)$.

SOLUTION The point $P = P(x, y, z)$ will lie on this plane iff

$$\overrightarrow{P_1P} \cdot (\overrightarrow{P_1P_2} \times \overrightarrow{P_1P_3}) = 0.$$

Here

$$\overrightarrow{P_1P} = x\mathbf{i} + (y - 1)\mathbf{j} + (z - 1)\mathbf{k}, \quad \overrightarrow{P_1P_2} = \mathbf{i} - \mathbf{j}, \quad \overrightarrow{P_1P_3} = \mathbf{i} - 4\mathbf{j} - 2\mathbf{k}.$$

As you can check,

$$\overrightarrow{P_1P_2} \times \overrightarrow{P_1P_3} = 2\mathbf{i} + 2\mathbf{j} - 3\mathbf{k}.$$

Thus,

$$\begin{aligned}\overrightarrow{P_1P} \cdot (\overrightarrow{P_1P_2} \times \overrightarrow{P_1P_3}) &= [x\mathbf{i} + (y - 1)\mathbf{j} + (z - 1)\mathbf{k}] \cdot [2\mathbf{i} + 2\mathbf{j} - 3\mathbf{k}] \\ &= 2x + 2(y - 1) - 3(z - 1) = 2x + 2y - 3z + 1.\end{aligned}$$

As an equation for the plane, we can use

$$2x + 2y - 3z + 1 = 0. \quad \square$$

The Distance from a Point to a Plane

In Figure 12.7.7, we have drawn a plane $p : Ax + By + Cz + D = 0$ and a point $P(x_1, y_1, z_1)$ not on p. The distance between the point P_1 and the plane p is given by the formula

(12.7.7)
$$d(P_1, p) = \frac{|Ax_1 + By_1 + Cz_1 + D|}{\sqrt{A^2 + B^2 + C^2}}.$$

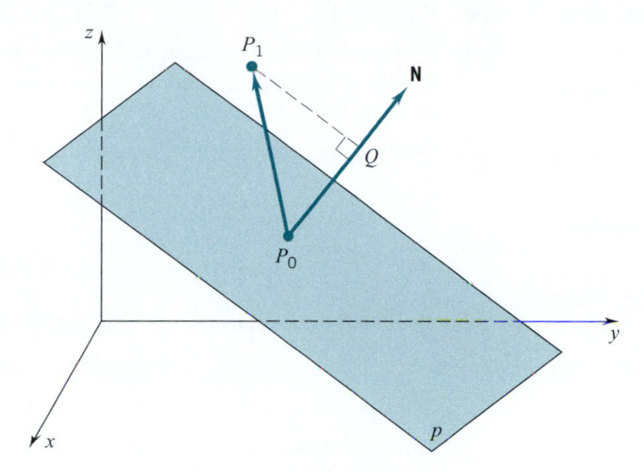

Figure 12.7.7

PROOF Pick any point $P_0(x_0, y_0, z_0)$ in the plane. As a normal to p we can take the vector

$$\mathbf{N} = A\mathbf{i} + B\mathbf{j} + C\mathbf{k}.$$

Then
$$\mathbf{u_N} = \frac{A\mathbf{i} + B\mathbf{j} + C\mathbf{k}}{\sqrt{A^2 + B^2 + C^2}}$$

is the corresponding unit normal. From Figure 12.7.7,

$$\begin{aligned} d(P_1, p) = d(P_0, Q) &= |\text{comp}_{\mathbf{N}}\, \overrightarrow{P_0P_1}| \\ &= |\overrightarrow{P_0P_1} \cdot \mathbf{u_N}| \\ &= \frac{|(x_1 - x_0)A + (y_1 - y_0)B + (z_1 - z_0)C|}{\sqrt{A^2 + B^2 + C^2}} \\ &= \frac{|Ax_1 + By_1 + Cz_1 - (Ax_0 + By_0 + Cz_0)|}{\sqrt{A^2 + B^2 + C^2}}. \end{aligned}$$

Since $P_0(x_0, y_0, z_0)$ lies on p,

$$Ax_0 + By_0 + Cz_0 = -D,$$

and we have

$$d(P_1, p) = \frac{|Ax_1 + By_1 + Cz_1 + D|}{\sqrt{A^2 + B^2 + C^2}}. \quad \square$$

EXERCISES 12.7

1. Which of the points $P(3,2,1), Q(2,3,1), R(1,4,1)$ lie on the plane

$$3(x-1)+4y-5(z+2)=0?$$

2. Which of the points $P(2,1,-2), Q(2,0,0), R(4,1,-1), S(0,-1,-3)$ lie on the plane $\mathbf{N}\cdot(\mathbf{r}-\mathbf{r}_0)=0$ if $\mathbf{N}=\mathbf{i}-3\mathbf{j}+\mathbf{k}$ and $\mathbf{r}_0=4\mathbf{i}+\mathbf{j}-\mathbf{k}$?

Find an equation for the plane which satisfies the given conditions.

3. Passes through the point $P(2,3,4)$ and is perpendicular to $\mathbf{i}-4\mathbf{j}+3\mathbf{k}$.

4. Passes through the point $P(1,-2,3)$ and is perpendicular to $\mathbf{j}+2\mathbf{k}$.

5. Passes through the point $P(2,1,1)$ and is parallel to the plane $3x-2y+5z-2=0$.

6. Passes through the point $P(3,-1,5)$ and is parallel to the plane $4x+2y-7z+5=0$.

7. Passes through the point $P(1,3,1)$ and contains the line $l: x=t,\ y=t,\ z=-2+t$.

8. Passes through the point $P(2,0,1)$ and contains the line $l: x=1-2t,\ y=1+4t,\ z=2+t$.

9. Passes through the point $P_0(x_0,y_0,z_0)$ and is perpendicular to $\overrightarrow{OP_0}$.

Find the unit normals to the plane.

10. $2x-3y+7z-3=0$. **11.** $2x-y+5z-10=0$.

12. Show that the plane $x/a+y/b+z/c=1$ intersects the coordinate axes at $x=a, y=b, z=c$. This is the equation of a plane in *intercept form*.

Write the equation of the plane in intercept form and find the points where it intersects the coordinate axes.

13. $4x+5y-6z=60$. **14.** $3x-y+4z+2=0$.

Find the angle between the planes.

15. $5(x-1)-3(y+2)+2z=0$,
$x+3(y-1)+2(z+4)=0$.

16. $2x-y+3z=5$, $5x+5y-z=1$.

17. $x-y+z-1=0$, $2x+y+3z+5=0$.

18. $4x+4y-2z=3$, $2x+y+z=-1$.

Determine whether the vectors are coplanar.

19. $4\mathbf{j}-\mathbf{k}$, $3\mathbf{i}+\mathbf{j}+2\mathbf{k}$, $\mathbf{0}$.

20. $\mathbf{i}$, $\mathbf{i}-2\mathbf{j}$, $3\mathbf{j}+\mathbf{k}$.

21. $\mathbf{i}+\mathbf{j}+\mathbf{k}$, $2\mathbf{i}-\mathbf{j}$, $3\mathbf{i}-\mathbf{j}-\mathbf{k}$.

22. $\mathbf{j}-\mathbf{k}$, $3\mathbf{i}-\mathbf{j}+2\mathbf{k}$, $3\mathbf{i}-2\mathbf{j}+3\mathbf{k}$.

Find the distance from the point P to the given plane.

23. $P(2,-1,3)$; $2x+4y-z+1=0$.

24. $P(3,-5,2)$; $8x-2y+z=5$.

25. $P(1,-3,5)$; $-3x+4z+5=0$.

26. $P(1,3,4)$; $x+y-2z=0$.

Find an equation in x,y,z for the plane that passes through the given points.

27. $P_1(1,0,1)$, $P_2(2,1,0)$, $P_3(1,1,1)$.

28. $P_1(1,1,1)$, $P_2(2,-2,-1)$, $P_3(0,2,1)$.

29. $P_1(3,-4,1)$, $P_2(3,2,1)$, $P_3(-1,1,-2)$.

30. $P_1(3,2,-1)$, $P_2(3,-2,4)$, $P_3(1,-1,3)$.

31. Write equations in symmetric form for the line that passes through $P_0(x_0,y_0,z_0)$ and is perpendicular to the plane $Ax+By+Cz+D=0$.

32. Find the distance between the parallel planes

$$Ax+By+Cz+D_1=0$$

and

$$Ax+By+Cz+D_2=0.$$

33. Show that the equations of a line in symmetric form

$$\frac{x-x_0}{d_1}=\frac{y-y_0}{d_2}=\frac{z-z_0}{d_3}$$

express the line as an intersection of two planes by finding equations for two such planes.

34. Find scalar parametric equations for the line formed by the two intersecting planes.
(a) $z=z_0$, $y=y_0$. (b) $x=x_0$, $z=z_0$.

In Exercises 35 and 36, find a set of scalar parametric equations for the line formed by the two intersecting planes.

35. $p_1: x+2y+3z=0$, $p_2: -3x+4y+z=0$.

36. $p_1: x+y+z+1=0$, $p_2: x-y+z+2=0$.

In Exercises 37 and 38, let l be the line determined by P_1, P_2, and let p be the plane determined by Q_1, Q_2, Q_3. Where, if anywhere, does l intersect p?

37. $P_1(1,-1,2)$, $P_2(-2,3,1)$;
$Q_1(2,0,-4)$, $Q_2(1,2,3)$, $Q_3(-1,2,1)$.

38. $P_1(4,-3,1)$, $P_2(2,-2,3)$;
$Q_1(2,0,-4)$, $Q_2(1,2,3)$, $Q_3(-1,2,1)$.

39. Let l_1, l_2 be lines that pass through the origin and have direction vectors

$$\mathbf{d}=\mathbf{i}+2\mathbf{j}+4\mathbf{k}, \quad \mathbf{D}=-\mathbf{i}-\mathbf{j}+3\mathbf{k}.$$

Find an equation for the plane that contains l_1 and l_2.

40. Show that two nonparallel lines $\mathbf{r}(t)=\mathbf{r}_0+t\mathbf{d}$ and $\mathbf{R}(t)=\mathbf{R}_0+t\mathbf{D}$ intersect iff the vectors $\mathbf{r}_0-\mathbf{R}_0, \mathbf{d}$, and $\mathbf{D}$ are coplanar.

41. Given that a plane contains the point P and has normal $\mathbf{N}$, describe the set of points Q on the plane for which $(\mathbf{N}+\overrightarrow{PQ})\perp(\mathbf{N}-\overrightarrow{PQ})$. HINT: Draw a figure.

42. Let $\mathbf{a},\mathbf{b},\mathbf{c}$ be three nonzero vectors such that the angle between any pair of them is $\frac{1}{2}\pi$. Can these vectors be coplanar?

43. Suppose that N $= A\mathbf{i} + B\mathbf{j} + C\mathbf{k}$ is a nonzero vector with its initial point on the plane $Ax + By + Cz + D = 0$. Take $P_1(x_1, y_1, z_1)$ as a point of space and set $\alpha = Ax_1 + By_1 + Cz_1 + D$. If $\alpha = 0$, then P_1 lies on the plane. What can you conclude about P_1 if α is positive? If α is negative?

44. Suppose that $\mathbf{a}, \mathbf{b}, \mathbf{c}$ are radius vectors the tips of which are not collinear. Give a geometric interpretation of the equation

$$\begin{vmatrix} x - a_1 & y - a_2 & z - a_3 \\ x - b_1 & y - b_2 & z - b_3 \\ x - c_1 & y - c_2 & z - c_3 \end{vmatrix} = 0.$$

45. Suppose that the points P, A, B, and C all lie on the same plane. Show that the vectors $\mathbf{a} = \overrightarrow{PA}$, $\mathbf{b} = \overrightarrow{PB}$, and $\mathbf{c} = \overrightarrow{PC}$ are coplanar vectors.

(*Sketching planes*). In Exercises 46–49, the equation of a plane p is given.

(a) Find the intercepts of p.

(b) Find the *traces* of p. (The traces of p are the lines of intersection of p with the coordinate planes.

(c) Find the unit normals.

(d) Sketch the plane.

46. $x + 2y + 3z - 6 = 0$.

47. $5x + 4y + 10z = 20$.

48. $3x + 2y - 6 = 0$.

49. $3x + 2z - 12 = 0$.

In Exercises 50–53, find an equation for the plane shown in the figure.

50.

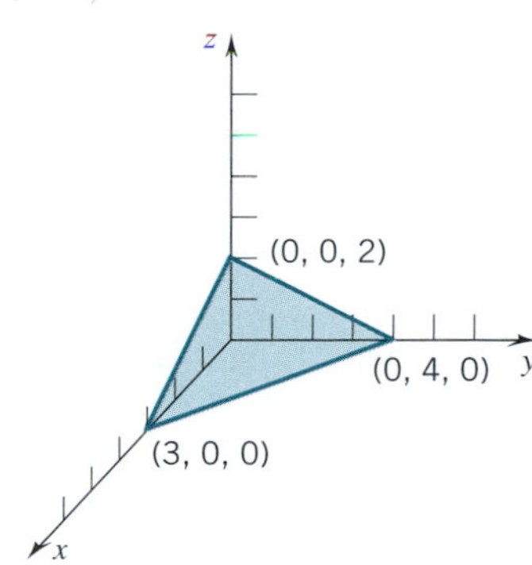

51.

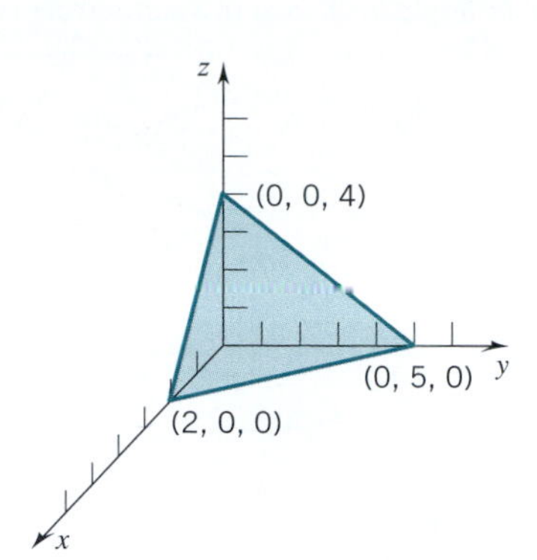

52.

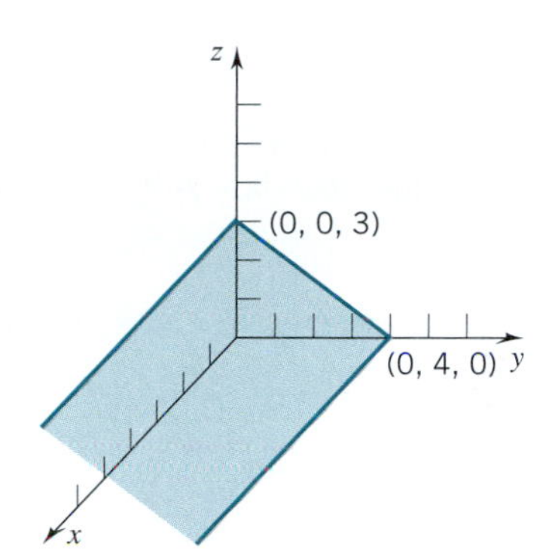

53.

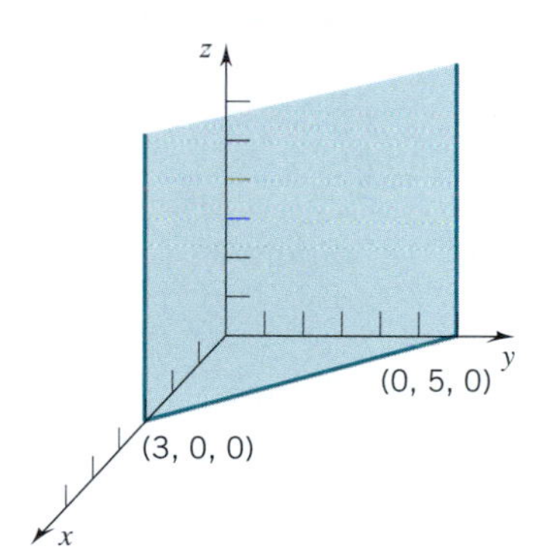

■ PROJECT 12.7 Some Geometry by Vector Methods

Try your hand at proving the following theorems by vector methods. Follow the hints if you like, but you may find it more interesting to disregard them and come up with proofs that are entirely you own.

Problem 1. The diagonals of a parallelogram are perpendicular iff the parallelogram is a rhombus.

HINT FOR PROOF With $\mathbf{a}$ and $\mathbf{b}$ as in Figure 1, the diagonals are $\mathbf{a} + \mathbf{b}$ and $\mathbf{a} - \mathbf{b}$.
Show that $(\mathbf{a} + \mathbf{b}) \cdot (\mathbf{a} - \mathbf{b}) = 0$ iff $\|\mathbf{a}\| = \|\mathbf{b}\|$. □

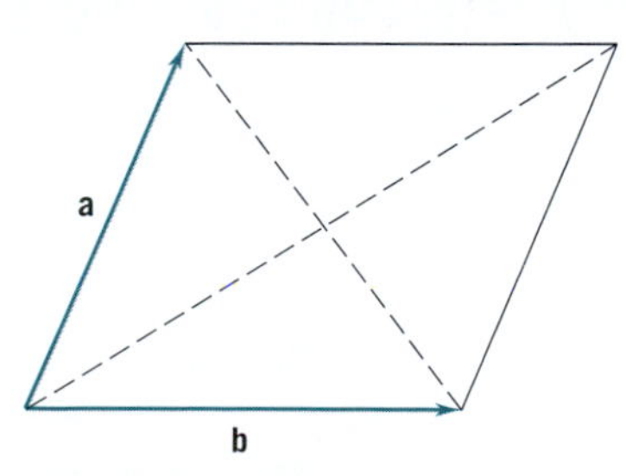

Figure 1

Problem 2. Every angle inscribed in a semicircle is a right angle.

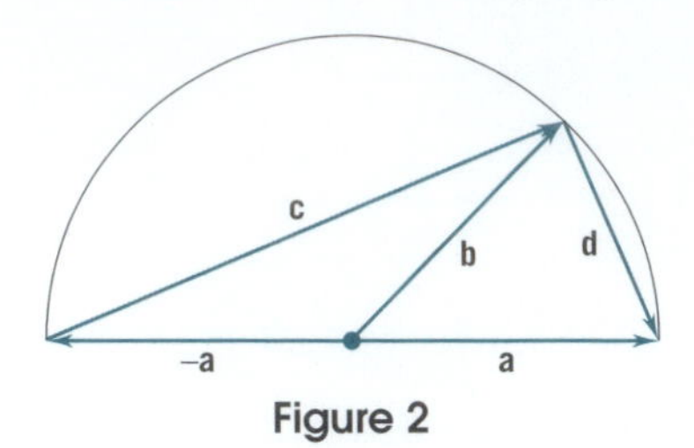

Figure 2

HINT FOR PROOF Take **c** and **d** as in Figure 2; express **c** and **d** in terms of **a** and **b**; then show that $\mathbf{c}\cdot\mathbf{d}=0$. ☐

Problem 3. In a parallelogram the sum of the squares of the lengths of the diagonals equals the sum of the squares of the lengths of the sides.

HINT FOR PROOF With **a** and **b** as in Figure 1, the diagonals are $\mathbf{a}+\mathbf{b}$ and $\mathbf{a}-\mathbf{b}$. Show that $\|\mathbf{a}+\mathbf{b}\|^2+\|\mathbf{a}-\mathbf{b}\|^2 = 2\|\mathbf{a}\|^2+2\|\mathbf{b}\|^2$. ☐

Problem 4. The three altitudes of a triangle intersect at one point.

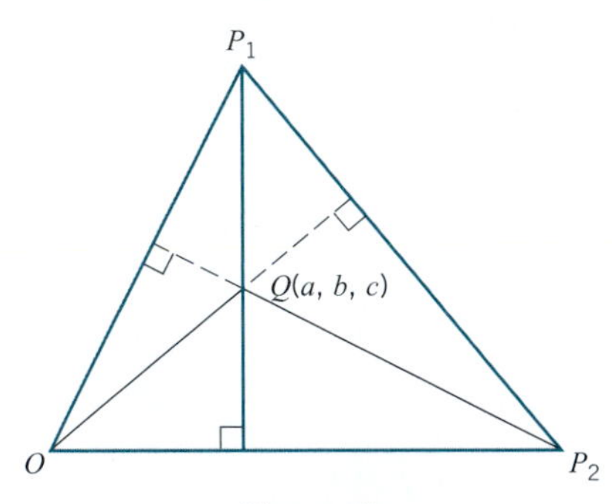

Figure 3

HINT FOR PROOF As in Figure 3 assume that the altitudes from P_1 and P_2 intersect at Q. Use the fact that $\overrightarrow{P_1Q}\perp\overrightarrow{OP_2}$ and $\overrightarrow{P_2Q}\perp\overrightarrow{OP_1}$ to show that $\overrightarrow{OQ}\perp\overrightarrow{P_1P_2}$. ☐

Problem 5. The three medians of a triangle intersect at one point.

HINT FOR PROOF With l_1, l_2, l_3 as in Figure 4,

$$l_1 : \mathbf{r}_1(t) = t(\mathbf{a}+\mathbf{b}),$$
$$l_2 : \mathbf{r}_2(u) = \tfrac{1}{2}\mathbf{b}+u(\mathbf{a}-\tfrac{1}{2}\mathbf{b}),$$
$$l_3 : \mathbf{r}_3(v) = \tfrac{1}{2}\mathbf{a}+v(\mathbf{b}-\tfrac{1}{2}\mathbf{a}).$$

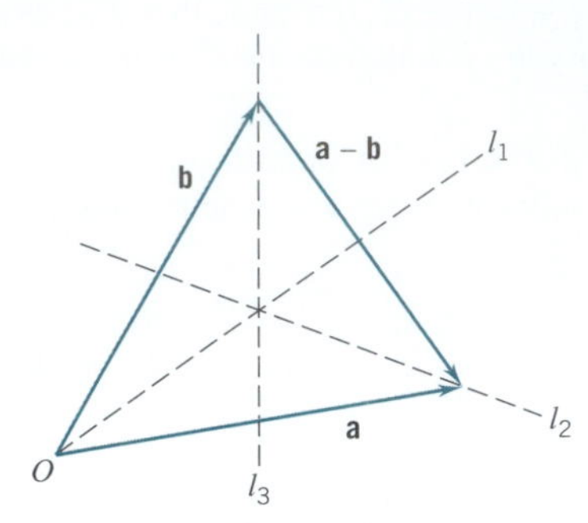

Figure 4

Show that l_1 intersects both l_2 and l_3 at the same point. ☐

Problem 6 (The Law of Sines). If a triangle has sides **a**, **b**, **c** and opposite angles A, B, C, then

$$\frac{\sin A}{\|\mathbf{a}\|}=\frac{\sin B}{\|\mathbf{b}\|}=\frac{\sin C}{\|\mathbf{c}\|}.$$

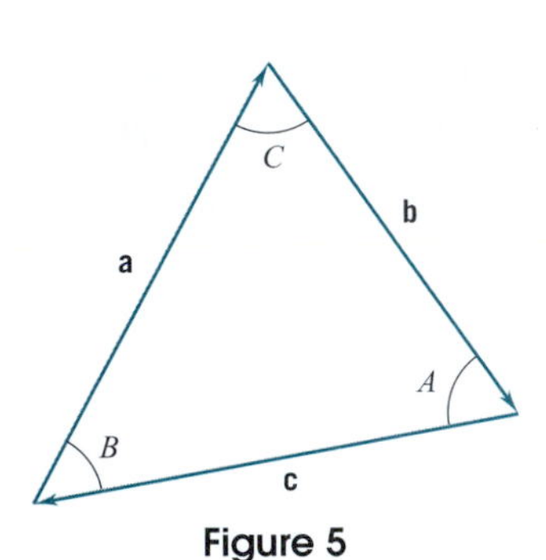

Figure 5

HINT FOR PROOF With **a**, **b**, **c** as in Figure 5, $\mathbf{a}+\mathbf{b}+\mathbf{c}=\mathbf{0}$. Observe then that $\mathbf{a}\times[\mathbf{a}+\mathbf{b}+\mathbf{c}]=\mathbf{0}$ and $\mathbf{b}\times[\mathbf{a}+\mathbf{b}+\mathbf{c}]=\mathbf{0}$. ☐

Problem 7. If two planes have a point in common, then they have a line in common.

HINT FOR PROOF As equations for the two planes, take $\mathbf{n}\cdot(\mathbf{r}-\mathbf{r}_0)=0$ and $\mathbf{N}\cdot(\mathbf{R}-\mathbf{R}_0)=0$. If the point $P(a_1,a_2,a_3)$ lies on both planes, then the vector $\mathbf{a}=a_1\mathbf{i}+a_2\mathbf{j}+a_3\mathbf{k}$ satisfies both equations. In that case we have $\mathbf{n}\cdot(\mathbf{a}-\mathbf{r}_0)=0$ and $\mathbf{N}\cdot(\mathbf{a}-\mathbf{R}_0)=0$. Consider the line $\mathbf{r}(t)=\mathbf{a}+t(\mathbf{n}\times\mathbf{N})$. ☐

■ CHAPTER HIGHLIGHTS

12.1 Cartesian Space Coordinates

distance formula (p. 708)
equation of sphere (p. 708)
midpoint formula (p. 709)

12.2 Displacements and Forces

12.3 Vectors

addition, multiplication by scalars (p. 714)
component (p.714)

parallel vectors (p. 716)
norm (p. 717)
unit vector (p. 718)
vectors in the plane (p. 719)
The zero vector $\mathbf{0}$ is parallel to every vector.

12.4 The Dot Product

$\mathbf{a} \cdot \mathbf{b} = a_1 b_1 + a_2 b_2 + a_3 b_3 = ||\mathbf{a}|| \, ||\mathbf{b}|| \cos\theta$
$\mathbf{a} \perp \mathbf{b}$ iff $\mathbf{a} \cdot \mathbf{b} = 0$
projections and components (p. 726)
directions angles, direction cosines (p. 728)
Schwarz's inequality (p. 729)

12.5 The Cross product

definition of $\mathbf{a} \times \mathbf{b}$ (p. 732)
properties of right-handed triples (p. 733)
properties of the cross product (p. 733)
distributive laws (p. 734)
scalar triple product (p. 734)
components of $\mathbf{a} \times \mathbf{b}$ (p. 735)
identities (p. 738)

12.6 Lines

position vector, radius vector (p. 741)
vector parametrization: $\mathbf{r}(t) = \mathbf{r}_0 + t\mathbf{d}$
direction vector, direction numbers (p. 742)
scalar parametric equations : $x(t) = x_0 + d_1 t,$
$y(t) = y_0 + d_2 t, \quad z(t) = z_0 + d_3 t$
symmetric form (p. 744)
distance from a point to a line (p. 747)

Two lines are parallel iff their direction vectors are parallel; two intersecting lines are perpendicular iff their direction vectors are perpendicular.

12.7 Planes

normal vector (p. 750)
scalar equation: $A(x - x_0) + B(y - y_0) + C(z - z_0) = 0$
vector equation of a plane (p. 752)
collinear vectors, coplanar vectors (p. 753)
parallel planes (p. 754)
angle between intersecting planes (p. 754)
plane determined by three noncollinear points (p. 756)
distance between a point and a plane (p. 757)

CHAPTER 13

VECTOR CALCULUS

■ 13.1 VECTOR FUNCTIONS

Introduction

If f_1, f_2, f_3 are real-valued functions defined on some interval I, then for each $t \in I$ we can form the vector

$$\mathbf{f}(t) = f_1(t)\,\mathbf{i} + f_2(t)\,\mathbf{j} + f_3(t)\,\mathbf{k}$$

and thereby create a *vector-valued function* **f**. For short we will call such a function a *vector function*. The real-valued functions f_1, f_2, f_3 are called the *components* of **f**. A point t is in the *domain* of a vector function **f** iff it is in the domain of each of its components.

For example, from the scalar functions

$$f_1(t) = x_0 + d_1 t, \quad f_2(t) = y_0 + d_2 t, \quad f_3(t) = z_0 + d_3 t,$$

we can form the vector function

$$\mathbf{f}(t) = (x_0 + d_1 t)\,\mathbf{i} + (y_0 + d_2 t)\,\mathbf{j} + (z_0 + d_3 t)\,\mathbf{k}.$$

The domain of **f** is the set of all real numbers. If d_1, d_2, d_3 are not all 0, then the radius vector $\mathbf{f}(t)$ traces out the line that passes through the point $P(x_0, y_0, z_0)$ and has direction numbers d_1, d_2, d_3. If d_1, d_2, d_3 are all 0, then we have the constant function

$$\mathbf{f}(t) = x_0\,\mathbf{i} + y_0\,\mathbf{j} + z_0\,\mathbf{k}.$$

Example 1 From the functions

$$f_1(t) = \cos t, \qquad f_2(t) = \sin t, \qquad f_3(t) = 0$$

we can form the vector function

$$\mathbf{f}(t) = \cos t\,\mathbf{i} + \sin t\,\mathbf{j}.$$

For each t

$$\|\mathbf{f}(t)\| = \sqrt{\cos^2 t + \sin^2 t} = 1.$$

Since the third component is zero, the radius vector $\mathbf{f}(t)$ lies in the xy-plane. As t increases, the tip of $\mathbf{f}(t)$ traces out the unit circle in a counterclockwise manner, effecting a complete revolution as t increases by 2π (Figure 13.1.1). ❑

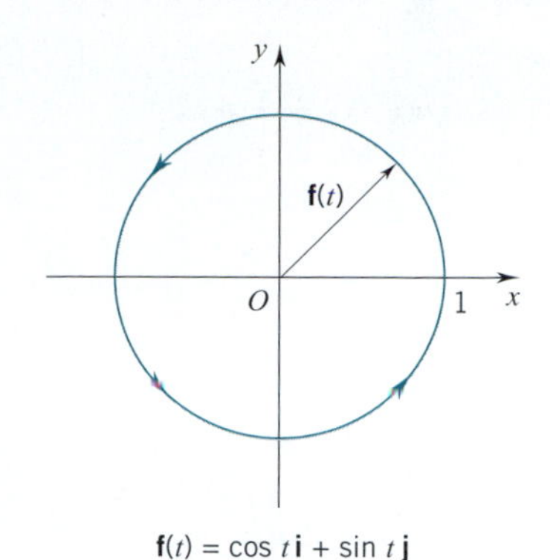

f(t) = cos t i + sin t j

Figure 13.1.1

Example 2 Each real-valued function f defined on an interval $[a, b]$ gives rise to a vector-valued function $\mathbf{f}$ in a natural way. Setting

$$f_1(t) = t, \quad f_2(t) = f(t), \quad f_3(t) = 0,$$

we obtain the vector function

$$\mathbf{f}(t) = t\,\mathbf{i} + f(t)\,\mathbf{j}.$$

As t ranges from a to b, the radius vector $\mathbf{f}(t)$ traces out the graph of f from left to right. See Figure 13.1.2. ❑

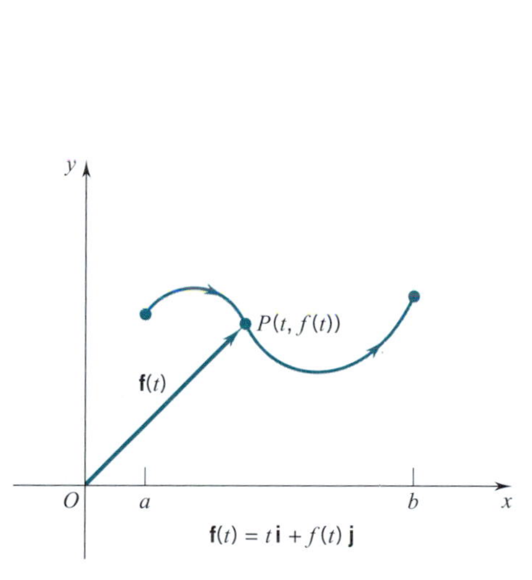

Figure 13.1.2

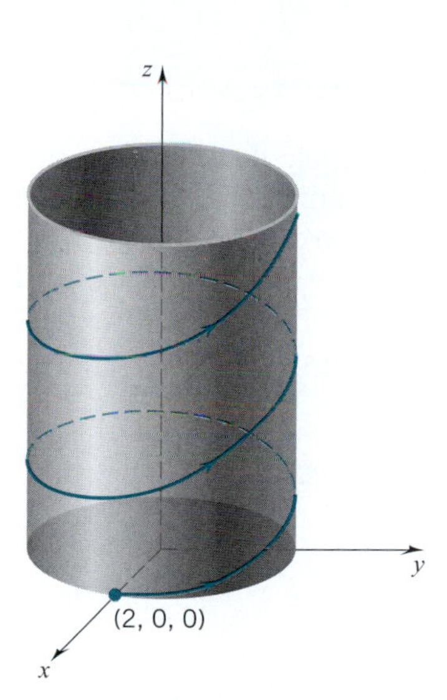

Figure 13.1.3

Example 3 From the functions

$$f_1(t) = 2\cos t, \qquad f_2(t) = 2\sin t, \qquad f_3(t) = t, \qquad t \geq 0,$$

we can form the vector function

$$\mathbf{f}(t) = 2\cos t\,\mathbf{i} + 2\sin t\,\mathbf{j} + t\,\mathbf{k}, \qquad t \geq 0.$$

At $t = 0$, the tip of the radius vector $\mathbf{f}(0)$ is at the point $(2, 0, 0)$. As t increases, the tip of $\mathbf{f}(t)$ spirals up the circular cylinder $x^2 + y^2 = 4$ (z arbitrary), effecting a complete turn on every t-interval of length 2π. This spiraling curve is called a *circular helix* (Figure 13.1.3). ❑

We return to the general case: a vector function

$$\mathbf{f}(t) = f_1(t)\,\mathbf{i} + f_2(t)\,\mathbf{j} + f_3(t)\,\mathbf{k}$$

defined on some t-interval I. As in the examples given, and under conditions to be spelled out later, as t ranges over the interval I, the tip of the radius vector $\mathbf{f}(t)$ traces out a curve C. The equations

$$x = f_1(t), \qquad y = f_2(t), \qquad z = f_3(t)$$

formed from the components of $\mathbf{f}$ serve as parametric equations for C. If one of the components is identically 0 on I, for example, if $\mathbf{f}$ has the form $\mathbf{f}(t) = f_1(t)\,\mathbf{i} + f_2(t)\,\mathbf{j}$, then C is a *plane curve*; otherwise C is a *space curve*.

The Limit Process

DEFINITION 13.1.1 LIMIT OF A VECTOR FUNCTION

$$\lim_{t\to t_0} \mathbf{f}(t) = \mathbf{L} \quad \text{iff} \quad \lim_{t\to t_0} \|\mathbf{f}(t) - \mathbf{L}\| = 0.$$

Note that for each t in the domain of $\mathbf{f}$, $\|\mathbf{f}(t) - \mathbf{L}\|$ is a real number, and therefore the limit on the right is the limit of a real-valued function. Thus we are still in familiar territory.

The first thing we show is that

(13.1.2) $$\text{if} \quad \lim_{t\to t_0} \mathbf{f}(t) = \mathbf{L}, \qquad \text{then} \qquad \lim_{t\to t_0} \|\mathbf{f}(t)\| = \|\mathbf{L}\|.$$

PROOF We know that

$$0 \le \Big|\, \|\mathbf{f}(t)\| - \|\mathbf{L}\| \,\Big| \le \|\mathbf{f}(t) - \mathbf{L}\|. \qquad \text{(Exercise 20, Section 12. 3)}$$

It follows from the pinching theorem that

$$\text{if} \quad \lim_{t\to t_0} \|\mathbf{f}(t) - \mathbf{L}\| = 0, \qquad \text{then} \qquad \lim_{t\to t_0} \Big|\, \|\mathbf{f}(t)\| - \|\mathbf{L}\| \,\Big| = 0. \quad \square$$

Remark The converse of (13.1.2) is false, as you can see by setting $\mathbf{f}(t) = \mathbf{k}$ and taking $\mathbf{L} = -\mathbf{k}$. $\square$

We can indicate that $\lim_{t\to t_0} \mathbf{f}(t) = \mathbf{L}$ by writing

$$\text{as } t \to t_0, \quad \mathbf{f}(t) \to \mathbf{L}.$$

We will state the limit rules in this form. As you will see below, there are no surprises.

THEOREM 13.1.3 LIMIT RULES

Let $\mathbf{f}$ and $\mathbf{g}$ be vector functions and let u be a real-valued function. Suppose that, as $t \to t_0$,

$$\mathbf{f}(t) \to \mathbf{L}, \quad \mathbf{g}(t) \to \mathbf{M}, \quad u(t) \to A.$$

Then

$$\mathbf{f}(t) + \mathbf{g}(t) \to \mathbf{L} + \mathbf{M}, \quad \alpha\mathbf{f}(t) \to \alpha\mathbf{L},$$

$$u(t)\mathbf{f}(t) \to A\mathbf{L}, \quad \mathbf{f}(t) \cdot \mathbf{g}(t) \to \mathbf{L} \cdot \mathbf{M}, \quad \mathbf{f}(t) \times \mathbf{g}(t) \to \mathbf{L} \times \mathbf{M}.$$

Each of these limit rules is easy to verify. We will verify the last one. To do this, we have to show that

$$\text{as } t \to t_0, \quad \|[\mathbf{f}(t) \times \mathbf{g}(t)] - [\mathbf{L} \times \mathbf{M}]\| \to 0.$$

We do this as follows:

$$\begin{aligned}
\|[\mathbf{f}(t) \times \mathbf{g}(t)] - [\mathbf{L} \times \mathbf{M}]\| &= \|[\mathbf{f}(t) \times \mathbf{g}(t)] - [\mathbf{L} \times \mathbf{g}(t)] + [\mathbf{L} \times \mathbf{g}(t)] - [\mathbf{L} \times \mathbf{M}]\| \\
&= \|[(\mathbf{f}(t) - \mathbf{L}) \times \mathbf{g}(t)] + [\mathbf{L} \times (\mathbf{g}(t) - \mathbf{M})]\| \qquad (12.5.4) \\
&\le \|(\mathbf{f}(t) - \mathbf{L}) \times \mathbf{g}(t)\| + \|\mathbf{L} \times (\mathbf{g}(t) - \mathbf{M})\| \\
&\qquad \text{(triangle inequality)} \\
&\le \|\mathbf{f}(t) - \mathbf{L}\|\ \|\mathbf{g}(t)\| + \|\mathbf{L}\|\ \|\mathbf{g}(t) - \mathbf{M}\|. \\
&\qquad \text{explain}
\end{aligned}$$

As $t \to t_0$, $\|\mathbf{g}(t)\| \to \|\mathbf{M}\|$. (This follows from 13.1.2.) Therefore, as $t \to t_0$,

$$\|\mathbf{f}(t) - \mathbf{L}\|\ \|\mathbf{g}(t)\| + \|\mathbf{L}\|\ \|\mathbf{g}(t) - \mathbf{M}\| \to (0)\|\mathbf{M}\| + \|\mathbf{L}\|(0) = 0.$$

Since $0 \le \|\mathbf{f}(t) \times \mathbf{g}(t)] - [\mathbf{L} \times \mathbf{M}]\|$, it follows from the pinching theorem that $\|[\mathbf{f}(t) \times \mathbf{g}(t)] - [\mathbf{L} \times \mathbf{M}]\| \to 0$.

The limit process can be carried out component by component. Let $\mathbf{f}(t) = f_1(t)\,\mathbf{i} + f_2(t)\,\mathbf{j} + f_3(t)\,\mathbf{k}$ and let $\mathbf{L} = L_1\,\mathbf{i} + L_2\,\mathbf{j} + L_3\,\mathbf{k}$. Then

(13.1.4)
$$\lim_{t \to t_0} \mathbf{f}(t) = \mathbf{L} \quad \text{iff}$$
$$\lim_{t \to t_0} f_1(t) = L_1, \quad \lim_{t \to t_0} f_2(t) = L_2, \quad \lim_{t \to t_0} f_3(t) = L_3.$$

PROOF

$$\begin{aligned}
\lim_{t \to t_0} \mathbf{f}(t) = \mathbf{L} \quad &\text{iff} \quad \lim_{t \to t_0} \|\mathbf{f}(t) - \mathbf{L}\| = 0 \\
&\text{iff} \quad \lim_{t \to t_0} \sqrt{[f_1(t) - L_1]^2 + [f_2(t) - L_2]^2 + [f_3(t) - L_3]^2} = 0 \\
&\text{iff} \quad \lim_{t \to t_0} f_1(t) = L_1, \quad \lim_{t \to t_0} f_2(t) = L_2, \quad \lim_{t \to t_0} f_3(t) = L_3. \quad \square
\end{aligned}$$

Example 4 Find $\lim_{t\to 0} \mathbf{f}(t)$ given that

$$\mathbf{f}(t) = \cos(t+\pi)\,\mathbf{i} + \sin(t+\pi)\,\mathbf{j} + e^{-t^2}\mathbf{k}.$$

SOLUTION

$$\begin{aligned}\lim_{t\to 0}\mathbf{f}(t) &= \lim_{t\to 0}[\cos(t+\pi)\,\mathbf{i} + \sin(t+\pi)\,\mathbf{j} + e^{-t^2}\mathbf{k}]\\ &= \left[\lim_{t\to 0}\cos(t+\pi)\right]\mathbf{i} + \left[\lim_{t\to 0}\sin(t+\pi)\right]\mathbf{j} + \left[\lim_{t\to 0}e^{-t^2}\right]\mathbf{k}\\ &= (-1)\,\mathbf{i} + (0)\,\mathbf{j} + (1)\,\mathbf{k} = -\mathbf{i} + \mathbf{k}. \quad \square\end{aligned}$$

Continuity and Differentiability

As you would expect, $\mathbf{f}$ is said to be *continuous* at t_0 if

$$\lim_{t\to t_0}\mathbf{f}(t) = \mathbf{f}(t_0).$$

Thus, by (13.1.4), $\mathbf{f}$ is continuous at t_0 iff each component of $\mathbf{f}$ is continuous at t_0.

The derivative of a vector function is defined as the limit of a *vector difference quotient*:

DEFINITION 13.1.5 DERIVATIVE OF A VECTOR FUNCTION

The vector function $\mathbf{f}$ is *differentiable* at t if

$$\lim_{h\to 0}\frac{\mathbf{f}(t+h)-\mathbf{f}(t)}{h} \quad \text{exists.}$$

If this limit exists, it is called the *derivative of* $\mathbf{f}$ at t and is denoted $\mathbf{f}'(t)$.

Differentiation can be carried out component by component; which is to say, if $\mathbf{f}(t) = f_1(t)\,\mathbf{i} + f_2(t)\,\mathbf{j} + f_3(t)\,\mathbf{k}$ is differentiable at t, then

$$\mathbf{f}'(t) = f_1'(t)\,\mathbf{i} + f_2'(t)\,\mathbf{j} + f_3'(t)\,\mathbf{k}.$$

PROOF

$$\begin{aligned}\mathbf{f}'(t) &= \lim_{h\to 0}\frac{\mathbf{f}(t+h)-\mathbf{f}(t)}{h}\\ &= \lim_{h\to 0}\left[\frac{f_1(t+h)-f_1(t)}{h}\,\mathbf{i} + \frac{f_2(t+h)-f_2(t)}{h}\,\mathbf{j} + \frac{f_3(t+h)-f_3(t)}{h}\,\mathbf{k}\right]\\ &= \left[\lim_{h\to 0}\frac{f_1(t+h)-f_1(t)}{h}\right]\mathbf{i} + \left[\lim_{h\to 0}\frac{f_2(t+h)-f_2(t)}{h}\right]\mathbf{j} +\\ &\qquad \left[\lim_{h\to 0}\frac{f_3(t+h)-f_3(t)}{h}\right]\mathbf{k}\\ &= f_1'(t)\,\mathbf{i} + f_2'(t)\,\mathbf{j} + f_3'(t)\,\mathbf{k}. \quad \square\end{aligned}$$

As with real-valued functions, if $\mathbf{f}$ is differentiable at t, then $\mathbf{f}$ is continuous at t. (Exercise 53.)

Interpretations of the vector derivative and applications of vector differentiation are introduced later in the chapter. Here we limit ourselves to computation.

Example 5 Given that $\mathbf{f}(t) = t\,\mathbf{i} + \sqrt{t+1}\,\mathbf{j} - e^t\,\mathbf{k}$, find:

(1) The domain of $\mathbf{f}$. (2) $\mathbf{f}(0)$.

(3) $\mathbf{f}'(t)$. (4) $\mathbf{f}'(0)$.

(5) $\|\mathbf{f}(t)\|$. (6) $\mathbf{f}(t) \cdot \mathbf{f}'(t)$.

SOLUTION

(1) For a number to be in the domain of $\mathbf{f}$, it is necessary only that it be in the domain of each of the components. The domain of $\mathbf{f}$ is $[-1, \infty)$.

(2) $\mathbf{f}(0) = 0\,\mathbf{i} + \sqrt{0+1}\,\mathbf{j} - e^0\,\mathbf{k} = \mathbf{j} - \mathbf{k}$.

(3) $\mathbf{f}'(t) = \mathbf{i} + \dfrac{1}{2\sqrt{t+1}}\,\mathbf{j} - e^t\,\mathbf{k}$.

(4) $\mathbf{f}'(0) = \mathbf{i} + \dfrac{1}{2\sqrt{0+1}}\,\mathbf{j} - e^0\,\mathbf{k} = i + \dfrac{1}{2}\mathbf{j} - \mathbf{k}$.

(5) $\|\mathbf{f}(t)\| = \sqrt{t^2 + (\sqrt{t+1})^2 + (-e^t)^2} = \sqrt{t^2 + t + 1 + e^{2t}}$.

(6) $\mathbf{f}(t) \cdot \mathbf{f}'(t) = (t\,\mathbf{i} + \sqrt{t+1}\,\mathbf{j} - e^t\,\mathbf{k}) \cdot \left(\mathbf{i} + \dfrac{1}{2\sqrt{t+1}}\,\mathbf{j} - e^t\,\mathbf{k}\right)$

$$= (t)(1) + (\sqrt{t+1})\left(\frac{1}{2\sqrt{t+1}}\right) + (-e^t)(-e^t) = t + \frac{1}{2} + e^{2t}. \quad \square$$

If $\mathbf{f}'$ is itself differentiable, we can calculate the second derivative $\mathbf{f}''$ and so on.

Example 6 Find $\mathbf{f}''(t)$ for $\mathbf{f}(t) = t\sin t\,\mathbf{i} + e^{-t}\,\mathbf{j} + t\,\mathbf{k}$.

SOLUTION

$$\mathbf{f}'(t) = (t\cos t + \sin t)\,\mathbf{i} - e^{-t}\mathbf{j} + \mathbf{k}.$$
$$\mathbf{f}''(t) = (-t\sin t + \cos t + \cos t)\,\mathbf{i} + e^{-t}\mathbf{j} = (2\cos t - t\sin t)\,\mathbf{i} + e^{-t}\mathbf{j}. \quad \square$$

Integration

Just as we can differentiate vector functions component by component, we can integrate component by component. For $\mathbf{f}(t) = f_1(t)\,\mathbf{i} + f_2(t)\,\mathbf{j} + f_3(t)\,\mathbf{k}$ continuous on $[a, b]$, we set

$$\int_a^b \mathbf{f}(t)\,dt = \left(\int_a^b f_1(t)\,dt\right)\mathbf{i} + \left(\int_a^b f_2(t)\,dt\right)\mathbf{j} + \left(\int_a^b f_3(t)\,dt\right)\mathbf{k}. \tag{13.1.6}$$

Example 7 Find $\displaystyle\int_0^1 \mathbf{f}(t)\,dt$ for $\mathbf{f}(t) = t\,\mathbf{i} + \sqrt{t+1}\,\mathbf{j} - e^t\,\mathbf{k}$.

SOLUTION

$$\int_0^1 \mathbf{f}(t)\,dt = \left(\int_0^1 t\,dt\right)\mathbf{i} + \left(\int_0^1 \sqrt{t+1}\,dt\right)\mathbf{j} + \left(\int_0^1 (-e^t)\,dt\right)\mathbf{k}$$

$$= \left[\tfrac{1}{2}t^2\right]_0^1\mathbf{i} + \left[\tfrac{2}{3}(t+1)^{3/2}\right]_0^1\mathbf{j} + \left[-e^t\right]_0^1\mathbf{k}$$

$$= \tfrac{1}{2}\,\mathbf{i} + \tfrac{2}{3}(2\sqrt{2}-1)\,\mathbf{j} + (1-e)\,\mathbf{k}. \quad \square$$

We can calculate indefinite integrals.

Example 8 Find $\mathbf{f}(t)$ given that

$$\mathbf{f}'(t) = 2\cos t\,\mathbf{i} - t\sin t^2\,\mathbf{j} + 2t\,\mathbf{k} \quad \text{and} \quad \mathbf{f}(0) = \mathbf{i} + 3\mathbf{k}.$$

SOLUTION By integrating $\mathbf{f}'(t)$, we find that

$$\mathbf{f}(t) = (2\sin t + C_1)\,\mathbf{i} + \left(\tfrac{1}{2}\cos t^2 + C_2\right)\mathbf{j} + (t^2 + C_3)\,\mathbf{k},$$

where C_1, C_2, C_3 are constants to be determined. Since

$$\mathbf{i} + 3\mathbf{k} = \mathbf{f}(0) = C_1\mathbf{i} + \left(\tfrac{1}{2} + C_2\right)\mathbf{j} + C_3\mathbf{k},$$

you can see that

$$C_1 = 1, \quad C_2 = -\tfrac{1}{2}, \quad C_3 = 3.$$

Thus
$$\mathbf{f}(t) = (2\sin t + 1)\,\mathbf{i} + \left(\tfrac{1}{2}\cos t^2 - \tfrac{1}{2}\right)\mathbf{j} + (t^2 + 3)\,\mathbf{k}. \quad \square$$

(Integration can also be carried out without direct reference to components. See Exercise 54.)

Properties of the Integral

It is easy to see that

(13.1.7)
$$\int_a^b [\mathbf{f}(t) + \mathbf{g}(t)]\,dt = \int_a^b \mathbf{f}(t)dt + \int_a^b \mathbf{g}(t)\,dt$$

and

(13.1.8)
$$\int_a^b [\alpha\mathbf{f}(t)]\,dt = \alpha\int_a^b \mathbf{f}(t)\,dt \qquad \text{for every constant scalar } \alpha.$$

It is also true that

(13.1.9)
$$\int_a^b [\mathbf{c}\cdot\mathbf{f}(t)]\,dt = \mathbf{c}\cdot\left(\int_a^b \mathbf{f}(t)\,dt\right) \qquad \text{for every constant vector } \mathbf{c}$$

and

(13.1.10)
$$\left\| \int_a^b \mathbf{f}(t)\,dt \right\| \leq \int_a^b \|\mathbf{f}(t)\|\,dt.$$

The proof of (13.1.9) is left as an exercise (Exercise 56). Here we prove (13.1.10). It is an important inequality.

PROOF Set $\mathbf{r} = \int_a^b \mathbf{f}(t)\,dt$ and note that

$$\|\mathbf{r}\|^2 = \mathbf{r}\cdot\mathbf{r} = \mathbf{r}\cdot \int_a^b \mathbf{f}(t)\,dt$$
$$= \int_a^b [\mathbf{r}\cdot\mathbf{f}(t)]\,dt \leq \int_a^b \|\mathbf{r}\|\,\|\mathbf{f}(t)\|\,dt = \|\mathbf{r}\| \int_a^b \|\mathbf{f}(t)\|\,dt.$$

by (13.1.9) ↑ ↑ by Schwarz's inequality (12.4.17)

If $\mathbf{r} \neq \mathbf{0}$, we can divide by $\|\mathbf{r}\|$ and conclude that

$$\|\mathbf{r}\| \leq \int_a^b \|\mathbf{f}(t)\|\,dt.$$

If $\mathbf{r} = \mathbf{0}$, the result is obvious in the first place. ❑

EXERCISES 13.1

Find the derivative.

1. $\mathbf{f}(t) = (1+2t)\,\mathbf{i} + (3-t)\,\mathbf{j} + (2+3t)\,\mathbf{k}$.
2. $\mathbf{f}(t) = 2\,\mathbf{i} - \cos t\,\mathbf{k}$.
3. $\mathbf{f}(t) = \sqrt{1-t}\,\mathbf{i} + \sqrt{1+t}\,\mathbf{j} + (1-t)^{-1}\,\mathbf{k}$.
4. $\mathbf{f}(t) = e^t\,\mathbf{i} + \ln t\,\mathbf{j} + \tan^{-1} t\,\mathbf{k}$.
5. $\mathbf{f}(t) = \sin t\,\mathbf{i} + \cos t\,\mathbf{j} + \tan t\,\mathbf{k}$.
6. $\mathbf{f}(t) = e^t(\mathbf{i} + t\,\mathbf{j} + t^2\mathbf{k})$.
7. $\mathbf{f}(t) = \ln(1-t)\,\mathbf{i} + \cos t\,\mathbf{j} + t^2\,\mathbf{k}$.
8. $\mathbf{f}(t) = \dfrac{t+1}{t-1}\,\mathbf{i} + t\,e^{2t}\,\mathbf{j} + \sec t\,\mathbf{k}$.

Calculate the second derivative

9. $\mathbf{f}(t) = 4t\,\mathbf{i} + 2t^3\,\mathbf{j} + (t^2+2t)\,\mathbf{k}$.
10. $\mathbf{f}(t) = t\sin t\,\mathbf{i} + t\cos t\,\mathbf{k}$.
11. $\mathbf{f}(t) = \cos 2t\,\mathbf{i} + \sin 2t\,\mathbf{j} + 4t\,\mathbf{k}$.
12. $\mathbf{f}(t) = \sqrt{t}\,\mathbf{i} + t\sqrt{t}\,\mathbf{j} + \ln t\,\mathbf{k}$.
13. Use a CAS to find $\mathbf{r}'(t_0)$.
    (a) $\mathbf{r}(t) = te^{-t^2}\,\mathbf{i} + t^2e^{-t}\,\mathbf{j}$, $\quad t_0 = 0$.
    (b) $\mathbf{r}(t) = \ln(\sin t)\,\mathbf{i} + \ln(\cos t)\,\mathbf{j} + (2\sin t - 3\cot t)\,\mathbf{k}$, $\quad t_0 = \pi/4$.
14. Use a CAS to find $\mathbf{r}''(t_0)$.
    (a) $\mathbf{r}(t) = t^2e^{-t}\,\mathbf{i} + te^{-t}\,\mathbf{j}$, $\quad t_0 = 1$.
    (b) $\mathbf{r}(t) = t\ln t\,\mathbf{i} + \dfrac{t}{\ln t}\,\mathbf{j} + \sqrt{\ln t}\,\mathbf{k}$, $t_0 = e$.

Carry out the intergration

15. $\int_1^2 \mathbf{f}(t)\,dt$ for $\mathbf{f}(t) = \mathbf{i} + 2t\,\mathbf{j}$.
16. $\int_0^\pi \mathbf{r}(t)\,dt$ for $\mathbf{r}(t) = \sin t\,\mathbf{i} + \cos t\,\mathbf{j} + t\,\mathbf{k}$.
17. $\int_0^1 \mathbf{g}(t)\,dt$ for $\mathbf{g}(t) = e^t\,\mathbf{i} + e^{-t}\,\mathbf{k}$.
18. $\int_0^1 \mathbf{h}(t)\,dt$ for $\mathbf{h}(t) = e^{-t}[t^2\,\mathbf{i} + \sqrt{2}t\,\mathbf{j} + \mathbf{k}]$.
19. $\int_0^1 \mathbf{f}(t)\,dt$ for $\mathbf{f}(t) = \dfrac{1}{1+t^2}\,\mathbf{i} + \sec^2 t\,\mathbf{j}$.
20. $\int_1^3 \mathbf{F}(t)\,dt$ for $\mathbf{F}(t) = \dfrac{1}{t}\,\mathbf{i} + \dfrac{\ln t}{t}\,\mathbf{j} + e^{-2t}\,\mathbf{k}$.

Find $\lim_{t\to 0} \mathbf{f}(t)$ if it exists.

21. $\mathbf{f}(t) = \dfrac{\sin t}{2t}\,\mathbf{i} + e^{2t}\,\mathbf{j} + \dfrac{t^2}{e^t}\,\mathbf{k}$.

22. $\mathbf{f}(t) = 3(t^2 - 1)\,\mathbf{i} + \cos t\,\mathbf{j} + \dfrac{t}{|t|}\,\mathbf{k}$.

23. $\mathbf{f}(t) = t^2\,\mathbf{i} + \dfrac{1 - \cos t}{3t}\,\mathbf{j} + \dfrac{t}{t+1}\,\mathbf{k}$.

24. $\mathbf{f}(t) = 3t\,\mathbf{i} + (t^2 + 1)\,\mathbf{j} + e^{2t}\,\mathbf{k}$.

C 25. Use a CAS to evaluate the definite integral.

(a) $\displaystyle\int_0^1 \mathbf{f}(t)\,dt$ for $\mathbf{f}(t) = te^t\,\mathbf{i} + te^{t^2}\,\mathbf{j}$.

(b) $\displaystyle\int_3^8 \mathbf{f}(t)\,dt$ for $\mathbf{f}(t) = \dfrac{t}{t+1}\,\mathbf{i} + \dfrac{t}{(t+1)^2}\,\mathbf{j} + \dfrac{t}{(t+1)^3}\,\mathbf{k}$.

C 26. Use a CAS to find the limit.

(a) $\displaystyle\lim_{t\to\pi/6}(\cos^2 t\,\mathbf{i} + \sin^2 t\,\mathbf{j} + \mathbf{k})$.

(b) $\displaystyle\lim_{t\to e^2}\left(t\ln t\,\mathbf{i} + \frac{\ln t}{t^2}\,\mathbf{j} + \sqrt{\ln t^2}\,\mathbf{k}\right)$.

Sketch the curve traced out by the vector-valued function and indicate the direction in which the curve is traversed as t increases.

27. $\mathbf{r}(t) = 2t\,\mathbf{i} + t^2\,\mathbf{j}$. $t \geq 0$.

28. $\mathbf{r}(t) = t^3\,\mathbf{i} + 2t\,\mathbf{j}$, $t \geq 0$.

29. $\mathbf{r}(t) = 2\sinh t\,\mathbf{i} + 2\cosh t\,\mathbf{j}$, $t \geq 0$.

30. $\mathbf{r}(t) = 3\cos t\,\mathbf{i} + 3\sin t\,\mathbf{k}$, $0 \leq t \leq 2\pi$.

31. $\mathbf{r}(t) = 2\cos t\,\mathbf{i} + 3\sin t\,\mathbf{j}$, $0 \leq t \leq 2\pi$.

32. $\mathbf{r}(t) = 2t\,\mathbf{i} + (5 - 2t)\,\mathbf{j} + 3t\,\mathbf{k}$, $t \geq 0$.

33. $\mathbf{r}(t) = (t^2 + 1)\,\mathbf{i} + t\,\mathbf{j} + 4\,\mathbf{k}$, $-2 \leq t \leq 2$.

34. $\mathbf{r}(t) = 2\cos t\,\mathbf{i} + 2\sin t\,\mathbf{j} + (2\pi - t)\,\mathbf{k}$, $0 \leq t \leq 2\pi$.

C Use a graphing utility to sketch the curve generated by the vector-valued function and indicate the direction in which the curve is traversed as t increases.

35. $\mathbf{r}(t) = 2\cos(t^2)\,\mathbf{i} + (2 - \sqrt{t})\,\mathbf{j}$.

36. $\mathbf{r}(t) = e^{\cos 2t}\,\mathbf{i} + e^{-\sin t}\,\mathbf{j}$.

37. $\mathbf{r}(t) = (2 - \sin 2t)\,\mathbf{i} + (3 + 2\cos t)\,\mathbf{j}$.

38. $\mathbf{r}(t) = (t - \sin t)\,\mathbf{i} + (1 - \cos t)\,\mathbf{j}$. (a cycloid)

Find a vector-valued function $\mathbf{f}$ that traces out the given curve in the indicated direction.

39. $4x^2 + 9y^2 = 36$ (a) Counterclockwise. (b) Clockwise.

40. $(x - 1)^2 + y^2 = 1$ (a) Counterclockwise. (b) Clockwise.

41. $y = x^2$ (a) From left to right. (b) From right to left.

42. $y = x^3$ (a) From left to right. (b) From right to left.

43. The directed line segment from $(1, 4, -2)$ to $(3, 9, 6)$.

44. The directed line segment from $(3, 2, -5)$ to $(7, 2, 9)$.

45. Set $\mathbf{f}(t) = t\,\mathbf{i} + f(t)\,\mathbf{j}$ and calculate

$$\mathbf{f}'(t_0), \quad \int_a^b \mathbf{f}(t)dt, \quad \int_a^b \mathbf{f}'(t)\,dt$$

given that

$$f'(t_0) = m, \quad f(a) = c, \quad f(b) = d, \quad \int_a^b f(t)dt = A.$$

Find $\mathbf{f}(t)$ from the following information.

46. $\mathbf{f}'(t) = t\,\mathbf{i} + t(1+t^2)^{-1/2}\,\mathbf{j} + t\,e^t\,\mathbf{k}$ and $\mathbf{f}(0) = \mathbf{i} + 2\mathbf{j} + 3\mathbf{k}$.

47. $\mathbf{f}'(t) = \mathbf{i} + t^2\mathbf{j}$ and $\mathbf{f}(0) = \mathbf{j} - \mathbf{k}$.

48. $\mathbf{f}'(t) = 2\mathbf{f}(t)$ and $\mathbf{f}(0) = \mathbf{i} - \mathbf{k}$.

49. $\mathbf{f}'(t) = \alpha\mathbf{f}(t)$ with α a real number and $\mathbf{f}(0) = \mathbf{c}$.

50. No ϵ, δ's have surface so far, but they are still there at the heart of the limit process. Give an ϵ, δ characterization of

$$\lim_{t\to t_0}\mathbf{f}(t) = \mathbf{L}.$$

51. (a) Show that, if $\mathbf{f}'(t) = \mathbf{0}$ for all t in an interval I, then $\mathbf{f}$ is a constant vector on I.

(b) Show that, if $\mathbf{f}'(t) = \mathbf{g}'(t)$ for all t in an interval I, then $\mathbf{f}$ and $\mathbf{g}$ differ by a constant vector on I.

52. Assume that, as $t \to t_0$, $\mathbf{f}(t) \to \mathbf{L}$ and $\mathbf{g}(t) \to \mathbf{M}$. Show that

$$\mathbf{f}(t) \cdot \mathbf{g}(t) \to \mathbf{L} \cdot \mathbf{M}.$$

53. Show that, if $\mathbf{f}$ is differentiable at t, then $\mathbf{f}$ is continuous at t.

54. A vector-valued function $\mathbf{G}$ is called an *antiderivative* for $\mathbf{f}$ on $[a, b]$ iff (i) $\mathbf{G}$ is continuous on $[a, b]$ and (ii) $\mathbf{G}'(t) = \mathbf{f}(t)$ for all $t \in (a, b)$. Show that:

(a) If $\mathbf{f}$ is continuous on $[a, b]$ and $\mathbf{G}$ is an antiderivative for $\mathbf{f}$ on $[a, b]$, then

$$\int_a^b \mathbf{f}(t)\,dt = \mathbf{G}(b) - \mathbf{G}(a).$$

(This is the vector version of the fundamental theorem of integral calculus.)

(b) If $\mathbf{f}$ is continuous on an interval I and $\mathbf{F}$ and $\mathbf{G}$ are antiderivatives for $\mathbf{f}$, then

$$\mathbf{F} = \mathbf{G} + \mathbf{C}$$

for some constant vector $\mathbf{C}$.

55. Is it always true that

$$\int_a^b [\mathbf{f}(t) \cdot \mathbf{g}(t)]\,dt = \left[\int_a^b \mathbf{f}(t)\,dt\right] \cdot \left[\int_a^b \mathbf{g}(t)\,dt\right]?$$

56. Prove that, if $\mathbf{f}$ is continuous on $[a, b]$, then for each constant vector $\mathbf{c}$,

$$\int_a^b [\mathbf{c} \cdot \mathbf{f}(t)]\,dt = \mathbf{c} \cdot \int_a^b \mathbf{f}(t)\,dt \quad \text{and}$$

$$\int_a^b [\mathbf{c} \times \mathbf{f}(t)]\,dt = \mathbf{c} \times \int_a^b \mathbf{f}(t)\,dt.$$

57. Let **f** be a differentiable vector-valued function. Show that if $\|\mathbf{f}(t)\| \neq 0$, then

$$\frac{d}{dt}(\|\mathbf{f}(t)\|) = \frac{\mathbf{f}(t)\cdot\mathbf{f}'(t)}{\|\mathbf{f}(t)\|}.$$

58. Let **f** be a differentiable vector-valued function. Show that where $\|\mathbf{f}(t)\| \neq 0$,

$$\frac{d}{dt}\left(\frac{\mathbf{f}(t)}{\|\mathbf{f}(t)\|}\right) = \frac{\mathbf{f}'(t)}{\|\mathbf{f}(t)\|} - \frac{\mathbf{f}(t)\cdot\mathbf{f}'(t)}{\|\mathbf{f}(t)\|^3}\mathbf{f}(t).$$

C In Exercise 59–62, use a graphing utility that can plot three-dimensional graphs to draw the curves.

59. The vector function

$$\mathbf{r}(t) = \cos at\,\mathbf{i} + \sin at\,\mathbf{j} + f(t)\,\mathbf{k}$$

describes the motion of an object on a circular cylinder.

(a) Set $f(t) = t$. Plot the helix for $a = 1, 2, 4, 5$ and $0 \le t \le 4\pi$.

(b) Set $a = 1$. Plot the helix for $f(t) = bt, b = 1, 2, 4, 5$ and $0 \le t \le 4\pi$.

(c) Describe the effect that the constants a and b have on the graph.

(d) Set $a = 1$ and experiment with other functions f. For example, try $f(t) = t^2$, $f(t) = e^t$, $f(t) = \ln(t + 1)$ and describe the effect on the curve.

60. The graph of the vector function

$$\mathbf{r}(t) = A\cos at\,\mathbf{i} + B\sin at\,\mathbf{j} + f(t)\,\mathbf{k}$$

is also a curve on a cylinder.

(a) Set $f(t) = t$ and $a = 1$. Plot the curve for the pairs $A = 2, B = 3; A = 4, B = 2; 0 \le t \le 4\pi$. What name would you give to this type of curve?

(b) Experiment with other values for the constants a, A, B and other functions f

61. Let $\mathbf{r}(t) = A\cos t\,\mathbf{i} + B\sin t\,\mathbf{j} + f(t)\,\mathbf{k}$.

(a) Set $A = B = 1$ and $f(t) = \sin bt$. Plot the curve for $b = 1, 2, 3, 4, 5; 0 \le t \le 2\pi$.

(b) Set $A = B = 1$ and $f(t) = \cos bt$. Plot the curve for $b = 1, 2, 3, 4, 5; 0 \le t \le 2\pi$.

(c) Set $f(t) = \sin bt$. Plot the curves for the pairs $A = 2, B = 3; A = 4, B = 2$.

(d) Describe the effects that the constants A, B, and b have on the graph.

62. The vector function

$$\mathbf{r}(t) = (A\cos at + B\cos bt)\,\mathbf{i} + A\sin at\,\mathbf{j} + B\sin bt\,\mathbf{k}.$$

describes the motion of an object on a torus. Sketch the curve generated by letting:

(a) $A = 2, B = 1, a = 1, b = 1$.

(b) $A = 1, B = 2, a = 2, b = 1$.

(c) Experiment with other values of the constants to see the effect on the graph.

■ 13.2 DIFFERENTIATION FORMULAS

Vector functions with a common domain can be combined in many ways to form new functions. From **f** and **g** we can form the sum $\mathbf{f} + \mathbf{g}$:

$$(\mathbf{f} + \mathbf{g})(t) = \mathbf{f}(t) + \mathbf{g}(t).$$

We can form scalar multiples $\alpha\mathbf{f}$ and thus linear combinations $\alpha\,\mathbf{f} + \beta\,\mathbf{g}$:

$$(\alpha\mathbf{f})(t) = \alpha\mathbf{f}(t), \quad (\alpha\mathbf{f} + \beta\mathbf{g})(t) = \alpha\mathbf{f}(t) + \beta\mathbf{g}(t).$$

We can form the dot product $\mathbf{f}\cdot\mathbf{g}$:

$$(\mathbf{f}\cdot\mathbf{g})(t) = \mathbf{f}(t)\cdot\mathbf{g}(t).$$

We can also form the cross product $\mathbf{f}\times\mathbf{g}$:

$$(\mathbf{f}\times\mathbf{g})(t) = \mathbf{f}(t)\times\mathbf{g}(t).$$

These operations on vector functions are simply the pointwise application of the algebraic operations on vectors that we introduced in Chapter 12.

There are two ways of bringing *scalar functions* (real-valued functions) into play. If a scalar function u has the same domain as $\mathbf{f}$, we can form the product $u\,\mathbf{f}$:

$$(u\,\mathbf{f})(t) = u(t)\mathbf{f}(t).$$

If $u(t)$ is in the domain of $\mathbf{f}$ for each t in some interval, then we can form the composition $\mathbf{f} \circ u$:

$$(\mathbf{f} \circ u)(t) = \mathbf{f}(u(t)).$$

For example, with $u(t) = t^2$ and $\mathbf{f}(t) = e^t\,\mathbf{i} + \sin 2t\,\mathbf{j}$,

$$(u\,\mathbf{f})(t) = u(t)\mathbf{f}(t) = t^2 e^t\,\mathbf{i} + t^2 \sin 2t\,\mathbf{j} \text{ and } (\mathbf{f} \circ u)(t) = \mathbf{f}(u(t)) = e^{t^2}\mathbf{i} + \sin 2t^2\,\mathbf{j}.$$

It follows from Theorem 13.1.3 that if $\mathbf{f}, \mathbf{g}$ and u are continuous on a common domain, then $\mathbf{f}+\mathbf{g}$, $\alpha\mathbf{f}$, $\mathbf{f} \cdot \mathbf{g}$, $\mathbf{f} \times \mathbf{g}$, and $u\mathbf{f}$ are all continuous on that same set. We have yet to show the continuity of $\mathbf{f} \circ u$. The verification of that is left to you (Exercise 36). What interests us here is that, if $\mathbf{f}, \mathbf{g}$ and u are differentiable, then the newly constructed functions are also differentiable and their derivatives satisfy the following rules:

(13.2.1)

(1) $(\mathbf{f} + \mathbf{g})'(t) = \mathbf{f}'(t) + \mathbf{g}'(t).$

(2) $(\alpha\mathbf{f})(t) = \alpha\mathbf{f}'(t)$ (α constant).

(3) $(u\mathbf{f})'(t) = u(t)\mathbf{f}'(t) + u'(t)\mathbf{f}(t).$

(4) $(\mathbf{f} \cdot \mathbf{g})'(t) = [\mathbf{f}(t) \cdot \mathbf{g}'(t)] + [\mathbf{f}'(t) \cdot \mathbf{g}(t)].$

(5) $(\mathbf{f} \times \mathbf{g})'(t) = [\mathbf{f}(t) \times \mathbf{g}'(t)] + [\mathbf{f}'(t) \times \mathbf{g}(t)].$

(6) $(\mathbf{f} \circ u)'(t) = \mathbf{f}'(u(t))u'(t) = u'(t)\mathbf{f}'(u(t))$ (the chain rule).

Rules (3),(4),(5) are all "product" rules and should remind you of the rule for differentiating the product of ordinary functions. Keep in mind, however, that the cross product is not commutative and therefore the order in Rule (5) is important.

In Rule (6) we first wrote the scalar part $u'(t)$ on the right so that the formula would look like the chain rule for ordinary functions. In general, $\mathbf{a}\alpha$ has the same meaning as $\alpha\mathbf{a}$.

Example 1 Taking $\mathbf{f}(t) = 2t^2\,\mathbf{i} - 3\mathbf{j}$, $g(t) = \mathbf{i} + t\,\mathbf{j} + t^2\,\mathbf{k}$, $u(t) = \frac{1}{3}t^3$

we have $\quad \mathbf{f}'(t) = 4t\,\mathbf{i}, \quad \mathbf{g}'(t) = \mathbf{j} + 2t\,\mathbf{k}, \quad u'(t) = t^2.$

Therefore

(a) $(\mathbf{f} + \mathbf{g})'(t) = \mathbf{f}'(t) + \mathbf{g}'(t) = 4t\,\mathbf{i} + (\mathbf{j} + 2t\,\mathbf{k}) = 4t\,\mathbf{i} + \mathbf{j} + 2t\,\mathbf{k};$

(b) $(2\mathbf{f})'(t) = 2\mathbf{f}'(t) = 2(4t\,\mathbf{i}) = 8t\,\mathbf{i};$

(c) $(u\mathbf{f})'(t) = u(t)\mathbf{f}'(t) + u'(t)\mathbf{f}(t) = \frac{1}{3}t^3(4t\,\mathbf{i}) + t^2(2t^2\mathbf{i} - 3\mathbf{j}) = \frac{10}{3}t^4\mathbf{i} - 3t^2\mathbf{j};$

(d) $(\mathbf{f} \cdot \mathbf{g})'(t) = [\mathbf{f}(t) \cdot \mathbf{g}'(t)] + [\mathbf{f}'(t) \cdot \mathbf{g}(t)]$

$$= [(2t^2\,\mathbf{i} - 3\mathbf{j}) \cdot (\mathbf{j} + 2t\,\mathbf{k})] + [4t\,\mathbf{i} \cdot (i + t\,\mathbf{j} + t^2\,\mathbf{k})] = -3 + 4t;$$

$$\begin{aligned}\text{(e) } (\mathbf{f}\times\mathbf{g})'(t) &= [\mathbf{f}(t)\times\mathbf{g}'(t)] + [\mathbf{f}'(t)\times\mathbf{g}(t)]\\ &= [(2t^2\,\mathbf{i} - 3\,\mathbf{j})\times(\mathbf{j} + 2t\,\mathbf{k})] + [4t\,\mathbf{i}\times(\mathbf{i} + t\,\mathbf{j} + t^2\,\mathbf{k})]\\ &= (2t^2\,\mathbf{k} - 4t^3\,\mathbf{j} - 6t\,\mathbf{i}) + (4t^2\,\mathbf{k} - 4t^3\,\mathbf{j}) = -6t\,\mathbf{i} - 8t^3\,\mathbf{j} + 6t^2\,\mathbf{k}\end{aligned}$$

while

$$\begin{aligned}(\mathbf{g}\times\mathbf{f})'(t) &= [\mathbf{g}(t)\times\mathbf{f}'(t)] + [\mathbf{g}'(t)\times\mathbf{f}(t)]\\ &= [(\mathbf{i} + t\,\mathbf{j} + t^2\,\mathbf{k})\times 4t\,\mathbf{i}] + [(\mathbf{j} + 2t\,\mathbf{k})\times(2t^2\,\mathbf{i} - 3\mathbf{j})]\\ &= (-4t^2\,\mathbf{k} + 4t^3\,\mathbf{j}) + (-2t^2\,\mathbf{k} + 4t^3\,\mathbf{j} + 6t\,\mathbf{i})\\ &= 6t\,\mathbf{i} + 8t^3\,\mathbf{j} - 6t^2\,\mathbf{k} = -(\mathbf{f}\times\mathbf{g})'(t);\end{aligned}$$

$$\begin{aligned}\text{(f) } (\mathbf{f}\circ u)'(t) &= \mathbf{f}'(u(t))u'(t)\\ &= [4u(t)\,\mathbf{i}]u'(t) = [4(\tfrac{1}{3}t^3)\,\mathbf{i}]t^2 = \tfrac{4}{3}t^5\,\mathbf{i}. \quad \square\end{aligned}$$

The differentiation formulas that we have given can all be derived component by component, and they can all be derived in a component-free manner. Take, for example, formula (3):

$$(u\mathbf{f})'(t) = u(t)\mathbf{f}'(t) + u'(t)\mathbf{f}(t).$$

COMPONENT-BY-COMPONENT DERIVATION Set

$$\mathbf{f}(t) = f_1(t)\,\mathbf{i} + f_2(t)\,\mathbf{j} + f_3(t)\,\mathbf{k}.$$

Then

$$(u\,\mathbf{f})(t) = u(t)\,\mathbf{f}(t) = u(t)\,f_1(t)\,\mathbf{i} + u(t)\,f_2(t)\,\mathbf{j} + u(t)\,f_3(t)\,\mathbf{k}$$

and

$$\begin{aligned}(u\,\mathbf{f})'(t) &= [u(t)\,f_1'(t) + u'(t)\,f_1(t)]\,\mathbf{i} + [u(t)\,f_2'(t) + u'(t)\,f_2(t)]\,\mathbf{j} +\\ &\qquad [u(t)\,f_3'(t) + u'(t)\,f_3(t)]\,\mathbf{k}\\ &= u(t)[f_1'(t)\,\mathbf{i} + f_2'(t)\,\mathbf{j} + f_3'(t)\,\mathbf{k}] + u'(t)[f_1(t)\,\mathbf{i} + f_2(t)\,\mathbf{j} + f_3(t)\,\mathbf{k}]\\ &= u(t)\mathbf{f}'(t) + u'(t)\mathbf{f}(t). \quad \square\end{aligned}$$

COMPONENT-FREE DERIVATION We find $(u\,\mathbf{f})'(t)$ by taking the limit as $h \to 0$ of the difference quotient

$$\frac{u(t+h)\mathbf{f}(t+h) - u(t)\mathbf{f}(t)}{h}.$$

By adding and subtracting $u(t+h)\mathbf{f}(t)$, we can rewrite this quotient as

$$\frac{u(t+h)\mathbf{f}(t+h) - u(t+h)\mathbf{f}(t) + u(t+h)\mathbf{f}(t) - u(t)\mathbf{f}(t)}{h},$$

which is equal to

$$u(t+h)\frac{\mathbf{f}(t+h) - \mathbf{f}(t)}{h} + \frac{u(t+h) - u(t)}{h}\mathbf{f}(t).$$

As $h \to 0$.

$$u(t+h) \to u(t), \qquad \text{(differentiable functions are continuous)}$$

$$\frac{\mathbf{f}(t+h)-\mathbf{f}(t)}{h} \to \mathbf{f}'(t), \qquad \text{(definition of derivative for vector functions)}$$

$$\frac{u(t+h)-u(t)}{h} \to u'(t). \qquad \text{(definition of derivative for scalar functions)}$$

It follows from the limit rules (Theorem 13.1.3) that

$$u(t+h)\frac{\mathbf{f}(t+h)-\mathbf{f}(t)}{h} \to u(t)\mathbf{f}'(t), \qquad \frac{u(t+h)-u(t)}{h}\mathbf{f}(t) \to u'(t)\mathbf{f}(t)$$

and therefore

$$u(t+h)\frac{\mathbf{f}(t+h)-\mathbf{f}(t)}{h} + \frac{u(t+h)-u(t)}{h}\mathbf{f}(t) \to u(t)\mathbf{f}'(t) + u'(t)\mathbf{f}(t). \quad \square$$

In Leibniz's notation the formulas take the following form:

(13.2.2)

$$\text{(1)}\ \frac{d}{dt}(\mathbf{f}+\mathbf{g}) = \frac{d\mathbf{f}}{dt} + \frac{d\mathbf{g}}{dt}.$$

$$\text{(2)}\ \frac{d}{dt}(\alpha\mathbf{f}) = \alpha\frac{d\mathbf{f}}{dt}. \qquad (\alpha \text{ constant})$$

$$\text{(3)}\ \frac{d}{dt}(u\mathbf{f}) = u\frac{d\mathbf{f}}{dt} + \frac{du}{dt}\mathbf{f}. \qquad (u = u(t))$$

$$\text{(4)}\ \frac{d}{dt}(\mathbf{f}\cdot\mathbf{g}) = \left(\mathbf{f}\cdot\frac{d\mathbf{g}}{dt}\right) + \left(\frac{d\mathbf{f}}{dt}\cdot\mathbf{g}\right).$$

$$\text{(5)}\ \frac{d}{dt}(\mathbf{f}\times\mathbf{g}) = \left(\mathbf{f}\times\frac{d\mathbf{g}}{dt}\right) + \left(\frac{d\mathbf{f}}{dt}\times\mathbf{g}\right).$$

$$\text{(6)}\ \frac{d\mathbf{f}}{dt} = \frac{d\mathbf{f}}{du}\frac{du}{dt}. \qquad \text{(chain rule)}$$

We conclude this section with two results that will prove useful as we go on.

Example 2 Let $\mathbf{r}$ be a differentiable vector function of t and set $r = \|\mathbf{r}\|$. Show that r is differentiable wherever it is not 0 and

(13.2.3)

$$\mathbf{r}\cdot\frac{d\mathbf{r}}{dt} = r\frac{dr}{dt}.$$

SOLUTION If $\mathbf{r}$ is differentiable, then $\mathbf{r}\cdot\mathbf{r} = \|\mathbf{r}\|^2 = r^2$ is differentiable. Let's assume now that $r \neq 0$. Since the square-root function is differentiable at all positive numbers and r^2 is positive, we can apply the square-root function to r^2 and conclude by the chain rule that r is itself differentiable.

To obtain the formula we differentiate the identity $\mathbf{r}\cdot\mathbf{r} = r^2$:

$$\mathbf{r}\cdot\frac{d\mathbf{r}}{dt} + \frac{d\mathbf{r}}{dt}\cdot\mathbf{r} = 2r\frac{dr}{dt}$$

$$2\mathbf{r}\cdot\frac{d\mathbf{r}}{dt} = 2r\frac{dr}{dt}$$

$$\mathbf{r}\cdot\frac{dr}{dt} = r\frac{dr}{dt}. \quad \square$$

Example 3 Let $\mathbf{r}$ be a differentiable vector function of t and set $r = \|\mathbf{r}\|$. Show that where $r \neq 0$

(13.2.4)
$$\frac{d}{dt}\left(\frac{\mathbf{r}}{r}\right) = \frac{1}{r^3}\left[\left(\mathbf{r} \times \frac{d\mathbf{r}}{dt}\right) \times \mathbf{r}\right].$$

SOLUTION This is a little tricky:

$$\begin{aligned}\frac{d}{dt}\left(\frac{\mathbf{r}}{r}\right) &= \frac{1}{r}\frac{d\mathbf{r}}{dt} - \frac{1}{r^2}\frac{dr}{dt}\mathbf{r}\\ &= \frac{1}{r^3}\left[r^2\frac{d\mathbf{r}}{dt} - r\frac{dr}{dt}\mathbf{r}\right]\\ &= \frac{1}{r^3}\left[(\mathbf{r}\cdot\mathbf{r})\frac{d\mathbf{r}}{dt} - \left(\mathbf{r}\cdot\frac{d\mathbf{r}}{dt}\right)\mathbf{r}\right] = \frac{1}{r^3}\left[\left(\mathbf{r}\times\frac{d\mathbf{r}}{dt}\right)\times\mathbf{r}\right]. \quad \square\end{aligned}$$

$(\mathbf{a} \times \mathbf{b}) \times \mathbf{c} = (\mathbf{c}\cdot\mathbf{a})\mathbf{b} - (\mathbf{c}\cdot\mathbf{b})\mathbf{a}$

EXERCISES 13.2

Find $\mathbf{f}'(t)$ and $\mathbf{f}''(t)$.

1. $\mathbf{f}(t) = \mathbf{a} + t\,\mathbf{b}$.
2. $\mathbf{f}(t) = \mathbf{a} + t\,\mathbf{b} + t^2\,\mathbf{c}$.
3. $\mathbf{f}(t) = e^{2t}\,\mathbf{i} - \sin t\,\mathbf{j}$.
4. $\mathbf{f}(t) = [(t^2\,\mathbf{i} - \mathbf{j})\cdot(\mathbf{i} - t^2\,\mathbf{j})]\,\mathbf{i}$.
5. $\mathbf{f}(t) = [(t^2\,\mathbf{i} - 2t\,\mathbf{j})\cdot(t\,\mathbf{i} + t^3\,\mathbf{j})]\,\mathbf{j}$.
6. $\mathbf{f}(t) = [(3t\,\mathbf{i} - t^2\,\mathbf{j} + \mathbf{k})\cdot(\mathbf{i} + t^3\,\mathbf{j} - 2t\,\mathbf{k})]\mathbf{k}$.
7. $\mathbf{f}(t) = (e^t\,\mathbf{i} + t\,\mathbf{k}) \times (t\,\mathbf{j} + e^{-t}\,\mathbf{k})$.
8. $\mathbf{f}(t) = (t\,\mathbf{i} - t^2\,\mathbf{j} + \mathbf{k}) \times (\mathbf{i} + t^3\,\mathbf{j} + 5t\,\mathbf{k})$.
9. $\mathbf{f}(t) = (\cos t\,\mathbf{i} + \sin t\,\mathbf{j} + \mathbf{k}) \times (\sin 2t\,\mathbf{i} + \cos 2t\,\mathbf{j} + t\,\mathbf{k})$.
10. $\mathbf{f}(t) = t\mathbf{g}(t^2)$.
11. $\mathbf{f}(t) = t\mathbf{g}(\sqrt{t})$.
12. $\mathbf{f}(t) = (e^{2t}\mathbf{i} + e^{-2t}\,\mathbf{j} + \mathbf{k}) \times (e^{2t}\,\mathbf{i} - e^{-2t}\,\mathbf{j} + \mathbf{k})$.

Find the indicated derivative.

13. $\dfrac{d}{dt}[e^{\cos t}\,\mathbf{i} + e^{\sin t}\,\mathbf{j}]$.
14. $\dfrac{d^2}{dt^2}[e^t\cos t\,\mathbf{i} + e^t\sin t\,\mathbf{j}]$.
15. $\dfrac{d^2}{dt^2}\left[(e^t\,\mathbf{i} + e^{-t}\,\mathbf{j})\cdot(e^t\,\mathbf{i} - e^{-t}\,\mathbf{j})\right]$.
16. $\dfrac{d}{dt}\left[(\ln t\,\mathbf{i} + t\,\mathbf{j} - (t^2+1)\,\mathbf{k}) \times \left(\dfrac{1}{t}\mathbf{i} + t^2\,\mathbf{j} - t\,\mathbf{k}\right)\right]$.
17. $\dfrac{d}{dt}[(\mathbf{a} + t\,\mathbf{b}) \times (\mathbf{c} + t\,\mathbf{d})]$.
18. $\dfrac{d}{dt}[(\mathbf{a} + t\,\mathbf{b}) \times (\mathbf{a} + t\,\mathbf{b} + t^2\,\mathbf{c})]$.
19. $\dfrac{d}{dt}[(\mathbf{a} + t\,\mathbf{b})\cdot(\mathbf{c} + t\,\mathbf{d})]$.
20. $\dfrac{d}{dt}[(\mathbf{a} + t\,\mathbf{b})\cdot(\mathbf{a} + t\,\mathbf{b} + t^2\,\mathbf{c})]$.

Find $\mathbf{r}(t)$ given that:

21. $\mathbf{r}'(t) = \mathbf{b}$ for all real t, $\quad \mathbf{r}(0) = \mathbf{a}$.
22. $\mathbf{r}''(t) = \mathbf{c}$ for all real t, $\quad \mathbf{r}'(0) = \mathbf{b}$, $\quad \mathbf{r}(0) = \mathbf{a}$.
23. $\mathbf{r}''(t) = \mathbf{a} + t\mathbf{b}$ for all real t, $\quad \mathbf{r}'(0) = \mathbf{c}$, $\quad \mathbf{r}(0) = \mathbf{d}$.
24. $\mathbf{r}''(t) = \cos 2t\,\mathbf{i} + \sin 2t\,\mathbf{j}$ for all real t, $\mathbf{r}'(0) = 2\,\mathbf{i} - \frac{1}{2}\,\mathbf{j}$, $\quad \mathbf{r}(0) = \frac{3}{4}\,\mathbf{i} + \mathbf{j}$.
25. Show that, if $\mathbf{r}(t) = \sin t\,\mathbf{i} + \cos t\,\mathbf{j}$, then $\mathbf{r}(t)$ and $\mathbf{r}''(t)$ are parallel. Is there a value of t for which $\mathbf{r}(t)$ and $\mathbf{r}''(t)$ have the same direction?
26. Show that, if $\mathbf{r}(t) = e^{kt}\mathbf{i} + e^{-kt}\mathbf{j}$, then $\mathbf{r}(t)$ and $\mathbf{r}''(t)$ have the same direction.
27. Calculate $\mathbf{r}(t)\cdot\mathbf{r}'(t)$ and $\mathbf{r}(t)\times\mathbf{r}'(t)$ for $\mathbf{r}(t) = \cos t\,\mathbf{i} + \sin t\,\mathbf{j}$.

Assume the rule for differentiating a cross product and show the following.

28. $(\mathbf{g} \times \mathbf{f})'(t) = -(\mathbf{f} \times \mathbf{g})'(t)$.
29. $\dfrac{d}{dt}[\mathbf{f}(t) \times \mathbf{f}'(t)] = \mathbf{f}(t) \times \mathbf{f}''(t)$.
30. $\dfrac{d}{dt}[u_1(t)\mathbf{r}_1(t) \times u_2(t)\mathbf{r}_2(t)] = u_1(t)u_2(t)\dfrac{d}{dt}[\mathbf{r}_1(t) \times \mathbf{r}_2(t)] + [\mathbf{r}_1(t) \times \mathbf{r}_2(t)]\,\dfrac{d}{dt}[u_1(t)u_2(t)]$.
31. Set $\mathbf{E}(t) = \mathbf{f}(t)\cdot[\mathbf{g}(t) \times \mathbf{h}(t)]$ and show that $\mathbf{E}'(t) = \mathbf{f}'(t)\cdot[\mathbf{g}(t)\times\mathbf{h}(t)] + \mathbf{f}(t)\cdot[\mathbf{g}'(t)\times\mathbf{h}(t)] + \mathbf{f}(t)\cdot[\mathbf{g}(t)\times\mathbf{h}'(t)]$.
32. Suppose that $\mathbf{f}(t)$ is parallel to $\mathbf{f}''(t)$ for all t. Prove that $\mathbf{f} \times \mathbf{f}'$ is constant. HINT: See Exercise 29.
33. Assume the rule for differentiating a dot product and show that

$$\|\mathbf{r}(t)\| \text{ is constant} \qquad \text{iff} \quad \mathbf{r}(t)\cdot\mathbf{r}'(t) = 0 \text{ identically.}$$

34. Derive the formula

$$(\mathbf{f} \cdot \mathbf{g})'(t) = [\mathbf{f}(t) \cdot \mathbf{g}'(t)] + [\mathbf{f}'(t) \cdot \mathbf{g}(t)]$$

(a) by appealing to components;

(b) without appealing to components.

35. Derive the formula

$$(\mathbf{f} \times \mathbf{g})'(t) = [\mathbf{f}(t) \times \mathbf{g}'(t)] + [\mathbf{f}'(t) \times \mathbf{g}(t)]$$

without appealing to components.

36. (a) Show that, if u is continuous at t_0 and $\mathbf{f}$ is continuous at $u(t_0)$, then the composition $\mathbf{f} \circ u$ is continuous at t_0.

(b) Derive the chain rule for vector functions:

$$\frac{d\mathbf{f}}{dt} = \frac{d\mathbf{f}}{du}\frac{du}{dt}.$$

■ 13.3 CURVES

Introduction

Earlier we explained how every linear vector function

$$\mathbf{r}(t) = \mathbf{r}_0 + t\,\mathbf{d} \quad \text{with} \quad \mathbf{d} \neq \mathbf{0}$$

parametrizes a line. More generally, every differentiable vector function parametrizes a curve.

Suppose that

$$\mathbf{r}(t) = x(t)\,\mathbf{i} + y(t)\,\mathbf{j} + z(t)\,\mathbf{k}$$

is differentiable on some interval I. (At the endpoints, if there are any, we require only continuity.) For each number $t \in I$, the tip of the radius vector $\mathbf{r}(t)$ is the point $P(x(t), y(t), z(t))$. As t ranges over I, the point P traces out some path C (Figure 13.3.1). We call C a *differentiable curve* and say that C is *parametrized* by $\mathbf{r}$ with parameter t. It is important to understand that a parametrized curve C is an *oriented* curve in the sense that as t increases on I, the tip of the radius vector traces out the curve C in a certain direction. For example, the curve parametrized by

$$\mathbf{r}(t) = \cos t\,\mathbf{i} + \sin t\,\mathbf{j}, \quad t \in [0, 2\pi]$$

is the unit circle traversed in the counterclockwise direction starting at the point $(1, 0)$ (Figure 13.3.2). The orientation of the elliptical helix parametrized by

$$\mathbf{r}(t) = t\,\mathbf{i} + 2\cos t\,\mathbf{j} + 3\sin t\,\mathbf{k}, \quad t \geq 0,$$

is indicated by the arrows drawn in Figure 13.3.3.

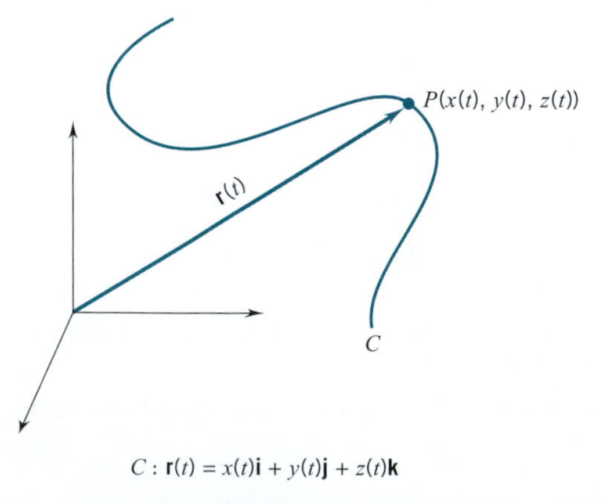

Figure 13.3.1

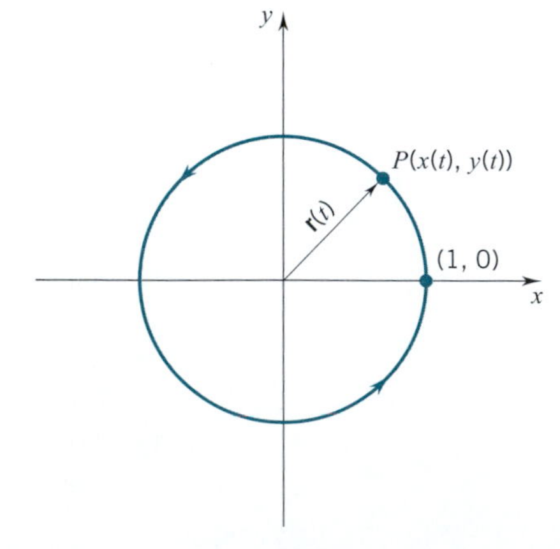

Figure 13.3.2

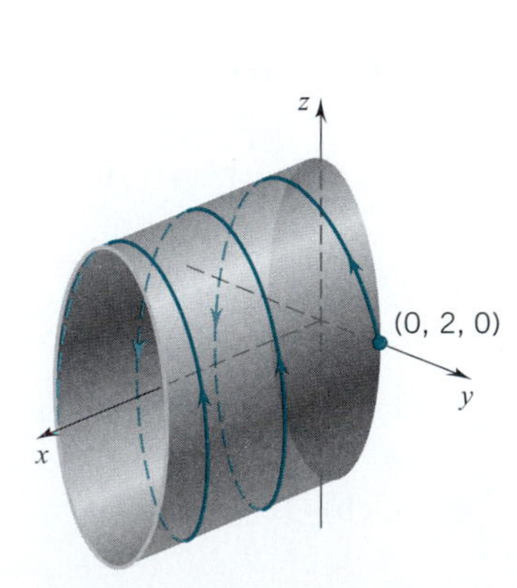

Figure 13.3.3

Tangent Vector, Tangent Line

Let's try to interpret the derivative

$$\mathbf{r}'(t) = x'(t)\,\mathbf{i} + y'(t)\,\mathbf{j} + z'(t)\,\mathbf{k}$$

geometrically. First of all,

$$\mathbf{r}'(t) = \lim_{h\to 0} \frac{\mathbf{r}(t+h) - \mathbf{r}(t)}{h}.$$

If $\mathbf{r}'(t) \neq \mathbf{0}$, then we can be sure that for $t + h$ close enough to t, the vector

$$\mathbf{r}(t+h) - \mathbf{r}(t)$$

will not be $\mathbf{0}$. (Explain.) Consequently, we can think of the vector $\mathbf{r}(t+h) - \mathbf{r}(t)$ as pictured in Figure 13.3.4.

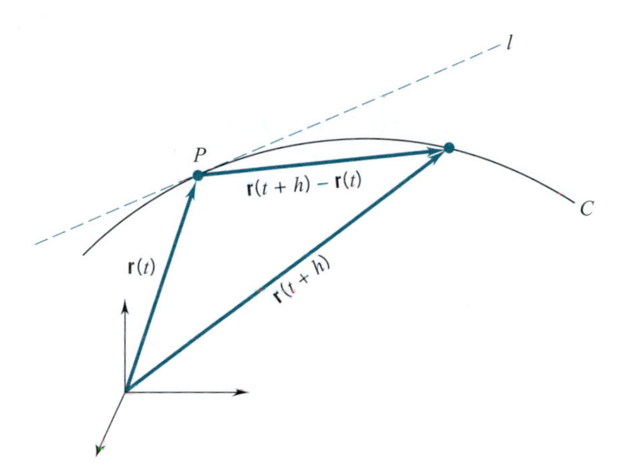

Figure 13.3.4

Let's agree that the line marked l in the figure corresponds to our intuitive notion of the tangent line at the point P. As h tends to zero, the vector

$$\mathbf{r}(t+h) - \mathbf{r}(t)$$

comes increasingly closer to serving as a direction vector for that tangent line. It may be tempting therefore to take the limiting case

$$\lim_{h\to 0} [\mathbf{r}(t+h) - \mathbf{r}(t)]$$

and call that a direction vector for the tangent line. The trouble is that this limit vector is $\mathbf{0}$ and $\mathbf{0}$ has no direction.

We can circumvent this difficulty by replacing $\mathbf{r}(t+h) - \mathbf{r}(t)$ by a vector which, for small h, has greater length: the difference quotient

$$\frac{\mathbf{r}(t+h) - \mathbf{r}(t)}{h}$$

For each real number $h \neq 0$, the vector $[\mathbf{r}(t+h) - \mathbf{r}(t)]/h$ is parallel to $\mathbf{r}(t+h) - \mathbf{r}(t)$, and therefore its limit,

$$\mathbf{r}'(t) = \lim_{h\to 0} \frac{\mathbf{r}(t+h) - \mathbf{r}(t)}{h},$$

which by assumption is not $\mathbf{0}$, can be taken as a direction vector for the tangent line. Hence the following definition.

DEFINITION 13.3.1 TANGENT VECTOR

Let

$$C : \mathbf{r}(t) = x(t)\,\mathbf{i} + y(t)\,\mathbf{j} + z(t)\,\mathbf{k}$$

be a differentiable curve. The vector $\mathbf{r}'(t)$, if not $\mathbf{0}$, is *tangent* to the curve C at the point $P(x(t), y(t), z(t))$.

Now the following question arises: Assuming that $\mathbf{r}'(t) \neq \mathbf{0}$, in which of the two possible directions does $\mathbf{r}'(t)$ point?

Figure 13.3.5 shows an oriented curve C and the vector $\mathbf{r}(t+h) - \mathbf{r}(t)$ with $h > 0$. In this case,

$$\frac{\mathbf{r}(t+h) - \mathbf{r}(t)}{h}$$

points in the same direction as $\mathbf{r}(t+h) - \mathbf{r}(t)$, which is the direction of increasing t.

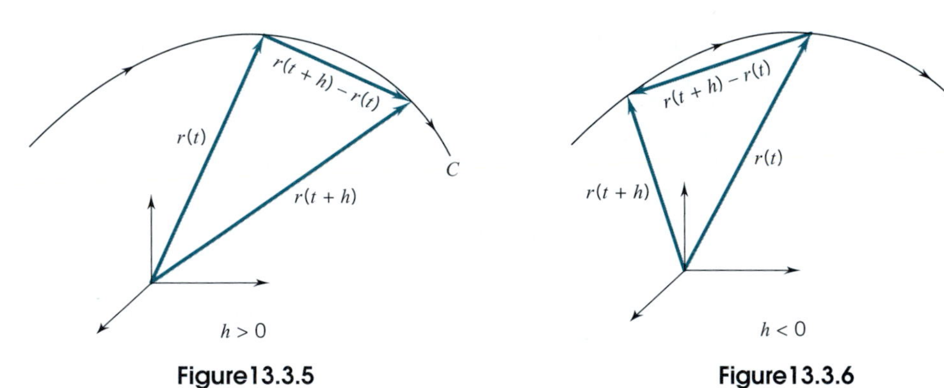

Figure 13.3.5

Figure 13.3.6

Figure 13.3.6 shows the vector $\mathbf{r}(t+h) - \mathbf{r}(t)$ for $h < 0$. In this case, dividing by h to form the difference quotient *reverses* the direction of the vector. Thus

$$\frac{\mathbf{r}(t+h) - \mathbf{r}(t)}{h}$$

also points in the direction of increasing t. We can now conclude that

$\mathbf{r}'(t)$ points in the direction of increasing t.

At each point of a line

$$l: \quad \mathbf{r}(t) = \mathbf{r}_0 + t\mathbf{d}$$

the tangent vector $\mathbf{r}'(t)$ is parallel to the line itself:

$$\mathbf{r}'(t) = \mathbf{d} \quad \text{and } \mathbf{d} \text{ is parallel to } l. \qquad \text{(Figure 13.3.7)}$$

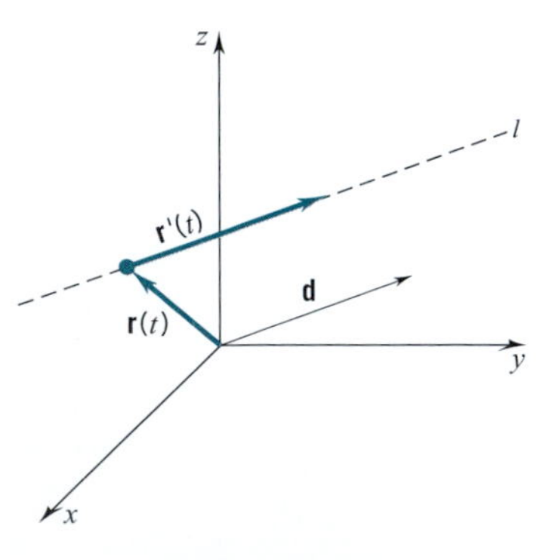

Figure 13.3.7

In the case of a circle

$$C: \quad \mathbf{r}(t) = a\cos t\,\mathbf{i} + a\sin t\,\mathbf{j}, \quad (a > 0)$$

the tangent vector $\mathbf{r}'(t)$ is perpendicular to the radius vector $\mathbf{r}(t)$:

$$\begin{aligned}\mathbf{r}'(t) \cdot \mathbf{r}(t) &= (-a\sin t\,\mathbf{i} + a\cos t\,\mathbf{j}) \cdot (a\cos t\,\mathbf{i} + a\sin t\,\mathbf{j}) \\ &= -a^2 \sin t \cos t + a^2 \cos t \sin t = 0.\end{aligned}$$

(Figure 13.3.8)

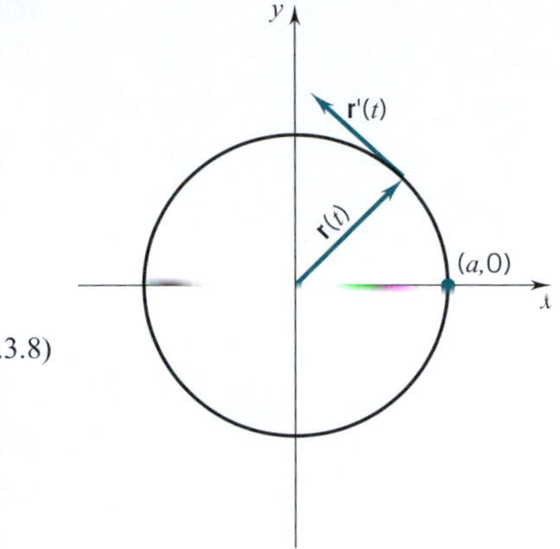

Figure 13.3.8

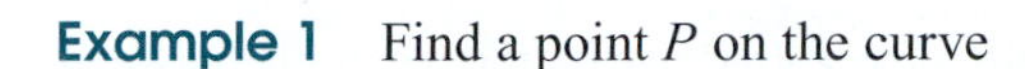

Example 1 Find a point P on the curve

$$\mathbf{r}(t) = (1 - 2t)\,\mathbf{i} + t^2\,\mathbf{j} + 2e^{2(t-1)}\,\mathbf{k}$$

at which the tangent vector $\mathbf{r}'(t)$ is parallel to the radius vector $\mathbf{r}(t)$.

SOLUTION $\mathbf{r}'(t) = -2\,\mathbf{i} + 2t\,\mathbf{j} + 4e^{2(t-1)}\,\mathbf{k}$.

For $\mathbf{r}'(t)$ to be parallel to $\mathbf{r}(t)$ there must exist a scalar α such that

$$\mathbf{r}(t) = \alpha\mathbf{r}'(t).$$

This vector equation holds iff

$$1 - 2t = -2\alpha, \quad t^2 = 2\alpha t, \quad 2e^{2(t-1)} = 4\alpha e^{2(t-1)}.$$

The last scalar equation requires that $\alpha = \frac{1}{2}$. The only value of t that satisfies all three equations with $\alpha = \frac{1}{2}$ is $t = 1$. Therefore the only point at which $\mathbf{r}'(t)$ is parallel to $\mathbf{r}(t)$ is the tip of $\mathbf{r}(1)$. This is the point $P(-1, 1, 2)$. ❑

If $\mathbf{r}'(t_0) \neq \mathbf{0}$, then $\mathbf{r}'(t_0)$ is tangent to the curve at the tip of $\mathbf{r}(t_0)$. The *tangent line* at this point can be parametrized by setting

(13.3.2)
$$\mathbf{R}(u) = \mathbf{r}(t_0) + u\,\mathbf{r}'(t_0).$$

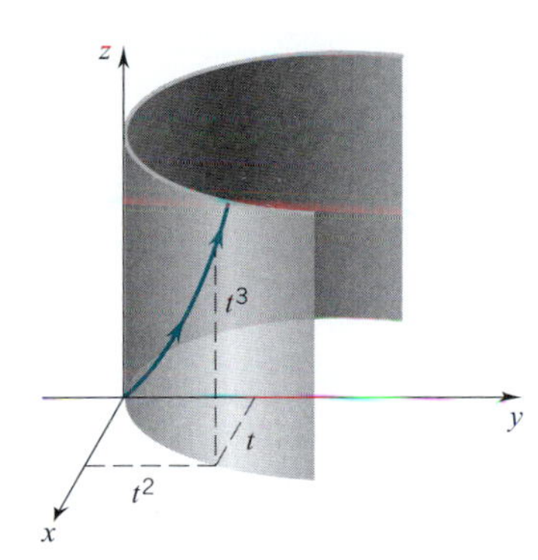

the twisted cubic, $t > 0$

Figure 13.3.9

Example 2 Find a vector tangent to the *twisted cubic*

$$\mathbf{r}(t) = t\,\mathbf{i} + t^2\mathbf{j} + t^3\mathbf{k}$$

(Figure 13.3.9)

at the point $P(2, 4, 8)$, and then parametrize the tangent line.

SOLUTION Here $\mathbf{r}'(t) = \mathbf{i} + 2t\,\mathbf{j} + 3t^2\,\mathbf{k}$.

Since $P(2, 4, 8)$ is the tip of $\mathbf{r}(2)$, the vector

$$\mathbf{r}'(2) = \mathbf{i} + 4\mathbf{j} + 12\mathbf{k}$$

is tangent to the curve at the point $P(2, 4, 8)$. The vector function

$$\mathbf{R}(u) = (2\mathbf{i} + 4\mathbf{j} + 8\mathbf{k}) + u(\mathbf{i} + 4\mathbf{j} + 12\mathbf{k})$$

parametrizes the tangent line. ❑

Intersecting Curves

Two curves

$$C_1: \ \mathbf{r}_1(t) = x_1(t)\,\mathbf{i} + y_1(t)\,\mathbf{j} + z_1(t)\,\mathbf{k}, \quad C_2: \ \mathbf{r}_2(u) = x_2(u)\,\mathbf{i} + y_2(u)\,\mathbf{j} + z_2(u)\,\mathbf{k}$$

intersect iff there are numbers t and u for which

$$\mathbf{r}_1(t) = \mathbf{r}_2(u).$$

If two curves C_1 and C_2 intersect at a point $\mathbf{r}_1(t_1) = \mathbf{r}_2(u_2)$, we define the angle between the curves at this point to be the angle between the corresponding tangent vectors $\mathbf{r}'_1(t_1)$ and $\mathbf{r}'_2(u_2)$.

Example 3 Show that the circles

$$C_1: \ \mathbf{r}_1(t) = \cos t\,\mathbf{i} + \sin t\,\mathbf{j}, \quad C_2: \ \mathbf{r}_2(u) = \cos u\,\mathbf{j} + \sin u\,\mathbf{k}$$

intersect at right angles at $P(0, 1, 0)$ and $Q(0, -1, 0)$.

SOLUTION Since $\mathbf{r}_1(\pi/2) = \mathbf{j} = \mathbf{r}_2(0)$, the curves meet at the tip of $\mathbf{j}$, which is $P(0, 1, 0)$. Also, since $\mathbf{r}_1(3\pi/2) = -\mathbf{j} = \mathbf{r}_2(\pi)$, the curves meet at the tip of $-\mathbf{j}$, which is $Q(0, -1, 0)$. Differentiation gives

$$\mathbf{r}'_1(t) = -\sin t\,\mathbf{i} + \cos t\,\mathbf{j} \quad \text{and} \quad \mathbf{r}'_2(u) = -\sin u\,\mathbf{j} + \cos u\,\mathbf{k}.$$

Since $\mathbf{r}'_1(\pi/2) = -\mathbf{i}$ and $\mathbf{r}'_2(0) = \mathbf{k}$, we have

$$\mathbf{r}'_1(\pi/2) \cdot \mathbf{r}'_2(0) = 0.$$

This tells us that the curves are perpendicular at $P(0, 1, 0)$. Since $\mathbf{r}'_1(3\pi/2) = \mathbf{i}$ and $\mathbf{r}'_2(\pi) = -\mathbf{k}$, we have

$$\mathbf{r}'_1(3\pi/2) \cdot \mathbf{r}'_2(\pi) = 0.$$

This tells us that the curves are perpendicular at $Q(0, -1, 0)$. The curves appear in Figure 13.3.10. ☐

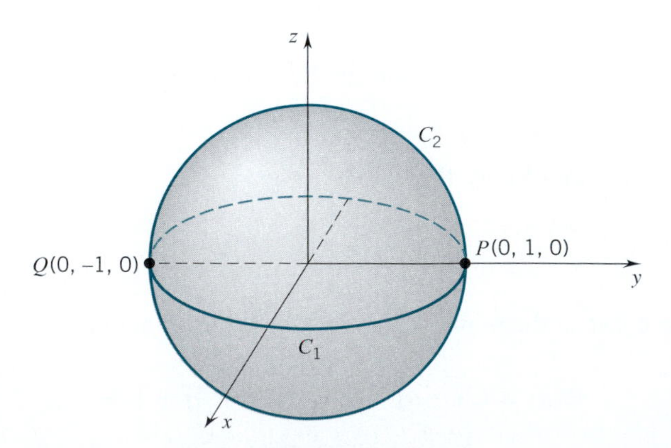

Figure 13.3.10

The Unit Tangent, the Principal Normal, the Osculating Plane

Suppose now that the curve

$$C: \ \mathbf{r}(t) = x(t)\,\mathbf{i} + y(t)\,\mathbf{j} + z(t)\,\mathbf{k}$$

is twice differentiable and $\mathbf{r}'(t)$ is never zero. Then at each point $P(x(t), y(t), z(t))$ of the curve, there is a *unit tangent vector*:

(13.3.3)
$$\mathbf{T}(t) = \frac{\mathbf{r}'(t)}{\|\mathbf{r}'(t)\|}.$$

Since $\|\mathbf{r}'(t)\| > 0$, $\mathbf{T}(t)$ points in the direction of $\mathbf{r}'(t)$ that is, in the direction of increasing t. Since $\|\mathbf{T}(t)\| = 1$, we have $\mathbf{T}(t) \cdot \mathbf{T}(t) = 1$. Differentiation gives

$$\mathbf{T}(t) \cdot \mathbf{T}'(t) + \mathbf{T}'(t) \cdot \mathbf{T}(t) = 0.$$

Since the dot product is commutative, we have

$$2[\mathbf{T}'(t) \cdot \mathbf{T}(t)] = 0 \qquad \text{and thus} \qquad \mathbf{T}'(t) \cdot \mathbf{T}(t) = 0.$$

At each point of the curve *the vector* $\mathbf{T}'(t)$ *is perpendicular to* $\mathbf{T}(t)$.

The vector $\mathbf{T}'(t)$ measures the rate of change of $\mathbf{T}(t)$ with respect to t. Since the norm of $\mathbf{T}(t)$ is constantly 1, $\mathbf{T}(t)$ can change only in direction. The vector $\mathbf{T}'(t)$ measures this change in direction.

If the unit tangent vector is not changing in direction (as in the case of a straight line), then $\mathbf{T}'(t) = \mathbf{0}$. If $\mathbf{T}'(t) \neq \mathbf{0}$, then we can form what is called the *principal normal vector*:

(13.3.4)
$$\mathbf{N}(t) = \frac{\mathbf{T}'(t)}{\|\mathbf{T}'(t)\|}.$$

This is the unit vector in the direction of $\mathbf{T}'(t)$. The *normal line* at P is the line through P parallel to the principal normal.

Figure 13.3.11 shows a curve on which we have marked several points. At each point we have drawn the unit tangent and the principal normal. The plane determined by these two vectors is called the *osculating plane* (literally, the "kissing plane"). This is the plane of greatest contact with the curve at the point in question. The principal normal points in the direction in which the curve is curving, that is, on the concave side of the curve.

If a curve is a plane curve but not a straight line, then the osculating plane is simply the plane of the curve. A straight line does not have an osculating plane. There is no principal normal vector $[\mathbf{T}'(t) = \mathbf{0}]$ and there is no single plane of greatest contact. Each straight line lies on an infinite number of planes.

Example 4 In Figure 13.3.12 you can see a curve spiraling up a circular cylinder. The curve is called a *circular helix* if the rate of climb is constant. The simplest parametrization for a circular helix takes the form

$$\mathbf{r}(t) = a\cos t\,\mathbf{i} + a\sin t\,\mathbf{j} + bt\,\mathbf{k} \quad \text{with } a > 0, \quad b > 0.$$

The first two components produce the rotational effect; the third component gives the rate of climb.

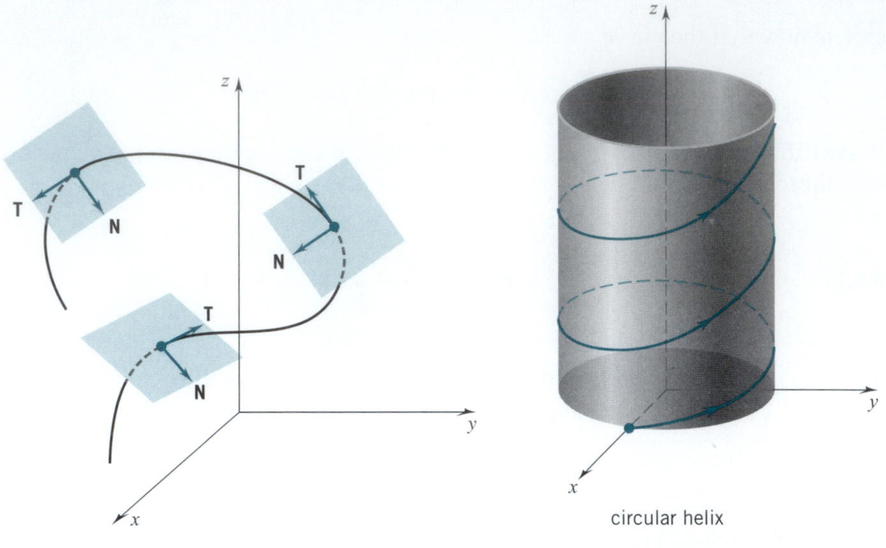

Figure 13.3.11 **Figure 13.3.12**

We will find the unit tangent, the principal normal, and then an equation for the osculating plane. Since

$$\mathbf{r}'(t) = -a\sin t\,\mathbf{i} + a\cos t\,\mathbf{j} + b\,\mathbf{k},$$

we have $$\|\mathbf{r}'(t)\| = \sqrt{a^2\sin^2 t + a^2\cos^2 t + b^2} = \sqrt{a^2+b^2}$$

and therefore

$$\mathbf{T}(t) = \frac{\mathbf{r}'(t)}{\|\mathbf{r}'(t)\|} = \frac{1}{\sqrt{a^2+b^2}}(-a\sin t\,\mathbf{i} + a\cos t\,\mathbf{j} + b\,\mathbf{k}).$$

This is the unit tangent vector.

The principal normal vector is the unit vector in the direction of $\mathbf{T}'(t)$. Since

$$\frac{d}{dt}(-a\sin t\,\mathbf{i} + a\cos t\,\mathbf{j} + b\,\mathbf{k}) = -a\cos t\,\mathbf{i} - a\sin t\,\mathbf{j} \quad \text{and} \quad a > 0,$$

you can see that

$$\mathbf{N}(t) = -\cos t\,\mathbf{i} - \sin t\,\mathbf{j}.$$

The principal normal is horizontal and points directly toward the z-axis.

Now let's find an equation for the osculating plane p at an arbitrary point $P(a\cos t, a\sin t, bt)$. The cross product $\mathbf{T}(t)\times\mathbf{N}(t)$ is perpendicular to p. Therefore, as a normal for p, we can take any nonzero scalar multiple of $\mathbf{T}(t)\times\mathbf{N}(t)$. In particular, we can take

$$(a\sin t\,\mathbf{i} - a\cos t\,\mathbf{j} - b\,\mathbf{k}) \times (\cos t\,\mathbf{i} + \sin t\,\mathbf{j}).$$

As you can check, this simplifies to

$$b\sin t\,\mathbf{i} - b\cos t\,\mathbf{j} + a\,\mathbf{k}.$$

The equation for the osculating plane at the point $P(a\cos t, a\sin t, bt)$ thus takes the form

$$b\sin t(x - a\cos t) - b\cos t(y - a\sin t) + a(z - bt) = 0. \qquad (12.7.1)$$

This simplifies to

$$(b\sin t)x - (b\cos t)y + az = abt.$$

To visualize how this osculating plane changes from point to point, think of a playground spiral slide or the threaded surface on a bolt. □

Reversing the Orientation of a Curve

As noted at the beginning of this section, a parametrized curve is an *oriented* curve; it is not just a set of points, but rather a succession of points traversed in a certain order.

We make a distinction between the curve

$$\mathbf{r} = \mathbf{r}(t), \qquad t \in [a, b]$$

and the curve

$$\mathbf{R}(u) = \mathbf{r}(a + b - u), \qquad u \in [a, b].$$

Both vector functions trace out the same set of points (check that out), but the order has been reversed. Whereas the first curve starts at $\mathbf{r}(a)$ and ends at $\mathbf{r}(b)$, the second curve starts at $\mathbf{r}(b)$ and ends at $\mathbf{r}(a)$:

$$\mathbf{R}(a) = \mathbf{r}(a + b - a) = \mathbf{r}(b), \qquad \mathbf{R}(b) = \mathbf{r}(a + b - b) = \mathbf{r}(a).$$

The vector function

$$\mathbf{r}(t) = \cos t\,\mathbf{i} + \sin t\,\mathbf{j}, \qquad t \in [0, 2\pi]$$

gives the unit circle traversed counterclockwise, while

$$\mathbf{R}(u) = \cos(2\pi - u)\,\mathbf{i} + \sin(2\pi - u)\,\mathbf{j}, \qquad u \in [0, 2\pi]$$

gives the unit circle traversed clockwise.

What happens to the unit tangent $\mathbf{T}$, the principal normal $\mathbf{N}$, and the osculating plane when we reverse the orientation of a curve? As you are asked to show in Exercise 43, $\mathbf{T}$ is replaced by $-\mathbf{T}$, but $\mathbf{N}$ remains the same. The osculating plane also remains the same.

Other Changes of Parameter

Not all changes of parameter change the succession of points. Suppose that

$$\mathbf{r} = \mathbf{r}(t), \qquad t \in I$$

is a differentiable curve, and let ϕ be a function that maps some interval J onto the interval I (domain J, range I). Now set

$$\mathbf{R}(u) = \mathbf{r}(\phi(u)) \quad \text{for all } u \in J.$$

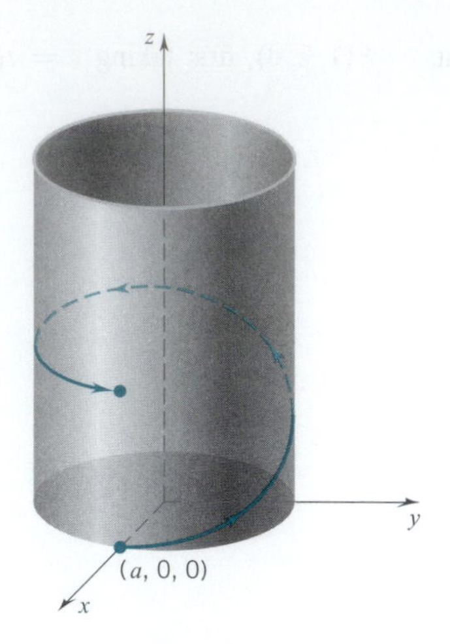

Figure 13.3.13

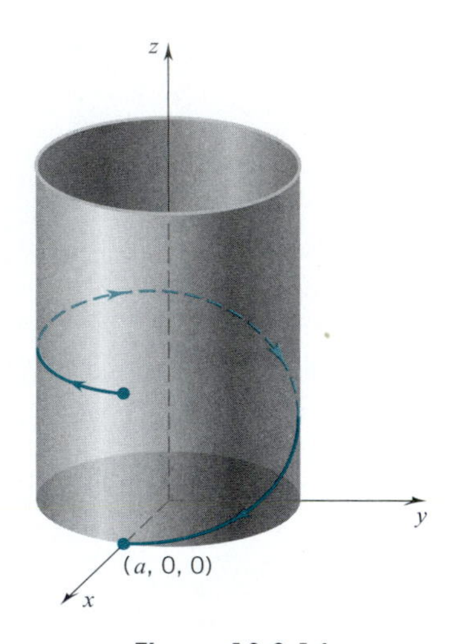

Figure 13.3.14

If $\phi'(u) > 0$ for all $u \in J$, then $\mathbf{r}$ and $\mathbf{R}$ are said to differ by an *orientation-preserving change of parameter*. In this case, $\mathbf{r}$ and $\mathbf{R}$ take on exactly the same values in exactly the same order. In other words, they produce exactly the same oriented curve (Exercise 44).

If, on the other hand, $\phi'(u) < 0$ for all $u \in J$. then the change in parameter is said to be *orientation-reversing*. In this case $\mathbf{r}$ and $\mathbf{R}$ still take on exactly the same values but in opposite order. The paths are the same, but they are traversed in opposite directions. You have already seen one example of this:

$$\mathbf{R}(u) = \mathbf{r}(a + b - u).$$

Example 5 The parametrization

$$\mathbf{r}(t) = a\cos t\,\mathbf{i} + a\sin t\,\mathbf{j} + bt\,\mathbf{k}. \quad t \in [0, 2\pi]$$

gives one "spiral" of the circular helix with the orientation indicated by the arrows (Figure 13.3.13). If we let $\phi(u) = \pi u$, $u \in [0, 2]$, then ϕ maps the interval $J = [0, 2]$ onto the interval $I = [0, 2\pi]$, and $\phi'(u) = \pi > 0$. Thus,

$$\mathbf{R}(u) = \mathbf{r}(\phi(u)) = a\cos(\pi u)\,\mathbf{i} + a\sin(\pi u)\,\mathbf{j} + b\pi u\,\mathbf{k}, \quad u \in [0, 2]$$

is precisely the same curve with the same orientation.

On the other hand, if we let $\psi(u) = (2 - u)\pi$, $u \in [0, 2]$, then ψ also maps the interval $J = [0, 2]$ onto the interval $I = [0, 2\pi]$, but $\psi'(u) = -\pi < 0$. Thus,

$$\mathbf{R}(u) = \mathbf{r}(\psi(u)) = a\cos(2 - u)\pi\,\mathbf{i} + a\sin(2 - u)\pi\,\mathbf{j} + b(2 - u)\pi\,\mathbf{k}$$

produces the same curve but with the opposite orientation. See Figure 13.3.14. ❑

EXERCISES 13.3

Find the tangent vector $\mathbf{r}'(t)$ at the indicated point and parametrize the tangent line.

1. $\mathbf{r}(t) = \cos \pi t\,\mathbf{i} + \sin \pi t\,\mathbf{j} + t\,\mathbf{k}$ at $t = 2$.

2. $\mathbf{r}(t) = e^t\,\mathbf{i} + e^{-t}\,\mathbf{j} - \ln t\,\mathbf{k}$ at $t = 1$.

3. $\mathbf{r}(t) = \mathbf{a} + t\,\mathbf{b} + t^2\,\mathbf{c}$ at $t = -1$.

4. $\mathbf{r}(t) = (t + 1)\,\mathbf{i} + (t^2 + 1)\,\mathbf{j} + (t^3 + 1)\,\mathbf{k}$ at $P(1, 1, 1)$.

5. $\mathbf{r}(t) = 2t^2\,\mathbf{i} + (1 - t)\,\mathbf{j} + (3 + 2t^2)\,\mathbf{k}$ at $P(2, 0, 5)$.

6. $\mathbf{r}(t) = 3t\,\mathbf{a} + \mathbf{b} - t^2\,\mathbf{c}$ at $t = 2$.

7. $\mathbf{r}(t) = 2\cos t\,\mathbf{i} + 3\sin t\,\mathbf{j} + t\,\mathbf{k}$; $t = \pi/4$.

8. $\mathbf{r}(t) = t\sin t\,\mathbf{i} + t\cos t\,\mathbf{j} + 2t\,\mathbf{k}$; $t = \pi/2$.

9. Show that $\mathbf{r}(t) = at\,\mathbf{i} + bt^2\,\mathbf{j}$ parametrizes a parabola. Find an equation in x and y for this parabola.

10. Show that $\mathbf{r}(t) = \frac{1}{2}a(e^{\omega t} + e^{-\omega t})\mathbf{i} + \frac{1}{2}a(e^{\omega t} - t^{-\omega t})\mathbf{j}$ parametrizes the right branch ($x > 0$) of the hyperbola $x^2 - y^2 = a^2$.

11. Find (a) the points on the curve $\mathbf{r}(t) = t\,\mathbf{i} + (1 + t^2)\,\mathbf{j}$ at which $\mathbf{r}(t)$ and $\mathbf{r}'(t)$ are perpendicular; (b) the points at which they have the same direction; (c) the points at which they have opposite directions.

12. Find the curve given that $\mathbf{r}'(t) = \alpha\mathbf{r}(t)$ for all real t and $\mathbf{r}(0) = \mathbf{i} + 2\mathbf{j} + 3\mathbf{k}$.

13. Suppose that $\mathbf{r}'(t)$ and $\mathbf{r}(t)$ are parallel for all t. Show that, if $\mathbf{r}'(t)$ is never $\mathbf{0}$, then the tangent line at each point passes through the origin.

C In Exercises 14–16, the given curves intersect at the indicated point. Find the angle of intersection. Express your answer in radians rounded to the nearest hundredth, and in degrees rounded to the nearest tenth.

14. $\mathbf{r}_1(t) = t\,\mathbf{i} + t^2\,\mathbf{j} + t^3\,\mathbf{k}$,
$\mathbf{r}_2(u) = \sin 2u\,\mathbf{i} + u\cos u\,\mathbf{j} + u\,\mathbf{k}$; $P(0, 0, 0)$.

15. $\mathbf{r}_1(t) = (e^t - 1)\,\mathbf{i} + 2\sin t\,\mathbf{j} + \ln(t + 1)\,\mathbf{k}$,
$\mathbf{r}_2(u) = (u + 1)\,\mathbf{i} + (u^2 - 1)\,\mathbf{j} + (u^3 + 1)\,\mathbf{k}$; $P(0, 0, 0)$.

16. $\mathbf{r}_1(t) = e^{-t}\,\mathbf{i} + \cos t\,\mathbf{j} + (t^2+4)\,\mathbf{k}$,
$\mathbf{r}_2(u) = (2+u)\,\mathbf{i} + u^4\,\mathbf{j} + 4u^2\,\mathbf{k}$; $\quad P(1,1,4)$.

17. Find the point at which the curves

$$\mathbf{r}_1(t) = e^t\,\mathbf{i} + 2\sin(t+\tfrac{1}{2}\pi)\,\mathbf{j} + (t^2-2)\,\mathbf{k},$$
$$\mathbf{r}_2(u) = u\,\mathbf{i} + 2\,\mathbf{j} + (u^2-3)\,\mathbf{k}$$

intersect and find the angle of intersection.

18. Consider the vector function $\mathbf{f}(t) = t\,\mathbf{i} + f(t)\,\mathbf{j}$ formed from a differentiable real-valued function f. The vector function $\mathbf{f}$ parametrizes the graph of f.

(a) Parametrize the tangent line at $P(t_0, f(t_0))$.

(b) Show that the parametrization obtained in part (a) can be reduced to the usual equation for the tangent line:

$$y - f(t_0) = f'(t_0)(x - t_0) \quad \text{if} \quad f'(t_0) \neq 0;$$
$$y = f(t_0) \quad \text{if}\ f'(t_0) = 0.$$

19. Define a vector function $\mathbf{r}$ on the interval $[0, 2\pi]$ that satisfies the initial condition $\mathbf{r}(0) = a\mathbf{i}$ and, as t increases to 2π, traces out the ellipse $b^2x^2 + a^2y^2 = a^2b^2$:

(a) Once in a counterclockwise manner.

(b) Once in a clockwise manner.

(c) Twice in a counterclockwise manner.

(d) Three times in a clockwise manner.

20. Repeat Exercise 19 given that $\mathbf{r}(0) = b\,\mathbf{j}$.

In Exercise 21–26, sketch the plane curve determined by the given vector-valued function $\mathbf{r}$ and indicate the orientation. Find $\mathbf{r}'(t)$ and draw the position vector and the tangent vector for the indicated value of t, placing the tangent vector at the tip of the position vector.

21. $\mathbf{r}(t) = \frac{1}{4}t^4\,\mathbf{i} + t^2\,\mathbf{j}$; $\quad t = 2$.

22. $\mathbf{r}(t) = 2t\,\mathbf{i} + (t^2+1)\,\mathbf{j}$; $\quad t = 4$.

23. $\mathbf{r}(t) = e^{2t}\,\mathbf{i} + e^{-4t}\,\mathbf{j}$; $\quad t = 0$.

24. $\mathbf{r}(t) = \sin t\,\mathbf{i} - 2\cos t\,\mathbf{j}$; $\quad t = \pi/3$.

25. $\mathbf{r}(t) = 2\cos t\,\mathbf{i} + 3\sin t\,\mathbf{j}$; $\quad t = \pi/6$.

26. $\mathbf{r}(t) = \sec t\,\mathbf{i} + \tan t\,\mathbf{j}$, $\quad |t| < \pi/2$; $\quad t = \pi/4$.

Find a vector parametrization for the curve.

27. $y^2 = x - 1, \quad y \geq 1$.

28. $r = 1 - \cos\theta, \quad \theta \in [0, 2\pi]$. (Polar coordinates)

29. $r = \sin 3\theta, \quad \theta \in [0, \pi]$. (Polar coordinates)

30. $y^4 = x^3, \quad y \leq 0$.

31. Find an equation in x and y for the curve $\mathbf{r}(t) = t^3\,\mathbf{i} + t^2\,\mathbf{j}$. Draw the curve. Does the curve have a tangent vector at the origin? If so, what is the unit tangent vector?

32. (a) Show that the curve

$$\mathbf{r}(t) = (t^2 - t + 1)\,\mathbf{i} + (t^3 - t + 2)\,\mathbf{j} + (\sin \pi t)\,\mathbf{k}$$

intersects itself at $P(1,2,0)$ by finding numbers $t_1 < t_2$ for which P is the tip of both $\mathbf{r}(t_1)$ and $\mathbf{r}(t_2)$.

(b) Find the unit tangents at $P(1,2,0)$, first taking $t = t_1$, then taking $t = t_2$.

33. Find the point(s) at which the twisted cubic

$$\mathbf{r}(t) = t\,\mathbf{i} + t^2\,\mathbf{j} + t^3\,\mathbf{k}$$

intersects the plane $4x + 2y + z = 24$. What is the angle of intersection between the curve and the normal to the plane?

34. (a) Find the unit tangent and the principal normal at an arbitrary point of the ellipse

$$\mathbf{r}(t) = a\cos t\,\mathbf{i} + b\sin t\,\mathbf{j}.$$

(b) Write vector equations for the tangent line and the normal line at the tip of $\mathbf{r}(\frac{1}{4}\pi)$.

Find the unit tangent vector, the principal normal vector, and an equation in x, y, z for the osculating plane at the point on the curve corresponding to the indicated value of t.

35. $\mathbf{r}(t) = \mathbf{i} + 2t\,\mathbf{j} + t^2\,\mathbf{k}$; $\quad t = 1$.

36. $\mathbf{r}(t) = t\,\mathbf{i} + t^2\,\mathbf{j} + 2t^2\,\mathbf{k}$; $\quad t = 1$.

37. $\mathbf{r}(t) = \cos 2t\,\mathbf{i} + \sin 2t\,\mathbf{j} + t\,\mathbf{k}$ at $t = \frac{1}{4}\pi$.

38. $\mathbf{r}(t) = t\,\mathbf{i} + 2t\,\mathbf{j} + t^2\,\mathbf{k}$ at $t = 2$.

39. $\mathbf{r}(t) = t\,\mathbf{i} + t^2\,\mathbf{j} + t^3\,\mathbf{k}$ at $t = 1$.

40. $\mathbf{r}(t) = \cos 3t\,\mathbf{i} + t\,\mathbf{j} - \sin 3t\,\mathbf{k}$ at $t = \frac{1}{3}\pi$.

41. $\mathbf{r}(t) = e^t \sin t\,\mathbf{i} + e^t\cos t\,\mathbf{j} + e^t\,\mathbf{k}$; $\quad t = 0$.

42. $\mathbf{r}(t) = (\cos t + t\sin t)\,\mathbf{i} + (\sin t - t\cos t)\,\mathbf{j} + 2k$; $\quad t = \frac{1}{4}\pi$.

43. Let $\mathbf{r} = \mathbf{r}(t), t \in [a, b]$ and set

$$\mathbf{R}(u) = \mathbf{r}(a + b - u), \quad u \in [a, b].$$

Show that this change of parameter changes the sign of the unit tangent but does not alter the principal normal.

HINT: Let P be the tip of $\mathbf{R}(u) = \mathbf{r}(a+b-u)$. At that point, $\mathbf{R}$ produces a unit tangent $\mathbf{T}_1(u)$ and a principal normal $\mathbf{N}_1(u)$. At that same point, $\mathbf{r}$ produces a unit tangent $\mathbf{T}(a + b - u)$ and a principal normal $\mathbf{N}(a + b - u)$.

44. Show that two vector functions that differ by a orientation-preserving change of parameter take on exactly the same values in exactly the same order. That is, set

$$\mathbf{r} = \mathbf{r}(t), \quad t \in I.$$

Assume that ϕ maps an interval J onto the interval I and that $\phi'(u) > 0$ for all $u \in J$. Set

$$\mathbf{R}(u) = \mathbf{r}(\phi(u)),$$

and show that $\mathbf{R}$ and $\mathbf{r}$ take on exactly the same values in exactly the same order.

45. Show that the unit tangent vector, the principal normal vector, and the osculating plane are left invariant (left unchanged) by every orientation-preserving change of parameter.

In Exercises 46 and 47, let $\mathbf{r}$ be the vector-valued function defined by

$$\mathbf{r}(t) = 2\cos t\,\mathbf{i} + 2\sin t\,\mathbf{j} + 4t\,\mathbf{k} \quad \text{for} \quad 0 \le t \le 2\pi.$$

The graph of $\mathbf{r}$ is one revolution on a circular helix, starting at the point $(2, 0, 0)$ and ending at the point $(2, 0, 8\pi)$.

46. Let $\varphi(u) = u^2$ for $0 \le u \le \sqrt{2\pi}$, and let

$$\mathbf{R}(u) = \mathbf{r}[\varphi(u)] = 2\cos u^2\mathbf{i} + 2\sin u^2\,\mathbf{j} + 4u^2\,\mathbf{k}.$$

(a) Show that φ determines a orientation-preserving change of parameter on $[0, \sqrt{2\pi}]$.

(b) Show that the unit tangent and principal normal vectors for $\mathbf{r}$ at the point $t = \frac{1}{4}\pi$ are the same as the unit tangent and principal normal vectors for $\mathbf{R}$ at $u = \frac{1}{2}\sqrt{\pi}$.

47. Let $\psi(v) = 2\pi - v^2$ for $0 \le v \le \sqrt{2\pi}$, and let

$$\mathbf{R}(v) = \mathbf{r}[\psi(v)] = 2\cos(2\pi - v^2)\,\mathbf{i} + 2\sin(2\pi - v^2)\,\mathbf{j} + 4(2\pi - v^2)\,\mathbf{k}.$$

(a) Show that ψ determines a orientation-reversing change of parameter on $[0, \sqrt{2\pi}]$.

(b) Show that the principal normal vector for $\mathbf{r}$ at $t = \pi/4$ is the same as the principal normal for $\mathbf{R}$ at $v = \frac{1}{2}\sqrt{7\pi}$, and show that the unit tangent vector for $\mathbf{r}$ at $\pi/4$ is the negative of the unit tangent vector for $\mathbf{R}$ at $v = \frac{1}{2}\sqrt{7\pi}$.

C **48.** Let $\mathbf{r}(t) = \sqrt{2}\cos t\,\mathbf{i} + \sqrt{2}\sin t\,\mathbf{j} + t\,\mathbf{k}, \quad 0 \le t \le 2\pi$.

(a) Find scalar parametric equations for the tangent line to the curve at the point $(1, 1, \pi/4)$.

(b) Use a CAS to draw the curve and the tangent line together. Experiment with the t-interval to find a good illustration of the curve and the tangent line.

(c) Are there points on the curve where the tangent line is parallel to the xy-plane? If so, find them.

C **49.** Let $\mathbf{r}(t) = \sqrt{2}\cos t\,\mathbf{i} + \sqrt{2}\sin t\,\mathbf{j} + \sin 5t\,\mathbf{k}, \quad 0 \le t \le 2\pi$.

(a) Find scalar parametric equations for the tangent line to the curve at the point $(1, 1, -\sqrt{2}/2)$.

(b) Use a CAS to draw the curve and the tangent line together. Experiment with the t-interval to find a good illustration of the curve and the tangent line.

(c) Are there points on the curve where the tangent line is parallel to the xy-plane? If so, find them

C **50.** Let $\mathbf{r}(t) = \sqrt{2}\cos t\,\mathbf{i} + \sqrt{2}\sin t\,\mathbf{j} + \ln t\,\mathbf{k}, \quad 0 \le t \le 2\pi$.

(a) Find the scalar parametric equations for the tangent line to the curve at the point $(1, 1, \ln(\pi/4))$.

(b) Use a CAS to draw the curve and the tangent line together. Experiment with the t-interval to find a good illustration of the curve and the tangent line.

■ 13.4 ARC LENGTH

In Section 9.8 we considered arc length in an intuitive manner and decided that the length of the path C traced out by a pair of continuously differentiable functions

$$x = x(t), \qquad y = y(t), \qquad t \in [a, b]$$

is given by the formula

$$L(C) = \int_a^b \sqrt{[x'(t)]^2 + [y'(t)]^2}\, dt.$$

Applied to a path C in space traced out by

$$x = x(t), \qquad y = y(t), \qquad z = z(t), \qquad t \in [a, b]$$

the formula becomes

$$L(C) = \int_a^b \sqrt{[x'(t)]^2 + [y'(t)]^2 + [z'(t)]^2}\, dt.$$

In vector notation, both formulas can be written

$$L(C) = \int_a^b \|\mathbf{r}'(t)\|\, dt.$$

We will prove this result in this form, but first we give a precise definition of arc length.

In Figure 13.4.1 we have sketched the path C traced out by a continuously differentiable vector function

$$\mathbf{r} = \mathbf{r}(t), \quad t \in [a, b].$$

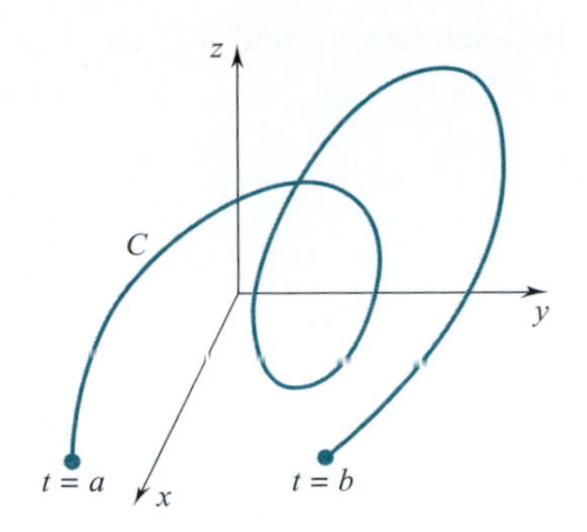

Figure 13.4.1

To decide what should be meant by the length of C, we approximate C by the union of a finite number of line segments.

Choosing a finite number of points in $[a, b]$,

$$a = t_0 < t_1 < \cdots < t_{i-1} < t_i < \cdots < t_{n-1} < t_n = b,$$

we obtain a finite number of points $P_0, P_1, \ldots, P_{i-1}, P_i, \ldots, P_{n-1}, P_n$, on C, where for each $k, 0 \leq k \leq n$, P_k denotes the point $P(x(t_k), y(t_k), z(t_k))$. Join these points consecutively by line segments and call the resulting path

$$\gamma = \overline{P_0P_1} \cup \cdots \cup \overline{P_{i-1}P_i} \cup \cdots \cup \overline{P_{n-1}P_n},$$

a *polygonal path* inscribed in C (Figure 13.4.2).

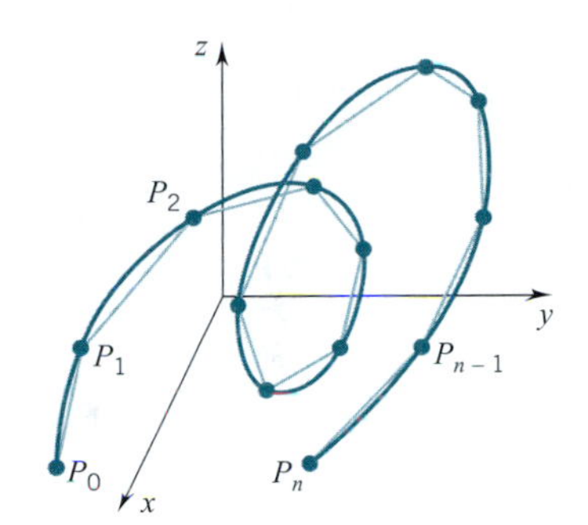

Figure 13.4.2

The length of such a polygonal path is the sum of the distances between consecutive vertices:

$$L(\gamma) = d(P_0, P_1) + \cdots + d(P_{i-1}, P_i) + \cdots + d(P_{n-1}, P_n).$$

The path γ serves as an approximation to the path C, and better approximations can be obtained by adding more vertices to γ. We now ask ourselves exactly what should we require of the number that we shall call the length of C. Certainly we require that

$$L(\gamma) \leq L(C) \quad \text{for each } \gamma \text{ inscribed in } C.$$

But that is not enough. There is another requirement that seems reasonable. If we can choose γ to approximate C as closely as we wish, then we should be able to choose γ so that $L(\gamma)$ approximates the length of C as closely as we wish; namely, for each positive number ϵ there should exist a polygonal path γ such that

$$L(C) - \epsilon < L(\gamma) \leq L(C).$$

In Section 10.1, we introduced the concept of least upper bound of a set of real numbers. Theorem 10.1.2 tells us that we can achieve the result we want by defining the length of C as the least upper bound of the set of all $L(\gamma)$. This is in fact what we do.

DEFINITION 13.4.1 ARC LENGTH

$$L(C) = \begin{cases} \text{the least upper bound of the set of all} \\ \text{lengths of polygonal paths inscribed in } C. \end{cases}$$

We are now ready to establish the arc length formula.

THEOREM 13.4.2 ARC LENGTH FORMULA

Let C be the path traced out by a continuously differentiable vector function

$$\mathbf{r} = \mathbf{r}(t), \qquad t \in [a.b].$$

The length of C is given by the formula

$$L(C) = \int_a^b \|\mathbf{r}'(t)\|\, dt.$$

PROOF First we show that

$$L(C) \leq \int_a^b \|\mathbf{r}'(t)\|\, dt.$$

To do this, we begin with an arbitrary partition P of $[a, b]$:

$$P = \{a = t_0, \ldots, t_{i-1}, t_i, \ldots, t_n = b\}.$$

Such a partition gives rise to a finite number of points of C:

$$\mathbf{r}(a) = \mathbf{r}(t_0), \ldots, \mathbf{r}(t_{i-1}), \mathbf{r}(t_i), \ldots, \mathbf{r}(t_n) = \mathbf{r}(b)$$

and thus to an inscribed polygonal path of total length

$$L_P = \sum_{i=1}^{n} \|\mathbf{r}(t_i) - \mathbf{r}(t_{i-1})\|. \qquad \text{(Figure 13.4.3)}$$

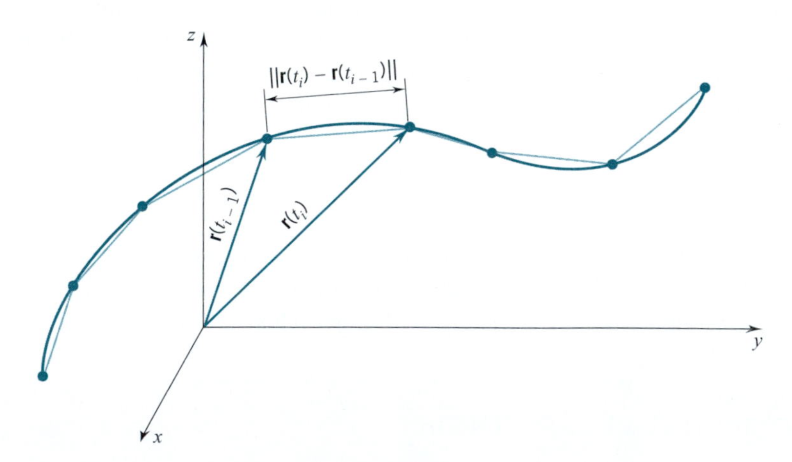

Figure 13.4.3

For each i,

$$\mathbf{r}(t_i) - \mathbf{r}(t_{i-1}) = \int_{t_{i-1}}^{t_i} \mathbf{r}'(t)\, dt.$$

This gives

$$\|\mathbf{r}(t_i) - \mathbf{r}(t_{i-1})\| = \left\| \int_{t_{i-1}}^{t_i} \mathbf{r}'(t)\, dt \right\| \overset{\text{by (13.1.10)}}{\leq} \int_{t_{i-1}}^{t_i} \|\mathbf{r}'(t)\|\, dt$$

and thus

$$L_P = \sum_{i=1}^{n} \|\mathbf{r}(t_i) - \mathbf{r}(t_{i-1})\| \leq \sum_{i=1}^{n} \int_{t_{i-1}}^{t_i} \|\mathbf{r}'(t)\|\, dt = \int_a^b \|\mathbf{r}'(t)\|\, dt.$$

Since the partition P is arbitrary, we know that the inequality

$$L_P \leq \int_a^b \|\mathbf{r}'(t)\|\, dt$$

must hold for all the L_P. This makes the integral on the right an upper bound for all the L_P. Since $L(C)$ is the *least* upper bound of all the L_P, we can conclude right now that

$$L(C) \leq \int_a^b \|\mathbf{r}'(t)\|\, dt.$$

The next step is to show that this inequality is actually an equation. To do this we need to know that arc length, *as we have defined it*, is additive. That is, with P, Q, and R as in Figure 13.4.4, we need to know that the arc length from P to Q plus the arc length from Q to R equals the arc length from P to R. It is clear arc length should have this property. We must prove that it does. A proof has been placed in a supplement to this section. For the moment we shall assume that arc length is additive and continue with the proof of the arc length formula

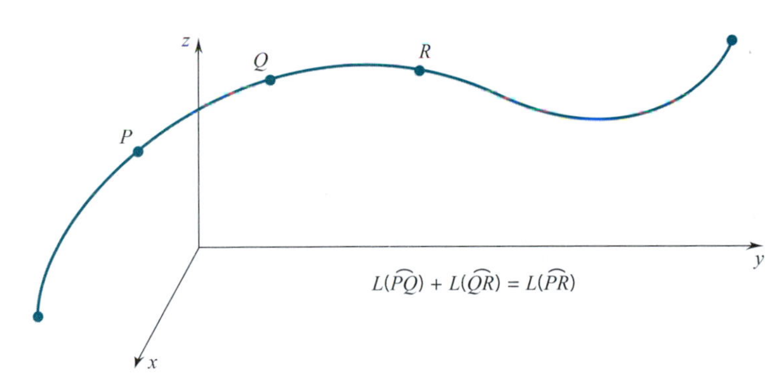

Figure 13.4.4

In Figure 13.4.5, we display the initial vector $\mathbf{r}(a)$, a general radius vector $\mathbf{r}(t)$, and a nearby vector $\mathbf{r}(t+h)$. Set

$$s(t) = \text{ length of the path from } \mathbf{r}(a) \text{ to } \mathbf{r}(t).$$

Then

$$s(a) = 0 \quad \text{and} \quad s(t+h) = \text{ length of the path from } \mathbf{r}(a) \text{ to } \mathbf{r}(t+h).$$

By the additivity of arc length (remember, we are assuming this for the moment),

$$s(t+h) - s(t) = \text{ length of the curve from } \mathbf{r}(t) \text{ to } \mathbf{r}(t+h).$$

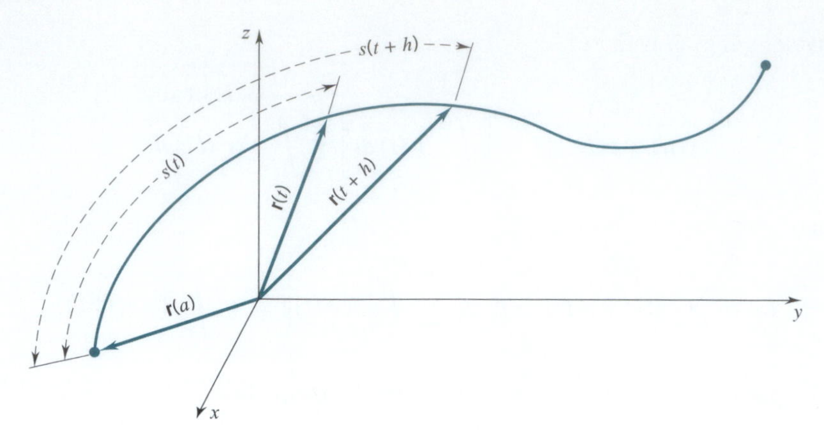

Figure 13.4.5

From that we have shown already, you can see that

$$\|\mathbf{r}(t+h)-\mathbf{r}(t)\| \leq s(t+h)-s(t) \leq \int_t^{t+h} \|\mathbf{r}'(u)\|\, du.$$

Dividing this inequality by h (which we are taking as positive), we get

$$\left\|\frac{\mathbf{r}(t+h)-\mathbf{r}(t)}{h}\right\| \leq \frac{s(t+h)-s(t)}{h} \leq \frac{1}{h}\int_t^{t+h} \|\mathbf{r}'(u)\|\, du.$$

As $h \to 0^+$, the left-hand side tends to $\|\mathbf{r}'(t)\|$ and, by the first mean-value theorem for integrals (Theorem 5.9.1), so does the right-hand side:

$$\frac{1}{h}\int_t^{t+h} \|\mathbf{r}'(u)\|\, du = \frac{1}{h}\|\mathbf{r}'(c_h)\|(t+h-t) = \|\mathbf{r}'(c_h)\| \to \|\mathbf{r}'(t)\|.$$

$c_h \in (t, t+h)$

Therefore

$$\lim_{h\to 0^+} \frac{s(t+h)-s(t)}{h} = \|\mathbf{r}'(t)\|.$$

By taking $h < 0$ and proceeding in a similar manner, one can show that

$$\lim_{h\to 0^-} \frac{s(t+h)-s(t)}{h} = \|\mathbf{r}'(t)\|.$$

Therefore, we can conclude that

$$\lim_{h\to 0} \frac{s(t+h)-s(t)}{h} = \|\mathbf{r}'(t)\|$$

and so $s'(t) = \|\mathbf{r}'(t)\|$. Integrating this equation from a to t, we get

$$s(t)-s(a) = \int_a^t s'(u)\, du = \int_a^t \|\mathbf{r}'(u)\|\, du.$$

Since $s(a) = 0$, it follows that

$$s(t) = \int_a^t \|\mathbf{r}'(u)\|\, du.$$

The total length of C is therefore

$$s(b) = \int_a^b \|\mathbf{r}'(t)\|\,dt. \quad \square$$

For convenience, and to follow custom, we shall speak of the length of a parametrized curve. You can take this to mean the length of the path traced out by the parametrizing vector function.

Example 1 Find the length of the curve

$$\mathbf{r}(t) = 2t^{3/2}\,\mathbf{i} + 4t\,\mathbf{j} \quad \text{from} \quad t = 0 \text{ to } t = 1.$$

SOLUTION

$$\mathbf{r}'(t) = 3t^{1/2}\,\mathbf{i} + 4\,\mathbf{j}.$$

$$\|\mathbf{r}'(t)\| = \sqrt{(3t^{1/2})^2 + 4^2} = \sqrt{9t + 16}.$$

$$L(C) = \int_0^1 \|\mathbf{r}'(t)\|\,dt = \int_0^1 \sqrt{9t+16}\,dt$$

$$= \left[\frac{1}{9}\left(\frac{2}{3}\right)(9t+16)^{3/2}\right]_0^1 = \frac{250}{27} - \frac{128}{27} = \frac{122}{27}. \quad \square$$

Example 2 Find the length of the curve

$$\mathbf{r}(t) = 2\cos t\,\mathbf{i} + 2\sin t\,\mathbf{j} + t^2\,\mathbf{k} \qquad \text{from } t = 0 \text{ to } t = \pi/2$$

and compare it to the straight-line distance between the endpoints of the curve. (See Figure 13.4.6.)

SOLUTION

$$\mathbf{r}'(t) = -2\sin t\,\mathbf{i} + 2\cos t\,\mathbf{j} + 2t\,\mathbf{k}.$$

$$\|\mathbf{r}'(t)\| = \sqrt{4\sin^2 t + 4\cos^2 t + 4t^2} = 2\sqrt{\sin^2 t + \cos^2 t + t^2} = 2\sqrt{1+t^2}.$$

$$L(C) = \int_0^{\pi/2} \|\mathbf{r}'(t)\|\,dt = \int_0^{\pi/2} 2\sqrt{1+t^2}\,dt$$

$$= \left[t\sqrt{1+t^2} + \ln\left(t + \sqrt{1+t^2}\right)\right]_0^{\pi/2} \qquad \text{(Formula 78)}$$

$$= \frac{\pi}{2}\sqrt{1+\frac{\pi^2}{4}} + \ln\left[\frac{\pi}{2} + \sqrt{1+\frac{\pi^2}{4}}\right] \cong 4.158.$$

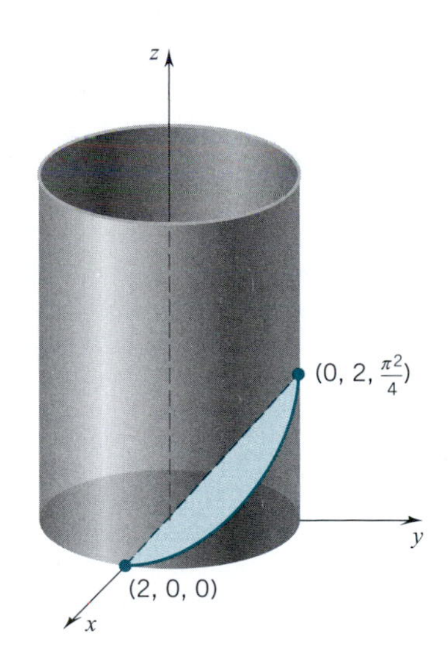

Figure 13.4.6

The curve begins at $\mathbf{r}(0) = 2\,\mathbf{i}$ and ends at $\mathbf{r}(\pi/2) = 2\,\mathbf{j} + (\pi^2/4)\,\mathbf{k}$. The straight-line distance between these two points is

$$\|\mathbf{r}(\pi/2) - \mathbf{r}(0)\| = \sqrt{2^2 + 2^2 + \frac{\pi^4}{16}} \cong 3.753.$$

The curve is about 11 percent longer than the straight-line distance between the endpoints of the curve. $\square$

EXERCISES 13.4

Find the length of the given curve.

1. $\mathbf{r}(t) = t\,\mathbf{i} + \frac{2}{3}t^{3/2}\,\mathbf{j}$ from $t = 0$ to $t = 8$.
2. $\mathbf{r}(t) = (\frac{1}{3}t^3 - t)\,\mathbf{i} + t^2\,\mathbf{j}$ from $t = 0$ to $t = 2$.
3. $\mathbf{r}(t) = a\cos t\,\mathbf{i} + a\sin t\,\mathbf{j} + bt\,\mathbf{k}$ from $t = 0$ to $t = 2\pi$.
4. $\mathbf{r}(t) = t\,\mathbf{i} + \frac{2}{3}\sqrt{2}\,t^{3/2}\,\mathbf{j} + \frac{1}{2}t^2\,\mathbf{k}$ from $t = 0$ to $t = 2$.
5. $\mathbf{r}(t) = t\,\mathbf{i} + \ln(\sec t)\,\mathbf{j} + 3\,\mathbf{k}$ from $t = 0$ to $t = \frac{1}{4}\pi$.
6. $\mathbf{r}(t) = \tan^{-1} t\,\mathbf{i} + \frac{1}{2}\ln(1 + t^2)\,\mathbf{j}$ from $t = 0$ to $t = 1$.
7. $\mathbf{r}(t) = t^3\,\mathbf{i} + t^2\,\mathbf{j}$ from $t = 0$ to $t = 1$.
8. $\mathbf{r}(t) = t\,\mathbf{i} + \mathbf{j} + (\frac{1}{6}t^3 + \frac{1}{2}t^{-1})\,\mathbf{k}$ from $t = 1$ to $t = 3$.
9. $\mathbf{r}(t) = e^t[\cos t\,\mathbf{i} + \sin t\,\mathbf{j}]$ from $t = 0$ to $t = \pi$.
10. $\mathbf{r}(t) = 3t\cos t\,\mathbf{i} + 3t\sin t\,\mathbf{j} + 4t\,\mathbf{k}$ from $t = 0$ to $t = 4$.
11. $\mathbf{r}(t) = 2t\,\mathbf{i} + (t^2 - 2)\,\mathbf{j} + (1 - t^2)\,\mathbf{k}$ from $t = 0$ to $t = 2$.
12. $\mathbf{r}(t) = t^2\,\mathbf{i} + (t^2 - 2)\,\mathbf{j} + (1 - t^2)\,\mathbf{k}$ from $t = 0$ to $t = 2$.
13. $\mathbf{r}(t) = (\ln t)\,\mathbf{i} + 2t\,\mathbf{j} + t^2\,\mathbf{k}$ from $t = 1$ to $t = e$.
14. $\mathbf{r}(t) = (t\sin t + \cos t)\,\mathbf{i} + (t\cos t - \sin t)\,\mathbf{j} + 2\,\mathbf{k}$ from $t = 0$ to $t = 2$.
15. $\mathbf{r}(t) = (\cos t + t\sin t)\,\mathbf{i} + (\sin t - t\cos t)\,\mathbf{j} + \frac{1}{2}\sqrt{3}t^2\,\mathbf{k}$ from $t = 0$ to $t = 2\pi$.
16. $\mathbf{r}(t) = \frac{2}{3}(1 + t)^{3/2}\,\mathbf{i} + \frac{2}{3}(1 - t)^{3/2}\,\mathbf{j} + \sqrt{2}t\,\mathbf{k}$ from $t = -\frac{1}{2}$ to $t = \frac{1}{2}$.
17. (*Important*) Let $\mathbf{r}(t) = x(t)\,\mathbf{i} + y(t)\,\mathbf{j} + z(t)\,\mathbf{k}$, $t \in [a, b]$ be a continuously differentiable curve. Show that, if s is the length of the curve from the tip of $\mathbf{r}(a)$ to the tip of $\mathbf{r}(t)$, then

$$\textbf{(13.4.3)}\qquad \frac{ds}{dt} = \sqrt{\left(\frac{dx}{dt}\right)^2 + \left(\frac{dy}{dt}\right)^2 + \left(\frac{dz}{dt}\right)^2}.$$

18. Use vector methods to show that, if $y = f(x)$ has a continuous first derivative, then the length of the graph from $x = a$ to $x = b$ is given by the integral

$$\int_a^b \sqrt{1 + [f'(x)]^2}\,dx.$$

19. (*Important*) Let $y = f(x)$, $x \in [a, b]$, be a continuously differentiable function. Show that, if s is the length of the graph from $(a, f(a))$ to $(x, f(x))$, then

$$\textbf{(13.4.4)}\qquad \frac{ds}{dx} = \sqrt{1 + \left(\frac{dy}{dx}\right)^2}.$$

20. Let C_1 be the curve

$$\mathbf{r}(t) = (t - \ln t)\,\mathbf{i} + (t + \ln t)\,\mathbf{j}.\quad 1 \le t \le e$$

and let C_2 be the graph of

$$y = e^x, \quad 0 \le x \le 1.$$

Find a relation between the length of C_1, and the length of C_2.

21. Show that the length of a continuously differentiable curve is left invariant (left unchanged) by an orientation-preserving (or an orientation-reversing) change of parameter. That is, set

$$\mathbf{r} = \mathbf{r}(t), \quad t \in [a, b].$$

Assume that ϕ maps $[c, d]$ onto $[a, b]$ and that ϕ' is positive (or negative) and continuous on $[a, b]$. Set $\mathbf{R}(u) = r(\phi(u))$ and show that the length of the curve as computed from $\mathbf{R}$ is the length of the curve as computed from $\mathbf{r}$.

22. Let $\mathbf{r}(t) = x(t)\,\mathbf{i} + y(t)\,\mathbf{j} + z(t)\,\mathbf{k}$ be a differentiable vector-valued function such that $\mathbf{r}'(t) \neq 0$ for all $t \ge 0$.

   (a) Show that the arc length function s defined by

$$s(t) = \int_0^t \sqrt{\left(\frac{dx}{dt}\right)^2 + \left(\frac{dy}{dt}\right)^2 + \left(\frac{dz}{dt}\right)^2}\,dt,\ t \ge 0$$

   has an inverse, $t = \varphi(s)$.

   (b) Let $\mathbf{R}(s) = \mathbf{r}[\varphi(s)]$. Show that $\|\mathbf{R}'(s)\| = 1$.

23. Consider the circular helix

$$\mathbf{r}(t) = 3\cos t\,\mathbf{i} + 3\sin t\,\mathbf{j} + 4t\,\mathbf{k} \text{ for } t \ge 0.$$

   (a) Determine the arc length s as a function of t by evaluating the integral

$$s = \int_0^t \sqrt{\left(\frac{dx}{dt}\right)^2 + \left(\frac{dy}{dt}\right)^2 + \left(\frac{dz}{dt}\right)^2}\,dt.$$

   (b) Use the relation found in (a) to express t as a function of s, $t = \varphi(s)$, and set

$$\mathbf{R}(s) = \mathbf{r}[\varphi(s)] = 3\cos\varphi(s)\,\mathbf{i} + 3\sin\varphi(s)\,\mathbf{j} + 4\varphi(s)\,\mathbf{k}.$$

   (c) Find the coordinates of the point Q on the helix such that the arc length from $P(3, 0, 0)$ to Q is 5π.

   (d) Show that $\|\mathbf{R}'(s)\| = 1$.

24. Repeat Exercise 23 (a), (b), (d) for the vector-valued function

$$\mathbf{r}(t) = (\sin t - t\cos t)\,\mathbf{i} + (\cos t + t\sin t)\,\mathbf{j} + \tfrac{1}{2}t^2\,\mathbf{k}$$

   for $t \ge 0$.

C In Exercise 25–28, use a CAS to estimate the length of the given curve.

25. $\mathbf{r}(t) = \frac{2}{5}t^{5/2}\,\mathbf{j} + t\,\mathbf{k}$ from $t = 0$ to $t = \frac{1}{2}$.
26. $\mathbf{r}(t) = t\,\mathbf{i} + \frac{1}{3}t^3\,\mathbf{j}$ from $t = 0$ to $t = 2$.

27. $\mathbf{r}(t) = 3\cos t\,\mathbf{i} + 4\sin t\,\mathbf{j} + 2\,\mathbf{k}$ from $t = 0$ to $t = 2\pi$.

28. $\mathbf{r}(t) = t\,\mathbf{i} + t^2\,\mathbf{j} + (\ln t)\,\mathbf{k}$ from $t = 0$ to $t = 4$.

29. Let $\mathbf{r}(t) = \cos t\,\mathbf{i} + \sin t\,\mathbf{j} + \sin 4t\,\mathbf{k}$, $0 \le t \le 2\pi$.
(a) Use a graphing utility to draw the curve.
(b) Use a CAS to estimate the length of the curve.

30. Let $\mathbf{r}(t) = \cos t\,\mathbf{i} + \ln(1+t)\,\mathbf{j} + \sin t\,\mathbf{k}$, $0 \le t \le 2\pi$.
(a) Use a graphing utility to draw the curve.
(b) Use a CAS to estimate the length of the curve.

*SUPPLEMENT TO SECTION 13.4

The Additivity of Arc Length We wish to show that with P, Q, R as in Figure 14.4.3,

$$L(\widehat{PQ}) + L(\widehat{QR}) = L(\widehat{PR}).$$

Let γ_1 be an arbitrary polygonal path inscribed in $\widehat{PQ}$ and γ_2 an arbitrary polygonal path inscribed in $\widehat{QR}$. Then $\gamma_1 \cup \gamma_2$ is a polygonal path inscribed in $\widehat{PR}$. Since

$$L(\gamma_1) + L(\gamma_2) = L(\gamma_1 \cup \gamma_2) \quad \text{and} \quad L(\gamma_1 \cup \gamma_2) \le L(\widehat{PR}),$$

we have

$$L(\gamma_1) + L(\gamma_2) \le L(\widehat{PR}) \quad \text{and thus} \quad L(\gamma_1) \le L(\widehat{PR}) - L(\gamma_2).$$

Since γ_1 is arbitrary, we can conclude that $L(\widehat{PR}) - L(\gamma_2)$ is an upper bound for the set of all lengths of polygonal paths inscribed in $\widehat{PQ}$. Since $L(\widehat{PQ})$is the *least* upper bound of this set, we have

$$L(\widehat{PQ}) \le L(\widehat{PR}) - L(\gamma_2).$$

It follows that

$$L(\gamma_2) \le L(\widehat{PR}) - L(\widehat{PQ}).$$

Arguing as we did with γ_1, we can conclude that

$$L(\widehat{QR}) \le L(\widehat{PR}) - L(\widehat{PQ}).$$

This gives

$$L(\widehat{PQ}) + L(\widehat{QR}) \le L(\widehat{PR}).$$

We now set out to prove that $L(\widehat{PR}) \le L(\widehat{PQ}) + L(\widehat{QR})$. To do this, we need only take $\gamma = \overline{T_0T_1} \cup \cdots \cup \overline{T_{n-1}T_n}$ as an arbitrary polygonal path inscribed in $\widehat{PR}$ and show that

$$L(\gamma) \le L(\widehat{PQ}) + L(\widehat{QR}).$$

If Q is one of the T_i, say $Q = T_k$, then

$$\gamma_1 = \overline{T_0T_1} \cup \cdots \cup \overline{T_{k-1}T_k} \quad \text{is inscribed in } \widehat{PQ}$$

and

$$\gamma_2 = \overline{T_kT_{k+1}} \cup \cdots \cup \overline{T_{n-1}T_n} \quad \text{is inscribed in } \widehat{QR}.$$

Moreover, $L(\gamma) = L(\gamma_1) + L(\gamma_2)$, so that

$$L(\gamma) \leq L(\widehat{PQ}) + L(\widehat{QR}).$$

If Q is none of the T_i, then Q lies between two consecutive points T_k and T_{k+1}. Set

$$\gamma' = \overline{T_0T_1} \cup \cdots \cup \overline{T_kQ} \cup \overline{QT_{k+1}} \cup \cdots \cup \overline{T_{n-1}T_n}.$$

Since
$$d(T_k, T_{k+1}) \leq d(T_k, Q) + d(Q, T_{k+1}),$$

we have
$$L(\gamma) \leq L(\gamma').$$

Proceed as before and you will see that

$$L(\gamma') \leq L(\widehat{PQ}) + L(\widehat{QR}),$$

and once again

$$L(\gamma) \leq L(\widehat{PQ}) + L(\widehat{QR}). \quad \square$$

■ 13.5 CURVILINEAR MOTION; CURVATURE

Curvilinear Motion from a Vector Viewpoint

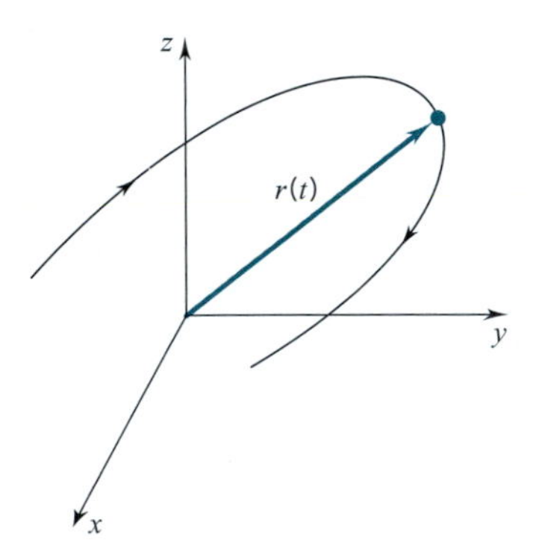

Figure 13.5.1

Here we use the theory we have developed for vector-valued functions to study the motion of an object moving in space. We can describe the position of a moving object at time t by a radius vector $\mathbf{r}(t)$. As t ranges over a time interval I, the object traces out some path

$$C: \mathbf{r}(t) = x(t)\,\mathbf{i} + y(t)\,\mathbf{j} + z(t)\,\mathbf{k}, \quad t \in I \qquad \text{(Figure 13.5.1)}$$

If $\mathbf{r}$ is twice differentiable, we can form $\mathbf{r}'(t)$ and $\mathbf{r}''(t)$. In this context these vectors have special names and special significance: $\mathbf{r}'(t)$ is called the *velocity* of the object at time t, and $\mathbf{r}''(t)$ is called the *acceleration*. In symbols, we have

(13.5.1)
$$\mathbf{r}'(t) = \mathbf{v}(t) \quad \text{and} \quad \mathbf{r}''(t) = \mathbf{v}'(t) = \mathbf{a}(t).$$

There should be nothing surprising about this. As before, velocity is the time rate of change of position and acceleration the time rate of change of velocity.

Since $\mathbf{v}(t) = \mathbf{r}'(t)$, the velocity vector, when not $\mathbf{0}$, is tangent to the path of the motion at the tip of $\mathbf{r}(t)$. (See Section 13.3.) The direction of the velocity vector at time t thus gives the direction of the motion at time t (Figure 13.5.2).

The magnitude of the velocity vector is called the *speed* of the object:

(13.5.2)
$$\|\mathbf{v}(t)\| = \text{ the speed at time } t.$$

The reasoning is as follows: during a time interval $[t_0, t]$ the object moves along its path from $\mathbf{r}(t_0)$ to $\mathbf{r}(t)$ for a total distance

$$s(t) = \int_{t_0}^{t} \|\mathbf{r}'(u)\|\, du. \qquad \text{(Section 13.4)}$$

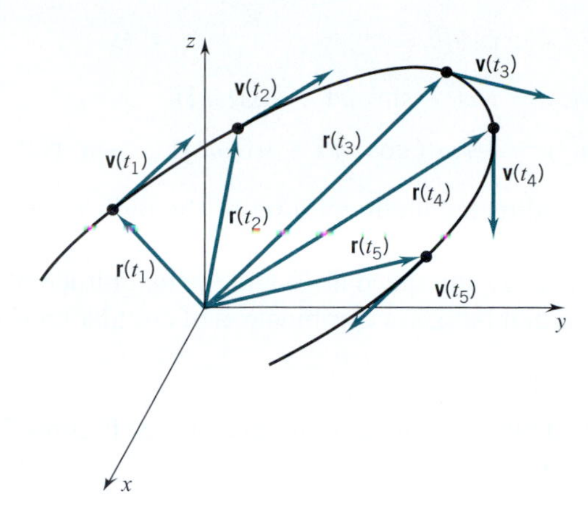

Figure 13.5.2

Differentiating with respect to t, we have

$$\frac{ds}{dt} = \|\mathbf{r}'(t)\|.$$

The magnitude of the velocity vector is thus *the rate of change of arc distance with respect to time*. This is why we call it the speed of the object.

Motion Along a Straight Line

The position at time t is given by a function of the form

$$\mathbf{r}(t) = \mathbf{r}_0 + f(t)\,\mathbf{d}, \quad \mathbf{d} \neq 0.$$

For convenience we take **d** as a unit vector.

The velocity and acceleration vectors are both directed along the line of the motion:

$$\mathbf{v}(t) = f'(t)\,\mathbf{d} \quad \text{and} \quad \mathbf{a}(t) = f''(t)\,\mathbf{d}.$$

The speed is $|f'(t)|$:

$$\|\mathbf{v}(t)\| = \|f'(t)\mathbf{d}\| = |f'(t)|\,\|\mathbf{d}\| = |f'(t)|,$$

and the magnitude of the acceleration is $|f''(t)|$:

$$\|\mathbf{a}(t)\| = \|f''(t)\mathbf{d}\| = |f''(t)|\,\|\mathbf{d}\| = |f''(t)|.$$

Circular Motion About the Origin

The position function can be written

$$\mathbf{r}(t) = r[\cos\theta(t)\,\mathbf{i} + \sin\theta(t)\,\mathbf{j}], \quad r > 0 \text{ constant}.$$

Here $\theta'(t)$ gives the time rate of change of the central angle θ. If $\theta'(t) > 0$, the motion is counterclockwise; if $\theta'(t) < 0$, the motion is clockwise. We call $\theta'(t)$ the *angular velocity* and $|\theta'(t)|$ the *angular speed*.

Uniform circular motion is circular motion with constant angular speed $\omega > 0$. The position function for uniform circular motion in the counterclockwise direction can be written

$$\mathbf{r}(t) = \mathbf{r}(\cos\omega t\,\mathbf{i} + \sin\omega t\,\mathbf{j}).$$

Differentiation gives

$$\mathbf{v}(t) = r\omega(-\sin \omega t\, \mathbf{i} + \cos \omega t\, \mathbf{j}),$$
$$\mathbf{a}(t) = -r\omega^2(\cos \omega t\, \mathbf{i} + \sin \omega t\, \mathbf{j}) = -\omega^2\, \mathbf{r}(t).$$

The acceleration is directed along the line of the radius vector toward the center of the circle and is therefore perpendicular to the velocity vector, which, as always, is tangential. As you can verify, the speed is $r\omega$ and the magnitude of acceleration is $r\omega^2$.

Motion along a circular helix is a combination of circular motion and motion along a straight line.

Example 1 An object moves along a circular helix (see Figure 13.3.12) with position at time t given by the function

$$\mathbf{r}(t) = a \cos \omega t\, \mathbf{i} + a \sin \omega t\, \mathbf{j} + b\omega t\, \mathbf{k} \quad (a > 0, b > 0, \omega > 0).$$

For each time t, find

(a) the velocity of the particle; **(b)** the speed; **(c)** the acceleration;
(d) the magnitude of the acceleration;
(e) the angle between the velocity vector and the acceleration vector.

SOLUTION

(a) Velocity: $\mathbf{v}(t) = \mathbf{r}'(t) = -a\omega \sin \omega t\, \mathbf{i} + a\omega \cos \omega t\, \mathbf{j} + b\omega\, \mathbf{k}$.

(b) Speed: $\|\mathbf{v}(t)\| = \sqrt{a^2\omega^2 \sin^2 \omega t + a^2\omega^2 \cos^2 \omega t + b^2\omega^2}$
$$= \sqrt{a^2\omega^2 + b^2\omega^2} = \omega\sqrt{a^2 + b^2}.$$

(The speed is thus constant.)

(c) Acceleration : $\mathbf{a}(t) = \mathbf{v}'(t) = -a\omega^2 \cos \omega t\, \mathbf{i} - a\omega^2 \sin \omega t\, \mathbf{j}$
$$= -a\omega^2(\cos \omega t\, \mathbf{i} + \sin \omega t\, \mathbf{j}).$$

(Since the speed is constant, the acceleration comes entirely from the change in direction.)

(d) Magnitude of the acceleration: $\|\mathbf{a}(t)\| = a\omega^2$.

(e) Angle between $\mathbf{v}(t)$ and $\mathbf{a}(t)$:

$$\cos\theta = \frac{\mathbf{v}(t)}{\|\mathbf{v}(t)\|} \cdot \frac{\mathbf{a}(t)}{\|\mathbf{a}(t)\|}$$
$$= \left[\frac{-a\omega \sin \omega t\, \mathbf{i} + a\omega \cos \omega t\, \mathbf{j} + b\omega\, \mathbf{k}}{\omega\sqrt{a^2 + b^2}}\right] \cdot \left[\frac{-a\omega^2(\cos \omega t\, \mathbf{i} + \sin \omega t\, \mathbf{j})}{a\omega^2}\right]$$
$$= \frac{a(\sin \omega t \cos \omega t - \cos \omega t \sin \omega t)}{\sqrt{a^2 + b^2}} = 0.$$

Therefore $\theta = \frac{1}{2}\pi$. (At each point the acceleration vector is perpendicular to the velocity vector.) □

The Curvature of a Plane Curve

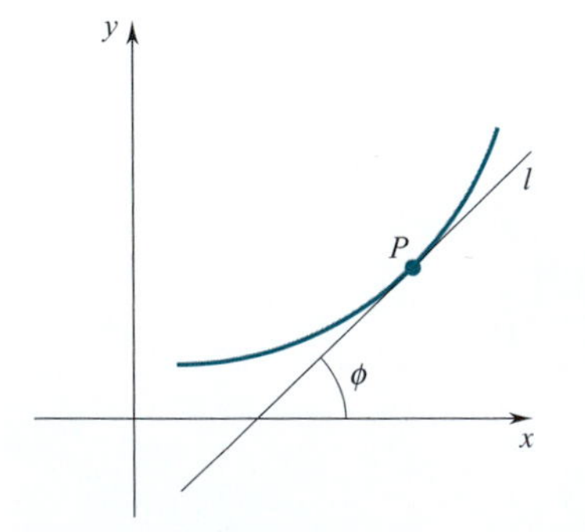

Figure 13.5.3

Figure 13.5.3 shows a plane curve which we assume to be twice differentiable. At the point P we have drawn the tangent line l. The angle that l makes with the x-axis has been labeled ϕ. As P moves along the curve, l changes and ϕ changes also. The magnitude κ† of the change in ϕ per unit of arc length is called the *curvature*:

† The symbol κ is the lower case Greek letter "kappa."

(13.5.3)
$$\kappa = \left|\frac{d\phi}{ds}\right|.$$

Calculating Curvature

If the curve is the graph of a twice differentiable function

$$y = y(x),$$

then the curvature can be calculated from the formula

(13.5.4)
$$\kappa = \frac{|y''|}{[1 + (y')^2]^{3/2}},$$

where the primes indicate differentiation with respect to x.

DERIVATION OF FORMULA 13.5.4 We know that $\tan\phi = y'$. Therefore,

$$\phi = \tan^{-1}(y').$$

Differentiating with respect to x, we have

$$\frac{d\phi}{dx} = \frac{1}{1 + (y')^2}\cdot\frac{d}{dx}(y') = \frac{y''}{1 + (y')^2}.$$

Since

$$\frac{d\phi}{dx} \underset{\text{chain rule}}{=} \frac{d\phi}{ds}\frac{ds}{dx} \underset{(13.4.4)}{=} \frac{d\phi}{ds}\sqrt{1 + (y')^2},$$

we have

$$\frac{d\phi}{ds}\sqrt{1 + (y')^2} = \frac{y''}{1 + (y')^2} \quad \text{and therefore} \quad \left|\frac{d\phi}{ds}\right| = \frac{|y''|}{[1 + (y')^2]^{3/2}}. \quad \square$$

If the curve is given parametrically by a twice differentiable vector function

$$\mathbf{r}(t) = x(t)\,\mathbf{i} + y(t)\,\mathbf{j},$$

then

(13.5.5)
$$\kappa = \frac{|x'y'' - y'x''|}{[(x')^2 + (y')^2]^{3/2}},$$

where the primes now indicate differentiation with respect to t.

We will derive the formula under the assumption that $x' \neq 0$. Actually, the formula holds provided that $(x')^2 + (y')^2 \neq 0$.

DERIVATION OF FORMULA 13.5.5

$$\frac{dy}{dx} = \frac{dy/dt}{dx/dt} = \frac{y'}{x'}.$$

Therefore, as you can verify,

$$\frac{d^2y}{dx^2} \underset{(9.7.5)}{=} \frac{(dx/dt)(d^2y/dt^2) - (dy/dt)(d^2x/dt^2)}{(dx/dt)^3} = \frac{x'y'' - y'x''}{(x')^3}.$$

Thus

$$\kappa = \underset{\uparrow_{(13.5.4)}}{\frac{|d^2y/dx^2|}{[1+(dy/dx)^2]^{3/2}}} = \left|\frac{x'y''-y'x''}{(x')^3}\right| \frac{1}{[1+(y'/x')^2]^{3/2}} = \frac{|x'y''-y'x''|}{[(x')^2+(y')^2]^{3/2}}. \qquad \square$$

Example 2 Since a straight line has a constant angle of inclination, we have

$$\frac{d\phi}{ds} = 0 \qquad \text{and thus} \qquad \kappa = 0.$$

> Along a straight line the curvature is constantly zero. $\square$

Example 3 For a circle of radius r,

$$\mathbf{r}(t) = r(\cos t\,\mathbf{i} + \sin t\,\mathbf{j}),$$

we have

$$x = r\cos t, \quad y = r\sin t.$$

Differentiation with respect to t gives

$$x' = -r\sin t, \quad x'' = -r\cos t; \quad y' = r\cos t, \quad y'' = -r\sin t.$$

Thus

$$\kappa = \frac{|x'y''-y'x''|}{[(x')^2+(y')^2]^{3/2}} = \frac{|(-r\sin t)(-r\sin t)-(r\cos t)(-r\cos t)|}{[(-r\sin t)^2+(r\cos t)^2]^{3/2}} = \frac{r^2}{r^3} = \frac{1}{r}.$$

> Along a circle of radius r the curvature is constantly $1/r$.

Hardly surprising. It is geometrically evident that along a circular path the change in direction takes place at a constant rate. Since a complete revolution entails a change of direction of 2π radians and this change is effected on a path of length $2\pi r$, the change in direction per unit of arc length is $2\pi/2\pi r = 1/r$. $\square$

To say that a curve $y = f(x)$ has slope m at a point P is to say that at the point P the curve is rising or falling at the rate of a line of slope m. To say that a plane curve C has curvature $1/r$ at a point P is to say that at the point P the curve is turning at the rate of a circle of radius r. (Figure 13.5.4). The smaller the circle, the tighter the turn and, thus, the greater the curvature.

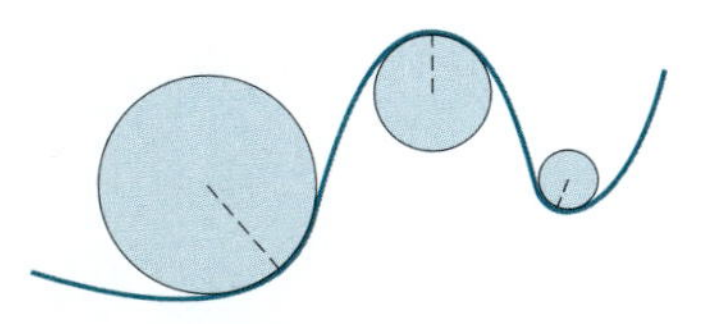

Figure 13.5.4

The reciprocal of the curvature,

$$\rho = \frac{1}{k}, \qquad \text{(for } k \neq 0\text{)}$$

is called the *radius of curvature*. The point at a distance ρ from the curve in the direction of the principal normal is called the *center of curvature*.

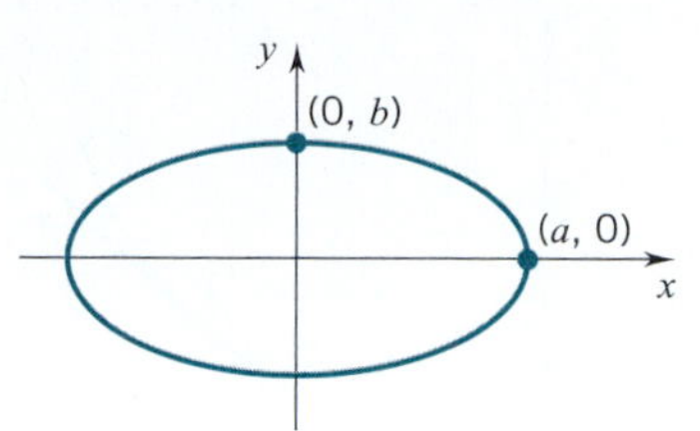

Figure 13.5.5

Example 4 Find the curvature at an arbitrary point (x, y) of the ellipse

$$\frac{x^2}{a^2} + \frac{y^2}{b^2} = 1. \qquad (a > b > 0). \qquad \text{(Figure 13.5.5)}$$

Determine the points of maximal curvature and the points of minimal curvature. What is the radius of curvature at each of these points?

SOLUTION We parametrize the ellipse by setting

$$\mathbf{r}(t) = a\cos t\,\mathbf{i} + b\sin t\,\mathbf{j}$$

and use the fact that

$$\kappa = \frac{|x'y'' - y'x''|}{[(x')^2 + (y')^2]^{3/2}}.$$

Here

$$x = a\cos t, \quad y = b\sin t,$$

and therefore

$$x' = -a\sin t, \quad x'' = -a\cos t; \quad y' = b\cos t, \quad y'' = -b\sin t.$$

Thus

$$k = \frac{|(-a\sin t)(-b\,\sin t) - (b\,\cos t)(-a\cos t)|}{[(-a\sin t)^2 + (b\,\cos t)^2]^{3/2}} = \frac{ab}{[a^2\sin^2 t + b^2\cos^2 t]^{3/2}}.$$

As you can check, the curvature at the point (x, y) can be written as

$$\kappa = \frac{a^4b^4}{(b^4x^2 + a^4y^2)^{3/2}}.$$

To find the points of maximal and minimal curvature, we go back to the parameter t. Observe that

$$\begin{aligned} a^2\sin^2 t + b^2\cos^2 t &= (a^2 - b^2)\sin^2 t + b^2(\sin^2 t + \cos^2 t) \\ &= (a^2 - b^2)\sin^2 t + b^2. \end{aligned}$$

Thus we have

$$\kappa = \frac{ab}{[(a^2 - b^2)\sin^2 t + b^2]^{3/2}}.$$

Since we have assumed that $a > b > 0$, the curvature is maximal when $\sin^2 t = 0$; that is, when $t = 0$ and when $t = \pi$. Thus the points of maximal curvature are the points $P(\pm a, 0)$, the ends of the major axis. The curvature at these points is a/b^2 and the radius of curvature is b^2/a. The curvature is minimal when $\sin^2 t = 1$; that is, when $t = \frac{1}{2}\pi$ and when $t = \frac{3}{2}\pi$. The points of minimal curvature are the points $P(0, \pm b)$, the ends of the minor axis. The curvature at these points is b/a^2 and the radius of curvature is a^2/b. □

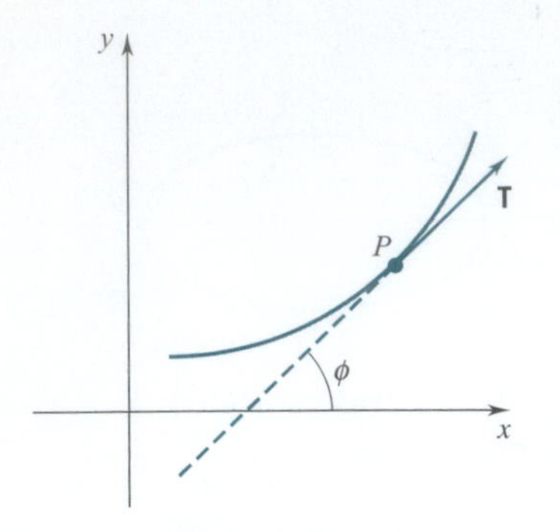

Figure 13.5.6

In Figure 13.5.6 you can see a plane curve. At the point P we have affixed the unit tangent vector $\mathbf{T}$. As P moves along the curve, $\mathbf{T}$ changes, not in length, but in direction. The curvature of the curve is the magnitude of the change in $\mathbf{T}$ per unit of arc length:

(13.5.6)
$$\kappa = \left\| \frac{d\mathbf{T}}{ds} \right\|.$$

PROOF Since $\mathbf{T}$ has length 1, we can write

$$\mathbf{T} = \cos\phi\,\mathbf{i} + \sin\phi\,\mathbf{j},$$

where ϕ is the angle between the tangent line and the x-axis. Differentiation with respect to s gives

$$\frac{d\mathbf{T}}{ds} = -\sin\phi\frac{d\phi}{ds}\mathbf{i} + \cos\phi\frac{d\phi}{ds}\mathbf{j} = \frac{d\phi}{ds}(-\sin\phi\mathbf{i} + \cos\phi\mathbf{j}).$$

Taking norms we have

$$\left\| \frac{d\mathbf{T}}{ds} \right\| = \left| \frac{d\phi}{ds} \right| \sqrt{\sin^2\phi + \cos^2\phi} = \left| \frac{d\phi}{ds} \right| = k$$

as asserted. ❑

This characterization of curvature generalizes to space curves.

The Curvature of a Space Curve

A space curve bends in two ways. It bends in the osculating plane (the plane of the unit tangent $\mathbf{T}$ and the principal normal $\mathbf{N}$) and it bends away from that plane. The first form of bending is measured by the rate at which the unit tangent $\mathbf{T}$ changes direction. The second form of bending is measured by the rate at which the vector $\mathbf{T} \times \mathbf{N}$ changes direction. We will concentrate here on the first form of bending, the bending in the osculating plane. The measure of this is called *curvature*.

What was a theorem on curvature in the case of a plane curve becomes a definition of curvature in the case of a space curve; namely, in the case of space curve, we *define* the curvature κ by setting

$$\kappa = \left\| \frac{d\mathbf{T}}{ds} \right\|.$$

If the space curve is given in terms of a parameter t, say

$$C: \quad \mathbf{r}(t) = x(t)\,\mathbf{i} + y(t)\,\mathbf{j} + z(t)\,\mathbf{k}, \qquad t \in [a, b],$$

then the curvature can be calculated from the formula

(13.5.7)
$$\kappa = \frac{\|d\mathbf{T}/dt\|}{ds/dt}.$$

We arrive at this by noting that

$$\frac{d\mathbf{T}}{ds}\frac{ds}{dt} = \frac{d\mathbf{T}}{dt},$$

then dividing through by ds/dt and taking norms. The assumption here is that $ds/dt = \|\mathbf{r}'(t)\|$ remains nonzero.

Example 5 Calculate the curvature of the circular helix

$$\mathbf{r}(t) = r \sin t\,\mathbf{i} + r\cos t\,\mathbf{j} + t\,\mathbf{k}. \qquad (r > 0)$$

SOLUTION We will use the Leibniz notation. We have:

$$\frac{d\mathbf{r}}{dt} = r\cos t\,\mathbf{i} - r\sin t\,\mathbf{j} + \mathbf{k}, \qquad \frac{ds}{dt} = \left\|\frac{d\mathbf{r}}{dt}\right\| = \sqrt{r^2+1},$$

$$\mathbf{T} = \frac{d\mathbf{r}/dt}{\|d\mathbf{r}/dt\|} = \frac{r\cos t\,\mathbf{i} - r\sin t\,\mathbf{j} + \mathbf{k}}{\sqrt{r^2+1}},$$

$$\frac{d\mathbf{T}}{dt} = \frac{-r\sin t\,\mathbf{i} - r\cos t\,\mathbf{j}}{\sqrt{r^2+1}}, \qquad \text{and} \qquad \left\|\frac{d\mathbf{T}}{dt}\right\| = \frac{r}{\sqrt{r^2+1}}.$$

Therefore,

$$\kappa = \frac{\|d\mathbf{T}/dt\|}{ds/dt} = \frac{r/\sqrt{r^2+1}}{\sqrt{r^2+1}} = \frac{r}{r^2+1}. \qquad \square$$

Components of Acceleration

In straight-line motion, acceleration is purely tangential; that is, the acceleration vector points in the direction of the motion. In uniform circular motion, the acceleration is normal; the acceleration vector is perpendicular to the tangent vector and points along the line of the normal vector toward the center of the circle.

In general, acceleration has two components, a tangential component and a normal component. To see this, let's suppose that the position of an object at time t is given by the vector function

$$\mathbf{r}(t) = x(t)\,\mathbf{i} + y(t)\,\mathbf{j} + z(t)\,\mathbf{k}.$$

Since

$$\mathbf{T} = \frac{d\mathbf{r}/dt}{\|d\mathbf{r}/dt\|} = \frac{\mathbf{v}}{ds/dt},$$

we have

$$\mathbf{v} = \frac{ds}{dt}\mathbf{T}.$$

Differentiation gives

$$\mathbf{a} = \frac{d^2s}{dt^2}\mathbf{T} + \frac{ds}{dt}\frac{d\mathbf{T}}{dt}.$$

Observe now that

$$\frac{d\mathbf{T}}{dt} \underset{(13.3.4)}{=} \left\|\frac{d\mathbf{T}}{dt}\right\|\mathbf{N} \underset{(13.5.7)}{=} \kappa\frac{ds}{dt}\mathbf{N}.$$

Substitution in the previous display gives

(13.5.8)

$$\boxed{\mathbf{a} = \frac{d^2s}{dt^2}\mathbf{T} + \kappa\left(\frac{ds}{dt}\right)^2\mathbf{N}.}$$

The acceleration vector lies in the osculating plane, the plane of **T** and **N**. The tangential component of acceleration,

$$a_{\mathrm{T}} = \frac{d^2s}{dt^2},$$

depends only on the change of speed of the object; if the speed is constant, the tangential component of acceleration is zero and the acceleration is directed entirely toward the center of curvature of the path. On the other hand, the normal component of acceleration,

$$a_{\mathrm{N}} = \kappa \left(\frac{ds}{dt}\right)^2,$$

depends both on the speed of the object and the curvature of the path. At a point where the curvature is zero, the normal component of acceleration is zero and the acceleration is directed entirely along the path of motion. If the curvature is not zero, then the normal component of acceleration is a multiple of the *square* of the speed. This means, for example, that if you are in a car going around a curve at 50 miles per hour, you will feel *four times* the normal component of acceleration that you would feel going around the same curve at 25 miles per hour.

We can use Equation (13.5.8) to obtain alternative formulas for $a_{\mathrm{T}}, a_{\mathrm{N}}$, and the curvature, κ.

If we take the dot product of **T** with **a**, we get

$$\mathbf{T} \cdot \mathbf{a} = a_{\mathrm{T}}(\mathbf{T} \cdot \mathbf{T}) + a_{N}(\mathbf{T} \cdot \mathbf{N}) = a_{\mathrm{T}}.$$

Therefore,

(13.5.9)

$$a_{\mathrm{T}} = \mathbf{T} \cdot \mathbf{a} = \frac{\mathbf{v} \cdot \mathbf{a}}{\|\mathbf{v}\|} = \frac{\mathbf{v} \cdot \mathbf{a}}{(ds/dt)}.$$

If we take the cross product of **T** with **a**, we get

$$\mathbf{T} \times \mathbf{a} = a_{\mathrm{T}}(\mathbf{T} \times \mathbf{T}) + a_{\mathrm{N}}(\mathbf{T} \times \mathbf{N}) = a_{\mathrm{N}}(\mathbf{T} \times \mathbf{N}),$$

and so

$$\|\mathbf{T} \times \mathbf{a}\| = a_{\mathrm{N}}\|\mathbf{T} \times \mathbf{N}\| = a_{\mathrm{N}}\|\mathbf{T}\|\|\mathbf{N}\| \sin(\pi/2) = a_{\mathrm{N}}.$$

Therefore

(13.5.10)

$$a_{\mathrm{N}} = \|\mathbf{T} \times \mathbf{a}\| = \frac{\|\mathbf{v} \times \mathbf{a}\|}{\|\mathbf{v}\|} = \frac{\|\mathbf{v} \times \mathbf{a}\|}{(ds/dt)}.$$

Since $a_{\mathrm{N}} = \kappa(ds/dt)^2$, it follows that

(13.5.11)

$$\kappa = \frac{\|\mathbf{v} \times \mathbf{a}\|}{(ds/dt)^3}.$$

Example 6 The position of a moving object at time t is given by

$$\mathbf{r}(t) = \ln t\,\mathbf{i} + 2t\,\mathbf{j} + t^2\,\mathbf{k}, \qquad t > 0.$$

Find the tangential and normal components of acceleration and the curvature of the path of the object at time $t = 1$.

SOLUTION $\mathbf{r}'(t) = \mathbf{v}(t) = \dfrac{1}{t}\,\mathbf{i} + 2\,\mathbf{j} + 2t\,\mathbf{k}$, and $\mathbf{r}''(t) = \mathbf{a}(t) = -\dfrac{1}{t^2}\,\mathbf{i} + 2\,\mathbf{k}$.

At $t = 1$, we have

$$\mathbf{v}(1) = \mathbf{i} + 2\mathbf{j} + 2\mathbf{k}, \qquad \|\mathbf{v}(1)\| = ds/dt = \sqrt{9} = 3, \qquad \text{and} \qquad \mathbf{a}(1) = -\mathbf{i} + 2\mathbf{k}.$$

Now,

$$a_{\mathbf{T}}(1) = \frac{\mathbf{v}\cdot\mathbf{a}}{(ds/dt)} = \frac{-1+4}{3} = 1,$$

$$\mathbf{a_N}(1) = \frac{\|\mathbf{v}\times\mathbf{a}\|}{(ds/dt)} = \frac{1}{3}\left\|\begin{vmatrix} \mathbf{i} & \mathbf{j} & \mathbf{k} \\ 1 & 2 & 2 \\ -1 & 0 & 2 \end{vmatrix}\right\| = \frac{1}{3}\,\|4\mathbf{i} - 4\mathbf{j} + 2\mathbf{k}\| = \frac{\sqrt{36}}{3} = 2,$$

and

$$k(1) = \frac{\|\mathbf{v}\times\mathbf{a}\|}{(ds/dt)^3} = \frac{\sqrt{36}}{27} = \frac{2}{9}. \quad \square$$

EXERCISES 13.5

1. A particle moves in a circle of radius r at constant speed v. Find the angular speed and the magnitude of the acceleration.
2. A particle moves so that

$$\mathbf{r}(t) = (a\cos\pi t + bt^2)\,\mathbf{i} + (a\sin\pi t - bt^2)\,\mathbf{j}.$$

Find the velocity, speed, acceleration, and the magnitude of the acceleration all at time $t = 1$.

3. A particle moves so that $\mathbf{r}(t) = at\,\mathbf{i} + b\sin at\,\mathbf{j}$. Show that the magnitude of the acceleration of the particle is proportional to its distance from the x-axis.
4. A particle moves so that $\mathbf{r}(t) = 2\,\mathbf{i} + t^2\,\mathbf{j} + (t-1)^2\,\mathbf{k}$. At what time is the speed a minimum?

Sketch the curve. Then compute and sketch the acceleration vector at the indicated points.

5. $\mathbf{r}(t) = (t/\pi)\,\mathbf{i} + \cos t\,\mathbf{j}$, $t \in [0, 2\pi]$; at $t = \frac{1}{4}\pi, \frac{1}{2}\pi, \pi$.
6. $\mathbf{r}(t) = t^3\,\mathbf{i} + t\,\mathbf{j}$, t real; at $t = -\frac{1}{2}, \frac{1}{2}, 1$.
7. $\mathbf{r}(t) = \sec t\,\mathbf{i} + \tan t\,\mathbf{j}$, $[-\frac{1}{4}\pi, \frac{1}{2}\pi)$; at $t = -\frac{1}{6}\pi, 0, \frac{1}{3}\pi$.
8. $\mathbf{r}(t) = \sin\pi t\,\mathbf{i} + t\,\mathbf{j}$, $t \in [0, 2]$; at $t = \frac{1}{2}, 1, \frac{5}{4}$.
9. An object moves so that

$$\mathbf{r}(t) = x_0\,\mathbf{i} + [\,y_0 + (\alpha\cos\theta)t]\,\mathbf{j} + [z_0 + (\alpha\sin\theta)t - 16t^2]\,\mathbf{k}, \quad t \geq 0.$$

Find (a) the initial position, (b) the initial velocity, (c) the initial speed, (d) the acceleration throughout the motion. Finally, (e) identify the curve.

10. A particle moves so that $\mathbf{r}(t) = 2\cos 2t\,\mathbf{i} + 3\cos t\,\mathbf{j}$.

(a) Show that the particle oscillates on an arc of the parabola $4y^2 - 9x = 18$. (b) Draw the path. (c) What are the acceleration vectors at the points of zero velocity? (d) Draw these vectors at the points in question.

11. Let $\mathbf{r}(t)$ be the position vector of a moving particle. Show that $\|\mathbf{r}(t)\|$ is constant iff $\mathbf{r}(t) \perp \mathbf{r}'(t)$.
12. Let $\mathbf{r}(t)$ be the position vector of a moving particle. Show that if the speed of the particle is constant, then the velocity vector is perpendicular to the acceleration vector.

Find the curvature of the given curve.

13. $y = e^{-x}$.
14. $y = x^3$.
15. $y = \sqrt{x}$.
16. $y = x - x^2$.
17. $y = \ln\sec x$.
18. $y = \tan x$.
19. $y = \sin x$.
20. $x^2 - y^2 = a^2$.

Find the radius of curvature at the indicated point.

21. $6y = x^3$; $(2, \frac{4}{3})$.
22. $2y = x^2$; $(0, 0)$.
23. $y^2 = 2x$; $(2, 2)$.
24. $y = 2\sin 2x$. $(\frac{1}{4}\pi, 2)$.
25. $y = \ln(x+1)$; $(2, \ln 3)$.
26. $y = \sec x$; $(\frac{1}{4}\pi, \sqrt{2})$.
27. Find the point of maximal curvature on the curve $y = \ln x$.
28. Find the curvature of the graph of $y = 3x - x^3$ at the point where the function takes on its local maximum value.

Express the curvature in terms of t.

29. $\mathbf{r}(t) = t\,\mathbf{i} + \frac{1}{2}t^2\,\mathbf{j}$.
30. $\mathbf{r}(t) = e^t\,\mathbf{i} + e^{-t}\,\mathbf{j}$.

31. $\mathbf{r}(t) = 2t\,\mathbf{i} + t^3\,\mathbf{j}$.

32. $\mathbf{r}(t) = t^2\,\mathbf{i} + t^3\,\mathbf{j}$.

33. $\mathbf{r}(t) = e^t(\cos t\,\mathbf{i} + \sin t\,\mathbf{j})$.

34. $\mathbf{r}(t) = 2\cos t\,\mathbf{i} + 3\sin t\,\mathbf{j}$.

35. $\mathbf{r}(t) = (t\cos t)\,\mathbf{i} + (t\sin t)\,\mathbf{j}$.

36. $\mathbf{r}(t) = (\cos t + t\sin t)\,\mathbf{i} + (\sin t - t\cos t)\,\mathbf{j},\ t > 0$.

37. Find the radius of curvature of the hyperbola $xy = 1$ at the points $(1, 1)$ and $(-1, -1)$.

38. Find the radius of curvature at the vertices of the hyperbola $x^2 - y^2 = 1$.

39. Find the curvature at each point (x, y) on the hyperbola $b^2x^2 - a^2y^2 = a^2b^2$.

HINT: Parametrize the hyperbola by setting $\mathbf{r}(t) = a\cosh t\,\mathbf{i} + b\sinh t\,\mathbf{j}$.

40. Find the curvature at the highest point of an arch of the cycloid

$$x(t) = r(t - \sin)t, \quad y(t) = r(1 - \cos t).$$

In Exercises 41–47, interpret $\mathbf{r}(t)$ as the position of a moving object at time t. Find the curvature of the path and determine the tangential and normal components of acceleration .

41. $\mathbf{r}(t) = e^t\cos t\,\mathbf{i} + e^t\sin t\,\mathbf{j} + e^t\,\mathbf{k}$.

42. $\mathbf{r}(t) = 2\cos t\,\mathbf{i} + t\,\mathbf{j} + \sin t\,\mathbf{k}$.

43. $\mathbf{r}(t) = \cos 2t\,\mathbf{i} + \sin 2t\,\mathbf{j} + \mathbf{k}$.

44. $\mathbf{r}(t) = 2t\,\mathbf{i} + t^2\,\mathbf{j} + \ln t\,\mathbf{k}$.

45. $\mathbf{r}(t) = 2t\,\mathbf{i} + t^2\,\mathbf{j} + \frac{1}{3}t^3\mathbf{k}$.

46. $\mathbf{r}(t) = (\cos t + t\sin t)\,\mathbf{i} + (\sin t - t\cos t)\,\mathbf{j} + \frac{1}{2}\sqrt{3}t^2\,\mathbf{k}$, from $t = 0$ to $t = 2\pi$.

47. $\mathbf{r}(t) = \frac{2}{3}(1 + t)^{3/2}\mathbf{i} + \frac{2}{3}(1 - t)^{3/2}\,\mathbf{j} + \sqrt{2}t\,\mathbf{k}$.

C In Exercises 48 and 49, let $\mathbf{r}(t) = 6t\,\mathbf{i} + 3t^2\,\mathbf{j} + 2t^3\,\mathbf{k}$

48. (a) Use a graphing utility to draw the graph.

(b) Use a CAS to find the maximum curvature of the curve.

49. Use a CAS to find the tangential and normal components of the acceleration.

50. Show that the curvature of a polar curve $r = f(\theta)$ is given by

$$\kappa = \frac{|[f(\theta)]^2 + 2[f'(\theta)]^2 - f(\theta)f''(\theta)|}{([f(\theta)]^2 + [f'(\theta)]^2)^{3/2}}.$$

51. Find the curvature of the logarithmic spiral $r = e^{a\theta}, a > 0$.

52. Find the curvature of the spiral of Archimedes $r = a\theta$, $a > 0$.

53. Find the curvature of the cardioid $r = a(1 - \cos\theta)$ in terms of r.

54. Find the curvature of the petal curve $r = a\sin 2\theta, a > 0$.

55. Let $s(\theta)$ be the arc distance from the highest point of the cycloidal arch

$$x(\theta) = R(\theta - \sin\theta),\ y(\theta) = R(1 - \cos\theta),\ \theta \in [0, 2\pi]$$

to the point $(x(\theta), y(\theta))$ of that same arch. Let $\rho(\theta)$ be the radius of curvature at the point $(x(\theta), y(\theta))$. (a) Calculate $s(\theta)$. (b) Calculate $\rho(\theta)$. (c) Then find an equation in s and ρ for that arch. (Such an equation is called a *natural equation* for the curve.)

56. Let $s(\theta)$ be the arc distance from the origin to the point $(x(\theta), y(\theta))$ along the exponential spiral $r = ae^{c\theta}$. (Take $a > 0, c > 0$.) Let $\rho(\theta)$ be the radius of curvature at that same point. Find an equation in s and ρ for that curve.

57. Let $s(\theta)$ be the arc distance from the point $(-2a, 0)$ to the point $(x(\theta), y(\theta))$ along the cardioid $r = a(1 - \cos\theta)$. (Take $a > 0$.) Let $\rho(\theta)$ be the radius of curvature at that same point. Find an equation is s and ρ for that curve.

(*The Frenet formulas*) The figure shows a space curve. At a point of the curve we have drawn the unit tangent. $\mathbf{T}$, the principal normal $\mathbf{N}$, and the vector $\mathbf{B} = \mathbf{T} \times \mathbf{N}$, which, being normal to both $\mathbf{T}$ and $\mathbf{N}$, is called the *binormal*. At each point of the curve, the vectors $\mathbf{T}, \mathbf{N}, \mathbf{B}$ form what is called the *Frenet frame* a set of mutually perpendicular unit vectors that, in the order given, form a local right-handed coordinate system.

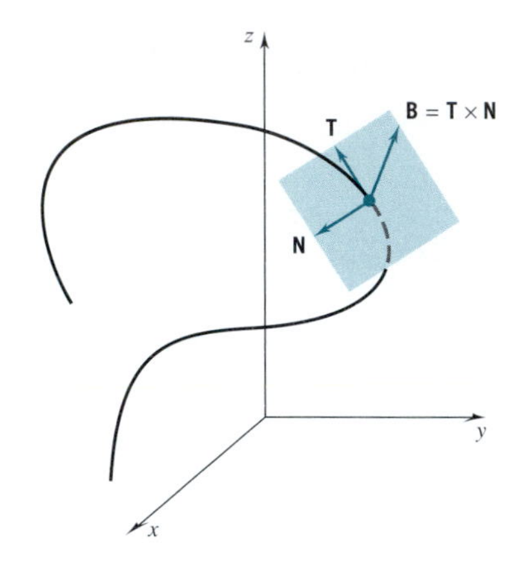

58. (a) Show that $d\mathbf{B}/ds$ is parallel to $\mathbf{N}$, and therefore there is a scalar τ for which

$$\frac{d\mathbf{B}}{ds} = \tau\mathbf{N}.$$

HINT: Since $\mathbf{B}$ has constant length one, $d\mathbf{B}/ds \perp \mathbf{B}$. Show that $d\,\mathbf{B}/ds \perp \mathbf{T}$ by carrying out the differentiation

$$\frac{d\,\mathbf{B}}{ds} = \frac{d}{ds}(\mathbf{T} \times \mathbf{N}).$$

(b) Now show that

$$\frac{d\,\mathbf{N}}{ds} = -\kappa\mathbf{T} - \tau\mathbf{B}.$$

HINT: Since $\mathbf{T}, \mathbf{N}, \mathbf{B}$ form a right-handed system of mutually perpendicular unit vectors, we can show that

$$\mathbf{N} \times \mathbf{B} = \mathbf{T} \quad \text{and} \quad \mathbf{B} \times \mathbf{T} = \mathbf{N}.$$

You can assume these relations.

(c) The scalar τ is called the *torsion* of the curve. Give a geometric interpretation to $|\tau|$.

■ PROJECT 13.5 Transition Curves

In the design of an automobile fender, engineers face the problem of connecting curved pieces in a smooth and elegant manner. In this context problems of the following sort arise.

The total curve of a fender is to be made up of two pieces each the graph of a cubic polynomial. The first piece p is to begin at the point $(1, 3)$ and end at the point $(3, 7)$. This requires that $p(1) = 3$ and $p(3) = 7$. The second piece q is to meet the first piece at the point $(3, 7)$ and end at the point $(9, -2)$. This requires that $q(3) = 7$ and $q(9) = -2$. The two pieces must meet smoothly. This requires that $p'(3) = q'(3)$ and $p''(3) = q''(3)$. Finally, the fender must be straight at the ends. This requires that $p''(1) = 0$ and $q''(9) = 0$. A curve that meets such requirements is called a *cubic spline*.

Problem 1. Let $p(x) = ax^3 + bx^2 + cx + d$ and $q(x) = \alpha x^3 + \beta x^2 + \gamma x + \delta$. Write the system of equations generated by the specified conditions and use a CAS to find the solution of the system.

Problem 2. Let

$$F(x) = \begin{cases} p(x), & 1 \le x \le 3 \\ q(x), & 3 \le x \le 9. \end{cases}$$

Show that F, F' and F'' are continuous on $[1, 9]$. Does F have continuous curvature? Sketch the graph of F using a graphing utility.

Problem 3. You are given the data in the table:

X	3	4	6
Y	10	15	35

a. Define a cubic polynomial p on $[3, 4]$ and a cubic polynomial q on $[4, 6]$ using the data in the table.

b. Write the 8×8 system of equations as in Problem 1.

c. Use a CAS to solve the system.

When engineers lay railroad track, they cannot allow any abrupt changes in curvature. To join a straight away that ends at a point P to a curved track that begins at a point Q, they need to lay some *transitional track* that has zero curvature at P and the curvature of the second piece at Q (see the figure).

Problem 4. Find an arc of the form $y = Cx^n, x \in [0, 1]$ that joins the arcs C_1 and C_2 in the figure without any discontinuities in curvature.

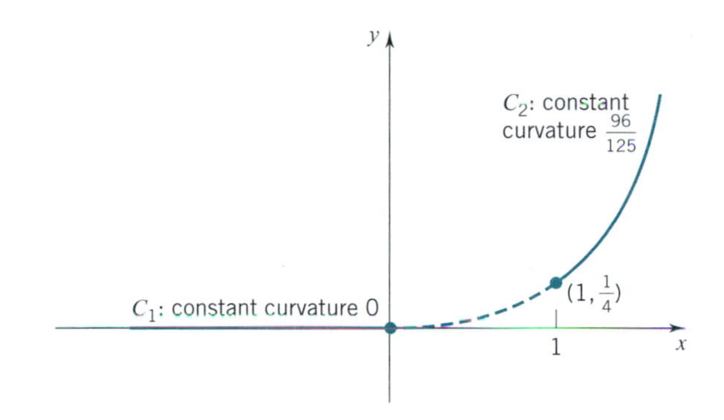

■ 13.6 VECTOR CALCULUS IN MECHANICS

The tools we have developed in the preceding sections have their premier application in Newtonian mechanics, the study of bodies in motion subject to Newton's laws. The heart of Newton's mechanics is his second law of motion:

$$\text{force} = \text{mass} \times \text{acceleration}.$$

We have worked with Newton's second law, but only in a very restricted context: motion along a coordinate line under the influence of a force directed along that same line. In that special setting, Newton's law was written as a scalar equation: $F = ma$. In general, objects do not move along straight lines (they move along curved paths) and the forces on them vary in direction. What happens to Newton's second law then? It becomes the vector equation

$$\mathbf{F} = m\,\mathbf{a}.$$

This is Newton's second law in its full glory.

An Introduction to Vector Mechanics

We are now ready to work with Newton's second law of motion in its vector form: $\mathbf{F} = m\,\mathbf{a}$. Since at each time t we have $\mathbf{a}(t) = \mathbf{r}''(t)$, Newton's law can be written

(13.6.1)
$$\boxed{\mathbf{F}(t) = m\,\mathbf{a}(t) = m\,\mathbf{r}''(t).}$$

This is a second-order differential equation in t. In Chapter 18 we give an introduction to the general theory of differential equations. Second-order differential equations of the type that we encounter here are treated in Sections 18.4 and 18.5. Our approach in this section is intuitive; we will search for solutions of (13.6.1) in particular situations.

When objects are moving, certain quantities (positions, velocities, and so on) are continually changing. This can make a situation difficult to grasp. In these circumstances it is particularly satisfying to find quantities that do not change. Such quantities are said to be *conserved*. (These conserved quantities are called the *constants of the motion*.) Mathematically we can determine whether a quantity is conserved by looking at its derivative with respect to time (the time derivative): *The quantity is conserved* (*is constant*) *iff its time derivative remains zero*. A *conservation law* is the assertion that in a given context a certain quantity does not change.

Momentum

We start with the idea of momentum. The *momentum* **p** of an object is the mass of the object times the velocity of the object:

$$\mathbf{P} = m\mathbf{v}.$$

To indicate the time dependence we write

(13.6.2)

$$\mathbf{p}(t) = m\mathbf{v}(t) = m\mathbf{r}'(t).$$

Assume that the mass of the object is constant. Then differentiation gives

$$\mathbf{p}'(t) = m\mathbf{r}''(t) = \mathbf{F}(t).$$

Thus, *the time derivative of the momentum of an object is the net force on the object*. If the net force on an object is continually zero, the momentum $\mathbf{p}(t)$ is constant. This is the law of *conservation of momentum*:

(13.6.3)

If the net force on an object is continually zero, then the momentum of the object is conserved.

Angular Momentum

The angular momentum of an object about any given point is a vector quantity that is intended to measure the extent to which the object is circling about that point. If the position of the object at time t is given by the radius vector $\mathbf{r}(t)$, then the object's *angular momentum about the origin* is defined by the formula

(13.6.4)

$$\mathbf{L}(t) = \mathbf{r}(t) \times \mathbf{p}(t) = \mathbf{r}(t) \times m\mathbf{v}(t).$$

At each time t of a motion, $\mathbf{L}(t)$ is perpendicular to $\mathbf{r}(t)$, perpendicular to $\mathbf{v}(t)$, and oriented so that $\mathbf{r}(t), \mathbf{v}(t), \mathbf{L}(t)$ form a right-handed triple. The magnitude of $\mathbf{L}(t)$ is given by the relation

$$\|\mathbf{L}(t)\| = \|\mathbf{r}(t)\| \; \|m\mathbf{v}(t)\| \sin\theta(t),$$

where $\theta(t)$ is the angle between $\mathbf{r}(t)$ and $\mathbf{v}(t)$. (All this, of course, comes from the definition of the cross product.)

If $\mathbf{r}(t)$ and $\mathbf{v}(t)$ are not zero, then we can express $\mathbf{v}(t)$ as a vector parallel to $\mathbf{r}(t)$ plus a vector perpendicular to $\mathbf{r}(t)$:

$$\mathbf{v}(t) = \mathbf{v}_{\parallel}(t) + \mathbf{v}_{\perp}(t). \qquad \text{(see Figure 13.6.1)}$$

Thus

$$\begin{aligned}\mathbf{L}(t) &= \mathbf{r}(t) \times m\mathbf{v}(t)\\ &= \mathbf{r}(t) \times m\,[\mathbf{v}_{\parallel}(t) + \mathbf{v}_{\perp}(t)]\\ &= \underbrace{[\mathbf{r}(t) \times m\mathbf{v}_{\parallel}(t)]}_{\mathbf{0}} + [\mathbf{r}(t) \times m\mathbf{v}_{\perp}(t)]\\ &= \mathbf{r}(t) \times m\mathbf{v}_{\perp}(t).\end{aligned}$$

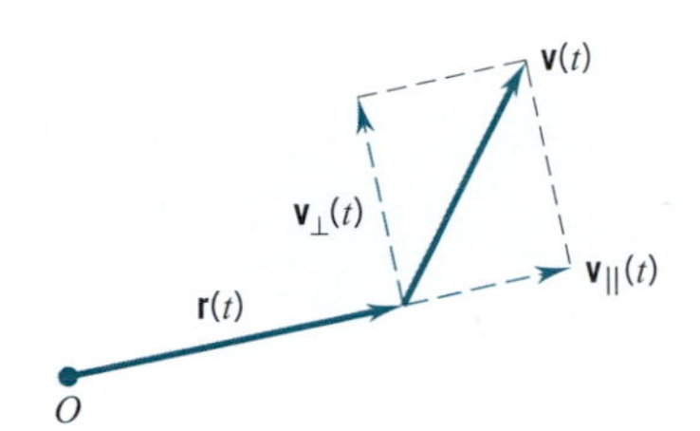

Figure 13.6.1

The component of velocity that is parallel to the radius vector contributes nothing to angular momentum. *The angular momentum comes entirely from the component of velocity that is perpendicular to the radius vector.*

Example 1 In uniform circular motion about the origin,

$$\mathbf{r}(t) = \mathbf{r}(\cos \omega t\,\mathbf{i} + \sin \omega t\,\mathbf{j}),$$

the velocity vector $\mathbf{v}(t)$ is always perpendicular to the radius vector $\mathbf{r}(t)$. In this case all of $\mathbf{v}(t)$ contributes to the angular momentum.

We can calculate $\mathbf{L}(t)$ as follows:

$$\begin{aligned}\mathbf{L}(t) &= \mathbf{r}(t) \times m\mathbf{v}(t)\\ &= [r(\cos \omega t\,\mathbf{i} + \sin \omega t\,\mathbf{j})] \times [mr(-\omega \sin \omega t\,\mathbf{i} + \omega \cos \omega t\,\mathbf{j})]\\ &= mr^2\omega(\cos^2 \omega t + \sin^2 \omega t)\,\mathbf{k} = mr^2\omega\,\mathbf{k}.\end{aligned}$$

The angular momentum is constant and is perpendicular to the xy-plane. If the motion is counterclockwise (if $\omega > 0$), then the angular momentum points up from the xy-plane. If the motion is clockwise (if $\omega < 0$), then the angular momentum points down from the xy-plane. (This is the right-handedness of the cross product coming in.) ❑

Example 2 In uniform straight-line motion with constant velocity $\mathbf{d}$,

$$\mathbf{r}(t) = \mathbf{r}_0 + t\,\mathbf{d},$$

we have

$$\mathbf{L}(t) = \mathbf{r}(t) \times m\mathbf{v}(t) = (\mathbf{r}_0 + t\,\mathbf{d}) \times m\mathbf{d} = m(\mathbf{r}_0 \times \mathbf{d}).$$

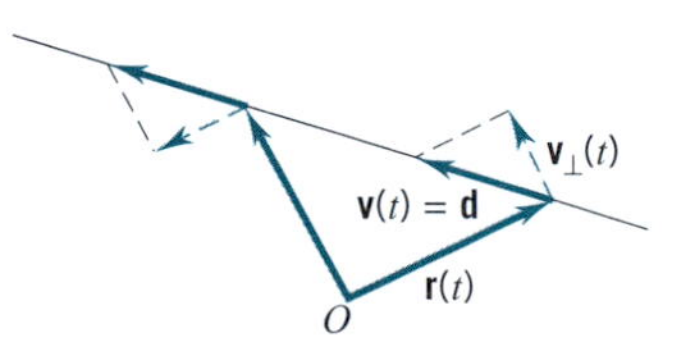

Figure 13.6.2

Here again the angular momentum is constant, but all is not quite so simple as it looks. In Figure 13.6.2 you can see the radius vector $\mathbf{r}(t)$, the velocity vector $\mathbf{v}(t) = \mathbf{d}$, and the part of the velocity vector that gives rise to angular momentum, the part perpendicular to $\mathbf{r}(t)$. As before, we have called this $\mathbf{v}_{\perp}(t)$. While $\mathbf{v}(t)$ is constant, $\mathbf{v}_{\perp}(t)$ is not constant. What happens here is that $\mathbf{r}(t)$ and $\mathbf{v}_{\perp}(t)$ vary in such a way that the cross product

$$\mathbf{L}(t) = \mathbf{r}(t) \times m\mathbf{v}_{\perp}(t)$$

remains constant. If the line of motion passes through the origin, then $\mathbf{v}(t)$ is parallel to $\mathbf{r}(t)$, $\mathbf{v}_{\perp}(t)$ is zero, and the angular momentum is zero. ❑

Torque

How the angular momentum of an object changes in time depends on the force acting on the object and on the position of the object relative to the origin: since $\mathbf{L}(t) = \mathbf{r}(t) \times m\mathbf{r}'(t)$,

$$\mathbf{L}'(t) = [\mathbf{r}(t) \times m\mathbf{r}''(t)] + \underbrace{[\mathbf{r}'(t) \times m\,\mathbf{r}'(t)]}_{\mathbf{0}} = \mathbf{r}(t) \times \mathbf{F}(t).$$

The cross product

(13.6.5)
$$\boldsymbol{\tau}(t) = \mathbf{r}(t) \times \mathbf{F}(t)$$

is called the *torque* about the origin.† See Project 12.5.

Since $\mathbf{L}'(t) = \boldsymbol{\tau}(t)$, we have the following conservation law:

(13.6.6) If the net torque on an object is continually zero, then the angular momentum of the object is conserved.

A force $\mathbf{F} = \mathbf{F}(t)$ is called a *central force* (*radial force*) if $\mathbf{F}(t)$ is always parallel to $\mathbf{r}(t)$. (Gravitational force, for example, is a central force.) For a central force, the cross product $\mathbf{r}(t) \times \mathbf{F}(t)$ is always zero. Thus a central force produces no torque about the origin. As you will see, this places severe restrictions on the kind of motion possible under a central force.

THEOREM 13.6.7

If an object moves under a central force and has constant angular momentum $\mathbf{L}$ different from zero, then:

1. The object is confined to the plane that passes through the origin and is perpendicular to $\mathbf{L}$.
2. The radius vector of the object sweeps out equal areas in equal times.

(This theorem plays an important role in astronomy and is embodied in Kepler's three celebrated laws of planetary motion. We will study Kepler's laws in Section *13.7.)

PROOF OF THEOREM 13.6.7 The first assertion is easy to verify: all the radius vectors pass through the origin and (by the very definition of angular momentum) they are all perpendicular to the constant vector $\mathbf{L}$.

To verify the second assertion, we introduce a right-handed coordinate system O–xyz setting the xy-plane as the plane of the motion, the positive z-axis pointing along $\mathbf{L}$. On the xy-plane we introduce polar coordinates r, θ. Thus, at time t, the object has some position $[r(t), \theta(t)]$.

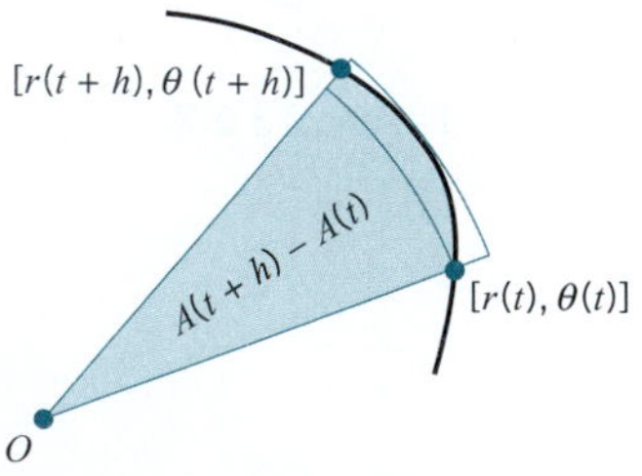

Let's denote by $A(t)$ the area swept out by the radius vector from some fixed time t_0 up to time t. Our task is to show that $A'(t)$ is constant.

The area swept out during the time interval $[t, t+h]$ is simply

$$A(t+h) - A(t).$$

Assuming that the motion takes place in the direction of increasing polar angle θ (see Figure 13.6.3), we have with obvious notation,

$[r(t+h), \theta(t+h)]$ $A(t+h) - A(t)$ $[r(t), \theta(t)]$ O

Figure 13.6.3

† The symbol τ is the Greek letter tau. The word *torque* comes from the Latin word *torquere*, "to twist."

$$\underbrace{\tfrac{1}{2}\min\,[r(t)]^2 \cdot [\theta(t+h)-\theta(t)]}_{\text{area of inner sector}} \le A(t+h)-A(t) \le \underbrace{\tfrac{1}{2}\max\,[r(t)]^2 \cdot [\theta(t+h)-\theta(t)]}_{\text{area of outer sector}}.$$

Divide through by h, take the limit as h tends to 0, and you will see that

$$(*) \qquad A'(t) = \tfrac{1}{2}[r(t)]^2\theta'(t).$$

Now

$$\begin{aligned}
\mathbf{r}(t) &= r(t)[\cos\theta(t)\,\mathbf{i} + \sin\theta(t)\,\mathbf{j}],\\
\mathbf{v}(t) &= r'(t)[\cos\theta(t)\,\mathbf{i} + \sin\theta(t)\,\mathbf{j}] + r(t)\theta'(t)[-\sin\theta(t)\,\mathbf{i} + \cos\theta(t)\,\mathbf{j}].\\
&= [r'(t)\cos\theta(t) - r(t)\theta'(t)\sin\theta(t)]\,\mathbf{i} + [r'(t)\sin\theta(t) + r(t)\theta'(t)\cos\theta(t)]\,\mathbf{j}.
\end{aligned}$$

A calculation that you can carry out yourself shows that

$$\mathbf{L} = \mathbf{r}(t) \times m\mathbf{v}(t) = mr^2(t)\theta'(t)\,\mathbf{k}.$$

Since $\mathbf{L}$ is constant, $r^2(t)\theta'(t)$ is constant. Thus, by $(*)$, $A'(t)$ is constant:

$$A'(t) = L/2m \qquad \text{where} \quad L = \|\mathbf{L}\|. \quad \square$$

Initial-Value Problems

In physics one tries to make predictions about the future on the basis of current information and a knowledge of the forces at work. In the case of an object in motion, the task can be to determine $\mathbf{r}(t)$ for all t given the force and some "initial conditions." Frequently the initial conditions give the position and velocity of the object at some time t_0. The problem then is to solve the differential equation

$$\mathbf{F} = m\mathbf{r}''$$

subject to conditions of the form

$$\mathbf{r}(t_0) = \mathbf{r}_0, \qquad \mathbf{v}(t_0) = \mathbf{v}_0.$$

Such problems are known as initial-value problems. We have considered initial-value problems in several different contexts earlier in the text.

By far the simplest problem of this sort concerns a *free particle*, an object on which there is no net force.

Example 3 At time t_0 a free particle has position $\mathbf{r}(t_0) = \mathbf{r}_0$ and velocity $\mathbf{v}(t_0) = \mathbf{v}_0$. Find $\mathbf{r}(t)$ for all t.

SOLUTION Since there is no net force on the object, the acceleration is zero and the velocity is constant. Since $\mathbf{v}(t_0) = \mathbf{v}_0$,

$$\mathbf{v}(t) = \mathbf{v}_0 \qquad \text{for all } t.$$

Integration with respect to t gives.

$$\mathbf{r}(t) = t\mathbf{v}_0 + \mathbf{c},$$

where $\mathbf{c}$, the constant of integration, is a vector that we can determine from the initial position. The initial position $\mathbf{r}(t_0) = \mathbf{r}_0$ gives

$$\mathbf{r}_0 = t_0\mathbf{v}_0 + \mathbf{c} \qquad \text{and therefore} \qquad \mathbf{c} = \mathbf{r}_0 - t_0\mathbf{v}_0.$$

Using this value for $\mathbf{c}$ in our equation for $\mathbf{r}(t)$, we have

$$\mathbf{r}(t) = t\mathbf{v}_0 + (\mathbf{r}_0 - t_0\mathbf{v}_0),$$

which we write as

$$\mathbf{r}(t) = \mathbf{r}_0 + (t - t_0)\mathbf{v}_0.$$

This is the equation of a straight line with direction vector $\mathbf{v}_0$. Free particles travel in straight lines with constant velocity. (We have tacitly assumed that $\mathbf{v}_0 \neq \mathbf{0}$. If $\mathbf{v}_0 = \mathbf{0}$, the particle remains at rest at $\mathbf{r}_0$.) ❑

Example 4 An object of mass m is subject to a force of the form

$$\mathbf{F}(t) = -m\omega^2\mathbf{r}(t) \qquad \text{with} \quad \omega > 0.$$

Find $\mathbf{r}(t)$ for all t given that

$$\mathbf{r}(0) = a\,\mathbf{i} \qquad \text{and} \qquad \mathbf{v}(0) = \omega a\,\mathbf{j} \qquad \text{with} \quad a > 0.$$

SOLUTION The force is a vector version of the restoring force exerted by a linear spring (Hooke's law). Since the force is central, the angular momentum of the object, $\mathbf{L}(t) = \mathbf{r}(t) \times m\mathbf{v}(t)$, is conserved. So $\mathbf{L}(t)$ is constantly equal to the value it had at time $t = 0$; for all t.

$$\mathbf{L}(t) = \mathbf{L}(0) = \mathbf{r}(0) \times m\mathbf{v}(0) = a\,\mathbf{i} \times m\omega a\,\mathbf{j} = ma^2\omega\,\mathbf{k}.$$

From our earlier discussion (Theorem 13.6.7) we can conclude that the motion takes place in the plane that passes through the origin and is perpendicular to $\mathbf{k}$. This is the xy-plane. Thus we can write

$$\mathbf{r}(t) = x(t)\,\mathbf{i} + y(t)\,\mathbf{j}.$$

Since $\mathbf{F}(t) = m\,\mathbf{r}''(t)$, the force equation can be written $\mathbf{r}''(t) = -\omega^2\mathbf{r}(t)$. In terms of components we have

$$x''(t) = -\omega^2 x(t), \qquad y''(t) = -\omega^2 y(t).$$

These are the equations of simple harmonic motion. We have already seen that functions of the form

$$x(t) = A_1 \sin(\omega t + \phi_1), \qquad y(t) = A_2 \sin(\omega t + \phi_2)$$

are solutions of these equations (Exercises 73 and 74, Section 3.6). In Section 18.4 we show that *all* solutions have this form. To evaluate the constants, we use the initial conditions. The condition $\mathbf{r}(0) = a\,\mathbf{i}$ means that $x(0) = a$ and $y(0) = 0$. So

$$(*) \qquad A_1 \sin\phi_1 = a \qquad A_2 \sin\phi_2 = 0.$$

The condition $\mathbf{v}(0) = \omega a\,\mathbf{j}$ means that $x'(0) = 0$ and $y'(0) = \omega a$. Since

$$x'(t) = \omega A_1 \cos(\omega t + \phi_1) \qquad \text{and} \qquad y'(t) = \omega A_2 \cos(\omega t + \phi_2),$$

we have

$$(**) \qquad \omega A_1 \cos\phi_1 = 0, \qquad \omega A_2 \cos\phi_2 = \omega a.$$

Conditions (*) and (**) are met by setting $A_1 = a, A_2 = a, \phi_1 = \frac{1}{2}\pi, \phi_2 = 0$. Thus

$$x(t) = a\sin(\omega t + \tfrac{1}{2}\pi) = a\cos\omega t, \quad y(t) = a\sin\omega t.$$

The vector equation reads

$$\mathbf{r}(t) = a\cos\omega t\,\mathbf{i} + a\sin\omega t\,\mathbf{j}.$$

The object moves in a circle of radius a about the origin with constant angular velocity ω. ❑

Example 5 A particle of charge q in a magnetic field $\mathbf{B}$ is subject to the force

$$\mathbf{F}(t) = \frac{q}{c}[\mathbf{v}(t) \times \mathbf{B}(t)],$$

where c is the speed of light and $\mathbf{v}$ is the velocity of the particle. Given that $\mathbf{r}(0) = \mathbf{r}_0$ and $\mathbf{v}(0) = \mathbf{v}_0$, find the path of the particle in the constant magnetic field $\mathbf{B}(t) = B_0\,\mathbf{k}$, $B_0 \neq 0$.

SOLUTION There is no conservation law that we can conveniently appeal to here. Neither momentum nor angular momentum is conserved: the force is not zero and it is not central. We start directly with Newton's $\mathbf{F} = m\mathbf{r}''$.

Since $\mathbf{r}'' = \mathbf{v}'$, we have

$$m\mathbf{v}'(t) = \frac{q}{c}[\mathbf{v}(t) \times B_0\mathbf{k}],$$

which we can write as

$$\mathbf{v}'(t) = \frac{qB_0}{mc}[\mathbf{v}(t) \times \mathbf{k}].$$

To simplify notation, we set $qB_0/mc = \omega$. We then have

$$\mathbf{v}'(t) = \omega[\mathbf{v}(t) \times \mathbf{k}].$$

Placing $\mathbf{v}(t) = v_1(t)\mathbf{i} + v_2(t)\mathbf{j} + v_3(t)\mathbf{k}$ in this last equation and working out the cross product, we find that

$$v_1'(t)\mathbf{i} + v_2'(t)\mathbf{j} + v_3'(t)\mathbf{k} = \omega[v_2(t)\mathbf{i} - v_1(t)\mathbf{j}].$$

This gives the scalar equations

$$v_1'(t) = \omega v_2(t), \qquad v_2'(t) = -\omega v_1(t), \qquad v_3'(t) = 0.$$

The last equation is trivial. It says that v_3 is constant:

$$v_3(t) = C.$$

The equations for v_1 and v_2 are linked together. We can get an equation that involves only v_1 by differentiating the first equation:

$$v_1''(t) = \omega v_2'(t) = -\omega^2 v_1(t).$$

As we know from our earlier work, this gives

$$v_1(t) = A_1\sin(\omega t + \phi_1).$$

Since $v_1'(t) = \omega v_2(t)$, we have

$$v_2(t) = \frac{v_1'(t)}{\omega} = \frac{A_1\omega}{\omega}\cos(\omega t + \phi_1) = A_1\cos(\omega t + \phi_1).$$

Therefore $$\mathbf{v}(t) = A_1\sin(\omega t + \phi_1)\,\mathbf{i} + A_1\cos(\omega t + \phi_1)\,\mathbf{j} + C\,\mathbf{k}.$$

A final integration with respect to t gives

$$\mathbf{r}(t) = \left[-\frac{A_1}{\omega}\cos(\omega t + \phi_1) + D_1\right]\mathbf{i} + \left[\frac{A_1}{\omega}\sin(\omega t + \phi_1) + D_2\right]\mathbf{j} + [Ct + D_3]\,\mathbf{k}$$

where $D_1,\ D_2,\ D_3$ are constants of integration. All six constants of integration — $A_1, \phi_1, C,\ D_1,\ D_2,\ D_3$— can be evaluated from the initial conditions. We will not pursue this. What is important here is that the path of the particle is a circular helix with axis parallel to **B**, in this case parallel to **k**. You should be able to see this from the equation for $\mathbf{r}(t)$: the z-component of **r** varies linearly with t from the value D_3, while the x and y components represent uniform motion with angular velocity ω in a circle of radius $|A_1/\omega|$ around the center (D_1, D_2). □

(Physicists express the behavior just found by saying that charged particles *spiral around* the magnetic field lines. Qualitatively, this behavior still holds even if the magnetic field lines are "bent," as is the case with the earth's magnetic field. Many charged particles become trapped by the earth's magnetic field. They keep spiraling around the magnetic field lines that run from pole to pole.)

EXERCISES 13.6

1. An object of mass m moves so that

$$\mathbf{r}(t) = \tfrac{1}{2}a(e^{\omega t} + e^{-\omega t})\,\mathbf{i} + \tfrac{1}{2}b(e^{\omega t} - e^{-\omega t})\,\mathbf{j}.$$

(a) What is the velocity at $t = 0$? (b) Show that the acceleration vector is a constant positive multiple of the radius vector. (This shows that the force is central and repelling.) (c) What does (b) imply about the angular momentum and the torque? Verify your answers by direct calculation.

2. (a) An object moves so that

$$\mathbf{r}(t) = a_1e^{bt}\,\mathbf{i} + a_2e^{bt}\,\mathbf{j} + a_3e^{bt}\,\mathbf{k}.$$

Show that, if $b > 0$, the object experiences a repelling central force.

(b) An object moves so that

$$\mathbf{r}(t) = \sin t\,\mathbf{i} + \cos t\,\mathbf{j} + (\sin t + \cos t)\,\mathbf{k}.$$

Show that the object experiences an attracting central force.

(c) Compute the angular momentum $\mathbf{L}(t)$ for the motion in part (*b*).

3. A constant force of magnitude α directed upward from the xy-plane is continually applied to an object of mass m. Given that the object starts at time 0 at the point $P(0, y_0, z_0)$ with initial velocity $2\mathbf{j}$, find: (a) the velocity of the object t seconds later; (b) the speed of the object t seconds later; (c) the momentum of the object t seconds later; (d) the path followed by the object, both in vector form and in Cartesian coordinates.

4. Show that, if the force on an object is always perpendicular to the velocity of the object, then the *speed* of the object is constant. (This tells us that the speed of a charged particle in a magnetic field is constant.)

5. Find the force required to propel a particle of mass m so that $\mathbf{r}(t) = t\mathbf{j} + t^2\mathbf{k}$.

6. Show that for an object of constant velocity, the angular momentum is constant.

7. At each point $P(x(t), y(t), z(t))$ of its motion, an object of mass m is subject to a force

$$\mathbf{F}(t) = m\pi^2[a\cos\pi t\,\mathbf{i} + b\sin\pi t\,\mathbf{j}]. \quad (a > 0, b > 0)$$

Given that $\mathbf{v}(0) = -\pi b\,\mathbf{j} + \mathbf{k}$ and $\mathbf{r}(0) = b\,\mathbf{j}$, find the following at time $t = 1$:

(a) The velocity. (b) The speed.
(c) The acceleration. (d) The momentum.
(e) The angular momentum. (f) The torque.

8. If an object of mass m moves with velocity $\mathbf{v}(t)$ subject to a force $\mathbf{F}(t)$, the scalar product

(13.6.8) $$\mathbf{F}(t)\cdot\mathbf{v}(t)$$

is called the *power* (expended by the force) and the number

(13.6.9) $$\boxed{\tfrac{1}{2}m[v(t)]^2}$$

is called the *kinetic energy* of the object. Show that the time rate of change of the kinetic energy of an object is the power expended on it:

(13.6.10) $$\boxed{\frac{d}{dt}\left(\tfrac{1}{2}m[v(t)]^2\right) = \mathbf{F}(t) \cdot \mathbf{v}(t).}$$

9. Two particles of equal mass m, one with constant velocity $\mathbf{v}$ and the other at rest, collide elastically (i.e., the kinetic energy of the system is preserved) and go off in different directions. Show that the two particles go off at right angles.

10. *(Elliptic harmonic motion)* Show that if the force on a particle of mass m is of the form

$$\mathbf{F}(t) = -m\omega^2\mathbf{r}(t),$$

then the path of the particle may be written

$$\mathbf{r}(t) = \cos\omega t\mathbf{A} + \sin\omega t\mathbf{B},$$

where $\mathbf{A}$ and $\mathbf{B}$ are constant vectors. Give the physical significance of $\mathbf{A}$ and $\mathbf{B}$ and specify conditions on $\mathbf{A}$ and $\mathbf{B}$ that restrict the particle to a circular path.

HINT: The solutions of the differential equation

$$x''(t) = -\omega^2 x(t)$$

can be written in the form

$$x(t) = A\cos\omega t + B\sin\omega t. \qquad \text{(Exercise 73, Section 3.6)}$$

11. A particle moves with constant acceleration $\mathbf{a}$. Show that the path of the particle lies entirely in some plane. Find a vector equation for this plane.

12. In Example 5 we stated that the path

$$\mathbf{r}(t) = \left[-\frac{A_1}{\omega}\cos(\omega t + \phi_1) + D_1\right]\mathbf{i} + \left[\frac{A_1}{\omega}\sin(\omega t + \phi_1) + D_2\right]\mathbf{j} + [Ct + D_3]\mathbf{k}$$

was a circular helix, Set $\omega = -1$ and show that, if $\mathbf{r}(0) = a\mathbf{i}$ and $\mathbf{v}(0) = a\,\mathbf{j} + b\,\mathbf{k}$, then the path takes the form

$$\mathbf{r}(t) = a\cos t\,\mathbf{i} + a\sin t\,\mathbf{j} + bt\,\mathbf{k},$$

the circular helix described in Section 13.3.

13. A charged particle in a time-independent electric field $\mathbf{E}$ experiences the force $q\mathbf{E}$, where q is the charge of the particle. Assume that the field has the constant value $\mathbf{E} = E_0\mathbf{k}$ and find the path of the particle given that $\mathbf{r}(0) = \mathbf{i}$ and $\mathbf{v}(0) = \mathbf{j}$.

14. *(Important)* A wheel is rotating about an axle with angular speed ω. Let $\boldsymbol{\omega}$ be the *angular velocity vector*, the vector of length ω that points along the axis of the wheel in such a direction that, observed from the tip of $\boldsymbol{\omega}$, the wheel rotates counterclockwise. Take the origin as the center of the wheel and let $\mathbf{r}$ be the vector from the origin to a point P on the rim of the wheel. Express the velocity $\mathbf{v}$ of P in terms of $\boldsymbol{\omega}$ and $\mathbf{r}$.

15. Solve the initial-value problem

$$\mathbf{F}(t) = m\mathbf{r}''(t) = t\mathbf{i} + t^2\mathbf{j}, \quad \mathbf{r}_0 = \mathbf{r}(0) = \mathbf{i},$$
$$\mathbf{v}_0 = \mathbf{v}(0) = \mathbf{k}.$$

16. Solve the initial-value problem

$$\mathbf{F}(t) = m\mathbf{r}''(t) = -m\beta^2 z(t)\,\mathbf{k}, \quad \mathbf{r}_0 = \mathbf{r}(0) = \mathbf{k},$$
$$\mathbf{v}_0 = \mathbf{v}(0) = 0.$$

[Here $z(t)$ is the z-component of $\mathbf{r}(t)$.]

17. An object of mass m moves subject to the force

$$\mathbf{F}(\mathbf{r}) = 4r^2\mathbf{r}$$

where $\mathbf{r} = \mathbf{r}(t)$ is the position of the object. Suppose $\mathbf{r}(0) = \mathbf{0}$ and $\mathbf{v}(0) = 2\mathbf{u}$, where $\mathbf{u}$ is a unit vector. Show that at each time t the speed v of the object satisfies the relation

$$v = \sqrt{4 + \frac{2}{m}r^4}.$$

HINT: Examine the quantity $\frac{1}{2}mv^2 - r^4$. (This is the *energy* of the object, a notion we will take up in Chapter 17.)

■ *13.7 PLANETARY MOTION

Tycho Brahe, Johannes Kepler

In the middle of the sixteenth century the arguments on planetary motion persisted. Was Copernicus right? Did the planets move in circles about the sun? Obviously not. Was not the earth the center of the universe?

In 1576, with the generous support of his king, the Danish astronomer Tycho Brahe built an elaborate astronomical observatory on the isle of Hveen and began his painstaking observations. For more than twenty years he looked through his telescopes and recorded what he saw. He was a meticulous observer, but he could draw no definite conclusions.

In 1599 the German astronomer-mathematician Johannes Kepler began his study of Brahe's voluminous tables. For a year and a half Brahe and Kepler worked together. Then Brahe died and Keper went on wrestling with the data. His persistence paid off. By 1619 Kepler had made three stupendous discoveries, known today as *Kepler's laws of planetary motion*:

> **I.** Each planet moves in a plane, not in a circle, but in an elliptical orbit with the sun at one focus.
>
> **II.** The radius vector from the sun to the planet sweeps out equal areas in equal times.
>
> **III.** The square of the period of the motion varies directly as the cube of the major semiaxis, and the constant of proportionality is the same for all the planets.

What Kepler formulated empirically, Newton was able to explain. Each of these laws, Newton showed, was deducible from his laws of motion and his law of gravitation.

Newton's Second Law of Motion for Extended Three-Dimensional Objects

Imagine an object that consists of n particles with masses $m_1, m_2, \ldots, m_n$ located at $\mathbf{r}_1, \mathbf{r}_2, \ldots \mathbf{r}_n$.† The total mass M of the object is the sum of the masses of the constituent particles:

$$M = m_1 + \cdots + m_n.$$

The center of mass of the object is by definition the point $\mathbf{R}_M$ where

$$M\mathbf{R}_M = m_1\mathbf{r}_1 + \cdots + m_n\mathbf{r}_n.$$

The total force $\mathbf{F}_{\text{TOT}}$ on the object is by definition the sum of all the forces that act on the particles that constitute the object:

$$\mathbf{F}_{\text{TOT}} = \mathbf{F}_1 + \cdots + \mathbf{F}_n.$$

Since $\mathbf{F}_1 = m_1\mathbf{r}_1'', \cdots, \mathbf{F}_n = m_n\mathbf{r}_n''$, we have

$$\mathbf{F}_{\text{TOT}} = m_1\mathbf{r}_1'' + \cdots + m_n\mathbf{r}_n'',$$

which we can write as

$$\mathbf{F}_{\text{TOT}} = M\mathbf{R}_M''.$$

The total force on an extended object is thus the total mass of the object times the acceleration of the center of mass.

We can simplify this still further. The forces that act between the constituent particles, the so-called internal forces, cancel in pairs: if particle 23 tugs at particle 71 in

† The case of a continuously distributed mass is taken up in Chapter 16.

a certain direction with a certain strength, then particle 71 tugs at particle 23 in the opposite direction with the same strength. (Newton's third law: To every action there is an equal reaction.) Therefore, in calculating the total force on our object, we can disregard the internal forces and simply add up the external forces. $\mathbf{F}_{\text{TOT}} = \mathbf{F}_{\text{TOT}}^{(\text{Ext})}$ and Newton's second law takes the form

(13.7.1)
$$\mathbf{F}_{\text{TOT}}^{(\text{Ext})} = M\,\mathbf{R}_M''.$$

The total external force on an extended three-dimensional object is thus the total mass of the object times the acceleration of the center of mass.

When an external force is applied to an extended object, we cannot predict the reaction of all the constituent particles, but we can predict the reaction of the center of mass. The center of mass will react to the force as if it were a particle with all the mass concentrated there. Suppose, for example, that a bomb is dropped from an airplane. The center of mass, "feeling" only the force of gravity (we are neglecting air resistance), falls in a parabolic arc toward the ground even if the bomb explodes at a thousand meters and individual pieces fly every which way. The forces of explosion are internal and do not affect the motion of the center of mass.

Some Preliminary Comments About the Planets

Roughly speaking, a planet is a massive object in the form of a ball with the center of mass at the center. In what follows, when we refer to the position of a planet, you are to understand that we really mean the position of the center of mass of the planet. Two other points require comment. First, we will write our equations as if the sun affected the motion of the planet, but not vice versa: we will assume that the sun stays put and that the planet moves. Really, each affects the other. Our viewpoint is justified by the immense difference in mass between the planets and the sun. In the case of the earth and the sun, for example, a reasonable analogy is to imagine a tug of war in space between someone who weighs three pounds and someone who weighs a million pounds: to a good approximation, the million-pound person does not move. The second point is that the planet is affected not only by the pull from the sun, but also by the gravitational pulls from the other planets and all the other celestial bodies. But these forces are much smaller, and they tend to cancel. We will ignore them. (Our results are only approximations, but they prove to be very good approximations.)

A Derivation of Kepler's Laws from Newton's Laws of Motion and His Law of Gravitation

The gravitational force exerted by the sun on a planet can be written in vector form as follows:

$$(*) \qquad \mathbf{F}(\mathbf{r}) = -G\frac{mM}{r^3}\mathbf{r}.$$

Here m is the mass of the planet, M is the mass of the sun, G is the gravitational constant, $\mathbf{r}$ is the vector from the center of the sun to the center of the planet, and r is the magnitude of $\mathbf{r}$. (Thus we are placing the sun at the origin of our coordinate system; namely, we are using what is known as a *heliocentric* coordinate system, from *hēlios*, the Greek word for sun.)

Let's make sure that equation $(*)$ conforms to our earlier characterization of gravitational force. First of all, because of the minus sign, the direction is toward the origin, where the sun is located. Taking norms we have

$$\|\mathbf{F}(\mathbf{r})\| = \frac{GmM}{r^3}\|\mathbf{r}\| = \frac{GmM}{r^2}$$
$$\uparrow \; \|\mathbf{r}\| = r$$

Thus the magnitude of the force is as expected: it does vary directly as the product of the masses and inversely as the square of the distance between them. So we are back in familiar territory.

We will derive Kepler's laws in a somewhat piecemeal manner. Since the force on the planet is a central force, **L**, the angular momentum of the planet, is conserved (Section 13.6). If **L** were zero, we would have

$$\|\mathbf{L}\| = L = m\|\mathbf{r}\| \; \|\mathbf{v}\| \sin\theta = 0.$$

This would mean that either $\mathbf{r} = \mathbf{0}$, $\mathbf{v} = \mathbf{0}$, or $\sin\theta = 0$. The first equation would place the planet at the center of the sun. The second equation could hold only if the planet stopped. The third equation could hold only if the planet moved directly toward the sun or directly away from the sun. We can be thankful that none of those things happen.

Since the planet moves under a central force and **L** is not zero, *the planet stays on the plane that passes through the center of the sun and is perpendicular to* **L**, and the radius vector does sweep out equal areas in equal times. (We know all this from Theorem 13.6.7.)

Now we go on to show that the path is an ellipse with the sun at one focus.† The equation of motion of a planet of mass m can be written

$$m\,\mathbf{a}(t) = -GmM\frac{\mathbf{r}(t)}{[r(t)]^3}.$$

Clearly m drops out of the equation. Setting $GM = \rho$ and suppressing the explicit dependence on t, we have

$$\mathbf{a} = -\rho\frac{\mathbf{r}}{r^3},$$

which gives

$$(**) \qquad \frac{\mathbf{r}}{r^3} = -\frac{1}{\rho}\mathbf{a} = -\frac{1}{\rho}\frac{d\mathbf{v}}{dt}.$$

From (13.2.4) we know that in general

$$\frac{d}{dt}\left(\frac{\mathbf{r}}{r}\right) = (\mathbf{r} \times \mathbf{v}) \times \frac{\mathbf{r}}{r^3}.$$

Since $\mathbf{L} = \mathbf{r} \times m\mathbf{v}$, we have

$$\frac{d}{dt}\left(\frac{\mathbf{r}}{r}\right) = \frac{\mathbf{L}}{m} \times \frac{\mathbf{r}}{r^3}.$$

† If some of the steps seem unanticipated to you, you should realize that we are discussing one of the most celebrated physical problems in history and there have been three hundred years to think of ingenious ways to deal with it. The argument we give here follows the lines of the excellent discussion that appears in Harry Pollard's *Celestial Mechanics*, Mathematical Association of America (1976).

Inserting (**) we see that

$$\frac{d}{dt}\left(\frac{\mathbf{r}}{r}\right) = \frac{\mathbf{L}}{m} \times \left(-\frac{1}{\rho}\frac{d\mathbf{v}}{dt}\right) = \frac{d\mathbf{v}}{dt} \times \frac{\mathbf{L}}{m\rho} = \frac{d}{dt}\left(\mathbf{v} \times \frac{\mathbf{L}}{m\rho}\right)$$

$\mathbf{L}, m, \rho$, are all constant ↑

and therefore
$$\frac{d}{dt}\left[\left(\mathbf{v} \times \frac{\mathbf{L}}{m\rho}\right) - \frac{\mathbf{r}}{r}\right] = \mathbf{0}.$$

Integration with respect to t gives

$$(***) \qquad \left(\mathbf{v} \times \frac{\mathbf{L}}{m\rho}\right) - \frac{\mathbf{r}}{r} = \mathbf{e}$$

where $\mathbf{e}$ is a constant vector that depends on the initial conditions. Dotting both sides with $\mathbf{r}$, we have

$$\mathbf{r} \cdot \left(\mathbf{v} \times \frac{\mathbf{L}}{m\rho}\right) - \frac{\mathbf{r} \cdot \mathbf{r}}{r} = \mathbf{r} \cdot \mathbf{e}.$$

Since

$$\mathbf{r} \cdot \left(\mathbf{v} \times \frac{\mathbf{L}}{m\rho}\right) = \frac{\mathbf{L}}{m\rho} \cdot (\mathbf{r} \times \mathbf{v}) = \frac{\mathbf{L}}{m\rho} \cdot \frac{\mathbf{L}}{m} = \frac{L^2}{m^2\rho} \qquad \text{and} \qquad \frac{\mathbf{r} \cdot \mathbf{r}}{r} = \frac{r^2}{r} = r,$$

(12.5.6) ↑

we find that
$$\frac{L^2}{m^2\rho} - r = \mathbf{r} \cdot \mathbf{e},$$

which we write as

$$r + (\mathbf{r} \cdot \mathbf{e}) = \frac{L^2}{m^2\rho}. \qquad \text{(orbit equation)}$$

If $\mathbf{e} = \mathbf{0}$, the orbit is a circle:

$$r = \frac{L^2}{m^2\rho}.$$

This is a possibility that requires very special initial conditions, conditions not met by any of the planets in our solar system. In our solar system, at least $\mathbf{e} \neq \mathbf{0}$.

Given that $\mathbf{e} \neq \mathbf{0}$, we can write the orbit equation as

(13.7.2)
$$\boxed{r(1 + e\cos\theta) = \frac{L^2}{m^2\rho}}$$

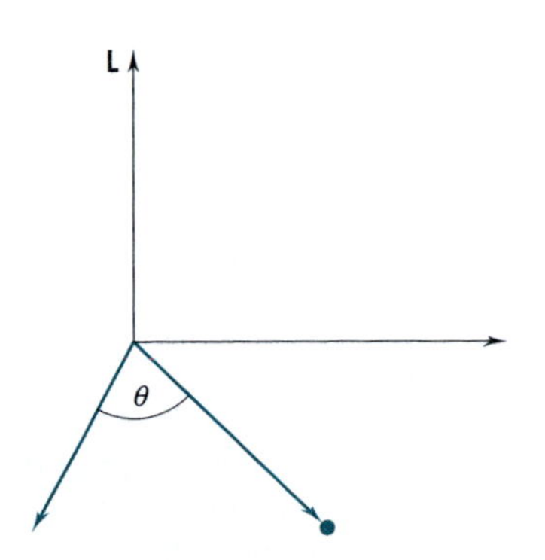

Figure 13.7.1

where $e = \|\mathbf{e}\|$ and θ is the angle between $\mathbf{r}$ and $\mathbf{e}$. We are almost through. Since $\mathbf{v} \times \mathbf{L}$ and $\mathbf{r}$ are both perpendicular to $\mathbf{L}$, we know from $(***)$ that $\mathbf{e}$ is perpendicular to $\mathbf{L}$. Therefore, in the plane of $\mathbf{e}$ and $\mathbf{r}$ (see Figure 13.7.1) we can interpret r and θ as the usual polar coordinates. The orbit equation (13.7.2) is then a polar equation. We refer you to Project 9.4, where it is shown that an equation of the form (13.7.2) represents a conic section with focus at the origin, which is to say *focus at the sun*. Accordingly, the conic section can be a parabola, a hyperbola, or an ellipse. The repetitiveness of planetary motion rules out the parabola and the hyperbola. *The orbit is an ellipse.*

Finally we will verify Kepler's third law: The square of the period of the motion varies directly as the cube of the major semi-axis, and the constant of proportionality is the same for all planets.

The elliptic orbit has an equation of the form

$$r(1 + e\cos\theta) = ed \qquad \text{with} \qquad 0 < e < 1, \quad d > 0.$$

Rewriting this equation in rectangular coordinates ($x = r\cos\theta,\ y = r\sin\theta$), we get

$$\frac{(x+c)^2}{a^2} + \frac{y^2}{a^2 - c^2} = 1, \qquad \text{where} \qquad a = \frac{ed}{1-e^2} \quad \text{and} \quad c = \frac{e^2 d}{1-e^2}.$$

By Section 9.2, the lengths of the major and minor semi-axes of the ellipse are given by

$$a = \frac{ed}{1-e^2} \qquad \text{and} \qquad b = \sqrt{a^2 - c^2} = a\sqrt{1-e^2}.$$

Denote the period of revolution by T. Since the radius vector sweeps out area at the constant rate of $L/2m$ (we know this from the proof of Theorem 13.6.7)

$$\left(\frac{L}{2m}\right)T = \text{ area of the ellipse} = \pi ab = \pi a^2\sqrt{1-e^2}.$$

Thus

$$T = \frac{2\pi m a^2\sqrt{1-e^2}}{L} \qquad \text{and} \qquad T^2 = \frac{4\pi^2 m^2 a^4(1-e^2)}{L^2}.$$

From (13.7.2) we know that

$$ed = \frac{L^2}{m^2\rho} = \frac{L^2}{m^2 GM} \qquad \text{and therefore} \qquad \frac{m^2}{L^2} = \frac{1}{edGM}.$$

It follows that

$$T^2 = \frac{4\pi^2 a^4(1-e^2)}{edGM} = \frac{4\pi^2 a^4(1-e^2)}{a(1-e^2)GM} = \frac{4\pi^2}{GM}a^3.$$

$\uparrow$ $ed = a(1-e^2)$

T^2 does vary directly with a^3, and the constant of proportionality $4\pi^2/GM$ is the same for all planets.

EXERCISES 13.7

1. Kepler's third law can be stated as follows: For each planet the square of the period of revolution varies directly as the cube of the planet's average distance from the sun, and the constant of proportionality is the same for all planets. A *year* on a given planet is the time taken by the planet to make one circuit around the sun. Thus, a year on a planet is the period of revolution of that planet. Given that on average Venus is 0.72 times as far from the sun as the earth, how does the length of a "Venus year" compare with the length of an "earth year"?

2. Verify by differentiation with respect to time t that if the acceleration of a planet is given by

$$\mathbf{a} = -\rho\frac{\mathbf{r}}{r^3}.$$

then the energy $E = \frac{1}{2}mv^2 - \dfrac{m\rho}{r}$ is constant.

3. Given that a planet moves in a plane, its motion can be described by rectangular coordinates (x, y) or polar coordinates $[r, \theta]$, with the origin at the sun. The kinetic energy of a planet is

$$\frac{1}{2}mv^2 = \frac{1}{2}m\left[\left(\frac{dx}{dt}\right)^2 + \left(\frac{dy}{dt}\right)^2\right].$$

Show that in polar coordinates

$$\frac{1}{2}mv^2 = \frac{1}{2}m\left[\left(\frac{dr}{dt}\right)^2 + r^2\left(\frac{d\theta}{dt}\right)^2\right].$$

4. We have seen that the energy of a planet

$$E = \frac{1}{2}mv^2 - \frac{m\rho}{r}$$

is constant (Exercise 2). Setting $dr/dt = \dot{r}, d\theta/dt = \dot{\theta}$, and using Exercise 3, we have

$$E = \frac{1}{2}m(\dot{r}^2 + r^2\dot{\theta}^2) - \frac{m\rho}{r}.$$

We also know that the angular momentum $\mathbf{L}$ is constant and that $L = mr^2\dot{\theta}$. Use this fact to verify that

$$E = \frac{L^2}{2m}\left\{\frac{1}{r^2} + \frac{1}{r^4}\left(\frac{dr}{d\theta}\right)^2\right\} - \frac{m\rho}{r}.$$

Since E is a constant, this is a differential equation for r as a function of θ.

5. Show that the function

$$r = \frac{a}{1 + e\cos\theta} \quad \text{with} \quad a = \frac{L^2}{m^2\rho} \quad \text{and} \quad e^2 = \frac{2Ea}{m\rho} + 1$$

satisfies the equation derived in Exercise 4.

■ CHAPTER HIGHLIGHTS

13.1 Vector Functions

basic definition: $\lim_{y\to t_0} \mathbf{f}(t) = \mathbf{L}$ iff $\lim_{t\to t_0} \|\mathbf{f}(t)\| - \|\mathbf{L}\| = 0$
theorem: if $\lim_{t\to t_0} \mathbf{f}(t) = \mathbf{L}$ then $\lim_{t\to t_0} \|\mathbf{f}(t)\| = \|\mathbf{L}\|$
limit rules (p. 765)
limits can be taken component by component (p. 765)
continuity and differentiability (p. 766) integration (p. 767)
properties of the integral (p. 768)

13.2 Differentiation Formulas

$$\frac{d}{dt}(\mathbf{f} + \mathbf{g}) = \frac{d\mathbf{f}}{dt} + \frac{d\mathbf{g}}{dt}.$$

$$\frac{d}{dt}(\alpha\mathbf{f}) = \alpha\frac{d\mathbf{f}}{dt} (\alpha \text{ constant}).$$

$$\frac{d}{dt}(u\mathbf{f}) = u\frac{d\mathbf{f}}{dt} + \frac{du}{dt}\mathbf{f} \quad (u = u(t)).$$

$$\frac{d}{dt}(\mathbf{f}\cdot\mathbf{g}) = \left(\mathbf{f}\cdot\frac{d\mathbf{g}}{dt}\right) + \left(\frac{d\mathbf{f}}{dt}\cdot g\right).$$

$$\frac{d}{dt}(\mathbf{f}\times\mathbf{g}) = \left(\mathbf{f}\times\frac{d\mathbf{g}}{dt}\right) + \left(\frac{d\mathbf{f}}{dt}\times g\right).$$

$$\frac{d\mathbf{f}}{dt} = \frac{d\mathbf{f}}{du}\frac{du}{dt} \quad \text{(chain rule)}.$$

13.3 Curves

tangent vector (p. 778) tangent line (p. 779)
unit tangent (p. 781) principal normal (p. 781)
normal line (p. 781) osculating plane (p. 781)
orientation-preserving (orientation-reversing) change of parameter (p. 784)

13.4 Arc Length

$$L(C) = \int_a^b \|\mathbf{r}'(t)\|\,dt \qquad \frac{ds}{dt} = \sqrt{\left(\frac{dx}{dt}\right)^2 + \left(\frac{dy}{dt}\right)^2 + \left(\frac{dz}{dt}\right)^2}$$

13.5 Curvature; Curvilinear Motion

velocity, acceleration (p. 794) speed (p. 794)
angular speed, uniform circular motion (p. 795)
curvature (p. 796) radius of curvature (p. 798)
components of acceleration (p. 801)
Frenet formulas (p. 804)

*13.6 Vector Calculus in Mechanics

conservation law (p. 806)
angular momentum (p. 806) torque (p. 808)
central force (radial force) (p. 808)
effect of a central force on an object with nonzero angular momentum (p. 806)
initial-value problems (p. 809) power, kinetic energy (p. 813)

*13.7 Planetary Motion

Kepler's three laws of planetary motion (p. 814)

■ 14.1 ELEMENTARY EXAMPLES

First a remark on notation. Points $P(x, y)$ of the xy-plane will be written (x, y) and points $P(x, y, z)$ of three-space will be written (x, y, z).

Let D be a nonempty subset of the xy-plane. A rule f that assigns a real number $f(x, y)$ to each point (x, y) in D is called a *real-valued function of two variables*. The set D is called the *domain of* f and the set of all values $f(x, y)$ is called the *range of* f.

Example 1 Take D as the entire xy-plane and to each point (x, y) assign the number

$$f(x, y) = xy. \quad \square$$

Example 2 Take D as the set of all points (x, y) with $y \neq 0$. To each such point assign the number

$$f(x, y) = \tan^{-1}\left(\frac{x}{y}\right). \quad \square$$

Example 3 Take D as the *open unit disc*: $D = \{(x, y) : x^2 + y^2 < 1\}$.
This set consists of all points inside the *unit circle* $x^2 + y^2 = 1$; the circle itself is omitted.† To each point (x, y) in D assign the number

$$f(x, y) = \frac{1}{\sqrt{1 - (x^2 + y^2)}}. \quad \square$$

Now let D be a nonempty subset of three-space. A rule f that assigns a real number $f(x, y, z)$ to each point (x, y, z) in D is called a *real-valued function of three variables*. The set D is called the *domain of* f and the set of all values $f(x, y, z)$ is called the *range of* f.

† The *closed unit disc* $D = \{(x, y) : x^2 + y^2 \leq 1\}$ consists of all points on or inside the unit circle; the circle is included.

Example 4 Take D as all of three-space and to each point (x, y, z) assign the number

$$f(x, y, z) = xyz. \quad \square$$

Example 5 Take D as the set of all points (x, y, z) with $z \neq x + y$. (Thus D consists of all points not on the plane $x + y - z = 0$.) To each point of D assign the number

$$f(x, y, z) = \cos\left(\frac{1}{x + y - z}\right). \quad \square$$

Example 6 Take D as the *open unit ball*: $D = \{(x, y, z) : x^2 + y^2 + z^2 < 1\}$. This set consists of all points inside the *unit sphere* $x^2 + y^2 + z^2 = 1$; the sphere itself is omitted. † To each point (x, y, z) in D assign the number

$$f(x, y, z) = \frac{1}{\sqrt{1 - (x^2 + y^2 + z^2)}}. \quad \square$$

Functions of several variables arise naturally in very elementary settings.

$f(x, y) = \sqrt{x^2 + y^2}$ gives the distance between (x, y) and the origin;

$f(x, y) = xy$ gives the area of a rectangle of dimensions x, y; and

$f(x, y) = 2(x + y)$ gives the perimeter.

$f(x, y, z) = \sqrt{x^2 + y^2 + z^2}$ gives the distance between (x, y, z) and the origin;

$f(x, y, z) = xyz$ gives the volume of a rectangular solid of dimensions x, y, z; and

$f(x, y, z) = 2(xy + xz + yz)$ gives the total surface area.

In general, a *real-valued function of n variables* is a rule f that assigns a real number $f(x_1, x_2, \ldots x_n)$ to each ordered n-tuple $(x_1, x_2, \ldots, x_n)$ in a subset D of n-dimensional space. For example, the function

$$f(x_1, x_2, \ldots, x_n) = \sqrt{x_1^2 + x_2^2 + \cdots + x_n^2}$$

gives the distance between the point $(x_1, x_2, \ldots, x_n)$ and the origin in n-dimensional space. In our study of functions of several variables we will restrict our attention to functions of two and three variables because, as you will see, graphs and other visualizations are possible.

Functions of several variables arise in many problems in science, engineering, economics, and so on. Indeed, the mathematical models for "real" problems are much more likely to involve functions of several variables than functions of a single variable; large-scale problems often involve functions of hundreds or even thousands of variables. Here are some simple examples.

An initial investment A_o compounded continuously at interest rate r grows over time t to have value.

$$A(t, r) = A_0 e^{rt} \qquad \text{(Section 7.6)}$$

Thus the principal A is a function of two variables r and t.

The magnitude of the gravitational force exerted by a body of mass M situated at the origin on a body of mass m at the point (x, y, z) is given by the function

$$F(x, y, z) = \frac{GmM}{x^2 + y^2 + z^2}. \qquad \text{(Section 13.6)}$$

† The *closed unit ball* $D = \{(x, y, z) : x^2 + y^2 + z^2 \leq 1\}$ consists of all points on or inside the unit sphere; the sphere is included.

The *ideal gas law* states that the pressure P of a gas depends on the volume V and on the temperature T according to the equation

$$P = \frac{cT}{V}$$

where c is a constant.

If a metal sphere is being heated by a flame, then the temperature T at a point $P(x, y, z)$ on the sphere is a function f of four variables: the position variables x, y, z and the time t,

$$T = f(x, y, z, t).$$

If the domain of a function of several variables is not explicitly given, it is to be understood that the domain is the maximal set of points for which the definition generates a real number. Thus, in the case of

$$f(x, y) = \frac{1}{x - y},$$

the domain is understood to be all points (x, y) with $y \neq x$, that is, all points of the plane not on the line $y = x$. In the case of

$$g(x,y,z) = \sin^{-1}(x + y + z),$$

the domain is understood to be all points (x, y, z) with $-1 \leq x + y + z \leq 1$. This set is the slab bounded by the parallel planes

$$x + y + z = -1 \qquad \text{and} \qquad x + y + z = 1.$$

To say that a function is *bounded* is to say that its range is bounded. Since the function

$$f(x,y) = \frac{1}{x - y}$$

takes on all values other than 0, its range is $(-\infty, 0) \cup (0, \infty)$. The function is unbounded. In the case of

$$g(x, y, z) = \sin^{-1}(x + y + z),$$

the range is the closed interval $[-\frac{1}{2}\pi, \frac{1}{2}\pi]$. This function is bounded, below by $-\frac{1}{2}\pi$ and above by $\frac{1}{2}\pi$.

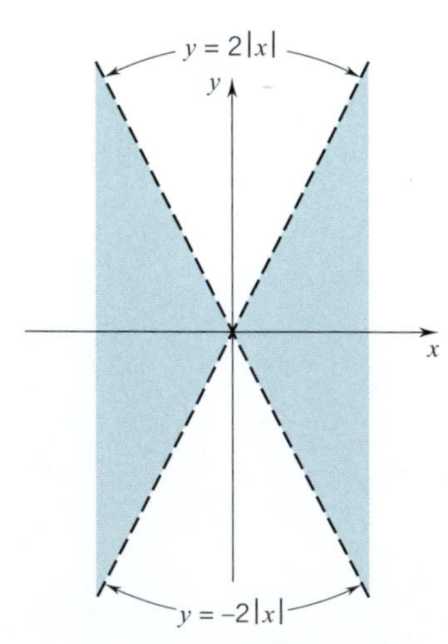

Figure 14.1.1

Example 7 Find the domain and range of the function $f(x,y) = \dfrac{1}{\sqrt{4x^2 - y^2}}$.

SOLUTION A point (x, y) is in the domain of f iff $4x^2 - y^2 > 0$. This occurs iff

$$y^2 < 4x^2$$

and thus iff

$$-2|x| < y < 2|x|.$$

The domain of f is the shaded region shown in Figure 14.1.1. It consists of all the points of the xy-plane that lie between the graph of $y = -2|x|$ and the graph of $y = 2|x|$. On this set $\sqrt{4x^2 - y^2}$ takes on all positive values, and so does its reciprocal $f(x,y)$. The range of f is $(0, \infty)$. ❑

EXERCISES 14.1

Find the domain and range of the function.

1. $f(x, y) = \sqrt{xy}$.

2. $f(x, y) = \sqrt{1 - xy}$.

3. $f(x, y) = \dfrac{1}{x + y}$.

4. $f(x, y) = \dfrac{1}{x^2 + y^2}$.

5. $f(x, y) = \dfrac{e^x - e^y}{e^x + e^y}$.

6. $f(x, y) = \dfrac{x^2}{x^2 + y^2}$.

7. $f(x, y) = \ln(xy)$.

8. $f(x, y) = \ln(1 - xy)$.

9. $f(x, y) = \dfrac{1}{\sqrt{y - x^2}}$.

10. $f(x,y) = \dfrac{\sqrt{9 - x^2}}{1 + \sqrt{1 - y^2}}$.

11. $f(x, y) = \sqrt{9 - x^2} - \sqrt{4 - y^2}$.

12. $f(x, y, z) = \cos x + \cos y + \cos z$.

13. $f(x, y, z) = \dfrac{x + y + z}{|x + y + z|}$.

14. $f(x, y, z) = \dfrac{z^2}{x^2 - y^2}$.

15. $f(x, y, z) = -\dfrac{z^2}{\sqrt{x^2 - y^2}}$.

16. $f(x, y, z) = \dfrac{z}{x - y}$.

17. $f(x, y) = \dfrac{2}{\sqrt{9 - (x^2 + y^2)}}$.

18. $f(x, y, z) = \ln(|x + 2y + 3z| + 1)$.

19. $f(x, y, z) = \ln(x + 2y + 3z)$.

20. $f(x, y, z) = e^{\sqrt{4-(x^2+y^2+z^2)}}$.

21. $f(x, y, z) = e^{-(x^2+y^2+z^2)}$.

22. $f(x, y, z) = \dfrac{\sqrt{1 - x^2} + \sqrt{4 - y^2}}{1 + \sqrt{9 - z^2}}$.

23. Let $f(x) = \sqrt{x}$, $g(x, y) = \sqrt{x}$, $h(x, y, z) = \sqrt{x}$. Determine the domain and range of each function and compare the results. Sketch the graphs of f and g.

24. Let $f(x, y) = \cos \pi x \sin \pi y$ and $g(x, y, z) = \cos \pi x \sin \pi y$. Determine the domain and range of each of these functions and compare the results.

C 25. Let $f(x, y) = \sqrt{1 - 4xy}$. Use a graphing utility to draw the graph of $g(x,y) = 1 - 4xy = 0$. Use this graph to help determine the domain of f.

C 26. Repeat Exercise 25 with $f(x,y) = \ln(x^2 + y - 2)$ and $g(x, y) = x^2 + y - 2$.

In Exercises 27–32, form the difference quotients

$$\frac{f(x + h, y) - f(x, y)}{h} \quad \text{and} \quad \frac{f(x, y + h) - f(x, y)}{h}, \quad (h \neq 0).$$

Then, assuming that x and y are fixed, calculate the limit as $h \to 0$. What is the connection between your results and derivatives?

27. $f(x,y) = 2x^2 - y$.

28. $f(x,y) = xy + 2y$.

29. $f(x,y) = 3x - xy + 2y^2$.

30. $f(x,y) = x \sin y$.

31. $f(x,y) = \cos(xy)$.

32. $f(x,y) = x^2 e^y$.

C 33. Use a CAS to find the difference quotients for $f(x,y) = \dfrac{x - 2y}{x + y}$.

(a) $\dfrac{f(x + h, y) - f(x, y)}{h}$.

(b) $\dfrac{f(x, y + h) - f(x, y)}{h}$.

(c) Calculate the limits as $h \to 0$.

C 34. Repeat Exercise 33 for the function $f(x, y) = x^2 - 4xy + y^2$.

35. Express each of the following as a function of two variables.

(a) The volume of a box with a square base of side length x and height y.

(b) The volume of a right circular cylinder whose radius is x and whose height is y.

(c) The area of the parallelogram whose sides are the vectors $2\mathbf{i}$ and $x\,\mathbf{i} + y\,\mathbf{j}$.

36. Express each of the following as a function of three variables.

(a) The surface area of a box with no top whose sides have lengths x, y and z.

(b) The angle between the vectors $\mathbf{i} + \mathbf{j}$ and $x\,\mathbf{i} + y\,\mathbf{j} + z\,\mathbf{k}$.

(c) The volume of the parallelepiped whose sides are the vectors $\mathbf{i}, \mathbf{i} + \mathbf{j}$, and $x\,\mathbf{i} + y\,\mathbf{j} + z\,\mathbf{k}$.

37. A closed box is to have a total surface area of 20 square feet. Express the volume V of the box as a function of the length l and the height h.

38. An open box is to contain a volume of 12 cubic meters. Given that the material for the sides of the box costs \$2 per square meter and the material for the bottom costs \$4 per square meter, express the total cost C of the box as a function of the length l and width w.

39. A petrochemical company is designing a cylindrical tank with hemispherical ends to be used in the transportation of its products. (See the figure.) Express the volume of the tank as a function of its radius r and the length h of the cylindrical portion.

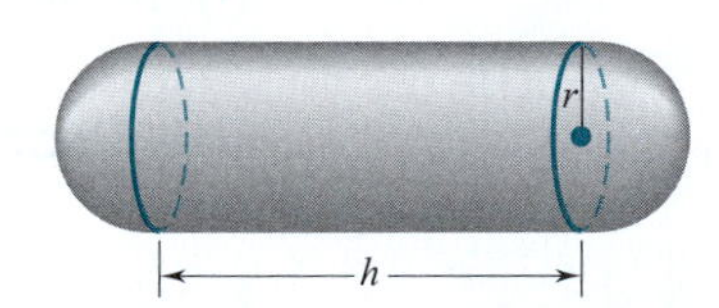

40. A 10-foot section of gutter is to be made from a 12-inch-wide strip of metal by folding up strips of length x on each side so that they make an angle θ with the bottom of the gutter. (See the figure.) Express the area of the trapezoidal cross section as a function of x and θ.

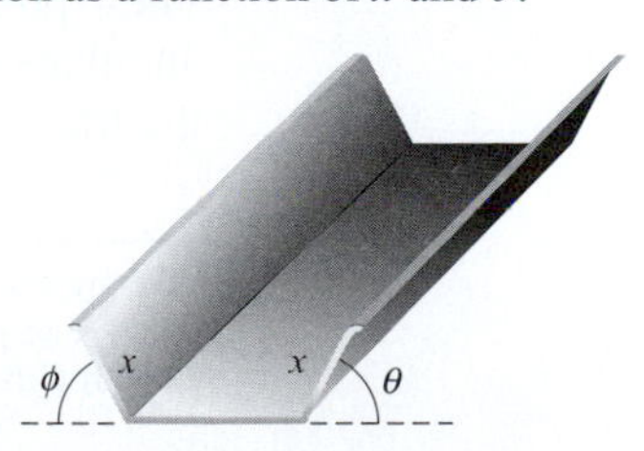

■ 14.2 A BRIEF CATALOG OF THE QUADRIC SURFACES; PROJECTIONS

As you know, the graph of an equation in x and y is typically a curve in the xy-plane. As you will see in this section, and in the sections which follow, the graph of an equation in three variables is, in general, a surface in three-space.

In this section we examine in a systematic manner the surfaces defined by equations of the form

$$Ax^2 + By^2 + Cz^2 + Dxy + Exz + Fyz + Hx + Iy + Jz + K = 0$$

where $A, B, C, \ldots, K$ are constants. Such surfaces are called *quadric surfaces.*

By suitable translations and rotations of the coordinate axes (see Section 9.1 and Appendix A.1), we can simplify such equations and thereby show that the nondegenerate† quadrics fall into nine distinct types:

1. The ellipsoid.
2. The hyperboloid of one sheet.
3. The hyperboloid of two sheets.
4. The elliptic cone.
5. The elliptic paraboloid.
6. The hyperbolic paraboloid.
7. The parabolic cylinder.
8. The elliptic cylinder.
9. The hyperbolic cylinder.

As you go on with calculus, you will encounter these surfaces time and time again. Here we give you a picture of each one, together with its equation in standard form and some information about its special properties. These are some of the things to look for:

(a) The *intercepts* (the points at which the surface intersects the coordinate axes).
(b) The *traces* (the intersections with the coordinate planes).
(c) The *sections* (the intersections with planes in general).
(d) The *center* (some quadrics have a center; some do not).
(e) *Symmetry*.
(f) *Boundedness, unboundedness.*

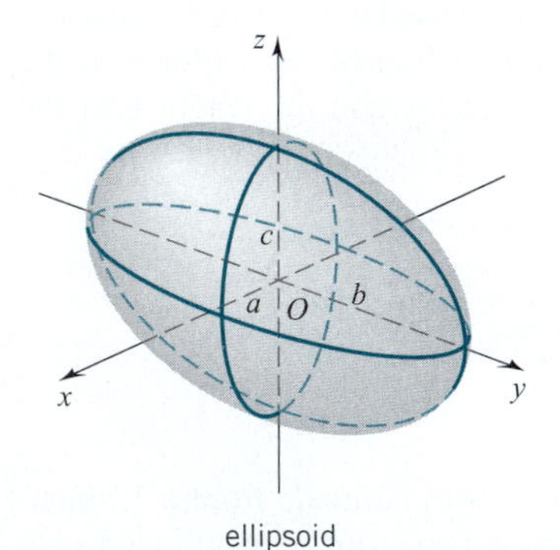

ellipsoid

Figure 14.2.1

The Ellipsoid

$$\frac{x^2}{a^2} + \frac{y^2}{b^2} + \frac{z^2}{c^2} = 1.\dagger\dagger \qquad \text{(Figure 14.2.1)}$$

This ellipsoid is centered at the origin and is symmetric about the three coordinate planes. It intersects the coordinate axes at six points: $(\pm a,\ 0,\ 0), (0,\ \pm b,\ 0), (0,\ 0,\ \pm c)$. These points are called the *vertices*. The surface is bounded, being contained in the rectangular solid: $|x| \le a,\ \ |y| \le b,\ \ |z| \le c$. All three traces are ellipses; thus, for example, the trace in the xy-plane (set $z = 0$) is the ellipse

†We are excluding such degenerate quadrics as $x^2 + y^2 + z^2 + 1 = 0$ and $x^2 + y^2 + z^2 = 0$. The first one has no points and the second consists of only one point, the origin.
†† Throughout this section we take a, b, c as positive constants.

$$\frac{x^2}{a^2} + \frac{y^2}{b^2} = 1.$$

Sections parallel to the coordinate planes are also ellipses; for example, taking $y = y_0$ we have

$$\frac{x^2}{a^2} + \frac{z^2}{c^2} = 1 - \frac{y_0^2}{b^2}.$$

This ellipse is the intersection of the ellipsoid with the plane $y = y_0$. The numbers a, b, c are called the *semiaxes* of the ellipsoid. If two of the semiaxes are equal, then we have an *ellipsoid of revolution.* (If, for example, $a = c$, then all sections parallel to the xz-plane are circles and the surface can be obtained by revolving the trace in the xy-plane about the y-axis.) If all three semiaxes are equal, the surface is a *sphere*.

The Hyperboloid of One Sheet

$$\frac{x^2}{a^2} + \frac{y^2}{b^2} - \frac{z^2}{c^2} = 1. \qquad \text{(Figure 14.2.2)}$$

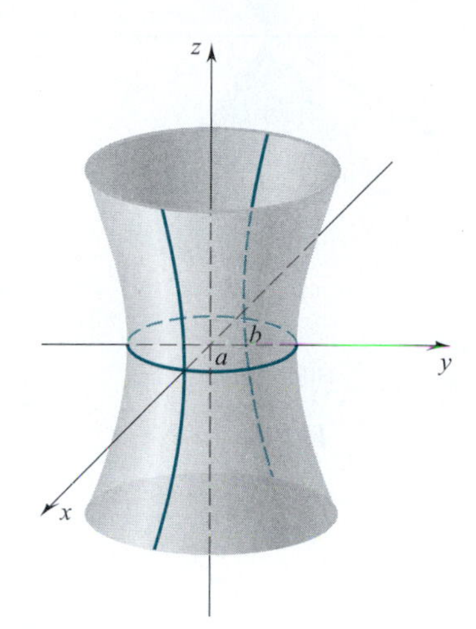

hyperboloid of one sheet

Figure 14.2.2

The surface is unbounded. It is centered at the origin and is symmetric about the three coordinate planes. The surface intersects the coordinate axes at four points: $(\pm a,\ 0,\ 0), (0,\ \pm b,\ 0)$. The trace in the xy-plane (set $z = 0$) is the ellipse

$$\frac{x^2}{a^2} + \frac{y^2}{b^2} = 1.$$

Sections parallel to the xy-plane are ellipses. The trace in the xz-plane (set $y = 0$) is the hyperbola

$$\frac{x^2}{b^2} - \frac{z^2}{c^2} = 1,$$

and the trace in the yz-plane (set $x = 0$) is the hyperbola

$$\frac{y^2}{b^2} - \frac{z^2}{c^2} = 1.$$

Sections parallel to the xz-plane or yz-plane are hyperbolas. If $a = b$, then sections parallel to the xy-plane are circles and we have a *hyperboloid of revolution*.

The Hyperboloid of Two Sheets

$$\frac{x^2}{a^2} + \frac{y^2}{b^2} - \frac{z^2}{c^2} = -1. \qquad \text{(Figure 14.2.3)}$$

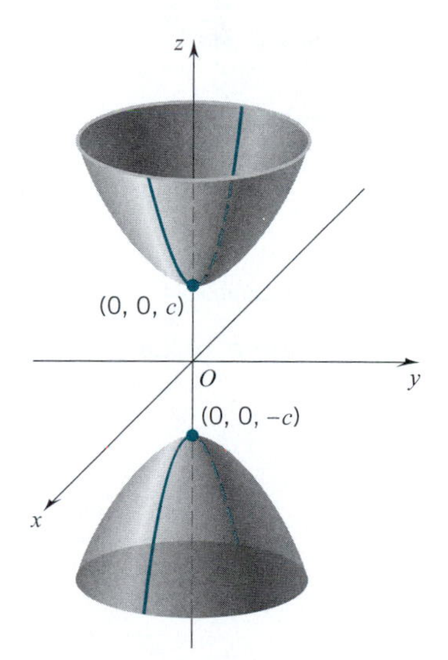

hyperboloid of two sheets

Figure 14.2.3

The surface intersects the coordinate axes only at the two vertics $(0,\ 0,\ \pm c)$. The surface consists of two parts: one for which $z \geq c$, another for which $z \leq -c$. We can see this by rewriting the equation as

$$\frac{x^2}{a^2} + \frac{y^2}{b^2} = \frac{z^2}{c^2} - 1.$$

The equation requires

$$\frac{z^2}{c^2} - 1 \geq 0, \quad z^2 \geq c^2, \quad |z| \geq c.$$

Each of the two parts is unbounded. Sections parallel to the xy-plane are ellipses: setting $z = z_0$ with $|z_0| \geq c$, we have

$$\frac{x^2}{a^2} + \frac{y^2}{b^2} = \frac{z_0^2}{c^2} - 1.$$

If $a = b$, then all sections parallel to the xy-plane are circles and we have a hyperboloid of revolution. Sections parallel to the other coordinate planes are hyperbolas; for example, setting $y = y_0$, we have

$$\frac{z^2}{c^2} - \frac{x^2}{a^2} = 1 + \frac{y_0^2}{b^2}.$$

The entire surface is symmetric about the three coordinate planes and is centered at the origin.

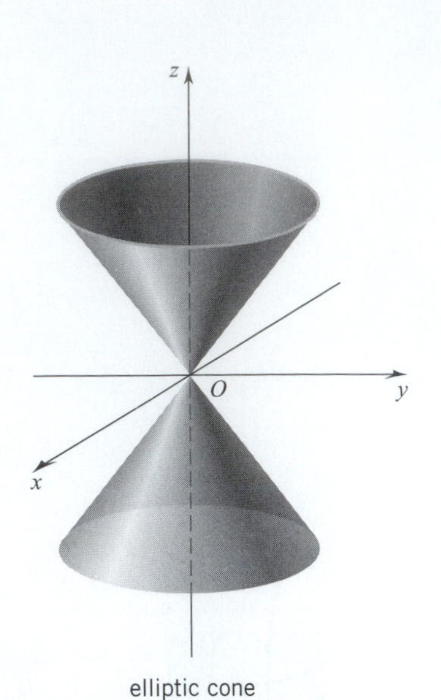

elliptic cone

Figure 14.2.4

The Elliptic Cone

$$\frac{x^2}{a^2} + \frac{y^2}{b^2} = z^2. \qquad \text{(Figure 14.2.4)}$$

The surface intersects the coordinate axes only at the origin. The surface is unbounded. Once again there is symmetry about the three coordinate planes. The trace in the xz-plane is a pair of intersecting lines: $z = \pm x/a$. The trace in the yz-plane is also a pair of intersecting lines: $z = \pm y/b$. The trace in the xy-plane is just the origin. Sections parallel to the xy-plane are ellipses. If $a = b$, these sections are circles and we have a surface of revolution, what is commonly called a *double circular cone* or simply a *cone*. The upper and lower portions of the cone are called *nappes*.

We come now to the *paraboloids*. The equations in standard form will involve x^2 and y^2, but then z instead of z^2.

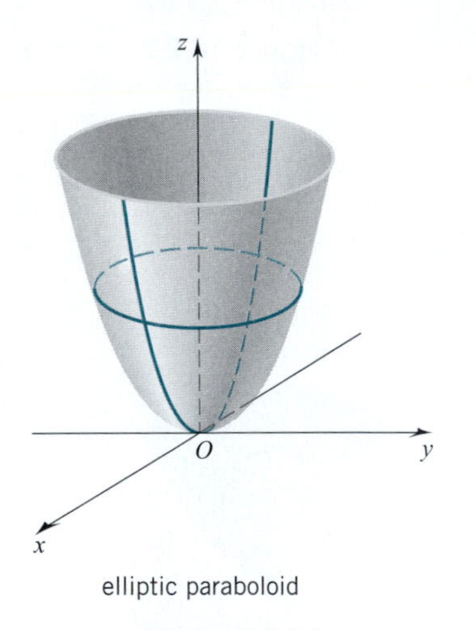

elliptic paraboloid

Figure 14.2.5

The Elliptic Paraboloid

$$\frac{x^2}{a^2} + \frac{y^2}{b^2} = z. \qquad \text{(Figure 14.2.5)}$$

The surface does not extend below the xy-plane; it is unbounded above. The origin is called the *vertex*. Sections parallel to the xy-plane are ellipses: sections parallel to the other coordinate planes are parabolas. Hence the term "elliptic paraboloid". The surface is symmetric about the xz-plane and about the yz-plane. It is also symmetric about the z-axis. If $a = b$, then the surface is a *paraboloid of revolution*.

The Hyperbolic Paraboloid

$$\frac{x^2}{a^2} - \frac{y^2}{b^2} = z. \qquad \text{(Figure 14.2.6)}$$

Here there is symmetry about the xz-plane and yz plane. Sections parallel to the xy-plane are hyperbolas; sections parallel to the other coordinate planes are parabolas. Hence the term "hyperbolic paraboloid." The origin is a minimum point for the trace in the xz-plane, but a maximum point for the trace in the yz-plane. The origin is called a *minimax* or *saddle point* of the surface. NOTE: The orientation of the coordinate axes was chosen to enhance the view of the surface.

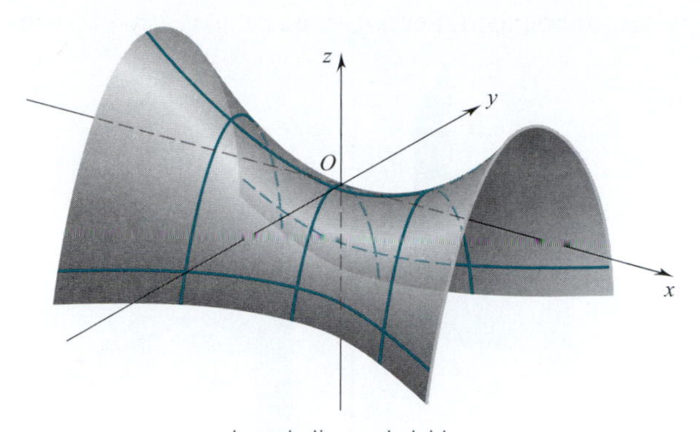

hyperbolic paraboloid

Figure 14.2.6

The rest of the quadric surfaces are *cylinders*. The term deserves definition. Take any plane curve C. All the lines through C that are perpendicular to the plane of C form a surface. Such a surface is called a *cylinder,* the cylinder with *base curve C*. The perpendicular lines are known as the *generators* of the cylinder.

If the base curve lies in the xy-plane (or in a plane parallel to the xy-plane), then the generators of the cylinder are parallel to the z-axis. In such a case the equation of the cylinder involves only x and y. The z-coordinate is left unrestricted; it can take on all values.

There are three basic types of quadric cylinders. We give you their equations in standard form: base curve in the xy-plane, generators parallel to the z-axis.

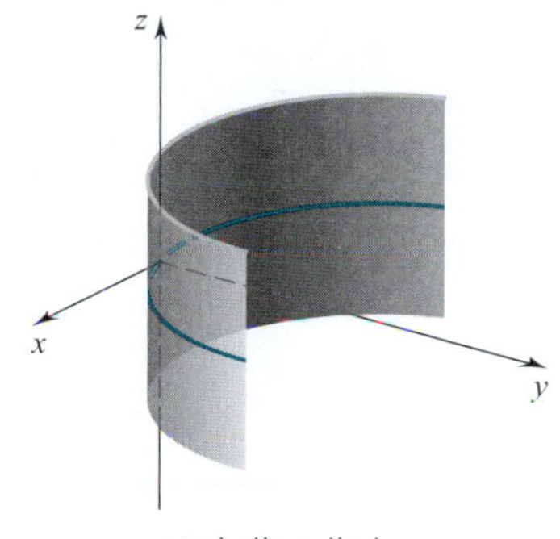

parabolic cylinder

Figure 14.2.7

The Parabolic Cylinder

$$x^2 = 4cy. \qquad \text{(Figure 14.2.7)}$$

This surface is formed by all lines that pass through the parabola $x^2 = 4cy$ and are perpendicular to the xy-plane.

The Elliptic Cylinder

$$\frac{x^2}{a^2} + \frac{y^2}{b^2} = 1. \qquad \text{(Figure 14.2.8)}$$

The surface is formed by all lines that pass through the ellipse

$$\frac{x^2}{a^2} + \frac{y^2}{b^2} = 1$$

and are perpendicular to the xy-plane. If $a = b$, we have the common *right circular cylinder*.

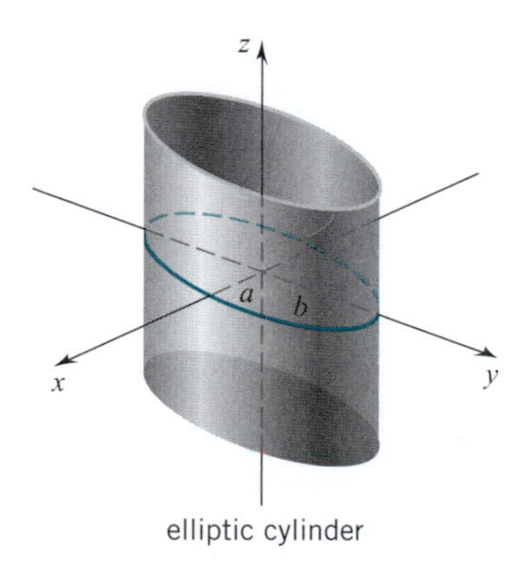

elliptic cylinder

Figure 14.2.8

The Hyperbolic Cylinder

$$\frac{x^2}{a^2} - \frac{y^2}{b^2} = 1. \qquad \text{(Figure 14.2.9)}$$

The surface has two parts, each generated by a branch of the hyperbola

$$\frac{x^2}{a^2} - \frac{y^2}{b^2} = 1.$$

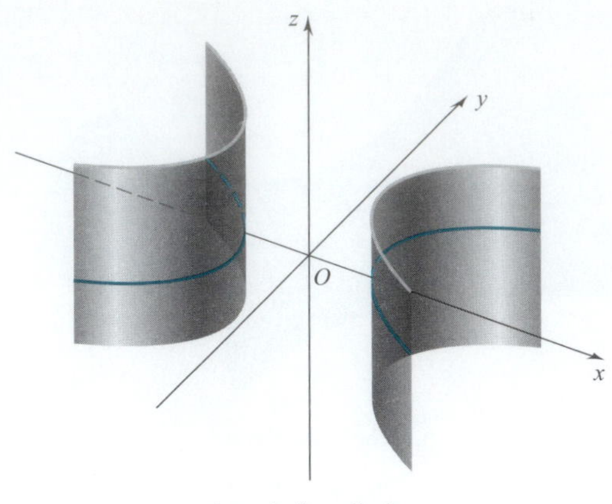

Figure 14.2.9

Projections

Suppose that $S_1 : z = f(x, y)$ and $S_2 : z = g(x, y)$ are surfaces in three-space that intersect in a space curve C. (See Figure 14.2.10.)

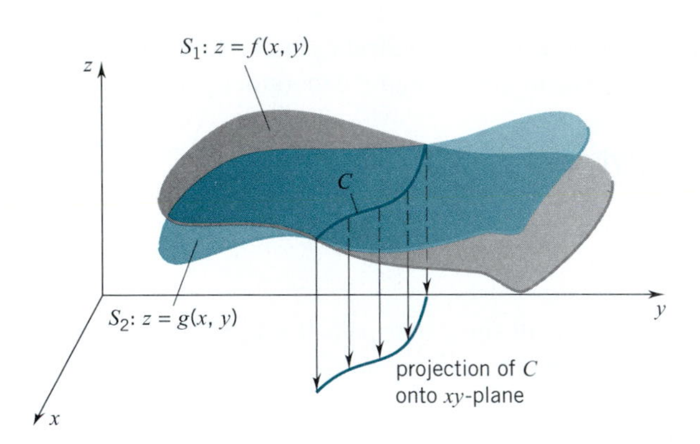

Figure 14.2.10

The curve C is the set of all points (x, y, z) with

$$z = f(x, y) \qquad \text{and} \qquad z = g(x, y).$$

The set of all points (x, y, z) with

$$f(x, y) = g(x, y) \qquad \text{(here z is unrestricted)}$$

is the vertical cylinder that passes through C.

The set of all points $(x, y, 0)$ with

$$f(x, y) = g(x, y) \qquad \text{(here } z = 0\text{)}$$

is called the *projection of C onto the xy-plane*. In Figure 14.2.10 it appears as the curve in the xy-plane that lies directly below C.

Example 1 The paraboloid of revolution $z = x^2 + y^2$ and the plane

$$z = 2y + 3$$

intersect in a curve C. See Figure 14.2.11. The projection of this curve onto the xy-plane is the set of all points $(x, y, 0)$ with

$$x^2 + y^2 = 2y + 3.$$

This equation can be written

$$x^2 + (y - 1)^2 = 4.$$

The projection of C onto the xy-plane is the circle of radius 2 centered at $(0, 1, 0)$. □

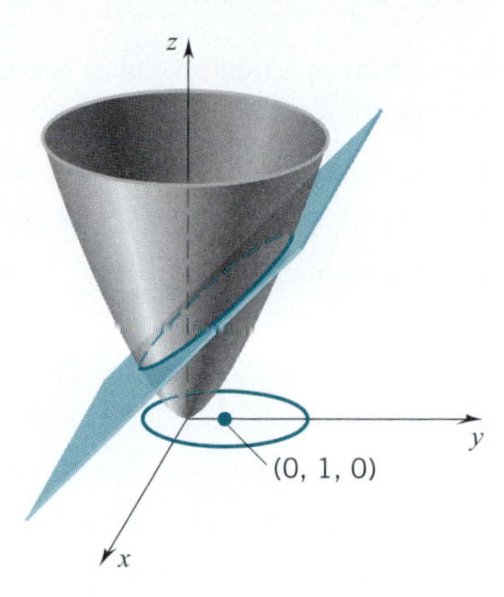

Figure 14.2.11

EXERCISES 14.2

Identify the surface.

1. $x^2 + 4y^2 - 16z^2 = 0.$
2. $x^2 + 4y^2 + 16z^2 - 12 = 0.$
3. $x - 4y^2 = 0.$
4. $x^2 - 4y^2 - 2z = 0.$
5. $5x^2 + 2y^2 - 6z^2 - 10 = 0.$
6. $2x^2 + 4y^2 - 1 = 0.$
7. $x^2 + y^2 + z^2 - 4 = 0.$
8. $5x^2 + 2y^2 - 6z^2 + 10 = 0.$
9. $x^2 + 2y^2 - 4z = 0.$
10. $2x^2 - 3y^2 - 6 = 0.$
11. $x - y^2 + 2z^2 = 0.$
12. $x - y^2 - 6z^2 = 0.$

Sketch the cylinder.

13. $25y^2 + 4z^2 - 100 = 0.$
14. $25x^2 + 4y^2 - 100 = 0.$
15. $y^2 - z = 0.$
16. $x^2 - y + 1 = 0.$
17. $y^2 + z = 0.$
18. $25x^2 - 9y^2 - 225 = 0.$
19. $x^2 + y^2 = 9.$
20. $\dfrac{x^2}{4} + \dfrac{y^2}{9} = 1.$
21. $y^2 - 4x^2 = 4.$
22. $z = x^2.$
23. $y = x^2 + 1.$
24. $(x - 1)^2 + (y - 1)^2 = 1.$

Identify the surface and find the traces. Then sketch the surface.

25. $9x^2 + 4y^2 - 36z = 0.$
26. $9x^2 + 4y^2 + 36z^2 - 36 = 0.$
27. $9x^2 + 4y^2 - 36z^2 = 0.$
28. $9x^2 + 4y^2 - 36z^2 - 36 = 0.$
29. $9x^2 + 4y^2 - 36z^2 + 36 = 0.$
30. $9x^2 - 4y^2 - 36z = 0.$
31. $9x^2 - 4y^2 - 36z^2 = 36.$
32. $9x^2 + 4z^2 - 36y^2 - 36 = 0.$
33. $4x^2 + 9z^2 - 36y = 0.$
34. $9x^2 + 4z^2 - 36y^2 = 0.$
35. $9y^2 - 4x^2 - 36z^2 - 36 = 0.$
36. $9y^2 + 4z^2 - 36x = 0.$
37. $x^2 + y^2 - 4z = 0.$
38. $36x^2 + 9y^2 + 4z^2 - 36 = 0.$
39. Identify all possibilities for the surface

$$z = Ax^2 + By^2$$

taking (a) $AB > 0$. (b) $AB < 0$. (c) $AB = 0$.

40. Find the planes of symmetry for the cylinder $x - 4y^2 = 0$.
41. Write an equation for the surface obtained by revolving the parabola $4z - y^2 = 0$ about the z-axis.
42. The hyperbola $c^2y^2 - b^2z^2 - b^2c^2 = 0$ is revolved about the z-axis. Find an equation for the resulting surface.
43. (a) The equation

$$\sqrt{x^2 + y^2} = kz \qquad \text{with } k > 0$$

represents the upper nappe of a cone, with vertex at the origin and the positive z-axis as the axis of symmetry. Describe the section in the plane $z = z_0$, $z_0 > 0$.

(b) Let S be one nappe of a cone, with vertex at the origin. Write an equation for S given that

(i) the negative z-axis is the axis of symmetry and the section in the plane $z = -2$ is a circle of radius 6,

(ii) the positive y-axis is the axis of symmetry and the section in the plane $y = 3$ is a circle of radius 1.

44. Form the elliptic paraboloid

$$x^2 + \frac{y^2}{b^2} = z.$$

(a) Describe the section in the plane $z = 1$.

(b) What happens to this section as b tends to infinity?

(c) What happens to the paraboloid as b tends to infinity?

The surfaces intersect in a space curve C. Determine the projection of C onto the xy-plane.

45. The planes $x + 2y + 3z = 6$ and $x + y - 2z = 6$.

46. The planes $x - 2y + z = 4$ and $3x + y - 2z = 1$.

47. The sphere $x^2 + y^2 + (z-1)^2 = \frac{3}{2}$ and the hyperboloid $x^2 + y^2 - z^2 = 1$.

48. The sphere $x^2 + y^2 + (z-2)^2 = 2$ and the cone $x^2 + y^2 = z^2$.

49. The paraboloids $x^2 + y^2 + z = 4$ and $x^2 + 3y^2 = z$.

50. The cylinder $y^2 + z - 4 = 0$ and the paraboloid $x^2 + 3y^2 = z$.

51. The cone $x^2 + y^2 = z^2$ and the plane $y + z = 2$.

52. The cone $x^2 + y^2 = z^2$ and the plane $y + 2z = 2$.

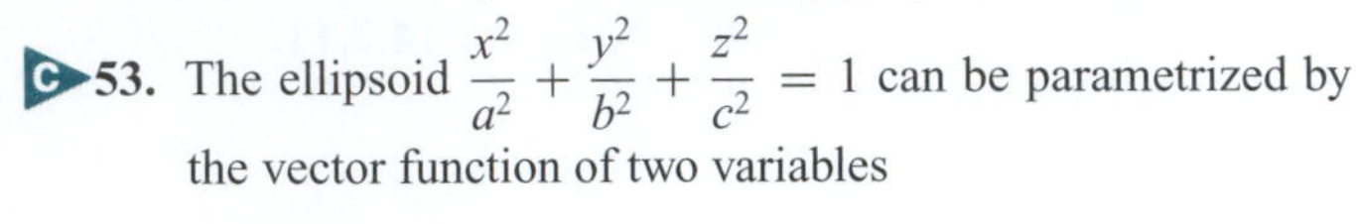

53. The ellipsoid $\dfrac{x^2}{a^2} + \dfrac{y^2}{b^2} + \dfrac{z^2}{c^2} = 1$ can be parametrized by the vector function of two variables

$$\mathbf{r}(u, v) = a\cos u \cos v\,\mathbf{i} + b\cos u \sin v\,\mathbf{j} + c\sin u\,\mathbf{k}.$$

(a) Verify that $\mathbf{r}$ parametrizes an ellipsoid.

(b) Use a graphing utility to draw the ellipsoid with $a = 3$, $b = 4$, $c = 2$.

(c) Experiment with other values of a, b, c to see how the ellipsoid changes shape. How would you choose a, b, c to obtain a sphere?

54. The hyperboloid of one sheet $\dfrac{x^2}{a^2} + \dfrac{y^2}{b^2} - \dfrac{z^2}{c^2} = 1$ can be parametrized by the vector function of two variables

$$\mathbf{r}(u, v) = a\sec u \cos v\,\mathbf{i} + b\sec u \sin v\,\mathbf{j} + c\tan u\,\mathbf{k}.$$

(a) Verify that $\mathbf{r}$ parametrizes a hyperboloid.

(b) Use a graphing utility to draw the hyperboloid with $a = 2, b = 3, c = 4$.

(c) Experiment with other values of a, b, c to see how the hyperboloid changes shape.

55. The elliptic cone $\dfrac{x^2}{a^2} + \dfrac{y^2}{b^2} = \dfrac{z^2}{c^2}$ can be parametrized by the vector function of two variables

$$\mathbf{r}(u, v) = av\cos u\,\mathbf{i} + bv\sin u\,\mathbf{j} + cv\,\mathbf{k}$$

(a) Verify that $\mathbf{r}$ parametrizes an elliptic cone.

(b) Use a graphing utility to draw the elliptic cone with $a = 1$, $b = 2$, $c = 3$.

(c) Experiment with other values of a, b, c to see how the cone changes shape. In particular, what effect does c have on the cone?

■ 14.3 GRAPHS; LEVEL CURVES AND LEVEL SURFACES

We begin with a function f of two variables defined on a subset D of the xy-plane. By the *graph of f* we mean the graph of the equation

$$z = f(x, y) \qquad (x, y) \in D.$$

Figure 14.3.1

Example 1 In the case of $f(x, y) = x^2 + y^2$, the domain is the entire plane. The graph of f is a paraboloid of revolution:

$$z = x^2 + y^2.$$

This surface can be generated by revolving the parabola

$$z = x^2 \qquad \text{(in the xz-plane)}$$

about the z-axis. See Figure 14.3.1. □

Example 2 Let a, b, and c be positive constants. The domain of the function

$$g(x, y) = c - ax - by$$

is also the entire xy-plane. The graph of g is the plane

$$z = c - ax - by \qquad \text{(Figure 14.3.2)}$$

with intercepts $x = c/a$, $y = c/b$, $z = c$. □

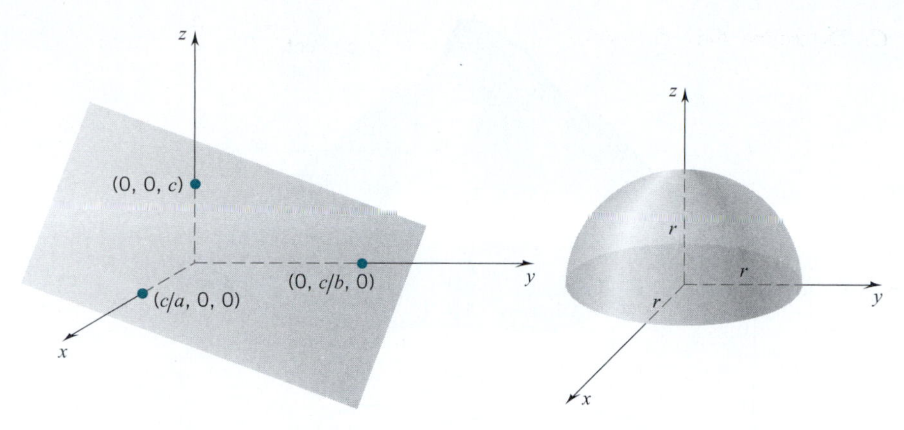

Figure 14.3.2

Figure 14.3.3

Example 3 The function $f(x,y) = \sqrt{r^2 - (x^2 + y^2)}$, $r > 0$ is defined only on the closed disc $x^2 + y^2 \leq r^2$. The graph is the surface

$$z = \sqrt{r^2 - (x^2 + y^2)}. \qquad \text{(Figure 14.3.3)}$$

This is the upper half of the sphere

$$x^2 + y^2 + z^2 = r^2. \quad \square$$

Example 4 The function $f(x,y) = xy$ is simple enough, but its graph, the surface $z = xy$, is quite difficult to draw. It is a "saddle-shaped" surface, a hyperbolic paraboloid: rotate the x and y axes by $\frac{1}{4}\pi$ radians and the equation takes the form

$$\frac{X^2}{2} - \frac{Y^2}{2} = z.\dagger$$

Try to visualize the surface in Figure 14.2.6 rotated $\frac{1}{4}\pi$ radians in the clockwise direction. $\square$

Level Curves

In practice, the graph of a function of two variables is difficult to visualize and difficult to draw. Moreover, if drawn, the drawing is often difficult to interpret. Computer-generated graphics can be very useful in resolving these matters. We will provide some illustration of this later in the section.

Here we discuss an approach which we take from the map maker. In mapping mountainous terrain, it is common practice to sketch curves joining points of constant elevation. The collection of such curves, called a topographic map, gives a good idea of the altitude variations in a region and suggests the shape of the mountains and valleys. (See Figure 14.3.5)

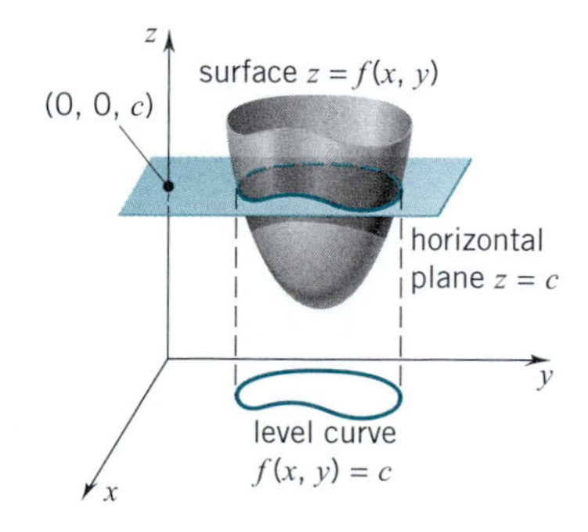

Figure 14.3.4

We can apply this technique to functions of two variables. Suppose that f is a nonconstant function defined on some portion of the xy-plane. If c is a value in the range of f, then we can sketch the curve $f(x, y) = c$. Such a curve is called a *level curve* for f. It can be obtained by intersecting the graph of f with the horizontal plane $z = c$ and then projecting that intersection onto the xy-plane. (See Figure 14.3.4)

The level curve $f(x,y) = c$ lies entirely in the domain of f, and on this curve f is constantly C.

† To see this, set $\alpha = \frac{1}{4}\pi$ in Formula A.1.2, Appendix A-1.

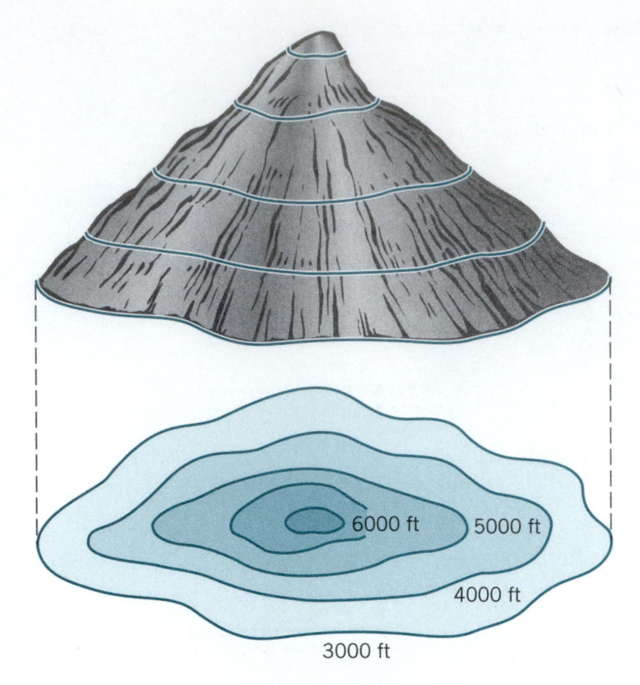

Figure 14.3.5

A collection of level curves, properly drawn and labeled, can lead to a good understanding of the overall behavior of a function.

Example 5 We begin with function $f(x, y) = x^2 + y^2$ (see Figure 14.3.1). The level curves are circles centered at the origin:

$$x^2 + y^2 = c, \qquad c \geq 0. \qquad \text{(Figure 14.3.6)}$$

The function has the value c on the circle of radius $\sqrt{c}$ centered at the origin. At the origin, the function has the value 0. ❑

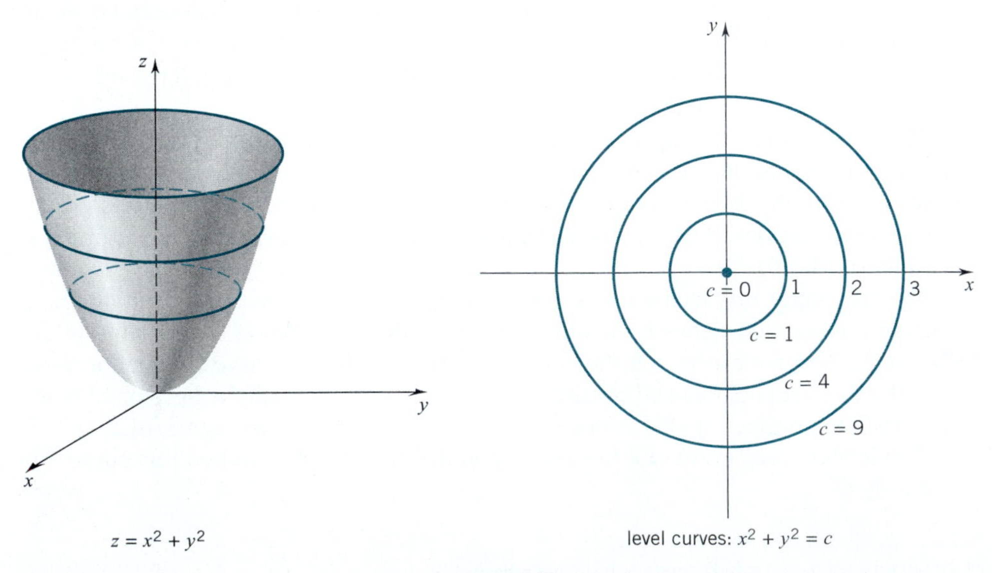

Figure 14.3.6

Example 6 The graph of the function $g(x, y) = 4 - x - y$ is a plane. The level curves are parallel lines of the form $4 - x - y = c$.

The surface and the level curves are indicated in Figure 14.3.7. ❑

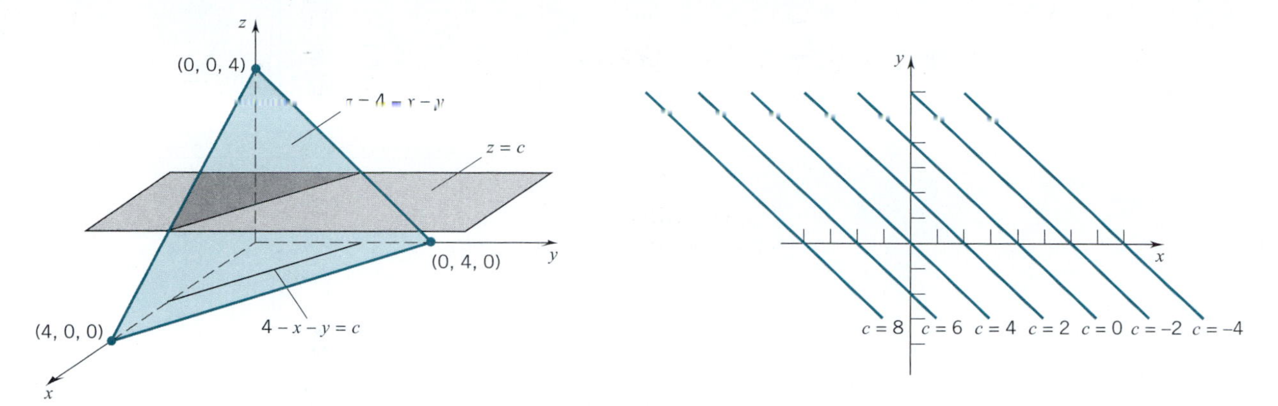

Figure 14.3.7

Example 7 Now we consider the function

$$h(x, y) = \begin{cases} \sqrt{x^2 + y^2}, & x \geq 0 \\ |y|, & x < 0. \end{cases}$$

For $x \geq 0, h(x, y)$ is the distance from (x, y) to the origin. For $x < 0, h(x, y)$ is the distance from (x, y) to the x-axis. The level curves are pictured in Figure 14.3.8. The 0-level curve is the nonpositive x-axis. The other level curves are horseshoe-shaped: pairs of horizontal rays capped on the right by semicircles. ❑

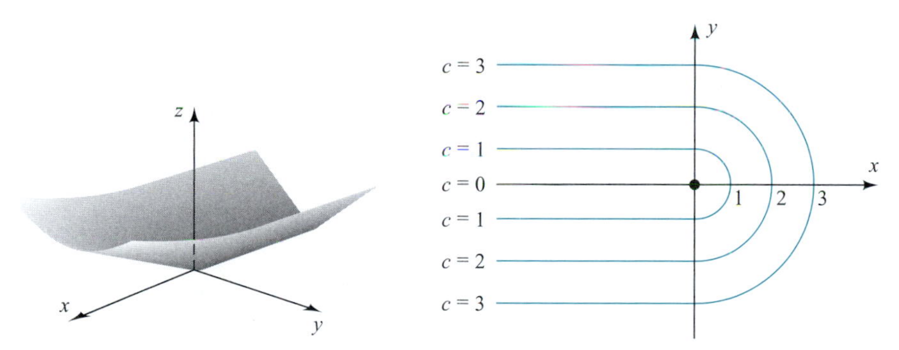

Figure 14.3.8

Example 8 Let's return to the function $f(x, y) = xy$. Earlier we noted that the graph is a "saddled-shaped" surface. You can visualize the surface from the few level curves sketched in Figure 14.3.9. The 0-level curve, $xy = 0$, consists of the two coordinate axes. The other level curves, $xy = c$ with $c \neq 0$, are hyperbolas.

Computer-Generated Graphs

The preceding examples illustrate how difficult it is to sketch an accurate graph of a function of two variables. But powerful help is at hand. Three-dimensional graphing programs for modern computers make it possible to visualize even quite complicated surfaces. These programs allow the user to view a surface from different perspectives, and they show level curves and the sections in various planes. Examples of computer-generated graphs are shown in Figure 14.3.10 and in the Exercises.

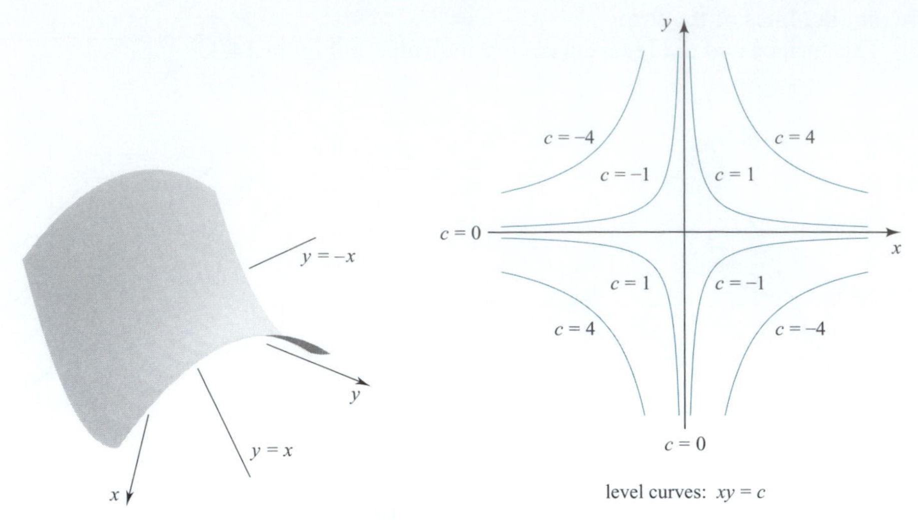

Figure 14.3.9

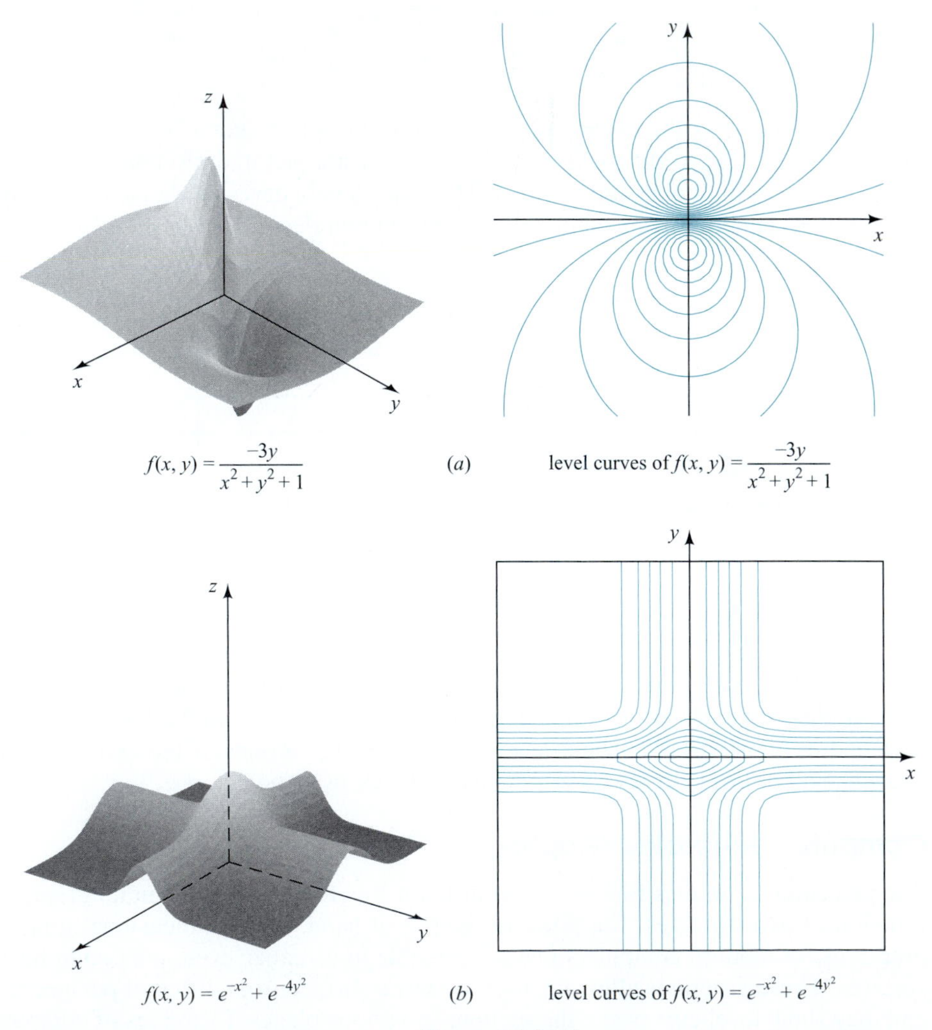

Figure 14.3.10

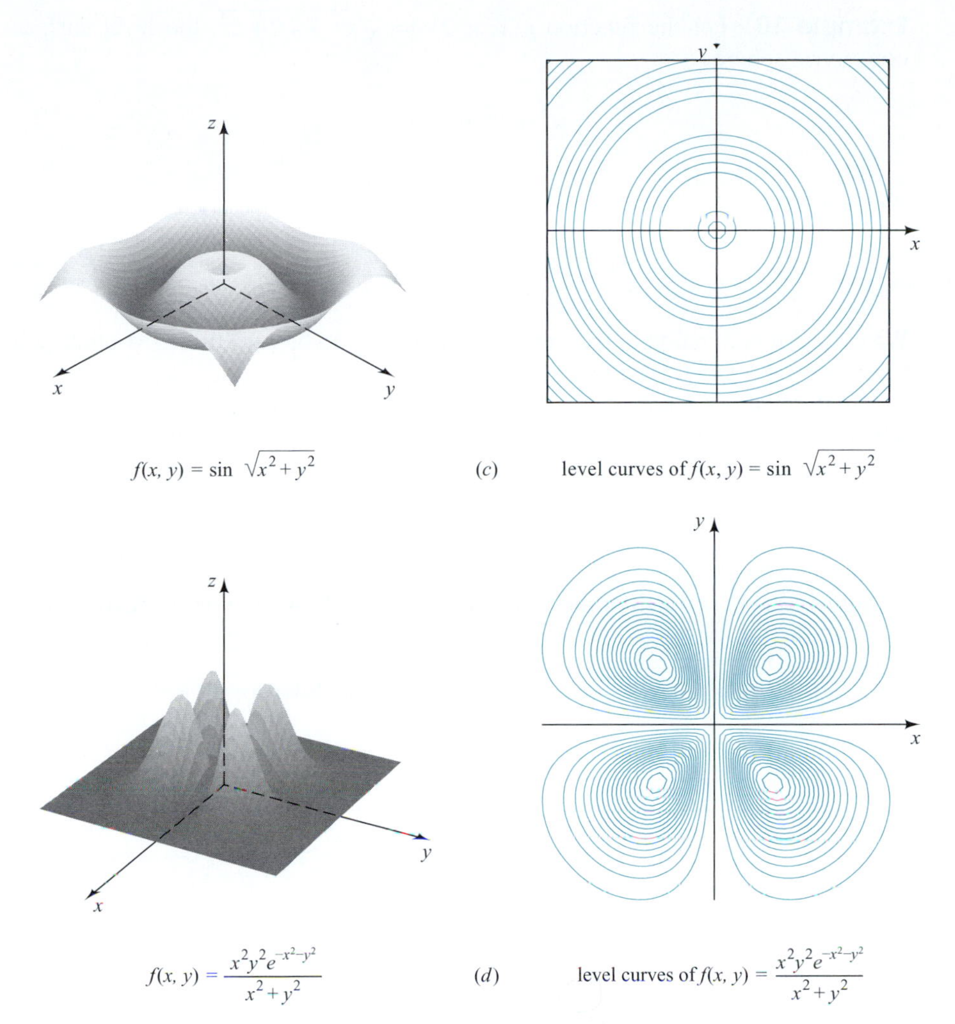

$f(x, y) = \sin\sqrt{x^2 + y^2}$ (c) level curves of $f(x, y) = \sin\sqrt{x^2 + y^2}$

$f(x, y) = \dfrac{x^2y^2e^{-x^2-y^2}}{x^2 + y^2}$ (d) level curves of $f(x, y) = \dfrac{x^2y^2e^{-x^2-y^2}}{x^2 + y^2}$

Figure 14.3.10 ***Continued***

Level Surfaces

While drawing graphs for functions of two variables is quite difficult, drawing graphs for functions of three variables is actually impossible. To draw such figures we would need four dimensions at our disposal; the domain itself is a portion of three-space.

One can try to visualize the behavior of a function of three variables, $w = f(x, y, z)$, by examining the *level surfaces* of f. These are the subsets of the domain of f with equations of the form

$$f(x, y, z) = c$$

where c is a value in the range of f.

Level surfaces are usually difficult to draw. Nevertheless, a knowledge of what they are can be helpful. Here we restrict ourselves to a few simple examples.

Example 9 For the function $f(x, y, z) = Ax + By + Cz$, the level surfaces are parallel planes

$$Ax + By + Cz = c. \quad \square$$

Example 10 For the function $g(x, y, z) = \sqrt{x^2 + y^2 + z^2}$, the level surfaces are concentric spheres

$$x^2 + y^2 + z^2 = c^2. \quad \square$$

Example 11 As our final example we take the function

$$f(x, y, z) = \frac{|z|}{x^2 + y^2}.$$

We extend this function to the origin by defining it to be zero there. At other points of the z-axis we leave f undefined.

In the first place note that f takes on only nonnegative values. Since f is zero only when $z = 0$, the 0-level surface is the xy-plane. To find the other level surfaces, we take $c > 0$ and set $f(x, y, z) = c$. This gives

$$\frac{|z|}{x^2 + y^2} = c \quad \text{and thus} \quad |z| = c(x^2 + y^2)$$

(Figure 14.3.11). Each of these surfaces is a double-paraboloid of revolution.† $\square$

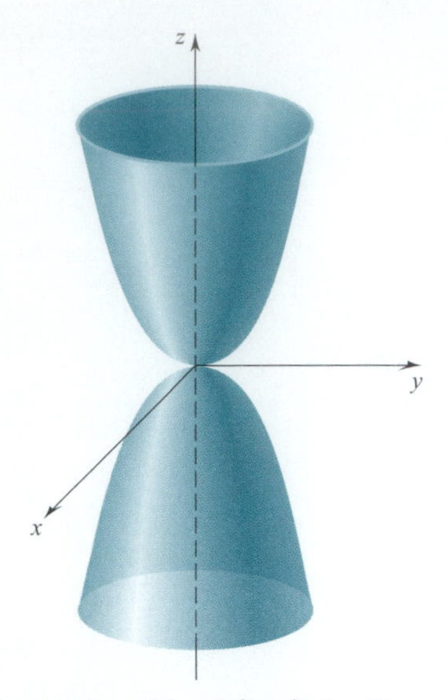

level surface: $|z| = c(x^2 + y^2)$, $(c > 0)$

Figure 14.3.11

† It is surface 5 of the last section together with its mirror image below the xy-plane.

EXERCISES 14.3

Identify the level curves $f(x, y) = c$ and sketch the curves corresponding to the indicated values of c.

1. $f(x, y) = x - y$; $c = -2, 0, 2$.

2. $f(x, y) = 2x - y$; $c = -2, 0, 2$.

3. $f(x, y) = x^2 - y$; $c = -1, 0, 1, 2$.

4. $f(x, y) = \dfrac{1}{x - y^2}$; $c = -2, -1, 1, 2$.

5. $f(x, y) = \dfrac{x}{x + y}$; $c = -1, 0, 1, 2$.

6. $f(x, y) = \dfrac{y}{x^2}$; $c = -1, 0, 1, 2$.

7. $f(x, y) = x^3 - y$; $c = -1, 0, 1, 2$.

8. $f(x, y) = e^{xy}$; $c = \frac{1}{2}, 1, 2, 3$.

9. $f(x, y) = x^2 - y^2$; $c = -2, -1, 0, 1, 2$.

10. $f(x, y) = x^2$; $c = 0, 1, 4, 9$.

11. $f(x, y) = y^2$; $c = 0, 1, 4, 9$.

12. $f(x, y) = x(y - 1)$; $c = -2, -1, 0, 1, 2$.

13. $f(x, y) = \ln(x^2 + y^2)$; $c = -1, 0, 1$.

14. $f(x, y) = \ln\left(\dfrac{y}{x^2}\right)$; $c = -2, -1, 0, 1, 2$.

15. $f(x, y) = \dfrac{\ln y}{x^2}$; $c = -2, -1, 0, 1, 2$.

16. $f(x, y) = x^2y^2$; $c = -4, -1, 0, 1, 4$.

17. $f(x, y) = \dfrac{x^2}{x^2 + y^2}$, $c = 0, \frac{1}{4}, \frac{1}{2}$.

18. $f(x, y) = \dfrac{\ln y}{x}$; $c = -2, -1, 0, 1, 2$.

Identify the c-level surface and sketch it.

19. $f(x, y, z) = x + 2y + 3z$, $c = 0$.

20. $f(x, y, z) = x^2 + y^2$, $c = 4$.

21. $f(x, y, z) = z(x^2 + y^2)^{-1/2}$, $c = 1$.

22. $f(x, y, z) = x^2/4 + y^2/6 + z^2/9$, $c = 1$.

23. $f(x, y, z) = 4x^2 + 9y^2 - 72z$, $c = 0$.

24. $f(x, y, z) = z^2 - 36x^2 - 9y^2$, $c = 1$.

25. Identify the c-level surfaces of

$$f(x,y,z) = x^2 + y^2 - z^2$$

taking (i) $c < 0$, (ii) $c = 0$, (iii) $c > 0$.

26. Identify the c-level surfaces of

$$f(x,y,z) = 9x^2 - 4y^2 + 36z^2$$

taking (i) $c < 0$, (ii) $c = 0$, (iii) $c > 0$.

Find an equation for the the level curve of f that contains the point P.

27. $f(x,y) = 1 - 4x^2 - y^2$; $P(0, 1)$.

28. $f(x,y) = (x^2 + y^2)\,e^{xy}$; $P(1, 0)$.

29. $f(x,y) = y^2 \tan^{-1} x$; $P(1, 2)$.

30. $f(x,y) = (x^2 + y)\ln[2 - x + e^y]$; $P(2, 1)$.

Find an equation for the level surface of f that contains the point P.

31. $f(x,y,z) = x^2 + 2y^2 - 2xyz; \quad P(-1,2,1)$.

32. $f(x,y,z) = \sqrt{x^2+y^2} - \ln z; \quad P(3,4,e)$.

C 33. Use a graphing utility to draw (a) the surfaces and (b) the default level curves.

(a) $f(x,y) = 3x + y^3$. (b) $f(x,y) = \dfrac{x^2+1}{y^2+4}$.

C 34. Use a graphing utility to draw the level surfaces corresponding to the values of c.

(a) $f(x,y,z) = x + 2y + 4z; \quad c = 0, 4, 8$.

(b) $f(x,y,z) = \dfrac{x+y}{1+z^2}; \quad c = -2, 0, 2$.

C 35. Use a CAS to find the level curve/surface at the point P.

(a) $f(x,y) = \dfrac{3x+2y+1}{4x^2+9}; \quad P(2,4)$.

(b) $f(x,y,z) = x^2 + 2y^2 - z^2; \quad P(2,-3,1)$.

C 36. Use a CAS to draw the surface and the level curves.

(a) $f(x,y) = (x^2-y^2)\,e^{(-x^2-y^2)}; \quad -2 \le x \le 2,$ $-2 \le y \le 2$.

(b) $f(x,y) = xy^3 - yx^3; \quad -5 \le x \le 5, \quad -5 \le y \le 5$.

37. The magnitude of the gravitational force exerted by a body of mass M situated at the origin on a body of mass m located at the point (x,y,z) is given by

$$F(x,y,z) = \frac{GmM}{x^2+y^2+z^2}$$

where G is the universal gravitational constant. If m and M are constants, describe the level surfaces of F. What is the physical significance of these surfaces?

38. The strength E of an electric field at a point (x,y,z) due to an infinitely long charged wire lying along the y-axis is given by

$$E(x,y,z) = \frac{k}{\sqrt{x^2+z^2}}$$

where k is a positive constant. Describe the level surfaces of E.

39. A thin metal plate is situated in the xy-plane. The temperature T(in °C) at the point (x,y) is inversely proportional to the square of its distance from the origin.

(a) Express T as a function of x and y.

(b) Describe the level curves and sketch a representative set. NOTE: The level curves of T are called *isothermals*; all points on an isothermal have the same temperature.

(c) Suppose the temperature at the point $(1,2)$ is 50°. What is the temperature at the point $(4,3)$?

40. The formula

$$V(x,y) = \frac{k}{\sqrt{r^2-x^2-y^2}},$$

where k and r are positive constants, gives the electric potential (in volts) at a point (x,y) in the xy-plane. Describe the level curves of V and sketch a representative set. NOTE: The level curves of V are called the *equipotential curves*; all points on an equipotential curve have the same electric potential.

In Exercises 41–46, a function f, together with a set of level curves for f, is given. Figures **A–F** are the surfaces $z = f(x,y)$ (in some order). Match f and its system of level curves with its graph $z = f(x,y)$.

41. $f(x,y) = y^2 - y^3$.

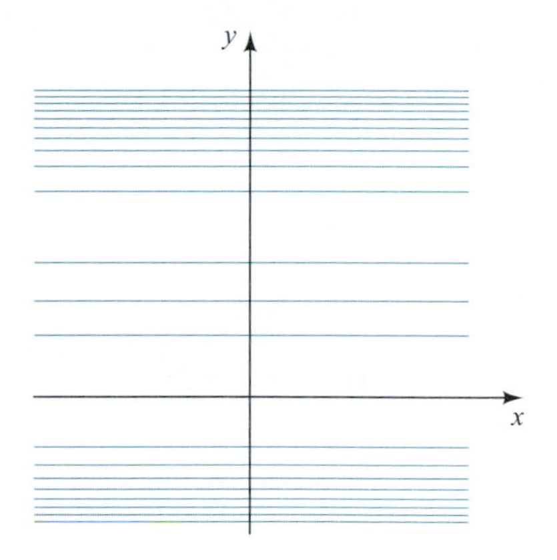

42. $f(x,y) = \sin x, 0 \le x \le 2\pi$.

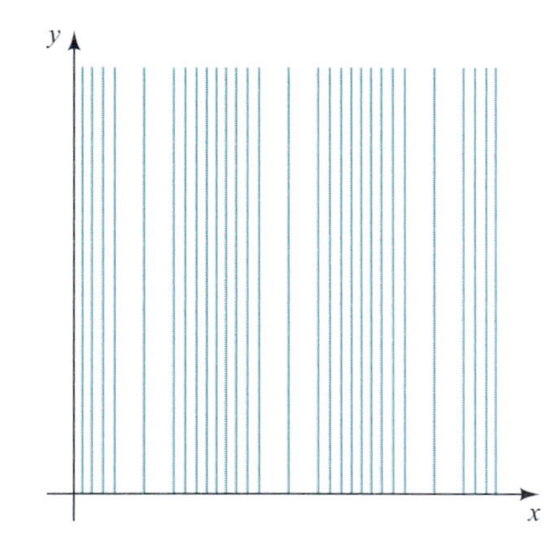

43. $f(x,y) = \cos\sqrt{x^2+y^2}, -10 \le x \le 10, -10 \le y \le 10$.

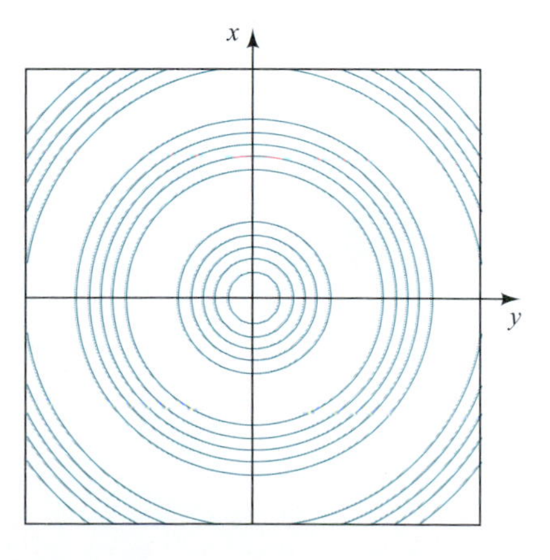

44. $f(x, y) = 2x^2 + 4y^2$.

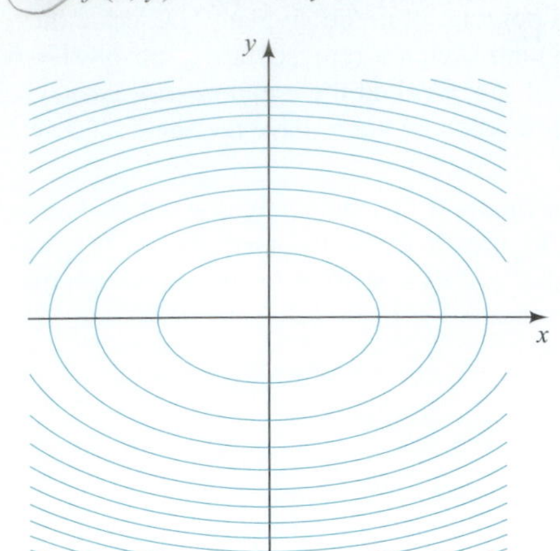

45. $f(x, y) = xye^{-(x^2+y^2)/2}$.

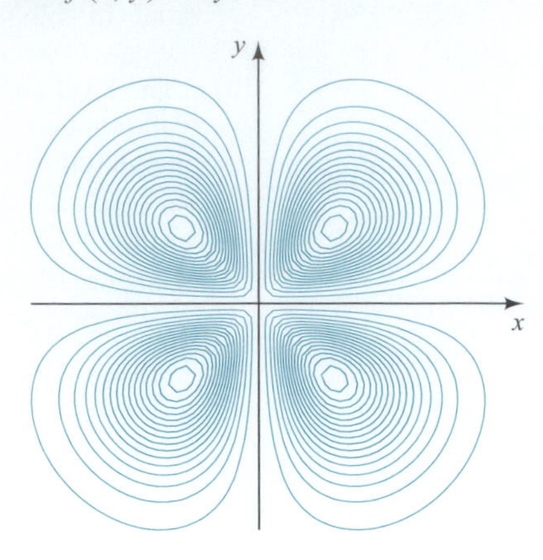

46. $f(x, y) = \sin x \sin y$.

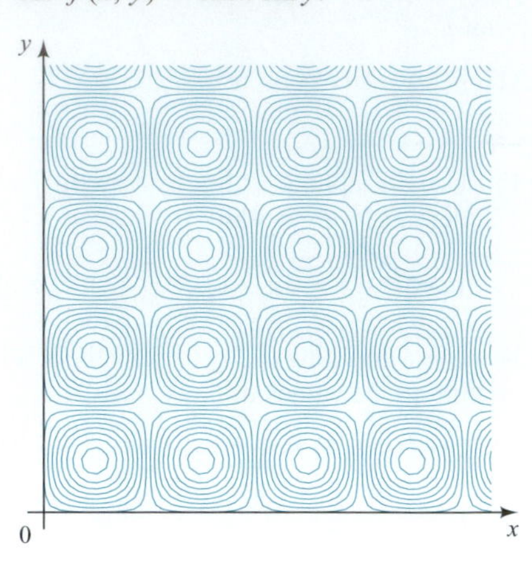

A

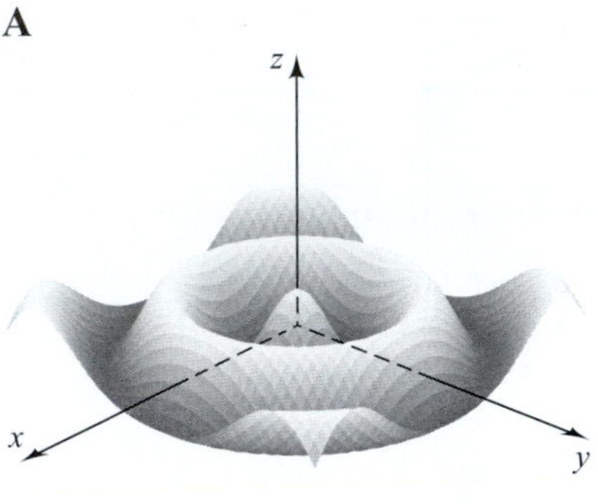

B

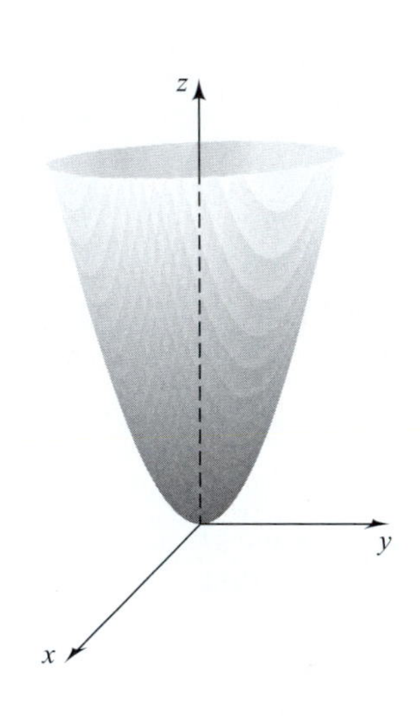

C

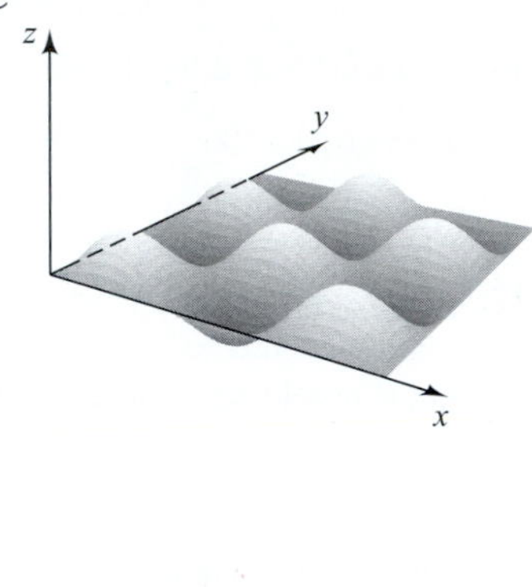

D

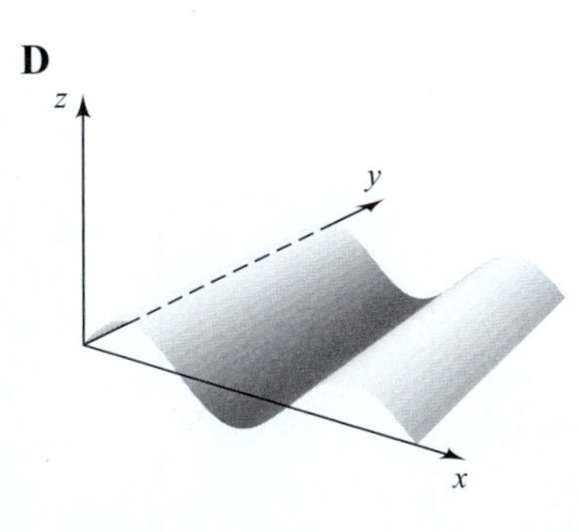

E

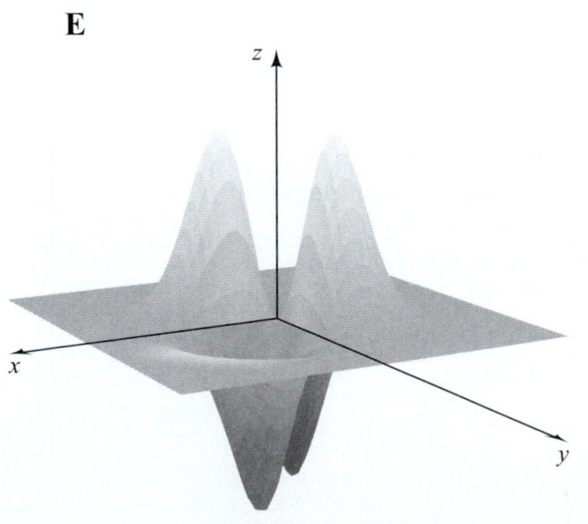

F

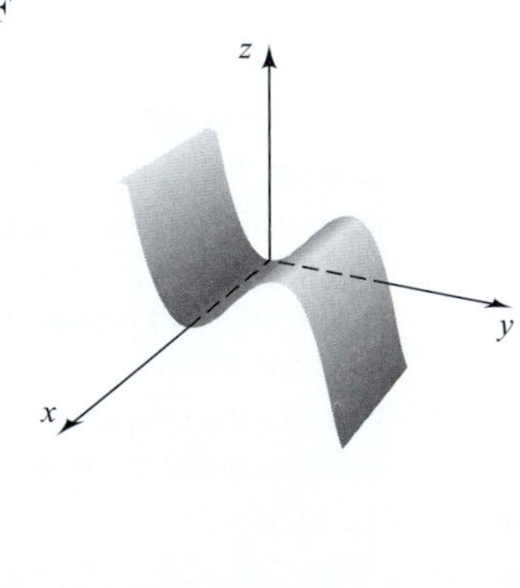

■ PROJECT 14.3 Level Curves and Surfaces

Computer systems such as Derive, Maple, and Mathematica are able to map the level curves of a function $f = f(x, y)$. In this project you are asked to map the level curves of a given function over a given rectangle and then you are asked to "visualize" the surface $z = f(x, y)$. For example, the level curves of

$$f(x,y) = x^2 y^2 e^{-(x^2+y^2)}$$

on the rectangle: $-3 \le x \le 3, -3 \le y \le 3$ are

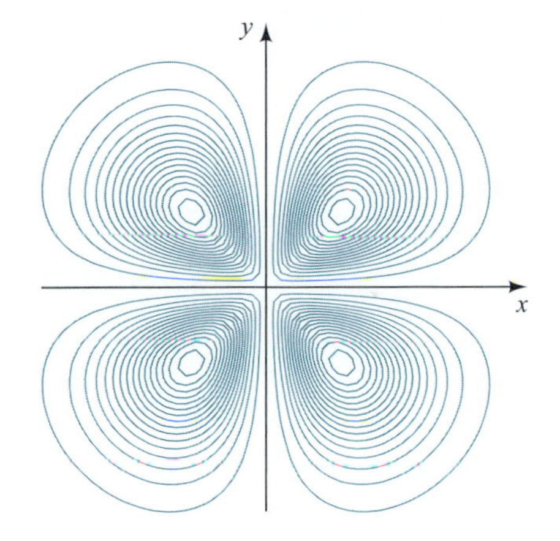

This map of level curves suggests that the surface $z = f(x, y)$ has either "peaks" or "pits" symmetrically placed in the four quadrants. The graph of the surface shown to the right confirms this conjecture.

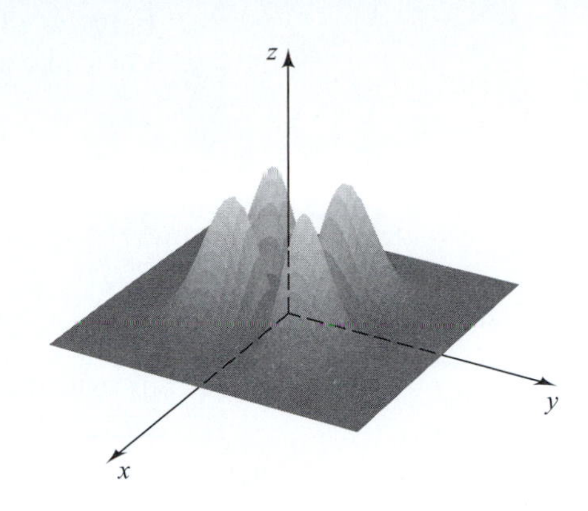

Problem 1. Make a map of the level curves of $f(x,y) = \dfrac{1}{x^2+y^2}$ over the rectangle: $-3 \le x \le 3, -3 \le y \le 3$. Try to visualize the graph of the surface from your map of the level curves. Then graph the surface $z = f(x,y)$ to confirm your visualization.

Problem 2. Repeat Problem 1 for the function $f(x,y) = \dfrac{2y}{x^2+y^2+1}$ over the rectangle $-5 \le x \le 5, -5 \le y \le 5$.

Problem 3. Repeat Problem 1 for the function $f(x,y) = \cos x \cos y \, e^{-(1/4)\sqrt{x^2+y^2}}$ over the rectangle $-2\pi \le x \le 2\pi$, $-2\pi \le y \le 2\pi$.

Problem 4. Repeat Problem 1 for the function $f(x,y) = \dfrac{-xy}{e^{x^2+y^2}}$ over the rectangle: $-2 \le x \le 2, -2 \le y \le 2$.

■ 14.4 PARTIAL DERIVATIVES

Functions of Two Variables

Let f be a function of x and y; for example

$$f(x, y) = 3x^2y - 5x \cos \pi y.$$

The *partial derivative of f with respect to x* is the function f_x obtained by differentiating f with respect to x, treating y as a constant. In this case

$$f_x(x, y) = 6xy - 5 \cos \pi y.$$

The *partial derivative of f with respect to y* is the function f_y obtained by differentiating f with respect to y, treating x as a constant. In this case

$$f_y(x, y) = 3x^2 + 5\pi x \sin \pi y.$$

These partial derivatives are formally defined as limits:

DEFINITION 14.4.1 PARTIAL DERIVATIVES OF $f(x, y)$

Let f be a function of two variables. The partial derivatives of f with respect to x and y are the functions f_x and f_y defined by setting

$$f_x(x,y) = \lim_{h\to 0} \frac{f(x+h,y) - f(x,y)}{h},$$

$$f_y(x,y) = \lim_{h\to 0} \frac{f(x,y+h) - f(x,y)}{h},$$

provided these limits exist.

Example 1 For $f(x,y) = x\tan^{-1} xy$, we have

$$f_x(x,y) = x\frac{y}{1+(xy)^2} + \tan^{-1} xy = \frac{xy}{1+x^2y^2} + \tan^{-1} xy$$

and

$$f_y(x,y) = x\frac{x}{1+(xy)^2} = \frac{x^2}{1+x^2y^2}. \quad \square$$

In the one-variable case, $f'(x_0)$ gives the rate of change with respect to x of $f(x)$ at $x = x_0$. In the two-variable case, *$f_x(x_0, y_0)$ gives the rate of change with respect to x of $f(x, y_0)$ at $x = x_0$, and $f_y(x_0, y_0)$ gives the rate of change with respect to y of $f(x_0, y)$ at $y = y_0$.*

Example 2 For the function $f(x,y) = e^{xy} + \ln(x^2 + y)$, we have

$$f_x(x,y) = ye^{xy} + \frac{2x}{x^2+y} \qquad \text{and} \qquad f_y(x,y) = xe^{xy} + \frac{1}{x^2+y}.$$

The number

$$f_x(2,1) = e^2 + \frac{4}{4+1} = e^2 + \frac{4}{5}$$

gives the rate of change with respect to x of the function

$$f(x,1) = e^x + \ln(x^2+1) \qquad \text{at } x = 2;$$

the number

$$f_y(2,1) = 2\,e^2 + \frac{1}{4+1} = 2\,e^2 + \frac{1}{5}$$

gives the rate of change with respect to y of the function

$$f(2,y) = e^{2y} + \ln(4+y) \qquad \text{at } y = 1. \quad \square$$

A Geometric Interpretation

In Figure 14.4.1 we have sketched a surface $z = f(x, y)$, which you can take as everywhere defined. Through the surface we have passed a plane $y = y_0$ parallel to the

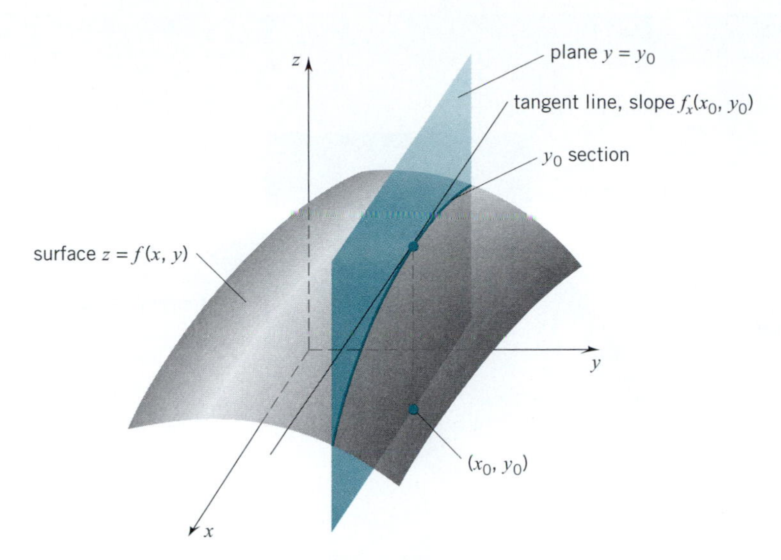

Figure 14.4.1

xz-plane. The plane $y = y_0$ intersects the surface in a curve, the y_0-section of the surface.

The y_0-section of the surface is the graph of the function

$$g(x) = f(x, y_0).$$

Differentiating with respect to x, we have

$$g'(x) = f_x(x, y_0)$$

and, in particular,

$$g'(x_0) = f_x(x_0, y_0).$$

The number $f_x(x_0, y_0)$ is thus the slope of the y_0-section of the surface $z = f(x, y)$ at the point $P(x_0, y_0, f(x_0, y_0))$.

The other partial derivative f_y can be given a similar interpretation. In Figure 14.4.2 you can see the same surface $z = f(x, y)$, this time sliced by a plane $x = x_0$ parallel to the yz-plane. The plane $x = x_0$ intersects the surface in a curve, the x_0-section of the surface.

The x_0-section of the surface is the graph of the function

$$h(y) = f(x_0, y).$$

Differentiating, this time with respect to y, we have

$$h'(y) = f_y(x_0, y)$$

and thus

$$h'(y_0) = f_y(x_0, y_0).$$

The number $f_y(x_0, y_0)$ is the slope of the x_0-section of the surface $z = f(x, y)$ at the point $P(x_0, y_0, f(x_0, y_0))$.

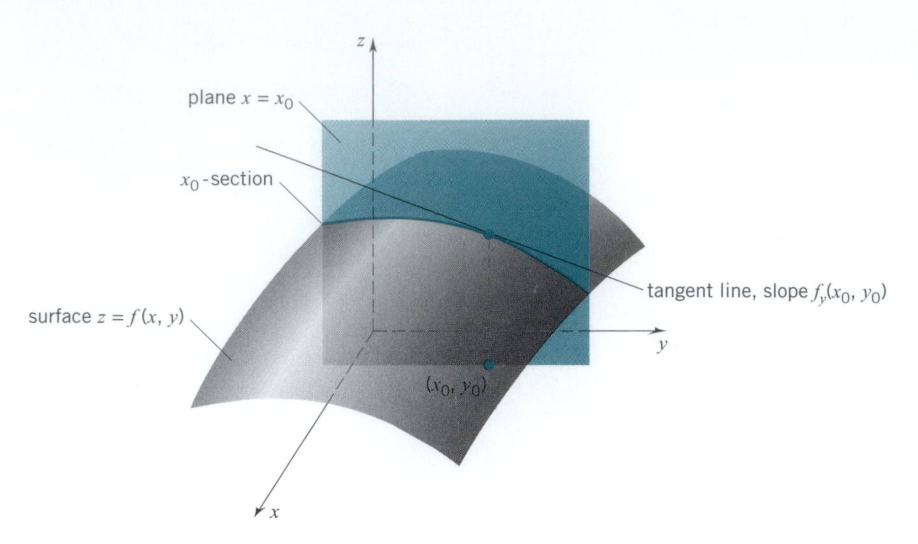

Figure 14.4.2

Functions of Three Variables

In the case of a function of three variables, you can look for three partial derivatives: the partial with respect to x, the partial with respect to y, and the partial with respect to z. These partials,

$$f_x(x, y, z), \quad f_y(x, y, z), \quad f_z(x, y, z),$$

are defined as follows.

DEFINITION 14.4.2 PARTIAL DERIVATIVES OF $f(x, y, z)$

Let f be a function of three variables. The partial derivatives of f with respect to x, y, and z are the functions f_x, f_y, and f_z defined by setting

$$f_x(x, y, z) = \lim_{h \to 0} \frac{f(x+h, y, z) - f(x, y, z)}{h},$$

$$f_y(x, y, z) = \lim_{h \to 0} \frac{f(x, y+h, z) - f(x, y, z)}{h},$$

$$f_z(x, y, z) = \lim_{h \to 0} \frac{f(x, y, z+h) - f(x, y, z)}{h},$$

provided these limits exist.

Each partial can be found by differentiating with respect to the subscript variable, treating the other two variables as constants.

Example 3 For the function $f(x, y, z) = xy^2z^3$ the partial derivatives are:

$$f_x(x, y, z) = y^2z^3, \quad f_y(x, y, z) = 2xyz^3, \quad f_z(x, y, z) = 3xy^2z^2.$$

In particular,

$$f_x(1, -2, -1) = -4, \quad f_y(1, -2, -1) = 4, \quad f_z(1, -2, -1) = 12. \quad \square$$

Example 4 For $g(x, y, z) = x^2 e^{y/z}$ we have

$$g_x(x,y,z) = 2x\, e^{y/z}, \quad g_y(x, y, z) = \frac{x^2}{z}\, e^{y/z}, \quad g_z(x, y, z) = -\frac{x^2 y}{z^2} e^{y/z}. \quad \square$$

Example 5 For a function of the form $f(x, y, z) = F(x,y)G(y,z)$ we can write

$$f_x(x, y, z) = F_x(x, y)G(y, z),$$
$$f_y(x, y, z) = F(x, y)G_y(y, z) + F_y(x, y)G(y, z),$$
$$f_z(x, y, z) = F(x, y)G_z(y, z). \quad \square$$

The number $f_x(x_0, y_0, z_0)$ gives the rate of change with respect to x of $f(x, y_0, z_0)$ at $x = x_0$; $f_y(x_0, y_0, z_0)$ gives the rate of change with respect to y of $f(x_0, y, z_0)$ at $y = y_0$; $f_z(x_0, y_0, z_0)$ gives the rate of change with respect to z of $f(x_0, y_0, z)$ at $z = z_0$.

Example 6 The function $f(x, y, z) = xy^2 - yz^2$ has partial derivatives

$$f_x(x, y, z) = y^2, \quad f_y(x, y, z) = 2xy - z^2, \quad f_z(x, y, z) = -2yz.$$

The number $f_x(1, 2, 3) = 4$ gives the rate of change with respect to x of the function

$$f(x, 2, 3) = 4x - 18 \qquad \text{at} \quad x = 1;$$

$f_y(1, 2, 3) = -5$ gives the rate of change with respect to y of the function

$$f(1, y, 3) = y^2 - 9y \qquad \text{at} \quad y = 2;$$

$f_z(1, 2, 3) = -12$ gives the rate of change with respect to z of the function

$$f(1, 2, z) = 4 - 2z^2 \qquad \text{at } z = 3. \quad \square$$

In general, if f is a function of n variables, $x_1, x_2, \ldots, x_n$, then the partial derivative of f with respect to the kth variable, x_k, is given by

$$f_{x_k}(x_1, x_2, \ldots, x_n) = \lim_{h \to 0} \frac{f(x_1, \ldots, x_{k-1}, x_k + h, x_{k+1}, \ldots x_n) - f(x_1, \ldots, x_k, \ldots, x_n)}{h}$$

provided the limit exists.

The geometric illustrations of the partial derivatives of a function of two variables given in Figures 14.4.1 and 14.4.2 require three-dimensional sketches. Corresponding illustrations for functions of three or more variables would require drawings in four or higher dimensions. Clearly, such drawings are not possible.

Other Notations

There is obviously no need to restrict ourselves to the variables x, y, z. Where more convenient we can use other letters.

Example 7 The volume of the frustum of a cone (Figure 14.4.3) is given by the function

$$V(R, r, h) = \tfrac{1}{3}\pi h(R^2 + Rr + r^2).$$

Find the rate of change of the volume with respect to each of its dimensions if the other dimensions are held constant. Determine the values of these rates of change when $R = 8, r = 4$, and $h = 6$.

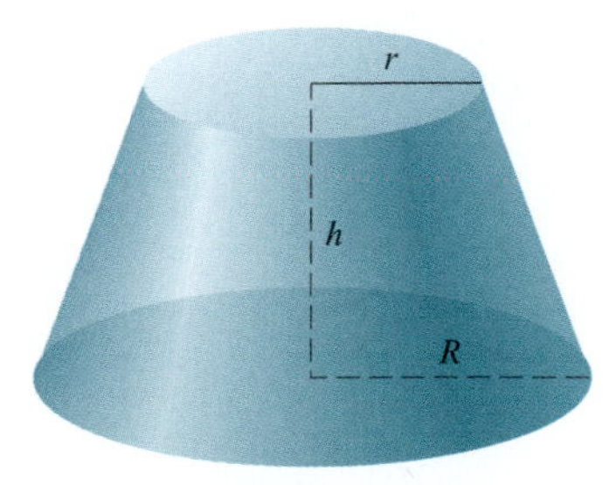

frustum of a cone

Figure 14.4.3

SOLUTION The partial derivatives of V with respect to R, r, and h are as follows:

$$V_R(R, r, h) = \tfrac{1}{3}\pi h(2R + r),$$
$$V_r(R, r, h) = \tfrac{1}{3}\pi h(R + 2r),$$
$$V_h(R, r, h) = \tfrac{1}{3}\pi(R^2 + Rr + r^2).$$

When $R = 8, r = 4$, and $h = 6$,

the rate of change of V with respect to R is $V_R(8, 4, 6) = 40\pi$,

the rate of change of V with respect to r is $V_r(8, 4, 6) = 32\pi$,

the rate of change of V with respect to h is $V_h(8, 4, 6) = \frac{112}{3}\pi$. ❑

The subscript notation is not the only one used in partial differentiation. A variant of Leibniz's double-d notation is also commonly used. In this notation the partials f_x, f_y, f_z are denoted by

$$\frac{\partial f}{\partial x}, \frac{\partial f}{\partial y}, \frac{\partial f}{\partial z}.$$

Thus, for

$$f(x, y, z) = x^3y^2z + \sin xy$$

we have

$$\frac{\partial f}{\partial x}(x, y, z) = 3x^2y^2z + y\cos xy, \quad \frac{\partial f}{\partial y}(x, y, z) = 2x^3yz + x\cos xy,$$
$$\frac{\partial f}{\partial z}(x, y, z) = x^3y^2,$$

or more simply,

$$\frac{\partial f}{\partial x} = 3x^2y^2z + y\cos xy, \quad \frac{\partial f}{\partial y} = 2x^3yz + x\cos xy, \quad \frac{\partial f}{\partial z} = x^3y^2.$$

We can also write

$$\frac{\partial}{\partial x}(x^3y^2z + \sin xy) = 3x^2y^2z + y\cos xy,$$

$$\frac{\partial}{\partial y}(x^3y^2z + \sin xy) = 2x^3yz + x\cos xy, \quad \frac{\partial}{\partial z}(x^3y^2z + \sin xy) = x^3y^2.$$

Of course, this notation is not restricted to the letters x, y, z. For instance, we can write

$$\frac{\partial}{\partial r}(r^2\cos\theta + e^{\theta r}) = 2r\cos\theta + \theta e^{\theta r},$$
$$\frac{\partial}{\partial \theta}(r^2\cos\theta + e^{\theta r}) = -r^2\sin\theta + r e^{\theta r}.$$

For the function

$$\rho = \sin 2\theta \, \cos 3\phi$$

we have

$$\frac{\partial \rho}{\partial \theta} = 2\cos 2\theta \, \cos 3\phi \qquad \text{and} \qquad \frac{\partial \rho}{\partial \phi} = -3\sin 2\theta \sin 3\phi.$$

EXERCISES 14.4

Calculate the partial derivatives.

1. $f(x, y) = 3x^2 - xy + y.$

2. $g(x, y) = x^2 e^{-y}.$

3. $\rho = \sin\phi \cos\theta.$

4. $\rho = \sin^2(\theta - \phi).$

5. $f(x, y) = e^{x-y} - e^{y-x}.$

6. $z = \sqrt{x^2 - 3y}.$

7. $g(x, y) = \dfrac{Ax + By}{Cx + Dy}.$

8. $u = \dfrac{e^z}{xy^2}.$

9. $u = xy + yz + zx.$

10. $z = Ax^2 + Bxy + Cy^2.$

11. $f(x, y, z) = z\sin(x - y).$

12. $g(u, v, w) = \ln(u^2 + vw - w^2).$

13. $\rho = e^{\theta+\phi}\cos(\theta - \phi).$

14. $f(x, y) = (x + y)\sin(x - y).$

15. $f(x, y) = x^2 y \sec xy.$

16. $g(x, y) = \tan^{-1}(2x + y).$

17. $h(x, y) = \dfrac{x}{x^2 + y^2}.$

18. $z = \ln\sqrt{x^2 + y^2}.$

19. $f(x, y) = \dfrac{x \sin y}{y \cos x}.$

20. $f(x, y, z) = e^{xy}\sin xz.$

21. $h(x, y) = [f(x)]^2 g(y).$

22. $h(x, y) = e^{f(x)g(y)}.$

23. $f(x, y, z) = z^{xy^2}.$

24. $h(x, y, z) = [f(x, y)]^3[g(x, z)]^2.$

25. $h(r, \theta, t) = r^2 e^{2t}\cos(\theta - t).$

26. $u = \ln(x/y) - ye^{xz}.$

27. $f(x, y, z) = z\tan^{-1}(y/x).$

28. $w = xy\sin z - yz\sin x.$

29. Find $f_x(0, e)$ and $f_y(0, e)$ given that $f(x, y) = e^x \ln y.$

30. Find $g_x(0, \frac{1}{4}\pi)$ and $g_y(0, \frac{1}{4}\pi)$ given that $g(x, y) = e^{-x}\sin(x + 2y).$

31. Find $f_x(1, 2)$ and $f_y(1, 2)$ given that $f(x, y) = \dfrac{x}{x + y}.$

32. Find $g_x(1, 2)$ and $g_y(1, 2)$ given that $g(x, y) = \dfrac{y}{x + y^2}.$

Find $f_x(x, y)$ and $f_y(x, y)$ by forming the appropriate difference quotient and taking the limit as h tends to zero.

33. $f(x, y) = x^2 y.$

34. $f(x, y) = y^2.$

35. $f(x, y) = \ln(x^2 y).$

36. $f(x, y) = \dfrac{1}{x + 4y}.$

37. $f(x, y) = \dfrac{1}{x - y}.$

38. $f(x, y) = e^{2x+3y}.$

Find $f_x(x, y, z)$, $f_y(x, y, z)$, and $f_z(x, y, z)$ by forming the appropriate difference quotient and taking the limit as h tends to zero.

39. $f(x, y, z) = xy^2 z.$

40. $f(x, y, z) = \dfrac{x^2 y}{z}.$

41. The intersection of a surface $z = f(x, y)$ with a plane $y = y_0$ is a curve C in space. The slope of the line tangent to C at the point $P(x_0, y_0, f(x_0, y_0))$ is $f_x(x_0, y_0)$. (See Figure 14.4.1.)

(a) Show that the equations for the tangent line can be written in the form

$$y = y_0, \quad z - z_0 = f_x(x_0, y_0)(x - x_0).$$

(b) Now let C be the curve formed by intersecting the surface $z = f(x, y)$ with the plane $x = x_0$. Derive equations for the line tangent to C at the point $P(x_0, y_0, f(x_0, y_0))$. (See Figure 14.4.2.)

In Exercises 42 and 43, let $z = x^2 + y^2$ and let C be the curve of intersection of the surface with the given plane. Find equations for the line tangent to C at the point P.

42. Plane $y = 3$; $P(1, 3, 10)$.

43. Plane $x = 2$; $P(2, 1, 5)$.

In Exercises 44 and 45, let

$$z = \frac{x^2}{y^2 - 3}$$

and let C be the curve of intersection of the surface with the given plane. Find equations for the line tangent to C at the point P.

44. Plane $x = 3$; $P(3, 2, 9)$.

45. Plane $y = 2$; $P(3, 2, 9)$.

46. The surface $z = \sqrt{4 - x^2 - y^2}$ is a hemisphere of radius 2 centered at the origin.

(a) Find equations for the line l_1 tangent to the curve of intersection of the hemisphere with plane $x = 1$ at the point $(1, 1, \sqrt{2})$.

(b) Find equations for the line l_2 tangent to the curve of intersection of the hemisphere with the plane $y = 1$ at the point $(1, 1, \sqrt{2})$.

(c) The tangent lines l_1 and l_2 determine a plane. Find an equation for this plane. As you might expect, this plane may be viewed as tangent to the surface at the point $(1, 1, \sqrt{2})$.

C **47.** Let $f(x, y) = 3x^2 - 6xy + 2y^3$. Use a graphing utility to draw the graph of f in a region around the point $P(1, 2)$.

(a) Use a CAS to find $m_x = \partial f/\partial x$ at P. Use a graphing utility to draw the graphs of $f(x, 2)$ and the line through P with slope m_x together.

(b) Use a CAS to find $m_y = \partial f/\partial y$ at P. Use a graphing utility to draw the graphs of $f(1, y)$ and the line through P with slope m_y together.

C **48.** Repeat Exercise 47 with $f(x, y) = \dfrac{x - y}{x^2 + y^2}$ and $P(1, 2)$.

In Exercises 49–52, show that the functions u and v satisfy

$$u_x(x, y) = v_y(x, y) \quad \text{and} \quad u_y(x, y) = -v_x(x, y).$$

These equations are called the *Cauchy-Riemann equations*. They arise in the study of functions of a complex variable and are of fundamental importance in that setting.

49. $u(x, y) = x^2 - y^2; \quad v(x, y) = 2xy.$

50. $u(x,y) = e^x \cos y; \quad v(x,y) = e^x \sin y.$

51. $u(x,y) = \frac{1}{2} \ln (x^2 + y^2); \quad v(x,y) = \tan^{-1} \dfrac{y}{x}.$

52. $u(x,y) = \dfrac{x}{x^2 + y^2}; \quad v(x,y) = \dfrac{-y}{x^2 + y^2}.$

53. Assume that f is a function defined on a set D in the xy-plane, and assume that the partial derivatives exist throughout D.

(a) Suppose that $f_x(x,y) = 0$ for all $(x,y) \in D$. What can you conclude about f?

(b) Suppose that $f_y(x,y) = 0$ for all $(x,y) \in D$. What can you conclude about f?

54. The law of cosines for a triangle can be written

$$a^2 = b^2 + c^2 - 2bc \cos \theta.$$

At time t_0 we have $b_0 = 10$ inches, $c_0 = 15$ inches, $\theta_0 = \frac{1}{3}\pi$ radians.

(a) Find a_0.

(b) Find the rate of change of a with respect to b at time t_0 given that c and θ remain constant.

(c) Using the rate found in part (b), calculate (by differentials) the approximate change in a if b is decreased by 1 inch.

(d) Find the rate of change of a with respect to θ at time t_0 given that b and c remain constant.

(e) Find the rate of change of c with respect to θ at time t_0 given that a and b remain constant.

55. The area of a triangle is given by the formula

$$A = \tfrac{1}{2} bc \sin \theta.$$

At time t_0 we have $b_0 = 10$ inches, $c_0 = 20$ inches, $\theta_0 = \frac{1}{3}\pi$ radians.

(a) Find the area of the triangle at time t_0.

(b) Find the rate of change of the area with respect to b at time t_0 given that c and θ remain constant.

(c) Find the rate of change of the area with respect to θ at time t_0 given that b and c remain constant.

(d) Using the rate found in part (c), calculate (by differentials) the approximate change in area if angle θ is increased by one degree.

(e) Find the rate of change of c with respect to b at time t_0 if the area and angle θ are to remain constant.

56. Let f be a function of x and y that satisfies a relation of the form

$$\frac{\partial f}{\partial x} = kf, \quad k \quad \text{a constant.}$$

Show that

$$f(x,y) = g(y)e^{kx},$$

where g is some function of y.

57. Let $z = f(x,y)$ be a function everywhere defined.

(a) Find a vector function that parametrizes the y_0-section of the graph. (See Figure 14.4.1.) Find a vector function that parametrizes the line tangent to this section at the point $P(x_0, y_0, f(x_0, y_0))$. See Exercise 41.

(b) Find a vector function that parametrizes the x_0-section of the graph. (See Figure 14.4.2.) Find a vector function that parametrizes the line tangent to this section at the point $P(x_0, y_0, f(x_0, y_0))$.

(c) Show that the equation of the plane determined by the tangent lines found in parts (a) and (b) can be written

$$z - f(x_0, y_0) = (x - x_0)\frac{\partial f}{\partial x}(x_0, y_0) + (y - y_0)\frac{\partial f}{\partial y}(x_0, y_0).$$

58. *(A chain rule)* Let f be a function of x and y, and g be a function of a single variable. Form the composition $h(x,y) = g(f(x,y))$ and show that

$$h_x(x,y) = g'(f(x,y)) f_x(x,y) \quad \text{and}$$
$$h_y(x,y) = g'(f(x,y)) f_y(x,y).$$

In Leibniz's notation, setting $u = f(x,y)$, we have

$$\frac{\partial h}{\partial x} = \frac{dg}{du}\frac{\partial u}{\partial x} \quad \text{and} \quad \frac{\partial h}{\partial y} = \frac{dg}{du}\frac{\partial u}{\partial y}.$$

59. Let g be a differentiable function of a single variable. Use Exercise 58 to verify the following results.

(a) If $w = g(ax + by)$, a, b constant, then

$$b\frac{\partial w}{\partial x} = a\frac{\partial w}{\partial y}.$$

(b) If m and n are nonzero integers and $w = g(x^m y^n)$, then

$$nx\frac{\partial w}{\partial x} = my\frac{\partial w}{\partial y}.$$

60. Given that $x = r\cos\theta$ and $y = r\sin\theta$, find

$$\frac{\partial x}{\partial r}\frac{\partial y}{\partial \theta} - \frac{\partial x}{\partial \theta}\frac{\partial y}{\partial r}.$$

61. For a gas confined in a container, the ideal gas law states that the pressure P is related to the volume V and the temperature T by an equation of the form

$$P = k\frac{T}{V}$$

where k is a positive constant. Show that

$$V\frac{\partial P}{\partial V} = -P \quad \text{and} \quad V\frac{\partial P}{\partial V} + T\frac{\partial P}{\partial T} = 0.$$

62. Three resistances R_1, R_2, R_3 connected in parallel in an electrical circuit produce a resistance R that is given by the formula

$$\frac{1}{R} = \frac{1}{R_1} + \frac{1}{R_2} + \frac{1}{R_3}.$$

Find $\partial R/\partial R_1$.

■ 14.5 OPEN AND CLOSED SETS

A *neighborhood* of a real number x_0 is by definition a set of the form $\{x : |x - x_0| < \delta\}$ where δ is a positive number. This is just an open interval centered at x_0 :

$$(x_0 - \delta, x_0 + \delta).$$

If we remove x_0 from the set, we obtain the set

$$(x_0 - \delta, x_0) \cup (x_0, x_0 + \delta).$$

Such a set is called a *deleted neighborhood* of x_0.

From your study of one-variable calculus you know that for a function to have a limit at x_0 it must be defined at least on a deleted neighborhood of x_0, and for it to be continuous or differentiable at x_0 it must be defined at least on a full neighborhood of x_0.

To pave the way for the calculus of functions of several variables, we will extend the notions of neighborhood and deleted neighborhood to higher dimensions and, in so doing, obtain access to other fruitful ideas.

Points in the domain of a function of several variables can be written in vector notation. In the two-variable case, set

$$\mathbf{x} = (x, y),$$

and, in the three-variable case, set

$$\mathbf{x} = (x, y, z).$$

The vector notation enables us to treat the two cases together.

In this section we introduce five important notions:

(1) Neighborhood of a point.

(2) Interior of a set.

(3) Boundary of a set.

(4) Open set.

(5) Closed Set.

For our purposes, the fundamental notion here is "neighborhood of a point." The other four notions can be derived from it.

DEFINITION 14.5.1 NEIGHBORHOOD OF A POINT

A *neighborhood* of a point $\mathbf{x}_0$ is a set of the form

$$\{\mathbf{x} : ||\mathbf{x} - \mathbf{x}_0|| < \delta\}$$

where δ is some number greater than zero.

In the plane, a neighborhood of $\mathbf{x}_0 = (x_0, y_0)$ consists of all the points inside a disc centered at (x_0, y_0). In three-space, a neighborhood of $\mathbf{x}_0 = (x_0, y_0, z_0)$. consists of all the points inside a ball (sphere) centered at (x_0, y_0, z_0). See Figure 14.5.1.

Remark On the real line, the norm of a number x is its absolute value, $||x|| = |x|$. Thus a neighborhood of a point x_0 on the real line is a set of the form $\{x : |x - x_0| < \delta\}$ for some $\delta > 0$ which is simply the open interval $(x_0 - \delta, x_0 + \delta)$.

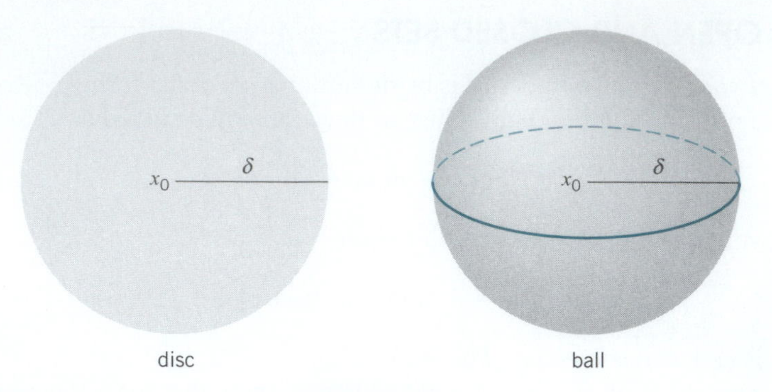

Figure 14.5.1

DEFINITION 14.5.2 THE INTERIOR OF A SET

A point $\mathbf{x}_0$ is said to be an *interior point* of the set S if the set S contains some neighborhood of $\mathbf{x}_0$. The set of all interior points of S is called the *interior* of S.

Example 1 Let Ω be the plane set shown in Figure 14.5.2. The point marked $\mathbf{x}_1$ is an interior point of Ω because Ω contains a neighborhood of $\mathbf{x}_1$. The point $\mathbf{x}_2$ is not an interior point of Ω because *no* neighborhood of $\mathbf{x}_2$ is completely contained in Ω. (Every neighborhood of $\mathbf{x}_2$ has points that lie outside of Ω.) □

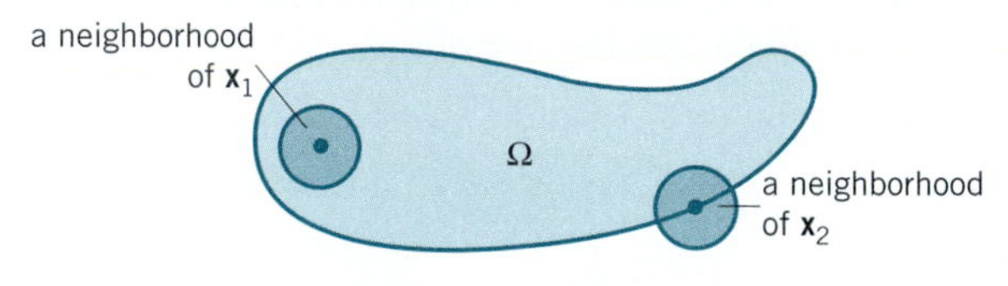

Figure 14.5.2

DEFINITION 14.5.3 THE BOUNDARY OF A SET

A point $\mathbf{x}_0$ is said to be a *boundary point* of the set S if every neighborhood of $\mathbf{x}_0$ contains points that are in S and points that are not in S. The set of all boundary points of S is called the *boundary* of S.

Example 2 The point marked $\mathbf{x}_2$ in Figure 14.5.2 is a boundary point of Ω: each neighborhood of $\mathbf{x}_2$ contains points in Ω and points not in Ω. □

DEFINITION 14.5.4 OPEN SET

A set S is said to be *open* if it contains a neighborhood of each of its points.

Equivalently:

(a) A set S is open iff each of its points is an interior point.

(b) A set S is open iff it contains no boundary points.

DEFINITION 14.5.5 CLOSED SET

A set S is said to be *closed* if it contains its boundary.

Here are some examples of sets that are open, sets that are closed, and sets that are neither open nor closed:

Two-Dimensional Examples

The sets

$$S_1 = \{(x,y) : 1 < x < 2, 1 < y < 2\},$$
$$S_2 = \{(x,y) : 3 \leq x \leq 4, 1 \leq y \leq 2\},$$
$$S_3 = \{(x,y) : 5 \leq x \leq 6, 1 < y < 2\}$$

are displayed if Figure 14.5.3. S_1 is the inside of the first square. S_1 is open because it contains a neighborhood of each of its points. S_2 is the inside of the second square together with the four bounding line segments. S_2 is closed because it contains its entire boundary. S_3 is the inside of the last square together with the two vertical bounding line segments. S_3 is not open because it contains part of its boundary, and it is not closed because it does not contain all of its boundary.

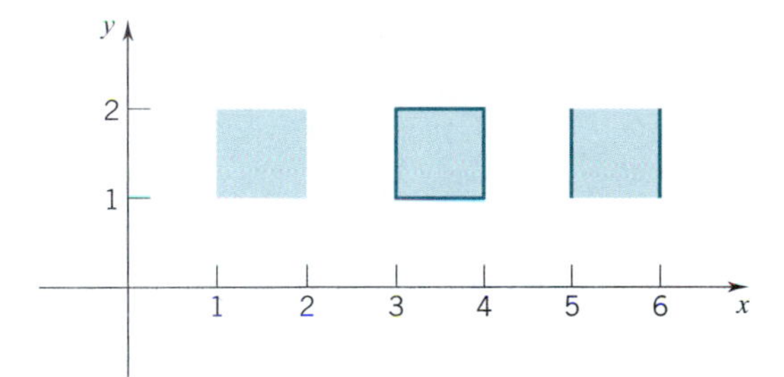

Figure 14.5.3

Three-Dimensional Examples

We now examine some three-dimensional sets:

$$S_1 = \{(x, y, z) : z > x^2 + y^2\},$$
$$S_2 = \{(x, y, z) : z \geq x^2 + y^2\},$$
$$S_3 = \left\{(x, y, z) : 1 \geq \frac{x^2 + y^2}{z}\right\}.$$

The boundary of each of these sets is the paraboloid of revolution

$$z = x^2 + y^2. \qquad \text{(Figure 14.5.4)}$$

The first set consists of all points above this surface. This set is open because, if a point is above this surface, then all points sufficiently close to it are also above this surface. Thus the set contains a neighborhood of each of its points. The second set is closed because it contains all of its boundary. The third set is neither open nor closed. It is not open because it contains some boundary points; for example, it contains the point (1, 1, 2). It is not closed because it fails to contain the boundary point (0, 0, 0).

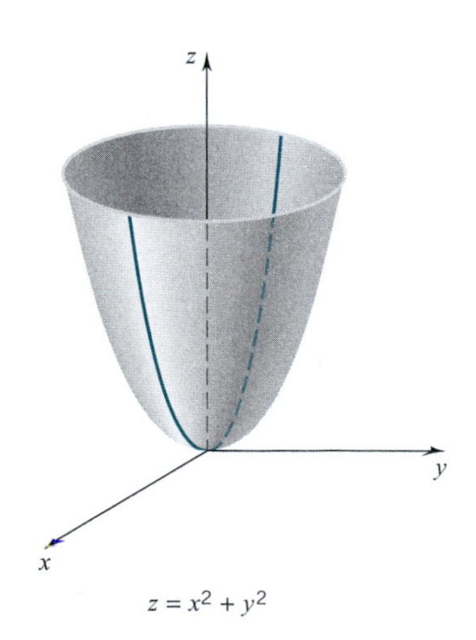

Figure 14.5.4

A Final Remark A neighborhood of $\mathbf{x}_0$ is a set of the form

$$\{\mathbf{x} : ||\mathbf{x} - \mathbf{x}_0|| < \delta\}.$$

If we remove $\mathbf{x}_0$ from the set, we have the set

$$\{\mathbf{x} : 0 < ||\mathbf{x} - \mathbf{x}_0|| < \delta\}.$$

Such a set is called a *deleted neighborhood* of $\mathbf{x}_0$. (We will use deleted neighborhoods in the next section to support the definition of limit. □

EXERCISES 14.5

Specify the interior and the boundary of the set. State whether the set is open, closed, or neither. Then sketch the set.

1. $\{(x, y) : 2 \le x \le 4, 1 \le y \le 3\}$.

2. $\{(x, y) : 2 < x < 4, 1 < y < 3\}$.

3. $\{(x, y) : 1 < x^2 + y^2 < 4\}$.

4. $\{(x, y) : 1 \le x^2 \le 4\}$.

5. $\{(x, y) : 1 < x^2 \le 4\}$.

6. $\{(x, y) : y < x^2\}$.

7. $\{(x, y) : y \le x^2\}$.

8. $\{(x, y, z) : 1 \le x \le 2, 1 \le y \le 2, 1 \le z < 2\}$.

9. $\{(x, y, z) : x^2 + y^2 \le 1, 0 \le z \le 4\}$.

10. $\{(x, y, z) : (x-1)^2 + (y-1)^2 + (z-1)^2 < \frac{1}{4}\}$.

11. Let $S = \{\mathbf{x}_1, \mathbf{x}_2, \ldots, \mathbf{x}_n\}$ be a nonempty, finite set of points. (a) What is the interior of S? (b) What is the boundary of S? (c) Is S open, closed, or neither?

All the notions introduced in this section can be applied to sets of real numbers: write x for $\mathbf{x}$ and $|x|$ for $||\mathbf{x}||$. As indicated earlier a *neighborhood* of a number x_0, a set of the form

$$\{x : |x - x_0| < \delta\} \quad \text{with } \delta > 0,$$

is just an open interval $(x_0 - \delta, x_0 + \delta)$. In Exercises 12–19, specify the interior and boundary of the given set of real numbers. State whether the set is open, closed, or neither.

12. $\{x : 1 < x < 3\}$.

13. $\{x : 1 \le x \le 3\}$.

14. $\{x : 1 \le x < 3\}$.

15. $\{x : x > 1\}$.

16. $\{x : x \le -1\}$.

17. $\{x : x < -1 \text{ or } x \ge 1\}$.

18. The set of positive integers: $\{1, 2, 3, \ldots, n, \ldots\}$.

19. The set of reciprocals: $\{1, 1/2, 1/3, \ldots, 1/n \ldots\}$.

20. Let $\emptyset$ be the empty set. Let X be the real line, the entire plane, or, in the three-dimensional case, all of three-space. For each subset A of X, let $X - A$ be the set of all points $\mathbf{x} \in X$ such that $\mathbf{x} \notin A$.

(a) Show that $\emptyset$ is both open and closed.

(b) Show that X is both open and closed. (It can be shown that $\emptyset$ and X are the only subsets of X that are both open and closed.)

(c) Let U be a subset of X. Show that U is open iff $X - U$ is closed.

(d) Let F be a subset of X. Show that F is closed iff $X - F$ is open.

■ 14.6 LIMITS AND CONTINUITY; EQUALITY OF MIXED PARTIALS

The Basic Notions

The limit process used in taking partial derivatives involved nothing new because in each instance all but one of the variables remained fixed. In this section we take up limits of the form

$$\lim_{(x,y)\to(x_0,y_0)} f(x, y) \qquad \text{and} \qquad \lim_{(x,y,z)\to(x_0,y_0,z_0)} f(x, y, z).$$

To avoid having to treat the two-and three-variable cases separately, we will write instead

$$\lim_{\mathbf{x}\to\mathbf{x}_0} f(\mathbf{x}).$$

This gives us both the two-variable case [set $\mathbf{x} = (x, y)$ and $\mathbf{x}_0 = (x_0, y_0)$] and the three-variable case [set $\mathbf{x} = (x, y, z)$ and $\mathbf{x}_0 = (x_0, y_0, z_0)$].

To take the limit of $f(\mathbf{x})$ as $\mathbf{x}$ tends to $\mathbf{x}_0$, we do not need f to be defined at $\mathbf{x}_0$ itself, but we do need f to be defined at points $\mathbf{x}$ close to $\mathbf{x}_0$. At this stage, we will assume that f is defined at all points $\mathbf{x}$ in some deleted neighborhood of $\mathbf{x}_0$ (f may or may not be defined at $\mathbf{x}_0$). This will guarantee that we can form $f(\mathbf{x})$ for all $\mathbf{x} \neq \mathbf{x}_0$ that are "sufficiently close" to $\mathbf{x}_0$. This approach is consistent with our approach to limits of functions of one variable in Chapter 2.

To say that

$$\lim_{\mathbf{x}\to\mathbf{x}_0} f(\mathbf{x}) = L$$

is to say that for $\mathbf{x}$ sufficiently close to $\mathbf{x}_0$ but different from $\mathbf{x}_0$, the number $f(\mathbf{x})$ is close to L; or, to put it another way, as $\|\mathbf{x} - \mathbf{x}_0\|$ tends to zero but remains different from zero, $|f(\mathbf{x}) - L|$ tends to zero. The $\epsilon - \delta$ definition is a direct generalization of the $\epsilon - \delta$ definition in the single-variable case.

DEFINITION 14.6.1 THE LIMIT OF A FUNCTION OF SEVERAL VARIABLES

Let f be a function defined at least on some deleted neighborhood of $\mathbf{x}_0$.

$$\lim_{\mathbf{x}\to\mathbf{x}_0} f(\mathbf{x}) = L$$

if for each $\epsilon > 0$ there exists a $\delta > 0$ such that

$$\text{if} \quad 0 < \|\mathbf{x} - \mathbf{x}_0\| < \delta \quad \text{then} \quad |f(\mathbf{x}) - L| < \epsilon.$$

Example 1 We will show that the function $f(x,y) = \dfrac{xy + y^3}{x^2 + y^2}$ does not have a limit at $(0,0)$. Note that f is not defined at $(0,0)$, but is defined for all $(x,y) \neq (0,0)$.

Along the obvious paths to $(0,0)$, the coordinate axes, the limiting value is 0:

Along the x-axis, $y = 0$; thus, $f(x,y) = f(x,0) = 0$ and $\lim_{x\to 0} f(x,0) = \lim_{x\to 0} 0 = 0$.

Along the y-axis, $x = 0$; thus, $f(x,y) = f(0,y) = y$ and $\lim_{y\to 0} f(0,y) = \lim_{y\to 0} y = 0$.

Along the line $y = 2x$, however, the limiting value is $\frac{2}{5}$:

$$f(x,y) = f(x,2x) = \frac{2x^2 + 8x^3}{x^2 + 4x^2} = \frac{2 + 8x}{5} \to \frac{2}{5} \quad \text{as } x \to 0.$$

There is nothing special about the line $y = 2x$ here. For example, as you can verify, $f(x,y) \to -\frac{1}{2}$ as $(x,y) \to (0,0)$ along the line $y = -x$. See Figure 14.6.1.

We have shown that not all paths to $(0,0)$ yield the same limiting value. It follows that f does not have a limit at $(0,0)$. ☐

Example 2 Show that the function $g(x,y) = \dfrac{x^2 y}{x^4 + y^2}$ has limiting value 0 as $(x,y) \to (0,0)$ along *any* line through the origin, but

$$\lim_{(x,y)\to(0,0)} g(x,y)$$

still does not exist. Note that the domain of g is all $(x,y) \neq (0,0)$.

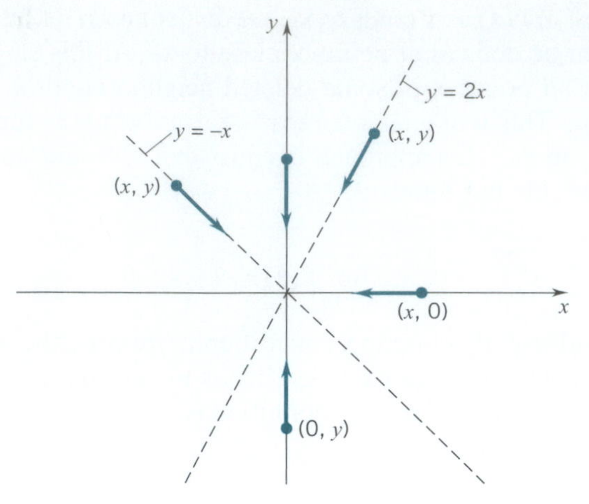

Figure 14.6.1

SOLUTION As in Example 1, it is easy to verify that $g(x,y) \to 0$ as $(x, y) \to (0,0)$. along the coordinate axes. If we let $y = mx$, then

$$g(x,y) = g(x,mx) = \frac{mx^3}{x^4 + m^2x^2} = \frac{mx}{x^2 + m^2} \qquad (x \neq 0)$$

and

$$\lim_{x\to 0} g(x,mx) = \lim_{x\to 0} \frac{mx}{x^2 + m^2} = 0.$$

Therefore, $g(x,y) \to 0$ as $(x,y) \to (0,0)$ along any line through the origin.

Now suppose that $(x,y) \to (0,0)$ along the parabola $y = x^2$. Then we have

$$g(x, y) = g(x,x^2) = \frac{x^4}{x^4 + x^4} = \frac{1}{2}$$

and

$$\lim_{x\to 0} g(x,x^2) = \lim_{x\to 0} \frac{1}{2} = \frac{1}{2}.$$

Thus, $g(x,y) \to \dfrac{1}{2}$ as $(x,y) \to (0,0)$ along $y = x^2$ (Figure 14.6.2).

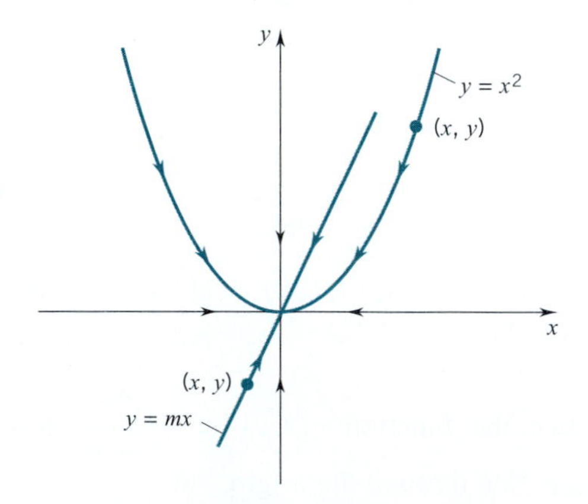

Figure 14.6.2

Since not all paths to $(0,0)$ yield the same limiting value, we conclude that g does not have a limit at $(0,0)$. □

As in the one-variable case, the limit (if it exists) is unique. Moreover, if

$$\lim_{\mathbf{x}\to\mathbf{x}_0} f(\mathbf{x}) = L \qquad \text{and} \qquad \lim_{\mathbf{x}\to\mathbf{x}_0} g(\mathbf{x}) = M,$$

then $\displaystyle\lim_{\mathbf{x}\to\mathbf{x}_0} [f(\mathbf{x}) + g(\mathbf{x})] = L + M, \quad \lim_{\mathbf{x}\to\mathbf{x}_0} [\alpha f(\mathbf{x})] = \alpha L, \quad \alpha$ a real number,

$$\lim_{\mathbf{x}\to\mathbf{x}_0} [f(\mathbf{x})g(\mathbf{x})] = LM, \qquad \text{and} \qquad \lim_{\mathbf{x}\to\mathbf{x}_0} [f(\mathbf{x})/g(\mathbf{x})] = L/M \quad \text{provided } M \neq 0.$$

These results are not hard to derive. You can do it simply by imitating the corresponding arguments in the one-variable case.

Suppose now that $\mathbf{x}_0$ is an interior point of the domain of f. To say that f is *continuous* at $\mathbf{x}_0$ is to say that

(14.6.2)

$$\lim_{\mathbf{x}\to\mathbf{x}_0} f(\mathbf{x}) = f(\mathbf{x}_0).$$

Another way to indicate that f is continuous at $\mathbf{x}_0$ is to write

(14.6.3)

$$\lim_{\mathbf{h}\to\mathbf{0}} f(\mathbf{x}_0 + \mathbf{h}) = f(\mathbf{x}_0).$$

For two variables we can write

$$\lim_{(x,y)\to(x_0,y_0)} f(x,y) = f(x_0,y_0)$$

and for three variables,

$$\lim_{(x,y,z)\to(x_0,y_0,z_0)} f(x,y,z) = f(x_0,y_0,z_0).$$

To say that f is *continuous on an open set S* is to say that f is continuous at all points of S.

Some Examples of Continuous Functions

Polynomials in several variables, for example,

$$P(x,y) = x^2y + 3x^3y^4 - x + 2y \qquad \text{and} \qquad Q(x,y,z) = 6x^3z - yz^3 + 2xyz$$

are everywhere continuous. In the two-variable case, that means continuity at each point of the xy-plane, and in the three-variable case, continuity at each point of three-space.

Rational functions (quotients of polynomials) are continuous everywhere except where the denominator is zero. Thus

$$f(x,y) = \frac{2x - y}{x^2 + y^2}$$

is continuous at each point of the xy-plane other than the origin $(0,0)$;

$$g(x,y) = \frac{x^4}{x - y}$$

is continuous except on the line $y = x$;

$$h(x, y) = \frac{1}{x^2 - y}$$

is continuous except on the parabola $y = x^2$;

$$F(x, y, z) = \frac{2x}{x^2 + y^2 + z^2}$$

is continuous at each point of three-space other than the origin $(0, 0, 0)$;

$$G(x, y, z) = \frac{x^5 - y}{ax + by + cz},$$

where a, b, c are constants, is continuous except on the plane $ax + by + cz = 0$.

You can construct more elaborate continuous functions by forming composites: take, for example,

$$f(x, y, z) = \tan^{-1}\left(\frac{xz^2}{x + y}\right), \quad g(x, y, z) = \sqrt{x^2 + y^4 + z^6}, \quad h(x, y, z) = \sin xyz.$$

The first function is continuous except along the vertical plane $x + y = 0$. The other two functions are continuous at each point of space. The continuity of such composites follows from a simple theorem that we now state and prove. In the theorem, g is a function of several variables, but f is a function of a single variable.

THEOREM 14.6.4 THE CONTINUITY OF COMPOSITE FUNCTIONS

If g is continuous at the point $\mathbf{x_0}$ and f is continuous at the number $g(\mathbf{x}_0)$, then the composition $f \circ g$ is continuous at the point $\mathbf{x}_0$.

PROOF We begin with $\epsilon > 0$. We must show that there exists a $\delta > 0$ such that

$$\text{if} \quad ||\mathbf{x} - \mathbf{x_0}|| < \delta, \qquad \text{then} \qquad |f(g(\mathbf{x})) - f(g(\mathbf{x_0}))| < \epsilon.$$

From the continuity of f at $g(\mathbf{x_0})$, we know that there exists a $\delta_1 > 0$ such that

$$\text{if} \quad |u - g(\mathbf{x_0})| < \delta_1, \qquad \text{then} \qquad |f(u) - f(g(\mathbf{x_0}))| < \epsilon.$$

From the continuity of g at $\mathbf{x_0}$, we know that there exists a $\delta > 0$ such that

$$\text{if} \quad ||\mathbf{x} - \mathbf{x_0}|| < \delta, \qquad \text{then} \qquad |g(\mathbf{x}) - g(\mathbf{x}_0)| < \delta_1.$$

This last δ obviously works; namely,

$$\text{if} \quad ||\mathbf{x} - \mathbf{x_0}|| < \delta, \qquad \text{then} \qquad |g(\mathbf{x}) - g(\mathbf{x}_0)| < \delta_1,$$

and so

$$|f(g(\mathbf{x})) - f(g(\mathbf{x_0}))| < \epsilon. \quad \square$$

Continuity in Each Variable Separately

A continuous function of several variables is continuous in each of its variables separately. In the two-variable case, this means that, if

$$\lim_{(x,y)\to(x_0,y_0)} f(x, y) = f(x_0, y_0),$$

then $\lim_{x \to x_0} f(x, y_0) = f(x_0, y_0)$ and $\lim_{y \to y_0} f(x_0, y) = f(x_0, y_0)$.

(This is not hard to prove.) The converse is false. *It is possible for a function to be continuous in each variable separately and yet fail to be continuous as a function of several variables.* You can see this in the next example.

Example 3 We set

$$f(x, y) = \begin{cases} \dfrac{2xy}{x^2 + y^2}, & (x, y) \neq (0, 0) \\ 0, & (x, y) = (0, 0). \end{cases}$$

Since $f(x, 0) = 0$ for all x and $f(0, y) = 0$ for all y,

we have $\lim_{x \to 0} f(x, 0) = 0 = f(0, 0)$ and $\lim_{y \to 0} f(0, y) = 0 = f(0, 0)$.

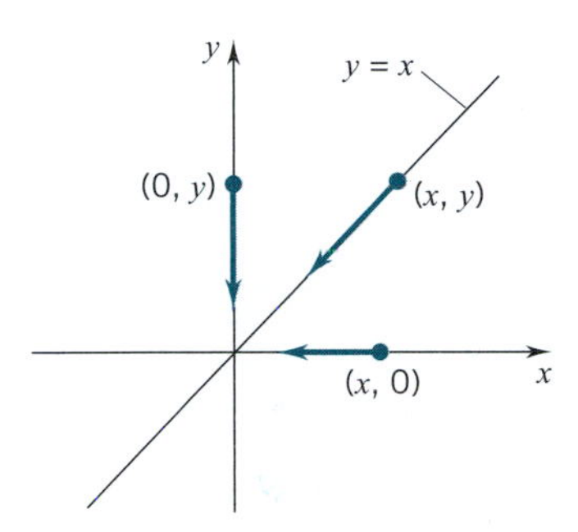

Figure 14.6.3

Thus, at the point $(0, 0)$, f is continuous in x and continuous in y. As a function of two variables, however, f is not continuous at $(0, 0)$. One way to see this is to note that we can approach $(0, 0)$ as closely as we wish by points of the form (t, t) with $t \neq 0$ (that is, along the line $y = x$.) (See Figure 14.6.3.) At such points f takes on the value 1:

$$f(t, t) = \frac{2t^2}{t^2 + t^2} = 1.$$

Hence, f cannot tend to $f(0, 0) = 0$ as required. ❑

Continuity and Partial Differentiability

For functions of a single variable the existence of the derivative guarantees continuity (Theorem 3.1.4). *For functions of several variables the existence of partial derivatives fails to guarantee continuity.*†

To show this, we can use the same function

$$f(x, y) = \begin{cases} \dfrac{2xy}{x^2 + y^2}, & (x, y) \neq (0, 0) \\ 0, & (x, y) = (0, 0). \end{cases}$$

Since both $f(x, 0)$ and $f(0, y)$ are constantly zero, both partials exist (and are zero) at $(0, 0)$, and yet, as you saw, the function is discontinuous at $(0, 0)$.

It is not hard to understand how a function can have partial derivatives and yet fail to be continuous. The existence of $\partial f/\partial x$ at (x_0, y_0) depends on the behavior of f only at points of the form $(x_0 + h, y_0)$. Similarly, the existence of $\partial f/\partial y$ at (x_0, y_0) depends on the behavior of f only at points of the form $(x_0, y_0 + k)$. On the other hand, continuity at (x_0, y_0) depends on the behavior of f at points of the more general form $(x_0 + h, y_0 + k)$. More briefly, we can put it this way: *the existence of a partial derivative depends on the behavior of the function along a line segment parallel to one of the axes, whereas continuity depends on the behavior of the function in all directions.*

† See. however, Exercise 32.

Derivatives of Higher Order; Equality of Mixed Partials

Suppose that f is a function of x and y with first partials f_x and f_y. These are again functions of x and y and may themselves possess partial derivatives: $(f_x)_x, (f_x)_y, (f_y)_x, (f_y)_y$. These functions are called the *second-order partials*. If $z = f(x, y)$, we use the following notations for second-order partials

$$(f_x)_x = f_{xx} = \frac{\partial}{\partial x}\left(\frac{\partial f}{\partial x}\right) = \frac{\partial^2 f}{\partial x^2} = \frac{\partial^2 z}{\partial x^2},$$

$$(f_x)_y = f_{xy} = \frac{\partial}{\partial y}\left(\frac{\partial f}{\partial x}\right) = \frac{\partial^2 f}{\partial y \partial x} = \frac{\partial^2 z}{\partial y \partial x},$$

$$(f_y)_x = f_{yx} = \frac{\partial}{\partial x}\left(\frac{\partial f}{\partial y}\right) = \frac{\partial^2 f}{\partial x \partial y} = \frac{\partial^2 z}{\partial x \partial y},$$

$$(f_y)_y = f_{yy} = \frac{\partial}{\partial y}\left(\frac{\partial f}{\partial y}\right) = \frac{\partial^2 f}{\partial y^2} = \frac{\partial^2 z}{\partial y^2}.$$

Note that there are two "mixed" partials: f_{xy} $(\partial^2 f/\partial y \partial x)$ and f_{yx} $(\partial^2 f/\partial x \partial y)$. The first of these is obtained by differentiating first respect to x and then with respect to y. The second is obtained by differentiating first with respect to y and then with respect to x.

Example 4 The function $f(x, y) = \sin x^2 y$ has first partials

$$\frac{\partial f}{\partial x} = 2xy\cos x^2 y \qquad \text{and} \qquad \frac{\partial f}{\partial y} = x^2 \cos x^2 y.$$

The second-order partials are

$$\frac{\partial^2 f}{\partial x^2} = -4x^2y^2 \sin x^2 y + 2y\cos x^2 y, \qquad \frac{\partial^2 f}{\partial y \partial x} = -2x^3 y \sin x^2 y + 2x \cos x^2 y,$$

$$\frac{\partial^2 f}{\partial x \partial y} = -2x^3 y \sin x^2 y + 2x \cos x^2 y, \qquad \frac{\partial^2 f}{\partial y^2} = -x^4 \sin x^2 y. \quad \square$$

Example 5 Setting $f(x, y) = \ln(x^2 + y^3)$, we have

$$\frac{\partial f}{\partial x} = \frac{2x}{x^2 + y^3} \qquad \text{and} \qquad \frac{\partial f}{\partial y} = \frac{3y^2}{x^2 + y^3}.$$

The second-order partials are

$$\frac{\partial^2 f}{\partial x^2} = \frac{(x^2 + y^3)2 - 2x(2x)}{(x^2 + y^3)^2} = \frac{2(y^3 - x^2)}{(x^2 + y^3)^2},$$

$$\frac{\partial^2 f}{\partial y \partial x} = \frac{-2x(3y^2)}{(x^2 + y^3)^2} = -\frac{6xy^2}{(x^2 + y^3)^2},$$

$$\frac{\partial^2 f}{\partial x \partial y} = \frac{-3y^2(2x)}{(x^2 + y^3)^2} = -\frac{6xy^2}{(x^2 + y^3)^2},$$

$$\frac{\partial^2 f}{\partial y^2} = \frac{(x^2 + y^3)6y - 3y^2(3y^2)}{(x^2 + y^3)^2} = \frac{3y(2x^2 - y^3)}{(x^2 + y^3)^2}. \quad \square$$

Perhaps you noticed that in both examples we had

$$\frac{\partial^2 f}{\partial y \partial x} = \frac{\partial^2 f}{\partial x \partial y}.$$

Since in neither case was f symmetric in x and y, this equality of the mixed partials was not due to symmetry. Actually it was due to continuity. It can be proved that

(14.6.5)

$$\frac{\partial^2 f}{\partial y \partial x} = \frac{\partial^2 f}{\partial x \partial y}$$

on every open set U on which f and its partials

$$\frac{\partial f}{\partial x}, \quad \frac{\partial f}{\partial y}, \quad \frac{\partial^2 f}{\partial y \partial x}, \quad \frac{\partial^2 f}{\partial x \partial y}$$

are all continuous.†

In the case of a function of three variables you can look for three first partials

$$\frac{\partial f}{\partial x}, \quad \frac{\partial f}{\partial y}, \quad \frac{\partial f}{\partial z},$$

and nine second partials

$$\frac{\partial^2 f}{\partial x^2}, \quad \frac{\partial^2 f}{\partial x \partial y}, \quad \frac{\partial^2 f}{\partial x \partial z}, \qquad \frac{\partial^2 f}{\partial y \partial x}, \quad \frac{\partial^2 f}{\partial y^2}, \quad \frac{\partial^2 f}{\partial y \partial z}, \qquad \frac{\partial^2 f}{\partial z \partial x}, \quad \frac{\partial^2 f}{\partial z \partial y}, \quad \frac{\partial^2 f}{\partial z^2}.$$

Here again, there is equality of the mixed partials

$$\frac{\partial^2 f}{\partial y \partial x} = \frac{\partial^2 f}{\partial x \partial y}, \qquad \frac{\partial^2 f}{\partial z \partial x} = \frac{\partial^2 f}{\partial x \partial z}, \qquad \frac{\partial^2 f}{\partial y \partial z} = \frac{\partial^2 f}{\partial z \partial y}$$

provided that f and its first and second partials are all continuous.

Example 6 For $f(x, y, z) = xe^y \sin \pi z$

we have

$$\frac{\partial f}{\partial x} = e^y \sin \pi z, \qquad \frac{\partial f}{\partial y} = xe^y \sin \pi z, \qquad \frac{\partial f}{\partial z} = \pi xe^y \cos \pi z.$$

$$\frac{\partial^2 f}{\partial x^2} = 0, \qquad \frac{\partial^2 f}{\partial y^2} = xe^y \sin \pi z, \qquad \frac{\partial^2 f}{\partial z^2} = -\pi^2 xe^y \sin \pi z.$$

$$\frac{\partial^2 f}{\partial y \partial x} = e^y \sin \pi z = \frac{\partial^2 f}{\partial x \partial y},$$

$$\frac{\partial^2 f}{\partial z \partial x} = \pi e^y \cos \pi z = \frac{\partial^2 f}{\partial x \partial z},$$

$$\frac{\partial^2 f}{\partial y \partial z} = \pi xe^y \cos \pi z = \frac{\partial^2 f}{\partial z \partial y}. \quad \square$$

† For a proof, consult a text on advanced calculus.

EXERCISES 14.6

Calculate the second-order partial derivatives. (Treat A, B, C, D as constants.)

1. $f(x, y) = Ax^2 + 2Bxy + Cy^2$.
2. $f(x, y) = Ax^3 + Bx^2y + Cxy^2$.
3. $f(x, y) = Ax + By + Ce^{xy}$.
4. $f(x, y) = x^2 \cos y + y^2 \sin x$.
5. $f(x, y, z) = (x + y^2 + z^3)^2$.
6. $f(x, y) = \sqrt{x + y^2}$.
7. $f(x, y) = \ln\left(\dfrac{x}{x+y}\right)$.
8. $f(x, y) = \dfrac{Ax + By}{Cx + Dy}$.
9. $f(x, y, z) = (x + y)(y + z)(z + x)$.
10. $f(x, y, z) = \tan^{-1} xyz$.
11. $f(x, y) = x^y$.
12. $f(x, y, z) = \sin(x + z^y)$.
13. $f(x, y) = xe^y + ye^x$.
14. $f(x, y) = \tan^{-1}(y/x)$.
15. $f(x, y) = \ln\sqrt{x^2 + y^2}$.
16. $f(x, y) = \sin(x^3y^2)$.
17. $f(x, y) = \cos^2(xy)$.
18. $f(x, y) = e^{xy^2}$.
19. $f(x, y, z) = xy \sin z - xz \sin y$.
20. $f(x, y, z) = xe^y + ye^z + ze^x$.
21. Show that

$$\text{if} \quad u = \frac{xy}{x+y},$$

$$\text{then} \quad x^2\frac{\partial^2 u}{\partial x^2} + 2xy\frac{\partial^2 u}{\partial x \partial y} + y^2\frac{\partial^2 u}{\partial y^2} = 0.$$

22. Verify that

$$\frac{\partial^2 f}{\partial y \partial x} = \frac{\partial^2 f}{\partial x \partial y}$$

for
(a) $f(x,y) = g(x) + h(y)$ with g and h differentiable.
(b) $f(x,y) = g(x)h(y)$ with g and h differentiable.
(c) $f(x,y)$ a polynomial in x and y. HINT: Check each term x^my^n separately.

23. Let f be a function of x and y with everywhere continuous second partials. Is it possible that
(a) $\dfrac{\partial f}{\partial x} = x + y$ and $\dfrac{\partial f}{\partial y} = y - x$?
(b) $\dfrac{\partial f}{\partial x} = xy$ and $\dfrac{\partial f}{\partial y} = xy$?

24. Let g be a twice-differentiable function of one variable and set

$$h(x,y) = g(x+y) + g(x-y).$$

Show that

$$\frac{\partial^2 h}{\partial x^2} = \frac{\partial^2 h}{\partial y^2}.$$

HINT: Use the chain rule of Exercise 58, Section 14.4.

25. Let f be a function of x and y with third-order partials

$$\frac{\partial^3 f}{\partial x^2 \partial y} = \frac{\partial}{\partial x}\left(\frac{\partial^2 f}{\partial x \partial y}\right) \quad \text{and} \quad \frac{\partial^3 f}{\partial y \partial x^2} = \frac{\partial}{\partial y}\left(\frac{\partial^2 f}{\partial x^2}\right).$$

Show that, if all the partials are continuous, then

$$\frac{\partial^3 f}{\partial x^2 \partial y} = \frac{\partial^2 f}{\partial y \partial x^2}.$$

26. Show that the following functions do not have a limit at $(0, 0)$:

(a) $f(x,y) = \dfrac{x^2 - y^2}{x^2 + y^2}$. (b) $f(x,y) = \dfrac{y^2}{x^2 + y^2}$.

In Exercises 27 and 28, evaluate the limit as (x,y) approaches the origin along:

(a) The x-axis.
(b) The y-axis.
(c) The line $y = mx$.
(d) The spiral $r = \theta, \theta > 0$.
(e) The differentiable curve $y = f(x)$, with $f(0) = 0$.
(f) The arc $r = \sin 3\theta$, $\frac{1}{6}\pi < \theta < \frac{1}{3}\pi$.
(g) The path $\mathbf{r}(t) = \dfrac{1}{t}\mathbf{i} + \dfrac{\sin t}{t}\mathbf{j}, t > 0$.

27. $\displaystyle\lim_{(x,y)\to(0,0)} \frac{xy}{x^2 + y^2}$.
28. $\displaystyle\lim_{(x,y)\to(0,0)} \frac{xy^2}{(x^2 + y^2)^{3/2}}$.
29. Set

$$g(x,y) = \begin{cases} \dfrac{x^2y^2}{x^4 + y^4}, & (x,y) \neq (0,0) \\ 0, & (x,y) = (0,0). \end{cases}$$

(a) Show that $\partial g/\partial x$ and $\partial g/\partial y$ both exist at $(0, 0)$. What are their values at $(0, 0)$?
(b) Show that $\displaystyle\lim_{(x,y)\to(0,0)} g(x,y)$ does not exist.

30. Set

$$f(x,y) = \frac{x - y^4}{x^3 - y^4}.$$

Determine whether or not f has a limit at $(1, 1)$.
HINT: Let (x, y) tend to $(1, 1)$ along the line $x = 1$ and along the line $y = 1$.

31. Set

$$f(x,y) = \begin{cases} \dfrac{xy(y^2 - x^2)}{x^2 + y^2}, & (x,y) \neq (0,0) \\ 0, & (x,y) = (0,0). \end{cases}$$

It can be shown that some of the second partials are discontinuous at $(0, 0)$. Show that

$$\frac{\partial^2 f}{\partial y \partial x}(0,0) \neq \frac{\partial^2 f}{\partial x \partial y}(0,0).$$

32. If a function of several variables has all first partials at a point, then it is continuous in each variable separately at

that point. Show, for example, that if f_x exists at (x_0, y_0), then f is continuous in x at (x_0, y_0).

33. Let f be a function of x and y which has continuous first and second partial derivatives throughout some set D in the plane. Suppose that $f_{xy}(x, y) = 0$ for all $(x, y) \in D$. What can you conclude about f?

34. Use a graphing utility to draw the graph of the function in Exercise 26(a) on the square $-2 \le x \le 2, -2 \le y \le 2$. Can you see that the limit as $(x, y) \to (0, 0)$ along the x-axis is 1 and the limit as $(x, y) \to (0, 0)$ along the y-axis is -1?

35. Use a graphing utility to draw the graph of the function in Exercise 27 on the square $-2 \le x \le 2, -2 \le y \le 2$. Can you see that the limit as $(x, y) \to (0, 0)$ along the coordinate axes is 0 and the limit as $(x, y) \to (0, 0)$ along the line $y = x$ is $\frac{1}{2}$?

36. Use a graphing utility to draw the graph of

$$f(x, y) = \frac{x^2 - y^2}{x^2 + y^2}$$

on the square $-2 \le x \le 2, -2 \le y \le 2$. From the graph, determine the limit of f as $(x, y) \to (0, 0)$ along the x-axis and as $(x, y) \to (0, 0)$ along the y-axis. Reverse the roles of x and y and see what happens.

■ PROJECT 14.6 Partial Differential Equations

The differential equations that we have studied so far are *ordinary differential equations.* They involve only ordinary derivatives, derivatives of functions of one variable. Here we examine some *partial differential equations,* the most prominent of which are equations that relate two or more of the partial derivatives of an unknown function of several variables.

Partial differential equations play an enormous role in science because the description of most natural phenomena is based on models that involve functions of several variables. For example, the partial differential equation known as the Schrödinger.† equation is viewed by many physicists as the cornerstone of quantum mechanics. Below we introduce two of the classical equations of physics having broad applications in science and engineering.

Problem 1. Show that the given function satisfies the corresponding partial differential equation

(a) $u = \dfrac{x^2y^2}{x + y}$; $x\dfrac{\partial u}{\partial x} + y\dfrac{\partial u}{\partial y} = 3u.$

(b) $u = x^2y + y^2z + z^2x$; $\dfrac{\partial u}{\partial x} + \dfrac{\partial u}{\partial y} + \dfrac{\partial u}{\partial z} = (x + y + z)^2.$

Laplace's Equation †† The partial differential equation

$$\frac{\partial^2 f}{\partial x^2} + \frac{\partial^2 f}{\partial y^2} = 0.$$

is known as *Laplace's equation in two dimensions.* It is used to describe potentials and steady-state temperature distributions in the plane. In three dimensions, Laplace's equation is

$$\frac{\partial^2 f}{\partial x^2} + \frac{\partial^2 f}{\partial y^2} + \frac{\partial^2 f}{\partial z^2} = 0.$$

It is satisfied by gravitational and electrostatic potentials and by steady-state temperature distributions in space. Functions that satisfy Laplace's equation are called *harmonic* functions.

Problem 2.

a. Show that the given functions satisfy Laplace's equation in two dimensions:

  (*i*) $f(x, y) = x^3 - 3xy^2$

  (*ii*) $f(x, y) = \cos x \sinh y + \sin x \cosh y$

  (*iii*) $f(x, y) = \ln\sqrt{x^2 + y^2}$.

b. Show that the given functions satisfy Laplace's equation in three dimensions:

  (*i*) $f(x, y, z) = \dfrac{1}{\sqrt{x^2 + y^2 + z^2}}$ (*ii*) $f(x, y, z) = e^{x+y}\cos\sqrt{2}\,z.$

The Wave Equation The partial differential equation

$$\frac{\partial^2 f}{\partial t^2} - c^2\frac{\partial^2 f}{\partial x^2} = 0,$$

where c is a positive constant, is known as the *wave equation.* It arises in the study of phenomena involving the propagation of waves in a continuous medium. For example, studies of water waves, sound waves, and light waves are all based on this equation. The wave equation is also used in the study of mechanical vibrations such as a vibrating string.

Problem 3. Show that the given functions satisfy the wave equation:

(i) $f(x, t) = (Ax + B)(Ct + D)$

(ii) $f(x, t) = \sin(x + ct)\cos(2x + 2ct)$

(iii) $f(x, t) = \ln(x + ct)$

(iv) $f(x, t) = (Ae^{kx} + Be^{-kx})(Ce^{ckt} + De^{-ckt})$

Problem 4. Let $f(x, t) = g(x + ct) + h(x - ct)$, where g and h are any two, twice differentiable functions. Show that f is a solution of the wave equation. [This is the most general form of a solution of the wave equation.]

† Introduced in 1926 by the Austrian theoretical physicist Ervin Schrödinger (1881–1961).

†† Named after the French mathematician Pierre-Simon Laplace (1749–1827). Laplace wrote two monumental works: one on celestial mechanics, the other on probability theory. He also made major contributions to the theory of differential equations.

■ CHAPTER HIGHLIGHTS

14.1 Elementary Examples

domain, range (p. 820) open (closed) unit disc (p. 820)
open (closed) unit ball (p. 821)

14.2 A Brief Catalog of the Quadric Surface: Projections

traces, sections (p. 824) ellipsoid (p. 824)
hyperboloid of one sheet (p. 825)
hyperboloid of two sheets (p. 825)
elliptic cone, nappe (p. 826) elliptic paraboloid (p. 826)
hyperbolic paraboloid (p. 826) cylinder (p. 827)
parabolic cylinder (p. 827) elliptic cylinder (p. 827)
hyperbolic cylinder (p. 827) projection (p. 828)

14.3 Graphs; Level Curves and Level Surfaces

level curve (p. 831) level surface (p. 835)

14.4 Partial Derivatives

for a function of two variables:
limit definition(p. 840)
geometric interpretation (Figs. 14.4.1 and 14.4.2)
for a function of three variables:
limit definitions (p. 842)

14.5 Open Sets and Closed Sets

neighborhood (p. 847) interior (p. 848)
boundary (p. 848)
open set (p. 848) closed set (p. 849)
deleted neighborhood (p. 850)
Some sets are neither open nor closed.

14.6 Limits and Continuity; Equality of Mixed Partials

limit of a function (p. 851) continuity at $\mathbf{x}_0$ (p. 853)

Points in the domain of a function of several variables can be written in vector notation. In the two-variable case, set $\mathbf{x}=(x, y)$, and, in the three-variable case, set $\mathbf{x} = (x, y, z)$. The vector notation enables us to treat the two cases together simply by writing $f(\mathbf{x})$.

A continuous function of several variables is continuous in each of its variables. The converse is false: it is possible for a function to be continuous in each variable separately and yet fail to be continuous as a function of several variables. (p. 855)

For functions of several variables the existence of partial derivatives fails to guarantee continuity. (p. 855)

The existence of a partial derivative depends on the behavior of the function along a line segment (two directions), whereas continuity depends on the behavior of the function in all directions. (p. 855)

equality of mixed partials (p. 857)

CHAPTER 15

GRADIENTS; EXTREME VALUES; DIFFERENTIALS

15.1 DIFFERENTIABILITY AND GRADIENT

The Notion of Differentiability

Our object here is to extend the notion of differentiability from real-valued functions of one variable to real-valued functions of several variables. Partial derivatives alone do not fulfill this role because they reflect behavior only along paths parallel to the coordinate axes.

In the one-variable case we formed the difference quotient

$$\frac{f(x+h)-f(x)}{h}$$

and called f differentiable at x provided that this quotient had a limit as h tended to zero. In the multivariable case we can still form the difference

$$f(\mathbf{x}+\mathbf{h})-f(\mathbf{x}),$$

but the "quotient"

$$\frac{f(\mathbf{x}+\mathbf{h})-f(\mathbf{x})}{\mathbf{h}}$$

is not defined because it makes no sense to divide a real number $[f(\mathbf{x}+\mathbf{h})-f(\mathbf{x})]$ by a vector $\mathbf{h}$.

We can get around this difficulty by going back to an idea introduced in the exercises to Section 3.9. We review the idea here.

Let g be a real-valued function of a single variable which is defined on some open interval containing 0. We say that $g(h)$ is *little-o(h)* (read "little oh of h") and write $g(h)=o(h)$ iff

$$\lim_{h\to 0}\frac{g(h)}{|h|}=0.$$

For a function of one variable, the following statements are equivalent:

$$\lim_{h\to 0} \frac{f(x+h)-f(x)}{h} = f'(x),$$

$$\lim_{h\to 0} \frac{[f(x+h)-f(x)]-f'(x)h}{h} = 0,$$

$$\lim_{h\to 0} \frac{[f(x+h)-f(x)]-f'(x)h}{|h|} = 0,$$

$$[f(x+h)-f(x)]-f'(x)h = o(h),$$

$$[f(x+h)-f(x)] = f'(x)h + o(h).$$

Thus, for a function of one variable, the derivative of f at x is the unique number $f'(x)$ such that

$$f(x+h)-f(x) = f'(x)h + o(h).$$

It is this view of the derivative that inspires the notion of differentiability in the multivariable case.

Paralleling the definition for a function of a single variable, let g be a function of several variables which is defined in some neighborhood of $\mathbf{0}$. We will say that $g(\mathbf{h})$ is $o(\mathbf{h})$ if

$$\lim_{h\to 0} \frac{g(\mathbf{h})}{||\mathbf{h}||} = 0.$$

We will denote by $o(\mathbf{h})$ any expression $g(\mathbf{h})$ that has this property.

Now let f be a function of several variables *defined at least in some neighborhood of* $\mathbf{x}$. [In the three-variable case, $\mathbf{x} = (x, y, z)$; in the two-variable case, $\mathbf{x} = (x, y)$.]

DEFINITION 15.1.1 DIFFERENTIABILITY

We say that f is *differentiable at* $\mathbf{x}$ if there exists a vector $\mathbf{y}$ such that

$$f(\mathbf{x}+\mathbf{h})-f(\mathbf{x}) = \mathbf{y}\cdot\mathbf{h} + o(\mathbf{h}).$$

It is not hard to show that, if such a vector $\mathbf{y}$ exists, it is unique (Exercise 41). We call this unique vector *the gradient of f at* $\mathbf{x}$ and denote it by $\nabla f(\mathbf{x})$: †

DEFINITION 15.1.2 GRADIENT

Let f be differentiable at $\mathbf{x}$. The *gradient of f at* $\mathbf{x}$ is the unique vector $\nabla f(\mathbf{x})$ such that

$$f(\mathbf{x}+\mathbf{h})-f(\mathbf{x}) = \nabla f(\mathbf{x})\cdot\mathbf{h} + o(\mathbf{h}).$$

† The symbol ∇, an inverted capital delta, is called a *nabla* and is read "del." The gradient of f is sometimes written grad f

The similarities between the one-variable case,

$$f(x+h) - f(x) = f'(x)h + o(h),$$

and the multivariable case,

$$f(\mathbf{x}+\mathbf{h}) - f(\mathbf{x}) = \nabla f(\mathbf{x}) \cdot \mathbf{h} + o(\mathbf{h}),$$

are obvious. We point to the differences. There are essentially two of them:

(1) While the derivative, $f'(x)$, is a number, the gradient $\nabla f(\mathbf{x})$ is a vector.

(2) While $f'(x)h$ is the ordinary product of two real numbers, $\nabla f(\mathbf{x}) \cdot \mathbf{h}$ is the dot product of two vectors.

Calculating Gradients

First we calculate some gradients by applying the definition directly. Then we give a theorem that makes such calculations much easier. Finally, we calculate some gradients with the aid of the theorem. As for notation, in the two-variable case we write

$$\nabla f(\mathbf{x}) = \nabla f(x,y) \qquad \text{and} \qquad \mathbf{h} = (h_1, h_2),$$

and in the three-variable case,

$$\nabla f(\mathbf{x}) = \nabla f(x,y,z) \qquad \text{and} \qquad \mathbf{h} = (h_1, h_2, h_3).$$

Example 1 For the function $f(x,y) = x^2 + y^2$, we have

$$\begin{aligned} f(\mathbf{x}+\mathbf{h}) - f(\mathbf{x}) &= f(x+h_1, y+h_2) - f(x,y) \\ &= \left[(x+h_1)^2 + (y+h_2)^2\right] - \left[x^2 + y^2\right] \\ &= [2xh_1 + 2yh_2] + \left[h_1^2 + h_2^2\right] \\ &= [2x\,\mathbf{i} + 2y\,\mathbf{j}] \cdot \mathbf{h} + \|\mathbf{h}\|^2. \end{aligned}$$

The remainder $\|\mathbf{h}\|^2$ is $o(\mathbf{h})$:

$$\frac{\|\mathbf{h}\|^2}{\|\mathbf{h}\|} = \|\mathbf{h}\| \to 0 \quad \text{as} \quad \mathbf{h} \to \mathbf{0}.$$

Thus,
$$\nabla f(\mathbf{x}) = \nabla f(x,y) = 2x\,\mathbf{i} + 2y\,\mathbf{j}. \quad \square$$

Example 2 For the function $f(x,y,z) = 2xy - 3z^2$, we have

$$\begin{aligned} f(\mathbf{x}+\mathbf{h}) - f(\mathbf{x}) &= f(x+h_1, y+h_2, z+h_3) - f(x,y,z) \\ &= 2(x+h_1)(y+h_2) - 3(z+h_3)^2 - \left[2xy - 3z^2\right] \\ &= 2xh_2 + 2yh_1 + 2h_1h_2 - 6zh_3 - 3h_3^2 \\ &= (2y\,\mathbf{i} + 2x\,\mathbf{j} - 6z\,\mathbf{k}) \cdot (h_1\,\mathbf{i} + h_2\,\mathbf{j} + h_3\,\mathbf{k}) + 2h_1h_2 - 3h_3^2 \\ &= (2y\,\mathbf{i} + 2x\,\mathbf{j} - 6z\,\mathbf{k}) \cdot \mathbf{h} + 2h_1h_2 - 3h_3^2. \end{aligned}$$

It remains to be shown that the remainder $g(\mathbf{h}) = 2h_1h_2 - 3h_3^2$ is $o(\mathbf{h})$. Since

$$g(\mathbf{h}) = (2h_2\,\mathbf{i} - 3h_3\,\mathbf{k}) \cdot (h_1\,\mathbf{i} + h_2\,\mathbf{j} + h_3\,\mathbf{k}) = (2h_2\,\mathbf{i} - 3h_3\,\mathbf{k}) \cdot \mathbf{h},$$

We can write

$$\frac{|g(\mathbf{h})|}{||\mathbf{h}||} = \frac{||2h_2\,\mathbf{i} - 3h_3\,\mathbf{k}||\;||\mathbf{h}||\;|\cos\theta|}{||\mathbf{h}||} \leq \frac{||2h_2\,\mathbf{i} - 3h_3\,\mathbf{k}||\;||\mathbf{h}||}{||\mathbf{h}||} = ||2h_2\,\mathbf{i} - 3h_3\,\mathbf{k}||.$$

Since $\mathbf{h} \to \mathbf{0}$ iff $h_1 \to 0, h_2 \to 0,$ and $h_3 \to 0$, it follows that $||2h_2\,\mathbf{i} - 3h_3\,\mathbf{k}|| \to 0$ as $\mathbf{h} \to 0$. Therefore, $g(\mathbf{h})/||\mathbf{h}|| \to 0$ as $\mathbf{h} \to \mathbf{0}$, and

$$\nabla f(\mathbf{x}) = 2y\,\mathbf{i} + 2x\,\mathbf{j} - 6z\,\mathbf{k}. \quad \square$$

Examples 1 and 2 illustrate the calculation of gradients directly from the definition of gradient. An easier way to calculate gradients is made possible by the theorem that follows

THEOREM 15.1.3

If f has continuous first partials in a neighborhood of $\mathbf{x}$, then f is differentiable at $\mathbf{x}$ and

$$\nabla f(\mathbf{x}) = \frac{\partial f}{\partial x}(\mathbf{x})\,\mathbf{i} + \frac{\partial f}{\partial y}(\mathbf{x})\,\mathbf{j} \qquad \text{(two variables)}$$

or

$$\nabla f(\mathbf{x}) = \frac{\partial f}{\partial x}(\mathbf{x})\,\mathbf{i} + \frac{\partial f}{\partial y}(\mathbf{x})\,\mathbf{j} + \frac{\partial f}{\partial z}(\mathbf{x})\,\mathbf{k} \qquad \text{(three variables)}$$

The proof is somewhat difficult. A proof of the two-variable case is given in a supplement at the end of this section.

Returning to Examples 1 and 2: for $f(x, y) = x^2 + y^2$, $\partial f/\partial x = 2x$ and $\partial f/\partial y = 2y$. Since these functions are continuous,

$$\nabla f(\mathbf{x}) = 2x\,\mathbf{i} + 2y\,\mathbf{j}.$$

For $f(x, y, z) = 2xy - 3z^2$, $\partial f/\partial x = 2y$, $\partial f/\partial y = 2x$ and $\partial f/\partial z = -6z$. Since these functions are continuous,

$$\nabla f(\mathbf{x}) = 2y\,\mathbf{i} + 2x\,\mathbf{j} - 6z\,\mathbf{k}.$$

Example 3 For $f(x, y) = x\,e^y - y\,e^x$, we have

$$\frac{\partial f}{\partial x}(x, y) = e^y - y\,e^x, \qquad \frac{\partial f}{\partial y}(x, y) = x\,e^y - e^x,$$

and therefore

$$\nabla f(x, y) = (e^y - y\,e^x)\,\mathbf{i} + (x\,e^y - e^x)\,\mathbf{j}. \quad \square$$

When there is no reason to emphasize the point of evaluation, we don't write

$$\nabla f(\mathbf{x}) \quad \text{or} \quad \nabla f(x, y) \quad \text{or} \quad \nabla f(x, y, z)$$

but simply ∇f. Thus for the function

$$f(x, y) = x\,e^y - y\,e^x$$

we write
$$\frac{\partial f}{\partial x} = e^y - y\,e^x, \quad \frac{\partial f}{\partial x} = x\,e^y - e^x$$

and
$$\nabla f = (e^y - y\,e^x)\,\mathbf{i} + (x\,e^y - e^x)\,\mathbf{j}.$$

Example 4 For $f(x, y, z) = \sin\,(xy^2z^3)$, we have

$$\frac{\partial f}{\partial x} = y^2z^3\cos(xy^2z^3), \quad \frac{\partial f}{\partial y} = 2xyz^3\cos(xy^2z^3), \quad \frac{\partial f}{\partial z} = 3xy^2z^2\cos(xy^2z^3)$$

and
$$\nabla f = yz^2\cos(xy^2z^3)[\,yz\,\mathbf{i} + 2xz\,\mathbf{j} + 3xy\,\mathbf{k}]. \quad \square$$

Example 5 Let f be the function defined by

$$f(x, y, z) = x\sin\pi y + y\cos\pi z.$$

Evaluate ∇f at $(0, 1, 2)$.

SOLUTION Here

$$\frac{\partial f}{\partial x} = \sin\pi y, \quad \frac{\partial f}{\partial y} = \pi x\cos\pi y + \cos\pi z, \quad \frac{\partial f}{\partial z} = -\pi y\sin\pi z.$$

At $(0, 1, 2)$,

$$\frac{\partial f}{\partial x} = 0, \quad \frac{\partial f}{\partial y} = 1, \quad \frac{\partial f}{\partial z} = 0 \quad \text{and thus} \quad \nabla f(0, 1, 2) = \mathbf{j}. \quad \square$$

Of special interest for later work are the powers of r where, as usual,

$$r = \|\mathbf{r}\| \quad \text{and} \quad \mathbf{r} = x\,\mathbf{i} + y\,\mathbf{j} + z\,\mathbf{k}.$$

We begin by showing that, for $r \neq 0$,

(15.1.4)
$$\boxed{\nabla r = \frac{\mathbf{r}}{r} \quad \text{and} \quad \nabla\left(\frac{1}{r}\right) = -\frac{\mathbf{r}}{r^3}.}$$

PROOF

$$\begin{aligned}
\nabla r &= \nabla(x^2 + y^2 + z^2)^{1/2} \\
&= \frac{\partial}{\partial x}(x^2 + y^2 + z^2)^{1/2}\,\mathbf{i} + \frac{\partial}{\partial y}(x^2 + y^2 + z^2)^{1/2}\,\mathbf{j} + \frac{\partial}{\partial z}(x^2 + y^2 + z^2)^{1/2}\,\mathbf{k} \\
&= \frac{x}{(x^2 + y^2 + z^2)^{1/2}}\,\mathbf{i} + \frac{y}{(x^2 + y^2 + z^2)^{1/2}}\,\mathbf{j} + \frac{z}{(x^2 + y^2 + z^2)^{1/2}}\,\mathbf{k} \\
&= \frac{1}{(x^2 + y^2 + z^2)^{1/2}}(x\,\mathbf{i} + y\,\mathbf{j} + z\,\mathbf{k}) = \frac{\mathbf{r}}{r}.
\end{aligned}$$

$$\nabla\left(\frac{1}{r}\right) = \nabla(x^2+y^2+z^2)^{-1/2}$$
$$= \frac{\partial}{\partial x}(x^2+y^2+z^2)^{-1/2}\,\mathbf{i} + \frac{\partial}{\partial y}(x^2+y^2+z^2)^{-1/2}\,\mathbf{j} + \frac{\partial}{\partial z}(x^2+y^2+z^2)^{-1/2}\,\mathbf{k}$$
$$= -\frac{x}{(x^2+y^2+z^2)^{3/2}}\,\mathbf{i} - \frac{y}{(x^2+y^2+z^2)^{3/2}}\,\mathbf{j} - \frac{z}{(x^2+y^2+z^2)^{3/2}}\,\mathbf{k}$$
$$= -\frac{1}{(x^2+y^2+z^2)^{3/2}}(x\,\mathbf{i}+y\,\mathbf{j}+z\,\mathbf{k}) = -\frac{\mathbf{r}}{r^3}. \quad \square$$

The formulas we just derived can be generalized. As you are asked to show in the Exercises, for each integer n and all $\mathbf{r} \neq \mathbf{0}$,

(15.1.5) $$\nabla r^n = nr^{n-2}\,\mathbf{r}.$$

(If n is positive and even, the result also holds at $\mathbf{r} = \mathbf{0}$.)

Differentiability Implies Continuity

As in the one-variable case, differentiability implies continuity; namely,

(15.1.6) if f is differentiable at $\mathbf{x}$, then f is continuous at $\mathbf{x}$.

To see this, write

$$f(\mathbf{x}+\mathbf{h}) - f(\mathbf{x}) = \nabla f(\mathbf{x}) \cdot \mathbf{h} + o(\mathbf{h})$$

and note that

$$|f(\mathbf{x}+\mathbf{h}) - f(\mathbf{x})| = |\nabla f(\mathbf{x}) \cdot \mathbf{h} + o(\mathbf{h})| \le |\nabla f(\mathbf{x}) \cdot \mathbf{h}| + |o(\mathbf{h})|. \qquad \text{(triangle inequality)}$$

As $\mathbf{h} \to \mathbf{0}$,

$$|\nabla f(\mathbf{x}) \cdot \mathbf{h}| \le \|\nabla f(\mathbf{x})\|\,\|\mathbf{h}\| \to 0 \qquad \text{and} \qquad |o(\mathbf{h})| \to 0.$$

Schwarz's inequality (for the $\le$); Exercise 42 (for $|o(\mathbf{h})| \to 0$)

It follows that

$$f(\mathbf{x}+\mathbf{h}) - f(\mathbf{x}) \to 0 \quad \text{and therefore} \quad f(\mathbf{x}+\mathbf{h}) \to f(\mathbf{x}). \quad \square$$

EXERCISES 15.1

Find the gradient.

1. $f(x,y) = 3x^2 - xy + y$.

2. $f(x,y) = Ax^2 + Bxy + Cy^2$.

3. $f(x,y) = xe^{xy}$.

4. $f(x,y) = \dfrac{x-y}{x^2+y^2}$.

5. $f(x,y) = 2xy^2 \sin(x^2+1)$.

6. $f(x,y) = \ln(x^2+y^2)$.

7. $f(x,y) = e^{x-y} - e^{y-x}$.

8. $f(x,y) = \dfrac{Ax+By}{Cx+Dy}$.

9. $f(x,y,z) = x^2y + y^2z + z^2x$.

10. $f(x,y,z) = \sqrt{x^2+y^2+z^2}$.

11. $f(x,y,z) = x^2ye^{-z}$.

12. $f(x,y,z) = xyz \ln(x+y+z)$.

13. $f(x,y,z) = e^{x+2y}\cos(z^2+1)$.

14. $f(x,y,z) = e^{yz^2/x^3}$.

15. $f(x,y,z) = \sin(2xy) + \ln(x^2z)$.

16. $f(x,y,z) = x^2y/z - 3xz^4$.

Find the gradient vector at the point P.

17. $f(x,y) = 2x^2 - 3xy + 4y^2$ at $P(2,3)$.

18. $f(x,y) = 2x(x-y)^{-1}$ at $P(3,1)$.

19. $f(x,y) = \ln(x^2+y^2)$ at $P(2,1)$.

20. $f(x,y) = x\tan^{-1}(y/x)$ at $P(1,1)$.

21. $f(x,y) = x\sin(xy)$ at $P(1,\pi/2)$.

22. $f(x,y) = xye^{-(x^2+y^2)}$ at $P(1,-1)$.

23. $f(x,y,z) = e^{-x}\sin(z+2y)$ at $P(0,\frac{1}{4}\pi,\frac{1}{4}\pi)$.

24. $f(x,y,z) = (x-y)\cos\pi z$ at $P(1,0,\frac{1}{2})$.

25. $f(x,y,z) = x - \sqrt{y^2+z^2}$ at $P(2,-3,4)$.

26. $f(x,y,z) = \cos(xyz^2)$ at $P(\pi,\frac{1}{4},-1)$.

C In Exercises 27 and 28, use a CAS to find the gradient of f at the point P.

27. (a) $f(x,y) = xy^2e^{-xy}$; $P(0,2)$.
(b) $f(x,y) = \sin(2x+y) - \cos(x-2y)$; $P(\pi/4,\pi/6)$.
(c) $f(x,y) = x - y\ln(x^2y)$; $P(1,e)$.

28. (a) $f(x,y,z) = \sqrt{x+y^2-z^3}$; $P(1,2,-3)$.
(b) $f(x,y,z) = \dfrac{xy}{x-y+z}$; $P(1,-2,3)$
(c) $f(x,y,z) = x\sin(z\ln y)$; $P(1,e^2,\pi/6)$.

Obtain the gradient directly from Definition 15.1.2

29. $f(x,y) = 3x^2 - xy + y$. **30.** $f(x,y) = \frac{1}{2}x^2 + 2xy + y^2$.

31. $f(x,y,z) = x^2y + y^2z + z^2x$.

32. $f(x,y,z) = 2x^2y - \dfrac{1}{z}$.

Find a function f with the gradient $\mathbf{F}$.

33. $\mathbf{F}(x,y) = 2xy\,\mathbf{i} + (1+x^2)\,\mathbf{j}$.

34. $\mathbf{F}(x,y) = (2xy+x)\,\mathbf{i} + (x^2+y)\,\mathbf{j}$.

35. $\mathbf{F}(x,y) = (x+\sin y)\,\mathbf{i} + (x\cos y - 2y)\,\mathbf{j}$.

36. $\mathbf{F}(x,y,z) = yz\,\mathbf{i} + (xz+2yz)\,\mathbf{j} + (xy+y^2)\,\mathbf{k}$.

37. Find (a) $\nabla(\ln r)$, (b) $\nabla(\sin r)$, (c) $\nabla(e^r)$, where $r = \sqrt{x^2+y^2+z^2}$.

38. Derive (15.1.5).

39. Let $f(x,y) = 1 + x^2 + y^2$.
(a) Find the points (x,y), if any, at which $\nabla f(x,y) = \mathbf{0}$.
(b) Sketch the graph of the surface $z = f(x,y)$.
(c) What can you say about the surface at the point(s) found in part (a)?

40. Repeat Exercise 39 for $f(x,y) = \sqrt{4-x^2-y^2}$.

41. (a) Show that, if $\mathbf{c}\cdot\mathbf{h}$ is $o(\mathbf{h})$, then $\mathbf{c} = \mathbf{0}$. HINT: First set $\mathbf{h} = h\,\mathbf{i}$, then set $\mathbf{h} = h\,\mathbf{j}$, then $\mathbf{h} = h\,\mathbf{k}$.
(b) Show that, if

$$f(\mathbf{x}+\mathbf{h}) - f(\mathbf{x}) = \mathbf{y}\cdot\mathbf{h} + o(\mathbf{h})$$

and

$$f(\mathbf{x}+\mathbf{h}) - f(\mathbf{x}) = \mathbf{z}\cdot\mathbf{h} + o(\mathbf{h}),$$

then

$$\mathbf{y} = \mathbf{z}.$$

42. Show that, if g is $o(\mathbf{h})$, then

$$\lim_{\mathbf{h}\to\mathbf{0}} g(\mathbf{h}) = 0.$$

43. Set

$$f(x,y) = \begin{cases} \dfrac{2xy}{x^2+y^2}, & (x,y) \neq (0,0) \\ 0, & (x,y) = (0,0). \end{cases}$$

(a) Show that f is not differentiable at $(0,0)$.
(b) In Section 14.6 you saw that the first partials $\partial f/\partial x$ and $\partial f/\partial y$ exist at $(0,0)$. Since these partials obviously exist at every other point of the plane, we can conclude from Theorem 15.1.3 that at least one of these partials is not continuous in a neighborhood of (0,0). Show that $\partial f/\partial x$ is discontinuous at $(0,0)$.

■ PROJECT 15.1 Points Where $\nabla f = 0$

In this project we find points (x,y) where $\nabla f = 0$. Then we investigate the behavior of the surface $z = f(x,y)$ at these points. In this context, recall the significance of $f'(x) = 0$ for a function f of a single variable.

For example, let

$$f(x,y) = \frac{-2y}{x^2+y^2+1}.$$

Then,

$$\frac{\partial f}{\partial x} = \frac{4xy}{(x^2+y^2+1)^2}, \quad \frac{\partial f}{\partial y} = \frac{2y^2-2x^2-2}{(x^2+y^2+1)^2}$$

and

$$\nabla f(x,y) = \frac{4xy}{(x^2+y^2+1)^2}\,\mathbf{i} + \frac{2y^2-2x^2-2}{(x^2+y^2+1)^2}\,\mathbf{j}.$$

Setting $\nabla f(x,y) = 0$, we get the pair of equations

$$\frac{4xy}{(x^2+y^2+1)^2} = 0. \quad \frac{2y^2-2x^2-2}{(x^2+y^2+1)^2} = 0$$

which are satisfied iff

$$4xy = 0 \quad \text{and} \quad 2y^2 - 2x^2 - 2 = 0$$

From the first equation, we get $x = 0$ or $y = 0$. Setting $x = 0$ in the second equation yields $y = \pm 1$; setting $y = 0$ in the second equation yields $x^2 = -1$, which has no solutions. Thus $\nabla f(x,y) = 0$ only at $(0,1)$ and $(0,-1)$. The figure shows the graph of the surface $z = f(x,y)$. Note that f has a maximum at $(0,-1)$ and a minimum at $(0,1)$.

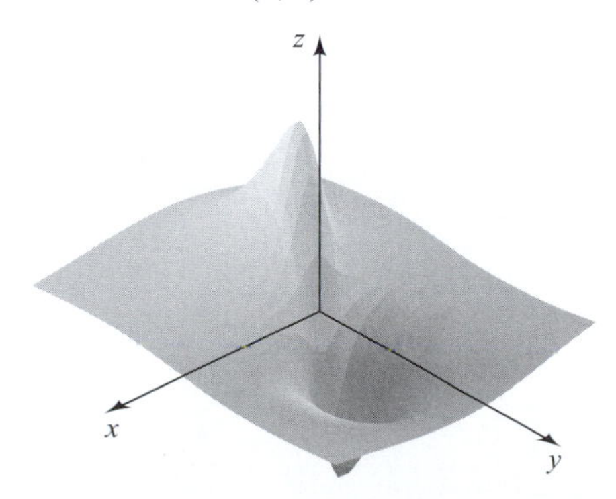

Problem 1. Let $f(x,y) = \dfrac{5x}{x^2+y^2+1}$, $-5 \le x \le 5$, $-5 \le y \le 5$.

(a) Find the points (x,y), if any, at which $\nabla f(x,y) = 0$.

(b) Use a CAS to draw the surface $z = f(x,y)$ and the level curves. Compare with the results of part (a).

(c) Investigate the behavior of f at the points found in part (a).

Problem 2. Repeat Problem 1 with $f(x,y) = (\sin x)(\sin y)$, $0 \le x \le 4\pi$, $0 \le y \le 4\pi$.

Problem 3. Repeat Problem 1 with $f(x,y) = 4xye^{-(x^2+y^2)}$, $-2 \le x \le 2$, $-2 \le y \le 2$.

Problem 4. Repeat Problem 1 with $f(x,y) = (x^2+4y^2)e^{1-(x^2+y^2)}$, $-2 \le x \le 2$, $-2 \le y \le 2$.

*SUPPLEMENT TO SECTION 15.1

PROOF OF THEOREM 15.1.3

We prove the theorem in the two-variable case. A similar argument yields a proof in the three-variable case, but there the details are more burdensome.

In the first place,

$$f(\mathbf{x}+\mathbf{h}) - f(\mathbf{x}) = f(x+h_1, y+h_2) - f(x,y).$$

Adding and subtracting $f(x, y+h_2)$, we have

$$(1) \qquad f(\mathbf{x}+\mathbf{h}) - f(\mathbf{x}) = [f(x+h_1,y+h_2) - f(x,y+h_2)] + [f(x,y+h_2) - f(x,y)].$$

By the mean-value theorem for functions of one variable, we know that there are numbers

$$0 < \theta_1 < 1 \qquad \text{and} \qquad 0 < \theta_2 < 1$$

such that

$$f(x+h_1,y+h_2) - f(x,y+h_2) = \frac{\partial f}{\partial x}(x+\theta_1 h_1, y+h_2)h_1$$

and

$$f(x,y+h_2) - f(x,y) = \frac{\partial f}{\partial y}(x, y+\theta_2 h_2)h_2. \qquad \text{(Exercise 40, Section 4.1)}$$

By the continuity of $\partial f/\partial x$,

$$\frac{\partial f}{\partial x}(x+\theta_1 h_1, y+h_2) = \frac{\partial f}{\partial x}(x,y) + \epsilon_1(\mathbf{h})$$

where $\qquad \epsilon_1(\mathbf{h}) \to 0 \quad \text{as} \quad \mathbf{h} \to 0.$†

---

†

$$\epsilon_1(\mathbf{h}) = \frac{\partial f}{\partial x}(x+\theta_1 h_1, y+h_2) - \frac{\partial f}{\partial x}(x,y) \to 0$$

since, by the continuity of $\partial f/\partial x$,

$$\frac{\partial f}{\partial x}(x+\theta_1 h_1, y+h_2) \to \frac{\partial f}{\partial x}(x,y).$$

By the continuity of $\partial f/\partial y$,

$$\frac{\partial f}{\partial x}(x, y+\theta_2 h_2) = \frac{\partial f}{\partial x}(x, y) + \epsilon_2(\mathbf{h})$$

where $$\epsilon_2(\mathbf{h}) \to 0 \quad \text{as} \quad \mathbf{h} \to 0.$$

Substituting these expressions in equation (1), we find that

$$\begin{aligned} f(\mathbf{x}+\mathbf{h}) - f(\mathbf{x}) &= \left[\frac{\partial f}{\partial x}(x, y) + \epsilon_1(\mathbf{h})\right] h_1 + \left[\frac{\partial f}{\partial y}(x, y) + \epsilon_2(\mathbf{h})\right] h_2 \\ &= \left[\frac{\partial f}{\partial x}(x, y) + \epsilon_1(\mathbf{h})\right] (\mathbf{i}\cdot\mathbf{h}) + \left[\frac{\partial f}{\partial y}(x, y) + \epsilon_2(\mathbf{h})\right] (\mathbf{j}\cdot\mathbf{h}) \\ &= \left[\frac{\partial f}{\partial x}(x, y)\,\mathbf{i} + \epsilon_1(\mathbf{h})\,\mathbf{i}\right]\cdot\mathbf{h} + \left[\frac{\partial f}{\partial y}(x, y)\,\mathbf{j} + \epsilon_2(\mathbf{h})\,\mathbf{j}\right]\cdot\mathbf{h} \\ &= \left[\frac{\partial f}{\partial x}(x, y)\,\mathbf{i} + \frac{\partial f}{\partial y}(x, y)\,\mathbf{j}\right]\cdot\mathbf{h} + \left[\epsilon_1(\mathbf{h})\,\mathbf{i} + \epsilon_2(\mathbf{h})\,\mathbf{j}\right]\cdot\mathbf{h}. \end{aligned}$$

To complete the proof of the theorem we need only show that

(2) $$[\epsilon_1(\mathbf{h})\,\mathbf{i} + \epsilon_2(\mathbf{h})\,\mathbf{j}]\cdot\mathbf{h} = o(\mathbf{h}).$$

From Schwarz's inequality, $|\mathbf{a}\cdot\mathbf{b}| \le \|\mathbf{a}\|\,\|\mathbf{b}\|$, we know that

$$|[\epsilon_1(\mathbf{h})\,\mathbf{i} + \epsilon_2(\mathbf{h})\,\mathbf{j}]\cdot\mathbf{h}| \le \|\epsilon_1(\mathbf{h})\,\mathbf{i} + \epsilon_2(\mathbf{h})\,\mathbf{j}\|\,\|\mathbf{h}\|.$$

It follows that

$$\frac{|[\epsilon_1(\mathbf{h})\,\mathbf{i} + \epsilon_2(\mathbf{h})\,\mathbf{j}]\cdot\mathbf{h}|}{\|\mathbf{h}\|} \le \|\epsilon_1(\mathbf{h})\,\mathbf{i}+\epsilon_2(\mathbf{h})\,\mathbf{j}\| \le \|\epsilon_1(\mathbf{h})\,\mathbf{i}\|+\|\epsilon_2(\mathbf{h})\,\mathbf{j}\| = |\epsilon_1(\mathbf{h})|+|\epsilon_2(\mathbf{h})|.$$

↑ by the triangle inequality

As $\mathbf{h} \to \mathbf{0}$, the expression on the right tends to 0. This shows that (2) holds and completes the proof of the theorem. ☐

■ 15.2 GRADIENTS AND DIRECTIONAL DERIVATIVES

Some Elementary Formulas

In many respects gradients behave just as derivatives do in the one-variable case. In particular, if $\nabla f(\mathbf{x})$ and $\nabla g(\mathbf{x})$ exist, then $\nabla[f(\mathbf{x})+g(\mathbf{x})]$, $\nabla[\alpha f(\mathbf{x})]$, and $\nabla[f(\mathbf{x})g(\mathbf{x})]$ all exist, and

(15.2.1)
$$\begin{aligned} \nabla[f(\mathbf{x})+g(\mathbf{x})] &= \nabla f(\mathbf{x}) + \nabla g(\mathbf{x}), \\ \nabla[\alpha f(\mathbf{x})] &= \alpha\nabla f(\mathbf{x}), \\ \nabla[f(\mathbf{x})g(\mathbf{x})] &= f(\mathbf{x})\,\nabla g(\mathbf{x}) + g(\mathbf{x})\,\nabla f(\mathbf{x}). \end{aligned}$$

The first two formulas are easy to derive. To derive the third formula, let's assume that $\nabla f(\mathbf{x})$ and $\nabla g(\mathbf{x})$ both exist. Our task is to show that

$$f(\mathbf{x}+\mathbf{h})g(\mathbf{x}+\mathbf{h}) - f(\mathbf{x})g(\mathbf{x}) = [f(\mathbf{x})\,\nabla g(\mathbf{x}) + g(\mathbf{x})\,\nabla f(\mathbf{x})]\cdot\mathbf{h} + o(\mathbf{h}).$$

We now sketch how this can be done. We leave it to you to justify each step. The key to proving a product rule is to add and subtract an appropriate expression (see, for instance, the proof of product rule Theorem 3.2.6). Starting from

$$f(\mathbf{x}+\mathbf{h})g(\mathbf{x}+\mathbf{h})-f(\mathbf{x})g(\mathbf{x}),$$

we add and subtract the term $f(\mathbf{x})g(\mathbf{x}+\mathbf{h})$. This gives

$$\begin{aligned}
&[f(\mathbf{x}+\mathbf{h})g(\mathbf{x}+\mathbf{h})-f(\mathbf{x})g(\mathbf{x}+\mathbf{h})]+[f(\mathbf{x})g(\mathbf{x}+\mathbf{h})-f(\mathbf{x})g(\mathbf{x})]\\
&=[f(\mathbf{x}+\mathbf{h})-f(\mathbf{x})]g(\mathbf{x}+\mathbf{h})+f(\mathbf{x})[g(\mathbf{x}+\mathbf{h})-g(\mathbf{x})]\\
&=[\nabla f(\mathbf{x})\cdot\mathbf{h}+o(\mathbf{h})]g(\mathbf{x}+\mathbf{h})+f(\mathbf{x})[\nabla g(\mathbf{x})\cdot\mathbf{h}+o(\mathbf{h})]\\
&=g(\mathbf{x}+\mathbf{h})\nabla f(\mathbf{x})\cdot\mathbf{h}+f(\mathbf{x})\nabla g(\mathbf{x})\cdot\mathbf{h}+o(\mathbf{h}) \qquad \text{(Exercise 30)}\\
&=g(\mathbf{x})\nabla f(\mathbf{x})\cdot\mathbf{h}+f(\mathbf{x})\nabla g(\mathbf{x})\cdot\mathbf{h}+[g(\mathbf{x}+\mathbf{h})-g(\mathbf{x})]\nabla f(\mathbf{x})\cdot\mathbf{h}+o(\mathbf{h})\\
&=[g(\mathbf{x})\nabla f(\mathbf{x})+f(\mathbf{x})\nabla g(\mathbf{x})]\cdot\mathbf{h}+o(\mathbf{h}). \qquad \text{(Exercise 30)} \quad \square
\end{aligned}$$

In Exercise 43, you are asked to derive this formula from Theorem 15.1.3.

Directional Derivatives

Here we take up an idea that generalizes the notion of partial derivative. Its connection with gradients will be made clear as we go on. We begin by recalling the definitions of the first partial derivatives:

(two variables)

$$\frac{\partial f}{\partial x}(x,y)=\lim_{h\to 0}\frac{f(x+h,y)-f(x,y)}{h},$$

$$\frac{\partial f}{\partial y}(x,y)=\lim_{h\to 0}\frac{f(x,y+h)-f(x,y)}{h},$$

(three variables)

$$\frac{\partial f}{\partial x}(x,y,z)=\lim_{h\to 0}\frac{f(x+h,y,z)-f(x,y,z)}{h},$$

$$\frac{\partial f}{\partial y}(x,y,z)=\lim_{h\to 0}\frac{f(x,y+h,z)-f(x,y,z)}{h},$$

$$\frac{\partial f}{\partial z}(x,y,z)=\lim_{h\to 0}\frac{f(x,y,z+h)-f(x,y,z)}{h}.$$

Expressed in vector notation, these definitions take the form

$(\mathbf{x}=(x,y))$

$$\frac{\partial f}{\partial x}(\mathbf{x})=\lim_{h\to 0}\frac{f(\mathbf{x}+h\,\mathbf{i})-f(\mathbf{x})}{h},$$

$$\frac{\partial f}{\partial y}(\mathbf{x})=\lim_{h\to 0}\frac{f(\mathbf{x}+h\,\mathbf{j})-f(\mathbf{x})}{h},$$

$(\mathbf{x}=(x,y,z))$

$$\frac{\partial f}{\partial x}(\mathbf{x})=\lim_{h\to 0}\frac{f(\mathbf{x}+h\,\mathbf{i})-f(\mathbf{x})}{h},$$

$$\frac{\partial f}{\partial y}(\mathbf{x})=\lim_{h\to 0}\frac{f(\mathbf{x}+h\,\mathbf{j})-f(\mathbf{x})}{h},$$

$$\frac{\partial f}{\partial z}(\mathbf{x})=\lim_{h\to 0}\frac{f(\mathbf{x}+h\,\mathbf{k})-f(\mathbf{x})}{h}.$$

Each partial is thus the limit of a quotient

$$\frac{f(\mathbf{x}+h\,\mathbf{u})-f(\mathbf{x})}{h}$$

where $\mathbf{u}$ is one of the unit coordinate vectors, $\mathbf{i}, \mathbf{j}, \mathbf{k}$. There is no reason to be so restrictive on $\mathbf{u}$. If f is defined in a neighborhood of $\mathbf{x}$, then, for small h, the difference quotient

$$\frac{f(\mathbf{x}+h\,\mathbf{u})-f(\mathbf{x})}{h}$$

makes sense for any unit vector $\mathbf{u}$.

DEFINITION 15.2.2 DIRECTIONAL DERIVATIVE

For each unit vector $\mathbf{u}$, the limit

$$f'_u(\mathbf{x}) = \lim_{h \to 0} \frac{f(\mathbf{x} + h\mathbf{u}) - f(\mathbf{x})}{h},$$

if it exists, is called the *directional derivative of f at $\mathbf{x}$ in the direction $\mathbf{u}$.*

It is important to recognize that each partial derivative $\partial f/\partial x$, $\partial f/\partial y$, $\partial f/\partial z$ is itself a directional derivative:

(15.2.3)
$$\frac{\partial f}{\partial x}(\mathbf{x}) = f'_{\mathbf{i}}(\mathbf{x}), \quad \frac{\partial f}{\partial y}(\mathbf{x}) = f'_{\mathbf{j}}(\mathbf{x}), \quad \frac{\partial f}{\partial z}(\mathbf{x}) = f'_{\mathbf{k}}(\mathbf{x}).$$

As you know, the partials of f give the rates of change of f in the $\mathbf{i}, \mathbf{j}, \mathbf{k}$ directions. The directional derivative $f'_{\mathbf{u}}$ gives *the rate of change of f in the direction* $\mathbf{u}$.

A geomentric interpretation of the directional derivative for a function f of two variables can be obtained from Figure 15.2.1. Fix a point (x, y) in the domain of f and let $\mathbf{u}$ be a unit vector with initial point (x, y) in the xy-plane. Let C be the curve of intersection of the surface $z = f(x, y)$ and the plane p which contains $\mathbf{u}$ and is perpendicular to the xy-plane. Then $f'_{\mathbf{u}}(\mathbf{x})$ is the slope of the line tangent to C at the point $(x, y, f(x, y))$.

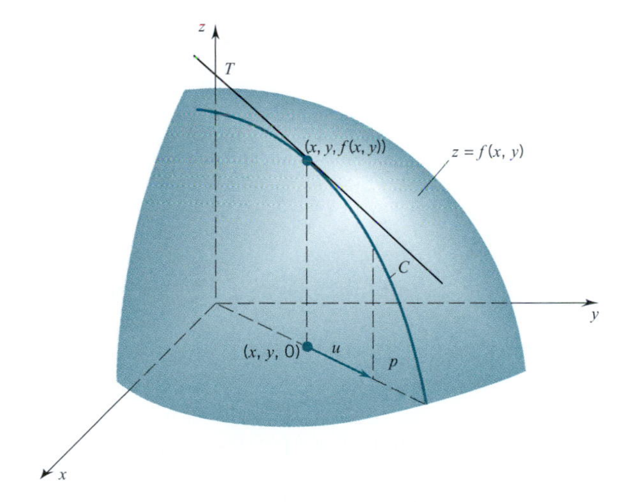

Figure 15.2.1

Remark The definition of the directional derivative in the direction $\mathbf{u}$ requires $\mathbf{u}$ to be a unit vector. However, we can extend the definition to arbitrary nonzero vectors by adopting the following convention: the directional derivative of f at $\mathbf{x}$ in the direction of a nonzero vector $\mathbf{a}$ is by definition $f'_{\mathbf{u}}(\mathbf{x})$ where $\mathbf{u} = \mathbf{a}/\|\mathbf{a}\|$ is the unit vector which has the same direction as $\mathbf{a}$. □

There is an important connection between the gradient at $\mathbf{x}$ and the directional derivatives at $\mathbf{x}$.

THEOREM 15.2.4

If f is differentiable at $\mathbf{x}$, then f has a directional derivative at $\mathbf{x}$ in every direction $\mathbf{u}$, where $\mathbf{u}$ is a unit vector, and

$$f'_{\mathbf{u}}(\mathbf{x}) = \nabla f(\mathbf{x}) \cdot \mathbf{u}.$$

PROOF We take $\mathbf{u}$ as a unit vector and assume that f is differentiable at $\mathbf{x}$. The differentiability at $\mathbf{x}$ tells us that $\nabla f(\mathbf{x})$ exists and

$$f(\mathbf{x} + h\,\mathbf{u}) - f(\mathbf{x}) = \nabla f\,(\mathbf{x}) \cdot h\,\mathbf{u} + o(h\,\mathbf{u}).$$

Division by h gives

$$\frac{f(\mathbf{x} + h\,\mathbf{u}) - f(\mathbf{x})}{h} = \nabla f(\mathbf{x}) \cdot \mathbf{u} + \frac{o(h\,\mathbf{u})}{h}.$$

Since
$$\left|\frac{o(h\,\mathbf{u})}{h}\right| = \frac{|o(h\,\mathbf{u})|}{|h|} = \frac{|o(h\,\mathbf{u})|}{||h\,\mathbf{u}||} \to 0,$$

you can see that

$$\frac{o(h\,\mathbf{u})}{h} \to 0 \qquad \text{and thus} \qquad \frac{f(\mathbf{x} + h\,\mathbf{u}) - f(\mathbf{x})}{h} \to \nabla f(\mathbf{x}) \cdot \mathbf{u}. \quad \square$$

Earlier (Theorem 15.1.3) you saw that, *if f has continuous first partials in a neighborhood of* $\mathbf{x}$, *then f is differentiable at* $\mathbf{x}$ *and*

$$\nabla f(\mathbf{x}) = \frac{\partial f}{\partial x}(\mathbf{x})\,\mathbf{i} + \frac{\partial f}{\partial y}(\mathbf{x})\,\mathbf{j}, \qquad [\mathbf{x} = (x, y)]$$

$$\nabla f(\mathbf{x}) = \frac{\partial f}{\partial x}(\mathbf{x})\,\mathbf{i} + \frac{\partial f}{\partial y}(\mathbf{x})\,\mathbf{j} + \frac{\partial f}{\partial z}(\mathbf{x})\,\mathbf{k}. \qquad [\mathbf{x} = (x, y, z)]$$

The next theorem shows that this formula for $\nabla f(\mathbf{x})$ holds wherever f is differentiable.

THEOREM 15.2.5

If f is differentiable at $\mathbf{x}$, then all the first partial derivatives of f exist at $\mathbf{x}$ and

$$\nabla f(\mathbf{x}) = \frac{\partial f}{\partial x}(\mathbf{x})\,\mathbf{i} + \frac{\partial f}{\partial y}(\mathbf{x})\,\mathbf{j} \qquad (\mathbf{x} = (x, y)),$$

$$\nabla f(\mathbf{x}) = \frac{\partial f}{\partial x}(\mathbf{x})\,\mathbf{i} + \frac{\partial f}{\partial y}(\mathbf{x})\,\mathbf{j} + \frac{\partial f}{\partial z}(\mathbf{x})\,\mathbf{k} \qquad (\mathbf{x} = (x, y, z)).$$

PROOF It is sufficient to prove the theorem for the case $\mathbf{x} = (x, y, z)$. Assume that f is differentiable at $\mathbf{x}$. Then $\nabla f(\mathbf{x})$ exists and we can write

$$\nabla f(\mathbf{x}) = [\nabla f(\mathbf{x}) \cdot \mathbf{i}]\,\mathbf{i} + [\nabla f(\mathbf{x}) \cdot \mathbf{j}]\,\mathbf{j} + [\nabla f(\mathbf{x}) \cdot \mathbf{k}]\,\mathbf{k}. \qquad (12.4.13)$$

The result follows from observing that

$$\begin{aligned}
&\quad\ \ (15.2.4)\quad (15.2.3)\\
\nabla f(\mathbf{x})\cdot\mathbf{i} &\overset{\downarrow}{=} f'_{\mathbf{i}}(\mathbf{x}) \overset{\downarrow}{=} \frac{\partial f}{\partial x}(\mathbf{x}),\\
\nabla f(\mathbf{x})\cdot\mathbf{j} &= f'_{\mathbf{j}}(\mathbf{x}) = \frac{\partial f}{\partial y}(\mathbf{x}),\\
\nabla f(\mathbf{x})\cdot\mathbf{k} &= f'_{\mathbf{k}}(\mathbf{x}) = \frac{\partial f}{\partial z}(\mathbf{x}). \quad \square
\end{aligned}$$

Example 1 Find the directional derivative of the function $f(x,y) = x^2 + y^2$ at the point $(1,2)$ in the direction of the vector $2\,\mathbf{i} - 3\,\mathbf{j}$.

SOLUTION In the first place, $2\,\mathbf{i} - 3\,\mathbf{j}$ is not a unit vector; its norm is $\sqrt{13}$. The unit vector in the direction of $2\,\mathbf{i} - 3\,\mathbf{j}$ is the vector

$$\mathbf{u} = \frac{1}{\sqrt{13}}[2\,\mathbf{i} - 3\,\mathbf{j}].$$

Next, $$\nabla f = 2x\,\mathbf{i} + 2y\,\mathbf{j},$$

and therefore $$\nabla f(1,2) = 2\,\mathbf{i} + 4\,\mathbf{j}.$$

By Theorem 15.2.4 we have

$$\begin{aligned}
f'_{\mathbf{u}}(1,2) &= \nabla f(1,2)\cdot\mathbf{u}\\
&= (2\mathbf{i} + 4\,\mathbf{j})\cdot\frac{1}{\sqrt{13}}[2\,\mathbf{i} - 3\,\mathbf{j}] = \frac{-8}{\sqrt{13}} \cong -2.219. \quad \square
\end{aligned}$$

Example 2 Find the directional derivative of the function

$$f(x,y,z) = 2xz^2\cos\pi y$$

at the point $P(1,2,-1)$ toward the point $Q(2,1,3)$.

SOLUTION The vector from P to Q is given by $\overrightarrow{PQ} = \mathbf{i} - \mathbf{j} + 4\,\mathbf{k}$. The unit vector in this direction is the vector

$$\mathbf{u} = \frac{1}{3\sqrt{2}}[\mathbf{i} - \mathbf{j} + 4\,\mathbf{k}].$$

Here, $$\frac{\partial f}{\partial x} = 2z^2\cos\pi y, \qquad \frac{\partial f}{\partial y} = -2\pi xz^2\sin\pi y, \qquad \frac{\partial f}{\partial z} = 4xz\cos\pi y,$$

so that $$\frac{\partial f}{\partial x}(1,2,-1) = 2, \qquad \frac{\partial f}{\partial y}(1,2,-1) = 0, \qquad \frac{\partial f}{\partial z}(1,2,-1) = -4.$$

Therefore, $$\nabla f(1,2,-1) = 2\,\mathbf{i} - 4\,\mathbf{k}$$

and $$f'_{\mathbf{u}}(1,2,-1) = \nabla f(1,2,-1) \cdot \mathbf{u} = (2\,\mathbf{i} - 4\,\mathbf{k}) \cdot \frac{1}{3\sqrt{2}}[\mathbf{i} - \mathbf{j} + 4\mathbf{k}]$$

$$= -\frac{14}{3\sqrt{2}} \cong -3.30. \quad \square$$

You know that for each unit vector $\mathbf{u}$

$$f'_{\mathbf{u}}(\mathbf{x}) = \nabla f(\mathbf{x}) \cdot \mathbf{u}$$

Since $$\nabla f(\mathbf{x}) \cdot \mathbf{u} = \text{comp}_{\mathbf{u}} \nabla f(\mathbf{x}), \qquad (12.4.11)$$

we have

(15.2.6) $$f'_{\mathbf{u}}(\mathbf{x}) = \text{comp}_{\mathbf{u}} \nabla f(\mathbf{x}).$$

Namely, the directional derivative in a direction $\mathbf{u}$ *is the component of the gradient vector in that direction.* (See Figure 15.2.2.)

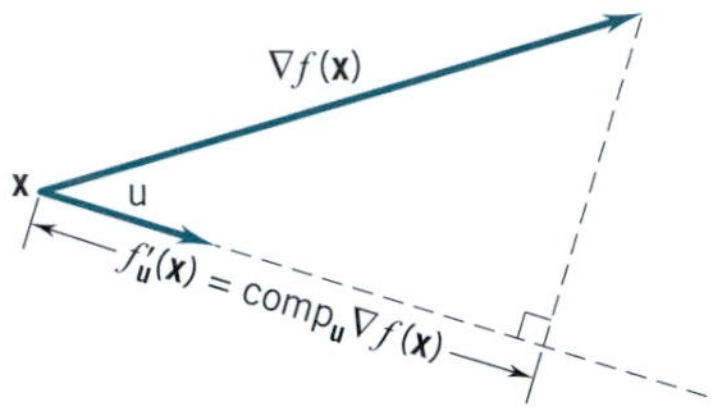

Figure 15.2.2

If $\nabla f(\mathbf{x}) \neq \mathbf{0}$, then

$$f'_{\mathbf{u}}(\mathbf{x}) = \nabla f(\mathbf{x}) \cdot \mathbf{u} = ||\nabla f(\mathbf{x})||\,||\mathbf{u}|| \cos\theta = ||\nabla f(\mathbf{x})|| \cos\theta$$

↑ (12.4.7) ↑ $||\mathbf{u}|| = 1$

where θ is the angle between $\nabla f(\mathbf{x})$ and $\mathbf{u}$. Since $-1 \leq \cos\theta \leq 1$, we have

$$-||\nabla f(\mathbf{x})|| \leq f'_{\mathbf{u}}(\mathbf{x}) \leq ||\nabla f(\mathbf{x})|| \quad \text{for all directions } \mathbf{u}.$$

If $\mathbf{u}$ points in the direction of $\nabla f(\mathbf{x})$, then

$$f'_{\mathbf{u}}(\mathbf{x}) = ||\nabla f(\mathbf{x})||, \qquad (\theta = 0, \cos\theta = 1)$$

and if $\mathbf{u}$ points in the direction of $-\nabla f(\mathbf{x})$, then

$$f'_{\mathbf{u}}(\mathbf{x}) = -||\nabla f(\mathbf{x})||. \qquad (\theta = \pi, \cos\theta = -1)$$

Since the directional derivative gives the rate of change of the function in that direction, it is clear that

(15.2.7) from each point $\mathbf{x}$ of the domain, a differentiable function f increases most rapidly in the direction of the gradient (the rate of change at $\mathbf{x}$ being $||\nabla f(\mathbf{x})||$); the function decreases most rapidly in the opposite direction (the rate of change at $\mathbf{x}$ being $-||\nabla f(\mathbf{x})||$).

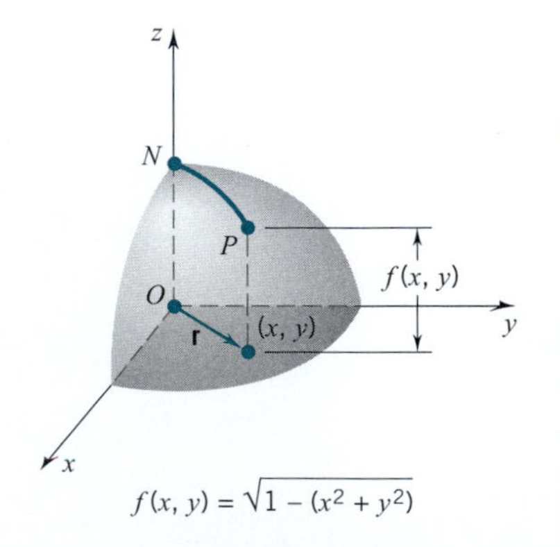

Figure 15.2.3

Example 3 The graph of the function $f(x,y) = \sqrt{1-(x^2+y^2)}$ is the upper half of the unit sphere $x^2 + y^2 + z^2 = 1$. The function is defined on the closed unit disc $x^2 + y^2 \leq 1$, but differentiable only on the open unit disc.

In Figure 15.2.3 we have marked a point (x, y) and drawn the corresponding radius vector $\mathbf{r} = x\,\mathbf{i} + y\,\mathbf{j}$. The gradient

$$\nabla f(x,y) = \frac{-x}{\sqrt{1-(x^2+y^2)}}\,\mathbf{i} + \frac{-y}{\sqrt{1-(x^2+y^2)}}\,\mathbf{j}$$

is a negative multiple of **r**:

$$\nabla f(x,y) = -\frac{1}{\sqrt{1-(x^2+y^2)}}(x\,\mathbf{i}+y\,\mathbf{j}) = -\frac{1}{\sqrt{1-(x^2+y^2)}}\,\mathbf{r}.$$

Since **r** points from the origin to (x,y), the gradient points from (x,y) to the origin. This means that f increases most rapidly toward the origin. This is borne out by the observation that along the hemispherical surface the path of steepest ascent from the point $P(x,y,f(x,y))$ is the "great circle route to the North Pole." □

Example 4 Suppose that the temperature at each point of a metal plate is given by the function

$$T(x,y) = e^x \cos\, y + e^y \cos x.$$

(a) In what direction does the temperature increase most rapidly at the point $(0,0)$? What is this rate of increase?

(b) In what direction does the temperature decrease most rapidly at $(0,0)$?

SOLUTION

$$\begin{aligned}\nabla T(x,y) &= \frac{\partial T}{\partial x}(x,y)\,\mathbf{i} + \frac{\partial T}{\partial y}(x,y)\,\mathbf{j}\\ &= (e^x\cos y - e^y \sin x)\,\mathbf{i} + (e^y\cos x - e^x \sin y)\,\mathbf{j}.\end{aligned}$$

(a) At $(0,0)$ the temperature increases most rapidly in the direction of the gradient

$$\nabla T(0,0) = \mathbf{i}+\mathbf{j}.$$

This rate of increase is

$$||\nabla T(0,0)|| = ||\mathbf{i}+\mathbf{j}|| = \sqrt{2}.$$

(b) The temperature decreases most rapidly in the direction of

$$-\nabla T(0,0) = -\mathbf{i}-\mathbf{j}. \quad □$$

Example 5 Suppose that the mass density (mass per unit volume) of a metal ball centered at the origin is given by the function

$$\lambda(x,y,z) = k\,e^{-(x^2+y^2+z^2)}, \quad k \text{ a positive constant.}$$

(a) In what direction does the density increase most rapidly at the point (x,y,z)? What is this rate of density increase?

(b) In what direction does the density decrease most rapidly?

(c) What are the rates of density change at (x,y,z) in the **i**, **j**, **k** directions?

SOLUTION The gradient

$$\begin{aligned}\nabla\lambda(x,y,z) &= \frac{\partial\lambda}{\partial x}(x,y,z)\,\mathbf{i} + \frac{\partial\lambda}{\partial y}(x,y,z)\,\mathbf{j} + \frac{\partial\lambda}{\partial z}(x,y,z)\,\mathbf{k}\\ &= -2k\,e^{-(x^2+y^2+z^2)}[x\,\mathbf{i}+y\,\mathbf{j}+z\,\mathbf{k}].\end{aligned}$$

Since $\lambda(x, y, z) = k\, e^{-(x^2+y^2+z^2)}$, we have

$$\nabla\lambda(x, y, z) = -2\lambda(x, y, z)\mathbf{r}.$$

From this, we see that the gradient points from (x, y, z) in the direction opposite to that of the radius vector.

(a) The density increases most rapidly toward the origin. The rate of increase is

$$||\nabla\lambda(x, y, z)|| = 2\lambda(x, y, z)||\mathbf{r}|| = 2\lambda(x, y, z)\sqrt{x^2 + y^2 + z^2}.$$

(b) The density decreases most rapidly directly away from the origin.

(c) The rates of density change in the $\mathbf{i}, \mathbf{j}, \mathbf{k}$ directions are given by the directional derivatives

$$\lambda'_{\mathbf{i}}(x, y, z) = \nabla\lambda(x, y, z) \cdot \mathbf{i} = -2x\, \lambda(x, y, z),$$
$$\lambda'_{\mathbf{j}}(x, y, z) = \nabla\lambda(x, y, z) \cdot \mathbf{j} = -2y\, \lambda(x, y, z),$$
$$\lambda'_{\mathbf{k}}(x, y, z) = \nabla\lambda(x, y, z) \cdot \mathbf{k} = -2z\, \lambda(x, y, z).$$

These are just the first partials of λ. ☐

Example 6 Suppose that the temperature at each point of a metal plate is given by the function

$$T(x, y) = 1 + x^2 - y^2.$$

Find the path followed by a heat-seeking particle that originates at $(-2, 1)$.

SOLUTION The particle moves in the direction of the gradient vector

$$\nabla T = 2x\,\mathbf{i} - 2y\,\mathbf{j}.$$

We want the curve

$$C : \mathbf{r}(t) = x(t)\,\mathbf{i} + y(t)\,\mathbf{j}$$

which begins at $(-2, 1)$ and at each point has its tangent vector in the direction of ∇T. We can satisfy the first condition by setting

$$x(0) = -2, \qquad y(0) = 1.$$

We can satisfy the second condition by setting

$$x'(t) = 2x(t), \qquad y'(t) = -2y(t). \qquad \text{(explain)}$$

These differential equations, together with initial conditions at $t = 0$, imply that

$$x(t) = -2\,e^{2t}, \qquad y(t) = e^{-2t}. \qquad \text{(Section 7.6)}$$

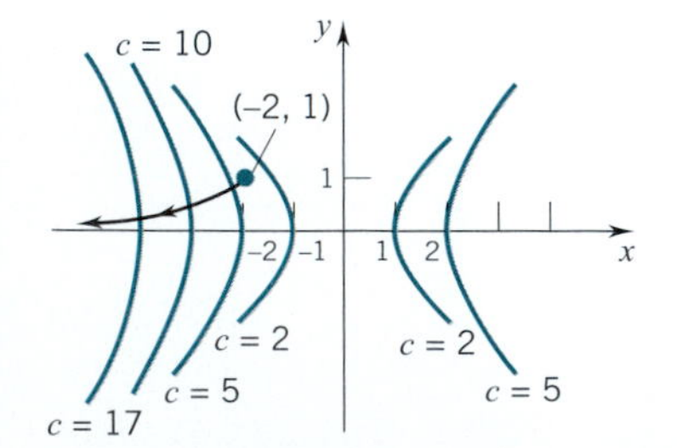

Figure 15.2.4

We can eliminate the parameter t by noting that

$$x(t)y(t) = (-2\,e^{2t})(e^{-2t}) = -2.$$

In terms of just x and y we have

$$xy = -2.$$

The particle moves from the point $(-2, 1)$ along the left branch of the hyperbola $xy = -2$ in the direction of decreasing x (see Figure 15.2.4).

The level curves, *isothermals* of the temperature distribution T, are also hyperbolas. As you can verify, the path of the object is perpendicular to each of the isothermals $x^2 - y^2 = c - 1$. ❑

Remark The pair of differential equations

$$x'(t) = 2x(t), \qquad y'(t) = -2y(t)$$

can be set as a single differential equation in x and y: the relation

$$\frac{y'(t)}{x'(t)} = -\frac{y(t)}{x(t)} \qquad \text{gives} \qquad \frac{dy}{dx} = -\frac{y}{x}.$$

This equation is readily solved directly (see Section 8.9):

$$\begin{aligned} \frac{1}{y}\frac{dy}{dx} &= -\frac{1}{x} \\ \ln|y| &= -\ln|x| + C \\ \ln|x| + \ln|y| &= C \\ \ln|xy| &= C \end{aligned}$$

Thus xy is constant:

$$xy = k.$$

Since the curve passes through the point $(-2, 1)$, $k = -2$ and once again we have the curve

$$xy = -2.$$

You will be called on to use this method of solution in some of the Exercises. ❑

EXERCISES 15.2

Find the directional derivative at the point P in the direction indicated.

1. $f(x,y) = x^2 + 3y^2$ at $P(1,1)$ in the direction of $\mathbf{i} - \mathbf{j}$.
2. $f(x,y) = x + \sin(x+y)$ at $P(0,0)$ in the direction of $2\mathbf{i} + \mathbf{j}$.
3. $f(x,y) = xe^y - ye^x$ at $P(1,0)$ in the direction of $3\mathbf{i} + 4\mathbf{j}$.
4. $f(x,y) = \dfrac{2x}{x-y}$ at $P(1,0)$ in the direction of $\mathbf{i} - \sqrt{3}\,\mathbf{j}$.
5. $f(x,y) = \dfrac{ax+by}{x+y}$ at $P(1,1)$ in the direction of $\mathbf{i} - \mathbf{j}$.
6. $f(x,y) = \dfrac{x+y}{cx+dy}$ at $P(1,1)$ in the direction of $c\mathbf{i} - d\mathbf{j}$.
7. $f(x,y) = \ln(x^2 + y^2)$ at $P(0,1)$ in the direction of $8\mathbf{i} + \mathbf{j}$.
8. $f(x,y) = x^2y + \tan y$ at $P(-1, \pi/4)$ in the direction of $\mathbf{i} - 2\mathbf{j}$.
9. $f(x,y,z) = xy + yz + zx$ at $P(1,-1,1)$ in the direction of $\mathbf{i} + 2\mathbf{j} + \mathbf{k}$.
10. $f(x,y,z) = x^2y + y^2z + z^2x$ at $P(1,0,1)$ in the direction of $3\mathbf{j} - \mathbf{k}$.
11. $f(x,y,z) = (x + y^2 + z^3)^2$ at $P(1,-1,1)$ in the direction of $\mathbf{i} + \mathbf{j}$.
12. $f(x,y,z) = Ax^2 + Bxyz + Cy^2$ at $P(1,2,1)$ in the direction of $A\mathbf{i} + B\mathbf{j} + C\mathbf{k}$.
13. $f(x,y,z) = x\tan^{-1}(y+z)$ at $P(1,0,1)$ in the direction of $\mathbf{i} + \mathbf{j} - \mathbf{k}$.
14. $f(x,y,z) = xy^2\cos z - 2yz^2\sin\pi x + 3zx^2$ at $P(0,-1,\pi)$ in the direction of $2\mathbf{i} - \mathbf{j} + 2\mathbf{k}$.
15. Find the directional derivative of $f(x,y) = \ln\sqrt{x^2+y^2}$ at $(x,y) \neq (0,0)$ toward the origin.
16. Find the directional derivative of $f(x,y) = (x-1)y^2e^{xy}$ at $(0,1)$ toward the point $(-1,3)$.
17. Find the directional derivative of $f(x,y) = Ax^2 + 2Bxy + Cy^2$ at (a,b) toward (b,a) (a) if $a > b$; (b) if $a < b$.
18. Find the directional derivative of $f(x,y,z) = z\ln(x/y)$ at $(1,1,2)$ toward the point $(2,2,1)$.
19. Find the directional derivative of $f(x,y,z) = xe^{y^2-z^2}$ at $(1,2,-2)$ in the direction of the path $\mathbf{r}(t) = t\,\mathbf{i} + 2\cos(t-1)\,\mathbf{j} - 2e^{t-1}\,\mathbf{k}$.

20. Find the directional derivative of $f(x,y,z) = x^2 + yz$ at $(1,-3,2)$ in the direction of the path $\mathbf{r}(t) = t^2\mathbf{i} + 3t\mathbf{j} + (1-t^3)\mathbf{k}$.

21 Find the directional derivative of $f(x,y,z) = x^2+2xyz-yz^2$ at $(1,1,2)$ in a direction parallel to the straight line

$$\frac{x-1}{2} = y - 1 = \frac{z-2}{-3}.$$

22. Find the directional derivative of $f(x,y,z) = e^x \cos \pi yz$ at $(0,1,\frac{1}{2})$ in a direction parallel to the line of intersection of the planes $x+y-z-5=0$ and $4x-y-z+2=0$.

In Exercises 23–26, find a unit vector in the direction in which f increases most rapidly at P and give the rate of change of f in that direction; find a unit vector in the direction in which f decreases most rapidly at P and give the rate of change of f in that direction.

23. $f(x,y) = y^2 e^{2x}$; $P(0,1)$.

24. $f(x,y) = x + \sin(x+2y)$; $P(0,0)$.

25. $f(x,y,z) = \sqrt{x^2+y^2+z^2}$; $P(1,-2,1)$.

26. $f(x,y,z) = x^2 z e^y + xz^2$; $P(1, \ln 2, 2)$.

27. Let $f = f(x)$ be a differentiable function of one variable. What is the gradient of f at x_0? What is the geometric significance of the direction of this gradient?

28. Suppose that f is differentiable at (x_0,y_0) and $\nabla f(x_o,y_o) \neq 0$. Compute the rate of change of f in the direction of the vector

$$\frac{\partial f}{\partial y}(x_0,y_0)\,\mathbf{i} - \frac{\partial f}{\partial x}(x_0,y_0)\,\mathbf{j}.$$

Give a geometric interpretation to your answer.

29. Let

$$f(x,y) = \sqrt{x^2+y^2}.$$

(a) Show that $\partial f/\partial x$ is not defined at $(0,0)$.

(b) Is f differentiable at $(0,0)$?

30. Verify that, if g is continuous at $\mathbf{x}$, then

(a) $g(\mathbf{x}+\mathbf{h})\,o(\mathbf{h}) = o(\mathbf{h})$ and

(b) $[g(\mathbf{x}+\mathbf{h}) - g(\mathbf{x})]\nabla f(\mathbf{x}) \cdot \mathbf{h} = o(\mathbf{h})$.

31. Given the density function $\lambda(x,y) = 48 - \frac{4}{3}x^2 - 3y^2$, find the rate of density change (a) at $(1,-1)$ in the direction of the most rapid density decrease; (b) at $(1,2)$ in the $\mathbf{i}$ direction;(c) at (2,2) away from the origin.

32. The intensity of light in a neighborhood of the point $(-2,1)$ is given by a function of the form $I(x,y) = A - 2x^2 - y^2$. Find the path followed by a light-seeking particle that originates at the center of the neighborhood.

33. Determine the path of steepest descent along the surface $z = x^2 + 3y^2$ from each of the following points: (a) $(1,1,4)$; (b) $(1,-2,13)$.

34. Determine the path of steepest ascent along the hyperbolic paraboloid $z = \frac{1}{2}x^2 - y^2$ from each of the following points: (a) $(-1,1,-\frac{1}{2})$; (b) $(1,0,\frac{1}{2})$.

35. Determine the path of steepest descent along the surface $z = a^2x^2 + b^2y^2$ from the point: (a,b,a^4+b^4).

36. The temperature in a neighborhood of the origin is given by a function of the form

$$T(x,y) = T_0 + e^y \sin x.$$

Find the path followed by a heat-fleeing particle that originates at the origin.

37. The temperature in a neighborhood of the point $(\frac{1}{4}\pi, 0)$ is given by the function

$$T(x,y) = \sqrt{2}\,e^{-y} \cos x.$$

Find the path followed by a heat-seeking particle that originates at the center of the neighborhood.

38. Determine the path of steepest descent along the surface $z = A + x + 2y - x^2 - 3y^2$ from the point $(0,0,A)$.

39. Set $f(x,y) = 3x^2 + y$.

(a) Find

$$\lim_{h\to 0} \frac{f(x(2+h), y(2+h)) - f(2,4)}{h}$$

given that $x(t) = t$ and $y(t) = t^2$. (These functions parametrize the parabola $y = x^2$.)

(b) Find

$$\lim_{h\to 0} \frac{f(x(4+h), y(4+h)) - f(2,4)}{h}$$

given that $x(t) = \frac{1}{4}(t+4)$ and $y(t) = t$. (These functions parametrize the line $y = 4x - 4$.)

(c) Compute the directional derivative of f at $(2,4)$ in the direction of $\mathbf{i} + 4\mathbf{j}$.

(d) Observe that $\mathbf{i} + 4\mathbf{j}$ is a direction vector for the line $y = 4x - 4$ and that this line is tangent to the parabola $y = x^2$ at $(2,4)$. Explain then why the computations in (a),(b), and (c) yield different values.

40. According to Newton's law of gravitation, the force exerted on a particle of mass m located at the point (x,y,z) by a particle of mass M located at the origin is given by

$$\mathbf{F}(x,y,z) = \frac{-GMm}{r^3}\mathbf{r}$$

where $\mathbf{r} = x\,\mathbf{i} + y\,\mathbf{j} + z\,\mathbf{k}$, $r = ||\mathbf{r}||$, and G is the gravitational constant. Show that $\mathbf{F}$ is the gradient of

$$f(x,y,z) = \frac{GMm}{r}.$$

41. Let $\mathbf{u}$ be a unit position vector (initial point at the origin) in the plane and let θ be the angle measured in the counterclockwise direction from the positive x-axis to $\mathbf{u}$. Let f be a differentiable function of two variables.

(a) Show that $f'_{\mathbf{u}}(x,y) = \dfrac{\partial f}{\partial x}\cos\theta + \dfrac{\partial f}{\partial y}\sin\theta$.

(b) Let $f(x,y) = x^3 + 2xy - xy^2$ and $\theta = 2\pi/3$. Find $f'_u(-1,2)$.

42. Refer to Exercise 41. Let $f(x,y) = x^2e^{2y}$ and $\theta = 5\pi/4$. Find $f'_{\mathbf{u}}(x,y)$ and $f'_{\mathbf{u}}(2, \ln 2)$.

43. Assume that $\nabla f(\mathbf{x})$ and $\nabla g(\mathbf{x})$ exist. Use Theorem 15.1.3 to derive the product rule

$$\nabla[f(\mathbf{x})g(\mathbf{x})] = f(\mathbf{x})\nabla g(\mathbf{x}) + g(\mathbf{x})\nabla f(\mathbf{x}).$$

44. Assume that $\nabla f(\mathbf{x})$ and $\nabla g(\mathbf{x})$ exist, and that $g(\mathbf{x}) \neq 0$. Derive the quotient rule

$$\nabla\left[\frac{f(\mathbf{x})}{g(\mathbf{x})}\right] = \frac{g(\mathbf{x})\nabla f(\mathbf{x}) - f(\mathbf{x})\nabla g(\mathbf{x})}{g^2(\mathbf{x})}.$$

45. Assume that $\nabla f(\mathbf{x})$ exists. Prove that, for each integer n,

$$\nabla f^n(\mathbf{x}) = nf^{n-1}(\mathbf{x})\nabla f(\mathbf{x}).$$

Does this result hold if n is replaced by an arbitrary real number?

■ 15.3 THE MEAN-VALUE THEOREM; CHAIN RULES

The Mean-Value Theorem

You have seen the important role played by the mean-value theorem in the calculus of functions of one variable. Here we take up the analogous result for functions of several variables. Let **a** and **b** be points (either in the plane or in three-space); by $\overline{\mathbf{ab}}$ we mean the line segment that joins point **a** to point **b**.

THEOREM 15.3.1 THE MEAN-VALUE THEOREM (SEVERAL VARIABLES)

If f is differentiable at each point of the line segment $\overline{\mathbf{ab}}$, then there exists on that line segment a point **c** between **a** and **b** such that

$$f(\mathbf{b}) - f(\mathbf{a}) = \nabla f(\mathbf{c}) \cdot (\mathbf{b} - \mathbf{a}).$$

PROOF As t ranges from 0 to 1, $\mathbf{a} + t(\mathbf{b} - \mathbf{a})$ traces out the line segment $\overline{\mathbf{ab}}$. The idea of the proof is to apply the one-variable mean-value theorem to the function

$$g(t) = f(\mathbf{a} + t[\mathbf{b} - \mathbf{a}]), \qquad t \in [0, 1].$$

To show that g is differentiable on the open interval $(0, 1)$ we take $t \in (0, 1)$ and form

$$\begin{aligned} g(t+h) - g(t) &= f(\mathbf{a} + (t+h)[\mathbf{b} - \mathbf{a}]) - f(\mathbf{a} + t[\mathbf{b} - \mathbf{a}]) \\ &= f(\mathbf{a} + t[\mathbf{b} - \mathbf{a}] + h[\mathbf{b} - \mathbf{a}]) - f(\mathbf{a} + t[\mathbf{b} - \mathbf{a}]) \\ &= \nabla f(\mathbf{a} + t[\mathbf{b} - \mathbf{a}]) \cdot h[\mathbf{b} - \mathbf{a}] + o(h[\mathbf{b} - \mathbf{a}]). \end{aligned}$$

Since $$\nabla f(\mathbf{a} + t[\mathbf{b} + \mathbf{a}]) \cdot h(\mathbf{b} - \mathbf{a}) = [\nabla f(\mathbf{a} + t[\mathbf{b} - \mathbf{a}]) \cdot (\mathbf{b} - \mathbf{a})]h$$

and the $o(h[\mathbf{b} - \mathbf{a}])$ term is obviously $o(h)$, we can write

$$g(t+h) - g(t) = [\nabla f(\mathbf{a} + t[\mathbf{b} - \mathbf{a}]) \cdot (\mathbf{b} - \mathbf{a})]h + o(h).$$

Dividing both sides by h, we see that g is differentiable and

$$g'(t) = \nabla f(\mathbf{a} + t[\mathbf{b} - \mathbf{a}]) \cdot (\mathbf{b} - \mathbf{a}).$$

The function g is clearly continuous at 0 and at 1. Applying the one-variable mean-value theorem to g, we can conclude that there exists a number t_0 between 0 and 1 such that

$$(*) \qquad g(1) - g(0) = g'(t_0)(1 - 0).$$

Since $g(1) = f(\mathbf{b}), g(0) = f(\mathbf{a})$, and $g'(t_0) = \nabla f(\mathbf{a} + t_0[\mathbf{b} - \mathbf{a}]) \cdot (\mathbf{b} - \mathbf{a})$, condition $(*)$ gives

$$f(\mathbf{b}) - f(\mathbf{a}) = \nabla f(\mathbf{a} + t_0[\mathbf{b} - \mathbf{a}]) \cdot (\mathbf{b} - \mathbf{a}).$$

Setting $\mathbf{c} = \mathbf{a} + t_0[\mathbf{b} - \mathbf{a}]$, we have

$$f(\mathbf{b}) - f(\mathbf{a}) = \nabla f(\mathbf{c}) \cdot (\mathbf{b} - \mathbf{a}). \quad \square$$

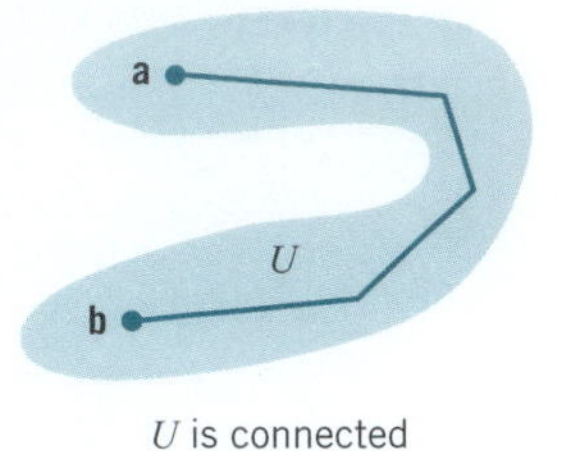

U is connected

Figure 15.3.1

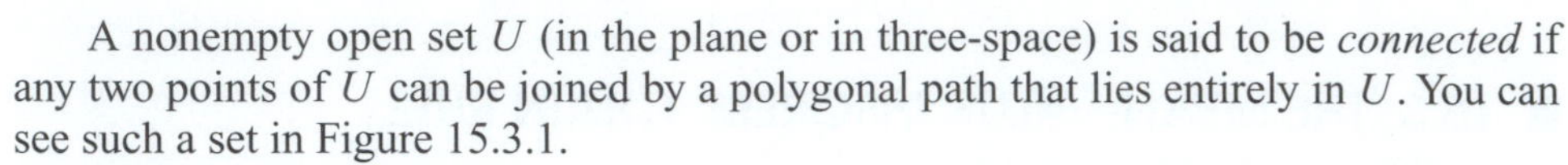

A nonempty open set U (in the plane or in three-space) is said to be *connected* if any two points of U can be joined by a polygonal path that lies entirely in U. You can see such a set in Figure 15.3.1.

The set shown in Figure 15.3.2 is the union of two disjoint open sets. The set is open but not connected: it is impossible to join **a** and **b** by a polygonal path that lies within the set.

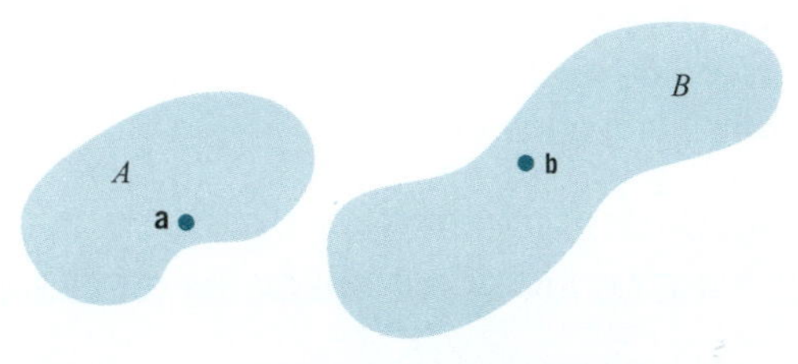

$U = A \cup B$ is not connected

Figure 15.3.2

In Chapter 4 you saw that, if $f'(x) = 0$ for all x in an open interval I, then f is constant on I. We have a similar result for functions of several variables.

THEOREM 15.3.2

Let U be an open connected set and let f be a differentiable function on U.

If $\nabla f(\mathbf{x}) = \mathbf{0}$ for all $\mathbf{x}$ in U, then f is constant on U.

PROOF Let **a** and **b** be any two points in U. Since U is connected, we can join these points by a polygonal path with vertices $\mathbf{a} = \mathbf{c}_0, \mathbf{c}_1, \mathbf{c}_2, \ldots, \mathbf{c}_{n-1}, \mathbf{c}_n = \mathbf{b}$ (see Figure 15.3.3). By the mean-value theorem (Theorem 15.3.1) there exist points

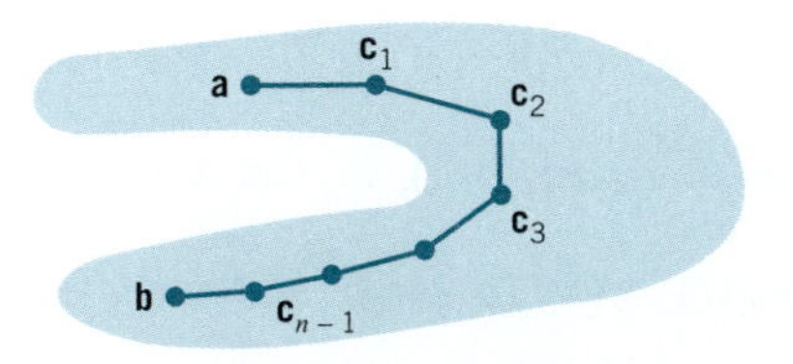

Figure 15.3.3

$$\begin{array}{lll}
\mathbf{c}_1^* \text{ between } \mathbf{c}_0 \text{ and } \mathbf{c}_1 & \text{such that} & f(\mathbf{c}_1) - f(\mathbf{c}_0) = \nabla f(\mathbf{c}_1^*) \cdot (\mathbf{c}_1 - \mathbf{c}_0), \\
\mathbf{c}_2^* \text{ between } \mathbf{c}_1 \text{ and } \mathbf{c}_2 & \text{such that} & f(\mathbf{c}_2) - f(\mathbf{c}_1) = \nabla f(\mathbf{c}_2^*) \cdot (\mathbf{c}_2 - \mathbf{c}_1), \\
\vdots & & \vdots \\
\mathbf{c}_n^* \text{ between } \mathbf{c}_{n-1} \text{ and } \mathbf{c}_n & \text{such that} & f(\mathbf{c}_n) - f(\mathbf{c}_{n-1}) = \nabla f(\mathbf{c}_n^*) \cdot (\mathbf{c}_n - \mathbf{c}_{n-1}).
\end{array}$$

If $\nabla f(\mathbf{x}) = \mathbf{0}$ for all $\mathbf{x}$ in U, then

$$f(\mathbf{c}_1) - f(\mathbf{c}_0) = 0, \;\; f(\mathbf{c}_2) - f(\mathbf{c}_1) = 0, \;\cdots, f(\mathbf{c}_n) - f(\mathbf{c}_{n-1}) = 0.$$

This shows that

$$f(\mathbf{a}) = f(\mathbf{c}_0) = f(\mathbf{c}_1) = f(\mathbf{c}_2) = \cdots = f(\mathbf{c}_{n-1}) = f(\mathbf{c}_n) = f(\mathbf{b}).$$

Since $\mathbf{a}$ and $\mathbf{b}$ are arbitrary points of U, f must be constant on U. ❑

THEOREM 15.3.3

Let U be an open connected set and let f and g be differentiable functions on U.

If $\nabla f(\mathbf{x}) = \nabla g(\mathbf{x})$ for all $\mathbf{x}$ in U, then f and g differ by a constant on U.

PROOF If $\nabla f(\mathbf{x}) = \nabla g(\mathbf{x})$ for all $\mathbf{x}$ in U, then

$$\nabla[f(\mathbf{x}) - g(\mathbf{x})] = \nabla f(\mathbf{x}) - \nabla g(\mathbf{x}) = 0 \quad \text{for all } \mathbf{x} \text{ in } U.$$

By Theorem 15. 3. 2, $f - g$ must be constant on U. ❑

The Chain Rule

You are by now thoroughly familiar with the chain rule for functions of a single variable (Theorem 3.5.7): If g is differentiable at x and f is differentiable at $g(x)$, then

$$\frac{d}{dx}(f[g(x)]) = f'[g(x)]g'(x).$$

Here we obtain generalizations of the chain rule for functions of several variables. As you will see, there are many versions of the chain rule in this setting.

A vector-valued function is said to be *continuous* provided that its components are continuous. If $f = f(x, y, z)$ is a scalar-valued function (a real-valued function), then its gradient ∇f is a vector-valued function. We say that f is *continuously differentiable* on an *open set* U if f is differentiable on U and ∇f is continuous on U.

If a curve $\mathbf{r}$ lies in the domain of f, then we can form the composition

$$(f \circ \mathbf{r})(t) = f(\mathbf{r}(t)).$$

The composition $f \circ \mathbf{r}$ is a real-valued function of the real variable t. The numbers $f(\mathbf{r}(t))$ are the values taken on by f *along the curve* $\mathbf{r}$. For example, let

$$f(x, y) = \tfrac{1}{3}(x^3 + y^3) \quad \text{and} \quad \mathbf{r}(t) = a\cos t\,\mathbf{i} + b\sin t\,\mathbf{j} \quad \text{(an ellipse)}.$$

Then, with $x(t) = a\cos t$ and $y(t) = b\sin t$, we have

$$f(\mathbf{r}(t)) = \tfrac{1}{3}(a^3\cos^3 t + b^3\sin^3 t).$$

The chain rule is a formula for calculating the derivative of the composition $f \circ \mathbf{r}$.

THEOREM 15.3.4 CHAIN RULE (ALONG A CURVE)

If f is continuously differentiable on an open set U and $\mathbf{r} = \mathbf{r}(t)$ is a differentiable curve that lies in U, then the composition $f \circ \mathbf{r}$ is differentiable and

$$\frac{d}{dt}[f(\mathbf{r}(t))] = \nabla f(\mathbf{r}(t)) \cdot \mathbf{r}'(t).$$

PROOF We will show that

$$\lim_{h\to 0} \frac{f(\mathbf{r}(t+h)) - f(\mathbf{r}(t))}{h} = \nabla f(\mathbf{r}(t)) \cdot \mathbf{r}'(t).$$

For $h \neq 0$ and sufficiently small, the line segment that joins $\mathbf{r}(t)$ to $\mathbf{r}(t+h)$ lies entirely in U. This we know because U is open and $\mathbf{r}$ is continuous. (See Figure 15.3.4.) For such h, the mean-value theorem assures us that there exists a point $\mathbf{c}(h)$ between $\mathbf{r}(t)$ and $\mathbf{r}(t+h)$ such that

$$f(\mathbf{r}(t+h)) - f(\mathbf{r}(t)) = \nabla f(\mathbf{c}(h)) \cdot [\mathbf{r}(t+h) - \mathbf{r}(t)].$$

Dividing both sides by h, we have

$$\frac{f(\mathbf{r}(t+h)) - f(\mathbf{r}(t))}{h} = \nabla f(\mathbf{c}(h)) \cdot \left[\frac{\mathbf{r}(t+h) - \mathbf{r}(t)}{h}\right].$$

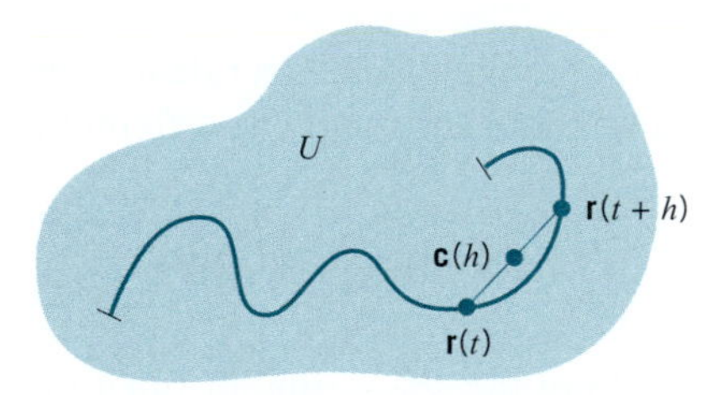

Figure 15.3.4

As h tends to zero, $\mathbf{c}(h)$ tends to $\mathbf{r}(t)$ and, by the continuity of ∇f,

$$\nabla f(\mathbf{c}(h)) \to \nabla f(\mathbf{r}(t)).$$

Since

$$\frac{\mathbf{r}(t+h) - \mathbf{r}(t)}{h} \to \mathbf{r}'(t),$$

the result follows. ☐

Example 1 Use the chain rule to find the rate of change of

$$f(x, y) = \tfrac{1}{3}(x^3 + y^3)$$

with respect to t along the ellipse $\mathbf{r}(t) = a\cos t\,\mathbf{i} + b\sin t\,\mathbf{j}$.

SOLUTION The rate of change of f with respect to t along the curve $\mathbf{r}$ is the derivative

$$\frac{d}{dt}[f(\mathbf{r}(t))].$$

By the chain rule (Theorem 15.3.4)

$$\frac{d}{dt}[f(\mathbf{r}(t))] = \nabla f(\mathbf{r}(t)) \cdot \mathbf{r}'(t).$$

Here

$$\nabla f = x^2\,\mathbf{i} + y^2\,\mathbf{j}.$$

With $x(t) = a\cos t$ and $y(t) = b\sin t$, we have

$$\nabla f(\mathbf{r}(t)) = a^2\cos^2 t\,\mathbf{i} + b^2\sin^2 t\,\mathbf{j}.$$

Since $\mathbf{r}'(t) = -a\sin t\,\mathbf{i} + b\cos t\,\mathbf{j}$, you can see that

$$\begin{aligned}\frac{d}{dt}[f(\mathbf{r}(t))] &= \nabla f(\mathbf{r}(t)) \cdot \mathbf{r}'(t)\\ &= (a^2\cos^2 t\,\mathbf{i} + b^2\sin^2 t\,\mathbf{j}) \cdot (-a\sin t\,\mathbf{i} + b\cos t\,\mathbf{j})\\ &= -a^3\sin t\cos^2 t + b^3\sin^2 t\cos t\\ &= \sin t\cos t(b^3\sin t - a^3\cos t).\end{aligned}$$ ❑

Remark Note that we could have obtained the same result without invoking Theorem 15.3.4 by first forming $f(\mathbf{r}(t))$ and then differentiating with respect to t. As you saw in the calculations carried out before the statement of theorem

$$f(\mathbf{r}(t)) = \tfrac{1}{3}(a^3\cos^3 t + b^3\sin^3 t).$$

This gives

$$\begin{aligned}\frac{d}{dt}[f(\mathbf{r}(t))] &= \tfrac{1}{3}[3a^3(\cos^2 t)(-\sin t) + 3b^3\sin^2 t\cos t]\\ &= \sin t\cos t(b^3\sin t - a^3\cos t).\end{aligned}$$ ❑

Example 2 Use the chain rule to find the rate of change of

$$f(x, y, z) = x^2 y + z\cos x$$

with respect to t along the twisted cubic $\mathbf{r}(t) = t\,\mathbf{i} + t^2\,\mathbf{j} + t^3\,\mathbf{k}$.

SOLUTION Once again we use the relation

$$\frac{d}{dt}[f(\mathbf{r}(t))] = \nabla f(\mathbf{r}(t)) \cdot \mathbf{r}'(t).$$

This time

$$\nabla f = (2xy - z\sin x)\,\mathbf{i} + x^2\,\mathbf{j} + \cos x\,\mathbf{k}.$$

With $x(t) = t, y(t) = t^2, z(t) = t^3$, we have

$$\nabla f(\mathbf{r}(t)) = (2t^3 - t^3\sin t)\,\mathbf{i} + t^2\,\mathbf{j} + \cos t\,\mathbf{k}.$$

Since $\mathbf{r}'(t) = \mathbf{i} + 2t\,\mathbf{j} + 3t^2\,\mathbf{k}$, we have

$$\begin{aligned}\frac{d}{dt}[(\mathbf{r}(t))] &= \nabla f(\mathbf{r}(t)) \cdot \mathbf{r}'(t)\\ &= [(2t^3 - t^3\sin t)\,\mathbf{i} + t^2\,\mathbf{j} + \cos t\,\mathbf{k}] \cdot [\mathbf{i} + 2t\,\mathbf{j} + 3t^2\,\mathbf{k}]\\ &= 2t^3 - t^3\sin t + 2t^3 + 3t^2\cos t\\ &= 4t^3 - t^3\sin t + 3t^2\cos t.\end{aligned}$$

You can check this answer by first forming $f(\mathbf{r}(t))$ and then differentiating. ❑

Another Formulation of Theorem 15.3.4

The chain rule for functions of one variable,

$$\frac{d}{dt}[u(x(t))] = u'(x(t))\,x'(t),$$

can be written

$$\frac{du}{dt} = \frac{du}{dx}\frac{dx}{dt}.$$

In a similar manner, the relation

$$\frac{d}{dt}[u(\mathbf{r}(t))] = \nabla u(\mathbf{r}(t)) \cdot \mathbf{r}'(t)$$

can be written

$$\frac{du}{dt} = \nabla u \cdot \frac{d\mathbf{r}}{dt}. \tag{1}$$

With $\quad \nabla u = \frac{\partial u}{\partial x}\mathbf{i} + \frac{\partial u}{\partial y}\mathbf{j} + \frac{\partial u}{\partial z}\mathbf{k} \quad$ and $\quad \frac{d\mathbf{r}}{dt} = \frac{dx}{dt}\mathbf{i} + \frac{dy}{dt}\mathbf{j} + \frac{dz}{dt}\mathbf{k},$

equation (1) takes the form

(15.3.5)
$$\boxed{\frac{du}{dt} = \frac{\partial u}{\partial x}\frac{dx}{dt} + \frac{\partial u}{\partial y}\frac{dy}{dt} + \frac{\partial u}{\partial z}\frac{dz}{dt}.}$$

In the two-variable case, the z-term drops out and we have

(15.3.6)
$$\boxed{\frac{du}{dt} = \frac{\partial u}{\partial x}\frac{dx}{dt} + \frac{\partial u}{\partial y}\frac{dy}{dt}.}$$

Example 3 Find du/dt given that $u = x^2 - y^2$ and $x = t^2 - 1, \quad y = 3\sin \pi t$.

SOLUTION Here we are in the two-variable case

$$\frac{du}{dt} = \frac{\partial u}{\partial x}\frac{dx}{dt} + \frac{\partial u}{\partial y}\frac{dy}{dt}.$$

Since $\quad \frac{\partial u}{\partial x} = 2x, \quad \frac{\partial u}{\partial y} = -2y \quad$ and $\quad \frac{dx}{dt} = 2t, \quad \frac{dy}{dt} = 3\pi \cos \pi t,$

We have
$$\begin{aligned}\frac{du}{dt} &= (2x)(2t) + (-\,2y)(3\pi \cos \pi t)\\ &= 2(t^2 - 1)(2t) + (-\,2)(3\sin \pi t)(3\pi \cos \pi t)\\ &= 4t^3 - 4t - 18\pi \sin \pi t \cos \pi t.\end{aligned}$$

You can obtain this same result by first writing u directly as a function of t and then differentiating:

$$u = x^2 - y^2 = (t^2 - 1)^2 - (3\sin \pi t)^2,$$

so that

$$\frac{du}{dt} = 2(t^2 - 1)2t - 2(3\sin\pi t)3\pi\cos\pi t = 4t^3 - 4t - 18\pi\sin\pi t\cos\pi t. \quad \square$$

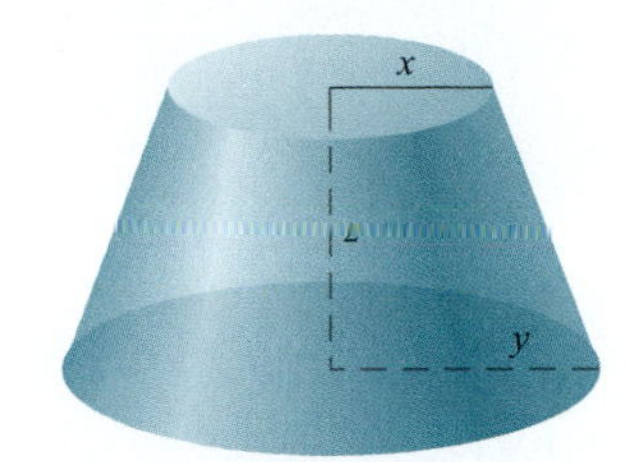

Figure 15.3.5

Example 4 A solid is in the shape of a frustum of a right circular cone (see Figure 15.3.5). If the upper radius x decreases at the rate of 2 inches per minute, the lower radius y increases at the rate of 3 inches per minute, and the height z decreases at the rate of 4 inches per minute, at what rate is the volume V changing at the instant the upper radius is 10 inches, the lower radius is 12 inches, and the height is 18 inches.

SOLUTION In Exercise 39 of Section 6.2, you were asked to derive the formula for the volume of a frustum of a cone.

With x, y, z as given,

$$V = \frac{1}{3}\pi z(x^2 + xy + y^2).$$

Here $\quad \dfrac{\partial V}{\partial x} = \dfrac{1}{3}\pi z(2x + y), \quad \dfrac{\partial V}{\partial y} = \dfrac{1}{3}\pi z(x + 2y), \quad \dfrac{\partial V}{\partial z} = \dfrac{1}{3}\pi(x^2 + xy + y^2).$

Since

$$\frac{dV}{dt} = \frac{\partial V}{\partial x}\frac{dx}{dt} + \frac{\partial V}{\partial y}\frac{dy}{dt} + \frac{\partial V}{\partial z}\frac{dz}{dt},$$

we have

$$\frac{dV}{dt} = \frac{1}{3}\pi z(2x + y)\frac{dx}{dt} + \frac{1}{3}\pi z(x + 2y)\frac{dy}{dt} + \frac{1}{3}\pi(x^2 + xy + y^2)\frac{dz}{dt}.$$

Set $\quad x = 10, \quad y = 12, \quad z = 18, \quad \dfrac{dx}{dt} = -2, \quad \dfrac{dy}{dt} = 3, \quad \dfrac{dz}{dt} = -4,$

and you will find that

$$\frac{dV}{dt} = -\frac{772}{3}\pi \cong -808.4.$$

The volume decreases at the rate of approximately 808 cubic inches per minute. $\square$

Other Chain Rules

In the setting of functions of several variables there are numerous versions of the chain rule. Some are stated here, others in the Exercises. They can all be deduced from Theorem 15.3.4 and its corollaries, (15.3.5) and (15.3.6).

If, for example,

$$u = u(x, y) \quad \text{where} \quad x = x(s, t) \quad \text{and} \quad y = y(s, t),$$

then

(15.3.7)
$$\frac{\partial u}{\partial s} = \frac{\partial u}{\partial x}\frac{\partial x}{\partial s} + \frac{\partial u}{\partial y}\frac{\partial y}{\partial s} \quad \text{and} \quad \frac{\partial u}{\partial t} = \frac{\partial u}{\partial x}\frac{\partial x}{\partial t} + \frac{\partial u}{\partial y}\frac{\partial y}{\partial t}.$$

To obtain the first equation, keep t fixed and differentiate u with respect to s according to Formula (15.3.6); to obtain the second equation, keep s fixed and differentiate u with respect to t.

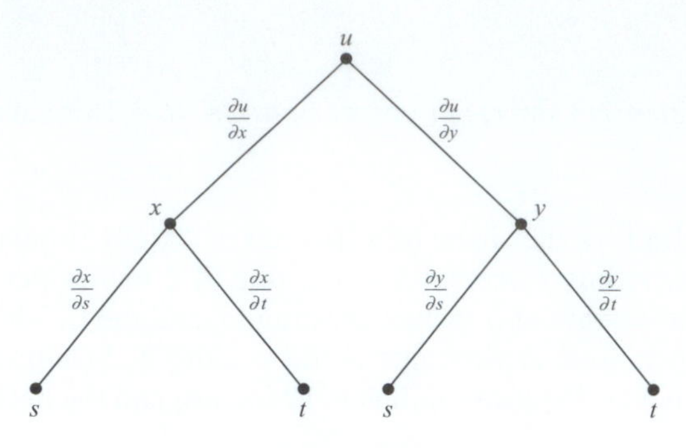

Figure 15.3.6

In Figure 15.3.6 we have drawn a *tree diagram* for Formula (15.3.7). We construct such a tree by branching at each stage from a function to all the variables that directly determine it. Each path starting at u and ending at a variable determines a product of (partial) derivatives. The partial derivative of u with respect to each variable is the sum of the products generated by all the direct paths to that variable.

Example 5 Let $u = x^2 - 2xy + 2y^3$, where $x = s^2 \ln t$ and $y = 2st^3$. Find $\partial u/\partial s$ and $\partial u/\partial t$.

SOLUTION Here u is a function of two variables, x and y, each of which is itself a function of two variables, s and t Thus (15.3.7) applies. Since

$$\frac{\partial u}{\partial x} = 2x - 2y, \quad \frac{\partial u}{\partial y} = -2x + 6y^2$$

and
$$\frac{\partial x}{\partial s} = 2s \ln t, \quad \frac{\partial y}{\partial s} = 2t^3, \quad \frac{\partial x}{\partial t} = \frac{s^2}{t}, \quad \frac{\partial y}{\partial t} = 6st^2,$$

we have
$$\frac{\partial u}{\partial s} = (2x - 2y)(2s \ln t) + (-2x + 6y^2)(2t^3)$$

and
$$\frac{\partial u}{\partial t} = (2x - 2y)\left(\frac{s^2}{t}\right) + (-2x + 6y^2)(6st^2).$$

These results can be expressed entirely in terms of s and t by replacing x by $s^2 \ln t$ and y by $2st^3$:

$$\frac{\partial u}{\partial s} = (2s^2 \ln t - 4st^3)(2s \ln t) + (-2s^2 \ln t + 24s^2t^6)(2t^3),$$

$$\frac{\partial u}{\partial t} = (2s^2 \ln t - 4st^3)\left(\frac{s^2}{t}\right) + (-2s^2 \ln t + 24s^2t^6)(6st^2). \quad \square$$

Now suppose that u is a function of three variables:

$$u = u(x, y, z) \quad \text{where} \quad x = x(s, t) \quad y = y(s, t), \quad z = z(s, t).$$

A tree diagram for the partials of u appears in Figure 15.3.7. The partials of u with respect to s and t can be read from the diagram:

(15.3.8)
$$\frac{\partial u}{\partial s} = \frac{\partial u}{\partial x}\frac{\partial x}{\partial s} + \frac{\partial u}{\partial y}\frac{\partial y}{\partial s} + \frac{\partial u}{\partial z}\frac{\partial z}{\partial s}, \quad \frac{\partial u}{\partial t} = \frac{\partial u}{\partial x}\frac{\partial x}{\partial t} + \frac{\partial u}{\partial y}\frac{\partial y}{\partial t} + \frac{\partial u}{\partial z}\frac{\partial z}{\partial t}.$$

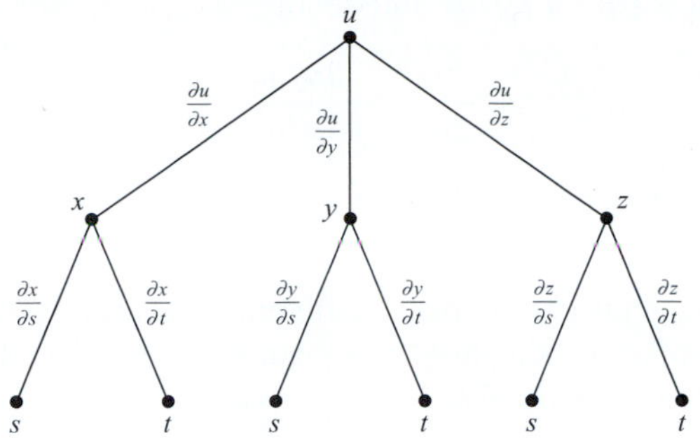

Figure 15.3.7

Example 6 Let $u = x^2y^3e^{xz}$, where $x = s^2 + t^2$, $y = 2st$, and $z = s \ln t$. Find $\partial u/\partial s$.

SOLUTION In this case, u is a function of three variables, x, y, z, each of which is a function of two variables, s and t. Thus (15.3.8) applies. To find $\partial u/\partial s$ we use

$$\frac{\partial u}{\partial s} = \frac{\partial u}{\partial x}\frac{\partial x}{\partial s} + \frac{\partial u}{\partial y}\frac{\partial y}{\partial s} + \frac{\partial u}{\partial z}\frac{\partial z}{\partial s}.$$

Here $$\frac{\partial u}{\partial x} = 2xy^3e^{xz} + x^2y^3ze^{xz}, \qquad \frac{\partial u}{\partial y} = 3x^2y^2e^{xz}, \qquad \frac{\partial u}{\partial z} = x^3y^3e^{xz}$$

and $$\frac{\partial x}{\partial s} = 2s, \qquad \frac{\partial y}{\partial s} = 2t, \qquad \frac{\partial z}{\partial s} = \ln t.$$

Therefore,

$$\frac{\partial u}{\partial s} = (2xy^3e^{xz} + x^2y^3ze^{xz})\,2s + (3x^2y^2e^{xz})\,2t + (x^3y^3e^{xz})\ln t.$$

The result can be expressed entirely in terms of s and t by substituting $s^2 + t^2$ for x, $2st$ for y, and $s \ln t$ for z. □

Implicit Differentiation

We return to implicit differentiation, a technique we introduced in Section 3.7.

Suppose that $u = u(x, y)$ is a continuously differentiable function. If y is a differentiable function of x that satisfies the equation $u(x, y) = 0$, then we can find the derivative of y without having to express y *explicitly* in terms of x. The process by which we do this is called *implicit differentiation*.

The process is based on (15.3.6.). To be able to apply that formula, we introduce a new variable t by setting $x = t$. We then have

$$u = u(x, y) \qquad \text{with} \quad x = t \qquad \text{and} \qquad y = y(t).$$

Formula (15.3.6) states that

$$\frac{du}{dt} = \frac{\partial u}{\partial x}\frac{dx}{dt} + \frac{\partial u}{\partial y}\frac{dy}{dt}.$$

Since $u(x(t), y(t)) = 0$ for all t under consideration, $du/dt = 0$ for such t. Since $x = t$, we have $dx/dt = 1$ and $dy/dt = dy/dx$. Therefore,

$$0 = \frac{\partial u}{\partial x} + \frac{\partial u}{\partial y}\frac{dy}{dx}.$$

If $\partial u/\partial y \neq 0$, we can solve for dy/dx and obtain

$$\frac{dy}{dx} = -\frac{\partial u/\partial x}{\partial u/\partial y}.$$

The result can be summarized as follows:

(15.3.9) If $u = u(x, y)$ is continuously differentiable, and y is a differentiable function of x that satisfies the equation $u(x, y) = 0$, then at all points (x, y) where $\partial u/\partial y \neq 0$,

$$\frac{dy}{dx} = -\frac{\partial u/\partial x}{\partial u/\partial y}.$$

Example 7 Suppose that y is a differentiable function of x that satisfies the equation

$$u(x, y) = 2x^2y - y^3 + 1 - x - 2y = 0.$$

Since

$$\frac{\partial u}{\partial x} = 4xy - 1 \quad \text{and} \quad \frac{\partial u}{\partial y} = 2x^2 - 3y^2 - 2,$$

we know that

$$\frac{dy}{dx} = -\frac{4xy - 1}{2x^2 - 3y^2 - 2}.$$

We obtained this result by a slightly different method in Section 3.7, Example 2 (a). □

This method of implicit differentiation can be applied to expressions involving more than two variables. For example,

(15.3.10) If $u = u(x, y, z)$ is continuously differentiable, and $z = z(x, y)$ is a differentiable function that satisfies the equation $u(x, y, z) = 0$, then at the points (x, y, z) where $\partial u/\partial z \neq 0$,

$$\frac{\partial z}{\partial x} = -\frac{\partial u/\partial x}{\partial u/\partial z} \quad \text{and} \quad \frac{\partial z}{\partial y} = -\frac{\partial u/\partial y}{\partial u/\partial z}.$$

PROOF To apply (15.3.8), we write

$$u = u(x, y, z) \quad \text{with} \quad x = s, \quad y = t, \quad z = z(s, t).$$

Then

$$\frac{\partial u}{\partial s} = \frac{\partial u}{\partial x}\frac{\partial x}{\partial s} + \frac{\partial u}{\partial y}\frac{\partial y}{\partial s} + \frac{\partial u}{\partial z}\frac{\partial z}{\partial s}.$$

Since $u(s, t, z(s, t)) = 0$, $\partial u/\partial s = 0$. Also, since

$$\frac{\partial x}{\partial s} = 1 \quad \text{and} \quad \frac{\partial y}{\partial s} = 0,$$

we have

$$0 = \frac{\partial u}{\partial x}(1) + \frac{\partial u}{\partial y}(0) + \frac{\partial u}{\partial z}\frac{\partial z}{\partial s} = \frac{\partial u}{\partial x} + \frac{\partial u}{\partial z}\frac{\partial z}{\partial s} = \frac{\partial u}{\partial x} + \frac{\partial u}{\partial z}\frac{\partial z}{\partial x}$$

$x = s$

At those points (x, y, z) where $\partial u/\partial z \neq 0$,

$$\frac{\partial z}{\partial x} = -\frac{\partial u/\partial x}{\partial u/\partial z}.$$

The formula for $\partial z/\partial y$ can be obtained in a similar manner. ❑

EXERCISES 15.3

1. Let $f(x, y) = x^3 - xy$. Set $\mathbf{a} = (0, 1)$ and $\mathbf{b} = (1, 3)$. Find a point $\mathbf{c}$ on the line segment $\overline{\mathbf{ab}}$ for which

$$f(\mathbf{b}) - f(\mathbf{a}) = \nabla f(\mathbf{c}) \cdot (\mathbf{b} - \mathbf{a}).$$

2. Let $f(x, y, z) = 4xz - y^2 + z^2$. Set $\mathbf{a} = (0, 1, 1)$ and $\mathbf{b} = (1, 3, 2)$. Find a point $\mathbf{c}$ on the line segment $\overline{\mathbf{ab}}$ for which

$$f(\mathbf{b}) - f(\mathbf{a}) = \nabla f(\mathbf{c}) \cdot (\mathbf{b} - \mathbf{a}).$$

3. (a) Find f if $\nabla f(x, y, z) = a_1\mathbf{i} + a_2\mathbf{j} + a_3\mathbf{k}$ for all (x, y, z).

(b) What can you conclude about f and g if $\nabla f(x, y, z) - \nabla g(x, y, z) = a_1\mathbf{i} + a_2\mathbf{j} + a_3\mathbf{k}$ for all (x, y, z)?

4. (*Rolle's theorem for functions of several variables*) Show that, if f is differentiable at each point of the line segment $\overline{\mathbf{ab}}$ and $f(\mathbf{a}) = f(\mathbf{b})$, then there exists a point $\mathbf{c}$ between $\mathbf{a}$ and $\mathbf{b}$ for which $\nabla f(\mathbf{c}) \perp (\mathbf{b} - \mathbf{a})$.

5. Let $U = \{\mathbf{x} : \|\mathbf{x}\| \neq 1\}$. Define f on U by setting

$$f(\mathbf{x}) = \begin{cases} 0, & \|\mathbf{x}\| < 1 \\ 1, & \|\mathbf{x}\| > 1. \end{cases}$$

(a) Note that $\nabla f(\mathbf{x}) = 0$ for all $\mathbf{x}$ in U, but f is not constant on U. Explain how this does not contradict Theorem 15.3.2.

(b) Define a function g on U different from f such that $\nabla f(\mathbf{x}) = \nabla g(\mathbf{x})$ for all $\mathbf{x}$ in U and $f - g$ is (i) constant on U, (ii) not constant on U.

6. A set of points is said to be *convex* provided that every pair of points in the set can be joined by a line segment that lies entirely within the set. Show that, if $\|\nabla f(\mathbf{x})\| \leq M$ for all $\mathbf{x}$ in some convex set Ω, then

$$|f(\mathbf{x}_1) - f(\mathbf{x}_2)| \leq M\|\mathbf{x}_1 - \mathbf{x}_2\| \qquad \text{for all } \mathbf{x}_1 \text{ and } \mathbf{x}_2 \text{ in } \Omega.$$

Find the rate of change of f with respect to t along the given curve.

7. $f(x, y) = x^2y, \quad \mathbf{r}(t) = e^t\,\mathbf{i} + e^{-t}\,\mathbf{j}.$

8. $f(x, y) = x - y, \quad \mathbf{r}(t) = at\,\mathbf{i} + b\cos at\,\mathbf{j}.$

9. $f(x, y) = \tan^{-1}(y^2 - x^2), \quad \mathbf{r}(t) = \sin t\,\mathbf{i} + \cos t\,\mathbf{j}.$

10. $f(x, y) = \ln(2x^2 + y^3), \quad \mathbf{r}(t) = e^{2t}\,\mathbf{i} + t^{1/3}\,\mathbf{j}.$

11. $f(x, y) = x\,e^y + y\,e^{-x}, \quad \mathbf{r}(t) = (\ln t)\,\mathbf{i} + t(\ln t)\,\mathbf{j}.$

12. $f(x, y, z) = \ln(x^2 + y^2 + z^2),$
$\mathbf{r}(t) = \sin t\,\mathbf{i} + \cos t\,\mathbf{j} + e^{2t}\,\mathbf{k}.$

13. $f(x, y, z) = xy - yz, \quad \mathbf{r}(t) = t\,\mathbf{i} + t^2\,\mathbf{j} + t^3\,\mathbf{k}.$

14. $f(x, y, z) = x^2 + y^2,$
$\mathbf{r}(t) = a\cos\omega t\,\mathbf{i} + b\sin\omega t\,\mathbf{j} + b\omega t\,\mathbf{k}.$

15. $f(x, y, z) = x^2 + y^2 + z,$
$\mathbf{r}(t) = a\cos\omega t\,\mathbf{i} + b\sin\omega t\,\mathbf{j} + b\omega t\,\mathbf{k}.$

16. $f(x, y, z) = y^2\sin(x + z), \quad \mathbf{r}(t) = 2t\,\mathbf{i} + \cos t\,\mathbf{j} + t^3\,\mathbf{k}.$

Find du/dt by applying (15.3.5) or (15.3.6).

17. $u = x^2 - 3xy + 2y^2; \quad x = \cos t, \quad y = \sin t.$

18. $u = x + 4\sqrt{xy} - 3y; \quad x = t^3, \quad y = t^{-1} \quad (t > 0).$

19. $u = e^x\sin y + e^y\sin x; \quad x = \frac{1}{2}t, \quad y = 2t.$

20. $u = 2x^2 - xy + y^2; \quad x = \cos 2t, \quad y = \sin t.$

21. $u = e^x\sin y; \quad x = t^2, \quad y = \pi t.$

22. $u = z\ln\left(\frac{y}{x}\right); \quad x = t^2 + 1, \quad y = \sqrt{t}, \quad z = t\,e^t$

23. $u = xy + yz + zx; \quad x = t^2, \quad y = t(1 - t), \quad z = (1 - t)^2.$

24. $u = x\sin\pi y - z\cos\pi x; \quad x = t^2, \quad y = 1 - t, \quad z = 1 - t^2.$

25. The radius of a right circular cone is increasing at the rate of 3 inches per second and the height is decreasing at the rate of 2 inches per second. At what rate is the volume of the cone changing at the instant the height is 20 inches and the radius is 14 inches?

26. The radius of a right circular cylinder is decreasing at the rate of 2 centimeters per second and the height is increasing at the rate of 3 centimeters per second. At what rate is the volume of the cylinder changing at the instant the radius is 13 centimeters and the height is 18 centimeters?

27. If the lengths of two sides of a triangle are x and y, and θ is the angle between the two sides, then the area A of the triangle is given by $A = \frac{1}{2}xy\sin\theta$. See the figure. If the sides are each increasing at the rate of 3 inches per second and θ is decreasing at the rate of 0.10 radian per second, how fast is the area changing at the instant $x = 1.5$ feet, $y = 2$ feet, and $\theta = 1$ radian?

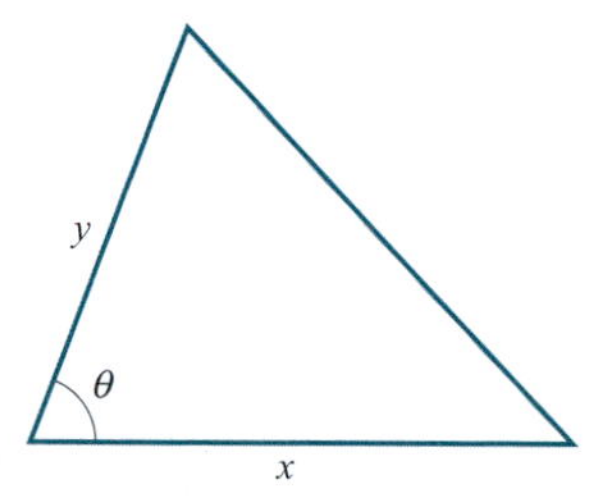

28. An object is moving along the curve of intersection of the paraboloid $z = x^2 + \frac{1}{4}y^2$ and the circular cylinder $x^2 + y^2 = 13$. If the x-coordinate is increasing at the rate of 5 centimeters per second, how fast is the z-coordinate changing at the instant when $x = 2$ centimeters and $y = 3$ centimeters?

Find $\partial u/\partial s$ and $\partial u/\partial t$.

29. $u = x^2 - xy$ where $x = s\cos t$, $y = t\sin s$.

30. $u = \sin(x-y) + \cos(x+y)$ where $x = st$, $y = s^2 - t^2$.

31. $u = x^2 \tan y$ where $x = s^2 t$, $y = s + t^2$.

32. $u = z^2 \sec xy$ where $x = 2st$, $y = s - t^2$, $z = s^2 t$.

33. $u = x^2 - xy + z^2$ where $x = s\cos t$, $y = \sin(t-s)$, $z = t\sin s$.

34. $u = x\,e^{yz^2}$ where $x = \ln st$, $y = t^3$, $z = s^2 + t^2$.

35. An object moves so that at time t it has position $\mathbf{r}(t) = x(t)\,\mathbf{i} + y(t)\,\mathbf{j} + z(t)\,\mathbf{k}$. Show that

$$\frac{d}{dt}[f(\mathbf{r}(t))]$$

is the directional derivative of f in the direction of the motion times the speed of the object.

36. (*Important*) Set $r = \|\mathbf{r}\|$ where $\mathbf{r} = x\,\mathbf{i} + y\,\mathbf{j} + z\,\mathbf{k}$. If f is a continuously differentiable function of r, then

(15.3.11) $$\nabla[f(r)] = f'(r)\frac{\mathbf{r}}{r}.$$

Derive this formula.

Calculate the following gradients taking $r = \|\mathbf{r}\|$.

37. (a) $\nabla(\sin r)$. (b) $\nabla(r\sin r)$.

38. (a) $\nabla(r\ln r)$. (b) $\nabla(e^{1-r^2})$.

39. (a) $\nabla\left(\dfrac{\sin r}{r}\right)$. (b) $\nabla\left(\dfrac{r}{\sin r}\right)$.

40. (a) Draw a tree diagram for du/dt given that $u = u(x,y)$ where $x = x(s), y = y(s)$, and $s = s(t)$.

(b) Calculate du/dt.

41. Set $u = u(x,y,z)$ where

$$x = x(w,t), \quad y = y(w,t), \quad z = z(w,t)$$
$$w = w(r,s), \quad t = t(r,s).$$

(a) Draw a tree diagram for the partials of u.

(b) Calculate $\partial u/\partial r$ *and* $\partial u/\partial s$.

42. Set $u = u(x,y,z,w)$ where

$$x = x(r,s,t), \quad y = y(s,t,v), \quad z = z(r,t),$$
$$w = w(r,s,t,v).$$

(a) Draw a tree diagram for the partials of u.

(b) Calculate $\partial u/\partial r$ *and* $\partial u/\partial v$.

Higher Derivatives

43. Let $u = u(x,y)$, where $x = x(t)$ and $y = y(t)$, and assume that these functions have continuous second derivatives. Show that

$$\frac{d^2u}{dt^2} = \frac{\partial^2 u}{\partial x^2}\left(\frac{dx}{dt}\right)^2 + 2\frac{\partial^2 u}{\partial x\partial y}\frac{dx}{dt}\frac{dy}{dt} + \frac{\partial^2 u}{\partial y^2}\left(\frac{dy}{dt}\right)^2 + \frac{\partial u}{\partial x}\frac{d^2x}{dt^2} + \frac{\partial u}{\partial y}\frac{d^2y}{dt^2}.$$

44. Let $u = u(x,y)$. where $x = x(s,t)$ and $y = y(s,t)$, and assume that all these functions have continuous second derivatives. Show that

$$\frac{\partial^2 u}{\partial s^2} = \frac{\partial^2 u}{\partial x^2}\left(\frac{\partial x}{\partial s}\right)^2 + 2\frac{\partial^2 u}{\partial x\partial y}\frac{\partial x}{\partial s}\frac{\partial y}{\partial s} + \frac{\partial^2 u}{\partial y^2}\left(\frac{\partial y}{\partial s}\right)^2 + \frac{\partial u}{\partial x}\frac{\partial^2 x}{\partial s^2} + \frac{\partial u}{\partial y}\frac{\partial^2 y}{\partial s^2}.$$

Polar Coordinates

45. Assume that $u = u(x,y)$ is differentiable.

(a) Show that the change of variables to polar coordinates $x = r\cos\theta$ and $y = r\sin\theta$ gives

$$\frac{\partial u}{\partial r} = \frac{\partial u}{\partial x}\cos\theta + \frac{\partial u}{\partial y}\sin\theta,$$
$$\frac{\partial u}{\partial \theta} = -\frac{\partial u}{\partial x}r\sin\theta + \frac{\partial u}{\partial y}r\cos\theta.$$

(b) Express

$$\left(\frac{\partial u}{\partial r}\right)^2 + \frac{1}{r^2}\left(\frac{\partial u}{\partial \theta}\right)^2$$

entirely in terms of $\partial u/\partial x$ and $\partial u/\partial y$.

46. Let w be a function of polar coordinates r and θ. Then w can be expressed in rectangular coordinates by using

$$x = r\cos\theta \quad \text{and} \quad y = r\sin\theta.$$

(a) Using the first part of Exercise 45, verify that

$$\frac{\partial w}{\partial x} = \frac{\partial w}{\partial r}\cos\theta - \frac{1}{r}\frac{\partial w}{\partial \theta}\sin\theta,$$
$$\frac{\partial w}{\partial y} = \frac{\partial w}{\partial r}\sin\theta + \frac{1}{r}\frac{\partial w}{\partial \theta}\cos\theta.$$

(b) Deduce from part (a) that

$$\frac{\partial r}{\partial x} = \cos\theta, \qquad \frac{\partial r}{\partial y} = \sin\theta;$$
$$\frac{\partial \theta}{\partial x} = -\frac{1}{r}\sin\theta, \qquad \frac{\partial \theta}{\partial y} = \frac{1}{r}\cos\theta.$$

(c) Find the fallacy in the following argument:

$$x = r\cos\theta, \qquad r = \frac{x}{\cos\theta}, \qquad \frac{\partial r}{\partial x} = \frac{1}{\cos\theta}.$$

47. (*The gradient in polar coordinates*) Let $u = u(x,y)$ be differentiable. Show that if u is written in terms of polar coordinates, then

$$\nabla u = \frac{\partial u}{\partial r}\,\mathbf{e}_r + \frac{1}{r}\frac{\partial u}{\partial \theta}\,\mathbf{e}_\theta$$

where

$$\mathbf{e}_r = \cos\theta\,\mathbf{i} + \sin\theta\,\mathbf{j} \quad \text{and} \quad \mathbf{e}_\theta = -\sin\theta\,\mathbf{i} + \cos\theta\,\mathbf{j}.$$

In Exercises 48 and 49, use the formula in Exercise 47 to express the gradient of the given function in polar coordinates.

48. $u(x,y) = x^2 + y^2$.

49. $u(x,y) = x^2 - xy + y^2$.

50. Let $u = u(x,y)$, where $x = r\cos\theta$ and $y = r\sin\theta$, and assume that u has continuous second partial derivatives. Derive a formula for $\partial^2 u/\partial r\partial\theta$.

51. (*The Laplace operator in polar coordinates*). Assume that $u = u(x,y)$ has continuous second partial derivatives. Show that

$$\frac{\partial^2 u}{\partial x^2} + \frac{\partial^2 u}{\partial y^2} = \frac{\partial^2 u}{\partial r^2} + \frac{1}{r^2}\frac{\partial^2 u}{\partial \theta^2} + \frac{1}{r}\frac{\partial u}{\partial r}.$$

Assume that y is a differentiable function of x that satisfies the given equation. Find dy/dx by implicit differentiation.

52. $x^2 - 2xy + y^4 = 4$.

53. $x\,e^y + y\,e^x - 2x^2y = 0$.

54. $x^{2/3} + y^{2/3} = a^{2/3}$ (a a constant).

55. $x\cos xy + y\cos x = 2$.

Assume that z is a differentiable function of (x,y) that satisfies the given equation. Find $\partial z/\partial x$ and $\partial z/\partial y$ by implicit differentiation.

56. $z^4 + x^2z^3 + y^2 + xy = 2$.

57. $\cos xyz + \ln(x^2 + y^2 + z^2) = 0$.

58. (*A chain rule for vector-valued functions*) Suppose that

$$\mathbf{u}(x,y) = u_1(x,y)\,\mathbf{i} + u_2(x,y)\,\mathbf{j}$$

$$\text{where } x = x(t), \quad y = y(t).$$

(a) Show that

(15.3.12)
$$\frac{d\mathbf{u}}{dt} = \frac{\partial \mathbf{u}}{\partial x}\frac{dx}{dt} + \frac{\partial \mathbf{u}}{\partial y}\frac{dy}{dt},$$

where

$$\frac{\partial \mathbf{u}}{\partial x} = \frac{\partial u_1}{\partial x}\mathbf{i} + \frac{\partial u_2}{\partial x}\mathbf{j} \quad \text{and} \quad \frac{\partial \mathbf{u}}{\partial y} = \frac{\partial u_1}{\partial y}\mathbf{i} + \frac{\partial u_2}{\partial y}\mathbf{j}.$$

(b) Let

$$\mathbf{u} = e^x\cos y\,\mathbf{i} + e^x\sin y\,\mathbf{j}$$

$$\text{where} \quad x = \tfrac{1}{2}t^2, \quad y = \pi t.$$

Calculate $d\mathbf{u}/dt$ (i) by applying (15.3.12), (ii) by forming $\mathbf{u}(t)$ directly.

59. Set

$$\mathbf{u}(x,y) = u_1(x,y)\,\mathbf{i} + u_2(x,y)\,\mathbf{j}$$

$$\text{where} \quad x = x(s,t), \quad y = y(s,t).$$

Find $\partial\mathbf{u}/\partial s$ and $\partial\mathbf{u}/\partial t$.

60. Set

$$\mathbf{u}(x,y,z) = u_1(x,y,z)\,\mathbf{i} + u_2(x,y,z)\,\mathbf{j} + u_3(x,y,z)\,\mathbf{k},$$

where

$$x = x(t), \quad y = y(t), \quad z = z(t).$$

Derive a formula for $d\mathbf{u}/dt$ analogous to (15.3.12).

*SUPPLEMENT TO SECTION 15.3

AN INTERMEDIATE-VALUE THEOREM

You have seen that a function of single variable that is continuous on an interval skips no values (the intermediate-value theorem, Theorem 2.6.1). There is an analogous result for functions of several variables. First, some terminology.

Open Regions, Closed Regions, Continuity

An open connected set is called an *open region*. If we start with an open region and adjoin to it the boundary, then we have what is called a *closed region*. (A closed region is therefore a closed set, the interior of which is an open region.)

Continuity on a closed region Ω requires continuity at the interior points *and* continuity at the boundary points. The meaning of this second requirement has to be explained.

Let the function f be defined on a closed region Ω, and let $\mathbf{x}_0 \in \Omega$. If $\mathbf{x}_0$ is an interior point of Ω, then all points $\mathbf{x}$ sufficiently close to $\mathbf{x}_0$ are in Ω and, by definition f is continuous at $\mathbf{x}_0$ iff

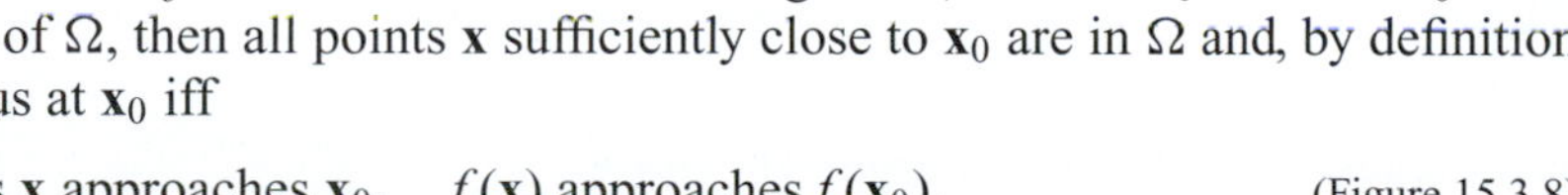

$$\text{as } \mathbf{x} \text{ approaches } \mathbf{x}_0, \quad f(\mathbf{x}) \text{ approaches } f(\mathbf{x}_0). \qquad \text{(Figure 15.3.8)}$$

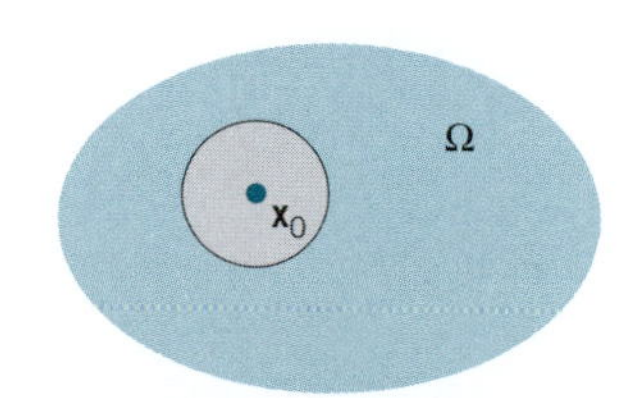

Figure 15.3.8

If $\mathbf{x}_0$ is a boundary point of Ω, then we have to modify the definition and say: f is continuous at $\mathbf{x}_0$ if

as $\mathbf{x}$ approaches $\mathbf{x}_0$ from *within* Ω, $f(\mathbf{x})$ approaches $f(\mathbf{x}_0)$. (Figure 15.3.9)

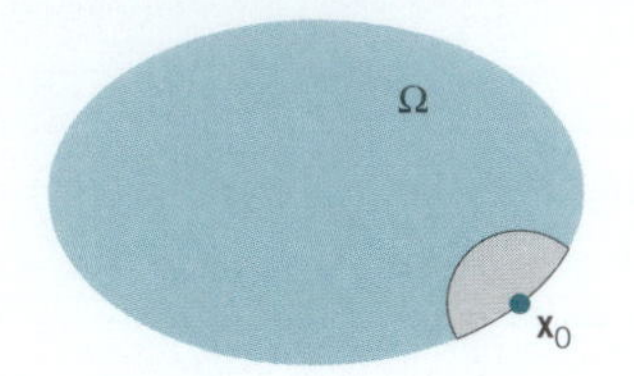

Figure 15.3.9

THEOREM 15.3.13 AN INTERMEDIATE-VALUE THEOREM

Suppose that f is continuous on a closed region Ω and A, B, C are real numbers such that $A < C < B$. If, somewhere in Ω, f takes on the value A and, somewhere in Ω, f takes on the value B, then, somewhere in Ω, f takes on the value C.

PROOF Let $\mathbf{a}$ and $\mathbf{b}$ be points in Ω for which

$$f(\mathbf{a}) = A \qquad \text{and} \qquad f(\mathbf{b}) = B.$$

We must show that there exists a point $\mathbf{c}$ in Ω for which $f(\mathbf{c}) = C$.

Let U be the interior of Ω and assume first that $\mathbf{a}$ and $\mathbf{b}$ are in U. Since U is connected, there is a polygonal path γ in U that joins $\mathbf{a}$ to $\mathbf{b}$. Let $\mathbf{r} = \mathbf{r}(t), a \leq t \leq b$, be a continuous parametrization of the path γ with $\mathbf{r}(a) = \mathbf{a}$ and $\mathbf{r}(b) = \mathbf{b}$. Since $\mathbf{r}$ is continuous, the composition

$$g(t) = f(\mathbf{r}(t))$$

is also continuous on $[a, b]$. Since

$$g(a) = f(\mathbf{r}(a)) = f(\mathbf{a}) = A \qquad \text{and} \qquad g(b) = f(\mathbf{r}(b)) = f(\mathbf{b}) = B,$$

we know from Theorem 2.6.1 that there is a number c in $[a, b]$ for which $g(c) = C$. Setting $\mathbf{c} = \mathbf{r}(c)$ we have $f(\mathbf{c}) = C$.

Now let $\mathbf{a}$ and $\mathbf{b}$ be *any* two points in Ω for which

$$f(\mathbf{a}) = A \qquad \text{and} \qquad f(\mathbf{b}) = B.$$

To take care of the possibility that one or both of these points lie on the boundary of Ω, we proceed as follows. Take ϵ small enough that

$$A + \epsilon < C < B - \epsilon.$$

By continuity there exist points $\mathbf{x}_1, \mathbf{x}_2$ in U, the interior of Ω, for which

$$f(\mathbf{x}_1) < A + \epsilon \qquad \text{and} \qquad B - \epsilon < f(\mathbf{x}_2).$$

Then

$$f(\mathbf{x}_1) < C < f(\mathbf{x}_2)$$

and the result follows by our argument above.

■ 15.4 THE GRADIENT AS A NORMAL; TANGENT LINES AND TANGENT PLANES

Functions of Two Variables

We begin with a nonconstant function $f = f(x, y)$ that is continuously differentiable. (*Remember:* That means f is differentiable and its gradient ∇f is continuous.) You have seen that at each point of the domain, the gradient vector, if not $\mathbf{0}$, points in the direction of the most rapid increase of f. Here we show that

(15.4.1) at each point of the domain, the gradient vector ∇f, if not $\mathbf{0}$, is perpendicular to the level curve of f that passes through that point.

PROOF We choose a point (x_0, y_0) in the domain and assume that $\nabla f(x_0, y_0) \neq \mathbf{0}$. The level curve through this point has equation

$$f(x, y) = c \quad \text{where} \quad c = f(x_0, y_0).$$

Under our assumptions on f, this curve can be parametrized in a neighborhood of (x_0, y_0) by a continuously differentiable vector function

$$\mathbf{r}(t) = x(t)\,\mathbf{i} + y(t)\,\mathbf{j}, \quad t \in I$$

with nonzero tangent vector $\mathbf{r}'(t)$. †

Now take t_0 such that

$$\mathbf{r}(t_0) = x_0\,\mathbf{i} + y_0\,\mathbf{j} = (x_0, y_0).$$

We will show that

$$\nabla f(\mathbf{r}(t_0)) \perp \mathbf{r}'(t_0).$$

Since f is constantly c on the curve, we have

$$f(\mathbf{r}(t)) = c \quad \text{for all } t \in I.$$

For such t

$$\frac{d}{dt}[f(\mathbf{r}(t))] = \nabla f(\mathbf{r}(t)) \cdot \mathbf{r}'(t) = 0.$$

↑ (Theorem 15.3.4)

In particular

$$\nabla f(\mathbf{r}(t_0)) \cdot \mathbf{r}'(t_0) = 0,$$

and thus

$$\nabla f(\mathbf{r}(t_0)) \perp \mathbf{r}'(t_0). \quad \square$$

Figure 15.4.1 illustrates the result.

† This follows from a result of advanced calculus known as the *implicit function theorem*.

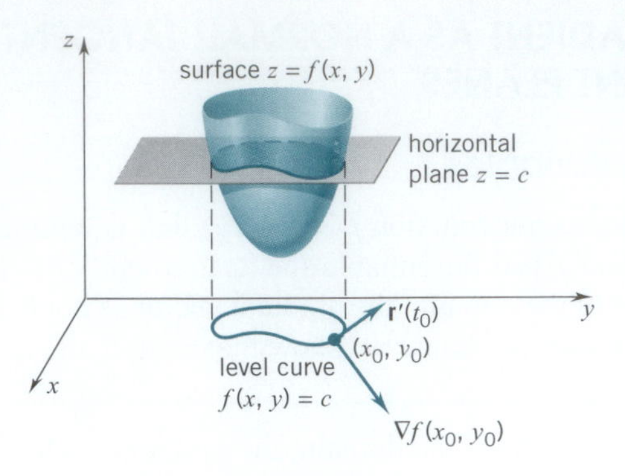

Figure 15.4.1

Example 1 For the function $f(x,y) = x^2+y^2$ the level curves are concentric circles:

$$x^2 + y^2 = c.$$

At each point $(x,y) \neq (0,0)$ the gradient vector

$$\nabla f(x,y) = 2x\,\mathbf{i} + 2y\,\mathbf{j} = 2\mathbf{r}$$

points away from the origin along the line of the radius vector and is thus perpendicular to the circle in question. At the origin the level curve is reduced to a point and the gradient is simply $\mathbf{0}$. See Figure 15.4.2. ❑

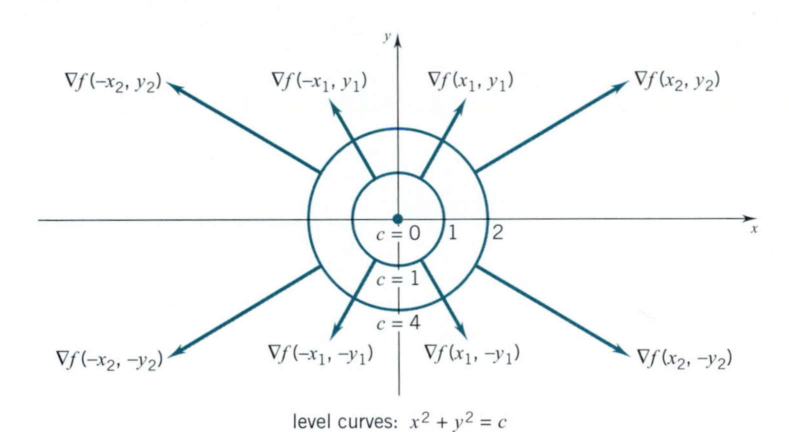

Figure 15.4.2

Consider now a curve in the xy-plane

$$C\colon\ f(x,y) = c.$$

As before, we assume that f is nonconstant and continuously differentiable. Let's suppose that (x_0, y_0) lies on the curve and $\nabla f(x_0, y_0) \neq \mathbf{0}$.

We can view C as the c-level curve of f and conclude from (15.4.1) that the gradient

(15.4.2)
$$\boxed{\nabla f(x_0, y_0) = \frac{\partial f}{\partial y}(x_0, y_0)\,\mathbf{i} + \frac{\partial f}{\partial y}(x_0, y_0)\,\mathbf{j}}$$

is perpendicular to C at (x_0, y_0). We call it a *normal vector.*

The vector

(15.4.3)
$$\mathbf{t}(x_0, y_0) = \frac{\partial f}{\partial y}(x_0, y_0)\,\mathbf{i} - \frac{\partial f}{\partial x}(x_0, y_0)\,\mathbf{j}$$

is perpendicular to the gradient:

$$\nabla f(x_0, y_0) \cdot \mathbf{t}(x_0, y_0) = \frac{\partial f}{\partial x}(x_0, y_0)\frac{\partial f}{\partial y}(x_0, y_0) - \frac{\partial f}{\partial y}(x_0, y_0)\frac{\partial f}{\partial x}(x_0, y_0) = 0.$$

It is therefore a *tangent vector*.

The line through (x_0, y_0) perpendicular to the gradient is the tangent line. To obtain an equation for the tangent line, we refer to Figure 15.4.3. A point (x, y) will lie on the tangent line iff

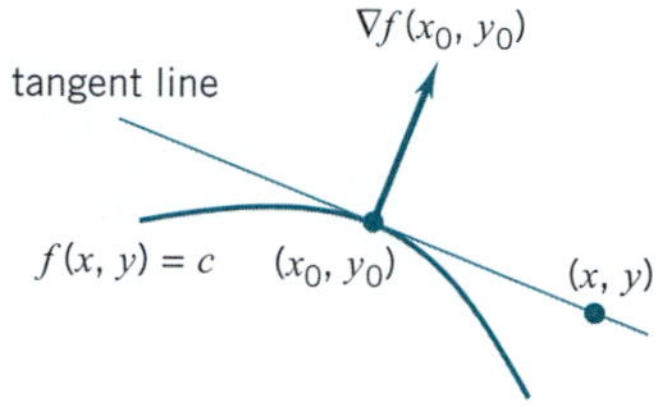

$$[(x - x_0)\,\mathbf{i} + (y - y_0)\,\mathbf{j}] \cdot \nabla f(x_0, y_0) = 0,$$

that is, iff

(15.4.4)
$$\frac{\partial f}{\partial x}(x_0, y_0)(x - x_0) + \frac{\partial f}{\partial y}(x_0, y_0)(y - y_0) = 0.$$

Figure 15.4.3

This is an equation for the *tangent line.*

The line through (x_0, y_0) perpendicular to the tangent vector $\mathbf{t}(x_0, y_0)$ is the normal line (Figure 15.4.4). A point (x, y) will lie on the normal line iff

$$[(x - x_0)\,\mathbf{i} + (y - y_0)\,\mathbf{j}] \cdot \mathbf{t}(x_0, y_0) = 0,$$

that is, iff

(15.4.5)
$$\frac{\partial f}{\partial y}(x_0, y_0)(x - x_0) - \frac{\partial f}{\partial x}(x_0, y_0)(y - y_0) = 0.$$

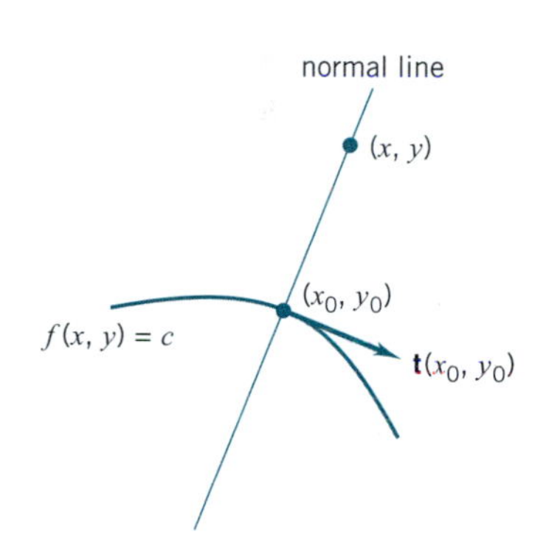

Figure 15.4.4

This is an equation for the *normal line.*

Example 2 Find a normal vector and a tangent vector to the plane curve

$$x^2 + 2y^3 = xy + 4$$

at the point (2,1). Then write equations for the tangent line and normal line to the curve at this point.

SOLUTION Set $f(x, y) = x^2 + 2y^3 - xy$. Then the given curve is the level curve $f(x, y) = 4$. Now the gradient of f is

$$\nabla f = (2x - y)\,\mathbf{i} + (6y^2 - x)\,\mathbf{j} \qquad \text{and} \qquad \nabla f(2, 1) = 3\,\mathbf{i} + 4\,\mathbf{j}.$$

Therefore, we have

normal vector: $\nabla f(2, 1) = 3\,\mathbf{i} + 4\,\mathbf{j}$, tangent vector: $\mathbf{t}(2, 1) = 4\,\mathbf{i} - 3\,\mathbf{j}$,

equation of tangent line: $3(x - 2) + 4(y - 1) = 0$ or $y = -\frac{3}{4}x + \frac{5}{2}$,

equation of normal line: $4(x - 2) - 3(y - 1) = 0$ or $y = \frac{4}{3}x - \frac{5}{3}$. □

Functions of Three Variables

Here, instead of level curves, we have level surfaces, but the results are similar. If $f = f(x, y, z)$ is nonconstant and continuously differentiable, then

(15.4.6) at each point of the domain, the gradient vector, if not $\mathbf{0}$, is perpendicular to the level surface that passes through that point.

PROOF We choose a point $\mathbf{x}_0 = (x_0, y_0, z_0)$ in the domain and assume that $\nabla f(x_0, y_0, z_0) \neq \mathbf{0}$. The level surface through this point has equation

$$f(x, y, z) = c \qquad \text{where} \qquad c = f(x_0, y_0, z_0).$$

We suppose now that

$$\mathbf{r}(t) = x(t)\,\mathbf{i} + y(t)\,\mathbf{j} + z(t)\,\mathbf{k}, \quad t \in I,$$

is a differentiable curve that lies on this surface and passes through the point $\mathbf{x}_0 = (x_0, y_0, z_0)$. We choose t_0 so that

$$\mathbf{r}(t_0) = \mathbf{x_0} = (x_0, y_0, z_0)$$

and suppose that $\mathbf{r}'(t_0) \neq \mathbf{0}$.

Since the curve lies on the given surface, we have

$$f(\mathbf{r}(t)) = c \quad \text{for all } t \in I.$$

For such t

$$\frac{d}{dt}[f(\mathbf{r}(t))] = \nabla f(\mathbf{r}(t)) \cdot \mathbf{r}'(t) = 0.$$

In particular,

$$\nabla f(\mathbf{r}(t_0)) \cdot \mathbf{r}'(t_0) = 0.$$

The gradient vector

$$\nabla f(\mathbf{r}(t_0)) = \nabla f(\mathbf{x}_0) = \nabla f(x_0, y_0, z_0)$$

is thus perpendicular to the curve in question.

This same argument applies to *every* differentiable curve that lies on this level surface and passes through the point $\mathbf{x}_0 = (x_0, y_0, z_0)$ with nonzero tangent vector. (See Figure 15.4.5.) Consequently, $\nabla f(\mathbf{x}_0)$ must be perpendicular to the surface itself. □

Example 3 For the function $f(x, y, z) = x^2 + y^2 + z^2$ the level surfaces are concentric spheres:

$$x^2 + y^2 + z^2 = c.$$

At each point $(x, y, z) \neq (0, 0, 0)$, the gradient vector

$$\nabla f(x, y, z) = 2x\,\mathbf{i} + 2y\,\mathbf{j} + 2z\,\mathbf{k} = 2\mathbf{r}$$

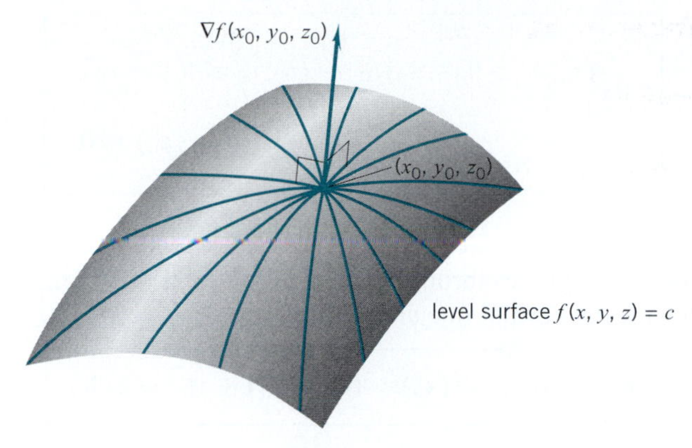

The gradient vector $\nabla f(x_0, y_0, z_0)$ is perpendicular to the level surface at (x_0, y_0, z_0)

Figure 15.4.5

points away from the origin along the line of the radius vector and is thus perpendicular to the sphere in question. At the origin the level surface is reduced to a point and the gradient is $\mathbf{0}$. ❑

The *tangent plane* for a surface

$$f(x, y, z) = c$$

at a point $\mathbf{x}_0 = (x_0, y_0, z_0)$ is the plane through $\mathbf{x}_0$ with normal $\nabla f(\mathbf{x}_0)$. See Figure 15.4.6.

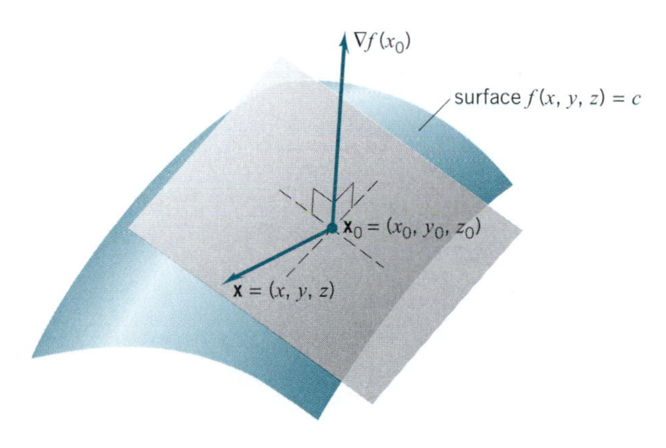

tangent plane to a surface $f(x, y, z) = c$

Figure 15.4.6

The tangent plane at a point $\mathbf{x}_0$ is the plane through $\mathbf{x}_0$ that best approximates the surface in a neighborhood of $\mathbf{x}_0$. (We return to this later.)

A point $\mathbf{x}$ lies on the tangent plane through $\mathbf{x}_0$ iff

(15.4.7) $$\boxed{\nabla f(\mathbf{x}_0) \cdot (\mathbf{x} - \mathbf{x}_0) = 0.}$$ (Figure 15.4.6)

This is an equation for the tangent plane in vector notation. In Cartesian coordinates the equation takes the form

(15.4.8)
$$\frac{\partial f}{\partial x}(x_0, y_0, z_0)(x - x_0) + \frac{\partial f}{\partial y}(x_0, y_0, z_0)(y - y_0) + \frac{\partial f}{\partial z}(x_0, y_0, z_0)(z - z_0) = 0.$$

The *normal line* to the surface $f(x, y, z) = c$ at a point $\mathbf{x}_0 = (x_0, y_0, z_0)$ on the surface is the line which passes through (x_0, y_0, z_0) parallel to $\nabla f(\mathbf{x}_0)$. Thus, $\nabla f(\mathbf{x}_0)$ is a direction vector for the normal line and

(15.4.9)
$$\mathbf{r}(t) = \mathbf{r}_0 + \nabla f(\mathbf{x}_0)t \quad (\mathbf{r}_0 = x_0\,\mathbf{i} + y_0\,\mathbf{j} + z_0\,\mathbf{k})$$

is a vector equation for the line. In scalar parametric form, equations for the normal line can be written

(15.4.10)
$$x = x_0 + \frac{\partial f}{\partial x}(x_0, y_0, z_0)\,t,$$
$$y = y_0 + \frac{\partial f}{\partial y}(x_0, y_0, z_0)\,t,$$
$$z = z_0 + \frac{\partial f}{\partial z}(x_0, y_0, z_0)\,t.$$

Example 4 Find an equation for the plane tangent to the surface

$$xy + yz + zx = 11 \quad \text{at the point } (1, 2, 3).$$

SOLUTION The surface is of the form

$$f(x, y, z) = c \quad \text{with} \quad f(x, y, z) = xy + yz + zx \quad \text{and} \quad c = 11.$$

Observe that

$$\frac{\partial f}{\partial x} = y + z, \qquad \frac{\partial f}{\partial y} = x + z, \qquad \frac{\partial f}{\partial z} = x + y.$$

At the point (1,2,3)

$$\frac{\partial f}{\partial x} = 5, \qquad \frac{\partial f}{\partial y} = 4, \qquad \frac{\partial f}{\partial z} = 3.$$

The equation for the tangent plane can therefore be written

$$5(x - 1) + 4(y - 2) + 3(z - 3) = 0.$$

This simplifies to

$$5x + 4y + 3z - 22 = 0. \quad \square$$

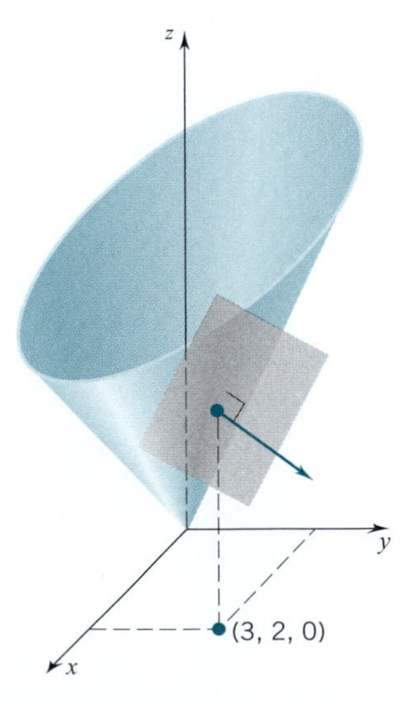

Figure 15.4.7

Example 5 Find an equation for the tangent plane and find scalar parametric equations for the normal line to the elliptic cone

$$x^2 + 4y^2 = z^2$$

at the point (3,2,5) on the cone (Figure 15.4.7).

SOLUTION The surface is of the form $f(x,y,z) = c$ with

$$f(x,y,z) = x^2 + 4y^2 - z^2 \qquad \text{and} \qquad c = 0.$$

The partial derivatives of f are

$$\frac{\partial f}{\partial x} = 2x, \quad \frac{\partial f}{\partial y} = 8y, \quad \frac{\partial f}{\partial z} = -2z$$

and
$$\nabla f = 2x\,\mathbf{i} + 8y\,\mathbf{j} - 2z\,\mathbf{k}.$$

Now, $\nabla f(3,2,5) = 6\mathbf{i} + 16\mathbf{j} - 10\mathbf{k}$ is normal to the cone at the point (3,2,5). Note that $\frac{1}{2}\nabla f(3,2,5) = 3\mathbf{i} + 8\mathbf{j} - 5\mathbf{k}$ is also normal to the cone and is a little easier to work with.

The equation for the tangent plane can be written

$$3(x-3) + 8(y-2) - 5(z-5) = 0,$$

which simplifies to

$$3x + 8y - 5z = 0.$$

Note that this plane passes through the origin, as we would expect. The following are scalar parametric equations for the normal line:

$$x = 3 + 3t, \quad y = 2 + 8t, \quad z = 5 - 5t. \quad \square$$

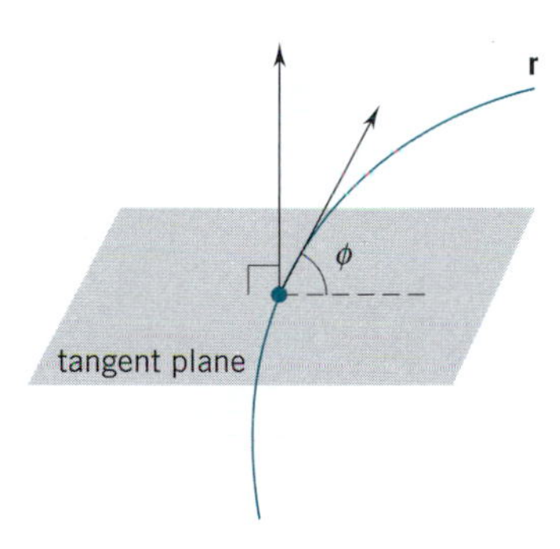

Figure 15.4.8

Example 6 The curve $\mathbf{r}(t) = \frac{1}{2}t^2\,\mathbf{i} + 4t^{-1}\,\mathbf{j} + (\frac{1}{2}t - t^2)\,\mathbf{k}$ intersects the hyperbolic paraboloid $x^2 - 4y^2 - 4z = 0$ at the point $(2, 2, -3)$. What is the angle of intersection?

SOLUTION We want the angle ϕ between the tangent vector of the curve and the tangent plane of the surface at the point of intersection (Figure 15.4.8).

A simple calculation shows that the curve passes through the point $(2,2,-3)$ at $t = 2$. Since

$$\mathbf{r}'(t) = t\,\mathbf{i} - 4t^{-2}\,\mathbf{j} + (\tfrac{1}{2} - 2t)\,\mathbf{k},$$

we have
$$\mathbf{r}'(2) = 2\,\mathbf{i} - \mathbf{j} - \tfrac{7}{2}\,\mathbf{k}.$$

Now set
$$f(x,y,z) = x^2 - 4y^2 - 4z.$$

This function has gradient $2x\,\mathbf{i} - 8y\,\mathbf{j} - 4\,\mathbf{k}$. At the point $(2,2,-3)$,

$$\nabla f = 4\,\mathbf{i} - 16\,\mathbf{j} - 4\,\mathbf{k}.$$

Now let θ be the angle between $\mathbf{r}'(2)$ and this gradient. By (12.4.9),

$$\cos\theta = \frac{\mathbf{r}'(2)}{\|\mathbf{r}'(2)\|} \cdot \frac{\nabla f}{\|\nabla f\|} = \frac{19}{414}\sqrt{138} \cong 0.539,$$

so that $\theta \cong 1.00$ radian. Since the gradient is normal to the tangent plane, the angle ϕ we want is

$$\phi = \tfrac{1}{2}\pi - \theta \cong 1.57 - 1.00 = 0.57 \text{ radian.} \quad \square$$

A surface of the form

$$z = g(x, y)$$

can be written in the form

$$f(x, y, z) = 0$$

by setting $\qquad f(x, y, z) = g(x, y) - z.$

If g is differentiable, so is f. Moreover,

$$\frac{\partial f}{\partial x}(x, y, z) = \frac{\partial g}{\partial x}(x, y), \quad \frac{\partial f}{\partial y}(x, y, z) = \frac{\partial g}{\partial y}(x, y), \quad \frac{\partial f}{\partial z}(x, y, z) = -1.$$

By (15.4.8), the tangent plane at (x_0, y_0, z_0) has equation

$$\frac{\partial g}{\partial x}(x_0, y_0)(x - x_0) + \frac{\partial g}{\partial y}(x_0, y_0)(y - y_0) + (-1)(z - z_0) = 0$$

which we can rewrite as

(15.4.11)
$$z - z_0 = \frac{\partial g}{\partial x}(x_0, y_0)(x - x_0) + \frac{\partial g}{\partial y}(x_0, y_0)(y - y_0).$$

If $\nabla g(x_0, y_0) = \mathbf{0}$, then both partials of g are zero at (x_0, y_0) and the equation reduces to

$$z = z_0.$$

In this case the tangent plane is *horizontal*.

Scalar parametric equations for the normal line to the surface $z = g(x, y)$ at the point (x_0, y_0, z_0) take the form

(15.4.12)
$$x = x_0 + \frac{\partial g}{\partial x}(x_0, y_0)\,t, \quad y = y_0 + \frac{\partial g}{\partial y}(x_0, y_0)\,t, \quad z = z_0 + (-1)\,t.$$

Example 7 Find an equation for the tangent plane and symmetric equations for the line normal to the surface.

$$z = \ln(x^2 + y^2)$$

at the point $(-2, 1, \ln 5)$ on the surface.

SOLUTION Set $g(x, y) = \ln(x^2 + y^2)$. The partial derivatives of g are

$$\frac{\partial g}{\partial x}(x, y) = \frac{2x}{x^2 + y^2}, \quad \frac{\partial g}{\partial y}(x, y) = \frac{2y}{x^2 + y^2}.$$

Where $x = -2$ and $y = 1$,

$$\frac{\partial g}{\partial x} = -\frac{4}{5}, \quad \frac{\partial g}{\partial y} = \frac{2}{5}.$$

Therefore, at the point $(-2, 1, \ln 5)$, the tangent plane has equation

$$z - \ln 5 = -\frac{4}{5}(x+2) + \frac{2}{5}(y-1).$$

The symmetric equations for the normal line are

$$\frac{x+2}{-\frac{4}{5}} = \frac{y-1}{\frac{2}{5}} = \frac{z-\ln 5}{-1}. \quad \square$$

Example 8 At what points on the surface $z = 3xy - x^3 - y^3$, is the tangent plane horizontal?

SOLUTION The function $g(x, y) = 3xy - x^3 - y^3$ has first partials

$$\frac{\partial g}{\partial x}(x, y) = 3y - 3x^2, \quad \frac{\partial g}{\partial y}(x, y) = 3x - 3y^2.$$

We set these partial derivatives equal to 0 and solve the resulting system of equations for x and y:

$$\begin{aligned} 3y - 3x^2 &= 0 \quad \text{simplifies to} \quad & y - x^2 &= 0 \\ 3x - 3y^2 &= 0 & x - y^2 &= 0. \end{aligned}$$

From the first equation, we get $y = x^2$. Substituting this into the second equation, we have

$$x - x^4 = 0,$$

and thus

$$x(1 - x^3) = 0.$$

Therefore, $x = 0$ (which implies that $y = 0$) or $x = 1$ (which implies that $y = 1$). Thus, the partials are both zero only at $(0, 0)$ and $(1, 1)$. The surface has a horizontal tangent plane only at the points $(0, 0, 0)$ and $(1, 1, 1)$. A graph of the surface and the level curves are shown in Figure 15.4.9 $\square$

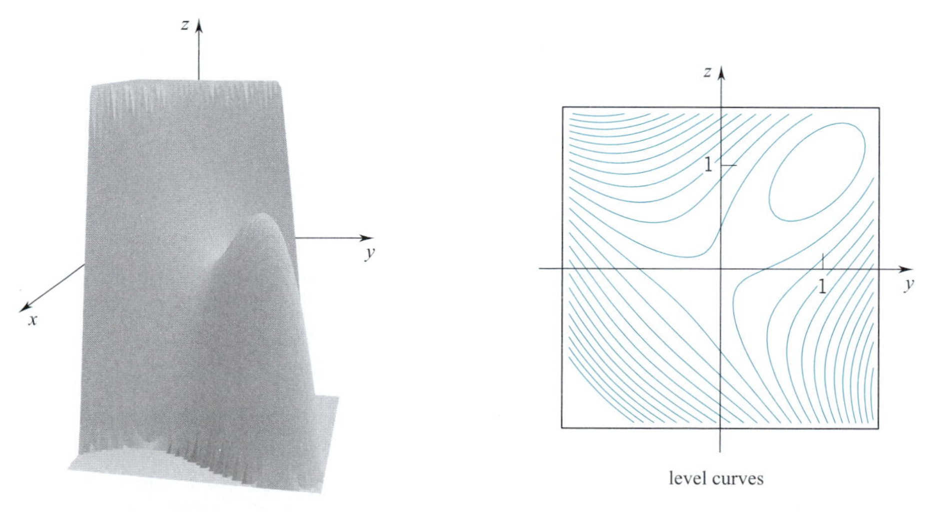

Figure 15.4.9

EXERCISES 15.4

Find a normal vector and a tangent vector at the point P. Write an equation for the tangent line and an equation for the normal line.

1. $x^2 + xy + y^2 = 3$; $P(-1, -1)$.

2. $(y - x)^2 = 2x$; $P(2, 4)$.

3. $(x^2 + y^2)^2 = 9(x^2 - y^2)$; $P(\sqrt{2}, 1)$.

4. $x^3 + y^3 = 9$; $P(1, 2)$.

5. $xy^2 - 2x^2 + y + 5x = 6$; $P(4, 2)$.

6. $x^5 + y^5 = 2x^3$; $P(1, 1)$.

7. $2x^3 - x^2y^2 = 3x - y - 7$; $P(1, -2)$.

8. $x^3 + y^2 + 2x = 6$; $P(-1, 3)$.

9. Find the slope of the curve $x^2y = a^2(a - y)$ at the point $(0, a)$.

Find an equation for the tangent plane and scalar parametric equations for the normal line at the point P.

10. $z = (x^2 + y^2)^2$; $P(1, 1, 4)$.

11. $x^3 + y^3 = 3xyz$; $P(1, 2, \frac{3}{2})$.

12. $xy^2 + 2z^2 = 12$; $P(1, 2, 2)$.

13. $z = axy$; $P(1, 1/a, 1)$.

14. $\sqrt{x} + \sqrt{y} + \sqrt{z} = 4$; $P(1, 4, 1)$.

15. $z = \sin x + \sin y + \sin(x + y)$; $P(0, 0, 0)$.

16. $z = x^2 + xy + y^2 - 6x + 2$; $P(4, -2, -10)$.

17. $b^2c^2x^2 - a^2c^2y^2 - a^2b^2z^2 = a^2b^2c^2$; $P(x_0, y_0, z_0)$.

18. $z = \sin(x \cos y)$; $P(1, \frac{1}{2}\pi, 0)$.

Find the point(s) on the surface at which the tangent plane is horizontal.

19. $xy + a^3x^{-1} + b^3y^{-1} - z = 0$.

20. $z = 4x + 2y - x^2 + xy - y^2$.

21. $z = xy$.

22. $x + y + z + xy - x^2 - y^2 = 0$.

23. $z - 2x^2 - 2xy + y^2 + 5x - 3y + 2 = 0$.

24. (a) Find the *upper unit normal* (the unit normal with positive **k** component) for the surface $z = xy$ at the point $(1, 1, 1)$.

(b) Find the *lower unit normal* (the unit normal with negative **k** component) for the surface $z = 1/x - 1/y$ at the point $(1, 1, 0)$.

25. Let $f = f(x, y, z)$ be continuously differentiable. Write equations in symmetric form for the line normal to the surface $f(x, y, z) = c$ at the point (x_0, y_0, z_0).

26. Show that in the case of a surface of the form $z = xf(x/y)$ with f continuously differentiable, all the tangent planes have a point in common.

27. Given that the surfaces $F(x, y, z) = 0$ and $G(x, y, z) = 0$ intersect at right angles in a curve γ, what condition must be satisfied by the partial derivatives of F and G on γ?

28. Show that, for all planes tangent to the surface $\sqrt{x} + \sqrt{y} + \sqrt{z} = \sqrt{a}$, the sum of the intercepts is the same.

29. Show that all pyramids formed by the coordinate planes and a plane tangent to the surface $xyz = a^3$ have the same volume. What is this volume?

30. Show that, for all planes tangent to the surface $x^{2/3} + y^{2/3} + z^{2/3} = a^{2/3}$, the sum of the squares of the intercepts is the same.

31. The curve $\mathbf{r}(t) = 2t\,\mathbf{i} + 3t^{-1}\,\mathbf{j} - 2t^2\,\mathbf{k}$ and the ellipsoid $x^2 + y^2 + 3z^2 = 25$ intersect at $(2, 3, -2)$. What is the angle of intersection?

32. Show that the curve $\mathbf{r}(t) = \frac{3}{2}(t^2 + 1)\,\mathbf{i} + (t^4 + 1)\,\mathbf{j} + t^3\mathbf{k}$ is perpendicular to the ellipsoid $x^2 + 2y^2 + 3z^2 = 20$ at the point $(3, 2, 1)$.

33. The surfaces $x^2y^2 + 2x + z^3 = 16$ and $3x^2 + y^2 - 2z = 9$ intersect in a curve that passes through the point $(2, 1, 2)$. What are the equations of the respective tangent planes for the two surfaces at this point?

34. Show that the sphere $x^2 + y^2 + z^2 - 8x - 8y - 6z + 24 = 0$ is tangent to the ellipsoid $x^2 + 3y^2 + 2z^2 = 9$ at the point $(2, 1, 1)$.

35. Show that the sphere $x^2 + y^2 + z^2 - 4y - 2z + 2 = 0$ is perpendicular to the paraboloid $3x^2 + 2y^2 - 2z = 1$ at the point $(1, 1, 2)$.

36. Show that the following surfaces are mutually perpendicular:

$$xy = az^2, \quad x^2 + y^2 + z^2 = b, \quad z^2 + 2x^2 = c(z^2 + 2y^2).$$

37. The surface $S : z = x^2 + 3y^2 + 2$ intersects the vertical plane $p : 3x + 4y + 6 = 0$ in a space curve C.

(a) Let C_1 be the projection of C onto the xy-plane. Find an equation for C_1.

(b) Find a parametrization $\mathbf{r}(t) = x(t)\,\mathbf{i} + y(t)\,\mathbf{j} + z(t)\,\mathbf{k}$ for C setting $x(t) = 4t - 2$.

(c) Find a parametrization $\mathbf{R}(s) = \mathbf{R}_0 + s\mathbf{d}$ for the line l tangent to C at the point $(2, -3, 33)$.

(d) Find an equation for the plane p_1 tangent to S at the point $(2, -3, 33)$.

(e) Find a parametrization $\mathbf{r}(t) = x(t)\,\mathbf{i} + y(t)\,\mathbf{j} + z(t)\,\mathbf{k}$ for the line l' formed by the intersection of p with p_1, taking $x(t) = t$. What is the relation between l and l'?

C 38. Let $f(x, y) = \dfrac{x^2 - 2y}{x^2 + y^2}$. The level curve $f(x, y) = 2/5$ passes through the point $P(2, 1)$. Use a CAS to find:

(a) A normal vector at P and scalar parametric equations for the normal line at P.

(b) Scalar parametric equations for the tangent line at P.

(c) Use a graphing utility to draw the level curve, normal line, and tangent line together.

C 39. Let $f(x, y, z) = x^2 + (y - 1)^2 + z^2$. The level surface $f(x, y, z) = 6$ passes through the point $P(1, 2, 2)$. Use a CAS to find:

(a) A normal vector at P and scalar parametric equations for the normal line at P.

(b) The equation of the tangent plane at P.

(c) Use a graphing utility to draw the level surface, normal line, and tangent plane together.

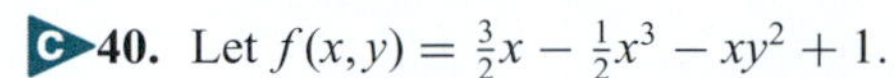

40. Let $f(x,y) = \frac{3}{2}x - \frac{1}{2}x^3 - xy^2 + 1$.

(a) Use a graphing utility to draw the surface $z = f(x,y)$. Choose a viewing window that reveals the features of the surface.

(b) Draw the level curves of the surface and use these to estimate the points where the surface may have a horizontal tangent plane.

(c) Calculate ∇f and find the points where $\nabla f(x,y) = \mathbf{0}$. Compare your answers with your estimates from part(b).

41. Repeat Exercise 40 with $f(x,y) = x^4 - y^4 - 2x^2 + 2y^2 + 2$.

42. Repeat Exercise 40 with $f(x,y) = 8xy\,e^{-(x^2+y^2)}$.

■ 15.5 LOCAL EXTREME VALUES

In Chapter 4 we discussed local extreme values for a function of one variable. Here we take up the same subject for functions of several variables. The ideas are very similar.

DEFINITION 15.5.1 LOCAL MAXIMUM AND LOCAL MINIMUM

Let f be a function of several variables and let $\mathbf{x}_0$ be an interior point of the domain:

f is said to take on a *local maximum* at $\mathbf{x}_0$ if

$$f(\mathbf{x}_0) \geq f(\mathbf{x}) \quad \text{for all } \mathbf{x} \text{ in some neighborhood of } \mathbf{x}_0;$$

f is said to take on a *local minimum* at $\mathbf{x}_0$ if

$$f(\mathbf{x}_0) \leq f(\mathbf{x}) \quad \text{for all } \mathbf{x} \text{ in some neighborhood of } \mathbf{x}_0.$$

As in the one-variable case, the local maxima and local minima together comprise the *local extreme values*.

In the one-variable case we know that if f takes on a local extreme value at x_0, then

$$\text{either} \quad f'(x_0) = 0 \quad \text{or} \quad f'(x_0) \text{ does not exist.}$$

We have a similar result for functions of several variables.

THEOREM 15.5.2

If f takes on a local extreme value at $\mathbf{x}_0$, then

$$\text{either} \qquad \nabla f(\mathbf{x}_0) = \mathbf{0} \quad \text{or} \quad \nabla f(\mathbf{x}_0) \text{ does not exist.}$$

PROOF We assume that f takes on a local extreme value at $\mathbf{x}_0$ and that f is differentiable at $\mathbf{x}_0$ [namely, that $\nabla f(\mathbf{x}_0)$ exists]. We need to show that $\nabla f(\mathbf{x}_0) = \mathbf{0}$. For simplicity we set $\mathbf{x}_0 = (x_0, y_0)$. The three-variable case is similar.

Since f takes on a local extreme value at (x_0, y_0), the function $g(x) = f(x, y_0)$ takes on a local a extreme value at x_0. Since f is differentiable at (x_0, y_0), g is differentiable at x_0 and therefore

$$g'(x_0) = \frac{\partial f}{\partial x}(x_0, y_0) = 0.$$

Similarly, the function $h(y)=f(x_0,y)$ takes on extreme value at y_0 and, being differentiable there, satisfies the relation

$$h'(y_0) = \frac{\partial f}{\partial y}(x_0, y_0) = 0.$$

The gradient is $\mathbf{0}$ since both partials are 0. ❑

Interior points of the domain at which the gradient is zero or the gradient does not exist are called *critical points*. By Theorem 15.5.2 these are the only points that can give rise to local extreme values.

Although the ideas introduced so far are completely general, their application to functions of more than two variables is generally laborious. We restrict ourselves mostly to functions of two variables. Not only are the computations less formidable, but also we can make use of our geometric intuition.

Two Variables

We suppose for the moment that $f = f(x, y)$ is defined on an open set and is continuously differentiable there. The graph of f is a surface

$$z = f(x, y).$$

Where f takes on a local maximum, the surface has a local high point. Where f takes on a local minimum, the surface has a local low point. Where f has either a local maximum or a local minimum, the gradient is $\mathbf{0}$ and therefore the tangent plane is horizontal. See Figure 15.5.1.

A zero gradient signals the possibility of a local extreme value; it does not guarantee it. For example, in the case of the saddle-shaped surface of Figure 15.5.2, there is a horizontal tangent plane at the origin and therefore the gradient is zero there, yet the origin gives neither a local maximum nor a local minimum.

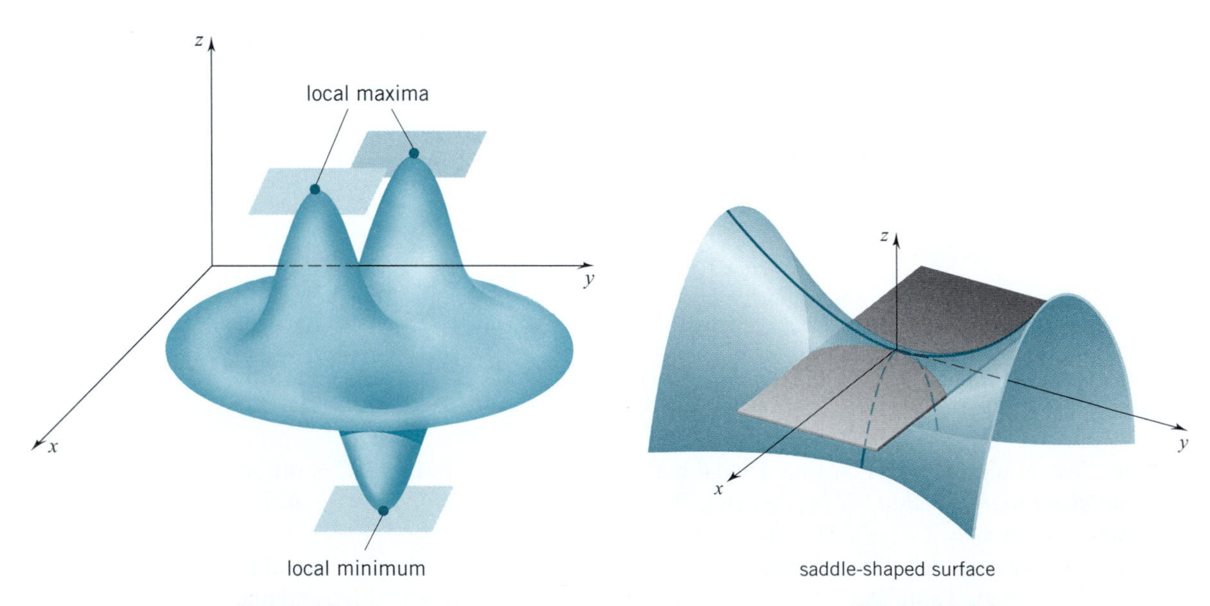

Figure 15.5.1 **Figure 15.5.2**

Critical points at which the gradient is zero are called *stationary points*. The stationary points that do not give rise to local extreme values are called *saddle points*.

Below we test some differentiable functions for extreme values. In each case, our first step is to find the stationary points.

Example 1 For the function $f(x,y) = 2x^2 + y^2 - xy - 7y$,

we have $$\nabla f(x,y) = (4x - y)\,\mathbf{i} + (2y - x - 7)\,\mathbf{j}.$$

To find the stationary points, we set $\nabla f(x,y) = \mathbf{0}$. This gives

$$4x - y = 0 \qquad \text{and} \qquad 2y - x - 7 = 0.$$

The only simultaneous solution to these equations is $x = 1, y = 4$. The point $(1,4)$ is therefore the only stationary point.

We now compare the value of f at $(1,4)$ with the values of f at nearby points $(1+h, 4+k)$:

$$\begin{aligned} f(1,4) &= 2 + 16 - 4 - 28 = -14,\\ f(1+h,4+k) &= 2(1+h)^2 + (4+k)^2 - (1+h)(4+k) - 7(4+k)\\ &= 2 + 4h + 2h^2 + 16 + 8k + k^2 - 4 - 4h - k - hk - 28 - 7k\\ &= 2h^2 + k^2 - hk - 14. \end{aligned}$$

The difference

$$\begin{aligned} f(1+h,4+k) - f(1,4) &= 2h^2 + k^2 - hk\\ &= h^2 + \left(h^2 - hk + k^2\right)\\ &\geq h^2 + \left(h^2 - 2|h||k| + k^2\right)\\ &= h^2 + (|h| - |k|)^2 \geq 0. \end{aligned}$$

Thus, $f(1+h,4+k) \geq f(1,4)$ for all small h and k (in fact for all real h and k). †

It follows that f takes on local minimum at $(1,4)$. This local minimum is -14. □

Example 2 In the case of $f(x,y) = y^2 - xy + 2x + y + 1$

we have $$\nabla f(x,y) = (2 - y)\,\mathbf{i} + (2y - x + 1)\,\mathbf{j}.$$

The gradient is $\mathbf{0}$ where

$$2 - y = \mathbf{0} \qquad \text{and} \qquad 2y - x + 1 = 0.$$

The only simultaneous solution to these equations is $x = 5, y = 2$. The point $(5,2)$ is the only stationary point.

We now compare the value of f at $(5,2)$ with the values of f at nearby points $(5+h, 2+k)$: S

$$\begin{aligned} f(5,2) &= 4 - 10 + 10 + 2 + 1 = 7,\\ f(5+h,2+k) &= (2+k)^2 - (5+h)(2+k) + 2(5+h) + (2+k) + 1\\ &= 4 + 4k + k^2 - 10 - 2h - 5k - hk + 10 + 2h + 2 + k + 1\\ &= k^2 - hk + 7. \end{aligned}$$

---

† Another way to see that $2h^2 + k^2 - hk$ is nonnegative is to complete the square:

$$2h^2 + k^2 - hk = \tfrac{1}{4}h^2 - hk + k^2 + \tfrac{7}{4}h^2 = (\tfrac{1}{2}h - k)^2 + \tfrac{7}{4}h^2 \geq 0.$$

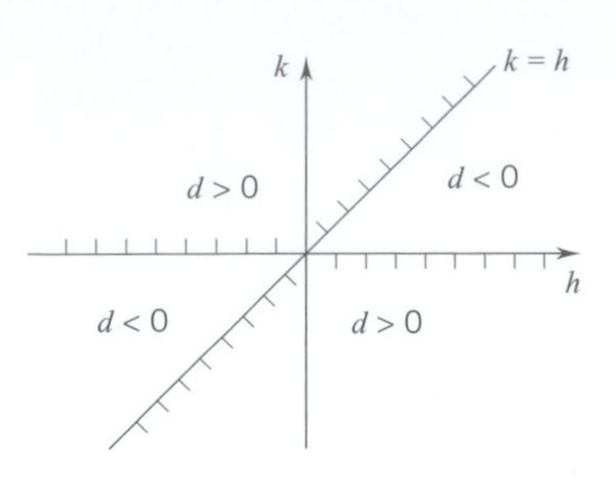

Figure 15.5.3

The difference

$$d = f(5+h, 2+k) - f(5,2) = k^2 - hk = k(k-h)$$

does not keep a constant sign for small h and k. See Figure 15.5.3. Therefore, $(5, 2)$ is a saddle point. ❑

Example 3 The function $f(x,y) = 1 + \sqrt{x^2+y^2}$ is everywhere defined and everywhere continuous. The graph is the upper nappe of a right circular cone. (See Figure 15.5.4). The number $f(0,0) = 1$ is obviously a local minimum.

Since the partials

$$\frac{\partial f}{\partial x} = \frac{x}{\sqrt{x^2+y^2}}, \quad \frac{\partial f}{\partial y} = \frac{y}{\sqrt{x^2+y^2}}$$

are not defined at $(0,0)$, the gradient is not defined at $(0,0)$ (Theorem 15.2.5). The point $(0,0)$ is thus a critical point, but not a stationary point. At $(0,0,1)$ the surface comes to a sharp point and there is no tangent plane. ❑

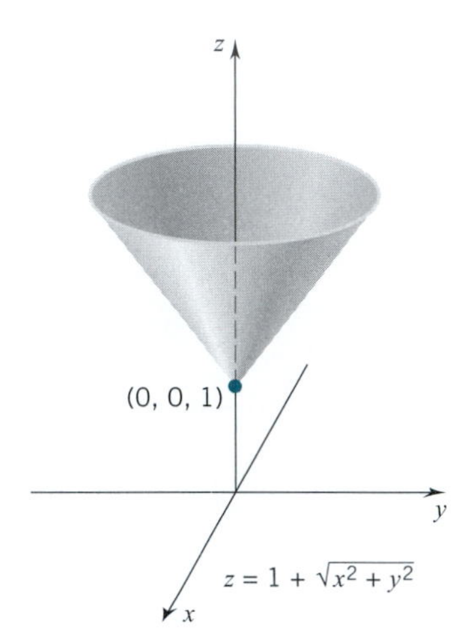

Figure 15.5.4

Second-Partials Test

Suppose that g is a function of one variable and $g'(x_0) = 0$. Then, according to the second-derivative test (Theorem 4.3.5), g takes on

$$\text{a local minimum at } x_0 \quad \text{if} \quad g''(x_0) > 0,$$
$$\text{a local maximum at } x_0 \quad \text{if} \quad g''(x_0) < 0.$$

We have a similar test for functions of two variables. As one might expect, the test is somewhat more complicated to state and definitely more difficult to prove. We will omit the proof. †

THEOREM 15.5.3 THE SECOND-PARTIALS TEST

Suppose that f has continuous second-order partial derivatives in a neighborhood of (x_0, y_0) and that $\nabla f(x_0, y_0) = \mathbf{0}$. Set

$$A = \frac{\partial^2 f}{\partial x^2}(x_0, y_0), \quad B = \frac{\partial^2 f}{\partial y \partial x}(x_0, y_0), \quad C = \frac{\partial^2 f}{\partial y^2}(x_0, y_0)$$

and form the *discriminant* $D = AC - B^2$

1. If $D < 0$, then (x_0, y_0) is a saddle point.
2. If $D > 0$, then f takes on

$$\text{a local minimum at } (x_0, y_0) \quad \text{if } A > 0,$$
$$\text{a local maximum at } (x_0, y_0) \quad \text{if } A < 0.$$

The test is geometrically evident for functions of the form

$$f(x,y) = \tfrac{1}{2}ax^2 + \tfrac{1}{2}cy^2. \qquad (a \neq 0, c \neq 0)$$

† You can find a proof in most texts on advanced calculus.

The graph of such a function is a paraboloid:

$$z = \tfrac{1}{2}ax^2 + \tfrac{1}{2}cy^2.$$

The gradient is $\mathbf{0}$ at the origin $(0, 0)$. Moreover

$$A = \frac{\partial^2 f}{\partial x^2}(0,0) = a, \quad B = \frac{\partial^2 f}{\partial y \partial x}(0,0) = 0, \quad C = \frac{\partial^2 f}{\partial y^2}(0,0) = c.$$

and $D = AC - B^2 = ac$. If $D < 0$, then a and c have opposite signs and we have a saddle point. (Figure 15.5.5).

Suppose now that $D > 0$. If $a > 0$, then $c > 0$ and the surface has a minimum point; if $a < 0$, then $c < 0$ and the surface has a maximum point.

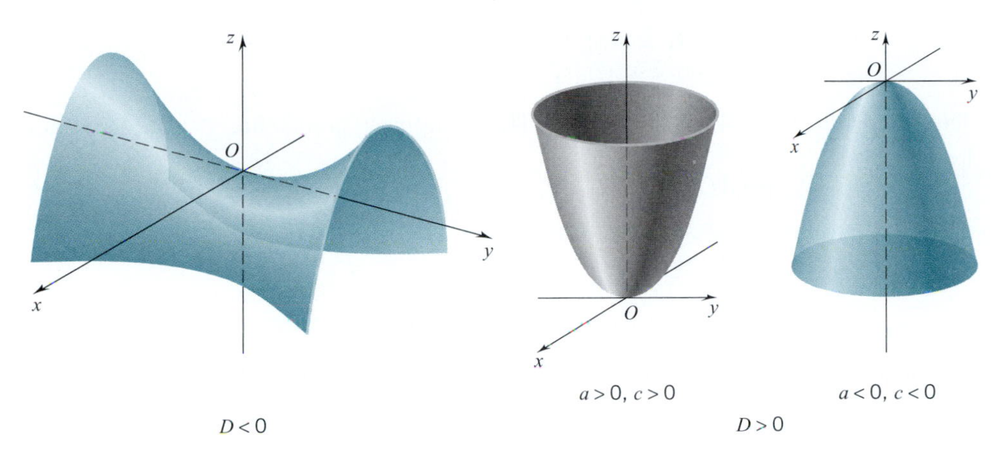

Figure 15.5.5

In the examples that follow we apply the second-partials test to a variety of functions.

Example 4 In Example 1, we saw that the point $(1, 4)$ is the only stationary point of the function

$$f(x, y) = 2x^2 + y^2 - xy - 7y.$$

The first partials of f are

$$\frac{\partial f}{\partial x} = 4x - y \quad \text{and} \quad \frac{\partial f}{\partial y} = 2y - x - 7,$$

and the second partials are constant:

$$\frac{\partial^2 f}{\partial x^2} = 4, \quad \frac{\partial^2 f}{\partial y \partial x} = -1, \quad \frac{\partial^2 f}{\partial y^2} = 2.$$

Thus, $A = 4, B = -1, C = 2$, and $D = AC - B^2 = 7 > 0$. Since $A > 0$, it follows from the second-partials test that

$$f(1, 4) = 2 + 16 - 4 - 28 = -14$$

is a local minimum. □

Example 5 The function $f(x, y) = -\frac{1}{4}x^4 + \frac{2}{3}x^3 + 4xy - y^2$ has partial derivatives

$$\frac{\partial f}{\partial x} = -x^3 + 2x^2 + 4y, \quad \frac{\partial f}{\partial y} = 4x - 2y.$$

Setting both partials equal to zero, we have

$$-x^3 + 2x^2 + 4y = 0, \quad 4x - 2y = 0.$$

We get $y = 2x$ from the second equation. Substituting $y = 2x$ into the first equation gives

$$-x^3 + 2x^2 + 8x = -x(x^2 - 2x - 8) = -x(x - 4)(x + 2) = 0.$$

Therefore $x = 0$, $x = 4$, and $x = -2$. The stationary points are $(0, 0)$, $(4, 8)$, and $(-2, -4)$.

The second partials are

$$\frac{\partial^2 f}{\partial x^2} = -3x^2 + 4x, \quad \frac{\partial^2 f}{\partial x \partial y} = 4, \quad \frac{\partial^2 f}{\partial y^2} = -2.$$

The data for the second partials test are given in the following table.

Point	A	B	C	D	Result
$(0, 0)$	0	4	−2	16	Saddle point
$(4, 8)$	−32	4	−2	48	Loc. max.
$(-2, -4)$	−20	4	−2	24	Loc. max.

A graph of the surface and the level curves are shown in Figure 15.5.6. ❑

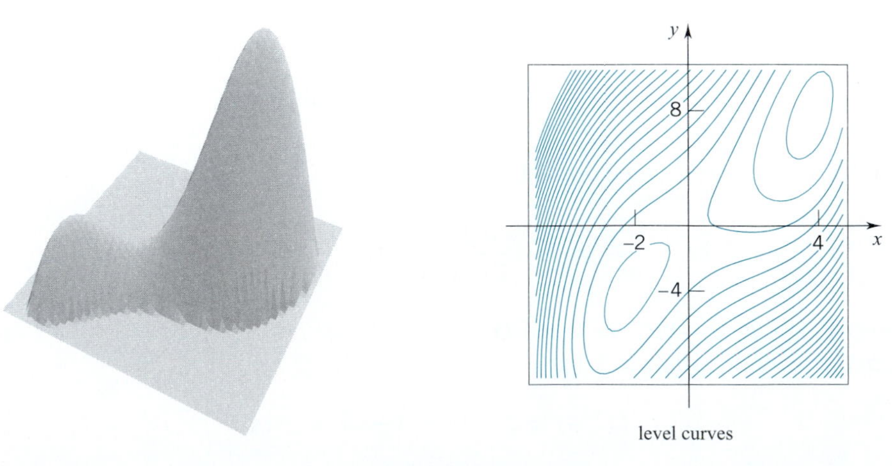

Figure 15.5.6

Example 6 For the function $f(x,y) = -xy\,e^{-(x^2+y^2)/2}$

we have
$$\frac{\partial f}{\partial x} = -y\,e^{-(x^2+y^2)/2} + x^2 y\,e^{-(x^2+y^2)/2} = y(x^2-1)\,e^{-(x^2+y^2)/2},$$
$$\frac{\partial f}{\partial y} = -x\,e^{-(x^2+y^2)/2} + xy^2\,e^{-(x^2+y^2)/2} = x(y^2-1)\,e^{-(x^2+y^2)/2},$$

and
$$\nabla f(x,y) = e^{-(x^2+y^2)/2}[y(x^2-1)\,\mathbf{i} + x(y^2-1)\,\mathbf{j}].$$

Since $e^{-(x^2+y^2)/2} \neq 0$ for all (x,y), $\nabla f(x,y) = \mathbf{0}$ iff
$$y(1-x^2) = 0 \quad \text{and} \quad x(y^2-1) = 0.$$

The simultaneous solutions to these equations are $x = y = 0$; $x = 1, y = \pm 1$; $x = -1$, $y = \pm 1$. Thus, $(0,0), (1,1), (1,-1), (-1,1)$, and $(-1,-1)$ are the stationary points.

You can verify that the second partial derivatives of f are
$$\frac{\partial^2 f}{\partial x^2} = xy(3-x^2)\,e^{-(x^2+y^2)/2}, \quad \frac{\partial^2 f}{\partial y^2} = xy(3-y^2)\,e^{-(x^2+y^2)/2},$$

and
$$\frac{\partial^2 f}{\partial y \partial x} = (x^2-1)(1-y^2)e^{-(x^2+y^2)/2}.$$

The data for the second-partials test are recorded in the following table.

Point	A	B	C	D	Result
(0,0)	0	−1	0	−1	Saddle point
(1,1)	$2\,e^{-1}$	0	$2\,e^{-1}$	$4\,e^{-2}$	Loc. min.
(1,−1)	$-2\,e^{-1}$	0	$-2\,e^{-1}$	$4\,e^{-2}$	Loc. max.
(−1,1)	$-2\,e^{-1}$	0	$-2\,e^{-1}$	$4\,e^{-2}$	Loc. max.
(−1,−1)	$2\,e^{-1}$	0	$2\,e^{-1}$	$4\,e^{-2}$	Loc. min.

A computer-generated graph of this function is shown in Figure 15.5.7. □

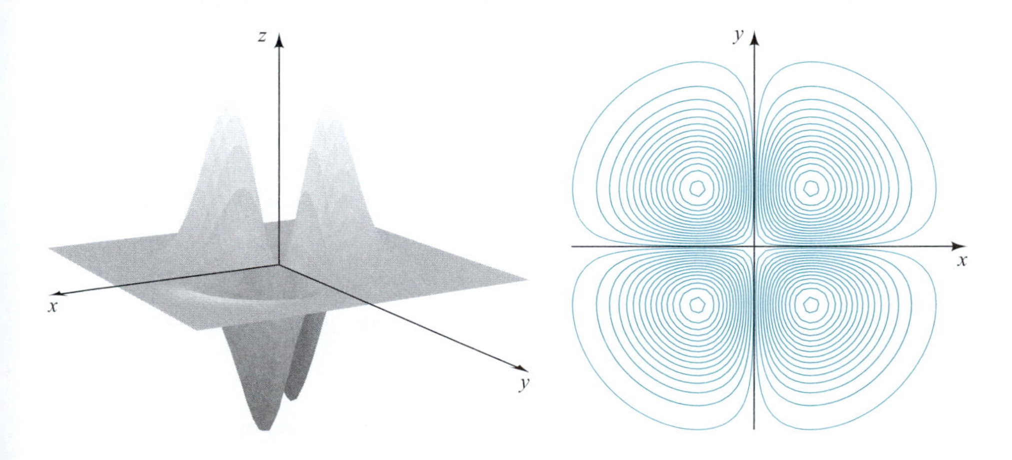

level curves of $f(x,y) = -xye^{-(x^2+y^2)/2}$

Figure 15.5.7

The second-derivative test for a function of one variable applies to points x_0 where $g'(x_0)=0$ but $g''(x_0)\neq 0$. If $g''(x_0)=0$, the second-derivative test provides no conclusive information. The second-partials test suffers from a similar limitation. It applies to points (x_0, y_0) where $\nabla f(x_0, y_0) = \mathbf{0}$ but $D \neq 0$. If $D = 0$, the second-partials test provides no information.

Consider, for example, the functions

$$f(x,y)=x^4+y^4, \quad g(x,y)=-(x^4+y^4), \quad h(x,y)=x^4-y^4.$$

Each of these functions has zero gradient at the origin, and, as you can check, in each case $D = 0$. Yet,

(1) for f, $(0,0)$ gives a local minimum;

(2) for g, $(0,0)$ gives a local maximum;

(3) for h, $(0,0)$ is a saddle point.

Statements (1) and (2) are obvious. To confirm (3), note that $h(0,0) = 0$, but in every neighborhood of $(0,0)$ the functions h takes on both positive and negative values:

$$h(x,0)>0 \quad \text{for } x\neq 0, \quad \text{while} \quad h(0,y)<0 \quad \text{for } y\neq 0. \quad \square$$

EXERCISES 15.5

Find the stationary points and use the method illustrated in Examples 1 and 2 to determine the local extreme values.

1. $f(x,y)=2x-x^2-y^2$.

2. $f(x,y)=2x+2y-x^2+y^2+5$.

3. $f(x,y)=x^2+xy+y^2+3x+1$.

4. $f(x,y)=x^3-3x+y$.

Find the stationary points and the local extreme values.

5. $f(x,y)=x^2+xy+y^2-6x+2$

6. $f(x,y)=x^2+2xy+3y^2+2x+10y+1$.

7. $f(x,y)=x^3-6xy+y^3$.

8. $f(x,y)=3x^2+xy-y^2+5x-5y+4$.

9. $f(x,y)=x^3+y^2-6xy+6x+3y-2$.

10. $f(x,y)=x^2-2xy+2y^2-3x+5y$.

11. $f(x,y)=x\sin y$.

12. $f(x,y)=y+x\sin y$.

13. $f(x,y)=(x+y)(xy+1)$.

14. $f(x,y)=xy^{-1}-yx^{-1}$.

15. $f(x,y)=xy+x^{-1}+8y^{-1}$.

16. $f(x,y)=x^2-2xy-y^2+1$.

17. $f(x,y)=xy+x^{-1}+y^{-1}$.

18. $f(x,y)=(x-y)(xy-1)$.

19. $f(x,y)=\dfrac{-2x}{x^2+y^2+1}$.

20. $f(x,y)=(x-3)\ln xy$.

21. $f(x,y)=x^4-2x^2+y^2-2$.

22. $f(x,y)=(x^2+y^2)e^{x^2-y^2}$.

23. $f(x,y)=\sin x\,\sin y, \quad 0<x<2\pi, \quad 0<y<2\pi$.

24. $f(x,y)=\cos x\,\cosh y, \quad -2\pi<x<2\pi$.

25. Let $f(x,y)=x^2+kxy+y^2$, k a constant.

(a) Show that f has a stationary point at $(0,0)$ no matter what value is assigned to k.

(b) For what values of k will f have a saddle point at $(0,0)$?

(c) For what values of k will f have a local minimum at $(0,0)$?

(d) For what values of k is the second-derivative test inconclusive?

26. Repeat Exercise 25 for the function

$$f(x,y)=x^2+kxy+4y^2, \quad k \text{ a constant.}$$

27. Find the point in the plane $2x-y+2z=16$ that is closest to the origin, and calculate the distance from the origin to the plane. Check your answer by using Formula (12.7.7).

28. Find the point in the plane $3x-4y+2z+32=0$ that is closest to the point $P(-1,2,4)$ and calculate the distance from P to the plane. Check your answer by using Formula (12.7.7).

29. Find the shortest distance from the point $(1,2,0)$ to the elliptic cone $z=\sqrt{x^2+2y^2}$. HINT: Minimize the square of the distance.

30. Find the maximum volume for a rectangular solid inscribed in the sphere.

$$x^2 + y^2 + z^2 = a^2.$$

If a continuous function of one variable has local maxima at $x = a$ and $x = b$, $a < b$, then it must have at least one local minimum at some number $c \in (a, b)$. The corresponding result holds with the roles of the local extrema interchanged. Exercises 31 and 32 illustrate that these properties do not hold in higher dimensions.

31. Let $f(x,y) = 4xy - x^4 - y^4 + 1$.

(a) Use a graphing utility to draw graphs of f and the level curves.

(b) Your graphs should show that f has local maxima at $(1, 1)$ and $(-1, -1)$, a saddle point at $(0, 0)$, and no local minima.

(c) Verify the conclusions of part (b) using the methods of this section.

32. Let $f(x,y) = x^4 - 2x^2 + y^2 + 1$.

(a) Use a graphing utility to draw graphs of f and the level curves.

(b) Your graphs should show that f has local minima at $(1, 0)$ and $(-1, 0)$, a saddle point at $(0, 0)$, and no local maxima.

(c) Verify the conclusions of part (b) using the methods of this section.

Use a graphing utility to draw graphs of f and the level curves. Locate the stationary points, if any, and state whether f has a local maximum, a local minimum, or a saddle point at each stationary point.

33. $f(x,y) = 3xy - x^3 - y^3 + 2.$

34. $f(x,y) = (x^2 + 2y^2)\,e^{-(x^2+y^2)}.$

35. $f(x,y) = \dfrac{-2x}{x^2 + y^2 + 1}.$

36. $f(x,y) = \sin x + \sin y - \cos(x + y)$, $0 \le x \le 3\pi, 0 \le y \le 3\pi.$

■ 15.6 ABSOLUTE EXTREME VALUES

In this section we consider absolute maxima and minima. As in the preceding section, the ideas here are similar to those discussed in Chapter 4.

DEFINITION 15.6.1 ABSOLUTE MAXIMUM AND ABSOLUTE MINIMUM

Let f be a function of several variables with domain D:

f takes on an *absolute maximum* at $\mathbf{x}_0$ if

$$f(\mathbf{x}_0) \ge f(\mathbf{x}) \qquad \text{for all } \mathbf{x} \in D;$$

f takes on an *absolute minimum* at $\mathbf{x}_0$ if

$$f(\mathbf{x}_0) \le f(\mathbf{x}) \qquad \text{for all } \mathbf{x} \in D.$$

In Chapter 2 we saw that a function of one variable that is continuous on a bounded closed interval takes on both an absolute maximum and an absolute minimum on that interval (Theorem 2.6.2). Before stating the general theorem, we need to extend the notion of bounded set of real numbers to include sets in higher dimensions.

DEFINITION 15.6.2 BOUNDED SET

A set S (of the real line, the plane, or three-space) is *bounded* if there exists a positive number R such that

$$\|\mathbf{x}\| \le R \qquad \text{for all} \quad \mathbf{x} \in S.$$

Thus a set S in the plane or three-space is bounded iff it can be contained in a ball of radius R.

THEOREM 15.6.3 EXTREME-VALUE THEOREM

If f is continuous on a bounded closed set D, then f takes on an absolute maximum value and an absolute minimum value.

Two Variables

In the search for local extreme values, the critical points are interior points of the domain: the stationary points and the points at which the gradient does not exist. In the search for absolute extreme values, we must also test the boundary points. This usually requires special methods. One approach is to try to parametrize the boundary by some vector function $\mathbf{r} = \mathbf{r}(t)$ and then work with the function of one variable $f(\mathbf{r}(t))$. This is the approach we take in this section. A more sophisticated approach, one which generalizes to functions of three or more variables, is given in the next section.

The general procedure for finding the absolute extrema of a function f continuous on a bounded closed set D is an extension of the procedure given in Section 4.4, namely:

1. Find the critical points in the interior of D.
2. Find the extreme points on the boundary of D.
3. Evaluate f at the points found in Steps 1 and 2.
4. The largest of the numbers found in Step 3 is the absolute maximum value of f and the smallest is the absolute minimum.

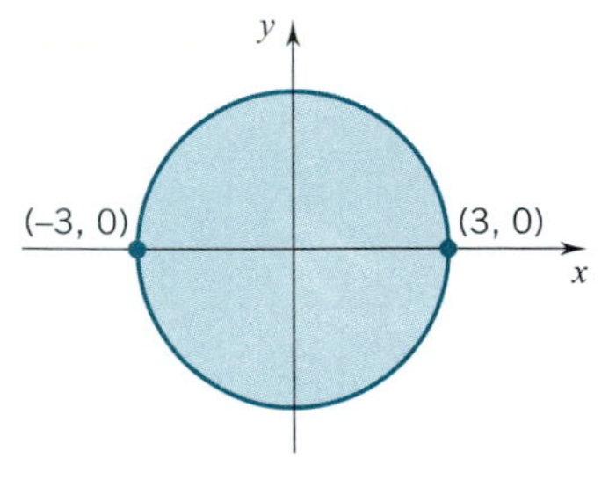

Figure 15.6.1

Example 1 Find the absolute extreme values of the function

$$f(x,y) = x^2 + y^2 - 2x - 2y + 4$$

on the closed disc $D = \{(x,y) : x^2 + y^2 \leq 9\}$. (Figure 15.6.1)

SOLUTION The region D is a bounded closed set and the function f, being continuous everywhere, is continuous on D. Therefore, we know that f takes on an absolute maximum and an absolute minimum on D.

First we find the critical points in the interior of D. The gradient of f,

$$\nabla f(x,y) = (2x-2)\,\mathbf{i} + (2y-2)\,\mathbf{j},$$

is defined everywhere. The gradient is $\mathbf{0}$ where

$$2x - 2 = 0 \qquad \text{and} \qquad 2y - 2 = 0.$$

The only simultaneous solution to these equations is $x = 1, y = 1$. The point $(1, 1)$ is the only stationary point of f and it is in the interior of D.

Now we look for extreme values on the boundary of D. The boundary can be parametrized by the equations

$$x = 3\cos t, \qquad y = 3\sin t, \qquad 0 \leq t \leq 2\pi.$$

The values of f on the boundary are given by the function

$$\begin{aligned} F(t) = f(\mathbf{r}(t)) &= 9\cos^2 t + 9\sin^2 t - 6\cos t - 6\sin t + 4 \\ &= 13 - 6\cos t - 6\sin t, \qquad 0 \leq t \leq 2\pi. \end{aligned}$$

Since F is a continuous function on a bounded closed interval, it has an absolute maximum and an absolute minimum. We find the absolute extreme values of F by the method of Section 4.4. To find the critical numbers, we differentiate:

$$F'(t) = 6\sin t - 6\cos t.$$

Setting $F'(t) = 0$, we get

$$\sin t = \cos t.$$

The solutions of this equation are $t = \pi/4$ and $t = 5\pi/4$.

The extreme values of f on the boundary are

$$\begin{aligned} F(0) &= F(2\pi) = f(3,0) = 7, \\ F(\pi/4) &= f(\tfrac{3}{2}\sqrt{2}, \tfrac{3}{2}\sqrt{2}) = 13 - 6\sqrt{2} \cong 4.51, \\ F(5\pi/4) &= f(-\tfrac{3}{2}\sqrt{2}, -\tfrac{3}{2}\sqrt{2}) = 13 + 6\sqrt{2} \cong 21.49. \end{aligned}$$

The value of f at the stationary point is $f(1,1) = 2$.

Thus, the absolute maximum of f on D is $13 + 6\sqrt{2}$ and the absolute minimum value of f is 2. ☐

Example 2 Find the absolute extreme values of the function

$$f(x,y) = 4xy - x^2 - y^2 - 6x$$

on the triangular region $D = \{(x,y) : 0 \le x \le 2, 0 \le y \le 3x\}$. (Figure 15.6.2)

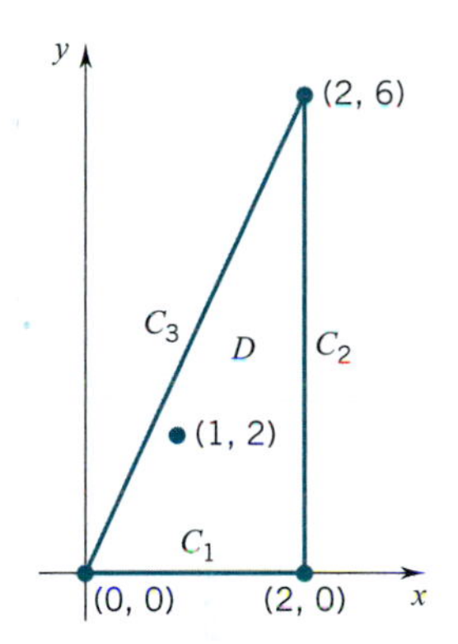

Figure 15.6.2

SOLUTION Since f is continuous and D is a bounded closed set, we know that f takes on an absolute maximum and an absolute minimum on D.

First we find the critical points of f in the interior of D. The gradient of f,

$$\nabla f = (4y - 2x - 6)\,\mathbf{i} + (4x - 2y)\,\mathbf{j},$$

is defined everywhere. The gradient is $\mathbf{0}$ iff

$$4y - 2x - 6 = 0 \quad \text{and} \quad 4x - 2y = 0.$$

Solving these equations simultaneously, we get $x = 1$, $y = 2$. The stationary point $(1,2)$ is in D.

Now we look for extreme values on the boundary by writing each side of the triangle in the form $\mathbf{r} = \mathbf{r}(t)$ and then analyzing $f(\mathbf{r}(t))$. With C_1, C_2, C_3 as in the figure, we have

$$\begin{aligned} C_1 &: \mathbf{r}_1(t) = t\,\mathbf{i}, && t \in [0,2]; \\ C_2 &: \mathbf{r}_2(t) = 2\,\mathbf{i} + t\,\mathbf{j}, && t \in [0,6]; \\ C_3 &: \mathbf{r}_3(t) = t\,\mathbf{i} + 3t\,\mathbf{j}, && t \in [0,2]. \end{aligned}$$

The values of f on these line segments are given by the functions

$$\begin{aligned} f_1(t) &= f(\mathbf{r}_1(t)) = -t^2 - 6t, && t \in [0,2]; \\ f_2(t) &= f(\mathbf{r}_2(t)) = -(t-4)^2, && t \in [0,6]; \\ f_3(t) &= f(\mathbf{r}_3(t)) = 2t^2 - 6t, && t \in [0,2]. \end{aligned}$$

As you can check, f_1 has no critical numbers in $(0, 2)$, f_2 has no critical numbers in $(0, 6)$, and f_3 has a critical number at $t = 3/2$. Evaluating these functions at the endpoints of their domains and at the critical number for f_3, we find that

$$\begin{aligned} f_1(0) &= f_3(0) = f(0,0) = 0, \\ f_1(2) &= f_2(0) = f(2,0) = -16, \\ f_2(6) &= f_3(2) = f(2,6) = -4, \\ f_3(\tfrac{3}{2}) &= f(\tfrac{3}{2}, \tfrac{9}{2}) = -\tfrac{9}{2}. \end{aligned}$$

The value of f at the stationary point is $f(1, 2) = -3$.

Thus the absolute maximum of f is 0 taken on at $(0, 0)$, and the absolute minimum is -16 taken on at $(2, 0)$. ❑

In some cases, physical or geometric considerations may allow us to conclude that an absolute maximum or an absolute minimum exists even if the function is not continuous, or the domain is not bounded or not closed.

Example 3 The rectangle $\{(x, y) : 0 \leq x \leq a, -b \leq y \leq b\}$ is a bounded closed subset of the plane. The function

$$f(x, y) = 1 + \sqrt{x^2 + y^2},$$

being everywhere continuous, is continuous on this rectangle. Thus we can be sure that f takes on both an absolute maximum and an absolute minimum on this set. The absolute maximum is taken on at the points $(a, -b)$ and (a, b), the points of the rectangle farthest away from the origin. (This should be clear from Figure 15.5.4.) The value at these points is $1 + \sqrt{a^2 + b^2}$. The absolute minimum is taken on at the origin $(0, 0)$. The value there is 1.

Now let's continue with the same function but apply it instead to the rectangle

$$\{(x, y) : 0 < x \leq a, -b \leq y \leq b\}.$$

This rectangle is bounded but not closed. On this set f takes on an absolute maximum (the same maximum as before and at the same points), but it takes on no absolute minimum (the origin is not in the set).

Finally, on the entire plane (which is closed but not bounded), f takes on an absolute minimum (1 at the origin) but no maximum. ❑

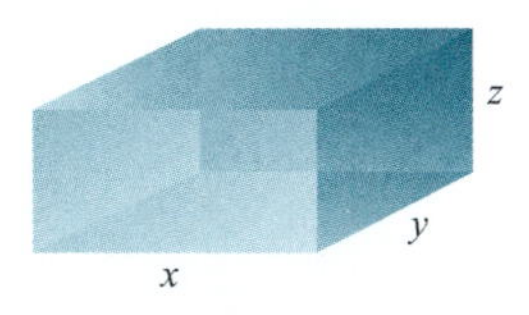

Figure 15.6.3

Example 4 A rectangular box without a top is to have a volume of 12 cubic feet. Find the dimensions of the box that will have minimum surface area.

SOLUTION Let the dimensions of the box be length x, width y, height z (Figure 15.6.3). Then the surface area is given by the expression

$$S = xy + 2xz + 2yz.$$

We can write S as a function of x and y by using the fact that the volume of the box is given by $xyz = 12$, so that

$$z = \frac{12}{xy}.$$

Thus, the expression for S can be written

$$S(x, y) = xy + \frac{24}{x} + \frac{24}{y}, \qquad \text{where} \quad x > 0, y > 0.$$

Note that the domain $D = \{(x, y) : x > 0, y > 0\}$ of S is open and unbounded.

The gradient of S,

$$\nabla S = \left(y - \frac{24}{x^2}\right)\mathbf{i} + \left(x - \frac{24}{y^2}\right)\mathbf{j},$$

is $\mathbf{0}$ iff

$$y - \frac{24}{x^2} = 0 \qquad \text{and} \qquad x - \frac{24}{y^2} = 0.$$

Solving these equations simultaneously, we get $x = y = 2\sqrt[3]{3}$. The point $(2\sqrt[3]{3}, 2\sqrt[3]{3})$ is the only stationary point.

The second partials are

$$\frac{\partial^2 S}{\partial x^2} = \frac{48}{x^3}, \qquad \frac{\partial^2 S}{\partial x \partial y} = 1, \qquad \frac{\partial^2 S}{\partial y^2} = \frac{48}{y^3}.$$

Evaluating these partials at $(2\sqrt[3]{3}, 2\sqrt[3]{3})$, we get

$$A = C = \tfrac{48}{24} = 2 \qquad \text{and} \qquad B = 1.$$

Thus $D = AC - B^2 = 4 - 1 = 3 > 0$. Since $A = 2 > 0$, it follows from the second-partials test that S has a local minimum at $(2\sqrt[3]{3}, 2\sqrt[3]{3})$. We can conclude that S actually has an absolute minimum at this point because $S(x, y) \to \infty$ as either $x \to 0$ or $y \to 0$, and $S(x, y) \to \infty$ as $x \to \infty$ or $y \to \infty$. Finally, the relation

$$xyz = 12$$

implies that $z = \sqrt[3]{3}$ when $x = y = 2\sqrt[3]{3}$, and so the dimensions that yield the minimum surface area are

$$\text{length } = 2\sqrt[3]{3}, \quad \text{width } = 2\sqrt[3]{3}, \quad \text{and} \quad \text{height } = \sqrt[3]{3}. \quad \square$$

EXERCISES 15.6

Find the absolute extreme values taken on by f on the set D.

1. $f(x, y) = 2x^2 + y^2 - 4x - 2y + 2$, $D = \{(x, y) : 0 \le x \le 2, 0 \le y \le 2x\}$.

2. $f(x, y) = 2 - 3x + 2y$, D the region enclosed by the triangle with vertices $(0, 0), (4, 0), (0, 6)$.

3. $f(x, y) = x^2 + xy + y^2 - 6x - 1$, $D = \{(x, y) : 0 \le x \le 5, -3 \le y \le 0\}$.

4. $f(x, y) = x^2 + 2xy + 3y^2$, $D = \{(x, y) : |x| \le 2, |y| \le 2\}$.

5. $f(x, y) = x^2 + y^2 + 3xy + 2$, $D = \{(x, y) : x^2 + y^2 \le 4\}$.

6. $f(x, y) = y(x - 3)$, $D = \{(x, y) : x^2 + y^2 \le 9\}$.

7. $f(x, y) = (x - 1)^2 + (y - 1)^2$, $D = \{(x, y) : x^2 + y^2 \le 4\}$.

8. $f(x, y) = 3 + x - y + xy$, D the region enclosed by $y = x^2$ and $y = 4$.

9. $f(x, y) = \dfrac{-2x}{x^2 + y^2 + 1}$, $D = \{(x, y) : |x| \le 2, |y| \le 2\}$.

10. $f(x, y) = \dfrac{-2x}{x^2 + y^2 + 1}$, $D = \{(x, y) : 0 \le x \le 2, -x \le y \le x\}$.

11. $f(x, y) = (4x - 2x^2)\cos y$, $D = \{(x, y) : 0 \le x \le 2, -\pi/4 \le y \le \pi/4\}$.

12. $f(x, y) = (x - 3)^2 + y^2$, $D = \{(x, y) : 0 \le x \le 4, x^2 \le y \le 4x\}$.

13. $f(x, y) = x^3 - 3xy - y^3$, $D = \{(x, y) : -2 \le x \le 2, x \le y \le 2\}$.

14. $f(x, y) = (x - 4)^2 + y^2$, $D = \{(x, y) : 0 \le x \le 2, x^3 \le y \le 4x\}$.

15. $f(x, y) = \dfrac{-2y}{x^2 + y^2 + 1}$, $D = \{(x, y) : x^2 + y^2 \le 4\}$.

16. $f(x, y) = x^2 + 4y^2 + x - 2y$, D the region enclosed by the ellipse $\frac{1}{4}x^2 + y^2 = 1$.

17. $f(x, y) = x^2 - 2xy + y^2$, $D = \{(x, y) : 0 \le x \le 6, 0 \le y \le 12 - 2x\}$.

18. $f(x, y) = \dfrac{1}{\sqrt{x^2 + y^2}}$, $D = \{(x, y) : 1 \le x \le 3, 1 \le y \le 4\}$.

19. Find positive numbers x, y, z such that $x + y + z = 18$ and xyz is a maximum. HINT: Maximize the function $f(x, y) = 18xy - x^2y - xy^2$ on the triangle formed by the positive x- and y-axes and the line $x + y = 18$.

20. Find positive numbers x, y, z such that $x + y + z = 30$ and xyz^2 is a maximum. HINT: Maximize the function $f(y, z) = 30yz^2 - y^2z^2 - yz^3$ on the triangle formed by the positive y- and z-axes and the line $y + z = 30$.

21. Find the maximum volume for a rectangular solid in the first octant ($x \geq 0, y \geq 0, z \geq 0$) with one vertex at the origin and opposite vertex on the plane $x + y + z = 1$.

22. Find the maximum volume for a rectangular solid in the first octant with one vertex at the origin and the opposite vertex on the plane

$$\frac{x}{a} + \frac{y}{b} + \frac{z}{c} = 1.$$

23. Define $f(x, y) = \frac{1}{4}x^2 - \frac{1}{9}y^2$ on the closed unit disc. Find

(a) the stationary points,

(b) the local extreme values,

(c) the absolute extreme values.

24. Suppose that the material to be used to construct the box of Example 4 costs \$3 per square foot for the sides and \$4 per square foot for the bottom. What dimensions will yield the minimum cost?

25. Describe the behavior of the function at the origin.

(a) $f(x, y) = x^2 - y^3$.

(b) $f(x, y) = 2\cos(x + y) + e^{xy}$.

26. Let n be an integer greater than 2 and set

$$f(x, y) = ax^n + cy^n, \quad \text{taking } ac \neq 0.$$

(a) Find the stationary points.

(b) Find the discriminant at each stationary point.

(c) Find the local and absolute extreme values given that

(i) $a > 0, c > 0$.

(ii) $a < 0, c < 0$.

(iii) $a > 0, c < 0$.

27. Find the point with the property that the sum of the squares of its distances from $P_1(x_1, y_1), P_2(x_2, y_2), P_3(x_3, y_3)$ is an absolute minimum.

28. Given that $0 < a < b$, find the absolute maximum value taken on by the function

$$f(x, y) = \frac{xy}{(a + x)(x + y)(b + y)}$$

on the open square $\{(x, y) : a < x < b, a < y < b\}$.

29. A pentagon is composed of a rectangle surmounted by an isosceles triangle (see the figure). Given that the perimeter of the pentagon has a fixed value P, find the dimensions for maximum area.

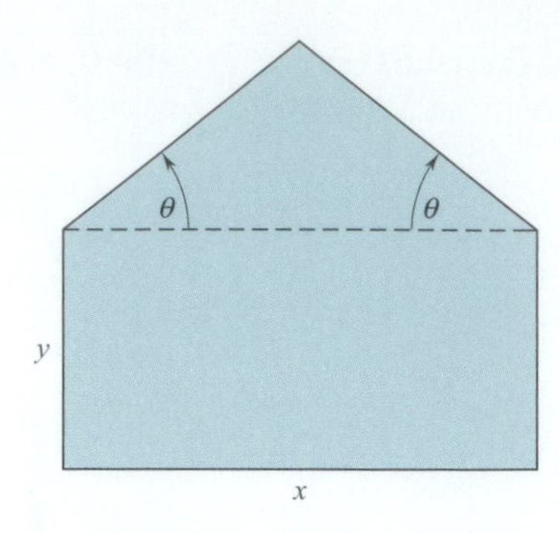

30. A bakery produces two types of bread, one at a cost of 50 cents per loaf, the other at a cost of 60 cents per loaf. Assume that if the first bread is sold at x cents a loaf and the second at y cents a loaf, then the number of loaves that can be sold each week is given by the formulas

$$N_1 = 250(y - x), \quad N_2 = 32{,}000 + 250(x - 2y).$$

Determine x and y for maximum profit.

31. Find the distance between the lines $x = \frac{1}{2}y = \frac{1}{3}z$ and $x = y - 2 = z$.

32. Find the absolute maximum value of the function

$$f(x, y) = \frac{(ax + by + c)^2}{x^2 + y^2 + 1}.$$

33. Find the dimensions of the most economical open-top rectangular crate 96 cubic meters in volume given that the base costs 30 cents per square meter and the sides cost 10 cents per square meter.

34. Let $f(x, y) = ax^2 + bxy + cy^2$, taking $abc \neq 0$.

(a) Find the discriminant D.

(b) Find the stationary points and local extreme values if $D \neq 0$.

(c) Suppose that $D = 0$. Find the stationary points and the local and absolute extreme values given that

(i) $a > 0, c > 0$.

(ii) $a < 0, c < 0$.

35. Show that a closed rectangular box of maximum volume having a prescribed surface area S is a cube.

36. If an open rectangular box has a prescribed surface area S, what dimensions yield the maximum volume?

37. (*The method of least squares*) In this exercise we illustrate an important method of fitting a curve to a collection of points. Consider three points

$$(x_1, y_1) = (0, 2), \quad (x_2, y_2) = (1, -5), \quad (x_3, y_3) = (2, 4).$$

(a) Find the line $y = mx + b$ that minimizes the sum of the squares of the vertical distances $d_i = |y_i - (mx_i + b)|$ from these points to the line.

(b) Find the parabola $y = \alpha x^2 + \beta$ that minimizes the sum of the squares of the vertical distances $d_i = |y_i - (\alpha x_i^2 + \beta)|$ from the points to the parabola.

38. Repeat Exercise 39 taking

$$(x_1, y_1) = (-1, 2),\ (x_2, y_2) = (0, -1),\ (x_3, y_3) = (1, 1).$$

39. According to U.S. Postal Service regulations, the length plus the girth (the perimeter of a cross section) of a package cannot exceed 108 inches. (See the figure.)

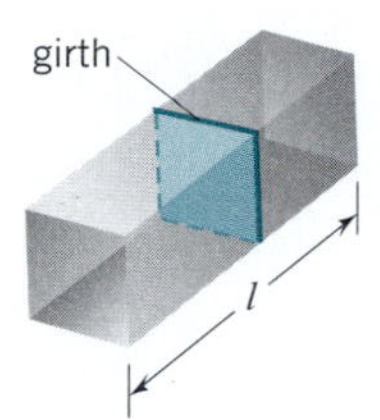

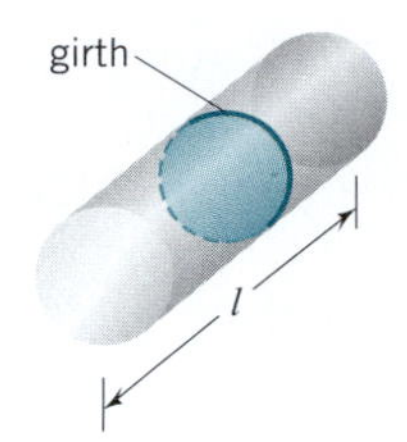

(a) Find the dimensions of the rectangular box of maximum volume that is acceptable for mailing.

(b) Find the dimensions of the cylindrical tube of maximum volume that is acceptable for mailing.

40. A petrochemical company is designing a cylindrical tank with hemispherical ends to be used in transporting its products. If the volume of the tank is to be 10,000 cubic meters, what dimensions should be used to minimize the amount of metal required?

41. A 10-foot section of gutter is to be made from a 12-inch wide strip of metal by folding up strips of length x on each side so that they make an angle θ with the bottom of the gutter. (See the figure.) Determine values for x and θ that will maximize the carrying capacity of the gutter.

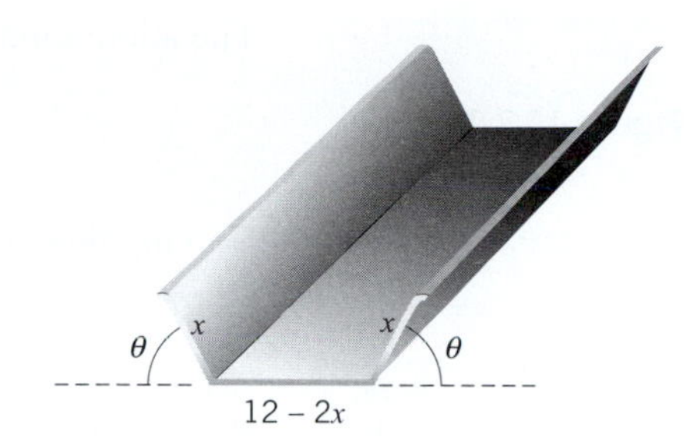

42. Find the volume of the largest rectangular box with edges parallel to the coordinate axes that can be inscribed in the ellipsoid

$$\frac{x^2}{a^2} + \frac{y^2}{b^2} + \frac{z^2}{c^2} = 1.$$

■ 15.7 MAXIMA AND MINIMA WITH SIDE CONDITIONS

When we ask for the distance from a point $P(x_0, y_0)$ to a line $l: Ax + By + C = 0$, we are asking for the minimum value of

$$f(x, y) = \sqrt{(x - x_0)^2 + (y - y_0)^2}$$

with (x, y) subject to the side condition† $Ax + By + C = 0$. When we ask for the distance from a point $P(x_0, y_0, z_0)$ to a plane $p : Ax + By + Cz + D = 0$, we are asking for a minimum value of

$$f(x, y, z) = \sqrt{(x - x_0)^2 + (y - y_0)^2 + (z - z_0)^2}$$

with (x, y, z) subject to the side condition $Ax + By + Cz + D = 0$.

We have already treated these particular problems by special techniques. Our interest here is to present techniques for handling problems of this sort in general. In the two-variable case, the problems will take the form of maximizing (or minimizing) some expression $f(x, y)$ subject to a side condition $g(x, y) = 0$. In the three-variable case, we will seek to maximize (or minimize) some expression $f(x, y, z)$ subject to a side condition $g(x, y, z) = 0$. We begin with two simple examples.

Example 1 Maximize the product xy subject to the side condition $x + y - 1 = 0$.

SOLUTION The condition $x + y - 1 = 0$ gives $y = 1 - x$. The original problem can therefore be solved simply by maximizing the product $h(x) = x(1 - x)$. The derivative $h'(x) = 1 - 2x$ is 0 only at $x = \frac{1}{2}$. Since $h''(x) = -2 < 0$, we know from the second-derivative test that $h(\frac{1}{2}) = \frac{1}{2}(1 - \frac{1}{2}) = \frac{1}{4}$ is the desired maximum. □

† Side conditions are often called *constraints*.

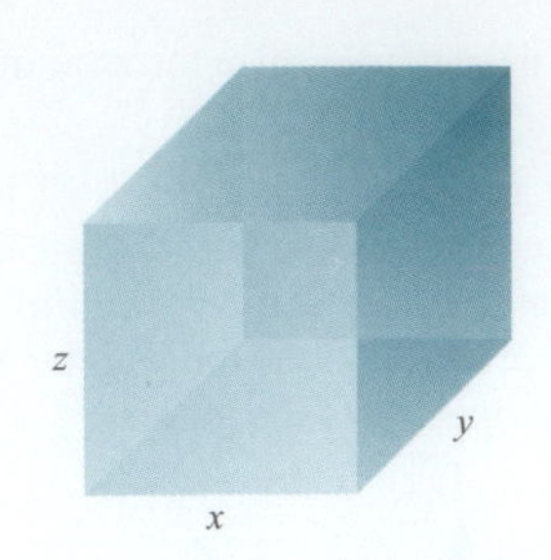

Figure 15.7.1

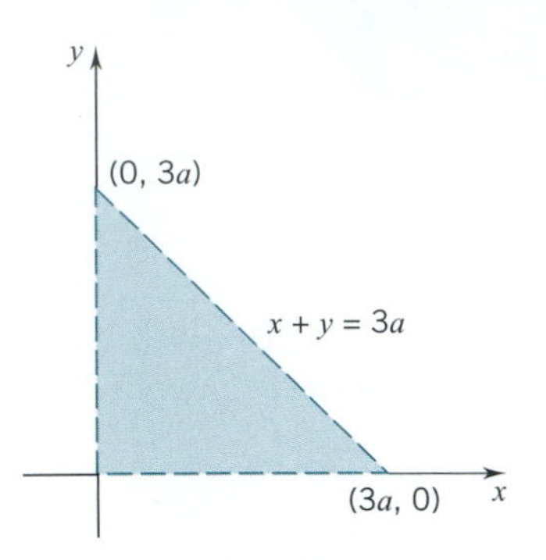

Figure 15.7.2

Example 2 Find the maximum volume of a rectangular solid given that the sum of the lengths of its edges is 12a.

SOLUTION We denote the dimensions of the solid by x, y, z. (Figure 15.7.1.) The volume is given by

$$V = xyz.$$

The stipulation on the edges requires that

$$4(x + y + z) = 12a.$$

Solving this last equation for z, we find that

$$z = 3a - (x + y).$$

Substituting this expression for z in the volume formula, we have

$$V = xy[3a - (x + y)].$$

Since $V = 0$ on the sides $(x = 0, y = 0, z = 3a - (x + y) = 0)$ of the triangle shown in Figure 15.7.2, we can conclude that the maximum value of V must be attained in the interior of the triangle.

The first partials are:

$$\frac{\partial V}{\partial x} = xy(-1) + y[3a - (x + y)] = 3ay - 2xy - y^2,$$
$$\frac{\partial V}{\partial y} = xy(-1) + x[3a - (x + y)] = 3ax - x^2 - 2xy.$$

Setting both partials equal to zero, we have

$$(3a - 2x - y)y = 0 \qquad \text{and} \qquad (3a - x - 2y)x = 0.$$

Since x and y are assumed positive, we can divide by x and y and get

$$3a - 2x - y = 0 \qquad \text{and} \qquad 3a - x - 2y = 0.$$

Solving these equations simultaneously, we find that $x = y = a$. The point (a, a), which does lie within the triangle, is the only stationary point. The value of V at that point is a^3. The conditions of the problem make it clear that this is a maximum. (If you are skeptical, you can confirm this by appealing to the second-partials test.) □

The last two problems were easy. They were easy in part because the side conditions were such that we could solve for one of the variables in terms of the other (s). In general this is not possible and a more sophisticated approach is required.

The Method of Lagrange

We begin with what looks like a detour. To avoid having to make separate statements for the two-and three-variable cases, we will use vector notation.

Throughout the discussion f will be a function of two or three variables which is continuously differentiable on some open set U. We take

$$C: \mathbf{r} = \mathbf{r}(t), \quad t \in I$$

to be a curve that lies entirely in U and has at each point a nonzero tangent vector $\mathbf{r}'(t)$. The basic result is this:

(15.7.1) if $\mathbf{x}_0$ maximizes (or minimizes) $f(\mathbf{x})$ on C, then $\nabla f(\mathbf{x}_0)$ is perpendicular to C at $\mathbf{x}_0$.

PROOF Assume that $\mathbf{x}_0$ maximizes (or minimizes) $f(\mathbf{x})$ on C. Choose t_0, so that

$$\mathbf{r}(t_0) = \mathbf{x}_0.$$

The composition $f(\mathbf{r}(t))$ has a maximum (or minimum) at t_0. Consequently, its derivative,

$$\frac{d}{dt}[f(\mathbf{r}(t))] = \nabla f(\mathbf{r}(t)) \cdot \mathbf{r}'(t),$$

must be zero at t_0 :

$$0 = \nabla f(\mathbf{r}(t_0)) \cdot \mathbf{r}'(t_0) = \nabla f(\mathbf{x}_0) \cdot \mathbf{r}'(t_0).$$

This shows that

$$\nabla f(\mathbf{x}_0) \perp \mathbf{r}'(t_0).$$

Since $\mathbf{r}'(t_0)$ is tangent to C at $\mathbf{x}_0$, $\nabla f(\mathbf{x}_0)$ is perpendicular to C at $\mathbf{x}_0$. ❑

We are now ready for side-condition problems. Suppose that g is a continuously differentiable function of two or three variables defined on a subset of the domain of f. Lagrange made the following observation: †

(15.7.2) if $\mathbf{x}_0$ maximizes (or minimizes) $f(\mathbf{x})$ subject to the side condition $g(\mathbf{x}) = 0$, then $\nabla f(\mathbf{x}_0)$ and $\nabla g(\mathbf{x}_0)$ are parallel. Thus, if $\nabla g(\mathbf{x}_0) \neq \mathbf{0}$, then there exists a scalar λ such that

$$\nabla f(\mathbf{x}_0) = \lambda \nabla g(\mathbf{x}_0).$$

Such a scalar λ has come to be called a *Lagrange multiplier.*

PROOF OF (15.7.2) Let's suppose that $\mathbf{x}_0$ maximizes (or minimizes) $f(\mathbf{x})$ subject to the side condition $g(\mathbf{x}) = 0$. If $\nabla g(\mathbf{x}_0) = \mathbf{0}$, the result is trivially true: every vector is parallel to the zero vector. We suppose therefore that $\nabla g(\mathbf{x}_0) \neq \mathbf{0}$.

In the two-variable case we have

$$\mathbf{x}_0 = (x_0, y_0) \qquad \text{and} \qquad \text{the side condition} \quad g(x, y) = 0.$$

The side condition defines a curve C that has a nonzero tangent vector at (x_0, y_0). †† Since (x_0, y_0) maximizes (or minimizes) $f(x, y)$ on C, we know from (15.7.1) that $\nabla f(x_0, y_0)$ is perpendicular to C at (x_0, y_0). By (15.4.2), $\nabla g(x_0, y_0)$ is also perpendicular to C at (x_0, y_0). The two gradients are therefore parallel. See Figure 15.7.3.

† Another contribution of the French mathematician Joseph Louis Lagrange.

†† $\mathbf{t}(x_0, y_0) = \dfrac{\partial g}{\partial y}(x_0, y_0)\,\mathbf{i} - \dfrac{\partial g}{\partial x}(x_0, y_0)\,\mathbf{j} \neq \mathbf{0}$.

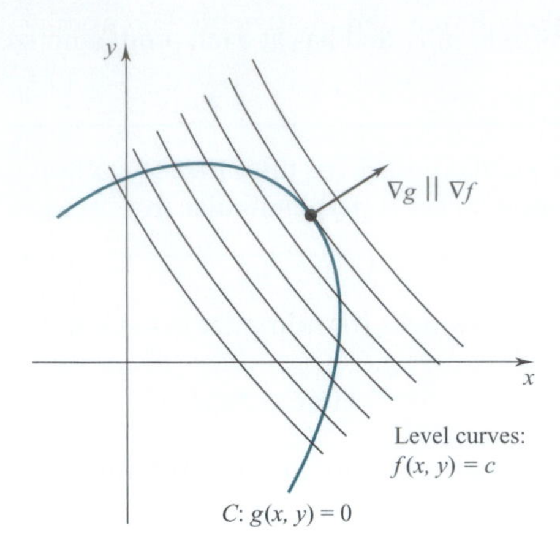

Figure 15.7.3

In the three-variable case we have

$$\mathbf{x}_0 = (x_0, y_0, z_0) \qquad \text{and} \qquad \text{the side condition} \quad g(x, y, z) = 0.$$

The side condition defines a surface Γ that lies in the domain of f. Now let C be a curve that lies on Γ and passes through (x_0, y_0, z_0) with nonzero tangent vector. We know that (x_0, y_0, z_0) maximizes (or minimizes) $f(x, y, z)$ on C. Consequently, $\nabla f(x_0, y_0, z_0)$ is perpendicular to C at (x_0, y_0, z_0). Since this is true for each such curve C, $\nabla f(x_0, y_0, z_0)$ must be perpendicular to Γ itself. But $\nabla g(x_0, y_0, z_0)$ is also perpendicular to Γ at (x_0, y_0, z_0) by (15.4.6). It follows that $\nabla f(x_0, y_0, z_0)$ and $\nabla g(x_0, y_0, z_0)$ are parallel. ❑

We come now to some problems that are susceptible to Lagrange's method. In each case ∇g is not $\mathbf{0}$ where g is 0 and therefore we can focus entirely on those points $\mathbf{x}$ that satisfy the Lagrange condition

$$\nabla f(\mathbf{x}) = \lambda \nabla g(\mathbf{x}). \tag{15.7.2}$$

Example 3 Maximize and minimize

$$f(x, y) = xy \quad \text{on the unit circle} \quad x^2 + y^2 = 1.$$

SOLUTION Since f is continuous and the unit circle is closed and bounded, it is clear that both a maximum and a minimum exist (see Section 15.6).

To apply Lagrange's method we set

$$g(x, y) = x^2 + y^2 - 1.$$

We want to maximize and minimize

$$f(x, y) = xy \quad \text{subject to the side condition } g(x, y) = 0.$$

The gradients are

$$\nabla f(x, y) = y\,\mathbf{i} + x\,\mathbf{j}, \quad \nabla g(x, y) = 2x\,\mathbf{i} + 2y\,\mathbf{j}.$$

Setting
$$\nabla f(x, y) = \lambda \nabla g(x, y),$$

we obtain
$$y = 2\lambda x, \quad x = 2\lambda y.$$

Multiplying the first equation by y and the second equation by x, we find that

$$y^2 = 2\lambda xy, \quad x^2 = 2\lambda xy$$

and thus
$$y^2 = x^2.$$

The side condition $x^2 + y^2 = 1$ now implies that $2x^2 = 1$ and therefore that $x = \pm\frac{1}{2}\sqrt{2}$. The only points that can give rise to an extreme value are

$$(\tfrac{1}{2}\sqrt{2}, \tfrac{1}{2}\sqrt{2}), \quad (\tfrac{1}{2}\sqrt{2}, -\tfrac{1}{2}\sqrt{2}), \quad (-\tfrac{1}{2}\sqrt{2}, \tfrac{1}{2}\sqrt{2}), \quad (-\tfrac{1}{2}\sqrt{2}, -\tfrac{1}{2}\sqrt{2}).$$

At the first and fourth points f takes on the value $\frac{1}{2}$. At the second and third points f takes on the value $-\frac{1}{2}$. Clearly then, $\frac{1}{2}$ is the maximum value and $-\frac{1}{2}$ the minimum value. ❑

Example 4 Find the minimum value taken on by the function

$$f(x, y) = x^2 + (y-2)^2 \quad \text{on the hyperbola} \quad x^2 - y^2 = 1.$$

SOLUTION Note that the expression $f(x, y) = x^2 + (y-2)^2$ gives the square of the distance between the points $(0, 2)$ and (x, y). Therefore, the problem asks us to minimize the square of the distance from the point $(0, 2)$ to the hyperbola, and this minimum value clearly exists. Note, also, that there is no maximum value. See Figure 15.7.4.

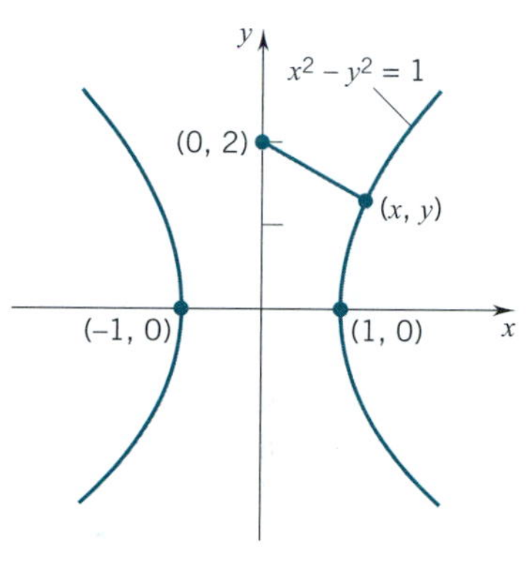

Figure 15.7.4

Now set
$$g(x, y) = x^2 - y^2 - 1.$$

We want to minimize

$$f(x, y) = x^2 + (y-2)^2 \quad \text{subject to the side condition} \quad g(x, y) = 0.$$

Here
$$\nabla f(x, y) = 2x\,\mathbf{i} + 2(y-2)\,\mathbf{j}, \quad \nabla g(x, y) = 2x\,\mathbf{i} - 2y\,\mathbf{j}.$$

The Lagrange condition $\nabla f(x, y) = \lambda \nabla g(x, y)$ gives

$$2x = 2\lambda x, \quad 2(y-2) = -2\lambda y,$$

which we can simplify to

$$x = \lambda x, \quad y - 2 = -\lambda y.$$

The side condition $x^2 - y^2 = 1$ shows that x cannot be zero. Dividing $x = \lambda x$ by x, we get $\lambda = 1$. This means that $y - 2 = -y$ and therefore $y = 1$. With $y = 1$, the side condition gives $x = \pm\sqrt{2}$. The points to be checked are therefore $(-\sqrt{2}, 1)$ and $(\sqrt{2}, 1)$. At each of these points f takes on the value 3. This is the desired minimum. ❑

REMARK The last problem could have been solved more simply by rewriting the side condition as $x^2 = 1 + y^2$ and eliminating x from $f(x, y)$ by substitution. Then it would have been simply a matter of minimizing the function $h(y) = 1 + y^2 + (y-2)^2 = 2y^2 - 4y + 5$. ❑

In the next example we use Lagrange's method to solve the problem of Example 4 of the previous section.

Example 5 A rectangular box without a top is to have a volume of 12 cubic feet. Find the dimensions of the box that will have minimum surface area.

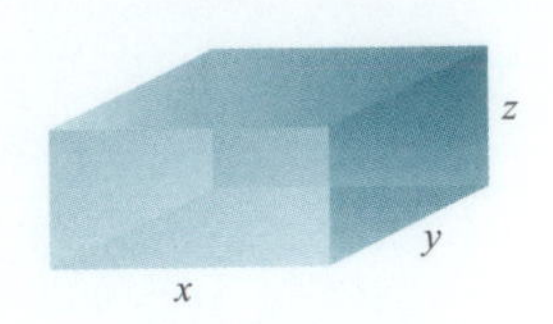

Figure 15.7.5

SOLUTION With the dimensions indicated in Figure 15.7.5, the surface area is given by the expression

$$S = xy + 2xz + 2yz.$$

We want to minimize S subject to the side condition $xyz = 12$ with $x > 0, y > 0, z > 0$.

We begin by setting

$$g(x, y, z) = xyz - 12$$

so that the side condition becomes $g(x, y, z) = 0$. We seek those triples (x, y, z) that simultaneously satisfy the Lagrange condition

$$\nabla f(x, y, z) = \nabla g(x, y, z) \qquad \text{and the side condition} \qquad g(x, y, z) = 0.$$

The gradients are

$$\nabla f = (y + 2z)\,\mathbf{i} + (x + 2z)\,\mathbf{j} + (2x + 2y)\,\mathbf{k}, \qquad \nabla g = yz\,\mathbf{i} + xz\,\mathbf{j} + xy\,\mathbf{k}.$$

The Lagrange condition gives

$$y + 2z = \lambda yz, \quad x + 2z = \lambda xz, \quad 2x + 2y = \lambda xy.$$

Multiplying the first equation by x, the second by $-y$, and adding the resulting equations, we get

$$2z(x - y) = 0.$$

Since $z \neq 0$, it follows that $y = x$. Replacing y by x in the third equation yields the equation

$$4x = \lambda x^2.$$

Since $x \neq 0$, we conclude that $x = 4/\lambda$, and since $y = x$, $y = 4/\lambda$. We can now solve either the first or second equation for z in terms of λ. The result is $z = 2/\lambda$.

Finally, substituting $x = y = 4/\lambda$ and $z = 2/\lambda$ into the side condition, we get

$$\left(\frac{4}{\lambda}\right)\left(\frac{4}{\lambda}\right)\left(\frac{2}{\lambda}\right) = 12 \qquad \text{which implies} \qquad \lambda^3 = \frac{8}{3} \qquad \text{so that} \qquad \lambda = \frac{2}{\sqrt[3]{3}}.$$

With $\lambda = 2/\sqrt[3]{3}$, we find that $x = y = 2\sqrt[3]{3}$ and $z = \sqrt[3]{3}$. The dimensions that minimize the surface area are

$$\text{length} = 2\sqrt[3]{3}, \qquad \text{width} = 2\sqrt[3]{3} \qquad \text{and} \qquad \text{height} = \sqrt[3]{3}$$

which is the result we obtained before. □

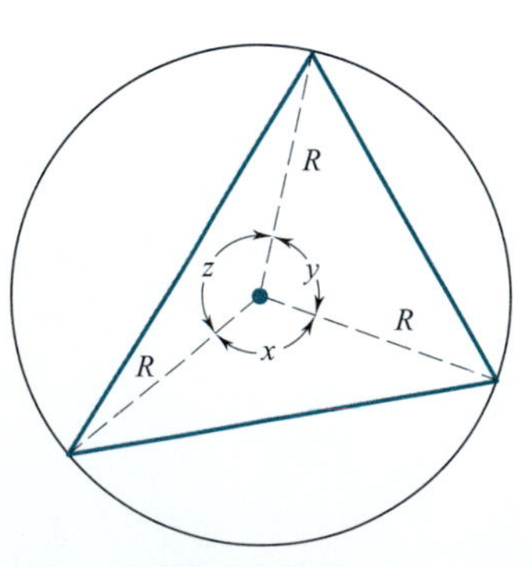

Figure 15.7.6

Example 6 Show that, of all the triangles inscribed in a fixed circle of radius R, the equilateral triangle has the largest perimeter.

SOLUTION It is intuitively clear that this maximum exists and geometrically clear that the triangle that offers this maximum contains the center of the circle in its interior or on its boundary. As in Figure 15.7.6, we denote by x, y, z the central angles that subtend the three sides. As you can verify by trigonometry, the perimeter of the triangle is given by the function.

$$f(x, y, z) = 2R(\sin \tfrac{1}{2}x + \sin \tfrac{1}{2}y + \sin \tfrac{1}{2}z).$$

As a side condition we have

$$g(x, y, z) = x + y + z - 2\pi = 0.$$

To maximize the perimeter we form the gradients

$$\nabla f(x, y, z) = R\left(\cos \tfrac{1}{2}x\,\mathbf{i} + \cos \tfrac{1}{2}y\,\mathbf{j} + \cos \tfrac{1}{2}z\,\mathbf{k}\right), \quad \nabla g(x, y, z) = \mathbf{i} + \mathbf{j} + \mathbf{k}.$$

The Lagrange condition $\nabla f(x, y, z) = \lambda \nabla g(x, y, z)$ gives

$$\lambda = R\cos \tfrac{1}{2}x, \quad \lambda = R\cos \tfrac{1}{2}y, \quad \lambda = R\cos \tfrac{1}{2}z,$$

and therefore

$$\cos \tfrac{1}{2}x = \cos \tfrac{1}{2}y = \cos \tfrac{1}{2}z.$$

With x, y, z all in $(0, \pi]$, we can conclude that $x = y = z$. Since the central angles are equal, the sides are equal. The triangle is therefore equilateral. □

An Application of the Cross Product

The Lagrange condition can be replaced by a cross-product equation: points that satisfy $\nabla f = \lambda \nabla g$ satisfy

(15.7.3)
$$\nabla f \times \nabla g = \mathbf{0}.$$

If f and g are functions of two variables, then $\nabla f \times \nabla g = 0$ has a simpler form:

$$\nabla f \times \nabla g = \begin{vmatrix} \mathbf{i} & \mathbf{j} & \mathbf{k} \\ f_x & f_y & 0 \\ g_x & g_y & 0 \end{vmatrix} = (f_x g_y - f_y g_x)\,\mathbf{k},$$

which implies that

(15.7.4)
$$\frac{\partial f}{\partial x}\frac{\partial g}{\partial y} - \frac{\partial f}{\partial y}\frac{\partial g}{\partial x} = 0.$$

Example 7 Maximize and minimize $f(x, y) = xy$ on the unit circle $x^2 + y^2 = 1$.

SOLUTION This problem was solved earlier by means of the Lagrange equation (see Example 3). This time we will use (15.7.4) instead.

As before, we set

$$g(x, y) = x^2 + y^2 - 1,$$

so that the side condition takes the form $g(x,y)=0$. Since

$$\frac{\partial f}{\partial x}=y,\quad \frac{\partial f}{\partial y}=x \quad \text{and} \quad \frac{\partial g}{\partial x}=2x,\quad \frac{\partial g}{\partial y}=2y,$$

(15.7.4) takes the form

$$y(2y)-x(2x)=0.$$

This gives $x^2=y^2$.

As before, the side condition $x^2+y^2=1$ implies that $2x^2=1$ and therefore $x=\pm\frac{1}{2}\sqrt{2}$. The points under consideration are

$$(\tfrac{1}{2}\sqrt{2},\tfrac{1}{2}\sqrt{2}),\quad (\tfrac{1}{2}\sqrt{2},-\tfrac{1}{2}\sqrt{2}),\quad (-\tfrac{1}{2}\sqrt{2},\tfrac{1}{2}\sqrt{2}),\quad (-\tfrac{1}{2}\sqrt{2},-\tfrac{1}{2}\sqrt{2}).$$

As we saw in Example 3, f takes on its maximum value $\frac{1}{2}$ at the first and fourth points, and its minimum value $-\frac{1}{2}$ at the second and third points. ❑

In three variables the computations demanded by the cross-product equation are often quite complicated, and it is usually easier to follow the method of Lagrange.

EXERCISES 15.7

1. Minimize x^2+y^2 on the hyperbola $xy=1$.
2. Maximize xy on the ellipse $b^2x^2+a^2y^2=a^2b^2$.
3. Minimize xy on the ellipse $b^2x^2+a^2y^2=a^2b^2$.
4. Minimize xy^2 on the unit circle $x^2+y^2=1$.
5. Maximize xy^2 on the ellipse $b^2x^2+a^2y^2=a^2b^2$.
6. Maximize $x+y$ on the curve $x^4+y^4=1$.
7. Maximize x^2+y^2 on the curve $x^4+7x^2y^2+y^4=1$.
8. Minimize xyz on the unit sphere $x^2+y^2+z^2=1$.
9. Maximize xyz on the ellipsoid $x^2/a^2+y^2/b^2+z^2/c^2=1$.
10. Minimize $x+2y+4z$ on the sphere $x^2+y^2+z^2=7$.
11. Maximize $2x+3y+5z$ on the sphere $x^2+y^2+z^2=19$.
12. Minimize $x^4+y^4+z^4$ on the plane $x+y+z=1$.
13. Maximize the volume of a rectangular solid in the first octant with one vertex at the origin and opposite vertex on the plane $x/a+y/b+z/c=1$. (Take $a>0, b>0, c>0$.)
14. Show that the square has the largest area of all the rectangles with a given perimeter.
15. Find the distance from the point $(0,1)$ to the parabola $x^2=4y$.
16. Find the distance from the point $(p,4p)$ to the parabola $y^2=2px$.
17. Find the points on the sphere $x^2+y^2+z^2=1$ that are closest to and farthest from the point $(2,1,2)$.
18. Let x, y, and z be the angles of a triangle. Determine the maximum value of $f(x,y,z)=\sin x\sin y\sin z$.
19. Maximize $f(x,y,z)=3x-2y+z$ on the sphere $x^2+y^2+z^2=14$.
20. A rectangular box has three of its faces on the coordinate planes and one vertex in the first octant on the paraboloid $z=4-x^2-y^2$. Determine the maximum volume of the box.
21. Use the method of Lagrange to find the distance from the origin to the plane with equation $Ax+By+Cz+D=0$.
22. Maximize the volume of a rectangular solid given that the sum of the areas of the six faces is $6a^2$.
23. Within a triangle there is a point P such that the sum of the squares of the distances from P to the sides of the triangle is a minimum. Find this minimum.
24. Show that of all the triangles inscribed in a fixed circle the equilateral one has the largest: (a) product of the lengths of the sides; (b) sum of the squares of the lengths of the sides.
25. The curve $x^3-y^3=1$ is asymptotic to the line $y=x$. Find the point(s) on the curve $x^3-y^3=1$ farthest from the line $y=x$.
26. A plane passes through the point $(a,b,c.)$ Find its intercepts with the coordinate axes if the volume of the solid bounded by the plane and the coordinate planes is to be a minimum.
27. Show that, of all the triangles with a given perimeter, the equilateral triangle has the largest area. HINT: Area$=\sqrt{s(s-a)(s-b)(s-c)}$, where s represents the semiperimeter $s=\frac{1}{2}(a+b+c)$.
28. Show that the rectangular box of maximum volume that can be inscribed in the sphere $x^2+y^2+z^2=a^2$ is a cube.

29. Determine the maximum value of $f(x,y) = (xy)^{1/2}$ given that x and y are nonnegative numbers and $x + y = k, k$ a constant. This result shows that if x and y are nonnegative numbers, then

$$(xy)^{1/2} \leq \frac{x+y}{2}.$$

(See Exercise 58, Section 1.3.)

30. (a) Determine the maximum value of $f(x,y,z) = (xyz)^{1/3}$ given that x, y, and z are nonnegative numbers and $x + y + z = k, k$ a constant.

(b) Use the result in part (a) to show that if x, y, and z are nonnegative numbers, then

$$(xyz)^{1/3} \leq \frac{x+y+z}{3}.$$

NOTE: $(xyz)^{1/3}$ is the *geometric mean* of x, y, z.

31. Let $x_1, x_2, \ldots, x_n$ be nonnegative numbers such that $x_1 + x_2 + \cdots + x_n = k, k$ a constant. Prove that

$$(x_1 x_2 \cdots x_n)^{1/n} \leq \frac{x_1 + x_2 + \cdots + x_n}{n}.$$

In words, the geometric mean of n nonnegative numbers cannot exceed the arithmetic mean of the numbers.

32. Assume that the Celsius temperature T at a point (x,y,z) on the sphere $x^2 + y^2 + z^2 = 1$ is given by

$$T(x,y,z) = 10xy^2z.$$

Find the point(s) on the sphere at which the temperature is greatest and the point(s) at which it is least. Give the temperature at each of these points.

33. A soft drink manufacturer wants to design an aluminum can in the shape of a right circular cylinder to hold a given volume V (measured in cubic inches.) If the objective is to minimize the amount of aluminum needed (top, sides, and bottom), what dimensions should be used?

Use the Lagrange method to give alternative solutions to the indicated exercises in Section 15.6.

34. Exercise 19.

35. Exercise 22.

36. Exercise 20.

37. Exercise 33.

38. Exercise 35.

39. Exercise 36.

40. Exercise 40.

41. Exercise 39.

42. Exercise 42.

43. A manufacturer can produce three distinct products in quantities Q_1, Q_2, Q_3, releate, and thereby derive a profit $p(Q_1, Q_2, Q_3) = 2Q_1, +8Q_2 + 24Q_3$. Find the values of Q_1, Q_2, Q_3 that maximize profit if production is subject to the constraint $Q_1^2 + 2Q_2^2 + 4Q_3^2 = 4.5 \times 10^9$.

44. Find the volume of the largest rectangular box that can be inscribed in the ellipsoid

$$4x^2 + 9y^2 + 36z^2 = 36$$

if the edges of the box are parallel to the coordinate axes.

■ PROJECT 15.7 Maxima and Minima with Two Side Conditions

The Lagrange method can be extended to problems with two side conditions as follows: If $\mathbf{x}_0$ is a maximum (or minimum) of $f(\mathbf{x})$ subject to the two side conditions $g(\mathbf{x}) = 0$ and $h(\mathbf{x}) = 0$, and if $\nabla g(\mathbf{x}_0), \nabla h(\mathbf{x}_0)$ are nonzero and nonparallel, then there exist scalars λ and μ such that

$$\nabla f(\mathbf{x}_0) = \lambda \nabla g(\mathbf{x}_0) + \mu \nabla h(\mathbf{x}_0).$$

Assume this result and solve the following problems.

Problem 1 Find the extreme values of

$$f(x,y,z) = xy + z^2$$

subject to the side conditions:

$$x^2 + y^2 + z^2 = 4 \qquad \text{and} \qquad y - x = 0.$$

Problem 2. The planes $x + 2y + 3z = 0$ and $2x + 3y + z = 4$ intersect in a straight line. Find the point on that line that is closest to the origin.

Problem 3. The plane $x + y - z + 1 = 0$ intersects the upper nappe of the cone $z^2 = x^2 + y^2$ in an ellipse. Find the points on this ellipse that are closest to and farthest from the origin.

■ 15.8 DIFFERENTIALS

We begin by reviewing the one-variable case. If f is differentiable at x, then for small h, the increment

$$\nabla f = f(x+h) - f(x)$$

can be approximated by the differential

$$df = f'(x)h.$$

The geometric interpretations of Δf and df are shown in Figure 15.8.1. We write

$$\Delta f \cong df.$$

How good is this approximation? It is good enough that the ratio

$$\frac{\Delta f - df}{|h|}$$

tends to 0 as h tends to 0.

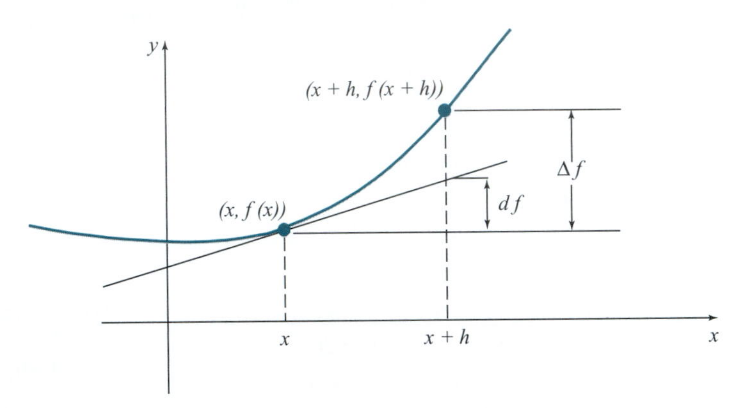

Figure 15.8.1

The differential of a function of several variables, defined in an analogous manner, plays a similar approximating role. Let's suppose that f, now a function of several variables, is differentiable at $\mathbf{x}$. The difference

(15.8.1) $$\Delta f = f(\mathbf{x}+\mathbf{h}) - f(\mathbf{x})$$

is called the *increment* of f, and the dot product

(15.8.2) $$df = \nabla f(\mathbf{x}) \cdot \mathbf{h}$$

is called the *differential* (more formally, the *total differential*). As in the one-variable case, for small $\mathbf{h}$, the differential and the increment are approximately equal:

(15.8.3) $$\Delta f \cong df.$$

How approximately equal are they? Enough so that the ratio

$$\frac{\Delta f - df}{\|\mathbf{h}\|}$$

tends to 0 as $\mathbf{h}$ tends to $\mathbf{0}$. How do we know this? We know that

$$f(\mathbf{x}+\mathbf{h}) - f(\mathbf{x}) = \nabla f(\mathbf{x}) \cdot \mathbf{h} + o(\mathbf{h}).$$

Therefore, $$[f(\mathbf{x}+\mathbf{h})-f(\mathbf{x})]-\nabla f(\mathbf{x})\cdot\mathbf{h}=o(\mathbf{h}),$$

and so

$$\frac{\overbrace{[f(\mathbf{x}+\mathbf{h})-f(\mathbf{x})]}^{\Delta f}-\overbrace{\nabla f(\mathbf{x})\cdot\mathbf{h}}^{df}}{\|\mathbf{h}\|}\to 0 \text{ as } \mathbf{h}\to\mathbf{0}. \qquad \text{(Section 15.1)}$$

In the two-variable case we set $\mathbf{x}=(x,y)$ and $\mathbf{h}=(\Delta x,\Delta y)$. The increment $\Delta f=f(\mathbf{x}+\mathbf{h})-f(\mathbf{x})$ then takes the form

$$\Delta f=f(x+\Delta x,y+\Delta y)-f(x,y),$$

and the differential $df=\nabla f(\mathbf{x})\cdot\mathbf{h}$ takes the form

$$df=\frac{\partial f}{\partial x}(x,y)\Delta x+\frac{\partial f}{\partial y}(x,y)\Delta y.$$

By suppressing the point of evaluation, we can write

(15.8.4) $$\boxed{df=\frac{\partial f}{\partial x}\Delta x+\frac{\partial f}{\partial y}\Delta y.}$$

The approximation $\Delta f\cong df$ is illustrated in Figure 15.8.2. There we have represented f as a surface $z=f(x,y)$. Through a point $P(x_0,y_0,f(x_0,y_0))$ we have drawn the tangent plane. *The difference* $df-\Delta f$ *is the vertical separation between this tangent plane and the surface as measured at the point* $(x_0+\Delta x,y_0+\Delta y)$.

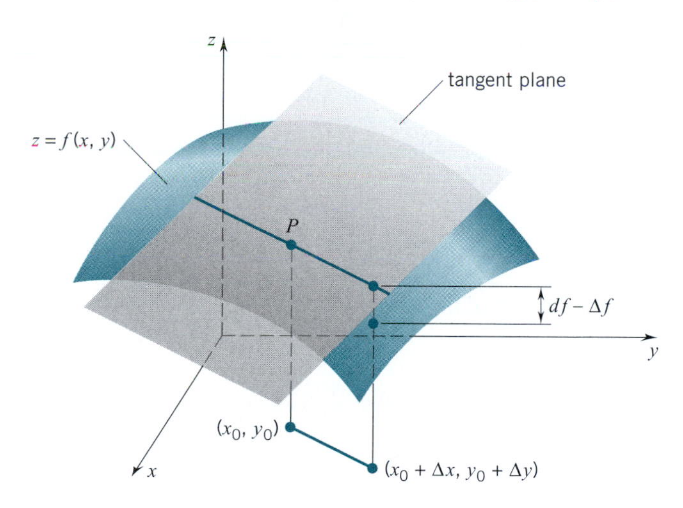

Figure 15.8.2

PROOF The tangent plane at P has equation

$$z-f(x_0,y_0)=\frac{\partial f}{\partial x}(x_0,y_0)(x-x_0)+\frac{\partial f}{\partial y}(x_0,y_0)(y-y_0).$$

The z-coordinate of this plane at the point $(x_0+\Delta x,y_0+\Delta y)$ is

$$f(x_0,y_0)+\frac{\partial f}{\partial x}(x_0,y_0)\Delta x+\frac{\partial f}{\partial y}(x_0,y_0)\Delta y. \qquad \text{(check this)}$$

The z-coordinate of the surface at this same point is

$$f(x_0 + \Delta x, y_0 + \Delta y).$$

The difference between these two,

$$\left[f(x_0, y_0) + \frac{\partial f}{\partial x}(x_0, y_0)\Delta x + \frac{\partial f}{\partial y}(x_0, y_0)\Delta y\right] - \left[f(x_0 + \Delta x, y_0 + \Delta y)\right],$$

can be written as

$$\left[\frac{\partial f}{\partial x}(x_0, y_0)\Delta x + \frac{\partial f}{\partial y}(x_0, y_0)\Delta y\right] - \left[f(x_0 + \Delta x, y_0 + \Delta y) - f(x_0, y_0)\right].$$

This is just $df - \Delta f$. ☐†

For the three-variable case we set $\mathbf{x} = (x, y, z)$ and $\mathbf{h} = (\Delta x, \Delta y, \Delta z)$. The increment becomes

$$\Delta f = f(x + \Delta x, y + \Delta y, z + \Delta z) - f(x, y, z),$$

and the approximating differential becomes

$$df = \frac{\partial f}{\partial x}(x, y, z)\Delta x + \frac{\partial f}{\partial y}(x, y, z)\Delta y + \frac{\partial f}{\partial z}(x, y, z)\,\Delta z.$$

Suppressing the point of evaluation we have

(15.8.5)
$$df = \frac{\partial f}{\partial x}\Delta x + \frac{\partial f}{\partial y}\Delta y + \frac{\partial f}{\partial z}\Delta z.$$

To illustrate the use of differentials, we begin with a rectangle of sides x and y. The area is given by

$$A(x, y) = xy.$$

An increase in the dimensions of the rectangle to $x + \Delta x$ and $y + \Delta y$ produces a change in area

$$\begin{aligned}\Delta A &= (x + \Delta x)(y + \Delta y) - xy \\ &= (xy + x\,\Delta y + y\,\Delta x + \Delta x\,\Delta y) - xy \\ &= x\,\Delta y + y\,\Delta x + \Delta x\,\Delta y.\end{aligned}$$

The differential estimate for this change in area is

$$dA = \frac{\partial A}{\partial x}\Delta x + \frac{\partial A}{\partial y}\Delta y = y\,\Delta x + x\,\Delta y. \qquad \text{(Figure 15.8.3)}$$

† As in the figure, we have been assuming that the tangent plane lies above the surface. If the tangent plane lies below the surface, then $df - \Delta f$ is negative. The vertical separation between the tangent plane and the surface is then $\Delta f - df$.

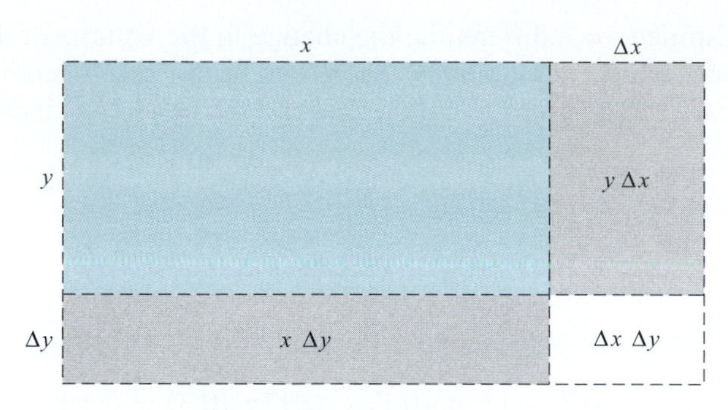

Figure 15.8.3

The error of our estimate, the difference between the actual change and the estimated change, is the difference $\Delta A - dA = \Delta x\,\Delta y$.

Example 1 Given that $f(x, y) = yx^{2/5} + x\sqrt{y}$, estimate by a differential the change in f from (32, 16) to (35, 18).

SOLUTION Since

$$\frac{\partial f}{\partial x} = \frac{2y}{5}\left(\frac{1}{x}\right)^{3/5} + \sqrt{y} \quad \text{and} \quad \frac{\partial f}{\partial y} = x^{2/5} + \frac{x}{2\sqrt{y}},$$

we have

$$df = \left[\frac{2y}{5}\left(\frac{1}{x}\right)^{3/5} + \sqrt{y}\right]\Delta x + \left[x^{2/5} + \frac{x}{2\sqrt{y}}\right]\Delta y.$$

At $x = 32, y = 16, \Delta x = 3, \Delta y = 2$, and

$$df = \left[\frac{32}{5}\left(\frac{1}{32}\right)^{3/5} + \sqrt{16}\right]3 + \left[32^{2/5} + \frac{32}{2\sqrt{16}}\right]2 = 30.4.$$

The change increases the value of f by approximately 30.4. □

Example 2 Use differentials to estimate $\sqrt{27}\,\sqrt[3]{1021}$.

SOLUTION We know $\sqrt{25}$ and $\sqrt[3]{1000}$. What we need is an estimate for the increase of

$$f(x, y) = \sqrt{x}\,\sqrt[3]{y} = x^{1/2}y^{1/3}$$

from $x = 25, y = 1000$ to $x = 27, y = 1021$. The differential is

$$df = \tfrac{1}{2}x^{-1/2}y^{1/3}\Delta x + \tfrac{1}{3}x^{1/2}y^{-2/3}\Delta y.$$

With $x = 25, y = 1000, \Delta x = 2, \Delta y = 21$, df becomes

$$(\tfrac{1}{2} \cdot 25^{-1/2} \cdot 1000^{1/3})2 + (\tfrac{1}{3} \cdot 25^{1/2} \cdot 1000^{-2/3})21 = 2.35.$$

The change increases the value of the function by about 2.35. It follows that

$$\sqrt{27}\,\sqrt[3]{1021} \cong \sqrt{25}\,\sqrt[3]{1000} + 2.35 = 52.35.$$

(Our calculator gives $\sqrt{27}\,\sqrt[3]{1021} \cong 52.323.$) □

Example 3 Estimate by a differential the change in the volume of the frustum of a right circular cone if the upper radius r is decreased from 3 to 2.7 centimeters, the base radius R is increased from 8 to 8.1 centimeters, and the height h is increased from 6 to 6.3 centimeters.

SOLUTION Since $V(r,R,h) = \frac{1}{3}\pi h(R^2 + Rr + r^2)$, we have

$$dV = \tfrac{1}{3}\pi h(R+2r)\Delta r + \tfrac{1}{3}\pi h(2R+r)\Delta R + \tfrac{1}{3}\pi(R^2 + Rr + r^2)\Delta h.$$

At $r = 3, R = 8, h = 6, \Delta r = -0.3, \Delta R = 0.1$, and $\Delta h = 0.3$,

$$dV = (28\pi)(-0.3) + (38\pi)(0.1) + \tfrac{1}{3}(97\pi)(0.3) = 5.1\,\pi \cong 16.02.$$

According to our differential estimate, the volume increases by about 16 cubic centimeters. ☐

EXERCISES 15.8

Find the differential df.

1. $f(x,y) = x^3y - x^2y^2$.
2. $f(x,y,z) = xy + yz + xz$.
3. $f(x,y) = x\cos y - y\cos x$.
4. $f(x,y,z) = x^2y\,e^{2z}$.
5. $f(x,y,z) = x - y\,\tan z$.
6. $f(x,y) = (x-y)\ln(x+y)$.
7. $f(x,y,z) = \dfrac{xy}{x^2+y^2+z^2}$.
8. $f(x,y) = \ln(x^2+y^2) + x\,e^{xy}$.
9. $f(x,y) = \sin(x+y) + \sin(x-y)$.
10. $f(x,y) = x\ln\left[\dfrac{1+y}{1-y}\right]$.
11. $f(x,y,z) = y^2e^{xz} + x\ln z$.
12. $f(x,y) = xy\,e^{-(x^2+y^2)}$.
13. Calculate Δu and du for $u = x^2 - 3xy + 2y^2$ at $x = 2$, $y = -3, \Delta x = -0.3, \Delta y = 0.2$.
14. Calculate du for $u = (x+y)\sqrt{x-y}$ at $x = 6, y = 2$, $\Delta x = \frac{1}{4}, \Delta y = -\frac{1}{2}$.
15. Calculate Δu and du for $u = x^2z - 2yz^2 + 3xyz$ at $x = 2$, $y = 1, z = 3, \Delta x = 0.1, \Delta y = 0.3, \Delta z = -0.2$.
16. Calculate du for
$$u = \frac{xy}{\sqrt{x^2+y^2+z^2}}$$
at $x = 1, y = 3, z = -2, \Delta x = \frac{1}{2}, \Delta y = \frac{1}{4}, \Delta z = -\frac{1}{4}$.

Use differentials to find the approximate value.

17. $\sqrt{125}\,\sqrt[4]{17}$.
18. $(1-\sqrt{10})(1+\sqrt{24})$.
19. $\sin\frac{6}{7}\pi\,\cos\frac{1}{5}\pi$.
20. $\sqrt{8}\tan\frac{5}{16}\pi$.

Use differentials to approximate the value of f at the point P.

21. $f(x,y) = x^2e^{xy}$; $P(2.9, 0.01)$.
22. $f(x,y,z) = x^2y\cos\pi z$; $P(2.12, 2.92, 3.02)$.
23. $f(x,y,z) = x\tan^{-1} yz$; $P(2.94, 1.1, 0.92)$.
24. $f(x,y) = \sqrt{x^2+y^2}$; $P(3.06, 3.88)$.
25. Given that $z = (x-y)(x+y)^{-1}$, use dz to find the approximate change in z if x is increased from 4 to $4\frac{1}{10}$ and y is increased from 2 to $2\frac{1}{10}$. What is the exact change?
26. Estimate by a differential the change in the volume of a right circular cylinder if the height is increased from 12 to 12.2 inches and the radius is decreased from 8 to 7.7 inches.
27. Estimate the change in the total surface area for the cylinder of Exercise 26.
28. Use a differential to estimate the change in $T = x^2\cos\pi z - y^2\sin\pi z$ from $x = 2, y = 2, z = 2$ to $x = 2.1, y = 1.9, z = 2.2$.
29. Estimate the surface area of a closed rectangular box whose dimensions are: length = 9.98 inches, width = 5.88 inches, height = 4.08 inches.
30. Estimate the volume of a right circular cone of base radius 7.2 centimeters and height 10.15 centimeters.
31. The dimensions of a closed rectangular box change from length = 12, width = 8, height = 6 to length = 12.02, width = 7.95, height = 6.03.
    (a) Use a differential to approximate the change in volume.
    (b) Calculate the exact change in volume.
32. Use the dimensions of the rectangular box in Exercise 31.
    (a) Approximate the change in the surface area using a differential.
    (b) Calculate the exact change in the surface area.
33. Suppose that $T(x,y,z) = 100 - x^2 - y^2 - z^2 + 2xyz$ gives the temperature T at the point $P(x,y,z)$ in space. Use differentials to approximate the temperature difference between the points $P(1,3,4)$ and $Q(1.15, 2.90, 4.10)$.
34. According to the ideal gas law, the relation between the pressure P, the temperature T, and the volume V of a confined gas is given by the equation $PV = kT$, where k is a

constant. If $P = 4$ pounds per square inch when $V = 81$ cubic inches and $T = 300$ K, approximate the change in pressure if the volume is decreased to 75 cubic inches and the temperature is increased to 325 K.

35. As illustrated in the following figure, the side x in the right triangle is increased by Δx and the angle θ is increased by $\Delta\theta$. Use a differential to approximate the change in the area of the triangle. Is the area more sensitive to a change in x or to a change in θ?

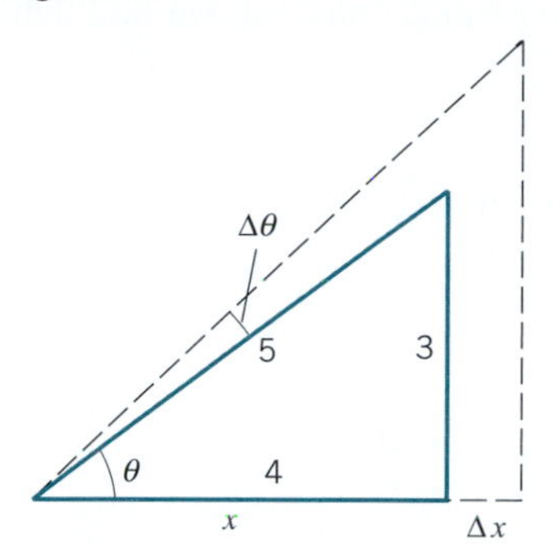

36. Use a differential to approximate the change in the area of the isosceles triangle shown in the figure if x changes by Δx and θ changes by $\Delta\theta$. Is the area more sensitive to changes in x or to changes in θ? HINT: The area of a triangle with sides a and b and included angle θ is given by $A = \frac{1}{2}ab\sin\theta$.

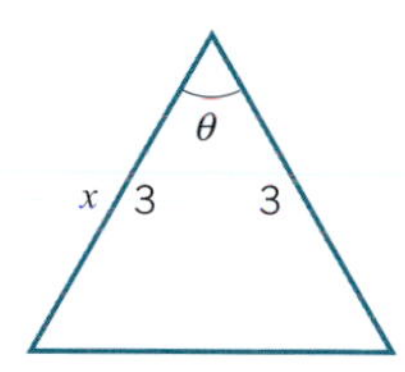

37. The radius of a right circular cylinder of height h is increased from r to $r + \Delta r$.

(a) Determine the exact change in h that will keep the volume constant. Then estimate this change in h by using a differential.

(b) Determine the exact change in h that will keep the total surface area constant. Then estimate this change in h by using a differential.

38. The dimensions of a rectangular box with a top are length $= 4$ feet, width $= 2$ feet, height $= 3$ feet. It has a coat of paint $\frac{1}{16}$ inch thick. Estimate the amount of paint (cubic inches) on the box.

Error Estimates

Let $u = u(x, y)$ be differentiable. If the variables x and y are known to be $x_0 \pm \Delta x$ and $y_0 \pm \Delta y$, then the maximum possible error in the calculated value of $u(x_0, y_0)$ is

$$\frac{\partial u}{\partial x}(x_0, y_0)(\pm\Delta x) + \frac{\partial u}{\partial y}(x_0, y_0)(\pm\Delta y).$$

39. The legs of a right triangle are measured to be 5 and 12 centimeters with a possible error of ± 15 millimeters in each measurement. What is the maximum possible error in the calculated value of (a) the hypotenuse and (b) the area of the triangle?

40. The radius of a right circular cone is measured to be 5 inches with a possible error of ± 0.2 inch and the height is measured to be 12 inches with a possible error of ± 0.3 inch. What is the maximum possible error in the calculated values of (a) the volume and (b) the lateral surface area of the cone?

41. The specific gravity of a solid is given by the formula $s = A(A - W)^{-1}$ where A is the weight in air and W is the weight in water. What is the maximum possible error in the calculated value of s if A is measured to be 9 pounds (within a tolerance of 0.01 pound) and W is measured to be 5 pounds (within a tolerance of 0.02 pound)?

42. The measurements of a closed rectangular box are length $=$ 5 feet, width $= 3$ feet, and height $= 3.5$ feet, with a possible error of $\pm\frac{1}{12}$ inch in each measurement. What is the maximum possible error in the calculated value of (a) the volume and (b) the surface area of the box? HINT: Extend the error estimate to three dimensions.

■ 15.9 RECONSTRUCTING A FUNCTION FROM ITS GRADIENT

This section has three parts. In Part 1 we show how to find $f(x, y)$ given its gradient

$$\nabla f(x, y) = \frac{\partial f}{\partial x}(x, y)\,\mathbf{i} + \frac{\partial f}{\partial y}(x, y)\,\mathbf{j}.$$

In Part 2 we show that, although all gradients $\nabla f(x, y)$ are expressions of the form

$$P(x, y)\,\mathbf{i} + Q(x, y)\,\mathbf{j}$$

(set $P = \partial f/\partial x$ and $Q = \partial f/\partial y$), not all such expressions are gradients. In Part 3 we tackle the problem of recognizing which expressions $P(x, y)\,\mathbf{i} + Q(x, y)\,\mathbf{j}$ are actually gradients.

Part 1

Example 1 Find f given that $\nabla f(x,y) = (4x^3y^3 - 3x^2)\,\mathbf{i} + (3x^4y^2 + \cos 2y)\,\mathbf{j}$.

SOLUTION The first partial derivatives of f are

$$\frac{\partial f}{\partial x}(x,y) = 4x^3y^3 - 3x^2, \quad \frac{\partial f}{\partial y}(x,y) = 3x^4y^2 + \cos 2y.$$

Integrating $\partial f/\partial x$ with respect to x, treating y as a constant, we find that

$$f(x,y) = x^4y^3 - x^3 + \phi(y)$$

where ϕ is an unknown function of y. The function ϕ plays the same role as the arbitrary constant C that arises when you integrate a function of one variable. Now, differentiation with respect to y gives

$$\frac{\partial f}{\partial y}(x,y) = 3x^4y^2 + \phi'(y).$$

Equating the two expressions for $\partial f/\partial y$, we have

$$3x^4y^2 + \phi'(y) = 3x^4y^2 + \cos 2y,$$

which implies that

$$\phi'(y) = \cos 2y \qquad \text{and thus} \qquad \phi(y) = \tfrac{1}{2}\sin 2y + C \qquad (C \text{ a constant.})$$

This means that

$$f(x,y) = x^4y^3 - x^3 + \tfrac{1}{2}\sin 2y + C. \quad \square$$

Remark The procedure for finding f just illustrated is symmetric in x and y. That is, rather than integrating $\partial f/\partial x$ with respect to x, we could have started by integrating $\partial f/\partial y$ with respect to y, with x held constant, followed by differentiating the result with respect to x:

$$\text{if} \quad \frac{\partial f}{\partial y}(x,y) = 3x^4y^2 + \cos 2y \qquad \text{then} \qquad f(x,y) = x^4y^3 + \tfrac{1}{2}\sin 2y + \psi(x),$$

where ψ is an unknown function of x. Now, differentiating with respect to x, we have

$$\frac{\partial f}{\partial x}(x,y) = 4x^3y^3 + \psi'(x).$$

Equating the two expressions for $\partial f/\partial x$ gives

$$4x^3y^3 + \psi'(x) = 4x^3y^3 - 3x^2.$$

Therefore,

$$\psi'(x) = -3x^2 \quad \text{which implies that} \quad \psi(x) = -x^3 + C \qquad (C \text{ a constant.})$$

Thus,

$$f(x,y) = x^4y^3 + \tfrac{1}{2}\sin 2y - x^3 + C,$$

as we found before. $\square$

Example 2 Find f given that

$$\nabla f(x,y) = \left(\sqrt{y} - \frac{y}{2\sqrt{x}} + 2x\right)\mathbf{i} + \left(\frac{x}{2\sqrt{y}} - \sqrt{x} + 1\right)\mathbf{j}.$$

SOLUTION Here we have

$$\frac{\partial f}{\partial x}(x,y) = \sqrt{y} - \frac{y}{2\sqrt{x}} + 2x, \quad \frac{\partial f}{\partial y}(x,y) = \frac{x}{2\sqrt{y}} - \sqrt{x} + 1.$$

Integrating $\partial f/\partial x$ with respect to x, we have

$$f(x,y) = x\sqrt{y} - y\sqrt{x} + x^2 + \phi(y)$$

with $\phi(y)$ independent of x. Differentiation with respect to y gives

$$\frac{\partial f}{\partial y}(x,y) = \frac{x}{2\sqrt{y}} - \sqrt{x} + \phi'(y).$$

The two equations for $\partial f/\partial y$ can be reconciled only by having

$$\phi'(y) = 1 \quad \text{and thus} \quad \phi(y) = y + C.$$

This means that

$$f(x,y) = x\sqrt{y} - y\sqrt{x} + x^2 + y + C. \quad \square$$

The function

$$f(x,y) = x\sqrt{y} - y\sqrt{x} + x^2 + y + C$$

is the *general solution* of the vector differential equation

$$\nabla f(x,y) = \left(\sqrt{y} - \frac{y}{2\sqrt{x}} + 2x\right)\mathbf{i} + \left(\frac{x}{2\sqrt{y}} - \sqrt{x} + 1\right)\mathbf{j}.$$

Each *particular solution* can be obtained by assigning a particular value to the constant C.

Part 2

The next example shows that not all linear combinations $P(x,y)\,\mathbf{i} + Q(x,y)\,\mathbf{j}$ are gradients.

Example 3 Show that $y\,\mathbf{i} - x\,\mathbf{j}$ is not a gradient.

SOLUTION Suppose on the contrary that it is a gradient. Then there exists a function f such that

$$\nabla f(x,y) = y\,\mathbf{i} - x\,\mathbf{j}.$$

This implies that

$$\frac{\partial f}{\partial x}(x,y) = y, \qquad \frac{\partial f}{\partial y}(x,y) = -x$$

$$\frac{\partial^2 f}{\partial y \partial x}(x,y) = 1, \qquad \frac{\partial^2 f}{\partial x \partial y}(x,y) = -1$$

and thus
$$\frac{\partial^2 f}{\partial y \partial x}(x,y) \neq \frac{\partial^2 f}{\partial x \partial y}(x,y).$$

This contradicts (14.6.5): the four partial derivatives under consideration are everywhere continuous and thus, according to (14.6.5), we must have

$$\frac{\partial^2 f}{\partial y \partial x}(x,y) = \frac{\partial^2 f}{\partial x \partial y}(x,y).$$

This contradiction shows that $y\,\mathbf{i} - x\,\mathbf{j}$ is not a gradient. □

Part 3

We come now to the problem of recognizing which linear combinations

$$P(x,y)\,\mathbf{i} + Q(x,y)\,\mathbf{j}$$

are actually gradients. But first we need to review some ideas and establish some terminology.

As indicated earlier (Section 15.3), an open set (in the plane or in three-space) is said to be *connected* if each pair of points of the set can be joined by a polygonal path that lies entirely within the set. An open connected set is called an *open region*. A curve

$$C: \quad \mathbf{r} = \mathbf{r}(t), \quad t \in [a,b]$$

is said to be *closed* if it begins and ends at the same point:

$$\mathbf{r}(a) = \mathbf{r}(b).$$

It is said to be *simple* if it does not intersect itself:

$$a < t_1 < t_2 < b \quad \text{implies} \quad \mathbf{r}(t_1) \neq \mathbf{r}(t_2).$$

These notions are illustrated in Figure 15.9.1.

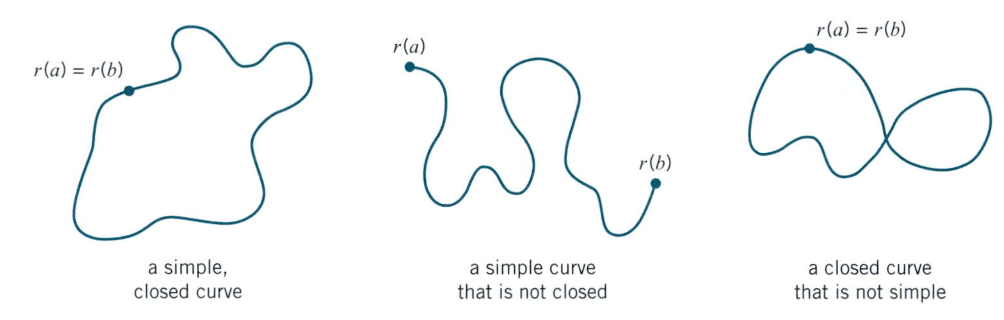

Figure 15.9.1

As is intuitively clear (Figure 15.9.2), a simple closed curve in the plane separates the plane into two disjoint open connected sets: a bounded inner region consisting of all points surrounded by the curve and an unbounded outer region consisting of all points not surrounded by the curve.†

†Add to this the assertion that the curve in question constitutes the total boundary of both regions and you have what is called the Jordan curve theorem, named after the French mathematician Camille Jordan (1838–1922). Jordan was the first to point out that, although all this is apparently obvious, it nevertheless requires proof. In recognition of Jordan, simple closed plane curves are now commonly known as *Jordan curves*.

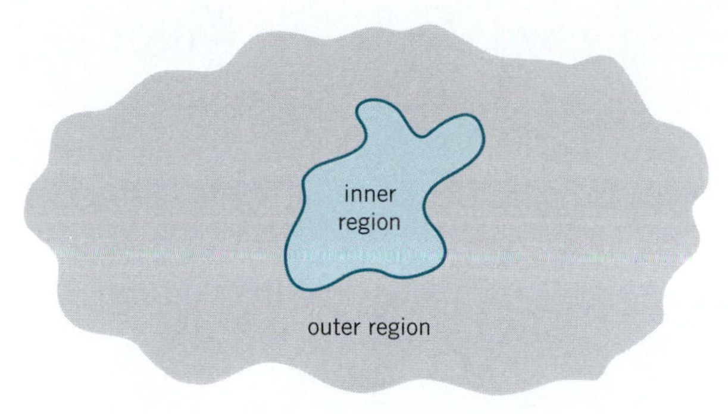

Figure 15.9.2

Finally, we come to a notion we will need in our work with gradients:

(15.9.1)

> Let Ω be an open region of the plane. Ω is said to be *simply connected* if, for every simple closed curve C in Ω, the inner region of C is contained in Ω.

The first two regions in Figure 15.9.3 are simply connected. The annular region is not; the annular region contains the simple closed curve drawn there, but it does not contain all of the inner region of that curve.

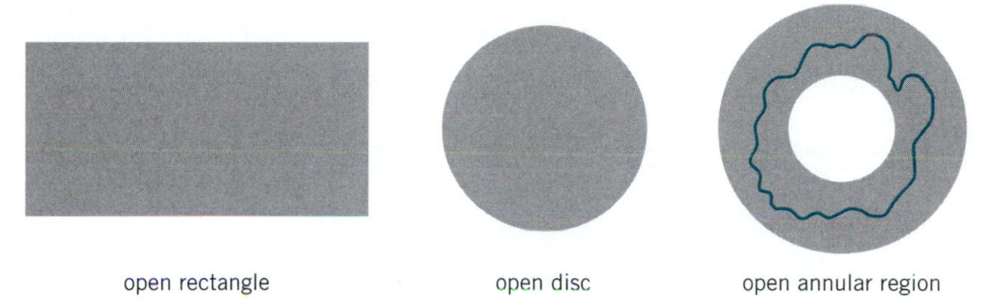

Figure 15.9.3

THEOREM 15.9.2

Let P and Q be functions of two variables, each continuously differentiable on a simply connected open region Ω. The linear combination $P(x,y)\,\mathbf{i} + Q(x,y)\,\mathbf{j}$ is a gradient on Ω iff

$$\frac{\partial P}{\partial y}(x,y) = \frac{\partial Q}{\partial x}(x,y) \qquad \text{for all } (x,y) \in \Omega.$$

A complete proof of this theorem for a general region Ω is complicated. We will prove the result under the additional assumption that Ω has the form of an open rectangle with sides parallel to the coordinate axes.

Suppose that $P(x,y)\,\mathbf{i} + Q(x,y)\,\mathbf{j}$ is a gradient on this open rectangle Ω, say,

$$\nabla f(x,y) = P(x,y)\,\mathbf{i} + Q(x,y)\,\mathbf{j}.$$

Since $$\nabla f(x,y) = \frac{\partial f}{\partial x}(x,y)\,\mathbf{i} + \frac{\partial f}{\partial y}(x,y)\,\mathbf{j},$$

we have $$P = \frac{\partial f}{\partial x} \quad \text{and} \quad Q = \frac{\partial f}{\partial y}.$$

Since P and Q have continuous first partials, f has continuous second partials. Thus, according to (14.6.5), the mixed partials are equal and we have

$$\frac{\partial P}{\partial y} = \frac{\partial^2 f}{\partial y \partial x} = \frac{\partial^2 f}{\partial x \partial y} = \frac{\partial Q}{\partial x}.$$

Conversely, suppose that

$$\frac{\partial P}{\partial y}(x,y) = \frac{\partial Q}{\partial x}(x,y) \quad \text{for all} \quad (x,y) \in \Omega.$$

We must show that $P(x,y)\,\mathbf{i} + Q(x,y)\,\mathbf{j}$ is a gradient on Ω. To do this, we choose a point (x_0, y_0) in Ω and form the function

$$f(x,y) = \int_{x_0}^{x} P(u, y_0)\, du + \int_{y_0}^{y} Q(x, v)\, dv, \quad (x,y) \in \Omega.$$

[If you want to visualize f, you can refer to Figure 15.9.4. The function P is being integrated along the horizontal line segment that joins (x_0, y_0) to (x, y_0), and Q is being integrated along the vertical line segment that joins (x, y_0) to (x, y). Our assumptions on Ω guarantee that these line segments remain in Ω.]

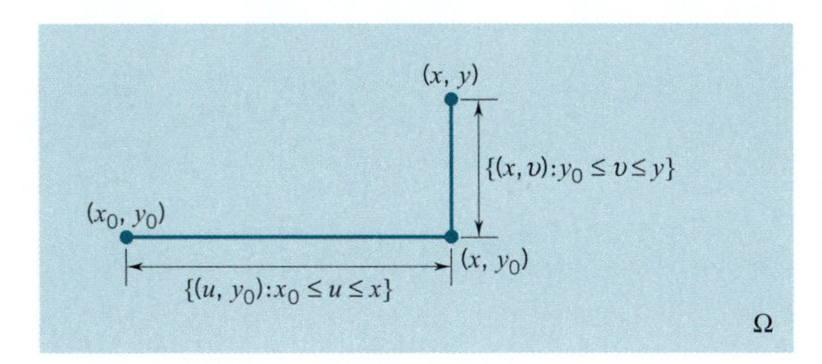

Figure 15.9.4

The first integral is independent of y. Hence

$$\frac{\partial f}{\partial y}(x,y) = \frac{\partial}{\partial y}\left(\int_{y_0}^{y} Q(x, v)\, dv\right) = Q(x,y).$$

The last equality holds because we are differentiating an integral with respect to its upper limit (Theorem 5.2.5). Differentiating f with respect to x we have

$$\frac{\partial f}{\partial x}(x,y) = \frac{\partial}{\partial x}\left(\int_{x_0}^{x} P(u, y_0)\, du\right) + \frac{\partial}{\partial x}\left(\int_{y_0}^{y} Q(x, v)\, dv\right).$$

The first term is $P(x, y_0)$, since once again we are differentiating with respect to the upper limit. In the second term the variable x appears in the integrand. It can be shown that, since Q and $\partial Q/\partial x$ are continuous,

$$\frac{\partial}{\partial x}\left(\int_{y_0}^{y} Q(x,v)\,dv\right) = \int_{y_0}^{y} \frac{\partial Q}{\partial x}(x,v)\,dv.\dagger$$

Anticipating this result, we have

$$\frac{\partial f}{\partial x}(x,y) = P(x,y_0) + \int_{y_0}^{y} \frac{\partial Q}{\partial x}(x,v)\,dv \underset{\text{explain}}{=} P(x,y_0) + \int_{y_0}^{y} \frac{\partial P}{\partial y}(x,v)\,dv$$

$$= P(x,y_0) + P(x,y) - P(x,y_0) = P(x,y).$$

We have now shown that

$$P(x,y) = \frac{\partial f}{\partial x}(x,y) \quad \text{and} \quad Q(x,y) = \frac{\partial f}{\partial y}(x,y) \quad \text{for all } (x,y) \in \Omega.$$

It follows that $P(x,y)\,\mathbf{i} + Q(x,y)\,\mathbf{j}$ is the gradient of f on Ω. ❑

Example 4 The vector functions

$$\mathbf{F}(x,y) = 2x\sin y\,\mathbf{i} + x^2\cos y\,\mathbf{j} \quad \text{and} \quad \mathbf{G}(x,y) = xy\,\mathbf{i} + \tfrac{1}{2}(x+1)^2 y^2\,\mathbf{j}$$

are both defined everywhere. The first vector function is the gradient of a function that is defined everywhere, since for $P(x,y) = 2x\sin y$ and $Q(x,y) = x^2\cos y$, we have

$$\frac{\partial P}{\partial y}(x,y) = 2x\cos y = \frac{\partial Q}{\partial x} \quad \text{for all } (x,y).$$

It is easy to verify that if $f(x,y) = x^2\sin y + C$ where C is a constant, then $\nabla f(x,y) = \mathbf{F}(x,y)$.

The vector function $\mathbf{G}$ is not a gradient: $P(x,y) = xy$, $Q(x,y) = \frac{1}{2}(x+1)^2y^2$, and

$$\frac{\partial P}{\partial y}(x,y) = x \quad \text{and} \quad \frac{\partial Q}{\partial x}(x,y) = (x+1)y^2.$$

Thus,

$$\frac{\partial P}{\partial y}(x,y) \neq \frac{\partial Q}{\partial x}(x,y). \quad ❑$$

Example 5 The vector function $\mathbf{F}$ defined on the *punctured disc* $0 < x^2 + y^2 < 1$ by setting

$$\mathbf{F}(x,y) = \frac{y}{x^2+y^2}\,\mathbf{i} - \frac{x}{x^2+y^2}\,\mathbf{j}$$

satisfies

$$\frac{\partial P}{\partial y}(x,y) = \frac{\partial Q}{\partial x}(x,y) \quad \text{on the punctured disc.}$$

(Check this out.) Nevertheless, as you will see in Exercise 28, Section 17.2, $\mathbf{F}$ is not a gradient on that set. The punctured disc is not simply connected and therefore Theorem 15.9.2 does not apply. ❑

---

†The validity of this equality is the subject of Exercise 60, Section 16.3.

EXERCISES 15.9

Determine whether or not the vector function is the gradient $\nabla f(x,y)$ of a function everywhere defined. If so, find all the functions with that gradient.

1. $xy^2\,\mathbf{i}+x^2y\,\mathbf{j}$.

2. $x\,\mathbf{i}+y\,\mathbf{j}$.

3. $y\,\mathbf{i}+x\,\mathbf{j}$.

4. $(x^2+y)\,\mathbf{i}+(y^3+x)\,\mathbf{j}$.

5. $(y^3+x)\,\mathbf{i}+(x^2+y)\,\mathbf{j}$.

6. $(y^2e^x-y)\,\mathbf{i}+(2y\,e^x-x)\,\mathbf{j}$.

7. $(\cos x-y\sin x)\,\mathbf{i}+\cos x\,\mathbf{j}$.

8. $(1+e^y)\,\mathbf{i}+(x\,e^y+y^2)\,\mathbf{j}$.

9. $e^x\cos y^2\,\mathbf{i}-2y\,e^x\sin y^2\mathbf{j}$.

10. $e^x\cos y\,\mathbf{i}+e^x\sin y\,\mathbf{j}$.

11. $y\,e^x(1+x)\,\mathbf{i}+(x\,e^x-e^{-y})\,\mathbf{j}$.

12. $(e^x+2xy)\,\mathbf{i}+(x^2+\sin y)\,\mathbf{j}$.

13. $(x\,e^{xy}+x^2)\,\mathbf{i}+(y\,e^{xy}-2y)\,\mathbf{j}$.

14. $(y\sin x+xy\cos x)\,\mathbf{i}+(x\sin x+2y+1)\,\mathbf{j}$.

15. $(1+y^2+xy^2)\,\mathbf{i}+(x^2y+y+2xy+1)\,\mathbf{j}$.

16. $\left[2\ln(3y)+\dfrac{1}{x}\right]\mathbf{i}+\left[\dfrac{2x}{y}+y^2\right]\mathbf{j}$.

In Exercises 17–20, find the most general function with the given gradient.

17. $\dfrac{x}{\sqrt{x^2+y^2}}\,\mathbf{i}+\dfrac{y}{\sqrt{x^2+y^2}}\,\mathbf{j}$.

18. $(x\tan y+\sec^2 x)\,\mathbf{i}+(\tfrac{1}{2}x^2\sec^2 y+\pi y)\,\mathbf{j}$.

19. $(x^2\sin^{-1}y)\,\mathbf{i}+\left(\dfrac{x^3}{3\sqrt{1-y^2}}-\ln y\right)\mathbf{j}$.

20. $\left(\dfrac{\tan^{-1}y}{\sqrt{1-x^2}}+\dfrac{x}{y}\right)\mathbf{i}+\left(\dfrac{\sin^{-1}x}{1+y^2}-\dfrac{x^2}{2y^2}+1\right)\mathbf{j}$.

C 21. Use a CAS to determine whether $\mathbf{F}$ is a gradient.

(a) $\mathbf{F}(x,y)=(y-2xy+y^2)\,\mathbf{i}+(x-x^2+2xy)\,\mathbf{j}$.

(b) $\mathbf{F}(x,y)=[2xy^2\cos(x^2-y)]\,\mathbf{i}+[-y^2\cos(x^2-y)+2y\sin(x^2-y)]\,\mathbf{j}$.

(c) $\mathbf{F}(x,y)=2xy(y-x)e^{-x^2y}\,\mathbf{i}+x^2(x-y)e^{-x^2y}\,\mathbf{j}$.

C 22. (a) Use a CAS to f to within an additive constant if

$$\nabla f=[(1-2xy[x-y])e^{-x^2y}]\,\mathbf{i}+[-(1+x^2[x-y])e^{-x^2y}]\,\mathbf{j}.$$

(b) Use a CAS to find f if $f\left(\frac{\pi}{3},\frac{\pi}{4}\right)=6$ and

$$\nabla f=[\cos(x+y)+\sin(x-y)]\,\mathbf{i}+[\cos(x+y)-\sin(x-y)]\,\mathbf{j}.$$

23. Find the general solution of the differential equation $\nabla f(x,y)=f(x,y)\,\mathbf{i}+f(x,y)\,\mathbf{j}$.

24. Given that g and its first and second partials are everywhere continuous, find the general solution of the differential equation $\nabla f(x,y)=e^{g(x,y)}[g_x(x,y)\,\mathbf{i}+g_y(x,y)\,\mathbf{j}]$.

Theorem 15.9.2 has a three-dimensional analog. In particular we can show that, if P,Q,R are continuously differentiable on an open rectangular box S, then the vector function

$$P(x,y,z)\,\mathbf{i}+Q(x,y,z)\,\mathbf{j}+R(x,y,z)\,\mathbf{k}$$

is a gradient on S iff

$$\frac{\partial P}{\partial y}=\frac{\partial Q}{\partial x},\quad \frac{\partial P}{\partial z}=\frac{\partial R}{\partial x},\quad \frac{\partial Q}{\partial z}=\frac{\partial R}{\partial y}\qquad \text{throughout S.}$$

25. (a) Verify that $2x\,\mathbf{i}+z\,\mathbf{j}+y\,\mathbf{k}$ is the gradient of a function f that is everywhere defined.

(b) Deduce from the relation $\partial f/\partial x=2x$ that $f(x,y,z)=x^2+g(y,z)$.

(c) Verify then that $\partial f/\partial y=z$ gives $g(y,z)=zy+h(z)$ and finally that $\partial f/\partial z=y$ gives $h(z)=C$.

(d) What is $f(x,y,z)$?

Determine whether the vector function is a gradient $\nabla f(x,y,z)$ and, if so, find all functions f with that gradient.

26. $yz\,\mathbf{i}+xz\,\mathbf{j}+xy\,\mathbf{k}$.

27. $(2x+y)\,\mathbf{i}+(2y+x+z)\,\mathbf{j}+(y-2z)\,\mathbf{k}$.

28. $(2x\sin 2y\cos z)\,\mathbf{i}+(2x^2\cos 2y\cos z)\,\mathbf{j}-(x^2\sin 2y\sin z)\,\mathbf{k}$.

29. $(y^2z^3+1)\,\mathbf{i}+(2xyz^3+y)\,\mathbf{j}+(3xy^2z^2+1)\,\mathbf{k}$.

30. $\left[\dfrac{y}{z}-e^z\right]\mathbf{i}+\left[\dfrac{x}{z}+1\right]\mathbf{j}-\left[xe^z+\dfrac{xy}{z^2}\right]\mathbf{k}$.

31. Verify that the gravitational force function

$$\mathbf{F}(\mathbf{r})=-G\frac{mM}{r^3}\,\mathbf{r}\qquad(\mathbf{r}=x\,\mathbf{i}+y\,\mathbf{j}+z\,\mathbf{k})$$

is a gradient.

32. Verify that every vector function of the form

$$\mathbf{h}(\mathbf{r})=kr^n\mathbf{r}\qquad(k\text{ constant, }n\text{ an integer})$$

is a gradient.

■ CHAPTER HIGHLIGHTS

15.1 Differentiability and Gradient

gradient of f at $\mathbf{x}$: $\nabla f(\mathbf{x})$ (p. 862)

$$\nabla f(x, y) = \frac{\partial f}{\partial x}\mathbf{i} + \frac{\partial f}{\partial y}\mathbf{j}$$

$$\nabla f(x, y, z) = \frac{\partial f}{\partial x}\mathbf{i} + \frac{\partial f}{\partial y}\mathbf{j} + \frac{\partial f}{\partial z}\mathbf{k}$$

If f is differentiable at $\mathbf{x}$, then f is continuous at $\mathbf{x}$.

$$\nabla r^n = nr^{n-2}\mathbf{r}.$$

15.2 Gradients and Directional Derivatives

directional derivative: $f'_{\mathbf{u}}(\mathbf{x}) = \nabla f(\mathbf{x}) \cdot \mathbf{u}$ (p. 872)

The directional derivative $f'_{\mathbf{u}}$ gives the rate of change of f in the direction of the unit vector $\mathbf{u}$.
The directional derivative in a direction $\mathbf{u}$ is the component of the gradient vector in that direction. (p. 874)
A differentiable function f increases most rapidly in the direction of the gradient (the rate of change is then $\|\nabla f(\mathbf{x})\|$) and it decreases most rapidly in the opposite direction (the rate of change is then $-\|\nabla f(\mathbf{x})\|$).

15.3 The Mean-Value Theorem; Chain Rules

The mean-value theorem (p. 879)
open, connected set (p. 880)
chain rule along a curve (p. 882)
tree diagram (p. 886)
intermediate-value theorem (p. 891)
open, closed regions (p. 891)

In the setting of functions of several variables, there are numerous versions of the chain rule. They can all be deduced from the chain rule along a curve.
If f is a continuously differentiable function of $r = \|\mathbf{r}\|$, then

$$\nabla[f(r)] = f'(r)\frac{\mathbf{r}}{r}.$$

15.4 The Gradient as a Normal; Tangent Lines and Tangent Planes

tangent and normal lines to a curve $f(x, y) = c$ (p. 895)
tangent plane to a surface $f(x, y, z) = c$ (p. 897)
upper and lower unit normals (p. 902)

At each point of the domain of a function, the gradient vector, if not $\mathbf{0}$, is perpendicular to the level curve (level surface) that passes through that point.

15.5 Local Extreme Values

local maximum and local minimum (p. 903)
critical points (p. 904)
stationary points, saddle points (p. 904)

If f has a local extreme value at $\mathbf{x}_0$, then either $\nabla f(\mathbf{x}_0) = \mathbf{0}$ or $\nabla f(\mathbf{x}_0)$ does not exist.

second-partials test, discriminant (p. 906)

15.6 Absolute Extreme Values

absolute maximum and absolute minimum (p. 911)
bounded set (p. 911)
extreme-value theorem (p. 912)
procedure for finding extreme values (p. 912)
method of least squares (p. 916)

15.7 Maxima and Minima with side Conditions

If $\mathbf{x}_0$ maximizes (or minimizes) $f(\mathbf{x})$ subject to the side condition $g(\mathbf{x}) = 0$, then $\nabla f(\mathbf{x}_0)$ and $\nabla g(\mathbf{x}_0)$ are parallel. Thus, if $\nabla g(\mathbf{x}_0) \neq \mathbf{0}$, then there exists a scalar λ, called a Lagrange multiplier, such that $\nabla f(\mathbf{x}_0) = \lambda \nabla g(\mathbf{x}_0)$.

15.8 Differentials

increment : $\Delta f = f(\mathbf{x} + \mathbf{h}) - f(\mathbf{x})$
differential : $df = \nabla f(\mathbf{x}) \cdot \mathbf{h}$

$$\Delta f \cong df \quad \text{in the sense that} \quad \frac{\Delta f - df}{\|\mathbf{h}\|} \to 0 \text{ as } \mathbf{h} \to \mathbf{0}$$

two variables: $df = \dfrac{\partial f}{\partial x}\Delta x + \dfrac{\partial f}{\partial y}\Delta y$

three variables : $df = \dfrac{\partial f}{\partial x}\Delta x + \dfrac{\partial f}{\partial y}\Delta y + \dfrac{\partial f}{\partial z}\Delta z$

15.9 Reconstructing a Function from its Gradient

finding f from ∇f (p. 932)
closed curve, simple curve (p. 934)
simply connected open region (p. 935)

necessary and sufficient conditions for a vector-valued function to be a gradient:

two variables (p. 935)
three variables (p. 938)

Let $\mathbf{a} = 3\,\mathbf{i} + 2\,\mathbf{j} - \mathbf{k}$, $\mathbf{b} = 5\,\mathbf{i} + 3\,\mathbf{j}$, $\mathbf{c} = -2\,\mathbf{i} + 4\,\mathbf{j} + \mathbf{k}$. Find the indicated vector or scalar.

1. $2\,\mathbf{a} - 3\,\mathbf{b}$

2. $\mathbf{a} \cdot (\mathbf{b} + \mathbf{c})$

3. $||\mathbf{a} + \mathbf{b}||$

4. A unit vector in the same direction as **a**.

5. The cosine of the angle between **b** and **c**.

6. $\mathbf{a} \times \mathbf{b}$

7. A unit vector perpendicular to **a** and **c**.

8. The volume of the parallelpiped with **a**, **b**, and **c** as sides.

9. Let $P(3, 2, -1)$, $Q(7, -5, 4)$, $R(5, 6, -3)$ be points in space.

(a) Find scalar parametric equations for the line that passes through R parallel to the line determined by P and Q.

(b) Find scalar parametric equations for the line that passes through R perpendicular to the line determined by P and Q.

(c) The lines found in parts (a) and (b) determine a plane. Find an equation for it.

10. Repeat Exercise 9 with $P(4, 2, 3)$, $Q(-2, 1, 4)$, $R(1, -1, -6)$.

11. (a) Are the points $P(3, 2, -1)$, $Q(7, -5, 4)$, $R(5, -1, 1)$ collinear?

(b) Are the points $P(3, 2, -1)$, $Q(7, -5, 4)$, $R(5, -1, 1)$, $S(1, 2, 0)$ coplanar?

Write an equation for the plane that satisfies the given conditions.

12. Contains the points $P(1, -2, 1)$, $Q(2, 0, 3)$, $R(0, 1, -1)$.

13. Contains the point $P(2, 1, -3)$ and is perpendicular to the line

$$\frac{x+1}{2} = \frac{y-1}{3} = -\frac{z}{4}$$

14. Contains the point $P(1, -2, -1)$ and is parallel to the plane $3x + 2y - z = 4$.

15. Contains the point $P(3, -1, 2)$ and the line $x = 2 + 2t$, $y = -1 + 3t$, $z = -2t$.

16. Find scalar parametric equations for the line of intersection of the planes $2x + y - 3z + 6 = 0$ and $x + 4y + 5z - 7 = 0$.

Find $\mathbf{f}'$ and $\mathbf{f}''$ for the vector-valued function **f**.

17. $\mathbf{f}(t) = e^{2t}\,\mathbf{i} + \ln(t^2 + 1)\,\mathbf{j}$

18. $\mathbf{f}(t) = e^t \cos t\,\mathbf{i} - \cos 2t\,\mathbf{j} + 3\,\mathbf{k}$

19. $\mathbf{f}(t) = \sinh 2t\,\mathbf{i} - te^{-t}\,\mathbf{j} + \cosh t\,\mathbf{k}$

Find the velocity, speed, and acceleration of the object with position **r**.

20. $\mathbf{r}(t) = \cos 2t\,\mathbf{i} + \sin 2t\,\mathbf{j} - t^2\mathbf{k}$

21. $\mathbf{r}(t) = 2t\,\mathbf{i} + \ln t\,\mathbf{j} - t^2\,\mathbf{k}$

22. $\mathbf{r}(t) = \cosh t\,\mathbf{i} + \sinh t\,\mathbf{j} + t\,\mathbf{k}$

23. Find $\mathbf{f}(t)$ if $\mathbf{f}'(t) = t^2\,\mathbf{i} + (e^{2t} + 1)\,\mathbf{j} + \sqrt{2t+1}\,\mathbf{k}$ and $\mathbf{f}(0) = \mathbf{i} - 3\,\mathbf{j} + 3\,\mathbf{k}$.

24. Find the position, velocity and speed of an object with initial velocity $\mathbf{v}_0 = \mathbf{k}$, initial position $\mathbf{r}_0 = \mathbf{i}$ and acceleration vector $\mathbf{a}(t) = -\cos t\,\mathbf{i} - \sin t\,\mathbf{j}$.

Sketch the curve represented by the vector-valued function and indicate the orientation.

25. $\mathbf{r}(t) = 2t^2\,\mathbf{i} - t\,\mathbf{j}$; $t \geq 0$

26. $\mathbf{r}(t) = e^{-t}\,\mathbf{i} + 2e^{2t}\,\mathbf{j}$; for all real t

27. $\mathbf{r}(t) = t\,\mathbf{i} + t\,\mathbf{j} + \sin t\,\mathbf{k}$; $t \geq 0$

Find the tangent vector and scalar parametric equations for the tangent line at the indicated point.

28. $\mathbf{r}(t) = t^2\,\mathbf{i} + (t+1)\,\mathbf{j} - t^3\,\mathbf{k}$; $P(1, 2, -1)$

29. $\mathbf{r}(t) = \cos 2t\,\mathbf{i} + \sin 2t\,\mathbf{j} + t\,\mathbf{k}$; $t = \pi/3$

Find the unit tangent vector and the principal unit normal vector for the curve **r**.

30. $\mathbf{r}(t) = 2t\,\mathbf{i} + \ln t\,\mathbf{j} - t^2\,\mathbf{k}$

31. $\mathbf{r}(t) = \cos t\,\mathbf{i} + \cos t\,\mathbf{j} - \sqrt{2}\sin t\,\mathbf{k}$

32. $\mathbf{r}(t) = e^t\,\mathbf{i} + e^{-t}\,\mathbf{j} - t\sqrt{2}\,\mathbf{k}$

Find the length of the curve **r**.

33. $\mathbf{r}(t) = 2t\,\mathbf{i} + \frac{2}{3}t^{3/2}\,\mathbf{j}$; from $t = 0$ to $t = 5$.

34. $\mathbf{r}(t) = e^t\,\mathbf{i} + e^{-t}\,\mathbf{j} - t\sqrt{2}\,\mathbf{k}$; from $t = 0$ to $t = \ln 3$.

35. $\mathbf{r}(t) = \sinh t\,\mathbf{i} + \cosh t\,\mathbf{j} + t\,\mathbf{k}$; from $t = 0$ to $t = 1$.

36. Find the curvature of the plane curves.

(a) $y = x^{3/2}$ (b) $y = \cos 2x$

37. Find the curvature of the plane curves.

(a) $x(t) = 2e^{-t}, y(t) = e^{-2t}$ (b) $\mathbf{r}(t) = \frac{1}{3}t^3\,\mathbf{i} + \frac{1}{2}t^2\,\mathbf{j}$

Interpret $\mathbf{r}(t)$ as the position of a moving object at time t. Find the curvature of the path and the tangential and normal components of acceleration.

38. $\mathbf{r}(t) = e^t\,\mathbf{i} + e^{-t}\,\mathbf{j} - t\sqrt{2}\,\mathbf{k}$

39. $\mathbf{r}(t) = \frac{4}{5}\cos t\,\mathbf{i} - \frac{3}{5}\cos t\,\mathbf{j} + (1 + \sin t)\,\mathbf{k}$

40. $\mathbf{r}(t) = \sinh t\,\mathbf{i} + \cosh t\,\mathbf{j} + t\,\mathbf{k}$

41. Find the domain and range of the function f.

(a) $f(x, y) = \ln(x^2 - y^2 - 1)$

(b) $f(x, y, z) = \sqrt{z - x^2 - y^2}$

42. Determine a function f whose value at (x, y) is:

(a) The volume of a circular cone of radius x and height y.

(b) The volume of a box whose length x is twice its width, and whose height is y.

(c) The cosine of the angle between the vectors $y\,\mathbf{i} + 2xy\,\mathbf{j}$ and $x\,\mathbf{i} + y\,\mathbf{j}$.

Identify the level curves $f(x,y) = c$ and sketch the curves corresponding to the given values of c.

43. $f(x,y) = \sqrt{x^2+y^2-4}$; $c = 0, \sqrt{5}$

44. $f(x,y) = \dfrac{y}{x^2} = c$; $c = -4, -1, 1, 4$

Identify the level surface $F(x,y,z) = c$ and sketch it.

45. $F(x,y,z) = 2x + y + 3z$; $c = 6$

46. $F(x,y,z) = 4x^2 + 9y^2 + 36z^2$; $c = 36$

Calculate the first-order partial derivatives

47. $z = x^2 \sin xy^2$

48. $f(x,y,z) = \dfrac{2xy}{x+y+z}$

49. $g(x,y,z) = \ln \sqrt{x^2+y^2+z^2}$

Calculate the indicated second-order partial derivatives of f.

50. $f(x,y) = x^3y^2 - 4xy^3 + 2x - y$; f_{xx}, f_{yx}.

51. $f(x,y,z) = 2x^2yz^3 + e^{xyz}$; f_{xx}, f_{zx}, f_{yz}.

52. The surface $z = \sqrt{20-2x^2-3y^2}$ is the top half of an ellipsoid centered at the origin.

(a) Find scalar parametric equations for the line l_1 tangent to the curve of intersection of the ellipsoid and the plane $x = 2$ at the point (2, 1, 3) on the surface.

(b) Find scalar parametric equations for the line l_2 tangent to the curve of intersection of the ellipsoid and the plane $y = 1$ at the point (2, 1, 3) on the surface.

(c) The tangent lines l_1 and l_2 determine a plane. Find an equation for this plane.

53. Find the gradient of the function f.

(a) $f(x,y) = 2x^2 - 4xy + y^3$ (b) $f(x,y) = \dfrac{xy}{x^2+y^2}$

54. Find the gradient of the function F

(a) $f(x,y,z) = \ln\sqrt{x^2+y^2+z^2}$

(b) $f(x,y,z) = x^2e^{-yz}\cos 2z$

Find the directional derivative at the point P in the direction indicated.

55. $f(x,y) = x^2 - 2xy$ at $(1,-2)$ in the direction of $\mathbf{a} = \mathbf{i} + 2\,\mathbf{j}$.

56. $f(x,y,z) = e^{x^2+y^2+z^2}$ at $(0,0,0)$ in the direction of the line $\dfrac{x}{2} = \dfrac{y}{-1} = \dfrac{z}{4}$.

57. Find the directional derivative of $f(x,y) = e^{2x}(\cos y - \sin y)$ at $(\frac{1}{2}, -\frac{1}{2}\pi)$ in the direction of the greatest increase of f.

58. Find the directional derivative of $F(x,y,z) = \sin(xyz)$ at $(\frac{1}{2}, \frac{1}{3}, \pi)$ in the direction of the greatest decrease of F.

Find an equation for the tangent plane and scalar parametric equations for the normal line to the surface at the point indicated.

59. $z = \sqrt{4-x^2-y^2}$ at $(1, -1, \sqrt{2})$.

60. $z^3 + xyz - 2 = 0$ at $(1,1,1)$.

61. $ye^{xy} + 2z^2 = 1$ at $(0,-1,1)$.

62. $f(x,y) = \ln(x^2+y^2)$ at $(1,-1,\ln 2)$.

Find the stationary points; determine the local maxima and minima and the saddle points.

63. $f(x,y) = x^2y - 2xy + 2y^2 - 15y - 2$

64. $f(x,y) = 3x^2 - 3xy^2 + y^3 + 3y^2$

65. $f(x,y) = (x-3)\ln xy$

Find the absolute maximum and minimum of f on the set D.

66. $f(x,y) = x^2 + y^2 - 2x + 2y + 2$; $D = \{(x,y){:}x^2+y^2 \le 4\}$.

67. $f(x,y) = 2x^2 - 4x + y^2 - 4y + 3$; D the closed triangular region bounded by the lines $x = 0$, $y = 3$, $y = x$.

68. Use differentials to calculate the approximate value.

(a) $e^{0.02}\sqrt{15.2+(1.01)^3}$ (b) $\sqrt[3]{64.5}\cos^2(28^\circ)$

69. Determine whether the given vector function is the gradient of a function f. If it is, find f.

(a) $\boldsymbol{R}(x,y) = (y^2e^{2x} + 4x + 4 - 2y)\,\mathbf{i} + (ye^{2x} - 2x^3 + 2y)\,\mathbf{j}$

(b) $\boldsymbol{S}(x,y) = \left(\dfrac{y}{x^2} + 4x^3 - 1 + 3y\sin 3x\right)\mathbf{i} + \left(3y^2 + 2 - \dfrac{1}{x} - \cos 3x\right)\mathbf{j}$

Find the extreme values of f subject to the side condition.

70. $f(x,y) = x^2y^2$; $\frac{1}{9}x^2 + \frac{1}{4}y^2 = 1$

71. $f(x,y,z) = xz + 2y$; $x^2 + y^2 + z^2 = 36$

72. Find the volume of the largest rectangular box that can be inscribed in the ellipsoid

$$4x^2 + 9y^2 + 36z^2 = 36$$

if the edges are parallel to the coordinate planes.

73. A closed rectangular box having a volume of 16 cubic feet will be constructed from three types of metal. The cost of the metal for the bottom is \$0. 50 per square foot, the metal for the sides costs \$0. 25 per square foot, and the cost for the top is \$0. 10 per square foot. Find the dimensions of the box that will minimize the cost of construction.

74. A metal silo in the shape of a right circular cylinder 22 feet high and 10 feet in diameter will be given a coat of paint 0.01 inches thick. The contractor for the job wants to estimate the number of gallons of paint that will be needed. Use differentials to obtain an estimate (there are 231 cubic inches in a gallon).

75. The temperature at a point $P(x,y)$ on the disc $x^2+y^2 \le 1$ is given by

$$T(x,y) = 2x^2 + y^2 - y.$$

Find the hottest and coldest points on the disc.

CHAPTER 16

DOUBLE AND TRIPLE INTEGRALS

We began with ordinary integrals

$$\int_a^b f(x)\,dx$$

which, with $b > a$, we can write as

$$\int_{[a,b]} f(x)\,dx.$$

Here we will study double integrals

$$\iint_{\Omega} f(x, y)\,dxdy$$

where Ω is a region in the xy-plane and, a little later, triple integrals

$$\iiint_{T} f(x, y, z)\,dxdydz$$

where T is a solid in three-dimensional space. Our first step is to introduce some new notation.

■ 16.1 MULTIPLE-SIGMA NOTATION

In an ordinary sequence $\{a_n\}$, each term a_i is indexed by a single integer. The sum of all the a_i from $i = 1$ to $i = m$ is then denoted by

$$\sum_{i=1}^{m} a_i.$$

When two indices are involved, say,

$$a_{ij} = 2^i 5^j, \qquad a_{ij} = \frac{2i}{5+j} \qquad \text{or} \qquad a_{ij} = (1+i)^j,$$

then we use double-sigma notation to denote the sum of all the doubly indexed terms. By

(16.1.1)
$$\sum_{i=1}^{m}\sum_{j=1}^{n} a_{ij}$$

we mean *the sum of all the* a_{ij} *where* i *ranges from* 1 *to* m *and* j *ranges from* 1 *to* n. For example,

$$\sum_{i=1}^{3}\sum_{j=1}^{2} 2^i 5^j = 2\cdot 5 + 2\cdot 5^2 + 2^2\cdot 5 + 2^2\cdot 5^2 + 2^3\cdot 5 + 2^3\cdot 5^2 = 420.$$

Since addition is associative and commutative, we can add the terms of (16.1.1) in any order we choose. Usually we set

(16.1.2)
$$\sum_{i=1}^{m}\sum_{j=1}^{n} a_{ij} = \sum_{i=1}^{m}\left(\sum_{j=1}^{n} a_{ij}\right).$$

We can expand the expression on the right by expanding first with respect to i and then with respect to j:

$$\begin{aligned}\sum_{i=1}^{m}\left(\sum_{j=1}^{n} a_{ij}\right) &= \sum_{j=1}^{n} a_{1j} + \sum_{j=1}^{n} a_{2j} + \cdots + \sum_{j=1}^{n} a_{mj}\\ &= (a_{11} + a_{12} + \cdots + a_{1n}) + (a_{21} + a_{22} + \cdots + a_{2n}) +\\ &\quad \cdots + (a_{m1} + a_{m2} + \cdots + a_{mn}),\end{aligned}$$

or we can expand first with respect to j and then with respect to i:

$$\begin{aligned}\sum_{i=1}^{m}\left(\sum_{j=1}^{n} a_{ij}\right) &= \sum_{i=1}^{m} (a_{i1} + a_{i2} + \cdots + a_{in})\\ &= (a_{11} + a_{12} + \cdots + a_{1n}) + (a_{21} + a_{22} + \cdots + a_{2n}) +\\ &\quad \cdots + (a_{m1} + a_{m2} + \cdots + a_{mn}).\end{aligned}$$

The results are the same.

For example, we can write

$$\begin{aligned}\sum_{i=1}^{3}\left(\sum_{j=1}^{2} a_{ij}\right) &= \sum_{j=1}^{2} a_{1j} + \sum_{j=1}^{2} a_{2j} + \sum_{j=1}^{2} a_{3j}\\ &= (a_{11} + a_{12}) + (a_{21} + a_{22}) + (a_{31} + a_{32}),\end{aligned}$$

or we can write

$$\sum_{i=1}^{3}\left(\sum_{j=1}^{2} a_{ij}\right) = \sum_{i=1}^{3}(a_{i1} + a_{i2}) = (a_{11} + a_{12}) + (a_{21} + a_{22}) + (a_{31} + a_{32}).$$

Since constants can be factored through single sums, they can also be factored through double sums; namely,

(16.1.3)
$$\sum_{i=1}^{m}\sum_{j=1}^{n} \alpha a_{ij} = \alpha \sum_{i=1}^{m}\sum_{j=1}^{n} a_{ij}.$$

Also,

(16.1.4)
$$\sum_{i=1}^{m}\sum_{j=1}^{n}(a_{ij} + b_{ij}) = \sum_{i=1}^{m}\sum_{j=1}^{n} a_{ij} + \sum_{i=1}^{m}\sum_{j=1}^{n} b_{ij}.$$

The easiest double sums to evaluate are those where each term a_{ij} appears as a product $b_i c_j$ in which each of the factors bears only one index. In that case, we can express the double sum as the product of two single sums:

(16.1.5)
$$\sum_{i=1}^{m}\sum_{j=1}^{n} b_i c_j = \left(\sum_{i=1}^{m} b_i\right)\left(\sum_{j=1}^{n} c_j\right).$$

PROOF Set

$$B = \sum_{i=1}^{m} b_i, \quad C = \sum_{j=1}^{n} c_j.$$

Then

$$\sum_{i=1}^{m}\sum_{j=1}^{n} b_i c_j = \sum_{i=1}^{m}\left(\sum_{j=1}^{n} b_i c_j\right) \overset{\dagger}{=} \sum_{i=1}^{m} b_i \left(\sum_{j=1}^{n} c_j\right) = \sum_{i=1}^{m} b_i C$$

$$= C\sum_{i=1}^{m} b_i = CB = BC. \quad \square$$

For example,

$$\sum_{i=1}^{3}\sum_{j=1}^{2} 2^i 5^j = \left(\sum_{i=1}^{3} 2^i\right)\left(\sum_{j=1}^{2} 5^j\right)$$

$$= (2 + 2^2 + 2^3)(5 + 5^2) = (14)(30) = 420.$$

† Since b_i is independent of j, it can be factored through the j-summation.

Triple-sigma notation is used when three indices are involved. The sum of all the a_{ijk} where i ranges from 1 to m, j from 1 to n, and k from 1 to q can be written

(16.1.6)
$$\sum_{i=1}^{m}\sum_{j=1}^{n}\sum_{k=1}^{q} a_{ijk}.$$

Multiple sums appear in the following sections. We introduced them here so as to avoid lengthy asides later.

EXERCISES 16.1

Evaluate the sum.

1. $\sum_{i=1}^{3}\sum_{j=1}^{3} 2^{i-1}3^{j+1}$.

2. $\sum_{i=1}^{4}\sum_{j=1}^{2} (1+i)^{j}$.

3. $\sum_{i=1}^{4}\sum_{j=1}^{3} (i^2+3i)(j-2)$.

4. $\sum_{i=1}^{3}\sum_{j=1}^{2}\sum_{k=1}^{3} \dfrac{2i}{k+j^2}$.

For Exercises 5–16, let

$$P_1 = \{x_0, x_1, \ldots, x_m\} \text{ be a partition of } [a_1, a_2],$$
$$P_2 = \{y_0, y_1, \ldots, y_n\} \text{ be a partition of } [b_1, b_2],$$
$$P_3 = \{z_0, z_1, \ldots, z_q\} \text{ be a partition of } [c_1, c_2],$$

and let

$$\Delta x_i = x_i - x_{i-1}, \quad \Delta y_j = y_j - y_{j-1}, \quad \Delta z_k = z_k - z_{k-1}.$$

Evaluate the sum.

5. $\sum_{i=1}^{m} \Delta x_i$.

6. $\sum_{j=1}^{n} \Delta y_j$.

7. $\sum_{i=1}^{m}\sum_{j=1}^{n} \Delta x_i \Delta y_j$.

8. $\sum_{j=1}^{n}\sum_{k=1}^{q} \Delta y_j \Delta z_k$.

9. $\sum_{i=1}^{m} (x_i + x_{i-1})\Delta x_i$.

10. $\sum_{j=1}^{n} \frac{1}{2}(y_j^2 + y_j y_{j-1} + y_{j-1}^2)\Delta y_j$.

11. $\sum_{i=1}^{m}\sum_{j=1}^{n} (x_i + x_{i-1})\Delta x_i \Delta y_j$.

12. $\sum_{i=1}^{m}\sum_{j=1}^{n} (y_j + y_{j-1})\Delta x_i \Delta y_j$.

13. $\sum_{i=1}^{m}\sum_{j=1}^{n} (2\Delta x_i - 3\Delta y_j)$.

14. $\sum_{i=1}^{m}\sum_{j=1}^{n} (3\Delta x_i - 2\Delta y_j)$.

15. $\sum_{i=1}^{m}\sum_{j=1}^{n}\sum_{k=1}^{q} \Delta x_i \Delta y_j \Delta z_k$.

16. $\sum_{i=1}^{m}\sum_{j=1}^{n}\sum_{k=1}^{q} (x_i + x_{i-1})\Delta x_i \Delta y_j \Delta z_k$.

17. Evaluate

$$\sum_{i=1}^{n}\sum_{j=1}^{n}\sum_{k=1}^{n} \delta_{ijk}\, a_{ijk} \quad \text{where} \quad \delta_{ijk} = \begin{cases} 1, & \text{if } i = j = k \\ 0, & \text{otherwise.} \end{cases}$$

18. Show that any sum written in double-sigma notation can be expressed in ordinary (that is, in single) sigma notation.

■ 16.2 DOUBLE INTEGRALS

The Double Integral over a Rectangle

We start with a function f continuous on a rectangle.

$$R: \quad a \le x \le b, \quad c \le y \le d.$$

See Figure 16.2.1. Our object is to define the double integral of f over R:

$$\iint_R f(x, y)\, dxdy.$$

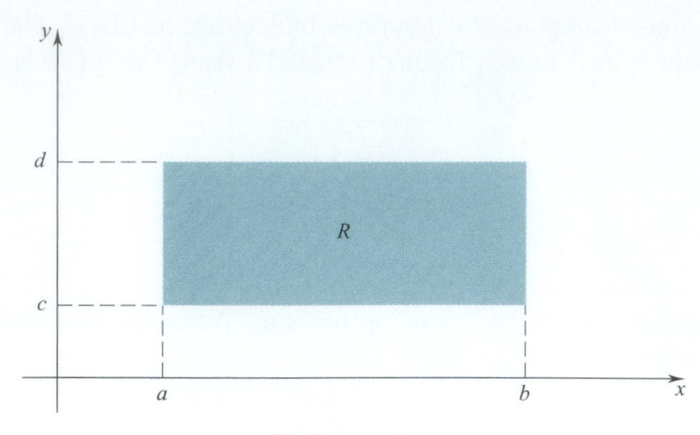

Figure 16.2.1

To define the integral

$$\int_a^b f(x)\,dx,$$

we introduced some auxiliary notions, namely: partition P of $[a, b]$, upper sum $U_f(P)$, and lower sum $L_f(P)$. We were then able to define

$$\int_a^b f(x)\,dx$$

as the unique number I that satisfies the inequality

$$L_f(P) \leq I \leq U_f(P) \quad \text{for all partitions } P \text{ of } [a, b].$$

We will follow exactly the same procedure to define the double integral

$$\iint\limits_R f(x, y)\,dxdy.$$

First we explain what we mean by a partition of the rectangle R. To do this, we begin with a partition

$$P_1 = \{x_0, x_1, \ldots, x_m\} \quad \text{of} \quad [a, b]$$

and a partition

$$P_2 = \{y_0, y_1, \ldots, y_n\} \quad \text{of} \quad [c, d].$$

The set

$$P = P_1 \times P_2 = \{(x_i, y_j) : x_i \in P_1, y_j \in P_2\}\dagger$$

is called a *partition of R* (see Figure 16.2.2); P consists of all the grid points (x_i, y_j).

† $P_1 \times P_2$ is the Cartesian product of P_1 and P_2; if A and B are sets, then the *Cartesian product* of A and B is the set $A \times B = \{(a, b) : a \in A \text{ and } b \in B\}$.

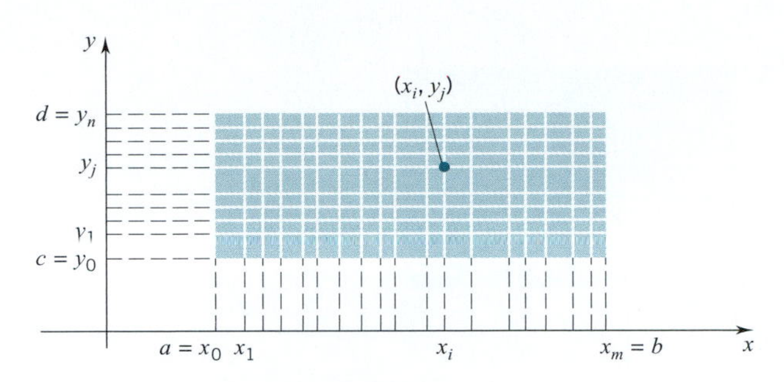

Figure 16.2.2

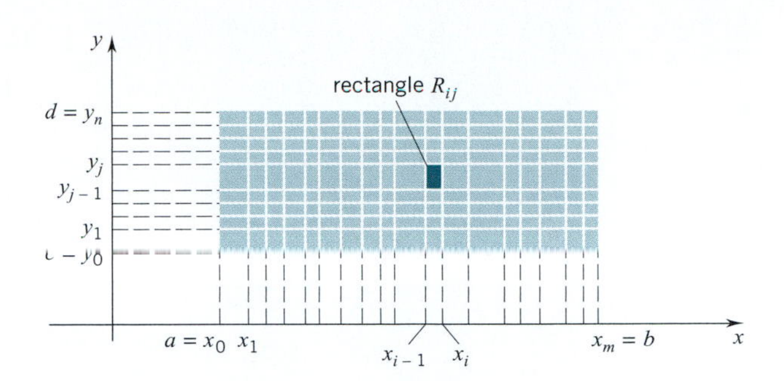

Figure 16.2.3

Using the partition P, we break up R into $m \times n$ nonoverlapping rectangles

$$R_{ij}: \quad x_{i-1} \le x \le x_i, \quad y_{j-1} \le y \le y_j, \qquad \text{(Figure 16.2.3)}$$

where $1 \le i \le m, 1 \le j \le n$. On each rectangle R_{ij}, the function f takes on a maximum value M_{ij} and a minimum value m_{ij}. We know this because f is continuous and R_{ij} is closed and bounded (Section 15.6). The sum of all the products

$$M_{ij}(\text{ area of } R_{ij}) = M_{ij}(x_i - x_{i-1})(y_j - y_{j-1}) = M_{ij}\,\Delta x_i \Delta y_j$$

is called the P *upper sum* for f:

(16.2.1)
$$U_f(P) = \sum_{i=1}^{m}\sum_{j=1}^{n} M_{ij}(\text{ area of } R_{ij}) = \sum_{i=1}^{m}\sum_{j=1}^{n} M_{ij}\,\Delta x_i\,\Delta y_j.$$

The sum of all the products

$$m_{ij}(\text{ area of } R_{ij}) = m_{ij}(x_i - x_{i-1})(y_j - y_{j-1}) = m_{ij}\,\Delta x_i\,\Delta y_j$$

is called the P *lower sum* for f:

(16.2.2)
$$L_f(P) = \sum_{i=1}^{m}\sum_{j=1}^{n} m_{ij}(\text{area of } R_{ij}) = \sum_{i=1}^{m}\sum_{j=1}^{n} m_{ij}\,\Delta x_i\,\Delta y_j.$$

Example 1 Consider the function $f(x, y) = x + y - 2$ on the rectangle

$$R: \quad 1 \le x \le 4, \quad 1 \le y \le 3.$$

As a partition of [1, 4] take

$$P_1 = \{1, 2, 3, 4\},$$

and as a partition of [1, 3] take

$$P_2 = \{1, \tfrac{3}{2}, 3\}.$$

The partition $P = P_1 \times P_2$ then breaks up the initial rectangle into the six rectangles marked in Figure 16.2.4. On each rectangle R_{ij}, the function f takes on its maximum value M_{ij} at the point (x_i, y_j), the corner farthest from the origin:

$$M_{ij} = f(x_i, y_j) = x_i + y_j - 2.$$

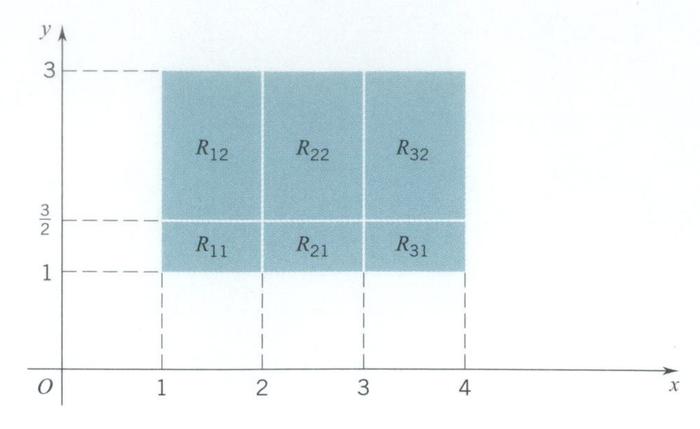

Figure 16.2.4

Thus
$$\begin{aligned}U_f(P) &= M_{11}(\text{area of } R_{11}) + M_{12}(\text{area of } R_{12}) + M_{21}(\text{area of } R_{21})\\ &\quad + M_{22}(\text{area of } R_{22}) + M_{31}(\text{area of } R_{31}) + M_{32}(\text{area of } R_{32})\\ &= \tfrac{3}{2}(\tfrac{1}{2}) + 3(\tfrac{3}{2}) + \tfrac{5}{2}(\tfrac{1}{2}) + 4(\tfrac{3}{2}) + \tfrac{7}{2}(\tfrac{1}{2}) + 5(\tfrac{3}{2}) = \tfrac{87}{4}.\end{aligned}$$

On each rectangle R_{ij}, f takes on its minimum value m_{ij} at the point (x_{i-1}, y_{j-1}), the corner closest to the origin:

$$m_{ij} = f(x_{i-1}, y_{j-1}) = x_{i-1} + y_{j-1} - 2.$$

Thus
$$\begin{aligned}L_f(P) &= m_{11}(\text{ area of } R_{11}) + m_{12}(\text{ area of } R_{12}) + m_{21}(\text{ area of } R_{21})\\ &\quad + m_{22}(\text{ area of } R_{22}) + m_{31}(\text{ area of } R_{31}) + m_{32}(\text{ area of } R_{32})\\ &= 0(\tfrac{1}{2}) + \tfrac{1}{2}(\tfrac{3}{2}) + 1(\tfrac{1}{2}) + \tfrac{3}{2}(\tfrac{3}{2}) + 2(\tfrac{1}{2}) + \tfrac{5}{2}(\tfrac{3}{2}) = \tfrac{33}{4}. \quad \square\end{aligned}$$

We return now to the general situation. As in the one-variable case, it can be shown that if f is continuous, then there exists one and only one number I that satisfies the inequality

$$L_f(P) \leq I \leq U_f(P) \quad \text{for all partitions } P \text{ of } R.$$

DEFINITION 16.2.3 THE DOUBLE INTEGRAL OVER A RECTANGLE R

Let f be continuous on a closed rectangle R. The unique number I that satisfies the inequality

$$L_f(P) \leq I \leq U_f(P) \quad \text{for all partitions } P \text{ of } R$$

is called the *double integral* of f over R, and is denoted by

$$\iint_R f(x, y)\, dxdy.\dagger$$

† The double integral $\iint_R f(x, y)\, dxdy$ can also be written $\iint_R f(x, y)\, dA$.

The Double Integral as a Volume

If f is continuous and nonnegative on the rectangle R, the equation

$$z = f(x, y)$$

represents a surface that lies above R. (See Figure 16.2.5.) In this case the double integral

$$\iint_R f(x, y)\, dxdy$$

gives the volume of the solid that is bounded below by R and bounded above by the surface $z = f(x, y)$.

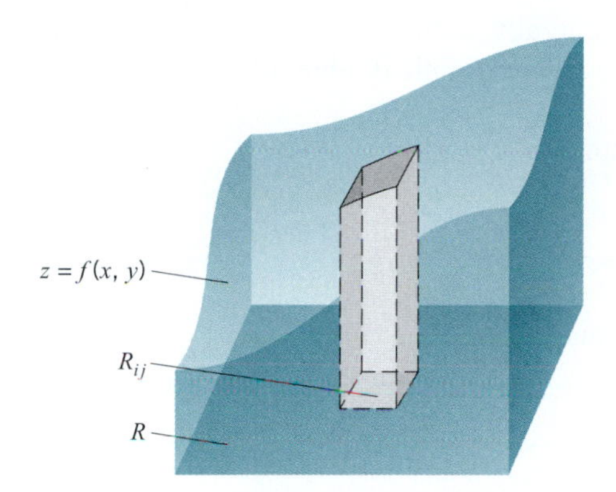

Figure 16.2.5

To see this, consider a partition P of R. P breaks up R into subrectangles R_{ij} and thus the entire solid T into parts T_{ij}. Since T_{ij} contains a rectangular solid with base R_{ij} and height m_{ij} (Figure 16.2.6), we must have

$$m_{ij}(\text{area of } R_{ij}) \leq \text{volume of } T_{ij}.$$

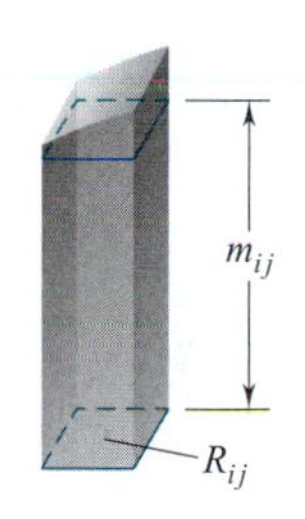

Figure 16.2.6

Since T_{ij} is contained in a rectangular solid with base R_{ij} and height M_{ij} (Figure 16.2.7), we must have

$$\text{volume of } T_{ij} \leq M_{ij}(\text{area of } R_{ij}).$$

In short, for each pair of indices i and j, we must have

$$m_{ij}(\text{area of } R_{ij}) \leq \text{ volume of } T_{ij} \leq M_{ij}(\text{area of } R_{ij}).$$

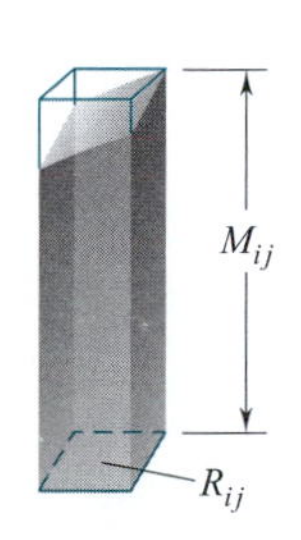

Figure 16.2.7

Adding up these inequalities, we can conclude that

$$L_f(P) \leq \text{ volume of } T \leq U_f(P).$$

Since P is arbitrary, the volume of T must be the double integral:

(16.2.4)
$$\boxed{\text{volume of } T = \iint_R f(x, y)\, dxdy.}$$

The double integral

$$\iint_R 1 \; dxdy = \iint_R dxdy$$

gives the volume of a solid of constant height 1 erected over R. In square units this is just the area of R:

(16.2.5)

$$\text{area of } R = \iint_R dxdy.$$

Some Computations

Double integrals are generally computed by techniques that we will take up later. It is possible, however, to evaluate simple double integrals directly from the definition.

Example 2 Evaluate

$$\iint_R \alpha \; dxdy$$

where α is a constant and R is the rectangle

$$R: \quad a \le x \le b, \quad c \le y \le d.$$

SOLUTION Here $f(x,y) = \alpha$ for all $(x,y) \in R$.

We begin with $P_1 = \{x_0, x_1, \ldots, x_m\}$ as an arbitrary partition of $[a,b]$ and $P_2 = \{y_0, y_1, \ldots, y_n\}$ as an arbitrary partition of $[c,d]$. This gives

$$P = P_1 \times P_2 = \{(x_i, y_j) : x_i \in P_1,\; y_j \in P_2\}$$

as an arbitrary partition of R. On each rectangle R_{ij}, f has constant value α. Therefore we have $M_{ij} = \alpha$ and $m_{ij} = \alpha$ throughout. Thus

$$U_f(P) = \sum_{i=1}^{m}\sum_{j=1}^{n} \alpha \Delta x_i \, \Delta y_j = \alpha \left(\sum_{i=1}^{m} \Delta x_i\right)\left(\sum_{j=1}^{n} \Delta y_j\right) = \alpha(b-a)(d-c).$$

Similarly, $$L_f(P) = \alpha(b-a)(d-c).$$

The inequality $$L_f(P) \le I \le U_f(P) \qquad \text{for all } P$$

forces $$\alpha(b-a)(d-c) \le I \le \alpha(b-a)(d-c).$$

The only number I that can satisfy this inequality is

$$I = \alpha(b-a)(d-c).$$

Therefore $$\iint_R f(x,y) \; dxdy = \alpha(b-a)(d-c). \quad \square$$

Remark If $\alpha > 0$,

$$\iint_R \alpha \, dxdy = \alpha(b-a)(d-c)$$

gives the volume of the rectangular solid of constant height α erected over the rectangle R (Figure 16.2.8). ❑

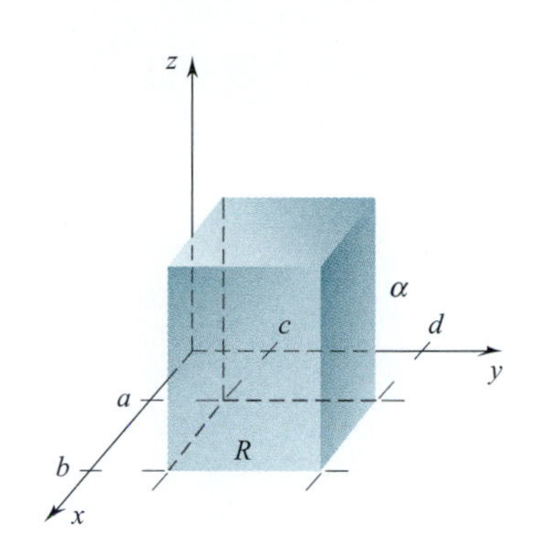

Figure 16.2.8

Example 3 Evaluate

$$\iint_R (x+y-2)\, dxdy,$$

where R is the rectangle: $1 \le x \le 4$, $1 \le y \le 3$. (This is a continuation of Example 1.)

SOLUTION With $P_1 = \{x_0, x_1, \ldots, x_m\}$ as an arbitrary partition of $[1,4]$ and $P_2 = \{y_0, y_1, \ldots, y_n\}$ as an arbitrary partition of $[1,3]$, we have

$$P_1 \times P_2 = \{(x_i, y_j) : x_i \in P_1, y_j \in P_2\}$$

as an arbitrary partition of R. On each rectangle $R_{ij} : x_{i-1} \le x \le x_i, y_{j-1} \le y \le y_j$, the function

$$f(x,y) = x + y - 2$$

has a maximum $M_{ij} = x_i + y_j - 2$ and a minimum $m_{ij} = x_{i-1} + y_{j-1} - 2$. Thus

$$L_f(P) = \sum_{i=1}^{m} \sum_{j=1}^{n} (x_{i-1} + y_{j-1} - 2)\Delta x_i \, \Delta y_j$$

$$U_f(P) = \sum_{i=1}^{m} \sum_{j=1}^{n} (x_i + y_j - 2)\, \Delta x_i \, \Delta y_j.$$

Now, for each pair of indices i and j

$$x_{i-1} + y_{j-1} - 2 \le \tfrac{1}{2}(x_i + x_{i-1}) + \tfrac{1}{2}(y_j + y_{j-1}) - 2 \le x_i + y_j - 2. \qquad \text{(explain)}$$

Adding up these inequalities, we have

$$L_f(P) \le \sum_{i=1}^{m} \sum_{j=1}^{n} [\tfrac{1}{2}(x_i + x_{i-1}) + \tfrac{1}{2}(y_j + y_{j-1}) - 2]\, \Delta x_i \, \Delta y_j \le U_f(P).$$

The double sum in the middle can be written

$$\sum_{i=1}^{m} \sum_{j=1}^{n} \tfrac{1}{2}(x_i + x_{i-1})\, \Delta x_i \, \Delta y_j + \sum_{i=1}^{m} \sum_{j=1}^{n} \tfrac{1}{2}(y_j + y_{j-1})\, \Delta x_i \, \Delta y_j - \sum_{i=1}^{m} \sum_{j=1}^{n} 2\Delta x_i \, \Delta y_j.$$

The first double sum reduces to

$$\begin{aligned} \sum_{i=1}^{m} \sum_{j=1}^{n} \tfrac{1}{2}(x_i^2 - x_{i-1}^2)\Delta y_j &= \tfrac{1}{2}\left(\sum_{i=1}^{m}(x_i^2 - x_{i-1}^2)\right)\left(\sum_{j=1}^{n}\Delta y_j\right) \\ &= \tfrac{1}{2}(16-1)(3-1) = 15. \end{aligned}$$

The second double sum reduces to

$$\sum_{i=1}^{m}\sum_{j=1}^{n}\tfrac{1}{2}\Delta x_i(y_j^2-y_{j-1}^2)=\tfrac{1}{2}\left(\sum_{i=1}^{m}\Delta x_i\right)\left(\sum_{j=1}^{n}(y_j^2-y_{j-1}^2)\right)$$
$$=\tfrac{1}{2}(4-1)(9-1)=12.$$

The third double sum reduces to

$$-\sum_{i=1}^{m}\sum_{j=1}^{n}2\Delta x_i\,\Delta y_j=-2\left(\sum_{i=1}^{m}\Delta x_i\right)\left(\sum_{j=1}^{n}\Delta y_j\right)=-2(4-1)(3-1)=-12.$$

The sum of these numbers, $15+12-12=15$, satisfies the inequality

$$L_f(P)\le 15\le U_f(P)\qquad\text{for arbitrary } P.$$

Thus,
$$\iint_R (x+y-2)\;dxdy=15.\quad\square$$

Remark Since $f(x,y)=x+y-2\ge 0$ on the rectangle R, the double integral gives the volume of the prism bounded above by the plane $z=x+y-2$ and below by R. See Figure 16.2.9. □

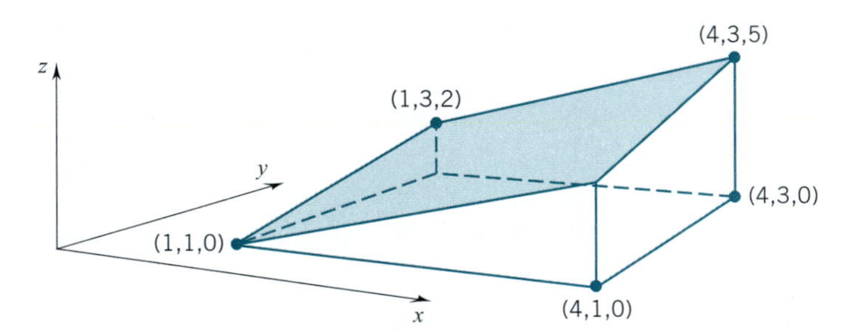

Figure 16.2.9

The Double Integral over a Region

We start with a closed and bounded set Ω in the xy-plane such as that depicted in Figure 16.2.10. We assume that Ω is a *basic region*, that is, we assume that Ω is a connected set (see Section 15.3) the total boundary of which consists of a finite number of continuous arcs of the form $y=\phi(x), x=\psi(y)$. See, for example, Figure 16.2.11.

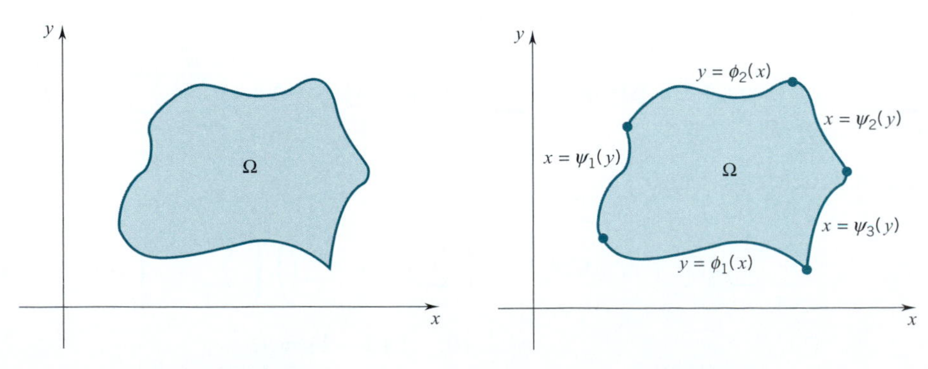

Figure 16.2.10

Figure 16.2.11

Now let f be a function continuous on Ω. We want to define the double integral

$$\iint_{\Omega} f(x,y)\ dxdy.$$

To do this, we enclose Ω by a rectangle R with sides parallel to the coordinate axes as in Figure 16.2.12. We extend f to all of R by setting f equal to 0 outside Ω. This extended function, which we continue to call f, is bounded on R, and it is continuous on all of R except possibly at the boundary of Ω. In spite of these possible discontinuities, it can be shown that f is still integrable on R; that is, there still exists a unique number I such that

$$L_f(P) \leq I \leq U_f(P) \qquad \text{for all partitions } P \text{ of } R.$$

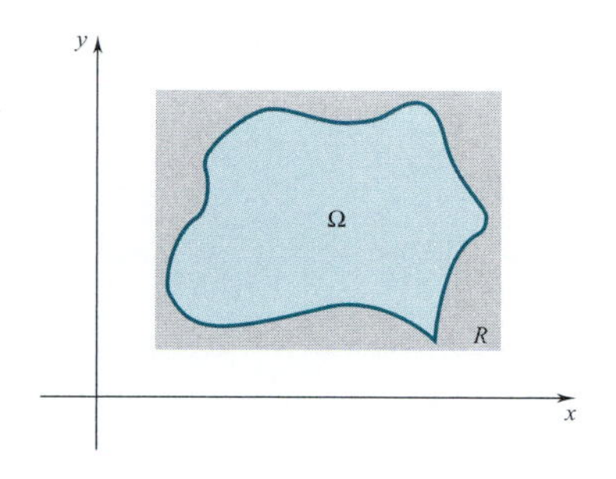

Figure 16.2.12

(We will not attempt to prove this.) This number I is by definition the double integral

$$\iint_{R} f(x,y)\ dxdy.$$

As you have probably guessed by now, we define the double integral over Ω by setting

(16.2.6)
$$\iint_{\Omega} f(x,y)\ dxdy = \iint_{R} f(x,y)\ dxdy.$$

If f is continuous and nonnegative over Ω, the extended f is nonnegative on all of R (Figure 16.2.13). The double integral gives the volume of the solid bounded above by the surface $z = f(x,y)$ and bounded below by the rectangle R. But since the surface has height 0 outside of Ω, the volume outside Ω is 0. It follows that

$$\iint_{\Omega} f(x,y)\ dxdy$$

gives the volume of the solid T bounded above by $z = f(x,y)$ and bounded below by Ω:

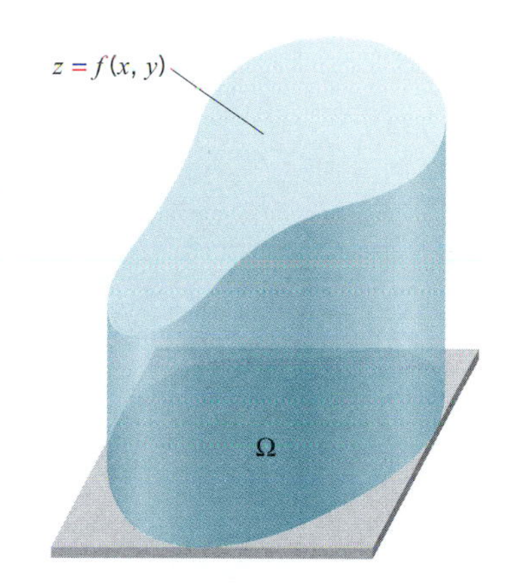

Figure 16.2.13

(16.2.7)
$$\text{volume of } T = \iint_{\Omega} f(x,y)\ dxdy.$$

The double integral

$$\iint_{\Omega} 1\ dxdy = \iint_{\Omega} dxdy$$

gives the volume of a solid of constant height 1 erected over Ω. In square units this is the area of Ω:

(16.2.8)
$$\text{area of } \Omega = \iint_{\Omega} dxdy.$$

Below we list four elementary properties of the double integral. They are all analogous to what you saw in the one-variable case. As specified above, the Ω referred to is a basic region. The functions f and g are assumed to be continuous on Ω.

I. Linearity: The double integral of a linear combination is the linear combination of the double integrals:

$$\iint_{\Omega} [\alpha f(x,y) + \beta g(x,y)]\ dxdy = \alpha \iint_{\Omega} f(x,y)\ dxdy + \beta \iint_{\Omega} g(x,y)\ dxdy.$$

II. Order: The double integral preserves order:

$$\text{if } f \geq 0 \text{ on } \Omega, \text{ then } \iint_{\Omega} f(x,y)\ dxdy \geq 0;$$

$$\text{if } f \leq g \text{ on } \Omega, \text{ then } \iint_{\Omega} f(x,y)\ dxdy \leq \iint_{\Omega} g(x,y)\ dxdy.$$

III. Additivity: If Ω is broken up into a finite number of nonoverlapping basic regions $\Omega_1, \ldots, \Omega_n$, then

$$\iint_{\Omega} f(x,y)\ dxdy = \iint_{\Omega_1} f(x,y)\ dxdy + \cdots + \iint_{\Omega_n} f(x,y)\ dxdy.$$

See, for example, Figure 16.2.14.

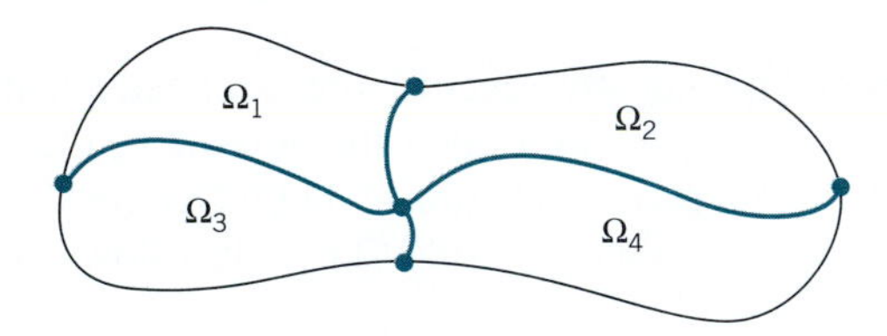

$$\iint_{\Omega} f(x,y)\ dxdy = \iint_{\Omega_1} f(x,y)\ dxdy + \iint_{\Omega_2} f(x,y)\ dxdy + \iint_{\Omega_3} f(x,y)\ dxdy + \iint_{\Omega_4} f(x,y)\ dxdy$$

Figure 16.2.14

IV. Mean-value condition: There is a point (x_0, y_0) in Ω for which

$$\iint_{\Omega} f(x,y)\ dxdy = f(x_0, y_0) \cdot (\text{ area of } \Omega).$$

We call $f(x_0, y_0)$ *the average value of f on Ω.*

This notion of average given in Property IV enables us to write

(16.2.9)
$$\iint_{\Omega} f(x,y)\ dxdy = \begin{pmatrix} \text{the average value} \\ \text{of } f \text{ on } \Omega \end{pmatrix} \cdot (\text{area of } \Omega).$$

This is a powerful, intuitive way of viewing the double integral. We will capitalize on it as we go on.

THEOREM 16.2.10 MEAN-VALUE THEOREM FOR DOUBLE INTEGRALS

Let f and g be functions continuous on the basic region Ω. If g is nonnegative on Ω, then there exists a point (x_0, y_0) in Ω for which

$$\iint_{\Omega} f(x,y)\, g(x,y)\, dxdy - f(x_0, y_0) \iint_{\Omega} g(x,y)\, dxdy.\dagger$$

We call $f(x_0, y_0)$ the *g-weighted average of f on Ω.*

PROOF Since f is continuous on Ω, and Ω is closed and bounded, we know that f takes on a minimum value m and a maximum value M. Since g is nonnegative on Ω,

$$mg(x,y) \le f(x,y)\, g(x,y) \le Mg(x,y) \quad \text{for all } (x,y) \text{ in } \Omega.$$

Therefore (by Property II)

$$\iint_{\Omega} m\, g(x,y)\, dxdy \le \iint_{\Omega} f(x,y)\, g(x,y)\, dxdy \le \iint_{\Omega} M\, g(x,y)\, dxdy,$$

and (by Property I)

$$(*) \qquad m \iint_{\Omega} g(x,y)\, dxdy \le \iint_{\Omega} f(x,y)\, g(x,y)\, dxdy \le M \iint_{\Omega} g(x,y)\, dxdy.$$

We know that $\iint_{\Omega} g(x,y)\, dxdy \ge 0$ (again, by Property II). If $\iint_{\Omega} g(x,y)\, dxdy = 0$, then by $(*)$ we have $\iint_{\Omega} f(x,y)\, g(x,y)\, dxdy = 0$ and the theorem holds for all choices of (x_0, y_0) in Ω. If $\iint_{\Omega} g(x,y)\, dxdy > 0$, then

$$m \le \frac{\iint_{\Omega} f(x,y)\, g(x,y)\, dxdy}{\iint_{\Omega} g(x,y)\, dxdy} \le M,$$

and, by the intermediate-value theorem (given in the supplement to Section 15.3), there exists (x_0, y_0) in Ω for which

$$f(x_0, y_0) = \frac{\iint_{\Omega} f(x,y)\, g(x,y)\, dxdy}{\iint_{\Omega} g(x,y)\, dxdy}.$$

Obviously, then,

$$f(x_0, y_0) \iint_{\Omega} g(x,y)\, dxdy = \iint_{\Omega} f(x,y)\, g(x,y)\, dxdy. \quad \square$$

† Property IV is this equation with g constantly 1.

EXERCISES 16.2

For Exercises 1–3, take

$$f(x,y) = x + 2y \quad \text{on} \quad R: 0 \le x \le 2. \quad 0 \le y \le 1.$$

and let P be the partition $P = P_1 \times P_2$.

1. Find $L_f(P)$ and $U_f(P)$ given that $P_1 = \{0, 1, \frac{3}{2}, 2\}$ and $P_2 = \{0, \frac{1}{2}, 1\}$.

2. Find $L_f(P)$ and $U_f(P)$ given that $P_1 = \{0, \frac{1}{2}, 1, \frac{3}{2}, 2\}$ and $P_2 = \{0, \frac{1}{4}, \frac{1}{2}, \frac{3}{4}, 1\}$.

3. (a) Find $L_f(P)$ and $U_f(P)$ given that

$$P_1 = \{x_0, x_1, \ldots, x_m\}$$

is an arbitrary partition of $[0, 2]$, and

$$P_2 = \{y_0, y_1, \ldots, y_n\}$$

is an arbitrary partition of $[0, 1]$.

(b) Use your answer to part (a) to evaluate the double integral

$$\iint_R (x + 2y)\, dxdy,$$

and give a geometric interpretation to your answer.

For Exercises 4–6, take

$$f(x,y) = x - y \quad \text{on} \quad R: 0 \le x \le 1, \quad 0 \le y \le 1.$$

and let P be the partition $P = P_1 \times P_2$.

4. Find $L_f(P)$ and $U_f(P)$ given that $P_1 = \{0, \frac{1}{2}, \frac{3}{4}, 1\}$ and $P_2 = \{0, \frac{1}{2}, 1\}$.

5. Find $L_f(P)$ and $U_f(P)$ given that $P_1 = \{0, \frac{1}{4}, \frac{1}{2}, \frac{3}{4}, 1\}$ and $P_2 = \{0, \frac{1}{3}, \frac{2}{3}, 1\}$.

6. (a) Find $L_f(P)$ and $U_f(P)$ given that

$$P_1 = \{x_0, x_1, \ldots, x_m\} \text{ and } P_2 = \{y_0, y_1, \ldots, y_n\}$$

are arbitrary partitions of $[0, 1]$.

(b) Use your answer to part (a) to evaluate the double integral

$$\iint_R (x - y)\, dxdy.$$

For Exercises 7–9, take $R: 0 \le x \le b, \quad 0 \le y \le d$ and let $P = P_1 \times P_2$, where

$$P_1 = \{x_0, x_1, \ldots, x_m\} \text{ is an arbitrary partition of } [0, b],$$
$$P_2 = \{y_0, y_1, \ldots, y_n\} \text{ is an arbitrary partition of } [0, d].$$

7. (a) Find $L_f(P)$ and $U_f(P)$ for $f(x,y) = 4xy$.

(b) Calculate

$$\iint_R 4xy\, dxdy.$$

HINT: $4x_{i-1}y_{j-1} \le (x_i + x_{i-1})(y_j + y_{j-1}) \le 4x_iy_j$.

8. (a) Find $L_f(P)$ and $U_f(P)$ for $f(x,y) = 3(x^2 + y^2)$.

(b) Calculate

$$\iint_R 3(x^2 + y^2)\, dxdy.$$

HINT: If $0 \le s \le t$, then $3s^2 \le t^2 + ts + s^2 \le 3t^2$.

9. (a) Find $L_f(P)$ and $U_f(P)$ for $f(x,y) = 3(x^2 - y^2)$.

(b) Calculate

$$\iint_R 3(x^2 - y^2)\, dxdy.$$

10. Let $f = f(x,y)$ be continuous on the rectangle R: $a \le x \le b, c \le y \le d$. Suppose that $L_f(P) = U_f(P)$ for some partition P of R. What can you conclude about f? What is

$$\iint_R f(x,y)\, dxdy?$$

11. Let $\phi = \phi(x)$ be continuous and nonnegative on the interval $[a, b]$, and set

$$\Omega = \{(x,y) : a \le x \le b, 0 \le y \le \phi(x)\}.$$

Compare

$$\iint_\Omega dxdy \quad \text{to} \quad \int_a^b \phi(x)\, dx.$$

12. Begin with a function f that is continuous on a closed bounded region Ω. Now surround Ω by a rectangle R as in Figure 16.2.12 and extend f to all of R by defining f to be 0 outside of Ω. Explain how the extended f can fail to be continuous on the boundary of Ω although the original function f, being continuous on all of Ω, was continuous on the boundary of Ω.

13. Suppose that f is continuous on a disc Ω centered at (x_0, y_0) and assume that

$$\iint_R f(x,y)\, dxdy = 0$$

for every rectangle R contained in Ω. Show that $f(x_0, y_0) = 0$.

14. Calculate the average value of $f(x,y) = x + 2y$ on the rectangle R: $0 \le x \le 2$, $0 \le y \le 1$. HINT: See Exercise 3.

15. Calculate the average value of $f(x,y) = 4xy$ on the rectangle $R: 0 \le x \le 2$, $0 \le y \le 3$. HINT: See Exercise 7.

16. Calculate the average value of $f(x,y) = x^2 + y^2$ on the rectangle R: $0 \le x \le b, 0 \le y \le d$. HINT: See Exercise 8.

17. Let f be continuous on a closed bounded region Ω and let (x_0, y_0) be a point in the interior of Ω. Let D_r be a closed disc with center (x_0, y_0) and radius r. Show that

$$\lim_{r\to 0} \frac{1}{\pi r^2} \iint_{D_r} f(x,y)\ dxdy = f(x_0, y_0).$$

18. Let $f(x,y) = \sin(x+y)$ on $R: 0 \le x \le 1$, $0 \le y \le 1$. Show that

$$0 \le \iint_R \sin(x+y)\ dxdy \le 1.$$

Sometimes it is possible to evaluate an integral by identifying it as the volume of an elementary solid that is known from geometry. Use this approach to evaluate the integrals in Exercises 19–21.

19. $\iint_\Omega \sqrt{4 - x^2 - y^2}\ dxdy$ where Ω is the quarter disk $x^2 + y^2 \le 4, x \ge 0, y \ge 0$.

20. $\iint_\Omega 8 - 4\sqrt{x^2 + y^2}\ dxdy$ where Ω is the disk $x^2 + y^2 \le 4$.

21. $\iint_\Omega (6 - 2x - 3y)\ dxdy$ where Ω is the triangular region bounded by the coordinate axes and the line $2x + 3y = 6$.

C▶22. Let $f(x,y) = 3y^2 - 2x$ on the rectangle R: $2 \le x \le 5$, $1 \le y \le 3$. Let P_1 be a regular partition of $[2,5]$ with $n = 100$ subintervals, let P_2 be a regular partition of $[1,3]$ with $m = 200$ subintervals, and let $P = P_1 \times P_2$.

(a) Use a CAS to find $L_f(P)$ and $U_f(P)$.

(b) Find $L_f(P)$ and $U_f(P)$ for several values of $n > 100$, $m > 200$.

(c) Estimate $\iint_R f(x,y)\ dxdy$.

■ 16.3 THE EVALUATION OF DOUBLE INTEGRALS BY REPEATED INTEGRALS

The Reduction Formulas

If an integral

$$\int_a^b f(x)\ dx$$

proves difficult to evaluate, it is not because of the interval $[a, b]$ but because of the integrand f. Difficulty in evaluating a double integral

$$\iint_\Omega f(x,y)\ dxdy$$

can come from two sources: from the integrand f or from the base region Ω. Even such a simple-looking integral as $\iint_\Omega 1\, dxdy$ is difficult to evaluate if Ω is complicated.

In this section we introduce a technique for evaluating double integrals of continuous functions over regions of Type I or Type II as depicted in Figure 16.3.1. In each case the region Ω is a basic region and so we know that the double integral exists. The fundamental idea of this section is that double integrals over such regions can each be reduced to a pair of ordinary integrals.

Type I Region. The *projection* of Ω onto the x-axis is a closed interval $[a, b]$ and Ω consists of all points (x, y) with

$$a \le x \le b \qquad \text{and} \qquad \phi_1(x) \le y \le \phi_2(x).$$

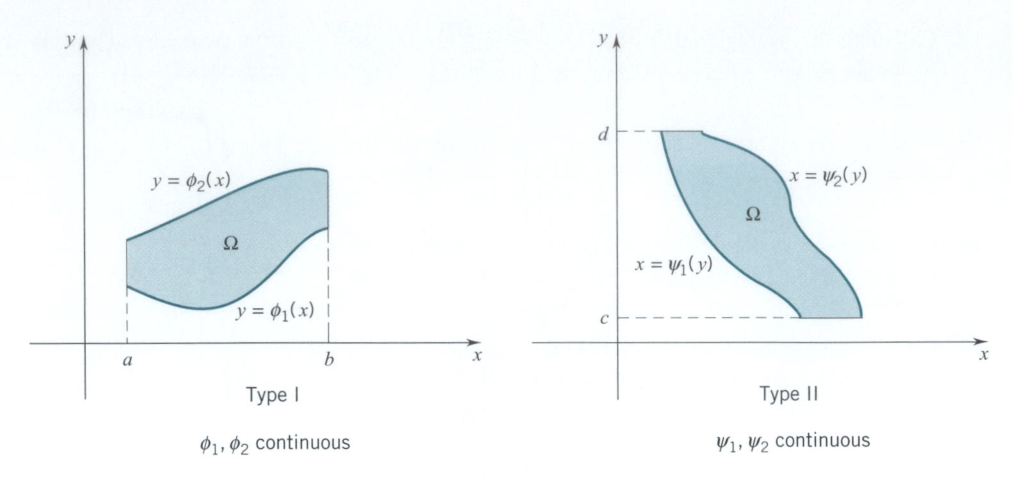

Figure 16.3.1

Then

(16.3.1)
$$\iint_{\Omega} f(x,y)\ dxdy = \int_a^b \left(\int_{\phi_1(x)}^{\phi_2(x)} f(x,y)\,dy\right)\,dx.$$

Here we first calculate

$$\int_{\phi_1(x)}^{\phi_2(x)} f(x,y)\,dy$$

by integrating $f(x,y)$ with respect to y from $y = \phi_1(x)$ to $y = \phi_2(x)$. The resulting expression is a function of x alone, which we then integrate with respect to x from $x = a$ to $x = b$.

Type II Region. The *projection* of Ω onto the y-axis is a closed interval $[c, d]$ and Ω consists of all points (x, y) with

$$c \leq y \leq d \qquad \text{and} \qquad \psi_1(y) \leq x \leq \psi_2(y).$$

Then

(16.3.2)
$$\iint_{\Omega} f(x,y)\ dxdy = \int_c^d \left(\int_{\psi_1(y)}^{\psi_2(y)} f(x,y)\ dx\right)\,dy.$$

This time we first calculate

$$\int_{\psi_1(y)}^{\psi_2(y)} f(x,y)\,dx$$

by integrating $f(x,y)$ with respect to x from $x = \psi_1(y)$ to $x = \psi_2(y)$. The resulting expression is a function of y alone, which we can then integrate with respect to y from $y = c$ to $y = d$.

The integrals on the right-hand sides of formulas (16.3.1) and (16.3.2) are called *repeated integrals*. These formulas are easy to understand geometrically.

The Reduction Formulas Viewed Geometrically

Suppose that f is nonnegative and Ω is a region of Type I. The double integral over Ω gives the volume of the solid T bounded above by the surface $z = f(x, y)$ and bounded below by the region Ω:

(1) $$\iint_{\Omega} f(x, y)\, dxdy = \text{volume of } T. \qquad \text{(Figure 16.3.2)}$$

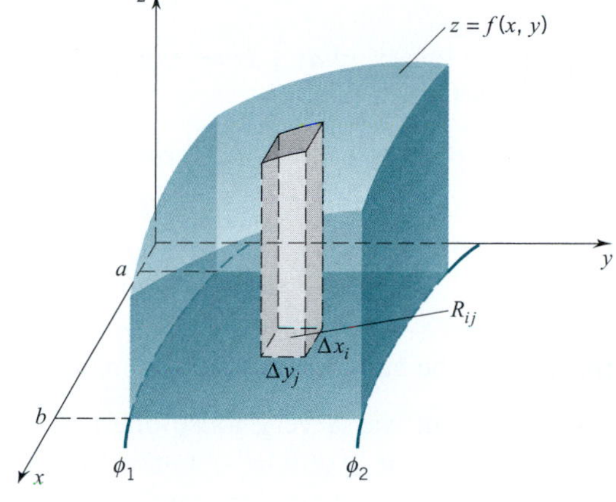

Figure 16.3.2

We can also compute the volume of T by the method of parallel cross sections (see Section 6.2). As in Figure 16.3.3, let $A(x)$ be the area of the cross section of T, all points of which have first coordinate x. Then by (6.2.1)

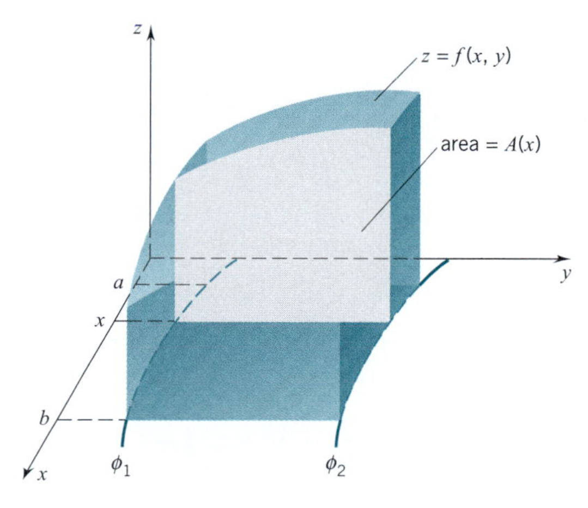

Figure 16.3.3

$$\int_a^b A(x)\,dx = \text{volume of } T.$$

Since
$$A(x) = \int_{\phi_1(x)}^{\phi_2(x)} f(x,y)\,dy,$$

we have

$$(2)\qquad \int_a^b \left(\int_{\phi_1(x)}^{\phi_2(x)} f(x,y)\,dy\right)dx = \text{volume of } T.$$

Combining (1) with (2), we have the first reduction formula

$$\iint_\Omega f(x,y)\,dxdy = \int_a^b \left(\int_{\phi_1(x)}^{\phi_2(x)} f(x,y)\,dy\right)dx.$$

The other reduction formula can be obtained in a similar manner.

Remark Note that our argument was a very loose one and certainly not a proof. How do we know, for example, that the "volume" obtained by double integration is the same as the "volume" obtained by the method of parallel cross sections? Intuitively it seems evident but, actually it is quite difficult to prove. The result is a special case of Fubini's theorem (after the Italian mathematician Guido Fubini (1879–1943) who proved a general version of the result in 1907). □

Computations

Example 1 Evaluate $\iint_\Omega (x^4 - 2y)\,dxdy$ with Ω as in Figure 16.3.4.

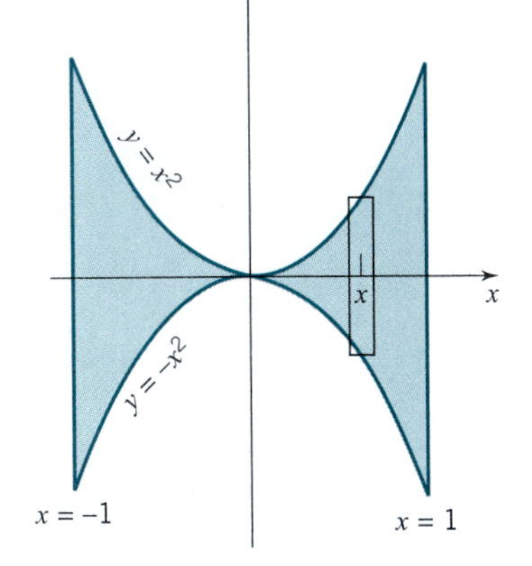

Figure 16.3.4

SOLUTION By projecting Ω onto the x-axis, we obtain the interval $[-1, 1]$. The region Ω consists of all points (x, y) with

$$-1 \le x \le 1 \qquad \text{and} \qquad -x^2 \le y \le x^2.$$

This is a region of Type I. By (16.3.1)

$$\begin{aligned}\iint_\Omega (x^4 - 2y)\,dxdy &= \int_{-1}^1 \left(\int_{-x^2}^{x^2} [x^4 - 2y]\,dy\right)dx\\ &= \int_{-1}^1 \left[x^4y - y^2\right]_{-x^2}^{x^2} dx\\ &= \int_{-1}^1 [(x^6 - x^4) - (-x^6 - x^4)]\,dx\\ &= \int_{-1}^1 2x^6\,dx = \left[\tfrac{2}{7}x^7\right]_{-1}^1 = \tfrac{4}{7}.\end{aligned}$$ □

Example 2 Evaluate $\iint_\Omega (xy - y^3)\,dxdy$ with Ω as in Figure 16.3.5.

Figure 16.3.5

SOLUTION By projecting Ω onto the y-axis, we obtain the interval $[0, 1]$. The region Ω consists of all points (x, y) with

$$0 \leq y \leq 1 \qquad \text{and} \qquad -1 \leq x \leq y.$$

This is a region of Type II. By (16.3.2)

$$\begin{aligned}\iint_{\Omega} (xy - y^3)\, dxdy &= \int_0^1 \left(\int_{-1}^{y} (xy - y^3)\, dx \right) dy \\ &= \int_0^1 \left[\tfrac{1}{2}x^2 y - xy^3 \right]_{-1}^{y} dy \\ &= \int_0^1 \left[\left(\tfrac{1}{2}y^3 - y^4\right) - \left(\tfrac{1}{2}y + y^3\right) \right] dy \\ &= \int_0^1 \left(-\tfrac{1}{2}y^3 - y^4 - \tfrac{1}{2}y\right) dy \\ &= \left[-\tfrac{1}{8}y^4 - \tfrac{1}{5}y^5 - \tfrac{1}{4}y^2 \right]_0^1 = -\tfrac{23}{40}.\end{aligned}$$

We can also project Ω onto the x-axis and express Ω as a region of Type I, but then the lower boundary is defined piecewise (see the figure) and the calculations are somewhat more complicated: setting

$$\phi(x) = \begin{cases} 0, & -1 \leq x \leq 0 \\ x, & 0 \leq x \leq 1, \end{cases}$$

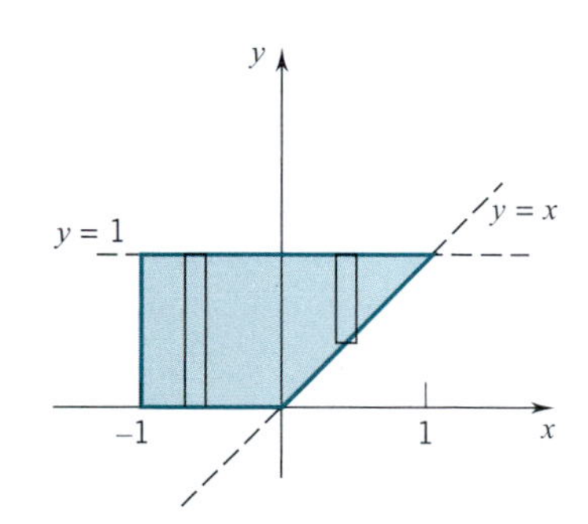

we have Ω as the set of all points (x, y) with

$$-1 \leq x \leq 1 \qquad \text{and} \qquad \phi(x) \leq y \leq 1;$$

thus

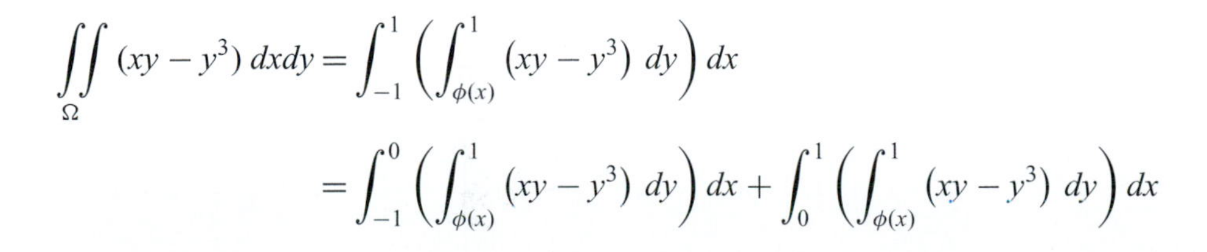

$$\begin{aligned}\iint_{\Omega} (xy - y^3)\, dxdy &= \int_{-1}^{1} \left(\int_{\phi(x)}^{1} (xy - y^3)\, dy \right) dx \\ &= \int_{-1}^{0} \left(\int_{\phi(x)}^{1} (xy - y^3)\, dy \right) dx + \int_{0}^{1} \left(\int_{\phi(x)}^{1} (xy - y^3)\, dy \right) dx\end{aligned}$$

$$= \int_{-1}^{0} \left(\int_{0}^{1} (xy - y^3)\, dy \right) dx + \int_{0}^{1} \left(\int_{x}^{1} (xy - y^3)\, dy \right) dx$$

as you can check ⟶

$$= (-\tfrac{1}{2}) + (-\tfrac{3}{40}) = -\tfrac{23}{40}. \quad \square$$

Repeated integrals

$$\int_a^b \left(\int_{\phi_1(x)}^{\phi_2(x)} f(x,y)\, dy \right) dx \qquad \text{and} \qquad \int_c^d \left(\int_{\psi_1(y)}^{\psi_2(y)} f(x,y)\, dx \right) dy$$

can be written in more compact form by omitting the large parentheses. From now on we will simply write

$$\int_a^b \int_{\phi_1(x)}^{\phi_2(x)} f(x,y)\, dy\, dx \qquad \text{and} \qquad \int_c^d \int_{\psi_1(y)}^{\psi_2(y)} f(x,y)\, dx\, dy$$

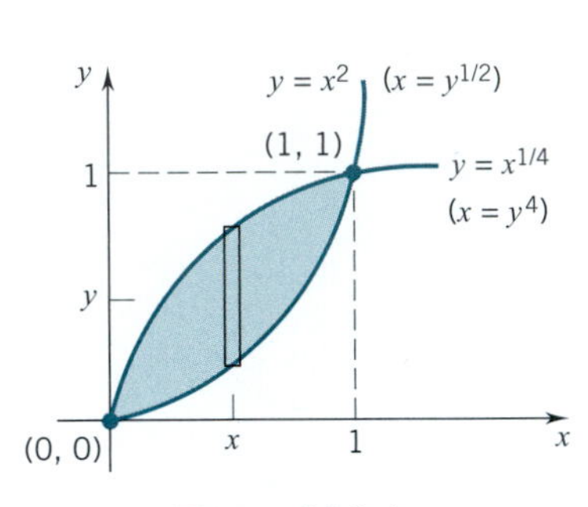

Figure 16.3.6

Example 3 Evaluate $\iint_\Omega (x^{1/2} - y^2)\, dxdy$ with Ω as in Figure 16.3.6.

SOLUTION The projection of Ω onto the x-axis is the closed interval $[0, 1]$ and Ω can be characterized as the set of all (x, y) with

$$0 \le x \le 1 \qquad \text{and} \qquad x^2 \le y \le x^{1/4}.$$

Thus

$$\iint_\Omega (x^{1/2} - y^2)\, dxdy = \int_0^1 \int_{x^2}^{x^{1/4}} (x^{1/2} - y^2)\, dy\, dx$$

$$= \int_0^1 \left[x^{1/2}y - \tfrac{1}{3}y^3 \right]_{x^2}^{x^{1/4}} dx$$

$$= \int_0^1 \left(\tfrac{2}{3}x^{3/4} - x^{5/2} + \tfrac{1}{3}x^6 \right) dx$$

$$= \left[\tfrac{8}{21}x^{7/4} - \tfrac{2}{7}x^{7/2} + \tfrac{1}{21}x^7 \right]_0^1 = \tfrac{8}{21} - \tfrac{2}{7} + \tfrac{1}{21} = \tfrac{1}{7}.$$

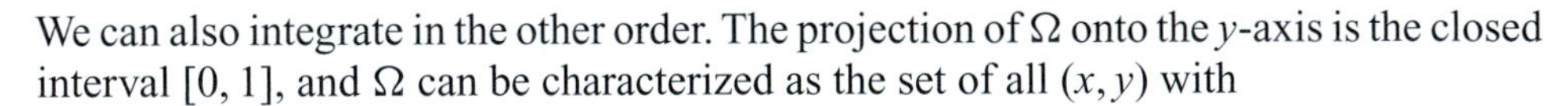

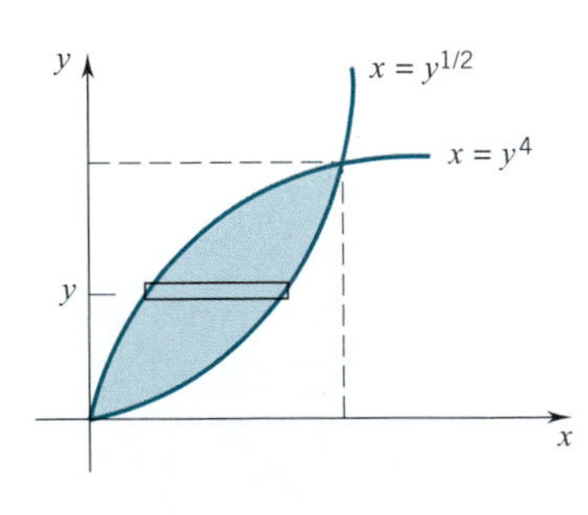

We can also integrate in the other order. The projection of Ω onto the y-axis is the closed interval $[0, 1]$, and Ω can be characterized as the set of all (x, y) with

$$0 \le y \le 1 \qquad \text{and} \qquad y^4 \le x \le y^{1/2}.$$

This gives the same result:

$$\iint_\Omega (x^{1/2} - y^2)\, dxdy = \int_0^1 \int_{y^4}^{y^{1/2}} (x^{1/2} - y^2)\, dx\, dy$$

$$= \int_0^1 \left[\tfrac{2}{3}x^{3/2} - y^2x \right]_{y^4}^{y^{1/2}} dy$$

$$= \int_0^1 \left(\tfrac{2}{3}y^{3/4} - y^{5/2} + \tfrac{1}{3}y^6 \right) dy$$

$$= \left[\tfrac{8}{21}y^{7/4} - \tfrac{2}{7}y^{7/2} + \tfrac{1}{21}y^7 \right]_0^1 = \tfrac{8}{21} - \tfrac{2}{7} + \tfrac{1}{21} = \tfrac{1}{7}. \quad \square$$

Example 4 Use double integration to calculate the area of the region Ω enclosed by

$$y = x^2 \qquad \text{and} \qquad x + y = 2.$$

SOLUTION The region Ω is pictured in Figure 16.3.7. Its area is given by the double integral

$$\iint_{\Omega} dxdy.$$

We project Ω onto the x-axis and write the boundaries as functions of x,

$$y = x^2 \qquad \text{and} \qquad y = 2 - x,$$

and view Ω as the set of all (x, y) with $-2 \leq x \leq 1$ and $x^2 \leq y \leq 2 - x$. This gives

$$\iint_{\Omega} dxdy = \int_{-2}^{1} \int_{x^2}^{2-x} dy\, dx = \int_{-2}^{1} (2 - x - x^2)\; dx = \left[2x - \tfrac{1}{2}x^2 - \tfrac{1}{3}x^3\right]_{-2}^{1}$$

$$= (2 - \tfrac{1}{2} - \tfrac{1}{3}) - (-4 - 2 + \tfrac{8}{3}) = \tfrac{9}{2}.$$

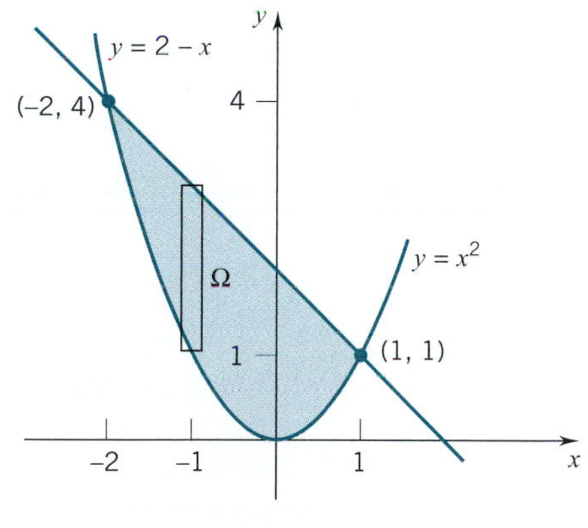

Figure 16.3.7

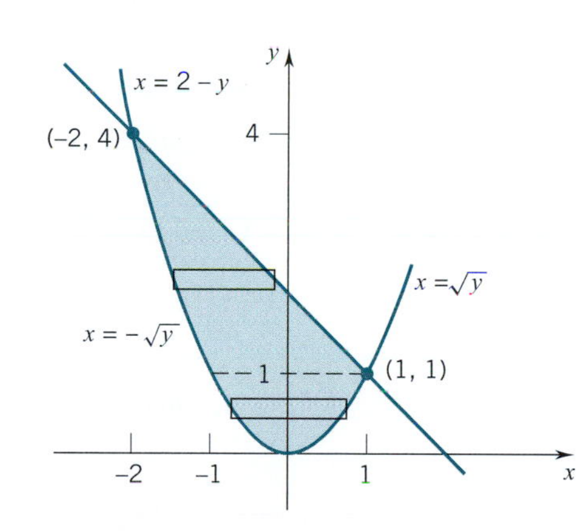

Figure 16.3.8

We can also project Ω onto the y-axis and write the boundaries as functions of y, but then the calculations become more complicated. As illustrated in Figure 16.3.8, Ω is the set of all (x, y) with

$$0 \leq y \leq 4 \qquad \text{and} \qquad -\sqrt{y} \leq x \leq \psi(y)$$

where

$$\psi(y) = \begin{cases} \sqrt{y}, & 0 \leq y \leq 1 \\ 2 - y, & 1 \leq y \leq 4. \end{cases}$$

Thus

$$\iint_{\Omega} dxdy = \int_{0}^{1} \int_{-\sqrt{y}}^{\sqrt{y}} dx\, dy + \int_{1}^{4} \int_{-\sqrt{y}}^{2-y} dx\, dy.$$

Carry out the calculations and you will get the same result as above. □

Symmetry in Double Integration

First we go back to the one-variable case (Section 5.8). Let's suppose that g is continuous on an interval that is symmetric about the origin, say $[-a, a]$.

$$\text{If } g \text{ is odd,} \quad \text{then} \quad \int_{-a}^{a} g(x)\, dx = 0.$$

$$\text{If } g \text{ is even,} \quad \text{then} \quad \int_{-a}^{a} g(x)\, dx = 2\int_{0}^{a} g(x)\, dx.$$

We have similar results for double integrals.

Suppose that Ω is symmetric about the y-axis.

$$\text{If } f \text{ is odd in } x\,[f(-x,y) = -f(x,y)], \quad \text{then} \quad \iint_{\Omega} f(x,y)\, dxdy = 0.$$

$$\text{If } f \text{ is even in } x\,[f(-x,y) = f(x,y)], \quad \text{then} \quad \iint_{\Omega} f(x,y)\, dxdy = 2\iint_{\substack{\text{right half}\\ \text{of } \Omega}} f(x,y)\, dxdy.$$

Suppose that Ω is symmetric about the x-axis.

$$\text{If } f \text{ is odd in } y\,[f(x,-y) = -f(x,y)], \quad \text{then} \quad \iint_{\Omega} f(x,y)\, dxdy = 0.$$

$$\text{If } f \text{ is even in } y\,[f(x,-y) = f(x,y)], \quad \text{then} \quad \iint_{\Omega} f(x,y)\, dxdy = 2\iint_{\substack{\text{upper half}\\ \text{of } \Omega}} f(x,y)\, dxdy.$$

As an example, take the region Ω of Figure 16.3.9. Suppose we wanted to calculate

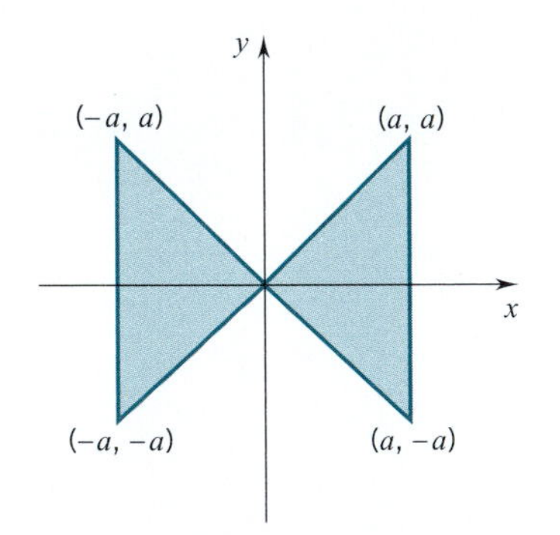

Figure 16.3.9

$$\iint_{\Omega} (2x - \sin x^2 y)\, dxdy.$$

First,

$$\iint_{\Omega} (2x - \sin x^2 y)\, dxdy = \iint_{\Omega} 2x\, dxdy - \iint_{\Omega} \sin x^2 y\, dxdy.$$

The symmetry of Ω about the y-axis gives

$$\iint_{\Omega} 2x\, dxdy = 0. \qquad \text{(the integrand is odd in } x\text{)}$$

The symmetry of Ω about the x-axis gives

$$\iint_{\Omega} \sin x^2 y\, dxdy = 0. \qquad \text{(the integrand is odd in } y\text{)}$$

Therefore $$\iint_{\Omega} (2x - \sin x^2 y)\, dxdy = 0.$$

Let's go back to Example 1 and reevaluate

$$\iint_{\Omega} (x^4 - 2y)\, dxdy,$$

this time capitalizing on the symmetry of Ω. Note that

$$\underbrace{\iint_{\Omega} 2y\, dxdy = 0}_{\text{symmetry about } x\text{-axis}} \quad \text{and} \quad \iint_{\Omega} x^4\, dxdy \overset{\text{symmetry about } y\text{-axis}}{=} 2 \iint_{\substack{\text{right half}\\ \text{of } \Omega}} x^4\, dxdy \overset{\text{symmetry about } x\text{-axis}}{=} 4 \iint_{\substack{\text{upper part}\\ \text{of right half}\\ \text{of } \Omega}} x^4\, dxdy.$$

Therefore $$\iint_{\Omega} (x^4 - 2y)\, dxdy = 4 \int_0^1 \int_0^{x^2} x^4\, dy\, dx = 4 \int_0^1 x^6\, dx = \tfrac{4}{7}.$$

Example 5 Calculate the volume within the cylinder $x^2 + y^2 = b^2$ between the planes $y + z = a$ and $z = 0$ given that $a \geq b > 0$.

SOLUTION See Figure 16.3.10. The solid in question is bounded below by the disc

$$\Omega: \quad 0 \leq x^2 + y^2 \leq b^2$$

and above by the plane

$$z = a - y.$$

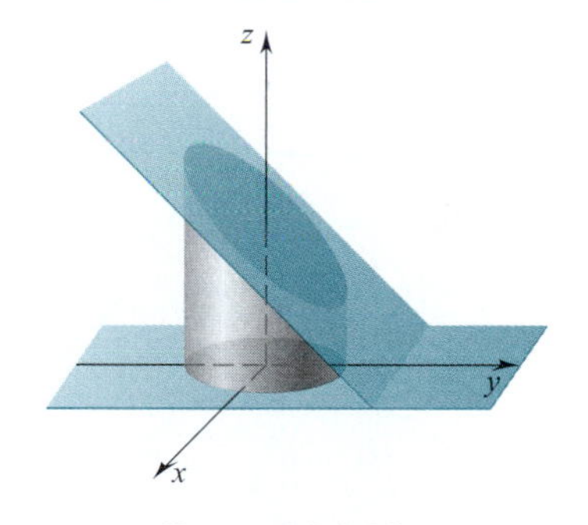

Figure 16.3.10

The volume is given by the double integral

$$\iint_{\Omega} (a - y)\, dxdy.$$

Since Ω is symmetric about the x-axis,

$$\iint_{\Omega} y\, dxdy = 0.$$

Thus

$$\iint_{\Omega} (a - y)\, dxdy = \iint_{\Omega} a\, dxdy = a \iint_{\Omega} dxdy = a\,(\text{area of } \Omega) = \pi ab^2. \quad \square$$

Concluding Remarks

When two orders of integration are possible, one order may be easy to carry out, while the other may present serious difficulties. Take as an example the double integral

$$\iint_{\Omega} \cos \tfrac{1}{2}\pi x^2 \, dxdy$$

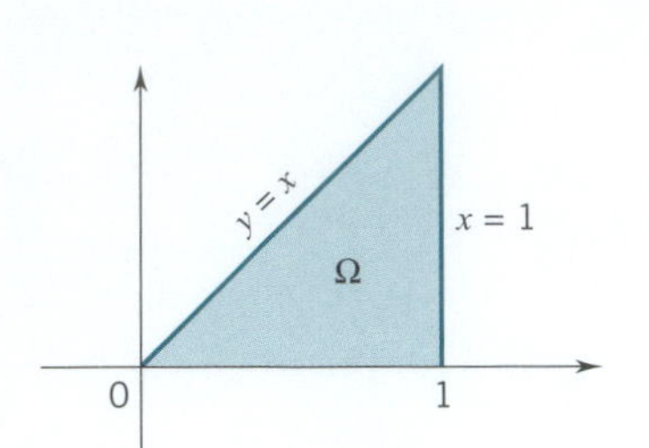

Figure 16.3.11

with Ω as in Figure 16.3.11. Projection onto the x-axis leads to

$$\int_0^1 \int_0^x \cos \tfrac{1}{2}\pi x^2 \, dy \, dx.$$

Projection onto the y-axis leads to

$$\int_0^1 \int_y^1 \cos \tfrac{1}{2}\pi x^2 \, dx \, dy.$$

The first expression is easy to evaluate:

$$\begin{aligned}
\int_0^1 \int_0^x \cos \tfrac{1}{2}\pi x^2 \, dy \, dx &= \int_0^1 \left(\int_0^x \cos \tfrac{1}{2}\pi x^2 \, dy \right) dx \\
&= \int_0^1 \left[y \cos \tfrac{1}{2}\pi x^2 \right]_0^x dx = \int_0^1 x \cos \tfrac{1}{2}\pi x^2 \, dx \\
&= \left[\tfrac{1}{\pi} \sin \tfrac{1}{2}\pi x^2 \right]_0^1 = \tfrac{1}{\pi}.
\end{aligned}$$

The second expression is not easy to evaluate:

$$\int_0^1 \int_y^1 \cos \tfrac{1}{2}\pi x^2 \, dx \, dy = \int_0^1 \left(\int_y^1 \cos \tfrac{1}{2}\pi x^2 \, dx \right) dy,$$

and $\cos \tfrac{1}{2}\pi x^2$ does not have an elementary antiderivative.

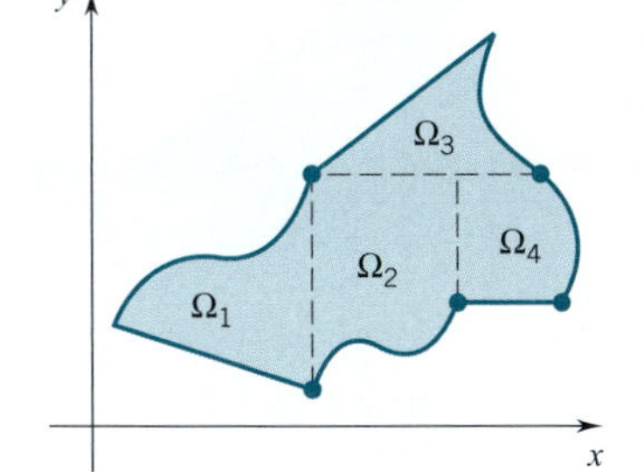

Figure 16.3.12

Finally, if Ω, the region of integration, is neither of Type I nor of Type II, it may be possible to break it up into a finite number of regions $\Omega_1, \ldots, \Omega_n$, each of which is of Type I or Type II. (See Figure 16.3.12.) Since the double integral is additive,

$$\iint_{\Omega_1} f(x,y) \, dxdy + \cdots + \iint_{\Omega_4} f(x,y) \, dxdy = \iint_{\Omega} f(x,y) \, dxdy.$$

Each of the integrals on the left can be evaluated by the methods of this section.

EXERCISES 16.3

Evaluate the integral taking $\Omega: 0 \le x \le 1, 0 \le y \le 3$.

1. $\displaystyle\iint_{\Omega} x^2 \, dxdy$.

2. $\displaystyle\iint_{\Omega} e^{x+y} \, dxdy$.

3. $\displaystyle\iint_{\Omega} xy^2 \, dxdy$.

Evaluate the integral taking $\Omega: 0 \le x \le 1, 0 \le y \le x$.

4. $\displaystyle\iint_{\Omega} x^3 y \, dxdy$.

5. $\displaystyle\iint_{\Omega} xy^3 \, dxdy$.

6. $\displaystyle\iint_{\Omega} x^2 y^2 \, dxdy$.

Evaluate the integral taking $\Omega: 0 \le x \le \frac{1}{2}\pi, 0 \le y \le \frac{1}{2}\pi$.

7. $\iint_{\Omega} \sin(x+y)\, dxdy$. **8.** $\iint_{\Omega} \cos(x+y)\, dxdy$.

9. $\iint_{\Omega} (1+xy)\, dxdy$.

Evaluate the double integral.

10. $\iint_{\Omega} (x+3y^3)\, dxdy, \quad \Omega: 0 \le x^2+y^2 \le 1$.

11. $\iint_{\Omega} \sqrt{xy}\, dxdy, \quad \Omega: 0 \le y \le 1, \quad y^2 \le x \le y$.

12. $\iint_{\Omega} y e^x\, dxdy, \quad \Omega: 0 \le y \le 1, \quad 0 \le x \le y^2$.

13. $\iint_{\Omega} (4-y^2)\, dxdy$, Ω the bounded region between $y^2 = 2x$ and $y^2 = 8-2x$.

14. $\iint_{\Omega} (x^4+y^2)\, dxdy$, Ω the bounded region between $y = x^3$ and $y = x^2$.

15. $\iint_{\Omega} (3xy^3 - y)\, dxdy$, Ω the region between $y = |x|$ and $y = -|x|$, $x \in [-1, 1]$.

16. $\iint_{\Omega} e^{-y^2/2}\, dxdy$, Ω the triangle formed by the y-axis, $2y = x$, $y = 1$.

17. $\iint_{\Omega} e^{x^2}\, dxdy$, Ω the triangle formed by the x-axis, $2y = x, x = 2$.

18. $\iint_{\Omega} (x+y)\, dxdy$, Ω the region between $y = x^3$ and $y = x^4$, $x \in [-1, 1]$.

Sketch the region Ω that gives rise to the repeated integral and change the order of integration.

19. $\int_0^1 \int_{x^4}^{x^2} f(x,y)\, dy\, dx$. **20.** $\int_0^1 \int_0^{y^2} f(x,y)\, dx\, dy$.

21. $\int_0^1 \int_{-y}^{y} f(x,y)\, dx\, dy$. **22.** $\int_{1/2}^1 \int_{x^3}^{x} f(x,y)\, dy\, dx$.

23. $\int_1^4 \int_x^{2x} f(x,y)\, dy\, dx$. **24.** $\int_1^3 \int_{-x}^{x^2} f(x,y)\, dy\, dx$.

Calculate by double integration the area of the bounded region determined by the given pairs of curves.

25. $x^2 = 4y, \quad 2y - x - 4 = 0$.

26. $y = x, \quad x = 4y - y^2$. **27.** $y = x, \quad 4y^3 = x^2$.

28. $x + y = 5, \quad xy = 6$.

Sketch the region Ω that gives rise to the repeated integral, change the order of integration, and then evaluate.

29. $\int_0^1 \int_{\sqrt{x}}^1 \sin\left(\frac{y^3+1}{2}\right) dy\, dx$.

30. $\int_{-1}^0 \int_{-\sqrt{y+1}}^{\sqrt{y+1}} x^2\, dx\, dy$. **31.** $\int_1^2 \int_0^{\ln y} e^{-x}\, dx\, dy$.

32. $\int_0^1 \int_{x^2}^1 \frac{x^3}{\sqrt{x^4+y^2}}\, dy\, dx$.

33. Find the area of the first quadrant region bounded by $xy = 2, \ y = 1, \ y = x + 1$.

34. Find the volume of the solid bounded above by $z = x + y$ and below by the triangle with vertices $(0,0,0)$, $(0,1,0)$, $(1,0,0)$.

35. Find the volume of the solid bounded by $\frac{1}{2}x + \frac{1}{3}y + \frac{1}{4}z = 1$ and the coordinate planes.

36. Find the volume of the solid bounded above by the plane $z = 2x + 3y$ and below by the unit square $0 \le x \le 1$, $0 \le y \le 1$.

37. Find the volume of the solid bounded above by $z = x^3y$ and below by the triangle with vertices $(0,0,0)$, $(2,0,0)$, $(0,1,0)$.

38. Find the volume under the paraboloid $z = x^2 + y^2$ within the cylinder $x^2 + y^2 \le 1, z \ge 0$.

39. Find the volume of the solid bounded above by the plane $z = 2x + 1$ and below by the disc $(x-1)^2 + y^2 \le 1$.

40. Find the volume of the solid bounded above by $z = 4 - y^2 - \frac{1}{4}x^2$ and below by the disc $(y-1)^2 + x^2 \le 1$.

41. Find the volume of the solid in the first octant bounded by $z = x^2+y^2$, the plane $x+y = 1$, and the coordinate planes.

42. Find the volume of the solid bounded by the circular cylinder $x^2+y^2 = 1$, the plane $z = 0$, and the plane $x + z = 1$.

43. Find the volume of the solid in the first octant bounded above by $z = x^2 + 3y^2$, below by the xy-plane, and on the sides by the cylinder $y = x^2$ and the plane $y = x$.

44. Find the volume of the solid bounded above by the surface $z = 1 + xy$ and below by the triangle with vertices $(1,1), (4,1)$, and $(3,2)$.

45. Find the volume of the solid in the first octant bounded by the two cylinders $x^2+y^2 = a^2$, $x^2+z^2 = a^2$.

46. Find the volume of the tetrahedron bounded by the coordinate planes and the plane

$$\frac{x}{a} + \frac{y}{b} + \frac{z}{c} = 1, \qquad a, b, c > 0.$$

Evaluate the double integral.

47. $\int_0^1 \int_y^1 e^{y/x}\, dx\, dy$. **48.** $\int_0^1 \int_0^{\cos^{-1} y} e^{\sin x}\, dx\, dy$.

49. $\int_0^1 \int_x^1 x^2 e^{y^4}\, dy\, dx$. **50.** $\int_0^1 \int_x^1 e^{y^2}\, dy\, dx$.

Calculate the average value of f over the region Ω.

51. $f(x,y) = x^2y;$ $\quad \Omega: \quad -1 \le x \le 1, \quad 0 \le y \le 4.$

52. $f(x,y) = xy;$ $\quad \Omega: \quad 0 \le x \le 1, \quad 0 \le y \le \sqrt{1-x^2}.$

53. $f(x,y) = \dfrac{1}{xy};$ $\quad \Omega$: $\ln 2 \le x \le 2\ln 2,$ $\ln 2 \le y \le 2\ln 2.$

54. $f(x,y) = e^{x+y};$ $\quad \Omega: \quad 0 \le x \le 1, \quad x-1 \le y \le x+1.$

55. (*Separated variables over a rectangle*) Let R be the rectangle $a \le x \le b,\ c \le y \le d$. Show that, if f is continuous on $[a,b]$ and g is continuous on $[c,d]$, then

(16.3.3)
$$\iint_R f(x)g(y)\,dxdy = \left[\int_a^b f(x)\,dx\right]\cdot\left[\int_c^b g(y)\,dy\right].$$

56. Let R be a rectangle symmetric about the x-axis, sides parallel to the coordinate axes. Show that, if f is odd with respect to y, then the double integral of f over R is 0.

57. Let R be a rectangle symmetric about the y-axis, sides parallel to the coordinate axes. Show that, if f is odd with respect to x, then the double integral of f over R is 0.

58. Given that $f(-x,-y) = -f(x,y)$ for all (x,y) in Ω, what form of symmetry in Ω will ensure that the double integral of f over Ω is zero?

59. Let Ω be the triangle with vertices $(0,0), (0,1), (1,1)$. Show that

$$\text{if} \quad \int_0^1 f(x)\,dx = 0 \quad \text{then} \quad \iint_\Omega f(x)f(y)\,dxdy = 0.$$

60. (*Differentiation under the integral sign*) If f and $\partial f/\partial x$ are continuous, then the function

$$H(t) = \int_a^b \frac{\partial f}{\partial x}(t,y)\,dy$$

can be shown to be continuous. Use the identity

$$\int_0^x \int_a^b \frac{\partial f}{\partial x}(t,y)\,dy\,dt = \int_a^b \int_0^x \frac{\partial f}{\partial x}(t,y)\,dt\,dy$$

to verify that

$$\frac{d}{dx}\left[\int_a^b f(x,y)\,dy\right] = \int_a^b \frac{\partial f}{\partial x}(x,y)\,dy.$$

61. We integrate over regions of Type I by means of the formula

$$\iint_\Omega f(x,y)\ dx\,dy = \int_a^b \int_{\phi_1(x)}^{\phi_2(x)} f(x,y)\ dy\,dx.$$

Here f is assumed to be continuous on Ω and ϕ_1, ϕ_2 are assumed to be continuous on $[a,b]$. Show that the function

$$F(x) = \int_{\phi_1(x)}^{\phi_2(x)} f(x,y)\,dy$$

is continuous on $[a,b]$.

C 62. Use a CAS to evaluate the double integral.

(a) $\displaystyle\int_{-1}^{3}\int_{2}^{5} x\,e^{-xy}\,dy\,dx.$ $\qquad$ (b) $\displaystyle\int_{3}^{7}\int_{1}^{4} \frac{xy}{x^2+y^2}\,dy\,dx.$

C 63. Use a graphing utility to draw the region Ω bounded by $y = x^2 - 2x + 2$ and $y = 1 + \sqrt{x-1}$. Find the area of Ω by evaluating the following integrals.

(a) $\displaystyle\int_{x=?}^{x=?}\int_{y=?}^{y=?} 1\,dy\,dx.$ $\qquad$ (b) $\displaystyle\int_{y=?}^{y=?}\int_{x=?}^{x=?} 1\,dx\,dy.$

C 64. Use a graphing utility to draw the region Ω bounded by $x - 2y = 0,\ x + y = 3$ and $y = 0$. Use a CAS to calculate

(a) $\displaystyle\int_{y=?}^{y=?}\int_{x=?}^{x=?} \sqrt{2x+y}\ dx\,dy.$

(b) $\displaystyle\int_{x=?}^{x=?}\int_{y=?}^{y=?} \sqrt{2x+y}\,dy\,dx.$

■ PROJECT 16.3 Numerical Methods for Double Integrals

The methods used to approximate ordinary definite integrals (see Section 8.7) can be extended to multiple integrals, but due the large number of calculations involved numerical integration methods for multiple integrals are much more time consuming. In this project, we explore numerical methods for the double integral

$$I = \iint_R f(x,y)\ dxdy,$$

where R is the rectangle: $a \le x \le b,\ c \le y \le d$. Let P_1 be a partition of $[a,b]$ into m subintervals of equal length $(b-a)/m$, and let P_2 be a partition of $[c,d]$ into n subintervals each of length $(d-c)/n$. Then $P = P_1 \times P_2$ partitions R into $m \times n$ rectangles R_{ij}, $1 \le i \le m, 1 \le j \le n$.

Midpoint Approximation. Let u_i be the midpoint of the x-subinterval $[x_{i-1}, x_i]$ and let v_j be the midpoint of the y-subinterval $[y_{j-1}, y_j]$. Then the midpoint approximation of I is given by the double sum.

$$I \cong M_{mn} = \frac{b-a}{m}\cdot\frac{d-c}{n}\sum_{i=1}^{m}\sum_{j=1}^{n} f(u_i, v_j).$$

The figure illustrates the case $m = 3, n = 2$.

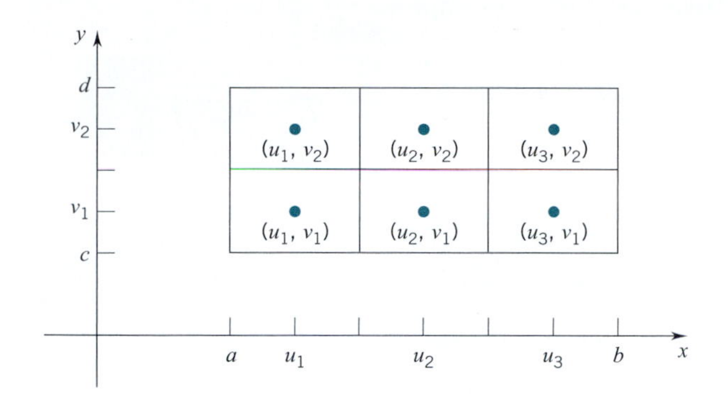

$$M_{32} = \frac{b-a}{3} \cdot \frac{d-c}{2}\,[f(u_1, v_1) + f(u_2, v_1) + f(u_3, v_1) + f(u_1, v_2) + f(u_2, v_2) + f(u_3, v_2)].$$

Problem 1. Let $I = \int_0^3 \int_0^2 (x + 2y)\, dx\, dy$.

a. Calculate M_{32}.

b. Use technology to calculate M_{mn} for larger values of m and n.

c. Calculate the exact value of I and compare it with your numerical approximations.

Problem 2. Let $I = \int_0^2 \int_0^2 4xy\, dx\, dy$.

a. Calculate M_{22}.

b. Use technology to calculate M_{mn} for larger values of m and n.

c. Calculate the exact value of I and compare it with your numerical approximations.

Problem 3. Let $I = \int_0^{\pi/2} \int_0^{\pi/2} \sin x \sin y\, dx\, dy$.

a. Calculate M_{22}.

b. Use technology to calculate M_{mn} for larger values of m and n.

c. Calculate the exact value of I and compare it with your numerical approximations.

Problem 4. Trapezoidal Rule: Derive a formula which extends the trapezoidal rule to the double integral over a rectangle. Denote your trapezoidal approximation of I by T_{mn}.

Problem 5. Let $I = \int_0^3 \int_0^2 (x + 2y)\, dx\, dy$.

a. Calculate T_{32}.

b. Use technology to calculate T_{mn} for larger values of m and n.

c. Compare your results here with the approximations in Problem 1 and with the exact value of I.

■ 16.4 THE DOUBLE INTEGRAL AS A LIMIT OF RIEMANN SUMS; POLAR COORDINATES

In the one-variable case we can write the integral as the limit of Riemann sums:

$$\int_a^b f(x)\, dx = \lim_{\max \Delta x \to 0} \sum_{i=1}^{n} f(x_i^*)\, \Delta x_i.$$

The same approach works with double integrals. To explain it, we need to explain what we mean by the *diameter of a set*.

Suppose that S is a bounded closed set (on the line, in the plane, or in three-space). For any two points P and Q of S, we can measure their separation, $d(P, Q)$. The maximal separation between points of S is called the *diameter of S*:

$$\text{diam } S = \max_{P, Q \in S} d(P, Q).\dagger$$

† If a set S is bounded, it is contained in some ball, which in turn has some finite diameter D. The set of all distances between points of S, being bounded above by D, has a least upper bound. This least upper bound is called the diameter of S:

$$\text{diam } S = \underset{P, Q \in S}{\text{lub}}\; d(P, Q).$$

It can be shown that, if S is bounded and closed, then this least upper bound is attained, in which case we can define

$$\text{diam } S = \max_{P, Q \in S} d(P, Q).$$

For a circle, a circular disc, a sphere, or a ball, this sense of diameter agrees with the usual one. Figure 16.4.1 gives some other examples.

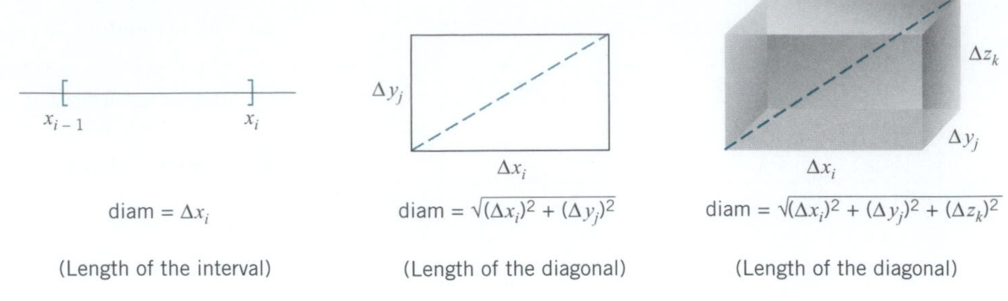

Figure 16.4.1

Now let's start with a basic region Ω and decompose it into a finite number of basic subregions $\Omega_1, \ldots, \Omega_N$. (See Figure 16.4.2.) If f is continuous on Ω, then f is continuous on each Ω_i. Now from each Ω_i we pick an arbitrary point (x_i^*, y_i^*) and form the *Riemann sum*

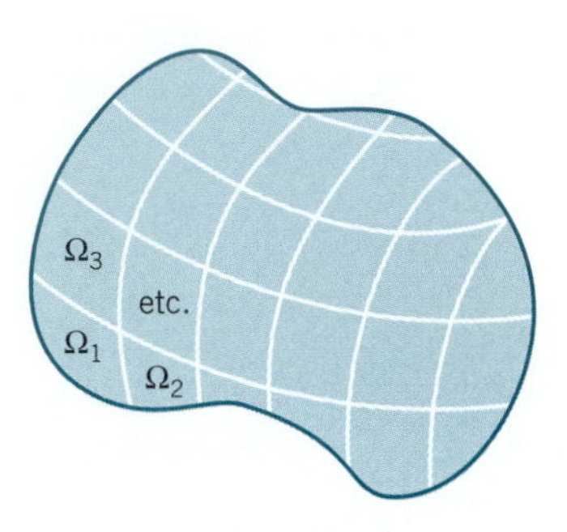

Figure 16.4.2

$$\sum_{i=1}^{N} f(x_i^*, y_i^*)(\text{area of } \Omega_i).$$

As you would expect, the double integral over Ω can be obtained as the limit of such sums; namely, given any $\epsilon > 0$, there exists $\delta > 0$ such that, if the diameters of the Ω_i are all less than δ, then

$$\left| \sum_{i=1}^{N} f(x_i^*, y_i^*)(\text{area of } \Omega_i) - \iint_{\Omega} f(x, y)\ dxdy \right| < \epsilon$$

no matter how the (x_i^*, y_i^*) are chosen within the Ω_i. We express this by writing

(16.4.1)
$$\iint_{\Omega} f(x, y)\ dxdy = \lim_{\text{diam } \Omega_i \to 0} \sum_{i=1}^{N} f(x_i^*, y_i^*)(\text{area of } \Omega_i).$$

Evaluating Double Integrals Using Polar Coordinates

Here we explain how to calculate double integrals

$$\iint_{\Omega} f(x, y)\ dxdy$$

using polar coordinates $[r, \theta]$. Throughout we take $r \geq 0$.

We will work with type of region shown in Figure 16.4.3. The region Ω is then the set of all points (x, y) that have polar coordinates $[r, \theta]$ in the set

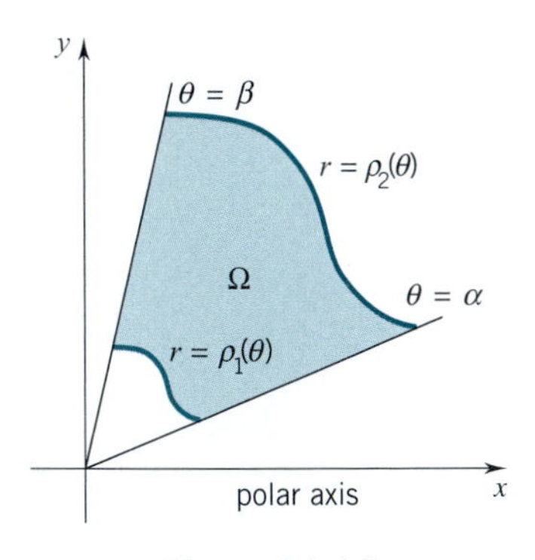

Figure 16.4.3

$$\Gamma: \quad \alpha \leq \theta \leq \beta, \quad \rho_1(\theta) \leq r \leq \rho_2(\theta),$$

where $\beta \leq \alpha + 2\pi$.

You already know how to calculate the area of Ω. By (9.5.2),

$$\text{area of } \Omega = \int_{\alpha}^{\beta} \tfrac{1}{2}\left([\rho_2(\theta)]^2 - [\rho_1(\theta)]^2 \right) d\theta.$$

We can write this as a double integral over Γ:

(16.4.2)
$$\text{area of } \Omega = \iint_{\Gamma} r\, drd\theta.$$

PROOF Simply note that

$$\tfrac{1}{2}\left([\rho_2(\theta)]^2 - [\rho_1(\theta)]^2\right) = \int_{\rho_1(\theta)}^{\rho_2(\theta)} r\, dr,$$

and therefore

$$\text{area of } \Omega = \int_{\alpha}^{\beta}\int_{\rho_1(\theta)}^{\rho_2(\theta)} r\, drd\theta = \iint_{\Gamma} r\, drd\theta. \quad \square$$

Now let's suppose that f is some function continuous at each point (x, y) of Ω. Then the composition

$$F(r, \theta) = f(r\cos\theta, r\sin\theta)$$

is continuous at each point $[r, \theta]$ of Γ. We will show that

(16.4.3)
$$\iint_{\Omega} f(x, y)\; dxdy = \iint_{\Gamma} f(r\cos\theta, r\sin\theta)\, r\, drd\theta. \qquad \text{(note the extra } r\text{)}$$

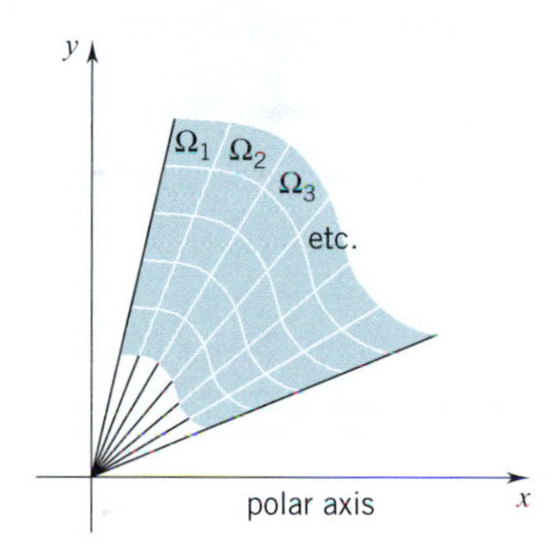

Figure 16.4.4

PROOF Our first step is to place a grid on Ω by using a finite number of rays $\theta = \theta_j$ and a finite number of continuous curves $r = \rho_k(\theta)$ in the manner of Figure 16.4.4. This grid decomposes Ω into a finite number of regions

$$\Omega_1, \ldots, \Omega_N$$

with polar coordinates in sets $\Gamma_1, \ldots, \Gamma_N$. Note that by (16.4.2)

$$\text{area of each } \Omega_i = \iint_{\Gamma_i} r\, drd\theta.$$

Writing $F(r, \theta)$ for $f(r\cos\theta, r\sin\theta)$, we have

$$\iint_{\Gamma} F(r, \theta)\, r\, drd\theta = \sum_{i=1}^{N} \iint_{\Gamma_i} F(r, \theta)\, r\, drd\theta \qquad \text{(additivity)}$$

$$= \sum_{i=1}^{N} F(r_i^*, \theta_i^*) \iint_{\Gamma_i} r\, drd\theta \qquad \text{(for some } [r_i^*, \theta_i^*] \in \Gamma_i \text{ (Theorem 16.2.10))}$$

$$= \sum_{i=1}^{N} F(r_i^*, \theta_i^*)(\text{area of } \Omega_i)$$

$$\overset{\downarrow}{=} \sum_{i=1}^{N} f(x_i^*, y_i^*)(\text{area of } \Omega_i). \qquad \text{(with } x_i^* = r_i^*\cos\theta_i^*, y_i^* = r_i^*\sin\theta_i^*\text{)}$$

This last expression is a Riemann sum for the double integral

$$\iint_{\Omega} f(x,y)\ dxdy.$$

As such, by (16.4.1), it differs from that integral by less than any preassigned positive ϵ provided only that the diameters of all the Ω_i are sufficiently small. This we can guarantee by making our grid sufficiently fine. ❑

Example 1 Use polar coordinates to evaluate $\displaystyle\iint_{\Omega} xy\, dxdy$, where Ω is the portion of the unit disc that lies in the first quadrant.

SOLUTION Here $\Gamma:\ 0 \le \theta \le \frac{1}{2}\pi,\quad 0 \le r \le 1.$ Therefore

$$\iint_{\Omega} xy\ dxdy = \iint_{\Gamma} (r\cos\theta)(r\sin\theta)r\,drd\theta$$

$$= \int_0^{\pi/2}\int_0^1 r^3\cos\theta\sin\theta\,dr\,d\theta = \tfrac{1}{8}. \quad ❑$$

check this

Example 2 Use polar coordinates to calculate the volume of a sphere of radius R.

SOLUTION In rectangular coordinates,

$$V = 2\iint_{\Omega} \sqrt{R^2 - (x^2 + y^2)}\ dxdy$$

where Ω is the disc of radius R centered at the origin. (Verify.) In polar coordinates, the disc of radius R centered at the origin is given by

$$\Gamma:\quad 0 \le \theta \le 2\pi,\quad 0 \le r \le R.$$

Therefore

$$V = 2\iint_{\Gamma} \sqrt{R^2 - r^2}\ r\,drd\theta = 2\int_0^{2\pi}\int_0^R \sqrt{R^2 - r^2}\ r\,dr\,d\theta = \tfrac{4}{3}\pi R^3. \quad ❑$$

check this

Example 3 Calculate the volume of the solid bounded above by the cone $z = 2 - \sqrt{x^2 + y^2}$ and bounded below by the disc $\Omega: (x-1)^2 + y^2 \le 1$. (See Figure 16.4.5.)

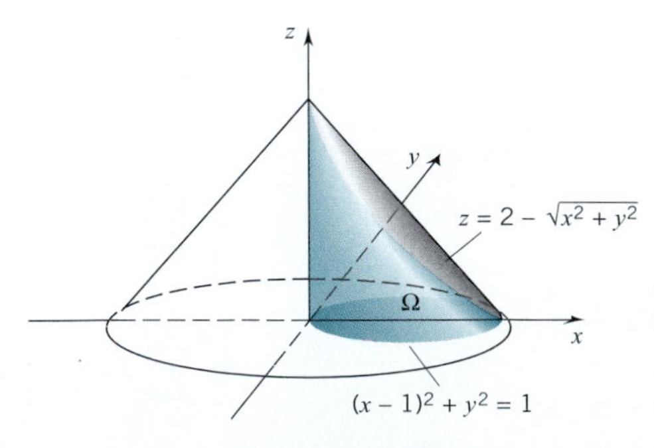

Figure 16.4.5

SOLUTION

$$V = \iint_{\Omega} (2 - \sqrt{x^2 - y^2})\, dxdy$$

$$= 2\iint_{\Omega} dxdy - \iint_{\Omega} \sqrt{x^2 + y^2}\, dxdy.$$

The first integral is $2 \times (\text{area of } \Omega) = 2\pi$. We evaluate the second integral by changing to polar coordinates.

The equation $(x - 1)^2 + y^2 = 1$ simplifies to $x^2 + y^2 = 2x$. In polar coordinates this becomes $r^2 = 2r\cos\theta$, which simplifies to $r = 2\cos\theta$. The disc Ω is the set of all points with polar coordinates in the set

$$\Gamma: -\tfrac{1}{2}\pi \le \theta \le \tfrac{1}{2}\pi, \quad 0 \le r \le 2\cos\theta.$$

Therefore $$\iint_{\Omega} \sqrt{x^2 + y^2}\, dxdy = \iint_{\Gamma} r^2 drd\theta = \int_{-\pi/2}^{\pi/2} \int_0^{2\cos\theta} r^2\, dr\, d\theta = \tfrac{32}{9}.$$

check this⟶↑

It follows that $$V = 2\pi - \tfrac{32}{9} \cong 2.73. \quad \square$$

Example 4 Evaluate $\displaystyle\iint_{\Omega} \frac{1}{(1 + x^2 + y^2)^{3/2}}\, dxdy$ where Ω is the triangle of Figure 16.4.6.

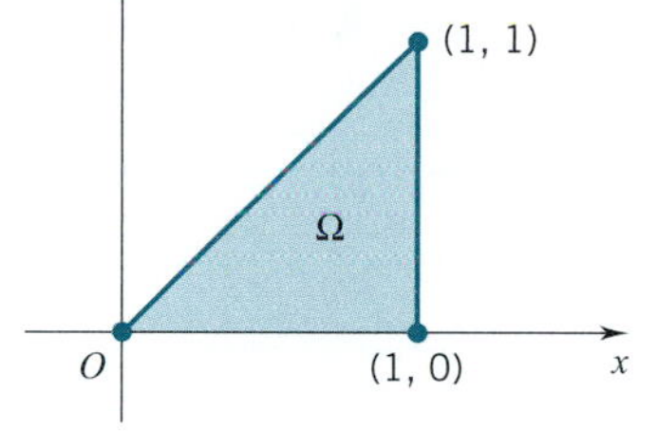

Figure 16.4.6

SOLUTION The vertical side of the triangle is part of the line $x = 1$. In polar coordinates this is $r\cos\theta = 1$, which can be written $r = \sec\theta$. Therefore

$$\iint_{\Omega} \frac{1}{(1 + x^2 + y^2)^{3/2}}\, dxdy = \iint_{\Gamma} \frac{r}{(1 + r^2)^{3/2}}\, drd\theta$$

where $\Gamma: 0 \le \theta \le \pi/4, \quad 0 \le r \le \sec\theta.$ (Figure 16.4.7)

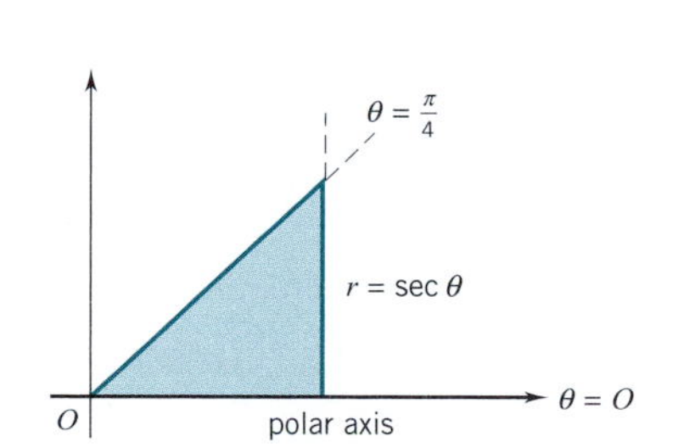

Figure 16.4.7

The double integral over Γ reduces to

$$\int_0^{\pi/4} \int_0^{\sec\theta} \frac{r}{(1 + r^2)^{3/2}}\, dr\, d\theta = \int_0^{\pi/4} \left[\frac{-1}{\sqrt{1 + r^2}}\right]_0^{\sec\theta} d\theta$$

$$= \int_0^{\pi/4} \left(1 - \frac{1}{\sqrt{1 + \sec^2\theta}}\right) d\theta.$$

For $\theta \in [0, \pi/4]$,

$$\frac{1}{\sqrt{1 + \sec^2\theta}} = \frac{\cos\theta}{\sqrt{\cos^2\theta + 1}} = \frac{\cos\theta}{\sqrt{2 - \sin^2\theta}}.$$

Therefore the integral can be written

$$\int_0^{\pi/4} \left(1 - \frac{\cos\theta}{\sqrt{2 - \sin^2\theta}}\right) d\theta = \left[\theta - \sin^{-1}\left(\frac{\sin\theta}{\sqrt{2}}\right)\right]_0^{\pi/4} = \frac{\pi}{4} - \frac{\pi}{6} = \frac{\pi}{12}. \quad \square$$

(7.7.4)⟶↑

The function $f(x) = e^{-x^2}$ has no elementary antiderivative. Nevertheless, by taking a circuitous route and then using polar coordinates, we can show that

(16.4.4)
$$\int_{-\infty}^{\infty} e^{-x^2}\, dx = \sqrt{\pi}.\ ^{\dagger}$$

PROOF The circular disc D_b: $x^2 + y^2 \leq b^2$ is the set of all (x, y) with polar coordinates $[r, \theta]$ in the set Γ: $0 \leq \theta \leq 2\pi, 0 \leq r \leq b$. Therefore

$$\iint_{D_b} e^{-(x^2+y^2)}\, dxdy = \iint_{\Gamma} e^{-r^2} r\, drd\theta = \int_0^{2\pi} \int_0^b e^{-r^2} r\, dr\, d\theta$$

$$= \int_0^{2\pi} \tfrac{1}{2}(1 - e^{-b^2})\, d\theta = \pi(1 - e^{-b^2}).$$

Let S_a be the square $-a \leq x \leq a, -a \leq y \leq a$. Since $D_a \subseteq S_a \subseteq D_{2a}$ and $e^{-(x^2+y^2)}$ is positive,

$$\iint_{D_a} e^{-(x^2+y^2)}\, dxdy \leq \iint_{S_a} e^{-(x^2+y^2)}\, dxdy \leq \iint_{D_{2a}} e^{-(x^2+y^2)}\, dxdy.$$

It follows that

$$\pi(1 - e^{-a^2}) \leq \iint_{S_a} e^{-(x^2+y^2)}\, dxdy \leq \pi(1 - e^{-4a^2}).$$

As $a \to \infty$, $\pi(1 - e^{-a^2}) \to \pi$ and $\pi(1 - e^{-4a^2}) \to \pi$. Therefore

$$\lim_{a\to\infty} \iint_{S_a} e^{-(x^2+y^2)}\, dxdy = \pi.$$

But
$$\iint_{S_a} e^{-(x^2+y^2)}\, dxdy = \int_{-a}^{a} \int_{-a}^{a} e^{-(x^2+y^2)}\, dx\, dy$$

$$= \int_{-a}^{a} \int_{-a}^{a} e^{-x^2} \cdot e^{-y^2}\, dx\, dy$$

$$= \left(\int_{-a}^{a} e^{-x^2}\, dx\right)\left(\int_{-a}^{a} e^{-y^2}\, dy\right) = \left(\int_{-a}^{a} e^{-x^2}\, dx\right)^2.$$

Therefore
$$\lim_{a\to\infty} \int_{-a}^{a} e^{-x^2}\, dx = \lim_{a\to\infty} \left(\iint_{S_a} e^{-(x^2+y^2)}\, dxdy\right)^{1/2} = \sqrt{\pi}. \quad \square$$

† This integral comes up frequently in probability theory and plays an important role in the branch of physics called "statistical mechanics." This integral was evaluated by numerical methods in Project 8.7

EXERCISES 16.4

Evaluate the iterated integral.

1. $\displaystyle\int_0^{\pi/2}\int_0^{\sin\theta} r\cos\theta\, dr\, d\theta.$

2. $\displaystyle\int_0^{\pi/4}\int_0^{\cos 2\theta} r\, dr\, d\theta.$

3. $\displaystyle\int_0^{\pi/2}\int_0^{3\sin\theta} r^2 dr\, d\theta.$

4. $\displaystyle\int_{-\pi/3}^{2\pi/3}\int_0^{2\cos\theta} r\sin\theta\, dr\, d\theta.$

5. Integrate $f(x,y)=\cos(x^2+y^2)$ over:
(a) the closed unit disc;
(b) the annular region $1\le x^2+y^2\le 4$.

6. Integrate $f(x,y)=\sin(\sqrt{x^2+y^2})$ over:
(a) the closed unit disc;
(b) the annular region $1\le x^2+y^2\le 4$.

7. Integrate $f(x,y)=x+y$ over:
(a) $0\le x^2+y^2\le 1,\ x\ge 0, y\ge 0$;
(b) $1\le x^2+y^2\le 4,\ x\ge 0, y\ge 0$.

8. Integrate $f(x,y)=\sqrt{x^2+y^2}$ over the triangle with vertices $(0,0),(1,0),(1,\sqrt{3})$.

Calculate by changing to polar coordinates.

9. $\displaystyle\int_{-1}^{1}\int_0^{\sqrt{1-y^2}}\sqrt{x^2+y^2}\,dx\,dy.$

10. $\displaystyle\int_0^{2}\int_0^{\sqrt{4-x^2}}\sqrt{x^2+y^2}\,dy\,dx.$

11. $\displaystyle\int_{1/2}^{1}\int_0^{\sqrt{1-x^2}} dy\,dx.$

12. $\displaystyle\int_0^{1/2}\int_0^{\sqrt{1-x^2}} xy\sqrt{x^2+y^2}\,dy\,dx.$

13. $\displaystyle\int_0^{1}\int_0^{\sqrt{1-x^2}} \sin\sqrt{x^2+y^2}\,dy\,dx.$

14. $\displaystyle\int_{-1}^{1}\int_{-\sqrt{1-y^2}}^{\sqrt{1-y^2}} e^{-(x^2+y^2)}\,dx\,dy.$

15. $\displaystyle\int_0^{2}\int_0^{\sqrt{2x-x^2}} x\,dy\,dx.$

16. $\displaystyle\int_0^{1}\int_{-\sqrt{x-x^2}}^{\sqrt{x-x^2}} (x^2+y^2)\,dy\,dx.$

In Exercises 17–22, use a double integral to find the area of the given region.

17. One leaf of the petal curve $r=3\sin 3\theta$.

18. The region enclosed by the cardioid $r=2(1-\cos\theta)$.

19. The region inside the circle $r=4\cos\theta$ but outside the circle $r=2$.

20. The region inside the large loop but outside the small loop of the limaçon $r=1+2\cos\theta$.

21. The region enclosed by the lemniscate $r^2=4\cos 2\theta$.

22. The region inside the circle $r=3\cos\theta$ but outside the cardioid $r=1+\cos\theta$.

23. Find the volume of the solid bounded above by the plane $z=y+b$, below by the xy-plane, and on the sides by the circular cylinder $x^2+y^2=b^2$.

24. Find the volume of the solid bounded below by the xy-plane and above by the paraboloid $z=1-(x^2+y^2)$.

25. Find the volume of the ellipsoid

$$x^2/4+y^2/4+z^2/3=1.$$

26. Find the volume of the solid bounded below by the xy-plane and above by the surface $x^2+y^2+z^6=5$.

27. Find the volume of the solid bounded below by the xy-plane, above by the spherical surface $x^2+y^2+z^2=4$, and on the sides by the cylinder $x^2+y^2=1$.

28. Find the volume of the solid bounded above by the surface $z=1-(x^2+y^2)$, below by the xy-plane, and on the sides by the cylinder $x^2+y^2-x=0$.

29. Find the volume of the solid bounded above by the plane $z=2x$ and below by the disc $(x-1)^2+y^2\le 1$.

30. Find the volume of the solid bounded above by the cone $z^2=x^2+y^2$ and below by the region Ω which lies inside the curve $x^2+y^2=2ax$.

31. Find the volume of the solid bounded below by the xy-plane, above by the ellipsoid of revolution $b^2x^2+b^2y^2+a^2z^2=a^2b^2$, and on the sides by the cylinder $x^2+y^2-ay=0$.

32. A cylindrical hole of radius r is drilled through the center of a sphere of radius R.
(a) Determine the volume of the material that has been removed from the sphere.
(b) Determine the volume of the ring-shaped solid that remains.

C 33. Use a graphing utility to draw the petal curve $r=2\cos 2\theta$. Then use a CAS to find the area of one petal by evaluating a double integral in polar coordinates.

C 34. Let $I=\displaystyle\iint_\Omega e^{x^2+y^2}\,dxdy$ where Ω is the annular region between the circles $x^2+y^2=4$ and $x^2+y^2=16$. Use a CAS to evaluate this integral after transforming it into a double integral in polar coordinates.

■ 16.5 SOME APPLICATIONS OF DOUBLE INTEGRATION

The Mass of a Plate

Suppose that a thin distribution of matter, called a *plate*, is laid out in the xy-plane in the form of a basic region Ω. If the mass density of the plate (the mass per unit area) is a constant λ, then the total mass M of the plate is simply the density λ times the area of the plate:

$$M = \lambda \times \text{the area of } \Omega.$$

If the density varies continuously from point to point, say $\lambda = \lambda(x, y)$, then the mass of the plate is the average density of the plate times the area of the plate:

$$M = \text{average density} \times \text{the area of } \Omega.$$

This is a double integral:

(16.5.1)
$$M = \iint_{\Omega} \lambda(x, y)\ dxdy.$$

The Center of Mass of a Plate

The center of mass x_M of a rod is a density-weighted average of position taken over the interval occupied by the rod:

$$x_M M = \int_a^b x\lambda(x)\ dx.$$

[This you have seen: (5.9.5).]

The coordinates of the center of mass of a plate (x_M, y_M) are determined by two density-weighted averages of position, each taken over the region occupied by the plate:

(16.5.2)
$$x_M M = \iint_{\Omega} x\lambda(x, y)\ dxdy, \quad y_M M = \iint_{\Omega} y\lambda(x, y)\ dxdy.$$

Example 1 A plate is in the form of a half-disc of radius a. Find the mass of the plate and the center of mass given that the mass density of the plate is directly proportional to the distance from the midpoint of the straight edge of the plate.

SOLUTION Place the plate over the region Ω: $-a \le x \le a,\ 0 \le y \le \sqrt{a^2 - x^2}$. See Figure 16.5.1. The mass density can then be written $\lambda(x, y) = k\sqrt{x^2 + y^2}$, where $k > 0$ is the constant of proportionality. Now

$$M = \iint_{\Omega} k\sqrt{x^2 + y^2}\ dxdy = \int_0^{\pi} \int_0^a (kr)\, r\, dr\, d\theta = k\left(\int_0^{\pi} 1\, d\theta\right)\left(\int_0^a r^2\, dr\right)$$

change to polar coordinates (↑ at the second equality)

$$= k(\pi)(\tfrac{1}{3}a^3) = \tfrac{1}{3}ka^3\pi,$$

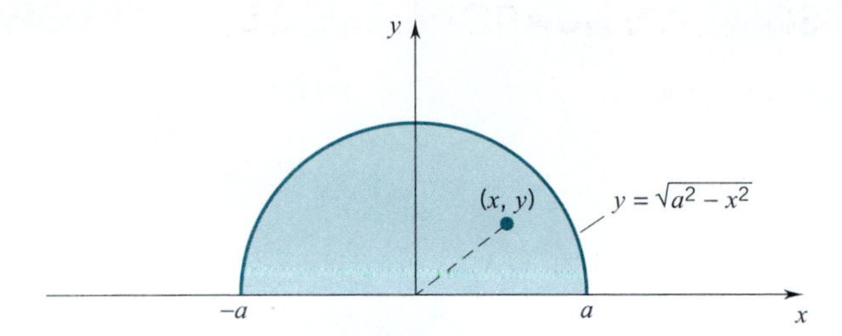

Figure 16.5.1

and
$$x_M M = \iint_{\Omega} x(k\sqrt{x^2+y^2})\ dxdy = 0$$

(Ω is symmetric with respect to the y-axis and the integrand is odd with respect to x). Thus $x_M = 0$. Also,

$$\begin{aligned} y_M M = \iint_{\Omega} y(k\sqrt{x^2+y^2})\ dxdy &= \int_0^{\pi}\int_0^a (r\sin\theta)(kr)\,r\,dr\,d\theta \\ &= k\left(\int_0^{\pi}\sin\theta\,d\theta\right)\left(\int_0^a r^3\,dr\right) \\ &= k(2)(\tfrac{1}{4}a^4) = \tfrac{1}{2}ka^4. \end{aligned}$$

Since $M = \frac{1}{3}ka^3\pi$, we have $y_M = (\frac{1}{2}ka^4)/(\frac{1}{3}ka^3\pi) = 3a/2\pi$. The center of mass of the plate is the point $(0, 3a/2\pi) \cong (0, 0.48a)$. ☐

Centroids

If the plate is homogeneous, then the mass density λ is constantly M/A where A is the area of the base region Ω. In this case the center of mass of the plate falls on the *centroid* of the base region (a notion with which you are already familiar). The centroid $(\bar{x}, \bar{y})$ depends only on the geometry of Ω:

$$\bar{x}M = \iint_{\Omega} x(M/A)\ dxdy = (M/A)\iint_{\Omega} x\ dxdy,$$
$$\bar{y}M = \iint_{\Omega} y(M/A)\ dxdy = (M/A)\iint_{\Omega} y\ dxdy.$$

Dividing by M and multiplying through by A, we have

(16.5.3)
$$\boxed{\bar{x}A = \iint_{\Omega} x\ dxdy, \quad \bar{y}A = \iint_{\Omega} y\ dxdy.}$$

Thus, in the sense of (16.2.9), $\bar{x}$ is the average x-coordinate on Ω and $\bar{y}$ is the average y-coordinate on Ω. The mass of the plate does not enter into this at all.

Example 2 Find the centroid of the region

$$\Omega: \quad a \le x \le b, \quad \phi_1(x) \le y \le \phi_2(x). \qquad \text{(Figure 16.5.2)}$$

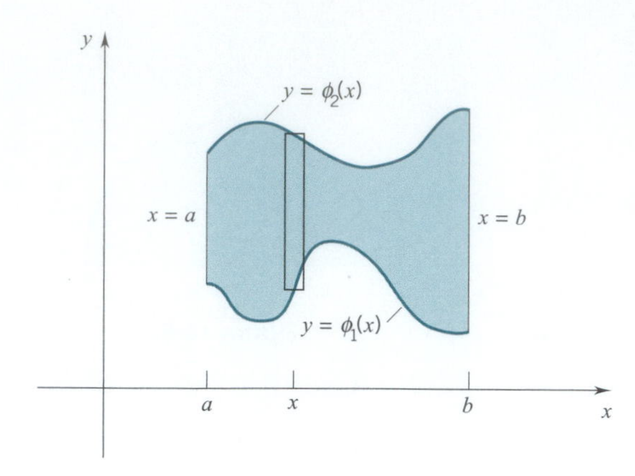

Figure 16.5.2

SOLUTION

$$\bar{x}A = \iint_{\Omega} x\ dxdy = \int_a^b \int_{\phi_1(x)}^{\phi_2(x)} x\,dy\,dx = \int_a^b x[\phi_2(x) - \phi_1(x)]\,dx;$$

$$\bar{y}A = \iint_{\Omega} y\ dxdy = \int_a^b \int_{\phi_1(x)}^{\phi_2(x)} y\,dy\,dx = \int_a^b \tfrac{1}{2}([\phi_2(x)]^2 - [\phi_1(x)]^2)\,dx.$$

These are the formulas for the centroid that we developed in Section 6.4. Having calculated many centroids there, we won't do so here. □

Kinetic Energy and Moment of Inertia

A particle of mass m at a distance r from a given line rotates about that line (called the *axis of rotation*) with angular speed ω. The speed v of the particle is then $r\omega$, and the kinetic energy is given by the formula

$$\text{KE} = \tfrac{1}{2}mv^2 = \tfrac{1}{2}mr^2\omega^2.$$

Imagine now a rigid body composed of a finite number of point masses m_i located at distances r_i from some fixed line. If the rigid body rotates about that line with angular speed ω, then all the point masses rotate about that same line with that same angular speed ω. The kinetic energy of the body can be obtained by adding up the kinetic energies of all the individual particles:

$$\text{KE} = \sum_i \tfrac{1}{2}m_i r_i^2\omega^2 = \tfrac{1}{2}\left(\sum_i m_i r_i^2\right)\omega^2.$$

The expression in parentheses is called the *moment of inertia* (or *rotational inertia*) of the body and is denoted by the letter I:

(16.5.4)

$$\boxed{I = \sum_i m_i r_i^2.}$$

For a rigid body in straight-line motion

$$\text{KE} = \tfrac{1}{2}Mv^2, \qquad \text{where } v \text{ is the speed of the body and } M = \sum m_i.$$

For a rigid body in rotational motion

$$\text{KE} = \tfrac{1}{2}I\omega^2, \qquad \text{where } \omega \text{ is the angular speed of the body.}$$

The Moment of Inertia of a Plate

Suppose that a plate in the shape of a basic region Ω rotates about a line. The moment of inertia of the plate about that axis of rotation is given by the formula

(16.5.5)
$$I = \iint_{\Omega} \lambda(x,y)[r(x,y)]^2 \, dxdy$$

where $\lambda = \lambda(x,y)$ is the mass density function and $r(x,y)$ is the distance from the axis to the point (x,y).

DERIVATION OF (16.5.5) Decompose the plate into N pieces in the form of basic regions $\Omega_1, \ldots, \Omega_N$. From each Ω_i choose a point (x_i^*, y_i^*) and view all the mass of the ith piece as concentrated there. The moment of inertia of this piece is then approximately

$$\underbrace{[\lambda(x_i^*, y_i^*)(\text{area of } \Omega_i)]}_{\text{approx. mass of piece}} \underbrace{[r(x_i^*, y_i^*)]^2}_{(\text{approx. distance})^2} = \lambda(x_i^*, y_i^*)[r(x_i^*, y_i^*)]^2(\text{area of } \Omega_i).$$

The sum of these approximations,

$$\sum_{i=1}^{N} \lambda(x_i^*, y_i^*)[r(x_i^*, y_i^*)]^2(\text{area of } \Omega_i),$$

is a Riemann sum for the double integral

$$\iint_{\Omega} \lambda(x,y)[r(x,y)]^2 \, dxdy.$$

As the maximum diameter of the Ω_i tends to zero, the Riemann sum tends to this integral. □

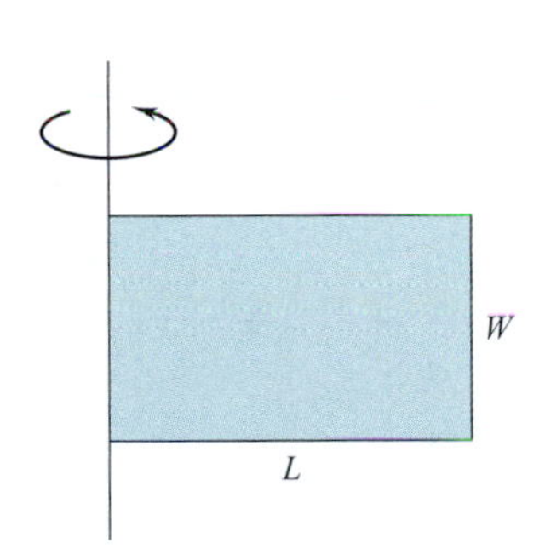

Figure 16.5.3

Example 3 A rectangular plate of mass M, length L, and width W rotates about the line shown in Figure 16.5.3. Find the moment of inertia of the plate about that line: **(a)** given that the plate has uniform mass density; **(b)** given that the mass density of the plate is directly proportional to the square of the distance from the rightmost side.

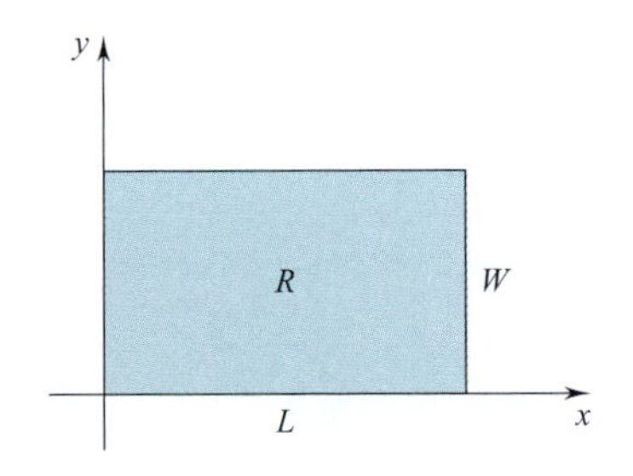

Figure 16.5.4

SOLUTION Coordinatize the plate as in Figure 16.5.4 and call the base region R.

(a) Here $\lambda(x,y) = M/LW$ and $r(x,y) = x$. Thus

$$I = \iint_R \frac{M}{LW}x^2 \, dxdy = \frac{M}{LW}\int_0^W \int_0^L x^2 \, dx \, dy$$

$$= \frac{M}{LW}W\int_0^L x^2 \, dx = \frac{M}{L}\left(\tfrac{1}{3}L^3\right) = \tfrac{1}{3}ML^2.$$

(b) In this case $\lambda(x,y) = k(L-x)^2$, but we still have $r(x,y) = x$. Therefore

$$I = \iint_R k(L-x)^2x^2\ dxdy = k\int_0^W \int_0^L (L-x)^2x^2\ dx\,dy$$

$$= kW\int_0^L (L^2x^2 - 2Lx^3 + x^4)\ dx = \tfrac{1}{30}kL^5W.$$

We can eliminate the constant of proportionality k by noting that

$$M = \iint_R k(L-x)^2\ dxdy = k\int_0^W \int_0^L (L-x)^2\ dx\,dy$$

$$= kW\Big[-\tfrac{1}{3}(L-x)^3\Big]_0^L = \tfrac{1}{3}kWL^3.$$

Therefore,

$$k = \frac{3M}{WL^3} \qquad \text{and} \qquad I = \tfrac{1}{30}\left(\frac{3M}{WL^3}\right)L^5W = \tfrac{1}{10}ML^2. \quad \square$$

Radius of Gyration

If the mass M of an object is all concentrated at a distance r from a given line, then the moment of inertia about that line is given by the product Mr^2.

Suppose now that we have a plate of mass M (actually any object of mass M will do here), and suppose that l is some line. The object has some moment of inertia I about l. Its *radius of gyration* about l is the distance K for which

$$I = MK^2.$$

Namely, the radius of gyration about l is the distance from l at which all the mass of the object would have to be concentrated to effect the same moment of inertia. The formula for radius of gyration is usually written

(16.5.6)
$$\boxed{K = \sqrt{I/M}.}$$

Example 4 A homogeneous circular plate of mass M and radius R rotates about an axle that passes through the center of the plate and is perpendicular to the plate. Calculate the moment of inertia and the radius of gyration.

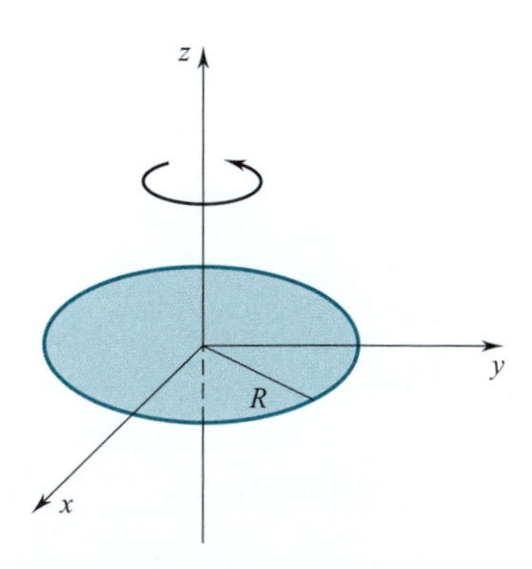

Figure 16.5.5

SOLUTION Take the axle as the z-axis and let the plate rest on the circular region $\Omega : x^2 + y^2 \le R^2$ (Figure 16.5.5). The density of the plate is $M/A = M/\pi R^2$ and $r(x,y) = \sqrt{x^2+y^2}$. Hence

$$I = \iint_\Omega \frac{M}{\pi R^2}(x^2+y^2)\ dxdy = \frac{M}{\pi R^2}\int_0^{2\pi}\int_0^R r^3\,dr\,d\theta = \tfrac{1}{2}MR^2.$$

The radius of gyration K is $K = \sqrt{I/M} = R/\sqrt{2}$.

The circular plate of radius R has the same moment of inertia about the central axle as a circular wire of the same mass with radius $R/\sqrt{2}$. $\square$

The Parallel Axis Theorem

Suppose we have an object of mass M and a line l_M that passes through the center of mass (x_M, y_M) of the object. The object has some moment of inertia about that line; call it I_M. If l is any line parallel to l_M, then the object has a certain moment of inertia about l; call that I. The parallel axis theorem states that

(16.5.7)
$$I = I_M + d^2 M$$

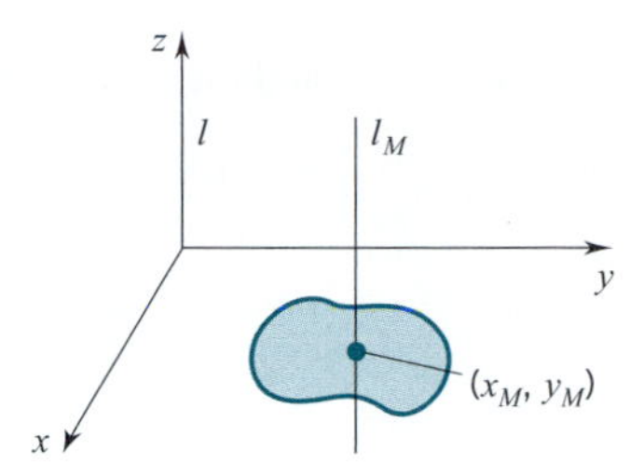

Figure 16.5.6

where d is the distance between the axes.

We prove the theorem under somewhat restrictive assumptions. Assume that the object is a plate of mass M in the shape of a basic region Ω, and assume that l_M is perpendicular to the plate. Call l the z-axis. Call the plane of the plate the xy-plane. (See Figure 16.5.6.) Denoting the points of Ω by (x, y) we have

$$\begin{aligned}
I - I_M &= \iint_\Omega \lambda(x,y)(x^2+y^2)\,dxdy - \iint_\Omega \lambda(x,y)[(x-x_M)^2+(y-y_M)^2]\,dxdy \\
&= \iint_\Omega \lambda(x,y)[2x_M x + 2y_M y - (x_M^2+y_M^2)]\,dxdy \\
&= 2x_M \iint_\Omega x\lambda(x,y)\,dxdy + 2y_M \iint_\Omega y\lambda(x,y)\,dxdy \\
&\quad -(x_M^2+y_M^2)\iint_\Omega \lambda(x,y)\,dxdy \\
&= 2x_M^2 M + 2y_M^2 M - (x_M^2+y_M^2)M = (x_M^2+y_M^2)M = d^2 M. \quad \square
\end{aligned}$$

An obvious consequence of the parallel axis theorem is that $I_M \leq I$ for all lines l parallel to l_M. To minimize the moment of inertia we must pass the axis of rotation through the center of mass.

EXERCISES 16.5

Find the mass and center of mass of the plate that occupies the region Ω and has the density function λ.

1. $\Omega: \quad -1 \leq x \leq 1, \quad 0 \leq y \leq 1, \quad \lambda(x,y) = x^2.$
2. $\Omega: \quad 0 \leq x \leq 1, \quad 0 \leq y \leq \sqrt{x}, \quad \lambda(x,y) = x + y.$
3. $\Omega: \quad 0 \leq x \leq 1, \quad x^2 \leq y \leq 1, \quad \lambda(x,y) = xy.$
4. $\Omega: \quad 0 \leq x \leq \pi, \quad 0 \leq y \leq \sin x, \quad \lambda(x,y) = y.$
5. $\Omega: \quad 0 \leq x \leq 8, \quad 0 \leq y \leq \sqrt[3]{x}, \quad \lambda(x,y) = y^2.$
6. $\Omega: \quad 0 \leq x \leq a, \quad 0 \leq y \leq \sqrt{a^2 - x^2}, \quad \lambda(x,y) = xy.$
7. Ω the triangle with vertices $(0,0)$, $(1,2)$, and $(1,3)$; $\lambda(x,y) = xy$.
8. Ω : the triangular region in the first quadrant bounded by the lines $x = 0$, $y = 0$, and $3x + 2y = 6$; $\lambda(x,y) = x + y$.
9. Ω : the region bounded by the cardioid $r = 1 + \cos\theta$; λ is the distance to the pole.
10. Ω: the region inside the circle $r = 2\sin\theta$ but outside the circle $r = 1$; $\lambda(x,y) = y$.

In the exercises that follow, I_x, I_y, I_z denote the moments of inertia about the x, y, z axes.

11. A rectangular plate of mass M, length L, and width W is placed on the xy-plane with center at the origin, long sides parallel to the x-axis. (We assume here that $L \geq W$.) Find I_x, I_y, I_z if the plate is homogeneous. Determine the corresponding radii of gyration K_x, K_y, K_z.
12. Verify that I_x, I_y, I_z are unchanged if the mass density of the plate of Exercise 11 varies directly as the distance from the leftmost side.
13. Determine the center of mass of the plate of Exercise 11 if the mass density varies as in Exercise 12.

14. Show that for any plate in the xy-plane

$$I_z = I_x + I_y.$$

How are the corresponding radii of gyration K_x, K_y, K_z related?

15. A homogeneous plate of mass M in the form of a quarter disc of radius R is placed in the xy-plane as in the figure. Find I_x, I_y, I_z and the corresponding radii of gyration.

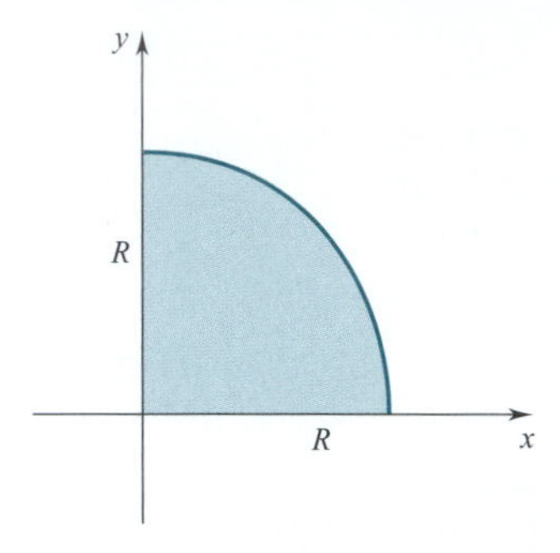

16. A plate in the xy-plane undergoes a rotation in that plane about its center of mass. Show that I_z remains unchanged.

17. A homogeneous disc of mass M and radius R is to be placed on the xy-plane so that it has moment of inertia I_0 about the z-axis. Where should the disc be placed?

18. A homogeneous plate of mass density λ occupies the region under the curve $y = f(x)$ from $x = a$ to $x = b$. Show that

$$I_x = \tfrac{1}{3}\lambda \int_a^b [f(x)]^3\, dx \qquad \text{and} \qquad I_y = \lambda \int_a^b x^2 f(x)\, dx.$$

19. A homogeneous plate of mass M in the form of an elliptical quadrant is placed on the xy-plane. (See the figure.) Find I_x, I_y, I_z.

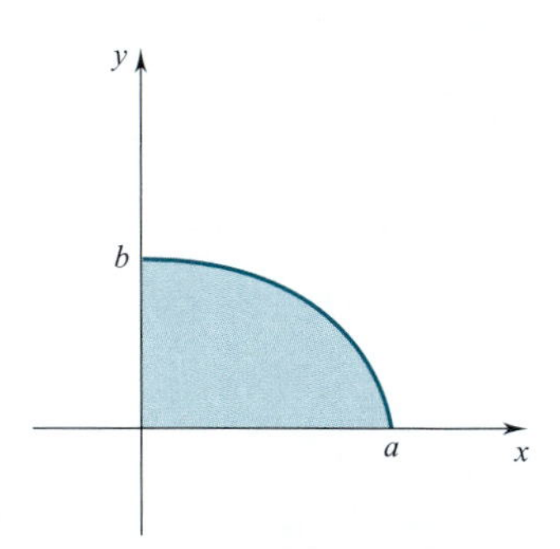

20. Find I_x, I_y, I_z for the plate in Exercise 2.

21. Find I_x, I_y, I_z for the plate in Exercise 3.

22. Find I_x, I_y, I_z for the plate in Exercise 5.

23. Find I_x, I_y, I_z for the plate in Exercise 9.

24. A plate of varying density occupies the region $\Omega = \Omega_1 \cup \Omega_2$ shown in the figure. Find the center of mass of the plate given that the Ω_1 piece has mass M_1 and center of mass (x_1, y_1), and the Ω_2 piece has mass M_2 and center of mass (x_2, y_2).

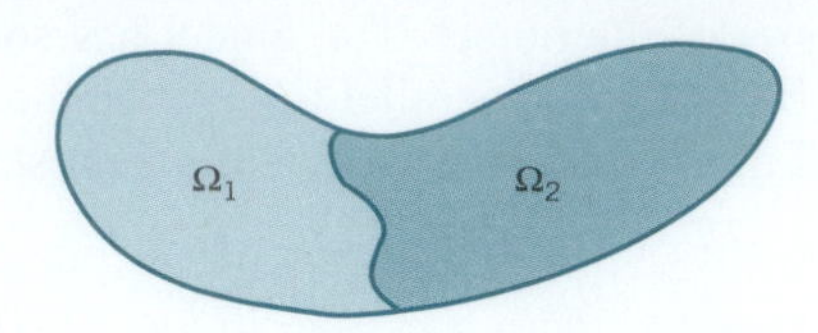

25. A homogeneous plate of mass M is in the form of a ring. (See the figure.) Calculate the moment of inertia of the plate:

(a) about a diameter;

(b) about a tangent to the inner circle;

(c) about a tangent to the outer circle.

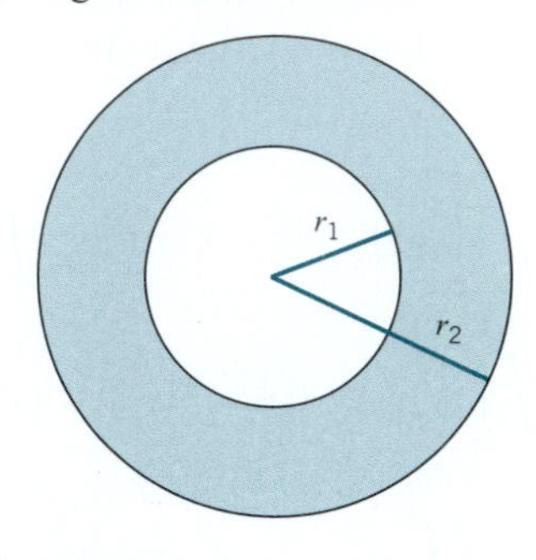

26. Find the moment of inertia of a homogeneous circular wire of mass M and radius r:

(a) about a diameter;

(b) about a tangent. HINT: Use the previous exercise.

27. The plate of Exercise 25 rotates about the axis that is perpendicular to the plate and passes through the center. Find the moment of inertia.

28. Prove the parallel axis theorem for the case where the line through the center of mass lies in the plane of the plate.

29. A plate of mass M has the form of a half disc Ω, $-R \le x \le R, 0 \le y \le \sqrt{R^2 - x^2}$. Find the center of mass given that the mass density varies directly as the distance from the curved boundary.

30. Find I_x, I_y, I_z for the plate of Exercise 29.

31. A plate of mass M is in the form of a disc of radius R. Given that the mass density of the plate varies directly as the distance from a point P on the boundary of the plate, locate the center of mass.

32. A plate of mass M is in the form of a right triangle of base b and height h. Given that the mass density of the plate varies directly as the square of the distance from the vertex of the right angle, locate the center of mass of the plate.

33. Use double integrals to justify an assumption we made about centroids in Chapter 6: Formula (6.4.1).

C **34.** A plate is in the form of a triangle with vertices $(0,0)$, $(0,1)$, $(2,1)$. The mass density of the plate is given by $\lambda(x,y) = x + y$. Use a CAS to find: (a) the center of mass of the plate, and (b) the moments of inertia I_x and I_y.

■ 16.6 TRIPLE INTEGRALS

Now that you are familiar with double integrals

$$\iint_{\Omega} f(x, y)\, dxdy,$$

you will find it easy to understand triple integrals

$$\iiint_{T} f(x, y, z)\, dxdydz.$$

Basically the only difference is that instead of working with functions of two variables continuous on a plane region Ω, we will be working with functions of three variables continuous on some portion T of three-space.

The Triple Integral over a Box

For double integration we began with a rectangle

$$R\colon a_1 \le x \le a_2, \quad b_1 \le y \le b_2.$$

For triple integration we begin with a *box* (a rectangular solid)

$$\Pi\colon a_1 \le x \le a_2, \quad b_1 \le y \le b_2, \quad c_1 \le z \le c_2. \qquad \text{(Figure 16.6.1)}$$

To partition this box, we first partition the edges. Taking

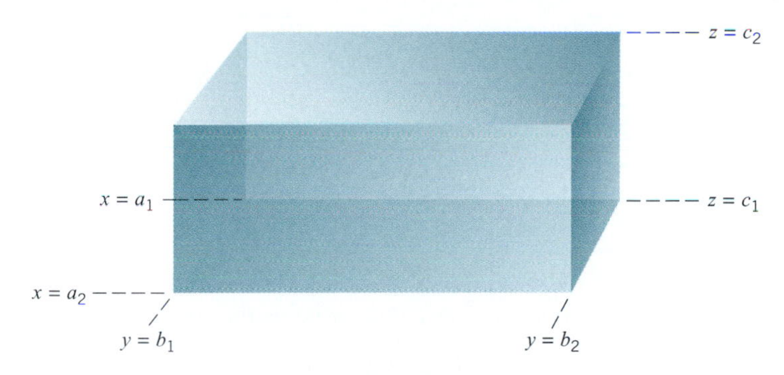

Figure 16.6.1

$$\begin{aligned} P_1 &= \{x_0, \ldots, x_m\} && \text{as a partition of } [a_1, a_2], \\ P_2 &= \{y_0, \ldots, y_n\} && \text{as a partition of } [b_1, b_2], \\ P_3 &= \{z_0, \ldots, z_q\} && \text{as a partition of } [c_1, c_2], \end{aligned}$$

we form the set

$$P = P_1 \times P_2 \times P_3 = \{(x_i, y_j, z_k) : x_i \in P_1, y_j \in P_2, z_k \in P_3\}^\dagger$$

and call this a *partition of* Π. The partition P breaks up Π into $m \times n \times q$ nonoverlapping boxes

$$\Pi_{ijk}\colon \quad x_{i-1} \le x \le x_i, \quad y_{j-1} \le y \le y_j, \quad z_{k-1} \le z \le z_k.$$

---

†$P_1 \times P_2 \times P_3$ is called the *Cartesian product* of P_1, P_2, and P_3.

A typical such box is pictured in Figured 16.6.2.

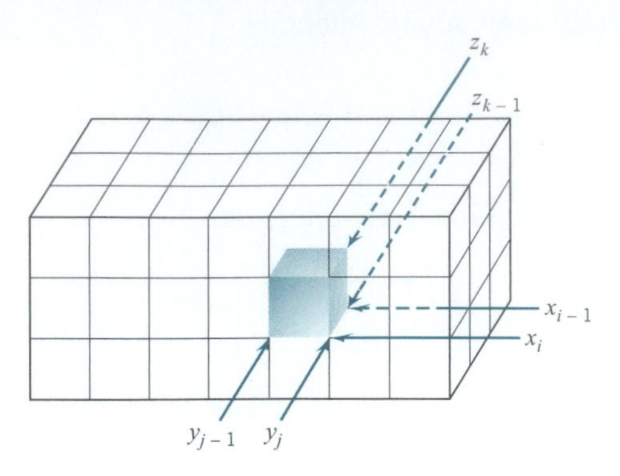

Figure 16.6.2

Let f be continuous on Π. Then, taking

$$M_{ijk} \text{ as the maximum value of } f \text{ on } \Pi_{ijk}$$

and

$$m_{ijk} \text{ as the minimum value of } f \text{ on } \Pi_{ijk},$$

we form the *upper sum*

$$U_f(P) = \sum_{i=1}^{m}\sum_{j=1}^{n}\sum_{k=1}^{q} M_{ijk}(\text{ volume of } \Pi_{ijk}) = \sum_{i=1}^{m}\sum_{j=1}^{n}\sum_{k=1}^{q} M_{ijk}\,\Delta x_i\;\Delta y_j\;\Delta z_k$$

and the *lower sum*

$$Ł_f(P) = \sum_{i=1}^{m}\sum_{j=1}^{n}\sum_{k=1}^{q} m_{ijk}(\text{ volume of } \Pi_{ijk}) = \sum_{i=1}^{m}\sum_{j=1}^{n}\sum_{k=1}^{q} m_{ijk}\,\Delta x_i\;\Delta y_j\;\Delta z_k.$$

As in the case of functions of one and two variables, it turns out that, with f continuous on Π, there is one and only one number I that satisfies the inequality

$$L_f(P) \le I \le u_f(P) \qquad \text{for all partitions } P \text{ of } \Pi.$$

DEFINITION 16.6.1 THE TRIPLE INTEGRAL OVER A BOX $\prod$

Let f be continuous on a closed box Π. The unique number I that satisfies the inequality

$$L_f(P) \le I \le U_f(P) \qquad \text{for all partitions } P \text{ of } \Pi$$

is called the *triple integral* of f over Π and is denoted by

$$\iiint_{\Pi} f(x, y, z)\; dxdydz.\dagger$$

† The triple integral can also be written $\int\int_{\Pi}\int f(x, y, z)\, dV$

The Triple Integral over a More General Solid

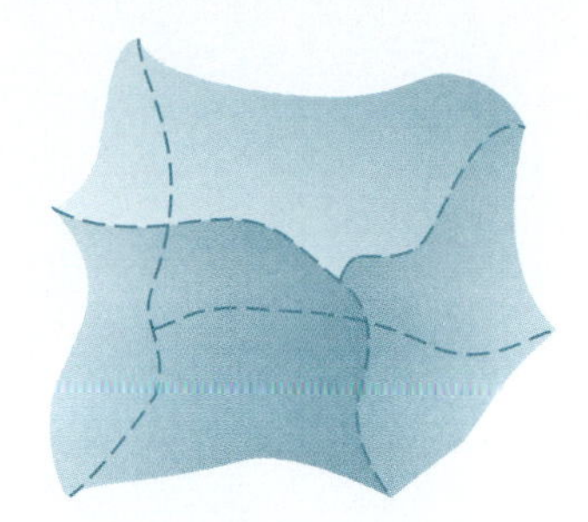

Figure 16.6.3

We start with a three-dimensional, bounded, closed, connected set T. We assume that T is a *basic solid;* that is, we assume that the boundary of T consists of a finite number of continuous surfaces $z = \alpha(x, y)$, $y = \beta(x, z)$, $x = \gamma(y, z)$. See, for example, Figure 16.6.3.

Now let's suppose that f is some function continuous on T. To define the triple integral of f over T, we first encase T in a rectangular box Π with sides parallel to the coordinate planes (Figure 16.6.4). We then extend f to all of Π by defining f to be zero outside of T. This extended function f is bounded on Π, and it is continuous on all of Π except possibly at the boundary of T. In spite of these possible discontinuities, f is still integrable over Π; that is, there still exists a unique number I such that

$$L_f(P) \leq I \leq U_f(P) \qquad \text{for all partitions } P \text{ of } \Pi$$

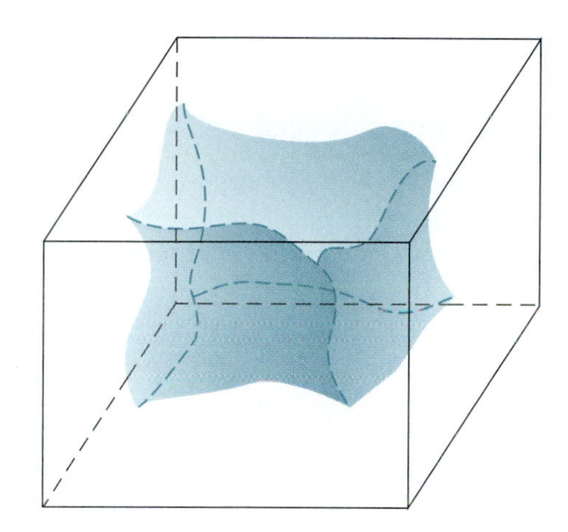

Figure 16.6.4

(We will not attempt to prove this.) The number I is by definition the triple integral

$$\iiint_{\Pi} f(x, y, z)\ dxdydz.$$

We define the triple integral over T by setting

(16.6.2)
$$\iiint_{T} f(x, y, z)\ dxdydz = \iiint_{\Pi} f(x, y, z)\ dxdydz.$$

Volume as a Triple Integral

The simplest triple integral of interest is the triple integral of the function that is constantly 1 on T. This gives the volume of T:

(16.6.3)
$$\text{volume of } T = \iiint_{T} dxdydz.$$

PROOF Set $f(x, y, z) = 1$ for all (x, y, z) in T. Encase T in a box Π. Define f to be zero outside of T. An arbitrary partition P of Π breaks up T into little boxes Π_{ijk}. Note that

$$L_f(P) = \text{ the sum of the volumes of all the } \Pi_{ijk} \text{ that are contained in } T$$

$$U_f(P) = \text{ the sum of the volumes of all the } \Pi_{ijk} \text{ that intersect } T.$$

It follows that

$$L_f(P) \leq \text{ the volume of } T \leq U_f(P).$$

The arbitrariness of P gives the formula. ❑

Remark As in the case of double integrals, there is no implied order in writing *dxdydz* in (16.6.2) and (16.6.3). The *dxdydz* represents an element of volume and could just as well have been written in any other order. An order of integration will be introduced when we evaluate a triple integral by means of a repeated integral. ❑

Some Properties of the Triple Integral

Below we give without proof the elementary properties of triple integrals analogous to what you saw in the one- and two-variable cases. Assume throughout that T is a basic solid. The functions f and g are assumed to be continuous on T.

I. Linearity:

$$\iiint_T [\alpha f(x, y, z) + \beta g(x, y, z)]\ dxdydz$$
$$= \alpha \iiint_T f(x, y, z)\ dxdydz + \beta \iiint_T g(x, y, z)\ dxdydz.$$

II. Order:

$$\text{if } f \geq 0 \text{ on } T, \quad \text{then} \quad \iiint_T f(x, y, z)\ dxdydz \geq 0;$$

$$\text{if } f \leq g \text{ on } T, \quad \text{then} \quad \iiint_T f(x, y, z)\ dxdydz \leq \iiint_T g(x, y, z)\ dxdydz.$$

III. Additivity: If T is broken up into a finite number of basic solids $T_1, \ldots, T_n$, then

$$\iiint_T f(x, y, z)\ dxdydz = \iiint_{T_1} f(x, y, z)\ dxdydz +$$
$$\cdots + \iiint_{T_n} f(x, y, z)\ dxdydz.$$

IV. Mean-value condition: There is a point (x_0, y_0, z_0) in T for which

$$\iiint_T f(x, y, z)\ dxdydz = f(x_0, y_0, z_0) \cdot (\text{ volume of } T).$$

We call $f(x_0, y_0, z_0)$ *the average value of* f *on* T.

The notion of average enables us to write

(16.6.4)
$$\iiint_T f(x,y,z)\ dxdydz = \left(\begin{array}{c}\text{the average value}\\ \text{of } f \text{ on } T\end{array}\right)\cdot(\text{ volume of } T).$$

We can also take weighted averages: if f and g are continuous and g is nonnegative on T, then there is a point (x_0, y_0, z_0) in T for which

(16.6.5)
$$\iiint_T f(x,y,z)\, g(x,y,z)\ dxdydz = f(x_0, y_0, z_0)\iiint_T g(x,y,z)\ dxdydz.$$

As you would expect, we call $f(x_0, y_0, z_0)$ *the g-weighted average of* f *on* T.

The formulas for mass, center of mass, and moments of inertia derived in the previous section for two-dimensional plates are easily extended to three-dimensional objects.

Suppose that T is an object in the form of a basic solid. If T has constant mass density λ (here density is mass per unit volume), then the mass of T is the density λ times the volume of T:

$$M = \lambda V.$$

If the mass density varies continuously over T, say, $\lambda = \lambda(x, y, z)$, then *the mass of* T *is the average density of* T *times the volume of* T. This is a triple integral

(16.6.6)
$$M = \iiint_T \lambda(x,y,z)\ dxdydz.$$

The coordinates of the center of mass (x_M, y_M, z_M) are density-weighted averages of position, each taken over the portion of space occupied by the solid.

(16.6.7)
$$x_M M = \iiint_T x\lambda(x,y,z)\ dxdydz, \qquad \text{etc.}$$

If the object T is homogeneous (constant mass density M/V), then the center of mass of T depends only on the geometry of T and falls on the centroid $(\bar{x}, \bar{y}, \bar{z})$ of the space occupied by T. The density is irrelevant. The coordinates of the centroid are simple averages over T:

(16.6.8)
$$\bar{x}V = \iiint_T x\ dxdydz, \qquad \text{etc.}$$

The moment of inertia of T about a line is given by the formula

(16.6.9)
$$I = \iiint_T \lambda(x,y,z)[r(x,y,z)]^2\ dxdydz.$$

Here $\lambda(x, y, z)$ is the mass density of T at (x, y, z) and $r(x, y, z)$ is the distance of (x, y, z) from the line in question. The moments of inertia about the x, y, z axes are again denoted by I_x, I_y, I_z.

All of this should be readily understandable. Techniques for evaluating triple integrals are introduced in the next three sections.

EXERCISES 16.6

1. Let $f(x,y)$ be a function continuous and nonnegative on a basic region Ω and set

$$T = \{(x, y, z) : (x, y) \in \Omega, \quad 0 \le z \le f(x, y)\}.$$

Compare

$$\iiint_T dxdydz \quad \text{to} \quad \iint_\Omega f(x, y)\, dxdy.$$

2. Set $f(x, y, z) = xyz$ on $\Pi : 0 \le x \le 1, \quad 0 \le y \le 1,$ $0 \le z \le 1$ and take P as the partition $P_1 \times P_2 \times P_3$.

(a) Find $L_f(P)$ and $U_f(P)$ given that

$$P_1 = \{x_0, \dots, x_m\}, \quad P_2 = \{y_0, \dots, y_n\},$$
$$P_3 = \{z_0, \dots, z_q\}$$

are all arbitrary partitions of [0, 1].

(b) Use your answer to (a) to calculate

$$\iiint_\Pi xyz\, dxdydz.$$

3. Let $f(x, y, z) = \alpha$, constant, over the rectangular solid Π : $a_1 \le x \le a_2, \; b_1 \le y \le b_2, \; c_1 \le z \le c_2$. Show that

$$\iiint_\Pi \alpha\, dxdydz = \alpha(a_2 - a_1)(b_2 - b_1)(c_2 - c_1).$$

4. Find the average value of $f(x, y, z) = xyz$ over the region Π in Exercise 2.

5. Calculate

$$\iiint_\Pi xy\, dxdydz \quad \text{where} \quad \Pi : 0 \le x \le a, \; 0 \le y \le b, \quad 0 \le z \le c.$$

6. Let T be a basic solid of varying mass density $\lambda = \lambda(x, y, z)$. The moment of inertia of T about the xy-plane is defined by setting

$$I_{xy} = \iiint_T \lambda(x, y, z) z^2\, dxdydz.$$

The other plane moments of inertia, I_{xz} and I_{yz}, have comparable definitions. Express I_x, I_y, I_z in terms of the plane moments of inertia.

7. A box Π_1 : $0 \le x \le 2a, 0 \le y \le 2b, 0 \le z \le 2c$ is cut away from a larger box $\Pi_0 : 0 \le x \le 2A, 0 \le y \le 2B,$ $0 \le z \le 2C$. Locate the centroid of the remaining solid.

8. Show that, if f is continuous and nonnegative on a basic solid T, then the triple integral of f over T is nonnegative.

9. Calculate the mass M of the cube Π: $0 \le x \le a,$ $0 \le y \le a, 0 \le z \le a$ given that the mass density varies directly with the distance from the face on the xy-plane.

10. Locate the center of mass of the cube of Exercise 9.

11. Find the moment of inertia I_z of the cube of Exercise 9.

C **12.** Let $f(x, y, z) = 3y^2 - 2x + z$ on the box $B : 2 \le x \le 5,$ $1 \le y \le 3, 3 \le z \le 4$. Let P_1 be a regular partition of $[2, 5]$ with $k = 100$ subintervals, let P_2 be a regular partition of $[1, 3]$ with $m = 200$ subintervals, let P_3 be a regular partition of $[3, 4]$ with $n = 150$ subintervals, and let $P = P_1 \times P_2 \times P_3$.

(a) Use a CAS to find $L_f(P)$ and $U_f(P)$.

(b) Investigate $L_f(P)$ and $U_f(P)$ for values of $k > 100,$ $m > 200$, and $n > 150$.

(c) Estimate $\int\int\int_B f(x, y, z)\, dxdydz$.

■ 16.7 REDUCTION TO REPEATED INTEGRALS

In this section we give no proofs. You can assume that all the solids that appear are basic solids and all the functions that you encounter are continuous.

In Figure 16.7.1 we have sketched a solid T. The projection of T onto the xy-plane has been labeled Ω_{xy}. The solid T is then the set of all (x, y, z) with

$$(x, y) \text{ in } \Omega_{xy} \qquad \text{and} \qquad \psi_1(x, y) \le z \le \psi_2(x, y).$$

The triple integral over T can be evaluated by setting

$$(*)\qquad \iiint_T f(x,y,z)\,dxdydz = \iint_{\Omega_{xy}} \left(\int_{\psi_1(x,y)}^{\psi_2(x,y)} f(x,y,z)\,dz\right) dxdy.$$

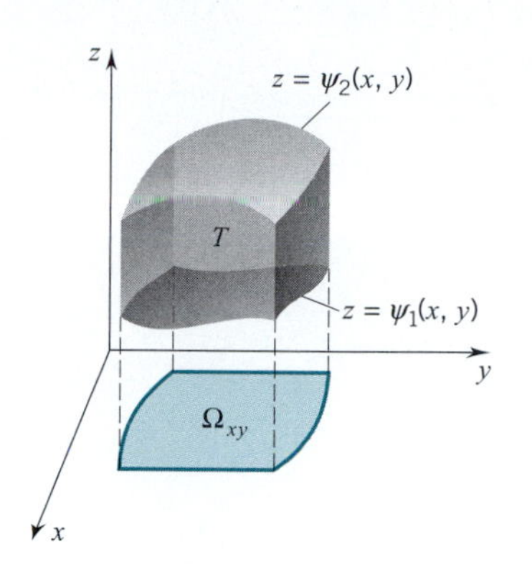

Figure 16.7.1

Moving to Figure 16.7.2, we see that in this case Ω_{xy} is the region

$$a_1 \le x \le a_2, \qquad \phi_1(x) \le y \le \phi_2(x),$$

and T itself is the set of all (x, y, z) with

$$a_1 \le x \le a_2, \qquad \phi_1(x) \le y \le \phi_2(x), \qquad \psi_1(x,y) \le z \le \psi_2(x,y).$$

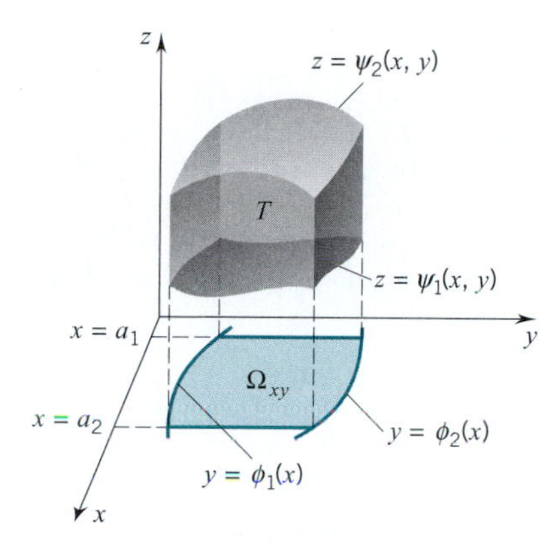

Figure 16.7.2

The triple integral over T can then be expressed by three ordinary integrals:

$$\iiint_T f(x,y,z)\,dxdydz = \int_{a_1}^{a_2} \left[\int_{\phi_1(x)}^{\phi_2(x)} \left(\int_{\psi_1(x,y)}^{\psi_2(x,y)} f(x,y,z)\,dz\right) dy\right] dx.$$

It is customary to omit the brackets and parentheses and write

(16.7.1)
$$\iiint_T f(x,y,z)\,dxdydz = \int_{a_1}^{a_2} \int_{\phi_1(x)}^{\phi_2(x)} \int_{\psi_1(x,y)}^{\psi_2(x,y)} f(x,y,z)\,dz\,dy\,dx.\ \dagger$$

Here we first integrate with respect to z [from $z = \psi_1(x,y)$ to $z = \psi_2(x,y)$], then with respect to y [from $y = \phi_1(x)$ to $y = \phi_2(x)$], and finally with respect to x [from $x = a_1$ to $x = a_2$].

There is nothing special about this order of integration. Other orders of integration are possible and in some cases more convenient. Suppose, for example, that the projection of T onto the xz-plane is a region of the form

$$\Omega_{xz}\colon a_1 \le z \le a_2, \qquad \phi_1(z) \le x \le \phi_2(z).$$

† This formula is formula (∗) taken one step further. Usually we skip the double-integral stage and go directly to three integrals.

If T is the set of all (x, y, z) with

$$a_1 \leq z \leq a_2, \quad \phi_1(z) \leq x \leq \phi_2(z), \quad \psi_1(x,z) \leq y \leq \psi_2(x,z),$$

then
$$\iiint_T f(x,y,z)\ dxdydz = \int_{a_1}^{a_2} \int_{\phi_1(z)}^{\phi_2(z)} \int_{\psi_1(x,z)}^{\psi_2(x,z)} f(x,y,z)\, dy\, dx\, dz.$$

In this case we integrate first with respect to y, then with respect to x, and finally with respect to z. Still four other orders of integration are possible.

Example 1 Evaluate the expression $\displaystyle\int_0^2 \int_0^x \int_0^{4-x^2} xyz\, dz\, dy\, dx.$

SOLUTION

$$\begin{aligned}
\int_0^2 \int_0^x \int_0^{4-x^2} xyz\, dz\, dy\, dx &= \int_0^2 \int_0^x \left(\int_0^{4-x^2} xyz\, dz \right) dy\, dx \\
&= \int_0^2 \int_0^x \left(\left[\tfrac{1}{2} xyz^2 \right]_0^{4-x^2} \right) dy\, dx \\
&= \tfrac{1}{2} \int_0^2 \int_0^x x(4-x^2)^2 y\, dy\, dx \\
&= \tfrac{1}{2} \int_0^2 \left(\int_0^x x(4-x^2)^2 y\, dy \right) dx \\
&= \tfrac{1}{2} \int_0^2 \left(\left[\tfrac{1}{2} x(4-x^2)^2 y^2 \right]_0^x \right) dx \\
&= \tfrac{1}{4} \int_0^2 x^3(4-x^2)^2\ dx = \tfrac{1}{4} \int_0^2 x^3(16 - 8x^2 + x^4)\, dx \\
&= \tfrac{1}{4} \left[4x^4 - \tfrac{8}{6}x^6 + \tfrac{1}{8}x^8 \right]_0^2 = \tfrac{8}{3}. \quad \square
\end{aligned}$$

Remark The solid determined by the limits of integration in Example 1 is the solid T in the first octant bounded by the parabolic cylinder $z = 4 - x^2$, the plane $z = 0$, the plane $y = x$, and the plane $y = 0$. This solid is shown in Figure 16.7.3. ❑

Example 2 Use triple integration to find the volume of the tetrahedron T shown in Figure 16.7.4, and find the coordinates of the centroid.

SOLUTION The volume of T is given by the triple integral

$$V = \iiint_T dxdydz.$$

To evaluate this triple integral, we can project T onto any one of the three coordinate planes. We will project onto the xy-plane. The base region is then the triangle

$$\Omega_{xy}: \quad 0 \leq x \leq 1, \quad 0 \leq y \leq 1 - x. \qquad \text{(Figure 16.7.5)}$$

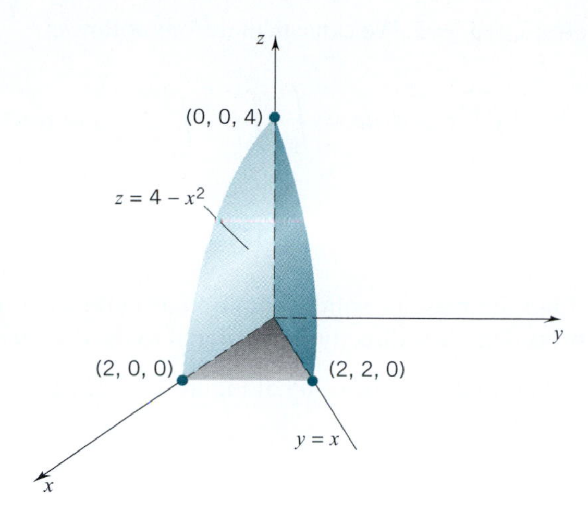

Figure 16.7.3

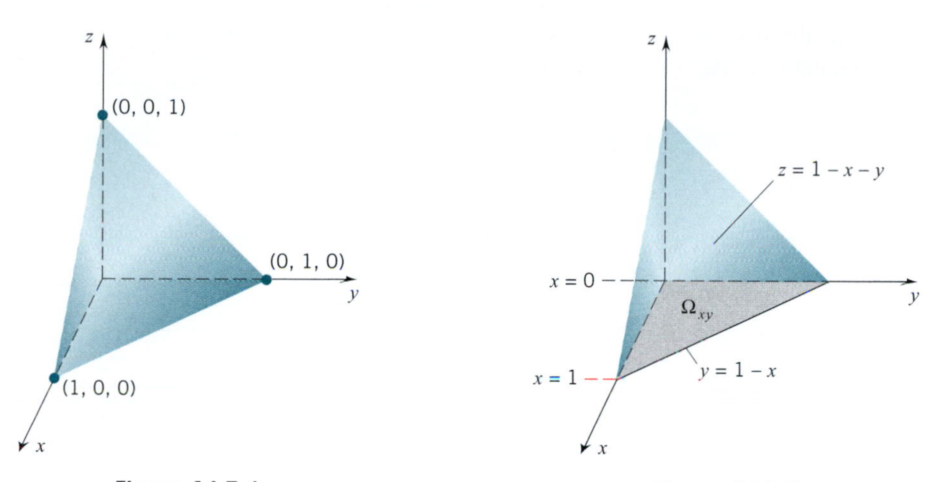

Figure 16.7.4

Figure 16.7.5

Since the inclined face is part of the plane $z = 1 - x - y$, we have T as the set of all (x, y, z) with

$$0 \leq x \leq 1, \quad 0 \leq y \leq 1 - x, \quad 0 \leq z \leq 1 - x - y.$$

It follows that

$$\begin{aligned} V = \iiint_T dxdydz &= \int_0^1 \int_0^{1-x} \int_0^{1-x-y} dz\, dy\, dx \\ &= \int_0^1 \int_0^{1-x} (1 - x - y)\, dy\, dx \\ &= \int_0^1 \left[(1-x)y - \tfrac{1}{2}y^2\right]_0^{1-x} dx \\ &= \int_0^1 \tfrac{1}{2}(1-x)^2\, dx = \left[-\tfrac{1}{6}(1-x)^3\right]_0^1 = \tfrac{1}{6}. \end{aligned}$$

By symmetry, $\bar{x} = \bar{y} = \bar{z}$. We can calculate $\bar{x}$ as follows:

$$\bar{x}\, V = \iiint_T x \; dxdydz = \int_0^1 \int_0^{1-x} \int_0^{1-x-y} x \, dz \, dy \, dx = \tfrac{1}{24}.$$

check this

Since $V = \frac{1}{6}$, we have $\bar{x} = \frac{1}{4}$. The centroid is the point $(\frac{1}{4}, \frac{1}{4}, \frac{1}{4})$. □

Example 3 Find the mass of solid right circular cylinder of radius r and height h given that the mass density is directly proportional to the distance from the lower base.

SOLUTION Call the solid T. in the setup of Figure 16.7.6, we can characterize T by the following inequalities:

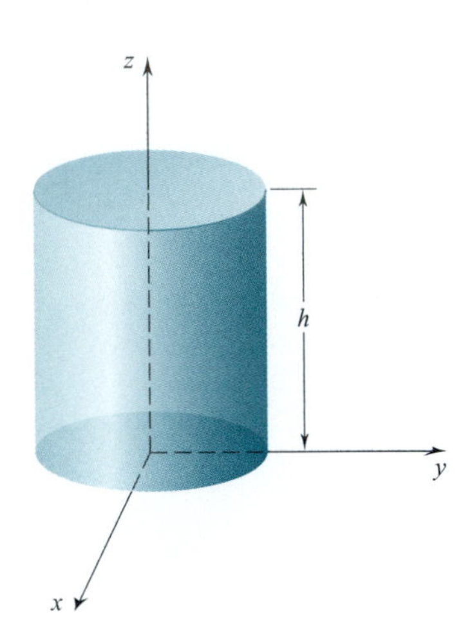

Figure 16.7.6

$$-r \le x \le r, \quad -\sqrt{r^2 - x^2} \le y \le \sqrt{r^2 - x^2}, \quad 0 \le z \le h.$$

The first two inequalities define the base region Ω_{xy}. Since the density varies directly with the distance from the lower base, we have $\lambda(x, y, z) = kz$, where $k > 0$ is the constant of proportionality. Then

$$\begin{aligned} M &= \iiint_T kz \; dxdydz \\ &= \int_{-r}^{r} \int_{-\sqrt{r^2-x^2}}^{\sqrt{r^2-x^2}} \int_0^h kz \, dz \, dy \, dx \\ &= \int_{-r}^{r} \int_{-\sqrt{r^2-x^2}}^{\sqrt{r^2-x^2}} \tfrac{1}{2} kh^2 \, dy \, dx \\ &= 4 \int_0^r \int_0^{\sqrt{r^2-x^2}} \tfrac{1}{2} kh^2 \, dy \, dx \qquad \text{(using the symmetry)} \\ &= 2kh^2 \int_0^r \sqrt{r^2 - x^2} \, dx. \end{aligned}$$

This integral can be evaluated by a trigonometric substitution (Section 8.4) or by Formula 87 in the Table of Integrals. Either way,

$$\int_0^r \sqrt{r^2 - x^2} \, dx = \tfrac{1}{4} \pi r^2.$$

It follows that

$$M = 2kh^2 (\tfrac{1}{4} \pi r^2) = \tfrac{1}{2} kh^2 r^2 \pi. \quad \square$$

Remark In Example 3 we would have profited by not skipping the double integral stage; namely, we could have written

$$\begin{aligned} M &= \iint_{\Omega_{xy}} \left(\int_0^h kz \, dz \right) \; dxdy = \iint_{\Omega_{xy}} \tfrac{1}{2} kh^2 \; dxdy \\ &= \tfrac{1}{2} kh^2 (\text{area of } \Omega_{xy}) = \tfrac{1}{2} kh^2 r^2 \pi. \quad \square \end{aligned}$$

Example 4 Integrate $f(x,y,z) = yz$ over that part of the first octant $x \geq 0$, $y \geq 0$, $z \geq 0$ that is contained in the ellipsoid

$$\frac{x^2}{a^2} + \frac{y^2}{b^2} + \frac{z^2}{c^2} = 1.$$

SOLUTION Call the solid T. The upper boundary of T has equation

$$z = \psi(x,y) = \frac{c}{ab}\sqrt{a^2b^2 - b^2x^2 - a^2y^2}.$$

This surface intersects the xy-plane in the curve

$$y = \phi(x) = \frac{b}{a}\sqrt{a^2 - x^2}.$$

We can take

$$\Omega_{xy}: \quad 0 \leq x \leq a, \quad 0 \leq y \leq \phi(x)$$

as the base region (see Figure 16.7.7) and characterize T as the set of all (x,y,z) with

$$0 \leq x \leq a, \quad 0 \leq y \leq \phi(x), \quad 0 \leq z \leq \psi(x,y).$$

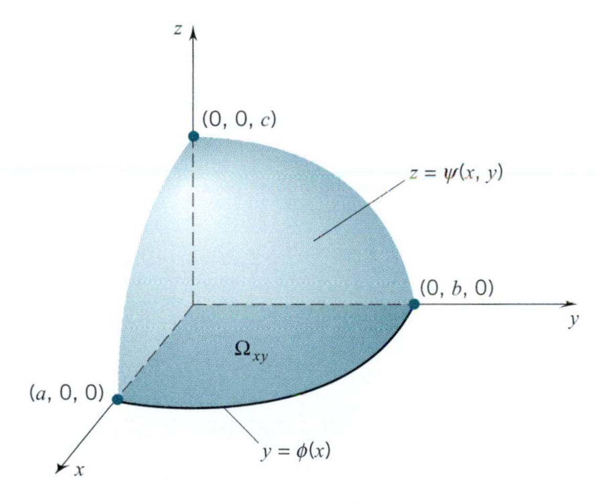

Figure 16.7.7

We can therefore calculate the triple integral by evaluating

$$\int_0^a \int_0^{\phi(x)} \int_0^{\psi(x,y)} yz\, dz\, dy\, dx.$$

A straightforward (but somewhat lengthy) computation that you can verify gives an answer of $\frac{1}{15}ab^2c^2$.

ANOTHER SOLUTION This time we carry out the integration in a different order. In Figure 16.7.8 you can see the same solid projected this time onto the yz-plane. In terms of y and z, the curved surface has equation

$$x = \Psi(y,z) = \frac{a}{bc}\sqrt{b^2c^2 - c^2y^2 - b^2z^2}.$$

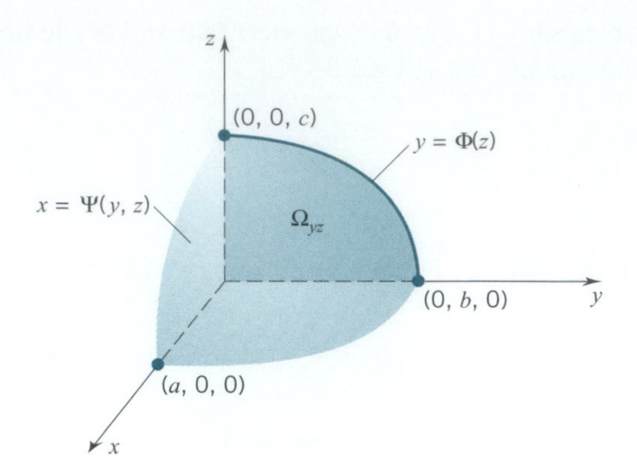

Figure 16.7.8

This surface intersects the yz-plane in the curve

$$y = \Phi(z) = \frac{b}{c}\sqrt{c^2 - z^2}.$$

We can take

$$\Omega_{yz}: \quad 0 \leq z \leq c, \quad 0 \leq y \leq \Phi(z)$$

as the base region and characterize T as the set of all (x, y, z) with

$$0 \leq z \leq c, \quad 0 \leq y \leq \Phi(z), \quad 0 \leq x \leq \Psi(y, z).$$

This leads to the repeated integral

$$\int_0^c \int_0^{\Phi(z)} \int_0^{\Psi(y,z)} yz \; dx \, dy \, dz,$$

which, as you can check, also gives $\frac{1}{15}ab^2c^2$. □

Example 5 Use triple integration to find the volume of the solid T bounded above by the parabolic cylinder $z = 4 - y^2$ and bounded below by the elliptic paraboloid $z = x^2 + 3y^2$.

SOLUTION Solving the two equations simultaneously, we have

$$4 - y^2 = x^2 + 3y^2 \qquad \text{and thus} \qquad x^2 + 4y^2 = 4.$$

This tells us that the two surfaces intersect in a space curve that lies on the elliptic cylinder $x^2 + 4y^2 = 4$. The projection of this intersection onto the xy-plane is the ellipse $x^2 + 4y^2 = 4$. (See Figure 16.7.9.)

The projection of T onto the xy-plane is the region

$$\Omega_{xy}: \quad -2 \leq x \leq 2, \quad -\tfrac{1}{2}\sqrt{4 - x^2} \leq y \leq \tfrac{1}{2}\sqrt{4 - x^2}.$$

The solid T is then the set of all (x, y, z) with

$$-2 \leq x \leq 2, \quad -\tfrac{1}{2}\sqrt{4 - x^2} \leq y \leq \tfrac{1}{2}\sqrt{4 - x^2}, \quad x^2 + 3y^2 \leq z \leq 4 - y^2.$$

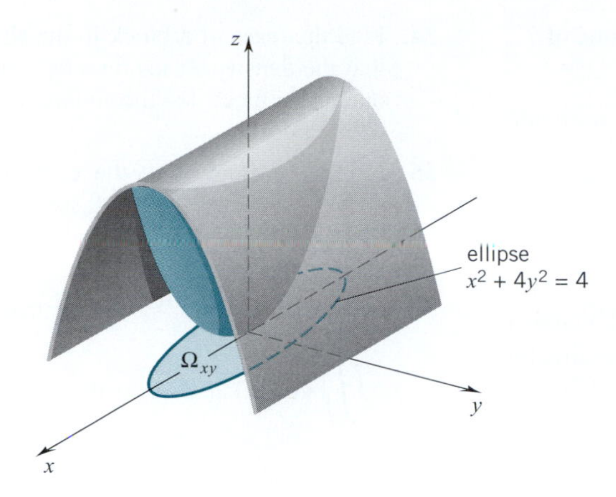

Figure 16.7.9

Its volume is given by

$$V = \int_{-2}^{2} \int_{-\frac{1}{2}\sqrt{4-x^2}}^{\frac{1}{2}\sqrt{4-x^2}} \int_{x^2+3y^2}^{4-y^2} dz\,dy\,dx$$

$$= 4\int_{0}^{2} \int_{0}^{\frac{1}{2}\sqrt{4-x^2}} \int_{x^2+3y^2}^{4-y^2} dz\,dy\,dx = 4\pi. \quad \square$$

explain ↑ ↑ check this

EXERCISES 16.7

Evaluate the repeated integral.

1. $\int_0^a \int_0^b \int_0^c dx\,dy\,dz.$

2. $\int_0^1 \int_0^x \int_0^y y\,dz\,dy\,dx.$

3. $\int_0^1 \int_1^{2y} \int_0^x (x + 2z)\,dz\,dx\,dy.$

4. $\int_0^1 \int_{1-x}^{1+x} \int_0^{xy} 4z\,dz\,dy\,dx.$

5. $\int_0^2 \int_{-1}^1 \int_1^3 (z - xy)\,dz\,dy\,dx.$

6. $\int_0^2 \int_{-1}^1 \int_1^3 (z - xz)\,dy\,dx\,dz.$

7. $\int_0^{\pi/2} \int_0^1 \int_0^{\sqrt{1-x^2}} x\cos z\,dy\,dx\,dz.$

8. $\int_{-1}^2 \int_1^{y-2} \int_e^{e^2} \frac{x+y}{z}\,dz\,dx\,dy.$

9. $\int_1^2 \int_y^{y^2} \int_0^{\ln x} y e^z\,dz\,dx\,dy.$

10. $\int_0^{\pi/2} \int_0^{\pi/2} \int_0^1 e^z \cos x \sin y\,dz\,dy\,dx.$

11. (*Separated variables over a box*) Set $\Pi : a_1 \le x \le a_2$, $b_1 \le y \le b_2, c_1 \le z \le c_2$. Show that, if f is continuous on $[a_1, a_2], g$ is continuous on $[b_1, b_2]$, and h is continuous on $[c_1, c_2]$, then

(16.7.2)
$$\iiint_{\Pi} f(x)\,g(y)\,h(z)\,dxdydz = \left(\int_{a_1}^{a_2} f(x)\,dx\right)\left(\int_{b_1}^{b_2} g(y)\,dy\right)\left(\int_{c_1}^{c_2} h(z)\,dz\right).$$

In Exercises 12 and 13, evaluate the triple integral, taking Π : $0 \le x \le 1,\ 0 \le y \le 2,\ 0 \le z \le 3.$

12. $\iiint_{\Pi} x^3y^2z\;dxdydz.$

13. $\iiint_{\Pi} x^2y^2z^2\;dxdydz.$

In Exercises 14–16, the mass density of a box $\Pi: 0 \le x \le a$, $0 \le y \le b,\ 0 \le z \le c$ varies directly with the product xyz.

14. Calculate the mass of Π.

15. Locate the center of mass.

16. Determine the moment of inertia of Π about: (a) the vertical line that passes through the point (a, b, c); (b) the vertical line that passes through the center of mass.

In Exercises 17–20, a homogeneous solid T of mass M consists of all points (x, y, z) with $0 \le x \le 1,\ 0 \le y \le 1,$ $0 \le z \le 1 - y.$

17. Sketch T. **18.** Find the volume of T.

19. Locate the center of mass.

20. Find the moments of inertia of T about the coordinate axes.

Express by repeated integrals. Do not evaluate.

21. The mass of a ball $x^2 + y^2 + z^2 \le r^2$ given that the density varies directly with the distance from the outer shell.

22. The mass of the solid bounded above by $z = 1$ and bounded below by $z = \sqrt{x^2 + y^2}$ given that the density varies directly with the distance from the origin. Identify the solid.

23. The volume of the solid bounded above by the parabolic cylinder $z = 1 - y^2$, below by the plane $2x + 3y + z + 10 = 0$, and on the sides by the circular cylinder $x^2 + y^2 - x = 0$.

24. The volume of the solid bounded above by the paraboloid $z = 4 - x^2 - y^2$ and bounded below by the parabolic cylinder $z = 2 + y^2$.

25. The mass of the solid bounded by the elliptic paraboloids $z = 4 - x^2 - \frac{1}{4}y^2$ and $z = 3x^2 + \frac{1}{4}y^2$ given that the density varies directly with the vertical distance from the lower surface.

26. The mass of the solid bounded by the paraboloid $x = z^2 + 2y^2$ and the parabolic cylinder $x = 4 - z^2$ given that the density varies directly with the distance from the z-axis.

Evaluate the triple integral.

27. $\iiint_T (x^2z + y)\, dxdydz$, where T is the solid bounded by the planes $x = 0, x = 1, y = 1, y = 3, z = 0, z = 2$.

28. $\iiint_T 2y\, e^x\, dxdydz$, where T is the solid given by $0 \le y \le 1,\ 0 \le x \le y,\ 0 \le z \le x + y$.

29. $\iiint_T x^2y^2z^2\, dxdydz$, where T is the solid bounded by the planes $z = y + 1,\ \ y + z = 1,\ \ x = 0,\ \ x = 1,\ \ z = 0$.

30. $\iiint_T xy\, dxdydz$, where T is the solid in the first octant bounded by the coordinate planes and the hemisphere $z = \sqrt{4 - x^2 - y^2}$.

31. $\iiint_T y^2\, dxdydz$, where T is the tetrahedron in the first octant bounded by the coordinate planes and the plane $2x + 3y + z = 6$.

32. $\iiint_T y^2\, dxdydz$, where T is the solid in the first octant bounded by the cylinders $x^2 + y = 1,\ z^2 + y = 1$.

33. Find the volume of the portion of the first octant bounded by the planes $z = x,\ y - x = 2$, and the cylinder $y = x^2$. Where is the centroid?

34. Find the mass of a block in the shape of a unit cube given that the density varies directly with: (a) the distance from one of the faces; (b) the square of the distance from one of the vertices.

35. Find the volume and the centroid of the solid bounded above by the cylindrical surface $x^2 + z = 4$, below by plane $x + z = 2$, and on the sides by the planes $y = 0$ and $y = 3$.

36. Show that, if $(\bar{x}, \bar{y}, \bar{z})$ is the centroid of a solid T, then

$$\iiint_T (x - \bar{x})\, dxdydz = 0,$$

$$\iiint_T (y - \bar{y})\, dxdydz = 0,$$

$$\iiint_T (z - \bar{z})\, dxdydz = 0.$$

37. Taking a, b, c as positive, find the volume of the tetrahedron with vertices $(0, 0, 0)$, $(a, 0, 0)$, $(0, b, 0)$, $(0, 0, c)$. Where is the centroid?

38. A homogeneous solid of mass M in the form and position of the tetrahedron of Figure 16.7.4 rotates about the z-axis. Find the moment of inertia I_z.

39. A homogeneous box of mass M has edges a, b, c. Calculate the moment of inertia

(a) about the edge c;

(b) about the line that passes through the center of the box and is parallel to the edge c;

(c) about the line that passes through the center of the face bc and is parallel to the edge c.

40. Where is the centroid of the solid bounded above by the plane $z = 1 + x + y$, below by the plane $z = -2$, and on the sides by the planes $x = 1,\ \ x = 2,\ \ y = 1,\ \ y = 2$?

41. Let T be the solid bounded above by the plane $z = y$, below by the xy-plane, and on the sides by the planes $x = 0,\ x = 1,\ y = 1$. Find the mass of T given that the density varies directly with the square of the distance from the origin. Where is the center of mass?

42. What can you conclude about T given that

$$\iiint_T f(x, y, z)\, dxdydz = 0$$

(a) for every continuous function f that is odd in x?

(b) for every continuous function f that is odd in y?

(c) for every continuous function f that is odd in z?

(d) for every continuous function f that satisfies the relation $f(-x, -y, -z) = -f(x, y, z)$?

43. (a) Integrate $f(x, y, z) = x + y^3 + z$ over the unit ball centered at the origin.

(b) Integrate $f(x, y, z) = a_1x + a_2y + a_3z + a_4$ over the unit ball centered at the origin.

44. Integrate $f(x,y,z) = x^2y^2$ over the solid bounded above by the cylinder $y^2 + z = 4$, below by the plane $y + z = 2$, and on the sides by the planes $x = 0$ and $x = 2$.

45. Use triple integrals to find the volume enclosed by the sphere $x^2 + y^2 + z^2 = a^2$.

46. Use triple integrals to find the volume enclosed by the ellipsoid

$$\frac{x^2}{a^2} + \frac{y^2}{b^2} + \frac{z^2}{c^2} = 1.$$

47. Find the mass of the solid in Example 5 given that the density varies directly with $|x|$.

48. Find the volume of the solid bounded by the paraboloids $z = 2 - x^2 - y^2$ and $z = x^2 + y^2$.

49. Find the mass and the center of mass of the solid of Exercise 35 given that the density varies directly with $1 + y$.

50. Let T be a solid with volume

$$V = \iiint_T dxdydz = \int_0^2 \int_0^{9-x^2} \int_0^{2-x} dz\,dy\,dx.$$

Sketch T and fill in the blanks.

(a) $V = \int_\square^\square \int_\square^\square \int_\square^\square dy\,dx\,dz.$

(b) $V = \int_\square^\square \int_\square^\square \int_\square^\square dy\,dz\,dx.$

(c) $V = \int_0^5 \int_\square^\square \int_\square^\square dz\,dx\,dy + \int_5^9 \int_\square^\square \int_\square^\square dz\,dx\,dy.$

51. Let T be a solid with volume

$$V = \iiint_T dxdydz = \int_0^3 \int_0^{6-x} \int_0^{2x} dz\,dy\,dx.$$

Sketch T and fill in the blanks.

(a) $V = \int_\square^\square \int_\square^\square \int_\square^\square dy\,dx\,dz.$

(b) $V = \int_\square^\square \int_\square^\square \int_\square^\square dy\,dz\,dx.$

(c) $V = \int_0^6 \int_\square^\square \int_\square^\square dx\,dy\,dz + \int_\square^\square \int_\square^\square \int_\square^\square dx\,dy\,dz.$

For the remaining exercises, let V be the volume of the solid T enclosed by the parabolic cylinder $y = 4 - z^2$ and the V-shaped cylinder $y = |x|$. Let $\Omega_{xy}, \Omega_{yz}, \Omega_{xz}$ be the projections of T onto the xy-, yz-, and xz-planes, respectively. Fill in the blanks.

52. (a) $V = \iint_{\Omega_{xy}} \square\, dxdy.$

(b) $V = \iint_{\Omega_{xy}} \left(\int_\square^\square dz\right) dxdy.$

(c) $V = \int_\square^\square \int_\square^\square \int_\square^\square dz\,dy\,dx.$

(d) $V = \int_\square^\square \int_\square^\square \int_\square^\square dz\,dx\,dy.$

53. (a) $V = \iint_{\Omega_{yz}} \square\, dydz.$

(b) $V = \iint_{\Omega_{yz}} \left(\int_\square^\square dx\right) dydz.$

(c) $V = \int_\square^\square \int_\square^\square \int_\square^\square dx\,dz\,dy.$

(d) $V = \int_\square^\square \int_\square^\square \int_\square^\square dx\,dy\,dz.$

54. (a) $V = \iint_{\Omega_{xz}} \square\, dxdz.$

(b) $V = \iint_{\Omega_{xz}} \left(\int_\square^\square dy\right) dxdz.$

(c) $V = \int_\square^\square \int_\square^\square \int_\square^\square dy\,dx\,dz.$

(d) $V = \int_{-2}^0 \int_\square^\square \int_\square^\square dy\,dz\,dx + \int_0^2 \int_\square^\square \int_\square^\square dy\,dz\,dx.$

C 55. Use a CAS to evaluate the triple integrals.

(a) $\int_2^4 \int_3^5 \int_1^2 \frac{\ln xy}{z}\, dz\,dy\,dx.$

(b) $\int_0^4 \int_1^2 \int_0^3 x\sqrt{yz}\, dz\,dy\,dx.$

C 56. Use a CAS to find the volume of the solid bounded by the plane $3x + 6y + 2z = 6$, the elliptic paraboloid $y = 36 - 9x^2 - 4z^2$, the yz-plane, and the xy-plane.

■ 16.8 CYLINDRICAL COORDINATES

Introduction to Cylindrical Coordinates

The cylindrical coordinates (r, θ, z) of a point P in xyz-space are shown geometrically in Figure 16.8.1. The first two coordinates, r and θ, are the usual plane polar coordinates,

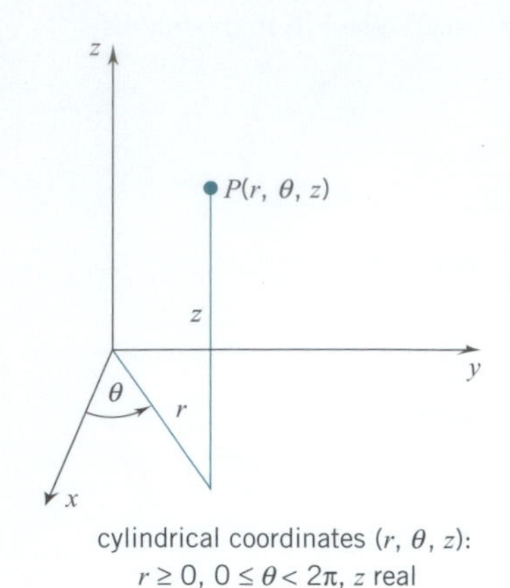

cylindrical coordinates (r, θ, z): $r \geq 0$, $0 \leq \theta < 2\pi$, z real

Figure 16.8.1

except that r is taken to be nonnegative and θ is restricted to the interval $[0, 2\pi]$. † The third coordinate is the third rectangular coordinate z.

In rectangular coordinates, the coordinate surfaces

$$x = x_0, \qquad y = y_0, \qquad z = z_0$$

are three mutually perpendicular planes. In cylindrical coordinates, the coordinate surfaces take the form

$$r = r_0, \qquad \theta = \theta_0, \qquad z = z_0. \qquad \text{(Figure 16.8.2)}$$

The surface $r = r_0$ is a right circular cylinder of radius r_0. The central axis of the cylinder is the z-axis. The surface $\theta = \theta_0$ is a vertical half-plane hinged at the z-axis. The plane stands at an angle of θ_0 radians from the positive x-axis. The last coordinate surface is the plane $z = z_0$.

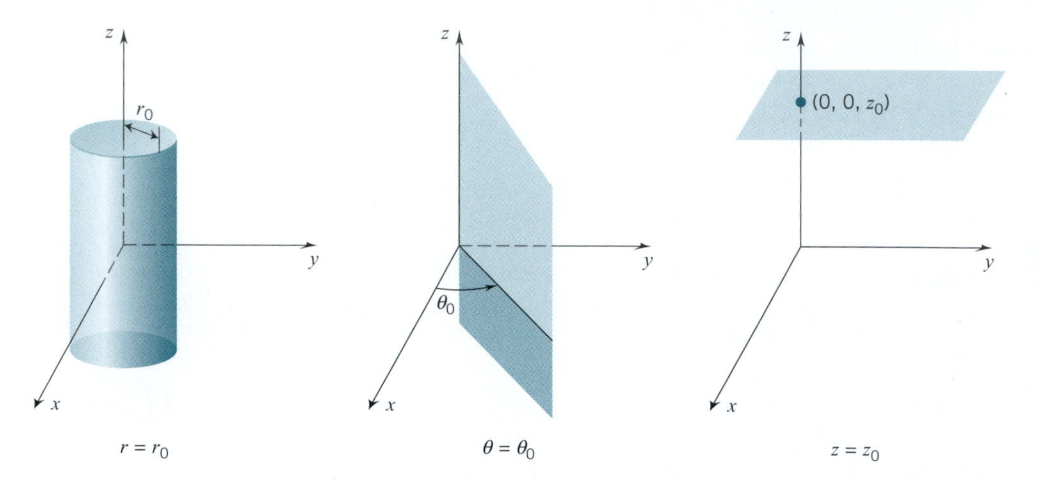

Figure 16.8.2

The point P with rectangular coordinates (x_0, y_0, z_0) lies on the plane $x = x_0$, on the plane $y = y_0$, and on the plane $z = z_0$. P is at the intersection of these three planes.

The point P with cylindrical coordinates (r_0, θ_0, z_0) lies on the cylinder $r = r_0$, on the vertical half-plane $\theta = \theta_0$, and on the horizontal plane $z = z_0$. P is at the intersection of these three surfaces.

Rectangular coordinates (x, y, z) can be obtained from cylindrical coordinates (r, θ, z) by means of the equations

$$x = r\cos\theta, \qquad y = r\sin\theta, \qquad z = z.$$

Conversely, with the obvious exclusions, cylindrical coordinates (r, θ, z) can be obtained from rectangular coordinates (x, y, z) by means of the equations

$$r = \sqrt{x^2 + y^2}, \qquad \tan\theta = \frac{y}{x}, \qquad z = z.$$

The solids in xyz-space easiest to describe in cylindrical coordinates are the *cylindrical wedges*. Such a wedge is pictured in Figure 16.8.3. The wedge consists of all points (x, y, z) that have cylindrical coordinates (r, θ, z) in the box

$$\Pi: \quad a_1 \leq r \leq a_2, \quad b_1 \leq \theta \leq b_2, \quad c_1 \leq z \leq c_2.$$

† By allowing θ to take on both 0 and 2π, we lose uniqueness but we gain flexibility and convenience.

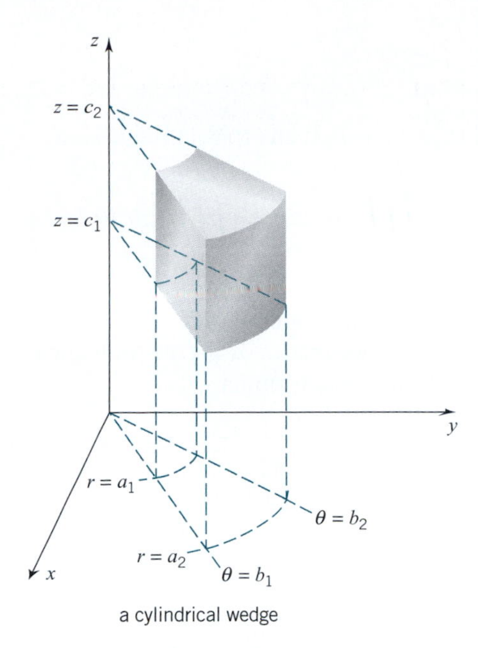

Figure 16.8.3

Evaluating Triple Integrals Using Cylindrical Coordinates

Suppose that T is some basic solid in xyz-space, not necessarily a wedge. If T is the set of all (x, y, z) with cylindrical coordinates in some basic solid S in $r\,\theta z$-space, then

(16.8.1)
$$\iiint_T f(x, y, z)\ dxdydz = \iiint_S f(r\cos\theta, r\sin\theta, z)\, r\, drd\theta dz.$$

DERIVATION OF (16.8.1) We will carry out the argument on the assumption that T is projectable onto some basic region Ω_{xy} of the xy-plane. (It is for such solids that the formula is most useful.) T has some lower boundary $z = \psi_1(x, y)$ and some upper boundary $z = \psi_2(x, y)$. T is then the set of all (x, y, z) with

$$(x, y) \in \Omega_{xy} \qquad \text{and} \qquad \psi_1(x, y) \leq z \leq \psi_2(x, y).$$

The region Ω_{xy} has polar coordinates in some set $\Omega_{r\theta}$ (which we assume is a basic region). Then S is the set of all (r, θ, z) with

$$[r, \theta] \in \Omega_{r\theta} \qquad \text{and} \qquad \psi_1(r\cos\theta,\ r\sin\theta) \leq z \leq \psi_2(r\cos\theta,\ r\sin\theta).$$

Therefore

$$\iiint_T f(x, y, z)\ dxdydz = \iint_{\Omega_{xy}} \left(\int_{\psi_1(x,y)}^{\psi_2(x,y)} f(x, y, z)\, dz \right) dxdy$$

$$\underset{(16.4.3)}{=} \iint_{\Omega_{r\theta}} \left(\int_{\psi_1(r\cos\theta,\ r\sin\theta)}^{\psi_2(r\cos\theta,\ r\sin\theta)} f(r\cos\theta,\ r\sin\theta, z)\, dz \right) r\, drd\theta$$

$$= \iiint_S f(r\cos\theta, r\sin\theta, z)\, r\, drd\theta dz. \qquad \square$$

Volume Formula

If $f(x, y, z) = 1$ for all (x, y, z) in T, then (16.8.1) reduces to

$$\iiint_T dxdydz = \iiint_S r\, drd\theta dz.$$

The triple integral on the left is the volume of T. In summary, if T is a basic solid in xyz-space and the cylindrical coordinates of T constitute a basic solid S in $r\theta z$-space, then the volume of T is given by the formula

(16.8.2)

$$V = \iiint_S r\, drd\theta dz.$$

Calculations

Cylindrical coordinates are particularly useful in cases where there is an axis of symmetry. The axis of symmetry is then taken as the z-axis.

Example 1 Use cylindrical coordinates to calculate

$$\iiint_T (x^2 + y^2)\ dxdydz$$

for $\quad T\colon -2 \le x \le 2, \quad -\sqrt{4 - x^2} \le y \le \sqrt{4 - x^2}, \quad 0 \le z \le 4 - x^2 - y^2.$

SOLUTION The solid is bounded above by the paraboloid of revolution $z = 4 - x^2 - y^2$ and below by the xy-plane (see Figure 16.8.4). Since the solid is symmetric about the z-axis, the integral has a simpler representation in cylindrical coordinates:

$$S\colon 0 \le r \le 2, \quad 0 \le \theta \le 2\pi, \quad 0 \le z \le 4 - r^2.$$

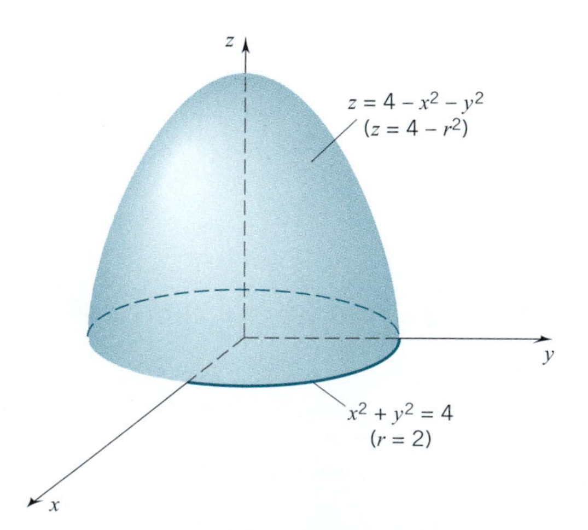

Figure 16.8.4

Now $\iiint_T (x^2+y^2)\,dxdydz = \iiint_S r^2\,r\,drd\theta dz = \int_0^{2\pi}\int_0^2\int_0^{4-r^2} r^3\,dz\,dr\,d\theta$

$$= \int_0^{2\pi}\int_0^2 \left[r^3 z\right]_0^{4-r^2} dr\,d\theta = \int_0^{2\pi}\int_0^2 (4r^3 - r^5)\,dr\,d\theta$$

$$= \int_0^{2\pi}\left[r^4 - \tfrac{1}{6}r^6\right]_0^2 d\theta$$

$$= \tfrac{16}{3}\int_0^{2\pi} d\theta = \frac{32\pi}{3}. \quad \square$$

Example 2 Find the mass of a solid right circular cylinder T of radius R and height h given that the density varies directly with the distance from the axis of the cylinder.

SOLUTION Place the cylinder T on the xy-plane so that the axis of T coincides with the z-axis. The density function then takes the form $\lambda(x,y,z) = k\sqrt{x^2+y^2}$, and T consists of all points (x,y,z) with cylindrical coordinates (r,θ,z) in the set

$$S:\quad 0 \le r \le R,\quad 0 \le \theta \le 2\pi,\quad 0 \le z \le h.$$

Therefore $M = \iiint_T k\sqrt{x^2+y^2}\,dxdydz = \iiint_S (kr)\,r\,drd\theta dz$

$$= k\int_0^R\int_0^{2\pi}\int_0^h r^2\,dz\,d\theta\,dr = \tfrac{2}{3}k\pi R^3 h. \quad \square$$

↑ check this

Example 3 Use cylindrical coordinates to find the volume of the solid T bounded above by the plane $z = y$ and below by the paraboloid $z = x^2 + y^2$.

SOLUTION In cylindrical coordinates the plane has equation $z = r\sin\theta$ and the paraboloid has equation $z = r^2$. Solving these two equations simultaneously, we have $r = \sin\theta$. This tells us that the two surfaces intersect in a space curve that lies along the circular cylinder $r = \sin\theta$. The projection of this intersection onto the xy-plane is the circle with polar equation $r = \sin\theta$. (See Figure 16.8.5.) The base region Ω_{xy} is thus the set of all (x,y) with polar coordinates in the set

$$0 \le \theta \le \pi,\quad 0 \le r \le \sin\theta.$$

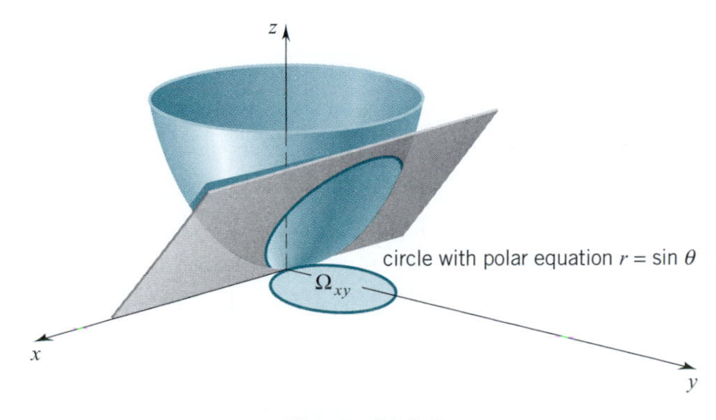

Figure 16.8.5

T itself is the set of all (x, y, z) with cylindrical coordinates in the set

$$S: \quad 0 \le \theta \le \pi, \quad 0 \le r \le \sin\theta, \quad r^2 \le z \le r\sin\theta. \qquad \text{(check this)}$$

Therefore,

$$\begin{aligned} V = \iiint_T dxdydz &= \iiint_S r\,drd\theta dz \\ &= \int_0^{\pi}\int_0^{\sin\theta}\int_{r^2}^{r\sin\theta} r\,dz\,dr\,d\theta \\ &= \int_0^{\pi}\int_0^{\sin\theta} (r^2\sin\theta - r^3)\,dr\,d\theta \\ &= \int_0^{\pi}\left[\tfrac{1}{3}r^3\sin\theta - \tfrac{1}{4}r^4\right]_0^{\sin\theta} d\theta \\ &= \tfrac{1}{12}\int_0^{\pi}\sin^4\theta\,d\theta = \tfrac{1}{12}(\tfrac{3}{8}\pi) = \tfrac{1}{32}\pi. \quad \square \end{aligned}$$

(verify this; Exercise 13, Section 8.3)

Example 4 Locate the centroid of the solid in Example 3.

SOLUTION Since T is symmetric about the yz-plane, we see that $\bar{x} = 0$. To get $\bar{y}$ we begin as usual:

$$\begin{aligned} \bar{y}V = \iiint_T y\,dxdydz &= \iiint_S (r\sin\theta)r\,drd\theta dz \\ &= \int_0^{\pi}\int_0^{\sin\theta}\int_{r^2}^{r\sin\theta} r^2\sin\theta dz\,dr\,d\theta \\ &= \int_0^{\pi}\int_0^{\sin\theta} (r^3\sin^2\theta - r^4\sin\theta)\,dr\,d\theta \\ &= \int_0^{\pi}\left[\tfrac{1}{4}r^4\sin^2\theta - \tfrac{1}{5}r^5\sin\theta\right]_0^{\sin\theta} d\theta \\ &= \tfrac{1}{20}\int_0^{\pi}\sin^6\theta\,d\theta = \tfrac{1}{20}(\tfrac{5}{16}\pi) = \tfrac{1}{64}\pi. \end{aligned}$$

(verify this; Exercise 13, Section 8.3)

Since $V = \frac{1}{32}\pi$, we have $\bar{y} = \frac{1}{2}$. Now for $\bar{z}$:

$$\begin{aligned} \bar{z}V &= \iiint_T z\ dxdydz \\ &= \iiint_S zr\,drd\theta dz \\ &= \int_0^{\pi}\int_0^{\sin\theta}\int_{r^2}^{r\sin\theta} zr\,dz\,dr\,d\theta = \cdots = \tfrac{1}{24}\int_0^{\pi}\sin^6\theta\,d\theta = \tfrac{1}{24}(\tfrac{5}{16}\pi) = \tfrac{5}{384}\pi. \end{aligned}$$

details are left to you

Division by $V = \frac{1}{32}\pi$ gives $\bar{z} = \frac{5}{12}$. The centroid is thus the point $(0, \frac{1}{2}, \frac{5}{12})$. $\square$

EXERCISE 16.8

Write the given equation in cylindrical coordinates and sketch the graph of the surface.

1. $x^2 + y^2 + z^2 = 9.$ **2.** $x^2 + y^2 = 4.$

3. $z = 2\sqrt{x^2 + y^2}.$ **4.** $x = 4z.$

5. $4x^2 + 4y^2 - z^2 = 0.$ **6.** $y^2 + z^2 = 8.$

The volume of a solid T is given by an integral in cylindrical coordinates. Sketch T and evaluate the integral

7. $\int_0^{\pi/2}\int_0^2\int_0^{4-r^2} r\,dz\,dr\,d\theta.$

8. $\int_0^{\pi/4}\int_0^1\int_0^{\sqrt{1-r^2}} r\,dz\,dr\,d\theta.$

9. $\int_0^{2\pi}\int_0^2\int_0^{r^2} r\,dz\,dr\,d\theta.$

10. $\int_0^3\int_0^{2\pi}\int_r^3 r\,dz\,d\theta\,dr.$

Evaluate using cylindrical coordinates.

11. $\iiint_T dxdydz;\quad T: 0 \le x \le 1,\ 0 \le y \le \sqrt{1-x^2},$ $0 \le z \le \sqrt{4-(x^2+y^2)}.$

12. $\iiint_T z^3 dxdydz;\quad T: -1 \le x \le 1,\ 0 \le y \le \sqrt{1-x^2},$ $\sqrt{x^2+y^2} \le z \le 1.$

13. $\iiint_T \frac{1}{\sqrt{x^2+y^2}}\,dxdydz;\quad T: 0 \le x \le \sqrt{9-y^2},\ 0 \le y \le 3,$ $0 \le z \le \sqrt{9-(x^2+y^2)}.$

14. $\iiint_T z\,dxdydz;\quad T: 0 \le x \le 1,\ 0 \le y \le \sqrt{1-x^2},$ $0 \le z \le \sqrt{1-x^2-y^2}.$

15. $\iiint_T \sin(x^2+y^2)\,dxdydz;\quad T: 0 \le x \le 1,$ $0 \le y \le \sqrt{1-x^2},\ 0 \le z \le 2.$

16. $\iiint_T \sqrt{x^2+y^2}\,dxdydz;\quad T: -1 \le x \le 1,$ $-\sqrt{1-x^2} \le y \le \sqrt{1-x^2},\ x^2+y^2 \le z \le 2-(x^2+y^2).$

17. Find the volume of the solid bounded above by the cone $z^2 = x^2 + y^2$, below by the xy-plane, and on the sides by the cylinder $x^2 + y^2 = 2ax$.

18. Find the volume of the solid bounded by the paraboloid of revolution $x^2 + y^2 = az$, the xy-plane, and the cylinder $x^2 + y^2 = 2ax$.

19. Find the volume of the solid bounded above by $z = a - \sqrt{x^2+y^2}$, below by the xy-plane, and on the sides by the cylinder $x^2 + y^2 = ax$.

20. Find the volume of the solid bounded above by the plane $2z = 4 + x$, below by the xy-plane, and on the sides by the cylinder $x^2 + y^2 = 2x$.

21. Find the volume of the solid bounded by the paraboloid $z = x^2 + y^2$ and the plane $z = x$.

22. Find the volume of the solid that is bounded above by $x^2 + y^2 + z^2 = 25$ and below by $z = \sqrt{x^2 + y^2} + 1$.

23. Find the volume of the "ice cream cone" bounded below by the half-cone $z = \sqrt{3(x^2+y^2)}$ and above by the unit sphere $x^2 + y^2 + z^2 = 1$.

24. Find the volume of the solid bounded by the hyperboloid $z^2 = a^2 + x^2 + y^2$ and the upper nappe of the cone $z^2 = 2(x^2 + y^2)$.

25. Find the volume of the solid that is bounded below by the xy-plane and lies inside the sphere $x^2 + y^2 + z^2 = 9$ but outside the cylinder $x^2 + y^2 = 1$.

26. Find the volume of the solid that lies between the cylinders $x^2 + y^2 = 1$ and $x^2 + y^2 = 4$, and is bounded above by the ellipsoid $x^2 + y^2 + 4z^2 = 36$ and below by the xy-plane.

In Exercises 27–29, let T be a solid right circular cylinder of base radius R and height h. Assume that the mass density varies directly with the distance from one of the bases.

27. Use cylindrical coordinates to find the mass M of T.

28. Locate the center of mass of T.

29. Find the moment of inertia of T about the axis of the cylinder.

30. Let T be a homogeneous right circular cylinder of mass M, base radius R, and height h. Find the moment of inertia of the cylinder about: (a) the central axis; (b) a line that lies in the plane of one of the bases and passes through the center of that base; (c) a line that passes through the center of the cylinder and is parallel to the bases.

In Exercises 31–34, let T be a homogeneous solid right circular cone of mass M, base radius R, and height h.

31. Use cylindrical coordinates to verify that the volume of the cone is given by the formula $V = \frac{1}{3}\pi R^2 h$.

32. Locate the center of mass.

33. Find the moment of inertia about the axis of the cone.

34. Find the moment of inertia about a line that passes through the vertex and is parallel to the base.

In Exercises 35–37, let T be the solid bounded above by the paraboloid $z = 1 - (x^2 + y^2)$ and bounded below by the xy-plane.

35. Use cylindrical coordinates to find the volume of T.

36. Find the mass of T if the density varies directly with the distance to the xy-plane.

37. Find the mass of T if the density varies directly with the square of the distance from the origin.

■ 16.9 THE TRIPLE INTEGRAL AS THE LIMIT OF RIEMANN SUMS; SPHERICAL COORDINATES

The Triple Integral as the Limit of Riemann Sums

You have seen how single integrals and double integrals can be obtained as limits of Riemann sums. The same holds true for triple integrals.

Start with a basic solid T in xyz-space and decompose it into a finite number of basic solids $T_1 \ldots, T_N$. If f is continuous on T, then f is continuous on each T_i. From each T_i pick an arbitrary point (x_i^*, y_i^*, z_i^*) and form the *Riemann sum*

$$\sum_{i=1}^{N} f(x_i^*, y_i^*, z_i^*)(\text{ volume of } T_i).$$

As you would expect, the triple integral over T is the limit of such sums; namely, given any $\epsilon > 0$, there exists a $\delta > 0$ such that, if the diameters of the T_i are all less than δ, then

$$\left| \sum_{i=1}^{N} f(x_i^*, y_i^*, z_i^*)(\text{ volume of } T_i) - \iiint_T f(x,y,z)\ dxdydz \right| < \epsilon$$

no matter how the (x_i^*, y_i^*, z_i^*) are chosen within the T_i. We express this by writing

(16.9.1) $$\iiint_T f(x,y,z)\ dxdydz = \lim_{\text{diam}\,T_i \to 0} \sum_{i=1}^{N} f(x_i^*, y_i^*, z_i^*)(\text{ volume of } T_i).$$

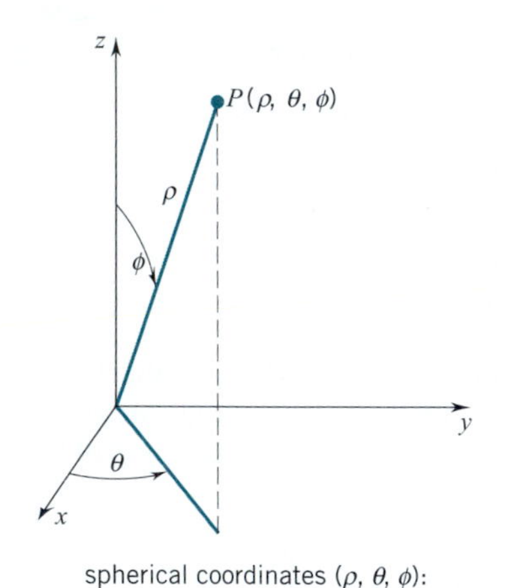

spherical coordinates (ρ, θ, ϕ): $\rho \geq 0$, $0 \leq \theta < 2\pi$, $0 \leq \phi \leq \pi$

Figure 16.9.1

Introduction to Spherical Coordinates

The spherical coordinates (ρ, θ, ϕ) of a point P in xyz-space are shown geometrically in Figure 16.9.1. The first coordinate, ρ, is the distance from P to the origin; thus $\rho \geq 0$. The second coordinate, the angle marked θ, is the second coordinate of cylindrical coordinates; θ ranges from 0 to 2π. We call θ the *longitude*. The third coordinate, the angle marked ϕ, ranges only from 0 to π. We call ϕ the *colatitude*, or more simply the *polar angle*. (The complement of ϕ would be the *latitude* on a globe.)

The coordinate surfaces

$$\rho = \rho_0, \quad \theta = \theta_0, \quad \phi = \phi_0$$

are shown in Figure 16.9.2. The surface $\rho = \rho_0$ is a sphere; the radius is ρ_0 and the center is the origin. The second surface, $\theta = \theta_0$, is the same as in cylindrical coordinates: the vertical half-plane hinged at the z-axis and standing at an angle of θ_0 radians from the positive x-axis. The surface $\phi = \phi_0$ requires detailed explanation. If $0 \leq \phi_0 \leq \frac{1}{2}\pi$ or $\frac{1}{2}\pi \leq \phi_0 < \pi$, the surface is a nappe of a cone; it is generated by revolving about the z-axis any ray that emerges from the origin at an angle of ϕ_0 radians from the positive z-axis. The surface $\phi = \frac{1}{2}\pi$ is the xy-plane. (The nappe of the cone has opened up completely.) The equation $\phi = 0$ gives the nonnegative z-axis, and the equation $\phi = \pi$ gives the nonpositive z-axis. (When $\phi = 0$ or $\phi = \pi$, the nappe of the cone has closed up completely.)

The point P with spherical coordinates $(\rho_0, \theta_0, \phi_0)$ is located at the intersection of the three surfaces $\rho = \rho_0$, $\theta = \theta_0$, $\phi = \phi_0$.

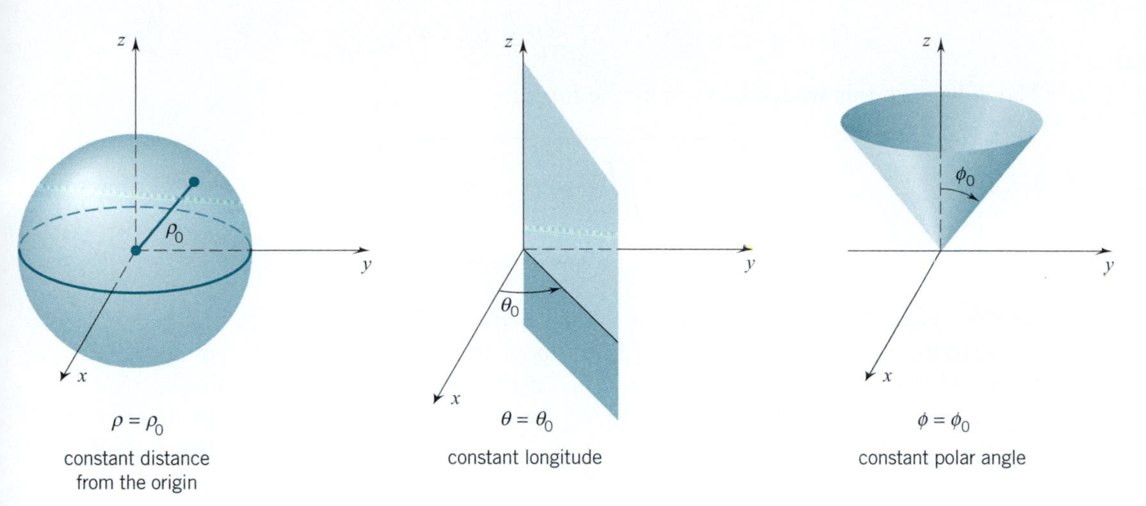

Figure 16.9.2

Rectangular coordinates (x, y, z) are related to spherical coordinates (ρ, θ, ϕ) by the following equations:

$$x = \rho \sin\phi \cos\theta, \quad y = \rho \sin\phi \sin\theta, \quad z = \rho\cos\phi.$$

You can verify these relations by referring to Figure 16.9.3. (Note that the factor $\rho \sin\phi$ appearing in the first two equations is the r of cylindrical coordinates: $r = \rho\sin\phi$.) Conversely, with obvious exclusions, we have

$$\rho = \sqrt{x^2 + y^2 + z^2}, \quad \tan\theta = \frac{y}{x}, \quad \cos\phi = \frac{z}{\sqrt{x^2 + y^2 + z^2}}.$$

The Volume of a Spherical Wedge

Figure 16.9.4 shows a *spherical wedge W* in *xyz*-space. The wedge W consists of all points (x, y, z) that have spherical coordinates in the box

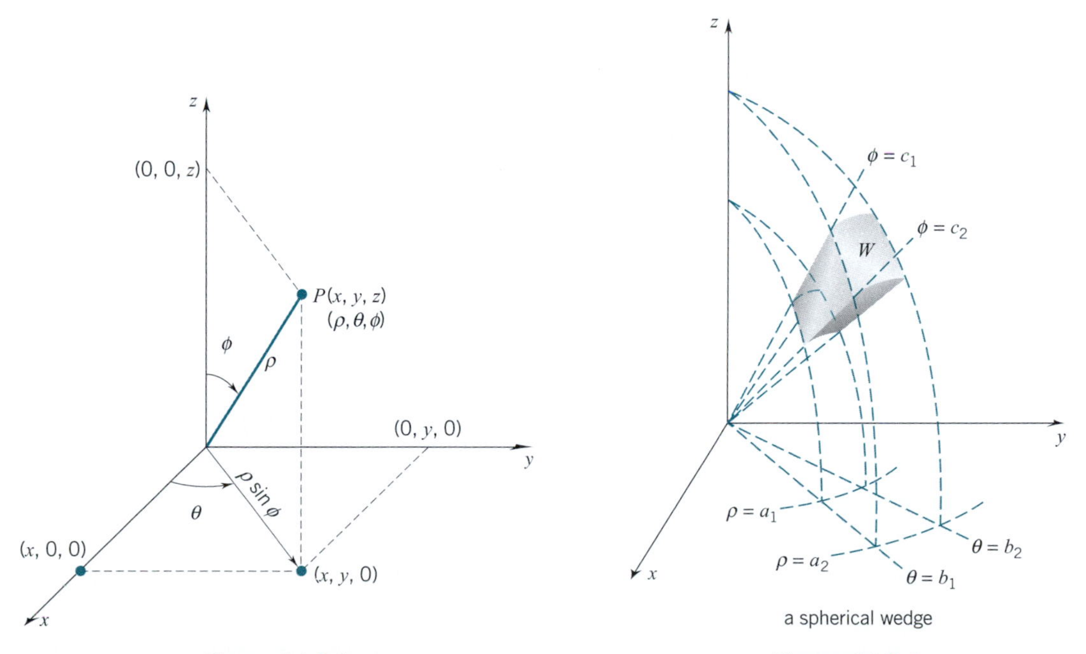

Figure 16.9.3

Figure 16.9.4

$$\Pi: \quad a_1 \le \rho \le a_2, \quad b_1 \le \theta \le b_2, \quad c_1 \le \phi \le c_2.$$

The volume of this wedge is given by the formula

(16.9.2) $$V = \iiint_{\Pi} \rho^2 \sin \phi d\rho d\theta d\phi.$$

PROOF Note first that W is part of a solid of revolution. One way to obtain W is to rotate the $\theta = b_1$ face of W, call it Ω, about the z-axis for $b_2 - b_1$ radians. (See Figure 16.9.4.) On that face ρ and $\alpha = \frac{1}{2}\pi - \phi$ play the role of polar coordinates. (See Figure 16.9.5.) In the setup of Figure 16.9.5 the face Ω is the set of all (X, z) with polar coordinates $[\rho, \alpha]$ in the set $\Gamma: a_1 \le \rho \le a_2, \frac{1}{2}\pi - c_2 \le \alpha \le \frac{1}{2}\pi - c_1$. The centroid of Ω is at a distance $\overline{X}$ from the z-axis where

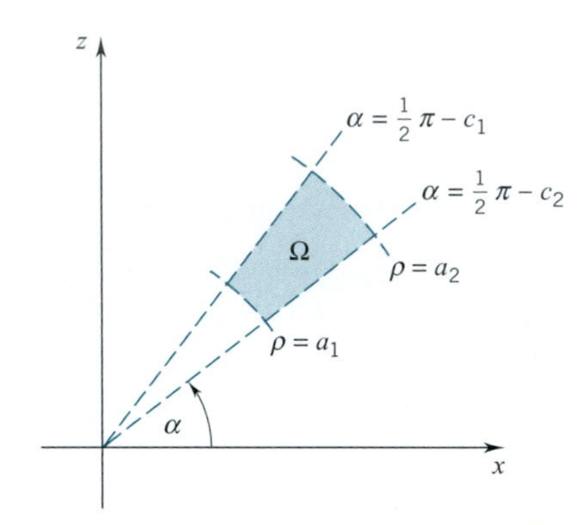

Figure 16.9.5

$$\overline{X}(\text{ area of } \Omega) = \iint_{\Omega} X\, dX\, dz = \iint_{\Gamma} \rho^2 \cos \alpha\, d\rho d\alpha$$

this follow from (16.4.3) together with the fact that here $[\rho, \alpha]$ play the role of polar coordinates

$$= \left(\int_{a_1}^{a_2} \rho^2 d\rho\right)\left(\int_{\frac{1}{2}\pi - c_2}^{\frac{1}{2}\pi - c_1} \cos \alpha\, d\alpha\right)$$

$$= \left(\int_{a_1}^{a_2} \rho^2 d\rho\right)\left(\int_{c_1}^{c_2} \sin \phi\, d\phi\right).$$

$\phi = \frac{1}{2}\pi - \alpha$

As the face Ω is rotated from $\theta = b_1$ to $\theta = b_2$, the centroid travels through a circular arc of length

$$s = (b_2 - b_1)\overline{X} = (b_2 - b_1)\frac{1}{\text{area of } \Omega}\left(\int_{a_1}^{a_2} \rho^2\, d\rho\right)\left(\int_{c_1}^{c_2} \sin \phi\, d\phi\right).$$

From Pappus's theorem (see the Remark on p. 354), we know that

$$\text{the volume of } W = s(\text{ area of } \Omega) = (b_2 - b_1)\left(\int_{a_1}^{a_2} \rho^2\, d\rho\right)\left(\int_{c_1}^{c_2} \sin \phi\, d\phi\right)$$

$$= \left(\int_{b_1}^{b_2} d\theta\right)\left(\int_{a_1}^{a_2} \rho^2\, d\rho\right)\left(\int_{c_1}^{c_2} \sin \phi\, d\phi\right)$$

$$= \int_{a_1}^{a_2}\int_{b_1}^{b_2}\int_{c_1}^{c_2} \rho^2 \sin\phi \, d\phi \, d\theta \, d\rho$$

$$= \iiint_{\Pi} \rho^2 \sin\phi \, d\rho d\theta d\phi. \quad \square$$

Evaluating Triple Integrals Using Spherical Coordinates

Suppose that T is a basic solid in xyz-space with spherical coordinates in some basic solid S of $\rho\theta\phi$-space. Then

(16.9.3)

$$\iiint_T f(x,y,z)\, dxdydz =$$
$$\iiint_S f(\rho\sin\phi\cos\theta,\ \rho\sin\phi\sin\theta,\ \rho\cos\phi)\,\rho^2\sin\phi\, d\rho d\theta d\phi.$$

DERIVATION OF (16.9.3) Assume first that T is a spherical wedge W. The solid S is then a box Π. Now decompose Π into N boxes $\Pi_1, \ldots, \Pi_N$. This induces a subdivision of W into N spherical wedges $W_1, \ldots, W_N$.

Writing $F(\rho, \theta, \phi)$ for $f(\rho\sin\phi\cos\theta,\ \rho\sin\phi\sin\theta,\ \rho\cos\phi)$ to save space, we have

$$\iiint_{\Pi} F(\rho, \theta, \phi)\,\rho^2\sin\phi\, d\rho d\theta d\phi = \sum_{i=1}^{N} \iiint_{\Pi_i} F(\rho, \theta, \phi)\,\rho^2\sin\phi\, d\rho d\theta d\phi$$

additivity

$$= \sum_{i=1}^{N} F(\rho_i^*, \theta_i^*, \phi_i^*) \iiint_{\Pi_i} \rho^2\sin\phi\, d\rho d\theta d\phi$$

for some $(\rho_i^*, \theta_i^*, \phi_i^*) \in \Pi_i$ (by 16.6.5)

$$= \sum_{i=1}^{N} F(\rho_i^*, \theta_i^*, \phi_i^*)(\text{volume of } W_i)$$

(16.9.2) applied to Π_i

$$= \sum_{i=1}^{N} f(x_i^*, y_i^*, z_i^*)(\text{volume of } W_i).$$

$x_i^* = \rho_i^*\sin\phi_i^*\cos\theta_i^*$, $y_i^* = \rho_i^*\sin\phi_i^*\sin\theta_i^*$, $z_i^* = \rho_i^*\cos\phi_i^*$

This last expression is a Riemann sum for

$$\iiint_W f(x,y,z)\, dxdydz$$

and, as such, by (16.9.1), will differ from that integral by less than any preassigned positive number ϵ provided only that the diameters of all the W_i are sufficiently small. This we can guarantee by making the diameters of all the Π_i sufficiently small.

This verifies the formula for the case where T is a spherical wedge. The more general case is left to you. HINT: Encase T in a wedge W and define f to be zero outside of T. $\square$

Volume Formula

If $f(x, y, z) = 1$ for all (x, y, z) in T, then the change of variables formula reduces to

$$\iiint_T dxdydz = \iiint_S \rho^2 \sin\phi \, d\rho d\theta d\phi.$$

The integral on the left is the volume of T. It follows that the volume of T is given by the formula

(16.9.4)
$$V = \iiint_S \rho^2 \sin\phi \, d\rho d\theta d\phi.$$

Calculations

Spherical coordinates are commonly used in applications where there is a center of symmetry. The center of symmetry is then taken as the origin.

Example 1 Calculate the mass M of a solid ball of radius 1 given that the density varies directly with the square of the distance from the center of the ball.

SOLUTION Center the ball at the origin. The ball, call it T, is now the set of all (x, y, z) with spherical coordinates (ρ, θ, ϕ) in the box

$$S: \quad 0 \le \rho \le 1, \quad 0 \le \theta \le 2\pi, \quad 0 \le \phi \le \pi. \qquad \text{(Figure 16.9.6)}$$

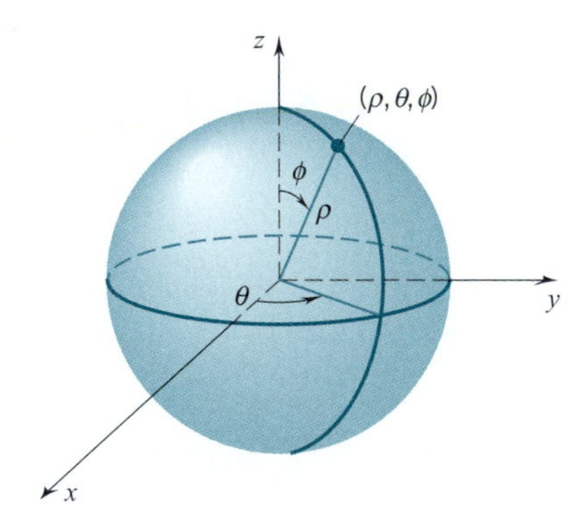

Figure 16.9.6

Therefore
$$M = \iiint_T k(x^2 + y^2 + z^2) \, dxdydz$$
$$= \iiint_S (k\rho^2)\rho^2 \sin\phi \, d\rho d\theta d\phi = k \int_0^{\pi}\int_0^{2\pi}\int_0^1 \rho^4 \sin\phi \, d\rho d\theta d\phi$$
$$= k\left(\int_0^{\pi} \sin\phi \, d\phi\right)\left(\int_0^{2\pi} d\theta\right)\left(\int_0^1 \rho^4 \, d\rho\right) = k(2)(2\pi)(\tfrac{1}{5}) = \tfrac{4}{5}k\pi. \quad \square$$

Example 2 Find the volume of the solid T that is bounded above by the cone $z^2 = x^2 + y^2$, below by the xy-plane, and on the sides by the hemisphere $z = \sqrt{4 - x^2 - y^2}$ (see Figure 16.9.7).

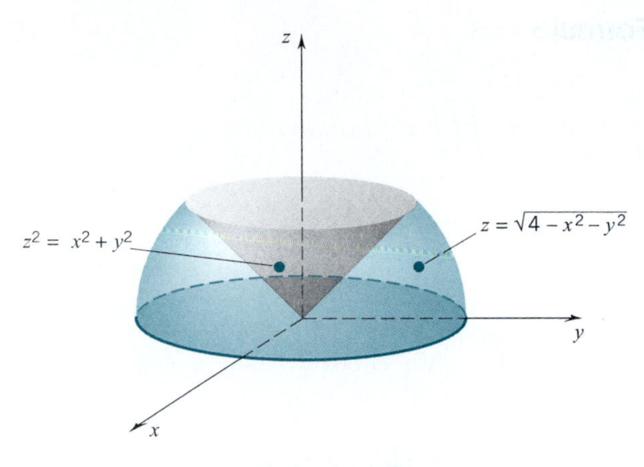

Figure16.9.7

SOLUTION In terms of spherical coordinates, the hemisphere is given by $\rho = 2$, $0 \le \phi \le \pi/2$. As you can verify, the hemisphere and the cone intersect in a circle which lies in the plane $z = \sqrt{2}$ and has its center on the z-axis. For points on this circle, the angle ϕ is $\pi/4$ (verify this). It follows that the solid T is the set of all (x, y, z) with spherical coordinates (ρ, θ, ϕ) in the set

$$S: \quad 0 \le \rho \le 2, \quad 0 \le \theta \le 2\pi, \quad \pi/4 \le \phi \le \pi/2.$$

Thus,

$$\begin{aligned} V &= \iiint_T dxdydz = \iiint_S \rho^2 \sin\phi \, d\rho d\phi d\theta \\ &= \int_{\pi/4}^{\pi/2}\int_0^{2\pi}\int_0^2 \rho^2 \sin\phi \, d\rho \, d\theta \, d\phi = \left(\int_{\pi/4}^{\pi/2} \sin\phi \, d\phi\right)\left(\int_0^{2\pi} d\theta\right)\left(\int_0^2 \rho^2 \, d\rho\right) \\ &= (\sqrt{2}/2)(2\pi)(\tfrac{8}{3}) = \frac{8\pi\sqrt{2}}{3} \cong 11.85. \quad \square \end{aligned}$$

Example 3 Find the volume of the solid T enclosed by the surface

$$(x^2 + y^2 + z^2)^2 = 2z(x^2 + y^2).$$

SOLUTION In spherical coordinates the bounding surface takes the form

$$\rho = 2\sin^2\phi\cos\phi. \qquad \text{(check this out)}$$

This equation places no restriction on θ; thus θ can range from 0 to 2π. Since ρ remains nonnegative, ϕ can range only from 0 to $\frac{1}{2}\pi$. Thus the solid T is the set of all (x, y, z) with spherical coordinates (ρ, θ, ϕ) in the set

$$S: \quad 0 \le \theta \le 2\pi, \quad 0 \le \phi \le \tfrac{1}{2}\pi, \quad 0 \le \rho \le 2\sin^2\phi\cos\phi.$$

The rest is straightforward:

$$V = \iiint_T dxdydz = \iiint_S \rho^2 \sin\phi \, d\rho d\theta d\phi$$

$$= \int_0^{2\pi} \int_0^{\pi/2} \int_0^{2\sin^2\phi\cos\phi} \rho^2 \sin\phi \, d\rho \, d\phi \, d\theta = \int_0^{2\pi} \int_0^{\pi/2} \tfrac{8}{3} \sin^7\phi \cos^3\phi \, d\phi \, d\theta$$

$$= \tfrac{8}{3} \left(\int_0^{2\pi} d\theta \right) \left(\int_0^{\pi/2} (\sin^7\phi \cos\phi - \sin^9\phi \cos\phi) \, d\phi \right)$$

$$= \tfrac{8}{3}(2\pi)(\tfrac{1}{40}) = \tfrac{2}{15}\pi \cong 0.42. \quad \square$$

EXERCISES 16.9

1. Find the spherical coordinates (ρ, θ, ϕ) of the point with rectangular coordinates $(1, 1, 1)$.
2. Find the rectangular coordinates of the point with spherical coordinates $(2, \frac{1}{6}\pi, \frac{1}{4}\pi)$.
3. Find the rectangular coordinates of the point with spherical coordinates $(3, \frac{1}{3}\pi, \frac{1}{6}\pi)$.
4. Find the spherical coordinates of the point with cylindrical coordinates $(2, \frac{2}{3}\pi, 6)$.
5. Find the spherical coordinates of the point with rectangular coordinates $(2, 2, \frac{2}{3}\sqrt{6})$.
6. Find the spherical coordinates of the point with rectangular coordinates $(2\sqrt{2}, -2\sqrt{2}, -4\sqrt{3})$.
7. Find the rectangular coordinates of the point with spherical coordinates $(3, \pi/2, 0)$.
8. Find the spherical coordinates of the point with cylindrical coordinates $(1/\sqrt{2}, \pi/4, 1/\sqrt{2})$.

Equations are given in spherical coordinates. Interpret each one geometrically.

9. $\rho \sin\phi = 1$.
10. $\sin\phi = 1$.
11. $\cos\phi = -\frac{1}{2}\sqrt{2}$.
12. $\tan\theta = 1$.
13. $\rho\cos\phi = 1$.
14. $\rho = \cos\phi$.

The volume of a solid T is given by an integral in spherical coordinates. Sketch T and evaluate the integral.

15. $\displaystyle\int_0^{2\pi} \int_0^{\pi} \int_0^{2} \rho^2 \sin\phi \, d\rho \, d\phi \, d\theta.$

16. $\displaystyle\int_0^{\pi/4} \int_0^{\pi/2} \int_0^{1} \rho^2 \sin\phi \, d\rho \, d\phi \, d\theta.$

17. $\displaystyle\int_{\pi/6}^{\pi/2} \int_0^{\pi/2} \int_0^{3} \rho^2 \sin\phi \, d\rho \, d\theta \, d\phi.$

18. $\displaystyle\int_0^{\pi/4} \int_0^{2\pi} \int_0^{\sec\varphi} \rho^2 \sin\phi \, d\rho \, d\theta \, d\phi.$

Evaluate using spherical coordinates.

19. $\displaystyle\iiint_T dxdydz;\quad T: 0 \le x \le 1,\ 0 \le y \le \sqrt{1-x^2},$ $\sqrt{x^2+y^2} \le z \le \sqrt{2-(x^2+y^2)}.$

20. $\displaystyle\iiint_T (x^2+y^2+z^2)\, dxdydz;$ $T: 0 \le x \le \sqrt{4-y^2},\ 0 \le y \le 2,$ $\sqrt{x^2+y^2} \le z \le \sqrt{4-x^2-y^2}.$

21. $\displaystyle\iiint_T z\sqrt{x^2+y^2+z^2}\, dxdydz;$ $T: 0 \le x \le \sqrt{9-y^2},\ 0 \le y \le 3,$ $0 \le z \le \sqrt{9-(x^2+y^2)}.$

22. $\displaystyle\iiint_T \frac{1}{(x^2+y^2+z^2)}\, dxdydz;$ $T: 0 \le x \le 1,\ 0 \le y \le \sqrt{1-x^2},$ $0 \le z \le \sqrt{1-x^2-y^2}.$

23. Derive the formula for the volume of a sphere of radius R using spherical coordinates.
24. Express cylindrical coordinates in terms of spherical coordinates.
25. A wedge is cut from a ball of radius R by two planes that meet in a diameter. Find the volume of the wedge if the angle between the planes is α radians.
26. Find the mass of a ball of radius R given that the density varies directly with the distance from the boundary.
27. Find the mass of a right circular cone of base radius r and height h given that the density varies directly with the distance from the vertex.
28. Use spherical coordinates to derive the formula for the volume of a right circular cone of base radius r and height h.

In Exercises 29 and 30, let T be a homogeneous ball of mass M and radius R.

29. Calculate the moment of inertia about: (a) a diameter; (b) a tangent line.

30. Locate the center of mass of the upper half given that the center of the ball is at the origin.

In Exercises 31 and 32, let T be a homogeneous solid bounded by two concentric spherical shells, an outer shell of radius R_2 and an inner shell of radius R_1.

31. (a) Calculate the moment of inertia about a diameter.
(b) Use your result in part (a) to determine the moment of inertia of a spherical shell of radius R and mass M about a diameter.
(c) What is the moment of inertia of that same shell about a tangent line?

32. (a) Locate the center of mass of the upper half of T given that the center of T is at the origin.
(b) Use your result in part (a) to locate the center of mass of a homogeneous hemispherical shell of radius R.

33. Find the volume of the solid common to the sphere $\rho = a$ and the cone $\phi = \alpha$. Take $\alpha \in (0, \frac{1}{2}\pi)$.

34. Let T be the solid bounded below by the half-cone $z = \sqrt{x^2 + y^2}$ and above by the spherical surface $x^2 + y^2 + z^2 = 1$. Use spherical coordinates to evaluate

$$\iiint_T e^{(x^2+y^2+z^2)^{3/2}}\, dxdydz.$$

35. (a) Find an equation in spherical coordinates for the sphere $x^2 + y^2 + (z - R)^2 = R^2$.
(b) Express the upper half of the ball $x^2 + y^2 + (z - R)^2 \leq R^2$ by inequalities in spherical coordinates.

36. Find the mass of the ball $\rho \leq 2R\cos\phi$ given that the density varies directly with (a) ρ; (b) $\rho\sin\phi$; (c) $\rho\cos^2\theta\sin\phi$.

37. Find the volume of the solid common to the spheres $\rho = 2\sqrt{2}\cos\phi$ and $\rho = 2$.

38. Find the volume of the solid enclosed by the surface $\rho = 1 - \cos\phi$.

39. Finish the argument in (16.9.3).

40. (*Gravitational attraction*) Let T be a basic solid and let (a, b, c) be a point not in T. Show that, if T has continuously varying mass density $\lambda = \lambda(x, y, z)$, then T attracts a point mass m at (a, b, c) with a force

$$\mathbf{F} = \iiint_T Gm\,\lambda(x, y, z)\,\mathbf{f}(x, y, z)\ dxdydz, \text{ where}$$

$$\mathbf{f}(x, y, z) = \frac{[(x - a)\,\mathbf{i} + (y - b)\,\mathbf{j} + (z - c)\,\mathbf{k}]}{\left[(x - a)^2 + (y - b)^2 + (z - c)^2\right]^{3/2}}.$$

[Assume that a point mass m_1 at P_1 attracts a point mass m_2 at P_2 with a force $\mathbf{F} = -(Gm_1m_2/r^3)\mathbf{r}$, where $\mathbf{r}$ is the vector $\overrightarrow{P_1P_2}$. Interpret the triple integral component by component.]

41. Let T be the upper half of the ball $x^2 + y^2 + (z - R)^2 \leq R^2$. Given that T is homogeneous and has mass M, find the gravitational force exerted by T on a point mass m located at the origin. (Note Exercise 40.)

42. A point mass m is placed on the axis of a homogeneous solid right circular cylinder at a distance α from the nearest base of the cylinder. Find the gravitational force exerted by the cylinder on the point mass given that the cylinder has base radius R, height h, and mass M. (Note Exercise 40.)

■ 16.10 JACOBIANS; CHANGING VARIABLES IN MULTIPLE INTEGRATION

During the course of the last few sections you have met several formulas for changing variables in multiple integration: to polar coordinates, to cylindrical coordinates, to spherical coordinates. The purpose of this section is to bring some unity into that material and provide a general description for other changes of variable.

We begin with a consideration of area. Figure 16.10.1 shows a basic region Γ in a plane that we are calling the uv-plane. (In this plane we denote the abscissa of a point by u and the ordinate by v.) Suppose that

$$x = x(u, v), \qquad y = y(u, v)$$

are continuously differentiable functions on Γ. As (u, v) ranges over Γ, the point $(x, y) = (x(u, v),\ y(u, v))$ generates a region Ω in the xy-plane. If the mapping

$$(u, v) \to (x, y)$$

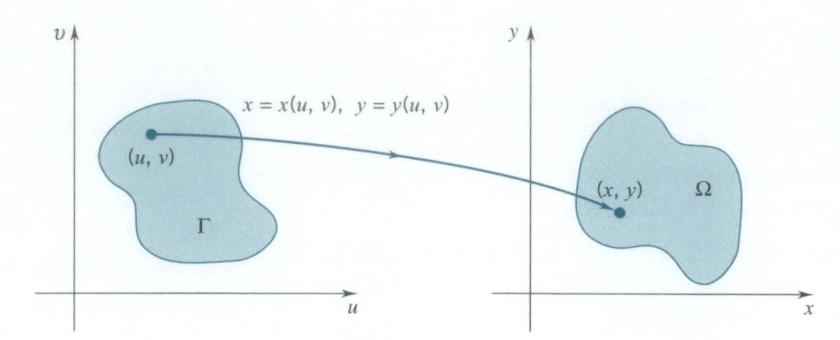

Figure 16.10.1

is one-to-one on the interior of Γ, and the *Jacobian J* given by the 2×2 determinant

$$J(u,v) = \begin{vmatrix} \dfrac{\partial x}{\partial u} & \dfrac{\partial y}{\partial u} \\ \dfrac{\partial x}{\partial v} & \dfrac{\partial y}{\partial v} \end{vmatrix} = \frac{\partial x}{\partial u}\frac{\partial y}{\partial v} - \frac{\partial x}{\partial v}\frac{\partial y}{\partial u},$$

is never zero on the interior of Γ, then

(16.10.1)
$$\text{area of } \Omega = \iint_{\Gamma} |J(u,v)|\, dudv.$$

It is very difficult to prove this assertion without making additional assumptions. A proof valid for all cases of practical interest is given in the supplement to Section 17.5. At this point we simply assume this area formula and go on from there.

Suppose now that we want to integrate some continuous function $f = f(x, y)$ over Ω. If this proves difficult to do directly, then we can change variables to u, v and try to integrate over Γ instead. It follows from (16.10.1) that

(16.10.2)
$$\iint_{\Omega} f(x,y)\; dxdy = \iint_{\Gamma} f(x(u,v), y(u,v))|J(u,v)|\, dudv.$$

The derivation of this formula from (16.10.1) follows the usual lines. Break up Γ into N little basic regions $\Gamma_1, \ldots, \Gamma_N$. These induce a decomposition of Ω into N little basic regions $\Omega_1, \ldots, \Omega_N$. We can then write

$$\iint_{\Gamma} f(x(u,v), y(u,v))|J(u,v)|\, dudv = \sum_{i=1}^{N} \iint_{\Gamma_i} f(x(u,v), y(u,v))|J(u,v)|\, dudv$$

additivity

$$= \sum_{i=1}^{N} f(x(u_i^*, v_i^*), y(u_i^*, v_i^*)) \iint_{\Gamma_i} |J(u,v)|\, dudv$$

Theorem 16.2.10 applied to Γ_i

$$= \sum_{i=1}^{N} f(x_i^*, y_i^*) \iint_{\Gamma_i} |J(u,v)|\, dudv$$

set $x_i^* = x(u_i^*, v_i^*), y_i^* = y(u_i^*, v_i^*)$

$$= \sum_{i=1}^{N} f(x_i^*, y_i^*)(\text{area of } \Omega_i).$$

(16.10.1) applied to Γ_i

This last expression is a Riemann sum for

$$\iint_{\Omega} f(x, y)\, dxdy$$

and tends to that integral as the maximum diameter of the Ω_i tends to zero. We can ensure this by letting the maximum diameter of the Γ_i tend to zero.

Example 1 Evaluate

$$\iint_{\Omega} (x+y)^2\, dxdy$$

where Ω is the parallelogram bounded by the lines

$$x+y=0, \qquad x+y=1, \qquad 2x-y=0, \qquad 2x-y=3.$$

SOLUTION The parallelogram is shown in Figure 16.10.2. The boundaries suggest that we set

$$u = x+y, \qquad v = 2x-y.$$

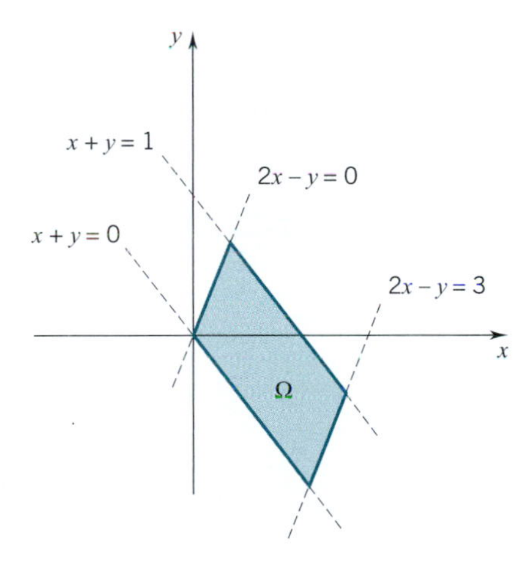

Figure 16.10.2

We want x and y in terms of u and v. Since

$$u+v = (x+y)+(2x-y) = 3x \quad \text{and} \quad 2u-v = (2x+2y)-(2x-y) = 3y,$$

we have

$$x = \frac{u+v}{3}, \qquad y = \frac{2u-v}{3}.$$

This transformation maps the rectangle Γ of Figure 16.10.3 onto Ω with Jacobian

$$J(u,v) = \begin{vmatrix} \dfrac{\partial}{\partial u}\left(\dfrac{u+v}{3}\right) & \dfrac{\partial}{\partial u}\left(\dfrac{2u-v}{3}\right) \\[2ex] \dfrac{\partial}{\partial v}\left(\dfrac{u+v}{3}\right) & \dfrac{\partial}{\partial v}\left(\dfrac{2u-v}{3}\right) \end{vmatrix} = \begin{vmatrix} \frac{1}{3} & \frac{2}{3} \\ \frac{1}{3} & -\frac{1}{3} \end{vmatrix} = -\tfrac{1}{3}.$$

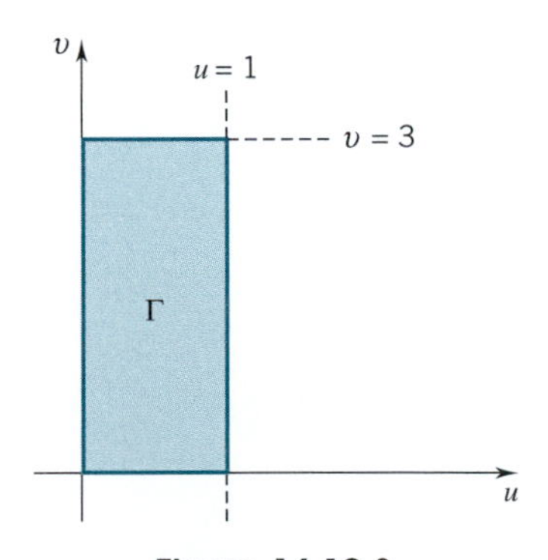

Figure 16.10.3

Therefore $$\iint_{\Omega} (x+y)^2\,dxdy = \iint_{\Omega} u^2|J(u,v)|\,dudv$$

$$= \tfrac{1}{3}\int_0^3\int_0^1 u^2\,dudv$$

$$= \tfrac{1}{3}\left(\int_0^3 dv\right)\left(\int_0^1 u^2\,du\right) = \tfrac{1}{3}(3)\tfrac{1}{3} = \tfrac{1}{3}. \quad \square$$

Example 2 Evaluate

$$\iint_{\Omega} xy\,dxdy$$

where Ω is the first-quadrant region bounded by the curves

$$x^2+y^2=4, \qquad x^2+y^2=9, \qquad x^2-y^2=1, \qquad x^2-y^2=4.$$

SOLUTION The region is shown in Figure 16.10.4. The boundaries suggest that we set

$$u = x^2+y^2, \qquad v = x^2-y^2.$$

We want x and y in terms of u and v. Since

$$u+v = 2x^2 \qquad \text{and} \qquad u-v = 2y^2,$$

we have $$x = \sqrt{\frac{u+v}{2}}, \quad y = \sqrt{\frac{u-v}{2}}.$$

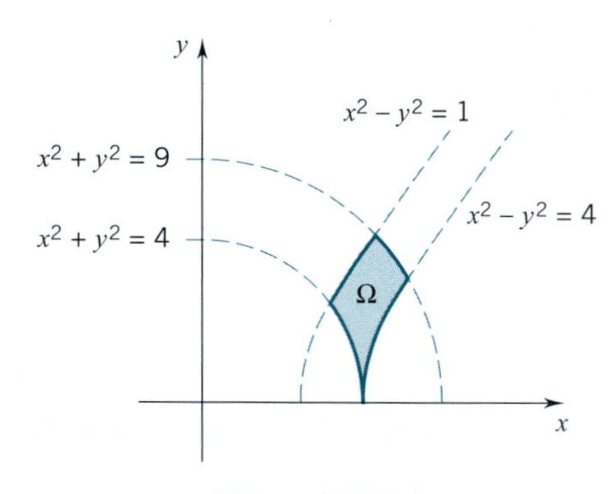

Figure 16.10.4

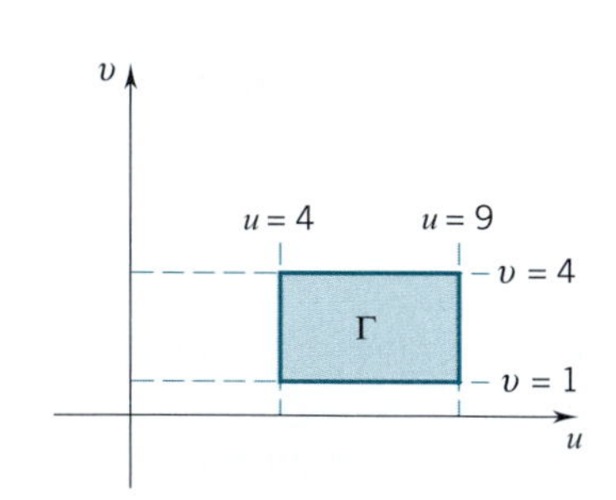

Figure 16.10.5

The transformation maps the rectangle Γ of Figure 16.10.5 onto Ω with Jacobian

$$J(u,v) = \begin{vmatrix} \dfrac{\partial}{\partial u}\left(\sqrt{\dfrac{u+v}{2}}\right) & \dfrac{\partial}{\partial u}\left(\sqrt{\dfrac{u-v}{2}}\right) \\ \dfrac{\partial}{\partial v}\left(\sqrt{\dfrac{u+v}{2}}\right) & \dfrac{\partial}{\partial v}\left(\sqrt{\dfrac{u-v}{2}}\right) \end{vmatrix} = -\frac{1}{4\sqrt{u^2-v^2}}.$$

check this

Therefore $$\iint_{\Omega} xy\ dxdy = \iint_{\Gamma} \left(\sqrt{\frac{u+v}{2}}\right)\left(\sqrt{\frac{u-v}{2}}\right)\left(\frac{1}{4\sqrt{u^2-v^2}}\right) dudv$$
$$= \iint_{\Gamma} \tfrac{1}{8}\, du\, dv = \tfrac{1}{8}(\text{area of } \Gamma) = \tfrac{15}{8}. \quad \square$$

In Section 16.4 you saw the formula for changing variables from rectangular coordinates (x, y) to polar coordinates $[r, \theta]$. The formula reads

$$\iint_{\Omega} f(x,y)\ dxdy = \iint_{\Gamma} f(r\cos\theta, r\sin\theta) r\ drd\theta. \qquad (16.4.3)$$

The factor r in the double integral over Γ is the Jacobian of the transformation $x = r\cos\theta,\ y = r\sin\theta$:

$$J(r,\theta) = \begin{vmatrix} \dfrac{\partial}{\partial r}(r\cos\theta) & \dfrac{\partial}{\partial r}(r\sin\theta) \\ \dfrac{\partial}{\partial \theta}(r\cos\theta) & \dfrac{\partial}{\partial \theta}(r\sin\theta) \end{vmatrix} = \begin{vmatrix} \cos\theta & \sin\theta \\ -r\sin\theta & r\cos\theta \end{vmatrix} = r(\cos^2\theta + \sin^2\theta) = r.$$

As you can see, (16.4.3) is a special case of (16.10.2).

When changing variables in a triple integral we make three coordinate changes:

$$x = x(u, v, w), \qquad y = y(u, v, w), \qquad z = z(u, v, w).$$

If these functions carry a basic solid Γ onto a solid T, then, under conditions analogous to the two-dimensional case,

$$\text{volume of } T = \iiint_{\Gamma} |J(u, v, w)|\ dudvdw$$

where now the Jacobian† is a three-by-three determinant:

$$J(u, v, w) = \begin{vmatrix} \dfrac{\partial x}{\partial u} & \dfrac{\partial y}{\partial u} & \dfrac{\partial z}{\partial u} \\ \dfrac{\partial x}{\partial v} & \dfrac{\partial y}{\partial v} & \dfrac{\partial z}{\partial v} \\ \dfrac{\partial x}{\partial w} & \dfrac{\partial y}{\partial w} & \dfrac{\partial z}{\partial w} \end{vmatrix}.$$

In this case the change of variables formula reads

$$\iiint_{T} f(x,y,z)\ dxdydz = \iiint_{\Gamma} f(x(u,v,w),\ y(u,v,w),\ z(u,v,w))|J(u,v,w)|\ dudvdw.$$

† The study of these functional determinants goes back to a memoir by the German mathematician C. G. Jacobi, 1804–1851.

EXERCISES 16.10

Find the Jacobian of the transformation.

1. $x = au + bv, \quad y = cu + dv.$ (linear transformation)

2. $x = u\cos\theta - v\sin\theta, \quad y = u\sin\theta + v\cos\theta.$ rotation by θ)

3. $x = uv, \quad y = u^2 + v^2.$

4. $x = u\ln v, \quad y = uv.$

5. $x = uv^2, \quad y = u^2 v.$

6. $x = u - \ln v, \quad y = \ln u + v.$

7. $x = au, \quad y = bv, \quad z = cw.$

8. $x = v + w, \quad y = u + w, \quad z = u + v.$

9. $x = r\cos\theta, \quad y = r\sin\theta, \quad z = z.$ (cylindrical coordinates)

10. $x = \rho\sin\phi\cos\theta, \quad y = \rho\sin\phi\sin\theta, \quad z = \rho\cos\phi.$ (spherical coordinates)

11. $x = (1 + w\cos v)\cos u, \quad y = (1 + w\cos v)\sin u,$
$z = w\sin v.$

12. Every linear transformation

$$x = au + bv, \quad y = cu + dv \quad \text{with} \quad ad - bc \neq 0$$

maps lines of the uv-plane onto lines of the xy-plane. Find the image of (a) a vertical line $u = u_0$; (b) a horizontal line $v = v_0$.

For Exercises 13–15, take Ω as the parallelogram bounded by

$$x + y = 0, \quad x + y = 1, \quad x - y = 0, \quad x - y = 2.$$

Evaluate.

13. $\displaystyle\iint_\Omega (x^2 - y^2)\, dx\, dy.$

14. $\displaystyle\iint_\Omega 4xy\, dx\, dy.$

15. $\displaystyle\iint_\Omega (x - y)\cos[\pi(x - y)]\, dx\, dy.$

For Exercises 16–18, take Ω as the parallelogram bounded by

$$x - y = 0, \quad x - y = \pi, \quad x + 2y = 0, \quad x + 2y = \tfrac{1}{2}\pi.$$

Evaluate.

16. $\displaystyle\iint_\Omega (x + y)\, dx\, dy.$

17. $\displaystyle\iint_\Omega \sin(x - y)\cos(x + 2y)\, dx\, dy.$

18. $\displaystyle\iint_\Omega \sin 3x\, dx\, dy.$

19. Let Ω be the first-quadrant region bounded by the curves $xy = 1$, $xy = 4$, $y = x$, $y = 4x$. (a) Determine the area of Ω and (b) locate the centroid.

20. Show that the ellipse $b^2x^2 + a^2y^2 = a^2b^2$ has area πab by setting $x = ar\cos\theta$, $y = br\sin\theta$.

21. A homogeneous plate in the xy-plane is in the form of a parallelogram. The parallelogram is bounded by the lines $x + y = 0, x + y = 1, 3x - 2y = 0, 3x - 2y = 2$. Calculate the moments of inertia of the plate about the three coordinate axes. Express your answers in terms of the mass of the plate.

22. Calculate the area of the region Ω bounded by the curves

$$x^2 - 2xy + y^2 + x + y = 0, \qquad x + y + 4 = 0.$$

HINT: Set $u = x - y, v = x + y$.

23. Calculate the area of the region Ω bounded by the curves

$$x^2 - 4xy + 4y^2 - 2x - y - 1 = 0, \qquad y = \tfrac{2}{5}.$$

24. Locate the centroid of the region in Exercise 22.

25. Calculate the area of the region Ω enclosed by the curve

$$11x^2 + 4\sqrt{3}xy + 7y^2 - 1 = 0.$$

HINT: Use a rotation $x = u\cos\theta - v\sin\theta$, $y = u\sin\theta + v\cos\theta$ such that the resulting uv-equation has no uv-term.

26. Evaluate

$$\int_{-\infty}^{\infty}\int_{-\infty}^{\infty} \frac{e^{-(x-y)^2}}{1 + (x + y)^2}\, dx\, dy$$

by integrating over the square $S_a: -a \leq x \leq a, -a \leq y \leq a$ and taking the limit as $a \to \infty$.
HINT: Set $u = x - y, v = x + y$ and see (16.4.4).

For Exercises 27–30, let T be the solid ellipsoid $x^2/a^2 + y^2/b^2 + z^2/c^2 \leq 1$.

27. Calculate the volume of T by setting

$$x = a\rho\sin\phi\cos\theta, \quad y = b\rho\sin\phi\sin\theta, \quad z = c\rho\cos\phi.$$

28. Locate the centroid of the upper half of T.

29. View the upper half of T as a homogeneous solid of mass M. Find the moments of inertia of this solid about the coordinate axes.

30. Evaluate

$$\iiint_T \left(\frac{x^2}{a^2} + \frac{y^2}{b^2} + \frac{z^2}{c^2}\right) dx\, dy\, dz.$$

■ PROJECT 16.10 Generalized Polar Coordinates

Recall the equations that transform polar coordinates to rectangular coordinates:

$$x = r\cos\theta, \quad y = r\sin\theta.$$

In this project, we investigate a generalization of these equations, and we apply the generalized polar coordinates to the problem of finding the area of regions enclosed by curves given in rectangular coordinates.

Let a, b, and α be fixed positive numbers, and let (x, y) be related to (r, θ) by the equations

$$(1) \qquad x = ar(\cos\theta)^{\alpha}, \quad y = br(\sin\theta)^{\alpha}.$$

Problem 1.

a. Show that the mapping defined by (1) carries the polar region $\Gamma: 0 \le r < \infty, 0 \le \theta \le \pi/2$ onto the first quadrant in the xy-plane. HINT: Find a point $[r, \theta]$ that maps onto (x, y) given that $x \ge 0$ and $y \ge 0$.

b. Show that the mapping is one-to-one on the interior of Γ.

Problem 2. Determine the Jacobian of the mapping defined by (1).

Problem 3. The curve given by the equation $x^{2/3} + y^{2/3} = a^{2/3}, a > 0$, is called an *astroid.*

a. Use a graphing utility to graph the astroid for several values of a.

b. Calculate the area enclosed by the astroid in the first quadrant by setting $x = ar\cos^3\theta, y = ar\sin^3\theta$.

c. What is the entire area enclosed by the astroid?

Problem 4. Consider the curve defined by the equation

$$\left(\frac{x}{a}\right)^{1/4} + \left(\frac{y}{b}\right)^{1/4} = 1.$$

a. Use a graphing utility to graph this curve in the cases $a = 3, b = 2$ and $a = 2, b = 3$.

b. Calculate the area enclosed by the curve in the first quadrant by setting $x = ar\cos^8\theta, y = br\sin^8\theta$.

■ CHAPTER HIGHLIGHTS

16.1 Multiple-Sigma Notation

16.2 Double Integrals

The Double Integral over a Rectangle:

partition (p. 946) upper sum, lower sum (p. 947)
double integral over a rectangle (p. 948)

The Double Integral over a Region:

basic region (p. 952)
double integral over a region (p. 953)
double integral as a volume (p. 953)
area as a double integral (p. 953)
properties of the double integral (p. 954)
average value over a region (p. 954)
weighted average over a region (p. 955)

16.3 The Evaluation of Double Integrals by Repeated Integrals

evaluation formulas (p. 958)
symmetry (p. 964)
separated variables over a rectangle (p. 968)

16.4 The Double Integral as the Limit of Riemann Sums; Polar Coordinates

diameter of set (p. 969) Riemann sum (p. 970)

$$\iint_{\Omega} f(x, y)\, dx\, dy = \lim_{\text{diam}\,\Omega_i \to 0} \sum_{i=1}^{N} f(x_i^*, y_i^*)(\text{ area of } \Omega_i)$$

area formula (p. 971)
evaluating double integrals using polar coordinates (p. 971)

16.5 Some Applications of Double Integration

mass of a plate (p. 976)
center of mass of a plate (p. 976)
centroid of a region (p. 977)
moment of inertia of a plate (p. 979)
radius of gyration (p. 980)
parallel axis theorem (p. 981)

16.6 Triple Integrals

box (p. 983) partition of a box (p. 983)
upper sum, lower sum (p. 984)
triple integral over a box (p. 984)
basic solid (p. 985)
triple integral over a basic solid (p. 985)
volume as a triple integral (p. 985)
properties of the triple integral (p. 986)
average value (p. 987) weighted average (p. 987)
mass (p. 987) center of mass (p. 987)
centroid (p. 987) moment of inertia (p. 987)

16.7 Reduction to Repeated Integrals

possible orders of integration (p. 989–990)
separated variables over a box (p. 995)

16.8 Cylindrical Coordinates

cylindrical coordinates (r, θ, z) (p. 997)
cylindrical wedge (p. 998)
evaluating triple integrals using cylindrical coordinates (p. 999)
volume in cylindrical coordinates (p. 1000)

16.9 The Triple Integral as the Limit of Riemann Sums; Spherical Coordinates

$$\iiint_T f(x,y,z)\,dx\,dy\,dz = \lim_{\operatorname{diam} T_i \to 0} \sum_{i=1}^{N} f(x_i^*, y_i^*, z_i^*)\,(\text{volume of } T_i)$$

spherical coordinates (ρ, θ, ϕ) (p. 1004)
spherical wedge (p. 1005)
volume of a spherical wedge (p. 1005)
evaluating triple integrals using spherical coordinates (p. 1007)
volume in spherical coordinates (p. 1008)

16.10 Jacobians; Changing Variables in Multiple Integration

Jacobian: two-dimensional case (p. 1012)
change of variables in double integration (p. 1012)
Jacobian: three-dimensional case (p. 1015)
change of variables in triple integration (p. 1015)

CHAPTER 17

LINE INTEGRALS AND SURFACE INTEGRALS

In this chapter we will study integration along curves and integration over surfaces. At the heart of this subject lie three great integration theorems: *Green's theorem, Gauss's theorem* (commonly known as the *divergence theorem*), and *Stokes's theorem*.

All three theorems are ultimately based on *The Fundamental Theorem of Integral Calculus*, and all can be cast in the same general form:

an integral over a set S = a related integral over the boundary of S.

A word about terminology. Suppose that S is some subset of the plane or of three-dimensional space. A function that assigns a scalar to each point of S (say, the temperature at that point or the mass density at that point) is known in science as a *scalar field*. A function that assigns a vector to each point of S (say, the wind velocity at that point or the gradient of a function f evaluated at that point) is called a *vector field*. We will be using this "field" language throughout.

17.1 LINE INTEGRALS

We are led to the definition of *line integral* by the notion of work.

The Work Done by a Varying Force over a Curved Path

The work done by a constant force **F** on an object that moves along a straight line is, by definition, the component of **F** in the direction of the displacement multiplied by the length of the displacement vector **d** (Project 12.4):

$$W = (\text{comp}_{\mathbf{d}}\, \mathbf{F})\, \|\mathbf{d}\|$$

We can write this more briefly as a dot product:

(17.1.1)
$$\boxed{W = \mathbf{F} \cdot \mathbf{d}}$$

This elementary notion of work is useful, but it is not sufficient. Consider, for example, an object that moves through a magnetic field or a gravitational field. The path of the motion is usually not a straight line but a curve, and the force, rather than remaining constant, tends to vary from point to point. What we want now is a notion of work that applies to this more general situation.

Let's suppose that an object moves along a curve

$$C: \ \mathbf{r}(u) = x(u)\,\mathbf{i} + y(u)\,\mathbf{j} + z(u)\,\mathbf{k}, \qquad u \in [a, b]$$

subject to a continuous force $\mathbf{F}$. (The vector field $\mathbf{F}$ may vary from point to point, not only in magnitude but also in direction.) We will suppose that the curve is *smooth*; namely, we will suppose that the tangent vector $\mathbf{r}'$ is continuous and never zero. What we want to do here is define the total work done by $\mathbf{F}$ along the curve C.

To decide how to do this, we begin by focusing our attention on what happens over a short parameter interval $[u, u+h]$. As an estimate for the work done over this interval we can use the dot product

$$\mathbf{F}(\mathbf{r}(u)) \cdot [\mathbf{r}(u+h) - \mathbf{r}(u)].$$

In making this estimate, we are evaluating the force vector $\mathbf{F}$ at $\mathbf{r}(u)$ and we are replacing the curved path from $\mathbf{r}(u)$ to $\mathbf{r}(u+h)$ by the line segment from $\mathbf{r}(u)$ to $\mathbf{r}(u+h)$. (See Figure 17.1.1.) If we set

$$\begin{aligned} W(u) &= \text{ total work done by } \mathbf{F} \text{ from } \mathbf{r}(a) \text{ to } \mathbf{r}(u), \text{ and} \\ W(u+h) &= \text{ total work done by } \mathbf{F} \text{ from } \mathbf{r}(a) \text{ to } \mathbf{r}(u+h), \end{aligned}$$

then the work done by $\mathbf{F}$ from $\mathbf{r}(u)$ to $\mathbf{r}(u+h)$ must be the difference

$$W(u+h) - W(u).$$

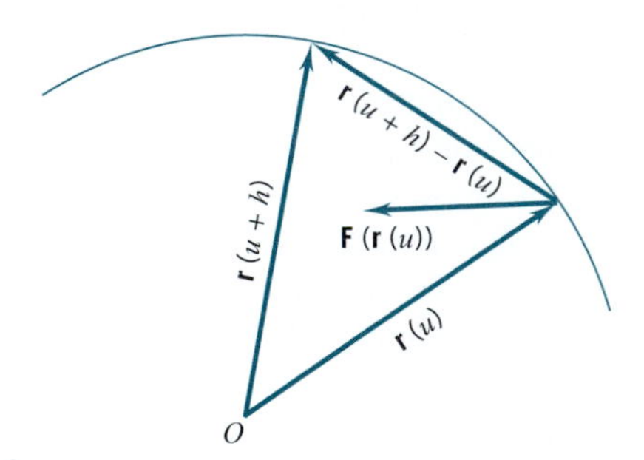

Figure 17.1.1

Bringing our estimate into play, we are led to the approximate equation

$$W(u+h) - W(u) \cong \mathbf{F}(\mathbf{r}(u)) \cdot [\mathbf{r}(u+h) - \mathbf{r}(u)],$$

which, upon division by h, becomes

$$\frac{W(u+h) - W(u)}{h} \cong \mathbf{F}(\mathbf{r}(u)) \cdot \frac{[\mathbf{r}(u+h) - \mathbf{r}(u)]}{h}.$$

The quotients here are average rates of change, and the equation is only an approximate one. The notion of work is made precise by requiring that both sides have exactly the same limit as h tends to zero; in other words, by requiring that

$$W'(u) = \mathbf{F}(\mathbf{r}(u)) \cdot \mathbf{r}'(u).$$

The rest is now determined. Since

$$W(a) = 0 \qquad \text{and} \qquad W(b) = \text{total work done by } \mathbf{F} \text{ on } C,$$

we have:

$$\text{total work done by } \mathbf{F} \text{ on } C = W(b) - W(a) = \int_a^b W'(u)\, du = \int_a^b [\mathbf{F}(\mathbf{r}(u)) \cdot \mathbf{r}'(u)]\, du.$$

In short, we have arrived at the following definition of work:

(17.1.2)

$$W = \int_a^b [\mathbf{F}(\mathbf{r}(u)) \cdot \mathbf{r}'(u)]\, du.$$

Example 1 Determine the work done by the force

$$\mathbf{F}(x, y, z) = xy\,\mathbf{i} + 2\,\mathbf{j} + 4z\,\mathbf{k}$$

along the circular helix C: $\mathbf{r}(u) = \cos u\,\mathbf{i} + \sin u\,\mathbf{j} + u\,\mathbf{k}$, from $u = 0$ to $u = 2\pi$.

SOLUTION We have: $x(u) = \cos u, y(u) = \sin u, z(u) = u$, and

$$\mathbf{F}(\mathbf{r}(u)) = \cos u \sin u\,\mathbf{i} + 2\,\mathbf{j} + 4u\,\mathbf{k},$$
$$\mathbf{r}'(u) = -\sin u\,\mathbf{i} + \cos u\,\mathbf{j} + \mathbf{k}.$$

Now,

$$\mathbf{F}(\mathbf{r}(u)) \cdot \mathbf{r}'(u) = -\sin^2 u \cos u + 2\cos u + 4u$$

and

$$W = \int_0^{2\pi} \mathbf{F}(\mathbf{r}(u)) \cdot \mathbf{r}'(u)\, du = \int_0^{2\pi} (-\sin^2 u \cos u + 2\cos u + 4u)\, du$$

$$= \left[-\frac{\sin^3 u}{3} + 2\sin u + 2u^2\right]_0^{2\pi} = 8\pi^2. \quad \square$$

Line Integrals

The integral on the right of (17.1.2) can be calculated not only for a force function $\mathbf{F}$ but for any vector field $\mathbf{h}$ continuous on C.

DEFINITION 17.1.3 LINE INTEGRAL

Let $\mathbf{h}(x, y, z) = h_1(x, y, z)\,\mathbf{i} + h_2(x, y, z)\,\mathbf{j} + h_3(x, y, z)\,\mathbf{k}$ be a vector field that is continuous on a smooth curve

$$C: \ \mathbf{r}(u) = x(u)\,\mathbf{i} + y(u)\,\mathbf{j} + z(u)\,\mathbf{k}, \qquad u \in [a, b].$$

The *line integral* of $\mathbf{h}$ over C is the number

$$\int_C \mathbf{h}(\mathbf{r}) \cdot \mathbf{dr} = \int_a^b [\mathbf{h}(\mathbf{r}(u)) \cdot \mathbf{r}'(u)]\, du.$$

Note also that, while we speak of integrating over C, we actually carry out the calculations over the parameter set $[a, b]$. If our definition of line integral is to make sense, the line integral as defined must be independent of the particular parametrization chosen for C. Within the limitations spelled out as follows, this is indeed the case:

THEOREM 17.1.4

Let $\mathbf{h}$ be a vector field that is continuous on a smooth curve C. The line integral

$$\int_C \mathbf{h}(\mathbf{r}) \cdot \mathbf{dr} = \int_a^b [\mathbf{h}(\mathbf{r}(u)) \cdot \mathbf{r}'(u)]\, du$$

is left invariant by every *orientation-preserving* change of parameter.†

PROOF Suppose that ϕ maps $[c, d]$ onto $[a, b]$ and that ϕ' is positive and continuous on $[c, d]$. We must show that the line integral over C as parametrized by

$$\mathbf{R}(w) = \mathbf{r}(\phi(w)), \qquad w \in [c, d]$$

equals the line integral over C as parametrized by $\mathbf{r}$. The argument is straightforward:

$$\begin{aligned}\int_c \mathbf{h}(\mathbf{R}) \cdot \mathbf{dR} &= \int_c^d [\mathbf{h}(\mathbf{R}(w)) \cdot \mathbf{R}'(w)]\, dw \\ &= \int_c^d [\mathbf{h}(\mathbf{r}(\phi(w))) \cdot \mathbf{r}'(\phi(w))\phi'(w)]\, dw \\ &= \int_c^d [\mathbf{h}(\mathbf{r}(\phi(w))) \cdot \mathbf{r}'(\phi(w))]\phi'(w)\, dw \\ &= \int_a^b [\mathbf{h}(\mathbf{r}(u)) \cdot \mathbf{r}'(u)]\, du = \int_C \mathbf{h}(\mathbf{r}) \cdot \mathbf{dr}. \quad \square\end{aligned}$$

Set $u = \phi(w), du = \phi'(w)\, dw$.
At $w = c, u = a$; at $w = d, u = b$.

Example 2 Calculate $\int_C \mathbf{h}(\mathbf{r}) \cdot \mathbf{dr}$ given that

$$\mathbf{h}(x, y) = xy\,\mathbf{i} + y^2\,\mathbf{j} \qquad \text{and} \qquad C\colon\ \mathbf{r}(u) = u\,\mathbf{i} + u^2\,\mathbf{j}, \qquad u \in [0, 1].$$

SOLUTION Here $x(u) = u, \quad y(u) = u^2$ and

$$\begin{aligned}\mathbf{h}(\mathbf{r}(u)) \cdot \mathbf{r}'(u) &= [x(u)y(u)\,\mathbf{i} + [y(u)]^2\,\mathbf{j}] \cdot [x'(u)\,\mathbf{i} + y'(u)\,\mathbf{j}] \\ &= x(u)y(u)x'(u) + [y(u)]^2 y'(u) \\ &= u(u^2)(1) + u^4(2u) = u^3 + 2u^5.\end{aligned}$$

It follows that $$\int_C \mathbf{h}(\mathbf{r}) \cdot \mathbf{dr} = \int_0^1 (u^3 + 2u^5)\, du = \left[\tfrac{1}{4}u^4 + \tfrac{1}{3}u^6\right]_0^1 = \tfrac{7}{12}. \quad \square$$

† Changes of parameter were explained in Section 13.3.

Example 3 Integrate the vector field $\mathbf{h}(x,y,z) = xy\,\mathbf{i} + yz\,\mathbf{j} + xz\,\mathbf{k}$ over the twisted cubic $\mathbf{r}(u) = u\,\mathbf{i} + u^2\,\mathbf{j} + u^3\,\mathbf{k}$ from $(-1, 1, -1)$ to $(1, 1, 1)$.

SOLUTION The path of integration begins at $u = -1$ and ends at $u = 1$. In this case

$$x(u) = u, \qquad y(u) = u^2, \qquad z(u) = u^3.$$

Therefore

$$\begin{aligned} \mathbf{h}(\mathbf{r}(u)) \cdot \mathbf{r}'(u) &= [x(u)y(u)\,\mathbf{i} + y(u)z(u)\,\mathbf{j} + x(u)z(u)\,\mathbf{k}] \cdot [x'(u)\,\mathbf{i} + y'(u)\,\mathbf{j} + z'(u)\,\mathbf{k}] \\ &= x(u)y(u)x'(u) + y(u)z(u)y'(u) + x(u)z(u)z'(u) \\ &= u(u^2)(1) + u^2(u^3)2u + u(u^3)3u^2 \\ &= u^3 + 5u^6 \end{aligned}$$

and $$\int_C \mathbf{h}(\mathbf{r}) \cdot d\mathbf{r} = \int_{-1}^{1} (u^3 + 5u^6)\,du = \left[\tfrac{1}{4}u^4 + \tfrac{5}{7}u^7\right]_{-1}^{1} = \tfrac{10}{7}. \quad \square$$

If a curve C is not smooth but is made up of a finite number of adjoining smooth pieces $C_1, C_2, \ldots, C_n$, then we define the integral over C as the sum of the integrals over the C_i:

(17.1.5) $$\int_C = \int_{C_1} + \int_{C_2} + \cdots + \int_{C_n}$$

A curve of this type is said to be *piecewise smooth*. See Figure 17.1.2.

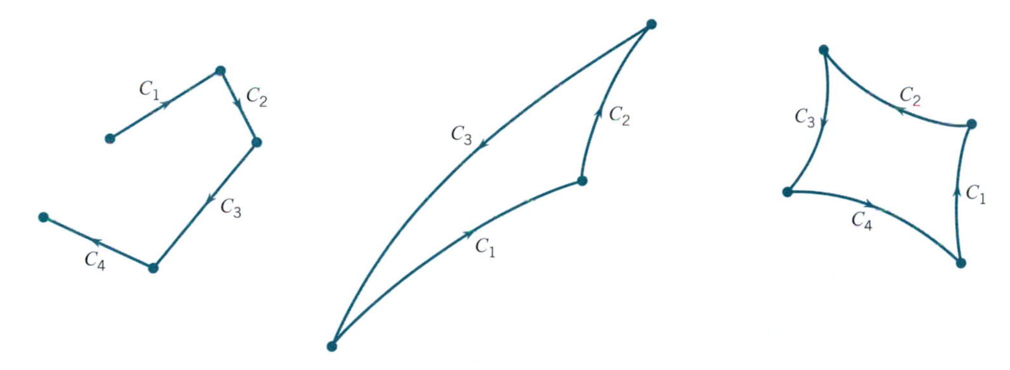

Figure 17.1.2

All polygonal paths are piecewise-smooth curves. In the next example we integrate over a triangle. We do this by integrating over each of the sides and then adding up the results. Observe that the directed line segment that begins at $\mathbf{a} = (a_1, a_2)$ and ends at $\mathbf{b} = (b_1, b_2)$ can be parametrized by setting

$$\mathbf{r}(u) = (1-u)\mathbf{a} + u\mathbf{b}, \qquad u \in [0, 1].$$

Example 4 Evaluate the line integral $\int_C \mathbf{h}(\mathbf{r}) \cdot d\mathbf{r}$ if $\mathbf{h}(x,y) = e^y\,\mathbf{i} - \sin \pi x\,\mathbf{j}$ and C is the triangle with vertices $(1,0), (0,1), (-1,0)$ traversed counterclockwise.

SOLUTION The path C is made up of three line segments:

$$\begin{aligned} C_1: &\quad \mathbf{r}(u) = (1-u)\,\mathbf{i} + u\,\mathbf{j}, \quad u \in [0,1], \\ C_2: &\quad \mathbf{r}(u) = (1-u)\,\mathbf{j} + u(-\mathbf{i}) = -u\,\mathbf{i} + (1-u)\,\mathbf{j}, \; u \in [0,1], \\ C_3: &\quad \mathbf{r}(u) = (1-u)(-\mathbf{i}) + u\,\mathbf{i} = (2u-1)\,\mathbf{i}, \; u \in [0,1]. \end{aligned}$$

See Figure 17.1.3.

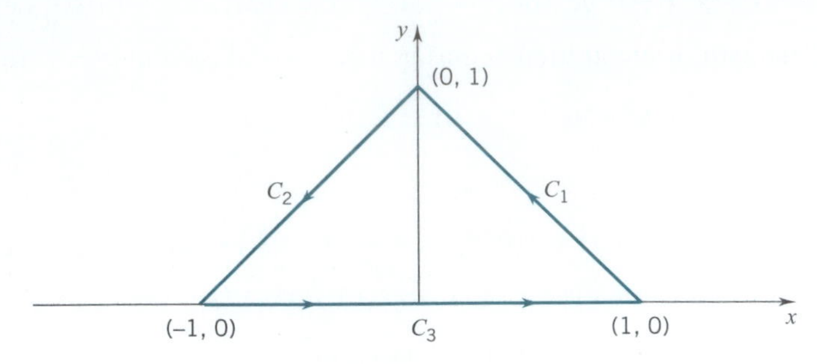

Figure 17.1.3

As you can verify,

$$\int_{C_1} \mathbf{h}(\mathbf{r}) \cdot d\mathbf{r} = \int_0^1 \left[e^{y(u)}x'(u) - \sin\left[\pi x(u)\right] y'(u)\right] du$$

$$= \int_0^1 \left[-e^u - \sin\left[\pi(1-u)\right]\right] du = 1 - e - \frac{2}{\pi};$$

$$\int_{C_2} \mathbf{h}(\mathbf{r}) \cdot d\mathbf{r} = \int_0^1 \left[e^{y(u)}x'(u) - \sin\left[\pi x(u)\right] y'(u)\right] du$$

$$= \int_0^1 \left[-e^{1-u} + \sin(-\pi u)\right] du = 1 - e - \frac{2}{\pi};$$

$$\int_{C_3} \mathbf{h}(\mathbf{r}) \cdot d\mathbf{r} = \int_0^1 \left[e^{y(u)}x'(u) - \sin\left[\pi x(u)\right] y'(u)\right] du = \int_0^1 2\, du = 2.$$

The integral over the entire triangle is the sum of these integrals:

$$\int_C \mathbf{h}(\mathbf{r}) \cdot d\mathbf{r} = \left(1 - e - \frac{2}{\pi}\right) + \left(1 - e - \frac{2}{\pi}\right) + 2 = 4 - 2e - \frac{4}{\pi} \cong -2.71. \quad \square$$

When we integrate over a parametrized curve, we integrate in the direction determined by the parametrization. If we integrate in the opposite direction, our answer is altered by a factor of -1. To be precise, let C be a piecewise-smooth curve and let $-C$ denote the same path traversed in the *opposite orientation*. (See Figure 17.1.4.) If C is parametrized by a vector function $\mathbf{r}$ defined on $[a, b]$, then $-C$ can be parametrized by setting

$$\mathbf{R}(w) = \mathbf{r}(a + b - w), \qquad w \in [a, b]. \qquad \text{(Section 13.3)}$$

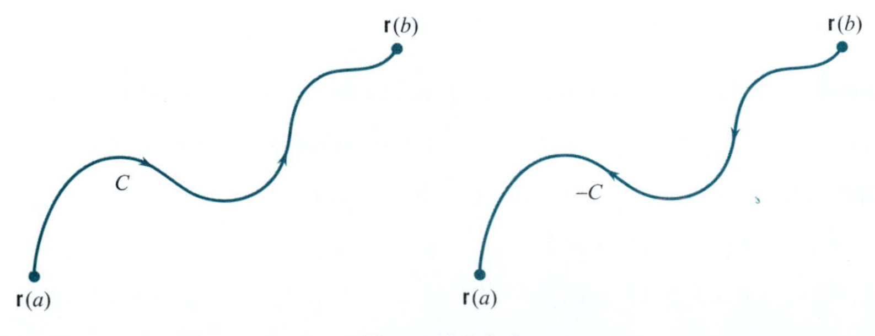

Figure 17.1.4

Our assertion is that

(17.1.6)
$$\int_{-C} \mathbf{h}(\mathbf{R}) \cdot d\mathbf{R} = -\int_C \mathbf{h}(\mathbf{r}) \cdot d\mathbf{r},$$

or, more briefly, that

(17.1.7)
$$\int_{-C} = -\int_C .$$

We leave the proof of this to you.

We were led to the definition of line integral by the notion of work. It follows from (17.1.2) that if a force $\mathbf{F}$ is continually applied to an object that moves over a piecewise-smooth curve C, then the work done by $\mathbf{F}$ is the line integral of $\mathbf{F}$ over C:

(17.1.8)
$$W = \int_C \mathbf{F}(\mathbf{r}) \cdot d\mathbf{r}.$$

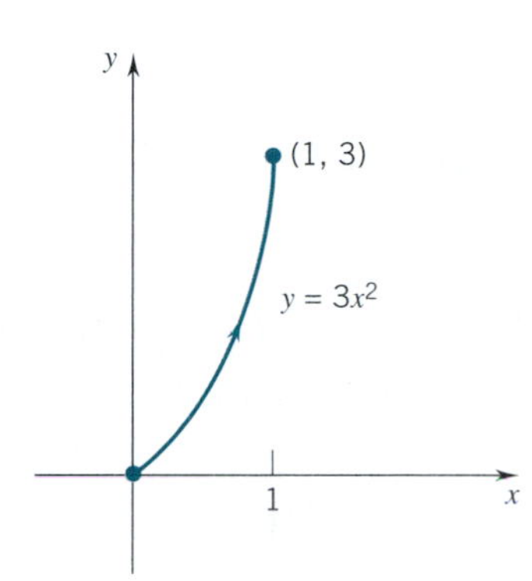

Figure 17.1.5

Example 5 An object, acted on by various forces, moves along the parabola $y = 3x^2$ from the origin to the point $(1, 3)$. (See Figure 17.1.5.) One of the forces acting on the object is $\mathbf{F}(x, y) = x^3\,\mathbf{i} + y\,\mathbf{j}$. Calculate the work done by $\mathbf{F}$.

SOLUTION We can parametrize the path by setting

$$C:\ \mathbf{r}(u) = u\,\mathbf{i} + 3u^2\,\mathbf{j}, \quad u \in [0, 1].$$

Here
$$x(u) = u, \quad y(u) = 3u^2$$

and
$$\mathbf{F}(\mathbf{r}\,(u)) \cdot \mathbf{r}'(u) = [x(u)]^3 x'(u) + y(u)y'(u) = u^3(1) + 3u^2(6u) = 19u^3.$$

It follows that

$$W = \int_C \mathbf{F}(\mathbf{r}) \cdot d\mathbf{r} = \int_0^1 19u^3\,du = \tfrac{19}{4}. \quad \square$$

If an object of mass m moves so that at time t it has position $\mathbf{r}(t)$, then, from Newton's second law, $\mathbf{F} = m\mathbf{a}$, the total force acting on the object at time t must be $m\mathbf{r}''(t)$.

Example 6 An object of mass m moves from time $t = 0$ to $t = 1$ so that its position a time t is given by the vector function

$$\mathbf{r}\,(t) = \alpha t^2\,\mathbf{i} + \sin\beta t\,\mathbf{j} + \cos\beta t\,\mathbf{k}, \qquad \alpha, \beta \text{ constant.}$$

Find the total force acting on the object at time t and calculate the total work done by this force.

SOLUTION Differentiation gives

$$\mathbf{r}'\,(t) = 2\alpha t\,\mathbf{i} + \beta\cos\beta t\,\mathbf{j} - \beta\sin\beta t\,\mathbf{k}, \qquad \mathbf{r}''(t) = 2\alpha\,\mathbf{i} - \beta^2\sin\beta t\,\mathbf{j} - \beta^2\cos\beta t\,\mathbf{k}.$$

The total force $\mathbf{F}(t)$ on the object at time t is therefore

$$\mathbf{F}(t) = m\,\mathbf{r}''(t) = m(2\alpha\,\mathbf{i} - \beta^2 \sin\beta t\,\mathbf{j} - \beta^2 \cos\beta t\,\mathbf{k}).$$

We can calculate the total work done by this force by integrating the force over the curve

$$C: \quad \mathbf{r}(t) = \alpha t^2\,\mathbf{i} + \sin\beta t\,\mathbf{j} + \cos\beta t\,\mathbf{k}, \qquad t \in [0, 1].$$

We leave it to you to verify that

$$W = \int_0^1 [m\,\mathbf{r}''(t) \cdot \mathbf{r}'(t)]\,dt = m\int_0^1 4\alpha^2 t\,dt = 2\alpha^2 m. \quad \square$$

EXERCISES 17.1

1. Integrate $\mathbf{h}(x,y) = y\,\mathbf{i} + x\,\mathbf{j}$ over the indicated path:
   (a) $\mathbf{r}(u) = u\,\mathbf{i} + u^2\,\mathbf{j}, \quad u \in [0,1]$.
   (b) $\mathbf{r}(u) = u^3\,\mathbf{i} - 2u\,\mathbf{j}, \quad u \in [0,1]$.
2. Integrate $\mathbf{h}(x,y) = x\,\mathbf{i} + y\,\mathbf{j}$ over the paths of Exercise 1.
3. Integrate $\mathbf{h}(x,y) = y\,\mathbf{i} + x\,\mathbf{j}$ over the unit circle traversed clockwise.
4. Integrate $\mathbf{h}(x,y) = xy^2\,\mathbf{i} + 2\,\mathbf{j}$ over the indicated path:
   (a) $\mathbf{r}(u) = e^u\,\mathbf{i} + e^{-u}\,\mathbf{j}, \quad u \in [0,1]$.
   (b) $\mathbf{r}(u) = (1-u)\,\mathbf{i}, \quad u \in [0,2]$.
5. Integrate $\mathbf{h}(x,y) = (x-y)\,\mathbf{i} + xy\,\mathbf{j}$ over the indicated path:
   (a) The line segment from (2, 3) to (1, 2).
   (b) The line segment from (1, 2) to (2, 3).
6. Integrate $\mathbf{h}(x,y) = x^{-1}y^{-2}\,\mathbf{i} + x^{-2}y^{-1}\,\mathbf{j}$ over the indicated path:
   (a) $\mathbf{r}(u) = \sqrt{u}\,\mathbf{i} + \sqrt{1+u}\,\mathbf{j}, \quad u \in [1,4]$.
   (b) The line segment from (1, 1) to (2, 2).
7. Integrate $\mathbf{h}(x,y) = y\,\mathbf{i} - x\,\mathbf{j}$ over the triangle with vertices $(-2,0), (2,0), (0,2)$ traversed counterclockwise.
8. Integrate $\mathbf{h}(x,y) = e^{x-y}\,\mathbf{i} + e^{x+y}\,\mathbf{j}$ over the line segment from $(-1,1)$ to $(1,2)$.
9. Integrate $\mathbf{h}(x,y) = (x+y)\,\mathbf{i} + (y^2 - x)\,\mathbf{j}$ over the closed curve that begins at $(-1, 0)$, goes along the x-axis to (1,0), and returns to $(-1, 0)$ by the upper part of the unit circle.
10. Integrate $\mathbf{h}(x,y) = 3x^2y\,\mathbf{i} + (x^3 + 2y)\,\mathbf{j}$ over the square with vertices $(0,0), (1,0), (1,1), (0,1)$ traversed counterclockwise.
11. Integrate $\mathbf{h}(x, y, z) = yz\,\mathbf{i} + x^2\,\mathbf{j} + xz\,\mathbf{k}$ over the indicated path:
    (a) The line segment from (0, 0, 0) to (1, 1, 1).
    (b) $\mathbf{r}(u) = u\,\mathbf{i} + u^2\,\mathbf{j} + u^3\,\mathbf{k}, \quad u \in [0,1]$.
12. Integrate $\mathbf{h}(x, y, z) = e^x\,\mathbf{i} + e^y\,\mathbf{j} + e^z\,\mathbf{k}$ over the paths of Exercise 11.
13. Integrate $\mathbf{h}(x, y, z) = \cos x\,\mathbf{i} + \sin y\,\mathbf{j} + yz\,\mathbf{k}$ over the indicated path:
    (a) The line segment from $(0, 0, 0)$ to $(2, 3, -1)$.
    (b) $\mathbf{r}(u) = u^2\,\mathbf{i} - u^3\,\mathbf{j} + u\,\mathbf{k}, \quad u \in [0,1]$.
14. Integrate $\mathbf{h}(x, y, z) = xy\,\mathbf{i} + x^2z\,\mathbf{j} + xyz\,\mathbf{k}$ over the indicated path:
    (a) The line segment from $(0, 0, 0)$ to $(2, -1, 1)$.
    (b) $\mathbf{r}(u) = e^u\,\mathbf{i} + e^{-u}\,\mathbf{j} + u\,\mathbf{k}, \quad u \in [0,1]$.
15. An object moves along the parabola $y = x^2$ from $(0,0)$ to $(2,4)$. One of the forces acting on the object is $\mathbf{F}(x, y) = (x + 2y)\,\mathbf{i} + (2x + y)\,\mathbf{j}$. Calculate the work done by $\mathbf{F}$.
16. An object moves along the polygonal path that connects $(0,0), (1,0), (1,1), (0,1)$ in the order indicated. One of the forces acting on the object is $\mathbf{F}(x,y) = x\cos y\,\mathbf{i} - y\sin x\,\mathbf{j}$. Calculate the work done by $\mathbf{F}$.
17. An object moves along the straight line from $(0,1,4)$ to $(1,0,-4)$. One of the forces acting on the object is $\mathbf{F}(x,y,z) = x\,\mathbf{i} + xy\,\mathbf{j} + xyz\,\mathbf{k}$. Calculate the work done by $\mathbf{F}$.
18. An object moves along the polygonal path that connects $(0,0,0), (1,0,0), (1,1,0), (1,1,1)$ in the order indicated. One of the forces acting on the object is $\mathbf{F}(x, y, z) = yz\,\mathbf{i} + xz\,\mathbf{j} + xy\,\mathbf{k}$. Calculate the work done by $\mathbf{F}$.
19. An object moves along the circular helix $\mathbf{r}(u) = \cos u\,\mathbf{i} + \sin u\,\mathbf{j} + u\,\mathbf{k}$ from $(1,0,0)$ to $(1,0,2\pi)$. One of the forces acting on the object is $\mathbf{F}(x,y,z) = x^2\,\mathbf{i} + xy\,\mathbf{j} + z^2\,\mathbf{k}$. Calculate the work done by $\mathbf{F}$.
20. A mass m, moving in a force field, traces out a circular arc at constant speed. Show that the force field does no work. Give a physical explanation for this.
21. Let $C: \mathbf{r} = \mathbf{r}(u), u \in [a,b]$ be a smooth curve and $\mathbf{q}$ a fixed vector. Show that

$$\int_C \mathbf{q} \cdot d\mathbf{r} = \mathbf{q} \cdot [\mathbf{r}(b) - \mathbf{r}(a)] \quad \text{and}$$

$$\int_C \mathbf{r} \cdot d\mathbf{r} = \frac{||\mathbf{r}(b)||^2 - ||\mathbf{r}(a)||^2}{2}.$$

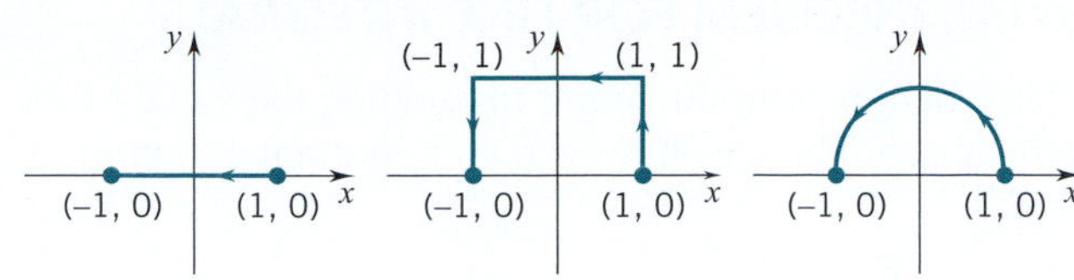

C_1: straight-line path C_2: rectangular path C_3: semicircular path

22. The preceding figure shows three paths from $(1, 0)$ to $(-1, 0)$. Calculate the line integral of

$$\mathbf{h}(x, y) = x^2\,\mathbf{i} + y\,\mathbf{j}$$

over (a) the straight-line path; (b) the rectangular path; (c) the semicircular path.

23. Let f be a continuous real-valued function of a real variable. Show that, if

$$\mathbf{f}(x, y, z) = f(x)\,\mathbf{i} \quad \text{and} \quad C: \mathbf{r}(u) = u\,\mathbf{i}, \quad u \in [a, b],$$

then $$\int_C \mathbf{f}(\mathbf{r}) \cdot d\mathbf{r} = \int_a^b f(u)\,du.$$

24. (*Linearity*) Show that, if **f** and **g** are continuous vector fields and C is piecewise smooth, then

(17.1.9)
$$\int_C [\alpha\,\mathbf{f}(\mathbf{r}) + \beta\,\mathbf{g}(\mathbf{r})] \cdot d\mathbf{r} = \alpha \int_C \mathbf{f}(\mathbf{r}) \cdot d\mathbf{r} + \beta \int_C \mathbf{g}(\mathbf{r}) \cdot d\mathbf{r}$$

for all real α, β.

25. The force $\mathbf{F}(x, y) = -\frac{1}{2}[\,y\,\mathbf{i} - x\,\mathbf{j}]$ is continually applied to an object that orbits an ellipse in standard position. Find a relation between the work done during each orbit and the area of the ellipse.

26. An object of mass m moves from time $t = 0$ to $t = 1$ so that its position at time t is given by the vector function

$$\mathbf{r}(t) = \alpha t\,\mathbf{i} + \beta t^2\,\mathbf{j}, \quad \alpha, \beta \quad \text{constant.}$$

Find the total force acting on the object at time t and calculate the work done by that force during the time interval $[0, 1]$.

27. Repeat Exercise 26 with $\mathbf{r}(t) = \alpha t\,\mathbf{i} + \beta t^2\,\mathbf{j} + \gamma t^3\,\mathbf{k}$.

28. (*Important*) The *circulation* of a vector field **v** around an oriented closed curve C is by definition the line integral

$$\int_C \mathbf{v}(\mathbf{r}) \cdot d\mathbf{r}.$$

Let **v** be the velocity field of a fluid in counterclockwise circular motion about the z-axis with constant angular speed ω.

(a) Verify that $\mathbf{v}(\mathbf{r}) = \omega\,\mathbf{k} \times \mathbf{r}$.

(b) Show that the circulation of **v** around any circle C in the xy-plane with center at the origin is $\pm 2\omega$ times the area of the circle.

29. Let **v** be the velocity field of a fluid that moves radially from the origin: $\mathbf{v} = f(x, y)\,\mathbf{r}$. What is the circulation of **v** around a circle C centered at the origin?

30. The force exerted on a charged particle at the point $(x, y) \neq (0, 0)$ in the xy-plane by an infinitely long uniformly charged wire lying along the z-axis is given by

$$\mathbf{F}(x, y) = \frac{c(x\,\mathbf{i} + y\,\mathbf{j})}{x^2 + y^2}, \qquad c > 0, \quad \text{constant.}$$

Find the work done by **F** in moving the particle along the indicated path:

(a) The line segment from $(1, 0)$ to $(1, 2)$.

(b) The line segment from $(0, 1)$ to $(1, 1)$.

31. An inverse-square force field is given by

$$\mathbf{F}(x, y, z) = \frac{c\,\mathbf{r}}{||\mathbf{r}||^3} = \frac{c(x\,\mathbf{i} + y\,\mathbf{j} + z\,\mathbf{k})}{(x^2 + y^2 + z^2)^{3/2}}, \qquad c > 0, \quad \text{constant.}$$

Find the work done by **F** in moving an object along the indicated path:

(a) The line segment from $(1, 0, 2)$ to $(1, 3, 2)$.

(b) From $(1, 0, 0)$ to $(0, \frac{5}{2}\sqrt{2}, \frac{5}{2}\sqrt{2})$ by the line segment from $(1, 0, 0)$ to $(5, 0, 0)$ and then to $(0, \frac{5}{2}\sqrt{2}, \frac{5}{2}\sqrt{2})$ along a path on the surface of the sphere $x^2 + y^2 + z^2 = 25$.

32. Suppose that $\mathbf{F}(\mathbf{r}) = c\,\mathbf{r}/||\mathbf{r}||^3$ is an inverse-square force field, where $\mathbf{r} = x\,\mathbf{i} + y\,\mathbf{j} + z\,\mathbf{k}$. Express the work done by **F** in moving an object from a point P_0 to a point P_1 in terms of the distances of these points to the origin.

33. An object moves along the curve $y = \alpha x(1 - x)$ from $(0, 0)$ to $(1, 0)$. One of the forces acting on the object is $\mathbf{F}(x, y) = (\,y^2 + 1)\,\mathbf{i} + (x + y)\,\mathbf{j}$. What value of α minimizes the work done by **F**?

34. Assume that the earth is located at the origin of a rectangular coordinate system. The gravitational force on an object at the point (x, y) is given by

$$\mathbf{F}(x, y) = \frac{-c\,\mathbf{r}}{||\mathbf{r}||^3} = \frac{-c(x\,\mathbf{i} + y\,\mathbf{j})}{(x^2 + y^2)^{3/2}}, \qquad c > 0, \quad \text{constant.}$$

Find the work done by **F** in moving an object from $(3, 0)$ to $(0, 4,)$ along the indicated path:

(a) The first quadrant part of the ellipse $\mathbf{r}(u) = 3\cos u\,\mathbf{i} + 4\sin u\,\mathbf{j}, \quad u \in [0, \pi/2]$.

(b) The line segment connecting the two points.

■ 17.2 THE FUNDAMENTAL THEOREM FOR LINE INTEGRALS

In general, if we integrate a vector field $\mathbf{h}$ from one point to another, the value of the line integral depends on the path chosen. There is, however, an important exception. If the vector field is a *gradient field*,

$$\mathbf{h} = \nabla f,$$

then the value of the line integral depends only on the endpoints of the path and not on the path itself. The details are spelled out in the following theorem.

THEOREM 17.2.1 THE FUNDAMENTAL THEOREM FOR LINE INTEGRALS

Let $C\colon \mathbf{r} = \mathbf{r}(u), u \in [a, b]$, be a piecewise-smooth curve that begins at $\mathbf{a} = \mathbf{r}(a)$ and ends at $\mathbf{b} = \mathbf{r}(b)$. If the scalar field f is continuously differentiable on an open set that contains the curve C, then

$$\int_C \nabla f(\mathbf{r}) \cdot \mathbf{dr} = f(\mathbf{b}) - f(\mathbf{a}).$$

PROOF If C is smooth,

$$\begin{aligned}\int_C \nabla f(\mathbf{r}) \cdot \mathbf{dr} &= \int_a^b [\nabla f(\mathbf{r}(u)) \cdot \mathbf{r}'(u)]\,du \\ &= \int_a^b \frac{d}{du}[f(\mathbf{r}(u))]\,du \quad \text{(chain rule (15.3.4))} \\ &= f(\mathbf{r}(b)) - f(\mathbf{r}(a)) \\ &= f(\mathbf{b}) - f(\mathbf{a}).\end{aligned}$$

If C is not smooth but only piecewise smooth, then we break up C into smooth pieces

$$C = C_1 \cup C_2 \cup \cdots \cup C_n.$$

With obvious notation,

$$\begin{aligned}\int_C \nabla f(\mathbf{r}) \cdot \mathbf{dr} &= \int_{C_1} \nabla f(\mathbf{r}) \cdot \mathbf{dr} + \int_{C_2} \nabla f(\mathbf{r}) \cdot \mathbf{dr} + \cdots + \int_{C_n} \nabla f(\mathbf{r}) \cdot \mathbf{dr} \\ &= [f(\mathbf{a}_1) - f(\mathbf{a}_0)] + [f(\mathbf{a}_2) - f(\mathbf{a}_1)] + \cdots + [f(\mathbf{a}_n) - f(\mathbf{a}_{n-1})] \\ &= f(\mathbf{a}_n) - f(\mathbf{a}_0) \\ &= f(\mathbf{b}) - f(\mathbf{a}). \quad \square\end{aligned}$$

The theorem we just proved has an important corollary:

(17.2.2) If the curve C is closed [that is, if $f(\mathbf{b}) = f(\mathbf{a})$], then

$$\int_C \nabla f(\mathbf{r}) \cdot \mathbf{dr} = 0.$$

Example 1 Integrate the vector field $\mathbf{h}(x, y) = y^2\,\mathbf{i} + (2xy - e^{2y})\,\mathbf{j}$ over the circular arc

$$C\colon \quad \mathbf{r}(u) = \cos u\,\mathbf{i} + \sin u\,\mathbf{j}, \qquad u \in [0, \tfrac{1}{2}\pi].$$

SOLUTION First we try to determine whether **h** is a gradient. We do this by applying Theorem 15.9.2.

Note that $\mathbf{h}(x,y)$ has the form $P(x,y)\,\mathbf{i} + Q(x,y)\,\mathbf{j}$ with

$$P(x,y) = y^2 \qquad \text{and} \qquad Q(x,y) = 2xy - e^{2y}.$$

Since P and Q are continuously differentiable everywhere and

$$\frac{\partial P}{\partial y} = 2y = \frac{\partial Q}{\partial x},$$

we can conclude that **h** is a gradient. Therefore, since the integral depends only on the endpoints of C and not on C itself, we can simplify the computations by integrating over the line segment C' that joins these same endpoints. (See Figure 17.2.1.)

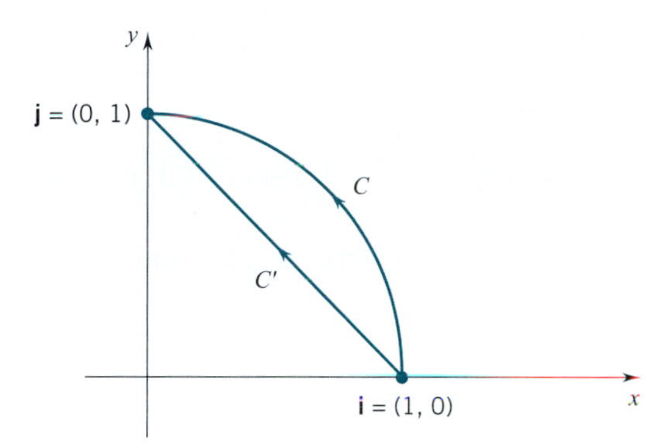

Figure 17.2.1

We parametrize C' by setting

$$\mathbf{r}(u) = (1-u)\,\mathbf{i} + u\,\mathbf{j}, \quad u \in [0,1].$$

We then have

$$\begin{aligned}
\int_C \mathbf{h}(\mathbf{r}) \cdot \mathbf{dr} &= \int_{C'} \mathbf{h}(\mathbf{r}) \cdot \mathbf{dr} \\
&= \int_0^1 [\mathbf{h}(\mathbf{r}(u)) \cdot \mathbf{r}'(u)]\, du \\
&= \int_0^1 \Big[[\,y(u)]^2 x'(u) + [2x(u)y(u) - e^{2y(u)}]y'(u)\Big]\, du \\
&= \int_0^1 \Big[u^2(-1) + [2(1-u)u - e^{2u}](1)\Big]\, du \\
&= \int_0^1 [2u - 3u^2 - e^{2u}]\, du = \Big[u^2 - u^3 - \tfrac{1}{2}e^{2u}\Big]_0^1 \\
&= \tfrac{1}{2} - \tfrac{1}{2}e^2.
\end{aligned}$$

ALTERNATIVE SOLUTION Once we recognize that $\mathbf{h}(x,y) = y^2\,\mathbf{i} + (2xy - e^{2y})\,\mathbf{j}$ is a gradient ∇f, we can try to determine $f(x,y)$ by the methods of Section 15.9. Since

$$\frac{\partial f}{\partial x} = y^2 \qquad \text{and} \qquad \frac{\partial f}{\partial y} = 2xy - e^{2y},$$

we have $\quad f(x, y) = xy^2 + \phi(y) \quad$ and therefore $\quad \dfrac{\partial f}{\partial y} = 2xy + \phi'(y).$

The two expressions for $\partial f/\partial y$ can be reconciled only if

$$\phi'(y) = -e^{2y} \qquad \text{and thus} \qquad \phi(y) = -\tfrac{1}{2}e^{2y} + K \qquad (K \text{ an arbitrary constant}).$$

Each function

$$f(x, y) = xy^2 - \tfrac{1}{2}e^{2y} + K$$

has gradient **h**. No matter what value we assign to K,

$$\int_C \mathbf{h}(\mathbf{r}) \cdot \mathbf{dr} = f(0, 1) - f(1, 0) = (-\tfrac{1}{2}e^2) - (-\tfrac{1}{2}) = \tfrac{1}{2} - \tfrac{1}{2}e^2. \quad \square$$

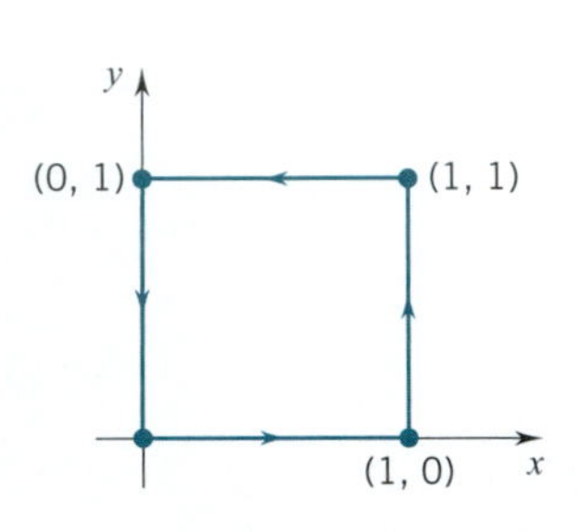

Figure 17.2.2

Example 2 Evaluate the line integral $\int_C \mathbf{h}(\mathbf{r}) \cdot \mathbf{dr}$, where C is the square with vertices $(0, 0), (1, 0), (1, 1), (0, 1)$ (Figure 17.2.2), and

$$\mathbf{h}(x, y) = (3x^2y + xy^2 - 1)\,\mathbf{i} + (x^3 + x^2y + 4y^3)\,\mathbf{j}.$$

SOLUTION First we try to determine whether **h** is a gradient. The functions

$$P(x, y) = 3x^2y + xy^2 - 1 \qquad \text{and} \qquad Q(x, y) = x^3 + x^2y + 4y^3$$

are continuously differentiable everywhere, and

$$\frac{\partial P}{\partial y} = 3x^2 + 2xy = \frac{\partial Q}{\partial x}.$$

Therefore **h** is the gradient of a function f. By (17.2.2),

$$\int_C \mathbf{h}(\mathbf{r}) \cdot \mathbf{dr} = \int_C \nabla f(\mathbf{r}) \cdot \mathbf{dr} = 0. \quad \square$$

Example 3 Evaluate the line integral $\int_C \mathbf{h}(\mathbf{r}) \cdot \mathbf{d\,r}$, where C is the unit circle

$$C\!: \ \mathbf{r}(u) = \cos u\,\mathbf{i} + \sin u\,\mathbf{j}, \quad u \in [0, 2\pi]$$

and

$$\mathbf{h}(x, y) = (y^2 + y)\,\mathbf{i} + (2xy - e^{2y})\,\mathbf{j}.$$

SOLUTION Although $(y^2 + y)\,\mathbf{i} + (2xy - e^{2y})\,\mathbf{j}$ is not a gradient $[\partial P/\partial y \neq \partial Q/\partial x]$, part of it,

$$y^2\,\mathbf{i} + (2xy - e^{2y})\,\mathbf{j},$$

is a gradient, as shown in Example 1. Therefore, we can write **h** as

$$\begin{aligned}\mathbf{h}(x, y) = (y^2 + y)\,\mathbf{i} + (2xy - e^{2y})\,\mathbf{j} &= [\,y^2\,\mathbf{i} + (2xy - e^{2y})\,\mathbf{j}] + y\,\mathbf{i}\\ &= \nabla f(x, y) + \mathbf{g}(x, y)\end{aligned}$$

where $\mathbf{g}(x, y) = y\,\mathbf{i}$. Now

$$\int_C \mathbf{h}(\mathbf{r}) \cdot \mathbf{dr} = \int_C \nabla f(\mathbf{r}) \cdot \mathbf{dr} + \int_C \mathbf{g}(\mathbf{r}) \cdot \mathbf{dr}.$$

Since we are integrating over a closed curve, the contribution of the gradient part is 0. The contribution of the remaining part is

$$\int_C \mathbf{g}(\mathbf{r}) \cdot d\mathbf{r} = \int_0^{2\pi} [\mathbf{g}(\mathbf{r}(u)) \cdot \mathbf{r}'(u)]\, du = \int_0^{2\pi} y(u)\, x'(u)\, du$$

$$= \int_0^{2\pi} -\sin^2 u\, du = -\pi.$$

Therefore $\int_C \mathbf{h}(\mathbf{r}) \cdot d\mathbf{r} = -\pi.$ ☐

EXERCISES 17.2

Determine whether **h** is a gradient, then calculate the line integral of **h** over the indicated curve.

1. $\mathbf{h}(x,y) = x\,\mathbf{i} + y\,\mathbf{j}$; $\mathbf{r}(u) = a\cos u\,\mathbf{i} + b\sin u\,\mathbf{j}$, $u \in [0, 2\pi]$.

2. $\mathbf{h}(x,y) = (x+y)\,\mathbf{i} + y\,\mathbf{j}$; the curve of Exercise 1.

3. $\mathbf{h}(x,y) = \cos \pi y\,\mathbf{i} - \pi x \sin \pi y\,\mathbf{j}$; $\mathbf{r}(u) = u^2\,\mathbf{i} - u^3\,\mathbf{j}$, $u \in [0, 1]$.

4. $\mathbf{h}(x,y) = (x^2 - y)\,\mathbf{i} + (y^2 - x)\,\mathbf{j}$; the curve of Exercise 1.

5. $\mathbf{h}(x,y) = xy^2\,\mathbf{i} + x^2y\,\mathbf{j}$; $\mathbf{r}(u) = u\sin \pi u\,\mathbf{i} + \cos \pi u^2\,\mathbf{j}$, $u \in [0, 1]$.

6. $\mathbf{h}(x,y) = (1+e^y)\,\mathbf{i} + (x\,e^y - x)\,\mathbf{j}$; the square with vertices $(-1,-1), (1,-1), (1,1), (-1,1)$ traversed counterclockwise.

7. $\mathbf{h}(x,y) = (2xy - y^2)\,\mathbf{i} + (x^2 - 2xy)\,\mathbf{j}$; $\mathbf{r}(u) = \cos u\,\mathbf{i} + \sin u\,\mathbf{j}$, $u \in [0, \pi]$.

8. $\mathbf{h}(x,y) = 3x(x^2+y^4)^{1/2}\,\mathbf{i} + 6y^3(x^2+y^4)^{1/2}\,\mathbf{j}$; the circular arc $y = (1-x^2)^{1/2}$ from $(1,0)$ to $(-1,0)$.

9. $\mathbf{h}(x,y) = 3x(x^2+y^4)^{1/2}\,\mathbf{i} + 6y^3(x^2+y^4)^{1/2}\,\mathbf{j}$; the arc $y = -(1-x^2)^{1/2}$ from $(-1,0)$ to $(1,0)$.

10. $\mathbf{h}(x,y) = 2xy\sinh x^2y\,\mathbf{i} + x^2\sinh x^2y\,\mathbf{j}$; the curve of Exercise 1.

11. $\mathbf{h}(x,y) = (2x\cosh y - y)\,\mathbf{i} + (x^2\sinh y - y)\,\mathbf{j}$; the square of Exercise 6.

Verify that the given vector field **h** is a gradient. Then calculate the line integral of **h** over the indicated curve C in two ways: (a) by calculating $\int_C \mathbf{h}(\mathbf{r}) \cdot d\mathbf{r}$ and (b) by finding f such that $\nabla f = \mathbf{h}$ and evaluating f at the endpoints of C.

12. $\mathbf{h}(x,y) = xy^2\,\mathbf{i} + yx^2\,\mathbf{j}$; $\mathbf{r}(u) = u\,\mathbf{i} + u^2\,\mathbf{j}$, $u \in [0, 2]$.

13. $\mathbf{h}(x,y) = (3x^2y^3 + 2x)\,\mathbf{i} + (3x^3y^2 - 4y)\,\mathbf{j}$; $\mathbf{r}(u) = u\,\mathbf{i} + e^u\,\mathbf{j}$, $u \in [0, 1]$.

14. $\mathbf{h}(x,y) = (2x\sin y - e^x)\,\mathbf{i} + (x^2\cos y)\,\mathbf{j}$; $\mathbf{r}(u) = \cos u\,\mathbf{i} + u\,\mathbf{j}$, $u \in [0, \pi]$.

15. $\mathbf{h}(x,y) = (e^{2y} - 2xy)\,\mathbf{i} + (2x\,e^{2y} - x^2 + 1)\,\mathbf{j}$; $\mathbf{r}(u) = u\,e^u\,\mathbf{i} + (1+u)\,\mathbf{j}$, $u \in [0, 1]$.

Use the three-dimensional analog of Theorem 15.9.2 given in Exercises 15.9 to show that the vector function **h** is a gradient. Then evaluate the line integral of **h** over the indicated curve.

16. $\mathbf{h}(x,y,z) = y^2z^3\,\mathbf{i} + 2xyz^3\,\mathbf{j} + 3xy^2z^2\mathbf{k}$; $\mathbf{r}(u) = u^2\,\mathbf{i} + u^4\,\mathbf{j} + u^6\,\mathbf{k}$, $u \in [0, 1]$.

17. $\mathbf{h}(x,y,z) = (2xz + \sin y)\,\mathbf{i} + x\cos y\,\mathbf{j} + x^2\,\mathbf{k}$; $\mathbf{r}(u) = \cos u\,\mathbf{i} + \sin u\,\mathbf{j} + u\,\mathbf{k}$, $u \in [0, 2\pi]$.

18. $\mathbf{h}(x,y,z) = \pi yz\cos \pi x\,\mathbf{i} + z\sin \pi x\,\mathbf{j} + y\sin \pi x\,\mathbf{k}$; $\mathbf{r}(u) = \cos u\,\mathbf{i} + \sin u\,\mathbf{j} + u\,\mathbf{k}$, $u \in [0, \pi/3]$.

19. $\mathbf{h}(x,y,z) = (2xy + z^2)\,\mathbf{i} + x^2\,\mathbf{j} + 2xz\,\mathbf{k}$; $\mathbf{r}(u) = 2u\,\mathbf{i} + (u^2+2)\,\mathbf{j} - u\,\mathbf{k}$, $u \in [0, 1]$.

20. $\mathbf{h}(x,y,z) = e^{-x}\ln y\,\mathbf{i} - \dfrac{e^{-x}}{y}\,\mathbf{j} + 3z^2\,\mathbf{k}$;
$\mathbf{r}(u) = (u+1)\,\mathbf{i} + e^{2u}\,\mathbf{j} + (u^2+1)\,\mathbf{k}$, $u \in [0, 1]$.

21. Calculate the work done by the force $\mathbf{F}(x,y) = (x+e^{2y})\,\mathbf{i} + (2y + 2x\,e^{2y})\,\mathbf{j}$ applied to an object that traverses the curve $\mathbf{r}(u) = 3\cos u\,\mathbf{i} + 4\sin u\,\mathbf{j}$, $u \in [0, 2\pi]$.

22. Calculate the work done by the force $\mathbf{F}(x,y,z) = (2x\ln y - yz)\,\mathbf{i} + [(x^2/y) - xz]\,\mathbf{j} - xy\,\mathbf{k}$ applied to an object that moves from the point $(1,2,1)$ to the point $(3,2,2)$.

23. If g is a continuously differentiable real-valued function defined on $[a,b]$, then by the fundamental theorem of integral calculus

$$\int_a^b g'(u)\, du = g(b) - g(a).$$

Show how this result is included in Theorem 17.2.1.

24. Let $\mathbf{r} = x\,\mathbf{i} + y\,\mathbf{j} + z\,\mathbf{k}$ and set $r = \|\mathbf{r}\|$. The central force field

$$\mathbf{F}(\mathbf{r}) = \frac{K}{r^n}\,\mathbf{r}, \qquad n \text{ a positive integer}$$

is a gradient field. Find f such that $\nabla f(\mathbf{r}) = \mathbf{F}(\mathbf{r})$ if: (a) $n = 2$; (b) $n \neq 2$.

25. Let $\mathbf{r} = x\,\mathbf{i} + y\,\mathbf{j} + z\,\mathbf{k}$ and set $r = \|\mathbf{r}\|$. Suppose that **F** is a vector field that is directed away from the origin with magnitude proportional to the square of the distance to the origin.

(a) Show that $\mathbf{F}(\mathbf{r}) = cr(x\,\mathbf{i} + y\,\mathbf{j} + z\,\mathbf{k})$, $c > 0$, constant.

(b) Show that **F** is a gradient field and find f such that $\mathbf{F}(\mathbf{r}) = \nabla f(\mathbf{r})$.

26. Let $\mathbf{r} = x\,\mathbf{i} + y\,\mathbf{j} + z\,\mathbf{k}$ and set $r = \|\mathbf{r}\|$. Suppose that $\mathbf{F}(\mathbf{r}) = g(r)\,\mathbf{r}$ where g is a continuous, real-valued function defined on $[0, \infty]$.

(a) Show that **F** is a gradient field

(b) Find f such that $\mathbf{F}(\mathbf{r}) = \nabla f(\mathbf{r})$.

27. Let $\mathbf{r} = x\,\mathbf{i} + y\,\mathbf{j} + z\,\mathbf{k}$ and set $r = ||\mathbf{r}||$. The function

$$\mathbf{F}(\mathbf{r}) = -\frac{mG}{r^3}\mathbf{r} \quad (G \text{ is the gravitational constant})$$

gives the gravitational force exerted by a unit mass at the origin on a mass m located at $\mathbf{r}$. What is the work done by $\mathbf{F}$ if m moves from $\mathbf{r}_1$ to $\mathbf{r}_2$?

28. Set

$$P(x,y) = \frac{y}{x^2+y^2} \quad \text{and} \quad Q(x,y) = -\frac{x}{x^2+y^2}$$

on *the punctured unit disc* $\Omega : 0 < x^2 + y^2 < 1$.

(a) Verify that P and Q are continuously differentiable on Ω and that

$$\frac{\partial P}{\partial y}(x,y) = \frac{\partial Q}{\partial x}(x,y) \quad \text{for all } (x,y) \in \Omega.$$

(b) Verify that, in spite of (a), the vector field $\mathbf{h}(x,y) = P(x,y)\,\mathbf{i} + Q(x,y)\,\mathbf{j}$ is not a gradient on Ω. HINT: Integrate $\mathbf{h}$ over a circle of radius less than 1 centered at the origin.

(c) Show that part (b) does not contradict Theorem 15.9.2.

29. The gravitational force acting on an object of mass m at a height z above the surface of the earth is given by

$$\mathbf{F}(x,y,z) = \frac{-mG\,r_0^2}{(r_0+z)^2}\mathbf{k},$$

where G is the gravitational constant and r_0 is the radius of the earth. Show that $\mathbf{F}$ is a gradient field and find f such that $\nabla f = \mathbf{F}$.

30. A rocket of mass m falls to the earth from a height of 300 miles. How much work is done by the gravitational force? Use Exercise 29 and assume that the radius of the earth is 4000 miles.

■ 17.3 WORK–ENERGY FORMULA; CONSERVATION OF MECHANICAL ENERGY

Suppose that a continuous force field $\mathbf{F} = \mathbf{F}(\mathbf{r})$ accelerates a mass m from $\mathbf{r} = \mathbf{a}$ to $\mathbf{r} = \mathbf{b}$ along some smooth curve C. The object undergoes a change in kinetic energy:

$$\tfrac{1}{2}m[v(\beta)]^2 - \tfrac{1}{2}m[v(\alpha)]^2.$$

The force does a certain amount of work W. How are these quantities related? They are equal:

(17.3.1)
$$W = \tfrac{1}{2}m[v(\beta)]^2 - \tfrac{1}{2}m[v(\alpha)]^2.$$

This relation is called the *work–energy formula*.

DERIVATION OF THE WORK–ENERGY FORMULA We parametrize the path of the motion by the time parameter t:

$$C: \quad \mathbf{r} = \mathbf{r}(t), \quad t \in [\alpha, \beta],$$

where $\mathbf{r}(\alpha) = \mathbf{a}$ and $\mathbf{r}(\beta) = \mathbf{b}$. The work done by $\mathbf{F}$ is given by the formula

$$W = \int_C \mathbf{F}(\mathbf{r}) \cdot d\mathbf{r} = \int_\alpha^\beta [\mathbf{F}(\mathbf{r}(t)) \cdot \mathbf{r}'(t)]\,dt.$$

From Newton's second law of motion, we know that at time t,

$$\mathbf{F}(\mathbf{r}(t)) = m\mathbf{a}(t) = m\mathbf{r}''(t).$$

It follows that

$$\begin{aligned}\mathbf{F}(\mathbf{r}(t)) \cdot \mathbf{r}'(t) &= m\,\mathbf{r}''(t) \cdot \mathbf{r}'(t)\\ &= \frac{d}{dt}\left[\frac{1}{2}m[\mathbf{r}'(t) \cdot \mathbf{r}'(t)]\right] = \frac{d}{dt}\left[\frac{1}{2}m||\mathbf{r}'(t)||^2\right] = \frac{d}{dt}\left[\frac{1}{2}m[v(t)]^2\right].\end{aligned}$$

Substituting this last expression into the work integral, we see that

$$W = \int_{\alpha}^{\beta} \frac{d}{dt}\left(\frac{1}{2}m[v(t)]^2\right) dt = \frac{1}{2}m[v(\beta)]^2 - \frac{1}{2}m[v(\alpha)]^2$$

as asserted. ❑

Conservative Force Fields

In general, if an object moves from one point to another, the work done (and hence the change in kinetic energy) depends on the path of the motion. There is, however, an important exception: if the force field is a gradient field,

$$\mathbf{F} = \nabla f,$$

then the work done (and hence the change in kinetic energy) depends only on the end-points of the path and not on the path itself. (This follows directly from the fundamental theorem for line integrals.) A force field that is a gradient field is called a *conservative field*.

Since the line integral over a closed path is zero, *the work done by a conservative field over a closed path is always zero. An object that passes through a given point with a certain kinetic energy returns to that same point with exactly the same kinetic energy.*

Potential Energy Functions

Suppose that $\mathbf{F}$ is a conservative force field. It is then a gradient field. Then $-\mathbf{F}$ is also a gradient field. The functions U for which $\nabla U = -\mathbf{F}$ are called *potential energy functions* for $\mathbf{F}$.

The Conservation of Mechanical Energy

Suppose that $\mathbf{F}$ is a conservative force field: $\mathbf{F} = -\nabla U$. In our derivation of the work–energy formula we showed that

$$\frac{d}{dt}\left(\tfrac{1}{2}m[v(t)]^2\right) = \mathbf{F}(\mathbf{r}(t)) \cdot \mathbf{r}'(t).$$

Since
$$\frac{d}{dt}[U(\mathbf{r}(t))] = \nabla U(\mathbf{r}(t)) \cdot \mathbf{r}'(t) = -\mathbf{F}(\mathbf{r}(t)) \cdot \mathbf{r}'(t),$$

we have
$$\frac{d}{dt}[\tfrac{1}{2}m[v(t)]^2 + U(\mathbf{r}(t))] = 0,$$

and therefore

$$\boxed{\underbrace{\tfrac{1}{2}m[v(t)]^2}_{\text{KE}} + \underbrace{U(\mathbf{r}(t))}_{\text{PE}} = \text{a constant.}}$$

As an object moves in a conservative force field, its kinetic energy can vary and its potential energy can vary, but the sum of these two quantities remains constant. We call this constant *the total mechanical energy*.

The total mechanical energy is usually denoted by the letter E. The law of conservation of mechanical energy can then be written

(17.3.2)
$$\tfrac{1}{2}mv^2 + U = E.$$

The conservation of energy is one of the cornerstones of physics. Here we have been talking about mechanical energy. There are other forms of energy and other energy conservation laws.

Differences in Potential Energy

Potential energy at a particular point has no physical significance. Only differences in potential energy are significant:

$$U(\mathbf{b}) - U(\mathbf{a}) = \int_C -\mathbf{F}(\mathbf{r}) \cdot \mathbf{dr}$$

is the work required to move from $\mathbf{r} = \mathbf{a}$ to $\mathbf{r} = \mathbf{b}$ *against* the force field $\mathbf{F}$.

Example 1 A planet moves in the gravitational field of the sun,

$$\mathbf{F}(\mathbf{r}) = -\rho m \frac{\mathbf{r}}{r^3},$$

where ρ is a constant and m is the mass of the planet. Show that the force field is conservative, find a potential energy function, and determine the total energy of the planet. How does the planet's speed vary with the planet's distance from the sun?

SOLUTION The field is conservative since

$$\mathbf{F}(\mathbf{r}) = -\rho m \frac{\mathbf{r}}{r^3} = \nabla\left(\frac{\rho m}{r}\right). \qquad \text{(check this out)}$$

As a potential energy function we can use

$$U(\mathbf{r}) = -\frac{\rho m}{r}.$$

The total energy of the planet is the constant

$$E = \tfrac{1}{2}mv^2 - \frac{\rho m}{r}.$$

(You met this quantity before: Exercises 2 and 4 of Section 13.7.)

Solving the energy equation for v, we have

$$v = \sqrt{\frac{2E}{m} + \frac{2\rho}{r}}.$$

As r decreases, $2\rho/r$ increases, and v increases; as r increases, $2\rho/r$ decreases, and v decreases. Thus every planet speeds up as it comes near the sun and slows down as it moves away. The same holds true for Halley's comet. The fact that it slows down as it gets farther away helps explain why it comes back. The simplicity of all this is a testimony to the power of the principle of energy conservation. ❑

EXERCISES 17.3

1. Let f be a continuous real-valued function of the real variable x. Show that the force field $\mathbf{F}(x,y,z) = f(x)\mathbf{i}$ is conservative and the potential functions for $\mathbf{F}$ are (except for notation) the antiderivatives of $-f$.

2. A particle with electric charge e and velocity $\mathbf{v}$ moves in a magnetic field $\mathbf{B}$, experiencing the force

$$\mathbf{F} = \frac{e}{c}[\mathbf{v} \times \mathbf{B}]. \qquad (c \text{ is the velocity of light})$$

F is not a gradient — it can't be, depending as it does on the *velocity* of the particle. Still, we can find a conserved quantity; the *kinetic energy* $\frac{1}{2}mv^2$. Show by differentiation with respect to t that this quantity is constant. (Assume Newton's second law.)

3. An object is subject to a constant force in the direction of $-\mathbf{k}$: $\mathbf{F} = -c\mathbf{k}$ with $c > 0$. Find a potential energy function for **F**, and use energy conservation to show that the speed of the object at time t_2 is related to that at time t_1 by the equation

$$v(t_2) = \sqrt{[v(t_1)]^2 + \frac{2c}{m}[z(t_1) - z(t_2)]}$$

where $z(t_1)$ and $z(t_2)$ are the z-coordinates of the object at times t_1 and t_2. (This analysis is sometimes used to model the behavior of an object in the gravitational field near the surface of the earth.)

4. (*Escape velocity*) An object is to be fired straight up from the surface of the earth. Assume that the only force acting on the object is the gravitational pull of the earth and determine the initial speed v_0 necessary to send the object off to infinity.

HINT: Appeal to conservation of energy and use the idea that the object is to arrive at infinity with zero speed.

5. (a) Justify the statement that a conservative force field **F** always acts so as to encourage motion toward regions of lower potential energy U.
   (b) Evaluate **F** at a point where U has a minimum.

6. A harmonic oscillator has a restoring force $\mathbf{F} = -\lambda x\,\mathbf{i}$. The associated potential is $U(x,y,z) = -\frac{1}{2}\lambda x^2$, and the constant total energy is

$$E = \tfrac{1}{2}mv^2 + U(x,y,z) = \tfrac{1}{2}mv^2 + \tfrac{1}{2}\lambda x^2.$$

Given that $x(0) = 2$ and $x'(0) = 1$, calculate the maximum speed of the oscillator and the maximum value of x.

7. The *equipotential surfaces* of a conservative field **F** are the surfaces where the potential energy is constant. Show that: (a) the speed of an object in such a field is constant on every equipotential surface; and (b) at each point of such a surface the force field is perpendicular to the surface.

8. Suppose a force field **F** is directed away from the origin with a magnitude that is inversely proportional to the distance from the origin. Show that **F** is a conservative field.

9. Let **F** be the inverse-square force field:

$$\mathbf{F}(x,y,z) = \frac{k(x\,\mathbf{i} + y\,\mathbf{j} + z\,\mathbf{k})}{(x^2 + y^2 + z^2)^{3/2}}$$

and let C be any curve on the unit sphere $x^2 + y^2 + z^2 = 1$. Show that the work done by **F** in moving an object along C is 0. Explain this result.

■ 17.4 ANOTHER NOTATION FOR LINE INTEGRALS; LINE INTEGRALS WITH RESPECT TO ARC LENGTH

If $\mathbf{h}(x,y,z) = P(x,y,z)\,\mathbf{i} + Q(x,y,z)\,\mathbf{j} + R(x,y,z)\,\mathbf{k}$, the line integral

$$\int_C \mathbf{h}(\mathbf{r}) \cdot \mathbf{dr} \qquad \text{can be written} \qquad \int_C P(x,y,z)\,dx + Q(x,y,z)\,dy + R(x,y,z)\,dz.$$

The notation arises as follows. With

$$C\colon \quad \mathbf{r}(u) = x(u)\,\mathbf{i} + y(u)\,\mathbf{j} + z(u)\,\mathbf{k}, \quad u \in [a,b],$$

the line integral

$$\int_C \mathbf{h}(\mathbf{r}) \cdot \mathbf{dr} = \int_a^b [\mathbf{h}(\mathbf{r}(u)) \cdot \mathbf{r}'(u)]\,du$$

expands to

$$\int_a^b \{P[x(u),y(u),z(u)]x'(u) + Q[x(u),y(u),z(u)]\,y'(u) + R[x(u),y(u),z(u)]z'(u)\}\,du.$$

Now set
$$\int_C P(x,y,z)\,dx = \int_a^b P[x(u),y(u),z(u)]\,x'(u)\,du,$$
$$\int_C Q(x,y,z)\,dy = \int_a^b Q[x(u),y(u),z(u)]\,y'(u)\,du,$$
$$\int_C R(x,y,z)\,dz = \int_a^b R[x(u),y(u),z(u)]\,z'(u)\,du.$$

Writing the sum of these integrals as

$$\int_C P(x,y,z)\,dx + Q(x,y,z)\,dy + R(x,y,z)\,dz,$$

we have
$$\int_C P(x,y,z)\,dx + Q(x,y,z)\,dy + R(x,y,z)\,dz = \int_C \mathbf{h}(\mathbf{r}) \cdot \mathbf{dr}.$$

If C lies in the xy-plane and $\mathbf{h}\,(x,y) = P(x,y)\,\mathbf{i} + Q(x,y)\,\mathbf{j}$, then the line integral reduces to

$$\int_C P(x,y)\,dx + Q(x,y)\,dy.$$

Example 1 Evaluate $\int_C x^2y\,dx + xy\,dy$, where C is

(a) The straight-line path connecting (1, 0) to (0, 1).
(b) The circular path $y = \sqrt{1-x^2}$ connecting $(1,0)$ to $(0,1)$.
(c) The polygonal path $(1,0)$, $(1,1)$, $(0,1)$.

SOLUTION The given integral is the line integral $\int_C \mathbf{h}(\mathbf{r}) \cdot \mathbf{dr}$, where

$$\mathbf{h}\,(x,y) = x^2y\,\mathbf{i} + xy\,\mathbf{j} \quad \text{and} \quad \mathbf{r} = \mathbf{r}\,(u) \;\text{ is a parametrization of } C.$$

(a) The straight-line path connecting $(1,0)$ to $(0,1)$ can be parametrized by

$$\mathbf{r}(u) = (1-u)\,\mathbf{i} + u\,\mathbf{j}, \quad 0 \le u \le 1.$$

Thus,
$$\begin{aligned}\int_C x^2y\,dx + xy\,dy &= \int_0^1 [x^2(u)y(u)x'(u) + x(u)y(u)y'(u)]\,du \\ &= \int_0^1 [(1-u)^2u(-1) + (1-u)u]\,du \\ &= \int_0^1 (u^2 - u^3)\,du = \left[\tfrac{1}{3}u^3 - \tfrac{1}{4}u^4\right]_0^1 = \tfrac{1}{12}.\end{aligned}$$

(b) We parametrize the circular arc $y = \sqrt{1-x^2}$, $0 \le x \le 1$ by setting

$$\mathbf{r}\,(u) = \cos u\,\mathbf{i} + \sin u\,\mathbf{j}, \quad 0 \le u \le \pi/2.$$

Thus,
$$\begin{aligned}\int_C x^2y\,dx + xy\,dy &= \int_0^{\pi/2} [x^2(u)y(u)x'(u) + x(u)y(u)y'(u)]\,du\\ &= \int_0^{\pi/2} [\cos^2 u\sin u\,(-\sin u) + \cos u\sin u\,(\cos u)]\,du\\ &= \int_0^{\pi/2} [-\cos^2 u\sin^2 u + \cos^2 u\sin u]\,du\\ &= -\int_0^{\pi/2} \tfrac{1}{4}\sin^2 2u\,du + \int_0^{\pi/2} \cos^2 u\sin u\,du\\ &= -\tfrac{1}{8}\int_0^{\pi/2} (1-\cos 4u)\,du + \left[-\tfrac{1}{3}\cos^3 u\right]_0^{\pi/2}\\ &= -\tfrac{1}{8}\left[u - \tfrac{1}{4}\sin 4u\right]_0^{\pi/2} + \tfrac{1}{3} = \tfrac{1}{3} - \tfrac{\pi}{16}.\end{aligned}$$

(c) The polygonal path $(1,0), (1,1), (0,1)$ is made up of the two segments

$$C_1:\ \mathbf{r}(u) = \mathbf{i} + u\,\mathbf{j},\ 0 \le u \le 1, \quad \text{and} \quad C_2:\ \mathbf{r}(u) = (1-u)\,\mathbf{i} + \mathbf{j},\ 0 \le u \le 1.$$

Now
$$\int_{C_1} x^2y\,dx + xy\,dy = \int_{C_1} xy\,dy = \int_0^1 x(u)\,y(u)\,y'(u)\,du = \int_0^1 u\,du = \tfrac{1}{2}$$

and
$$\begin{aligned}\int_{C_2} x^2y\,dx + xy\,dy = \int_{C_2} x^2y\,dx &= \int_0^1 x^2(u)\,y(u)\,x'(u)\,du\\ &= \int_0^1 -(1-u)^2\,du = -\tfrac{1}{3}.\end{aligned}$$

Therefore,

$$\int_C x^2y\,dx + xy\,dy = \int_{C_1} x^2y\,dx + xy\,dy + \int_{C_2} x^2y\,dx + xy\,dy = \tfrac{1}{2} - \tfrac{1}{3} = \tfrac{1}{6}. \quad \square$$

Line Integrals with Respect to Arc Length

Suppose that f is a scalar field continuous on a piecewise-smooth curve

$$C: \quad \mathbf{r}(u) = x(u)\,\mathbf{i} + y(u)\,\mathbf{j} + z(u)\,\mathbf{k}, \quad u \in [a,b].$$

If $s(u)$ is the length of the curve from the tip of $\mathbf{r}(a)$ to the tip of $\mathbf{r}(u)$, then, as you have seen,

$$s'(u) = \|\mathbf{r}'(u)\| = \sqrt{[x'(u)]^2 + [y'(u)]^2 + [z'(u)]^2}.$$

The line integral of f over C *with respect to arc length s* is defined by setting

(17.4.1)
$$\int_C f(\mathbf{r})\,ds = \int_a^b f(\mathbf{r}(u))\,s'(u)\,du.$$

In the xyz-notation we have

$$\int_C f(x,y,z)\,ds = \int_a^b f(x(u),y(u),z(u))\,s'(u)\,du,$$

which, in the two-dimensional case, becomes

$$\int_C f(x,y)\,ds = \int_a^b f(x(u),y(u))\,s'(u)\,du.$$

Suppose now that C represents a thin wire (a material curve) of varying mass density $\lambda = \lambda(\mathbf{r})$. (Here mass density is mass per unit length.) The *length* of the wire can be written

(17.4.2)
$$L = \int_C ds.$$

The *mass* of the wire is given by

(17.4.3)
$$M = \int_C \lambda(\mathbf{r})\,ds,$$

and the *center of mass* $\mathbf{r}_M$ can be obtained from the vector equation

(17.4.4)
$$\mathbf{r}_M M = \int_C \mathbf{r}\,\lambda(\mathbf{r})\,ds.$$

The equivalent scalar equations read

$$x_M M = \int_C x\,\lambda(\mathbf{r})\,ds, \quad y_M M = \int_C y\,\lambda(\mathbf{r})\,ds, \quad z_M M = \int_C z\,\lambda(\mathbf{r})\,ds.$$

Finally, the *moment of inertia* about an axis is given by the formula

(17.4.5)
$$I = \int_C \lambda(\mathbf{r})[R(\mathbf{r})]^2\,ds$$

where $R(\mathbf{r})$ is the distance from the axis to the tip of $\mathbf{r}$.

Example 2 The mass density of a semicircular wire of radius a varies directly as the distance from the diameter that joins the two endpoints of the wire. (a) Find the mass of the wire. (b) Locate the center of mass. (c) Determine the moment of inertia of the wire about the diameter.

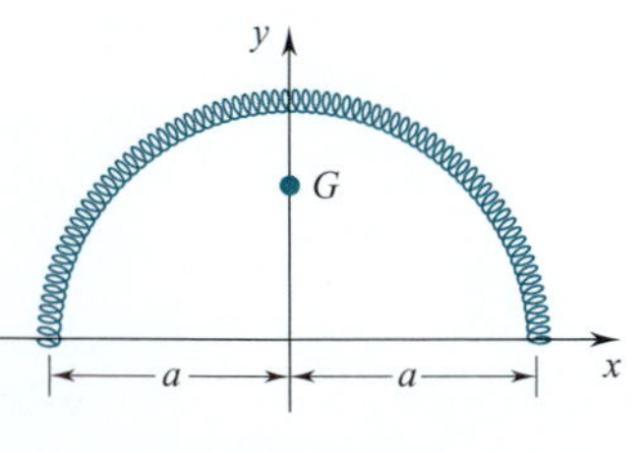

Figure 17.4.1

SOLUTION Placed as in Figure 17.4.1, the wire can be parametrized by

$$\mathbf{r}(u) = a\cos u\,\mathbf{i} + a\sin u\,\mathbf{j}, \quad u \in [0,\pi]$$

and the mass density function can be written $\lambda(x,y) = ky$. Since $\mathbf{r}'(u) = -a\sin u\,\mathbf{i} + a\cos u\,\mathbf{j}$, we have

$$s'(u) = ||\mathbf{r}'(u)|| = a.$$

Therefore $$M = \int_C \lambda(x,y)\,ds = \int_C ky\,ds$$

$$= \int_0^\pi ky(u)\,s'(u)\,du$$

$$= \int_0^\pi k(a\sin u)\,a\,du = ka^2\int_0^\pi \sin u\,du = 2ka^2.$$

By the symmetry of the configuration, $x_M = 0$. To find y_M we have to integrate:

$$y_M M = \int_C y\lambda(x,y)\,ds = \int_C ky^2\,ds$$

$$= \int_0^\pi k[y(u)]^2 s'(u)\,du$$

$$= \int_0^\pi k(a\sin u)^2 a\,du = ka^3\int_0^\pi \sin^2 u\,du = \tfrac{1}{2}ka^3\pi.$$

Since $M = 2ka^2$, we have $y_M = (\frac{1}{2}ka^3\pi)/(2ka^2) = \frac{1}{4}a\pi$. The center of mass G lies on the perpendicular bisector of the wire at a distance $\frac{1}{4}a\pi$ from the diameter. (See Figure 17.4.1.) Note that in this instance, the center of mass does not lie on the wire.

Now let's find the moment of inertia about the diameter:

$$I = \int_C \lambda(x,y)[R(x,y)]^2\,ds = \int_C (ky)y^2\,ds$$

$$= \int_C k[y(u)]^3 s'(u)\,du$$

$$= \int_0^\pi k(a\sin u)^3 a\,du$$

$$= ka^4\int_0^\pi \sin^3 u\,du = \tfrac{4}{3}ka^4.$$

It is customary to express I in terms of M. With $M = 2ka^2$, we have

$$I = \tfrac{2}{3}(2ka^2)a^2 = \tfrac{2}{3}Ma^2. \quad \square$$

EXERCISES 17.4

In Exercises 1–4, evaluate

$$\int_C (x - 2y)\,dx + 2x\,dy$$

along the given path C from $(0,0)$ to $(1,2)$.

1. The straight-line path.
2. The parabolic path $y = 2x^2$.
3. The polygonal path $(0,0), (1,0), (1,2)$.
4. The polygonal path $(0,0), (0,2), (1,2)$.

In Exercises 5–8, evaluate

$$\int_C y\,dx + xy\,dy$$

along the given path C from $(0,0)$ to $(2,1)$.

5. The parabolic path $x = 2y^2$.
6. The straight-line path.
7. The polygonal path $(0,0), (0,1), (2,1)$.
8. The cubic path $x = 2y^3$.

In Exercises 9–12, evaluate

$$\int_C y^2\,dx + (xy - x^2)\,dy$$

along the given path C from (0, 0) to (2, 4).

9. The straight-line path.

10. The parabolic path $y = x^2$.

11. The parabolic path $y^2 = 8x$.

12. The polygonal path $(0,0), (2,0), (2,4)$.

In Exercises 13–16, evaluate

$$\int_C (y^2 + 2x + 1)\,dx + (2xy + 4y - 1)\,dy$$

along the given path C from $(0,0)$ to $(1,1)$.

13. The straight-line path.

14. The parabolic path $y = x^2$.

15. The cubic path $y = x^3$.

16. The polygonal path $(0,0), (4,0), (4,2), (1,1)$.

In Exercises 17–20, evaluate

$$\int_C y\,dx + 2z\,dy + x\,dz$$

along the given path C from $(0,0,0)$ to $(1,1,1)$.

17. The straight-line path.

18. $\mathbf{r}(u) = u\,\mathbf{i} + u^2\,\mathbf{j} + u^3\,\mathbf{k}$.

19. The polygonal path $(0,0,0), (0,0,1), (0,1,1), (1,1,1)$.

20. The polygonal path $(0,0,0), (1,0,0), (1,1,0), (1,1,1)$.

In Exercises 21–24, evaluate

$$\int_C xy\,dx + 2z\,dy + (y + z)\,dz$$

along the given path C from $(0,0,0)$ to $(2,2,8)$.

21. The straight-line path.

22. The polygonal path $(0,0,0), (2,0,0), (2,2,0), (2,2,8)$.

23. The parabolic path $\mathbf{r}(u) = u\,\mathbf{i} + u\,\mathbf{j} + 2u^2\,\mathbf{k}$.

24. The polygonal path $(0,0,0), (2,2,2), (2,2,8)$.

25. Evaluate $\int_C x^2y\,dx + y\,dy + xz\,dz$, where C is the curve of intersection of the cylinder $y - 2z^2 = 1$ and the plane $z = x + 1$ from $(0,3,1)$ to $(1,9,2)$.

26. Evaluate $\int_C y\,dx + yz\,dy + z(x-1)\,dz$, where C is the curve of intersection of the sphere $x^2 + y^2 + z^2 = 4$ and the cylinder $(x-1)^2 + y^2 = 1$ from $(2,0,0)$ to $(0,0,2)$.

27. Let $\mathbf{h}$ be the vector field

$$\mathbf{h}(x, y) = (x^2 + 6xy - 2y^2)\,\mathbf{i} + (3x^2 - 4xy + 2y)\,\mathbf{j}.$$

(a) Show that $\mathbf{h}$ is a gradient field.

(b) What is the value of

$$\int_C (x^2 + 6xy - 2y^2)\,dx + (3x^2 - 4xy + 2y)\,dy$$

along any piecewise-smooth curve from $(3,0)$ to $(0,4)$?

(c) What is the value of

$$\int_C (x^2 + 6xy - 2y^2)\,dx + (3x^2 - 4xy + 2y)\,dy$$

where C is any piecewise-smooth curve from (4, 0) to (0, 3)?

28. Let $\mathbf{h}$ be the vector field

$$\mathbf{h}(x,y,z) = (2xy + z^2)\,\mathbf{i} + (x^2 - 2yz)\,\mathbf{j} + (2xz - y^2)\,\mathbf{k}.$$

(a) Show that $\mathbf{h}$ is a gradient field.

(b) What is the value of

$$\int_C (2xy + z^2)\,dx + (x^2 - 2yz)\,dy + (2xz - y^2)\,dz$$

along any piecewise-smooth curve from $(1,0,1)$ to $(3,2,-1)$?

(c) What is the value of

$$\int_C (2xy + z^2)\,dx + (x^2 - 2yz)\,dy + (2xz - y^2)\,dz$$

where C is any piecewise-smooth curve from $(3,2,-1)$ to (1, 0, 1)?

29. A wire in the shape of the quarter-circle

$$C\colon\ \mathbf{r}(u) = a(\cos u\,\mathbf{i} + \sin u\,\mathbf{j}),\quad u \in [0, \tfrac{1}{2}\pi]$$

has varying mass density $\lambda(x,y) = k(x + y)$ where k is a positive constant.

(a) Find the total mass of the wire and locate the center of mass.

(b) What is the moment of inertia of the wire about the x-axis?

30. Find the moment of inertia of a homogeneous circular wire of radius a and mass M about (a) a diameter; (b) the axis through the center that is perpendicular to the plane of the wire.

31. Find the moment of inertia of the wire of Exercise 29 about

(a) the z-axis; (b) the line $y = x$.

32. A wire of constant mass density k has the form

$$\mathbf{r}(u) = (1 - \cos u)\,\mathbf{i} + (u - \sin u)\,\mathbf{j},\quad u \in [0, 2\pi].$$

(a) Determine the mass of the wire.

(b) Locate the center of mass.

33. A homogeneous wire of mass M winds around the z-axis as

$$C: \quad \mathbf{r}(u) = a\cos u\,\mathbf{i} + a\sin u\,\mathbf{j} + bu\,\mathbf{k}, \qquad u \in [0, 2\pi].$$

(a) Find the length of the wire.

(b) Locate the center of mass.

(c) Determine the moments of inertia of the wire about the coordinate axes.

34. A homogeneous wire of mass M is of the form

$$C: \quad \mathbf{r}(u) = u\,\mathbf{i} + u^2\,\mathbf{j} + \tfrac{2}{3}u^3\,\mathbf{k}, \qquad u \in [0, a].$$

(a) Find the length of the wire.

(b) Locate the center of mass.

(c) Determine the moment of inertia of the wire about the z-axis.

35. Calculate the mass of the wire of Exercise 33 if the mass density varies directly as the square of the distance from the origin.

36. Show that

(17.4.6) $$\int_C \mathbf{h}(\mathbf{r}) \cdot \mathbf{dr} = \int_C [\mathbf{h}(\mathbf{r}) \cdot \mathbf{T}(\mathbf{r})]\,ds$$

where $\mathbf{T}$ is the unit tangent vector.

■ 17.5 GREEN'S THEOREM

Green's theorem is the first of the three integration theorems heralded at the beginning of this chapter.

A *Jordan curve*, as you may recall from a footnote in Section 15.9, is a plane curve that is both closed and simple. Thus circles, ellipses, and triangles are Jordan curves; figure eights are not.

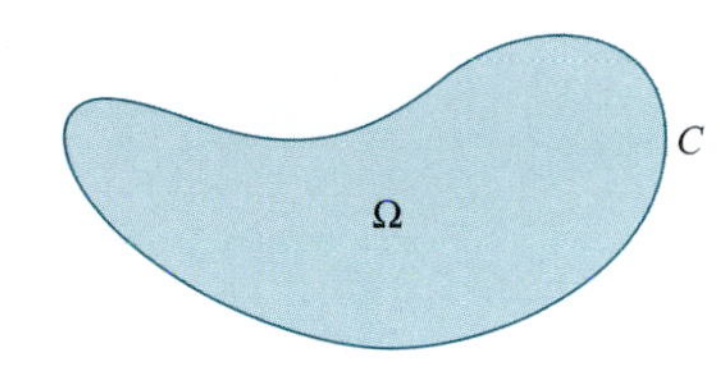

Figure 17.5.1

Figure 17.5.1 depicts a closed region Ω, the total boundary of which is a Jordan curve C. Such a region is called a *Jordan region*. We know how to integrate over Ω, and if the boundary C is piecewise smooth, we know how to integrate over C. Green's theorem expresses a double integral over Ω as a line integral over C.

THEOREM 17.5.1 GREEN'S THEOREM †

Let Ω be a Jordan region with a piecewise-smooth boundary C. If P and Q are scalar fields that are continuously differentiable on an open set that contains Ω, then

$$\iint_\Omega \left[\frac{\partial Q}{\partial x}(x,y) - \frac{\partial P}{\partial y}(x,y)\right] dxdy = \oint_C P(x,y)\,dx + Q(x,y)\,dy$$

where the integral on the right is the line integral taken over C in the counterclockwise direction.††

We will prove the theorem only for special cases. First of all let's assume that Ω is an *elementary region*, a region that is both of Type I and Type II as defined in Section 16.3. For simplicity we take Ω as in Figure 17.5.2.

With Ω being of Type I, we can show that

(1) $$\oint_C P(x,y)\,dx = \iint_\Omega -\frac{\partial P}{\partial y}(x,y)\,dxdy.$$

† The result was established in 1828 by the English mathematician George Green (1793–1841).

†† Counterclockwise as viewed from $z > 0$ in a right-handed coordinate system. The integral over C in the clockwise direction is denoted $\oint$.

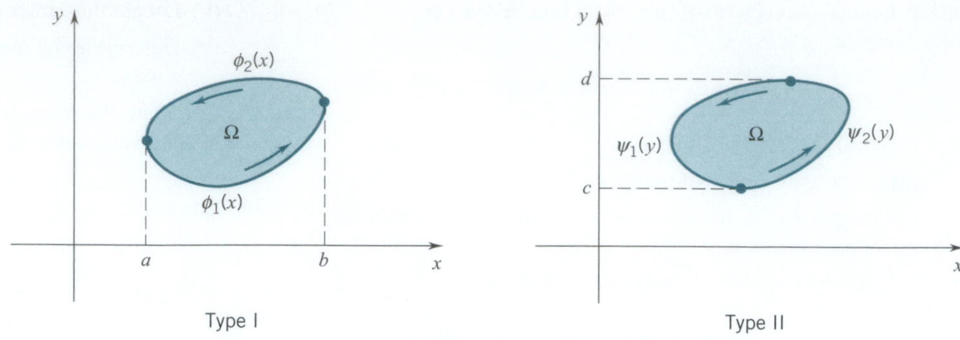

Ω is an elementary region: it is both of Type I and Type II

Figure 17.5.2

In the first place

$$\iint_{\Omega} -\frac{\partial P}{\partial y}(x,y)\,dxdy = -\int_a^b \int_{\phi_1(x)}^{\phi_2(x)} \frac{\partial P}{\partial y}(x,y)\,dy\,dx$$

$$= -\int_a^b \{P[x,\phi_2(x)] - P[x,\phi_1(x)]\}\,dx$$

by the fundamental theorem of integral calculus ↑

$$(*) \qquad = \int_a^b P[x,\phi_1(x)]\,dx - \int_a^b P[x,\phi_2(x)]\,dx.$$

The graph of ϕ_1 parametrized from left to right is the curve

$$C_1: \quad \mathbf{r}_1(u) = u\,\mathbf{i} + \phi_1(u)\,\mathbf{j}, \quad u \in [a,b];$$

the graph of ϕ_2, also parametrized from left to right, is the curve

$$C_2: \quad \mathbf{r}_2(u) = u\,\mathbf{i} + \phi_2(u)\,\mathbf{j}, \quad u \in [a,b].$$

Since C traversed counterclockwise consists of C_1 followed by $-C_2$ (C_2 traversed from right to left), you can see that

$$\oint_C P(x,y)\,dx = \int_{C_1} P(x,y)\,dx - \int_{C_2} P(x,y)\,dx$$

$$= \int_a^b P[u,\phi_1(u)]\,du - \int_a^b P[u,\phi_2(u)]\,du.$$

Since u is a dummy variable, it can be replaced by x. Comparison with $(*)$ proves (1).

We leave it to you to verify that

$$(2) \qquad \oint_C Q(x,y)\,dy = \iint_{\Omega} \frac{\partial Q}{\partial x}(x,y)\,dxdy$$

by using the fact that Ω is of Type II. This completes the proof of the theorem for Ω as in Figure 17.5.2.

A slight modification of this argument applies to elementary regions which are bordered entirely or in part by line segments parallel to the coordinate axes.

Figure 17.5.3 shows a Jordan region that is not elementary but can be broken up into two elementary regions. (See Figure 17.5.4.) Green's theorem applied to the elementary parts tells us that

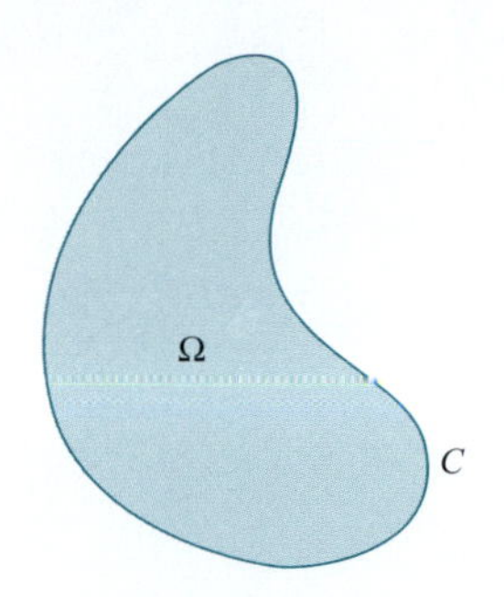

Figure 17.5.3

$$\iint_{\Omega_1} \left(\frac{\partial Q}{\partial x} - \frac{\partial P}{\partial y}\right) dxdy = \oint_{\text{bdry of } \Omega_1} P\,dx + Q\,dy,$$

$$\iint_{\Omega_2} \left(\frac{\partial Q}{\partial x} - \frac{\partial P}{\partial y}\right) dxdy = \oint_{\text{bdry of } \Omega_2} P\,dx + Q\,dy.$$

We now add these equations. The sum of the double integrals is, by additivity, the double integral over Ω. The sum of the line integrals is the integral over C (see the figure) plus the integrals over the crosscut. Since the crosscut is traversed twice and in opposite directions, the total contribution of the crosscut is zero and therefore Green's theorem holds:

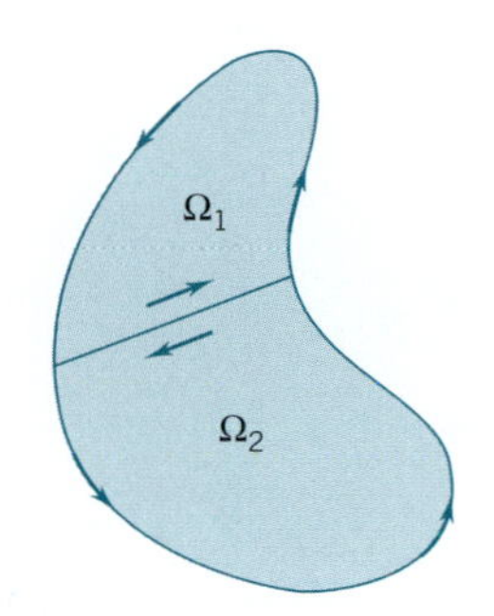

Figure 17.5.4

$$\iint_{\Omega} \left(\frac{\partial Q}{\partial x} - \frac{\partial P}{\partial y}\right) dxdy = \oint_C P\,dx + Q\,dy.$$

This same argument can be extended to a Jordan region Ω that breaks up into n elementary regions $\Omega_1, \ldots, \Omega_n$. (Figure 17.5.5 gives an example with $n = 4$.) The double integrals over the Ω_i add up to the double integral over Ω, and, since the line integrals over the crosscuts cancel, the line integrals over the boundaries of the Ω_i add up to the line integral over C. (This is as far as we will carry the proof of Green's theorem. It is far enough to cover all the Jordan regions encountered in practice.)

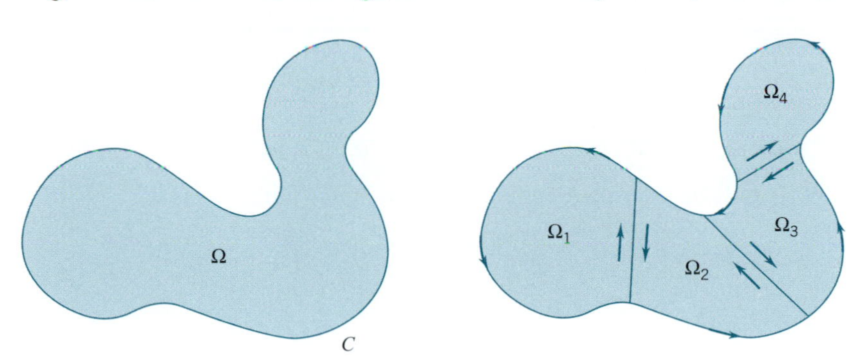

Figure 17.5.5

Example 1 Use Green's theorem to evaluate

$$\oint_C (3x^2 + y)\,dx + (2x + y^3)\,dy$$

where C is the circle $x^2 + y^2 = a^2$. (Figure 17.5.6)

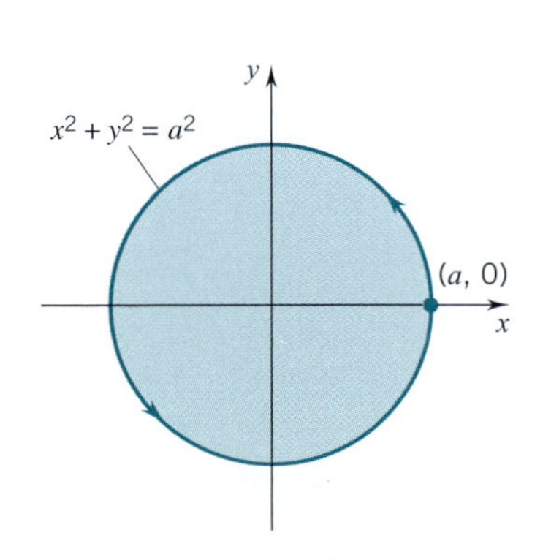

Figure 17.5.6

SOLUTION Let Ω be the closed disc $0 \leq x^2 + y^2 \leq a^2$. With

$$P(x,y) = 3x^2 + y \qquad \text{and} \qquad Q(x,y) = 2x + y^3,$$

we have $\quad \dfrac{\partial P}{\partial y} = 1, \quad \dfrac{\partial Q}{\partial x} = 2, \quad$ and $\quad \dfrac{\partial Q}{\partial x} - \dfrac{\partial P}{\partial y} = 2 - 1 = 1.$

By Green's theorem

$$\oint_C (3x^2 + y)\,dx + (2x + y^3)\,dy = \iint_\Omega 1\,dxdy = \text{area of } \Omega = \pi a^2. \quad \square$$

Remark The line integral in Example 1 could have been calculated directly as follows: The circle $x^2 + y^2 = a^2$ can be parametrized counterclockwise by setting

$$x = a\cos u, \quad y = a\sin u, \quad 0 \le u \le 2\pi.$$

Thus

$$\oint_C (3x^2 + y)\,dx + (2x + y^3)\,dy$$

$$= \int_0^{2\pi} [(3a^2\cos^2 u + a\sin u)(-a\sin u) + (2a\cos u + a^3\sin^3 u)(a\cos u)]\,du$$

$$= \int_0^{2\pi} [-3a^3\cos^2 u\sin u - a^2\sin^2 u + 2a^2\cos^2 u + a^4\sin^3 u\ \cos u]\,du,$$

which, as you can verify, also yields πa^2. In this case, at least, Green's theorem gives us a more direct route to the answer. $\square$

Example 2 Use Green's theorem to evaluate

$$\oint_C (1 + 10xy + y^2)\,dx + (6xy + 5x^2)\,dy$$

where C is the square with vertices $(0, 0), (a, 0), (a, a), (0, a)$. (Figure 17.5.7)

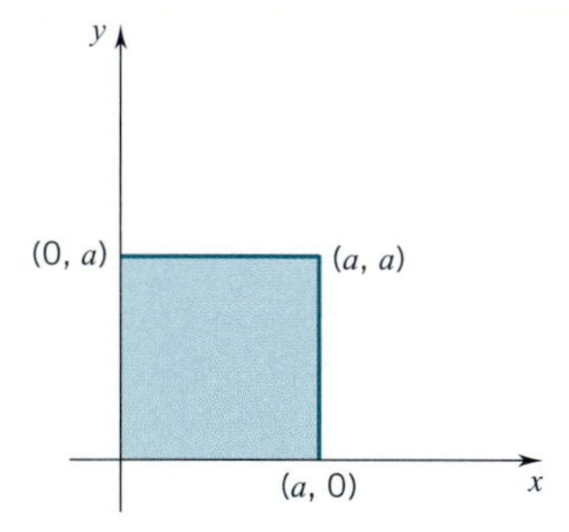

Figure 17.5.7

SOLUTION Let Ω be the square region enclosed by C. With

$$P(x, y) = 1 + 10xy + y^2 \quad \text{and} \quad Q(x, y) = 6xy + 5x^2,$$

we have $\quad \dfrac{\partial P}{\partial y} = 10x + 2y, \quad \dfrac{\partial Q}{\partial x} = 6y + 10x, \quad \dfrac{\partial Q}{\partial x} - \dfrac{\partial P}{\partial y} = 4y.$

By Green's theorem,

$$\begin{aligned}\oint_C (1 + 10xy + y^2)\,dx + (6xy + 5x^2)\,dy &= \iint_\Omega 4y\,dxdy \\ &= \int_0^a \int_0^a 4y\,dx\,dy \\ &= \left(\int_0^a dx\right)\left(\int_0^a 4y\,dy\right) \\ &= (a)(2a^2) = 2a^3.\end{aligned}$$

ALTERNATIVE SOLUTION By (16.5.3),

$$\iint_{\Omega} y\,dxdy = \bar{y}\,(\text{area of } \Omega)$$

where $\bar{y}$ is the y-coordinate of the centroid of Ω. Since $\bar{y} = \frac{1}{2}a$, it is evident that

$$\iint_{\Omega} 4y\,dxdy = 4\bar{y}\,(\text{area of } \Omega) = 4(\tfrac{1}{2}a)a^2 = 2a^3. \quad \square$$

Example 3 Use Green's theorem to evaluate

$$\oint_C e^x \sin y\,dx + e^x \cos y\,dy$$

where C is the closed curve consisting of the semicircle $y = \sqrt{1 - x^2}$ and the interval $[-1, 1]$. (See Figure 17.5.8.)

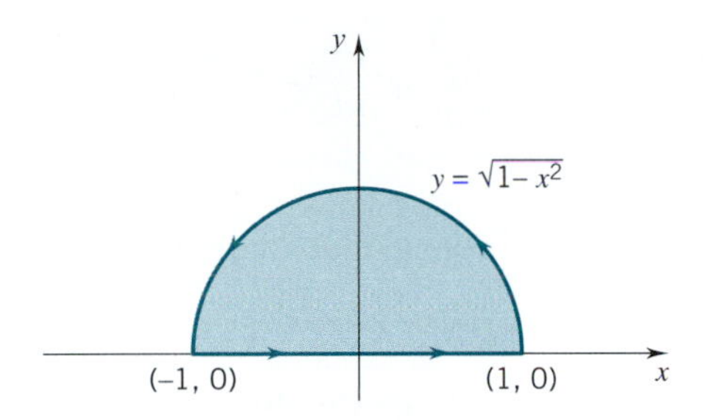

Figure 17.5.8

SOLUTION The curve bounds the closed semicircular disc $\Omega : x^2 + y^2 \leq 1,\ y \geq 0.$

Here $\quad \dfrac{\partial P}{\partial y} = e^x \cos y, \quad \dfrac{\partial Q}{\partial x} = e^x \cos y \quad$ and $\quad \dfrac{\partial Q}{\partial x} - \dfrac{\partial P}{\partial y} = 0.$

By Green's theorem,

$$\oint_C e^x \sin y\,dx + e^x \cos y\,dy = \iint_{\Omega} 0\,dxdy = 0.$$

To see the power of Green's theorem, try to evaluate this line integral directly. $\square$

The preceding examples illustrate the use of Green's theorem to convert a line integral over the boundary of a Jordan region Ω into a double integral over Ω. In some instances, the theorem can be used in reverse. That is, it may be possible to find the value of a double integral over a Jordan region by evaluating a line integral over its boundary. For example, Green's theorem enables us to calculate the area of a Jordan region by integrating over the boundary of the region.

(17.5.2) The area of a Jordan region with boundary C is given by each of the following integrals :

$$\oint_C -y\,dx, \quad \oint_C x\,dy, \quad \tfrac{1}{2}\oint_C -y\,dx + x\,dy.$$

PROOF Let Ω be the region enclosed by C. In the first integral

$$P(x,y) = -y, \quad Q(x,y) = 0.$$

Therefore
$$\frac{\partial P}{\partial y} = -1, \qquad \frac{\partial Q}{\partial x} = 0, \qquad \text{and} \qquad \frac{\partial Q}{\partial x} - \frac{\partial P}{\partial y} = 1.$$

Thus by Green's theorem

$$\oint_C -y\,dx = \iint_\Omega 1\,dxdy = \text{ area of } \Omega.$$

That the second integral also gives the area of Ω can be verified in a similar manner.

We can see the validity of the third formula by observing that

$$\oint_C -y\,dx + \oint_C x\,dy = \text{ twice the area of } \Omega. \quad \square$$

Example 4 Show that the area of the region Ω enclosed by the ellipse

$$\frac{x^2}{a^2} + \frac{y^2}{b^2} = 1 \qquad \text{(Figure 17.5.9)}$$

is πab.

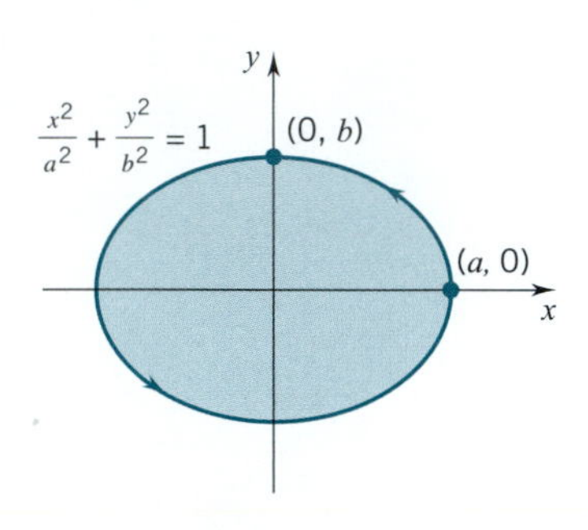

Figure 17.5.9

SOLUTION The ellipse is oriented counterclockwise by the parametrization

$$x = a\cos u, \qquad y = b\sin u, \qquad 0 \le u \le 2\pi.$$

Although the third integral in (17.5.2) appears to be more complicated than either of the other two integrals, it is actually simpler to use in this case:

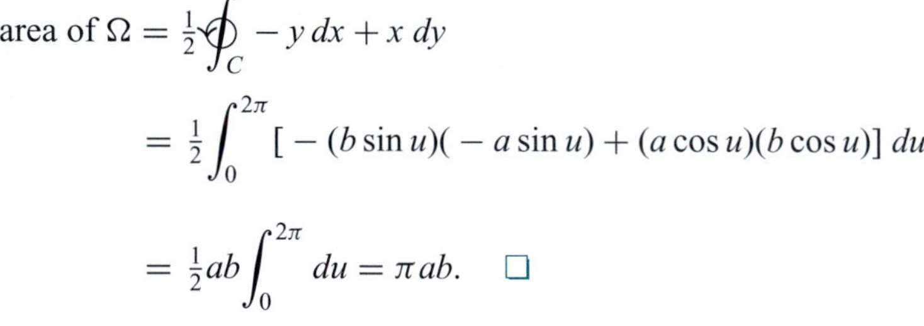

$$\begin{aligned} \text{area of } \Omega &= \tfrac{1}{2}\oint_C -y\,dx + x\,dy \\ &= \tfrac{1}{2}\int_0^{2\pi} [-(b\sin u)(-a\sin u) + (a\cos u)(b\cos u)]\,du \\ &= \tfrac{1}{2}ab\int_0^{2\pi} du = \pi ab. \quad \square \end{aligned}$$

Example 5 Let Ω be a Jordan region of area A with a piecewise-smooth boundary C. Show that the coordinates of the centroid of Ω are given by

$$\bar{x}A = \tfrac{1}{2}\oint_C x^2\,dy, \qquad \bar{y}A = -\tfrac{1}{2}\oint_C y^2\,dx.$$

SOLUTION

$$\tfrac{1}{2}\oint_C x^2\,dy = \tfrac{1}{2}\iint_\Omega 2x\,dxdy = \iint_\Omega x\,dxdy = \bar{x}A,$$

by Green's theorem

$$-\tfrac{1}{2}\oint_C y^2\,dx = -\tfrac{1}{2}\iint_\Omega (-2y)\,dxdy = \iint_\Omega y\,dxdy = \bar{y}A. \quad \square$$

Regions Bounded by Two or More Jordan Curves

(All the curves that appear here are assumed to be piecewise smooth.)

Figure 17.5.10 shows an annular region Ω. The region is not a Jordan region: the boundary consists of two Jordan curves C_1 and C_2. We cannot apply Green's theorem to Ω directly, but we can break up Ω into two Jordan regions as in Figure 17.5.11 and then apply Green's theorem to each piece. With Ω_1 and Ω_2 as in Figure 17.5.11,

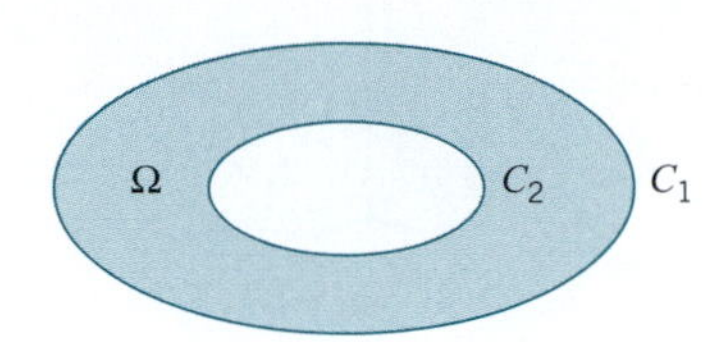

Figure 17.5.10

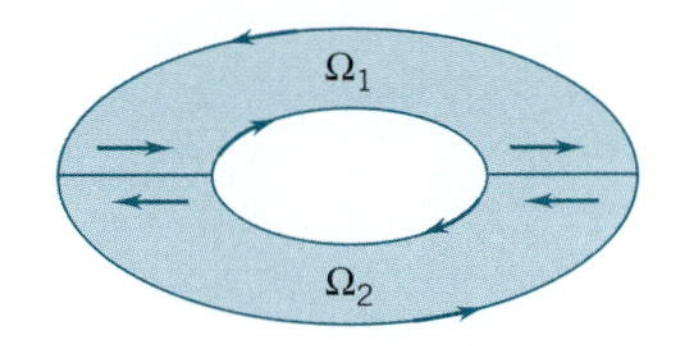

Figure 17.5.11

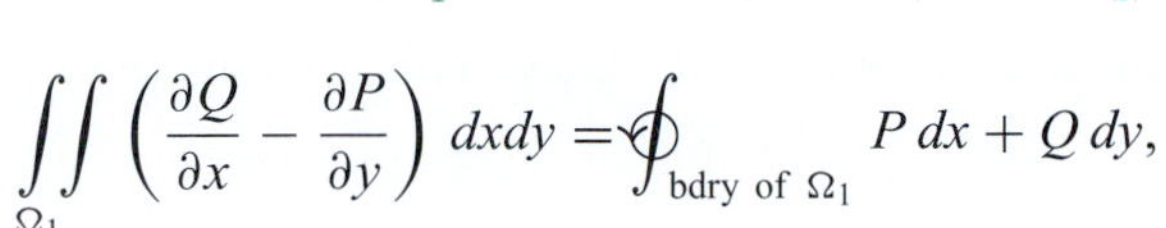

$$\iint_{\Omega_1} \left(\frac{\partial Q}{\partial x} - \frac{\partial P}{\partial y}\right) dxdy = \oint_{\text{bdry of } \Omega_1} P\,dx + Q\,dy,$$

$$\iint_{\Omega_2} \left(\frac{\partial Q}{\partial x} - \frac{\partial P}{\partial y}\right) dxdy = \oint_{\text{bdry of } \Omega_2} P\,dx + Q\,dy.$$

When we add the double integrals, we get the double integral over Ω. When we add the line integrals, the integrals over the crosscuts cancel and we are left with the *counterclockwise integral* over C_1 and the *clockwise integral* over C_2. (See the figure.) Thus, for the annular region,

(17.5.3)
$$\iint_{\Omega} \left(\frac{\partial Q}{\partial x} - \frac{\partial P}{\partial y}\right) dxdy = \oint_{C_1} P\,dx + Q\,dy + \oint_{C_2} P\,dx + Q\,dy.$$

If $\partial Q/\partial x = \partial P/\partial y$ throughout Ω, then the double integral on the left is 0, and the sum of the integrals on the right is also 0. Therefore we see that

(17.5.4)
if $\partial Q/\partial x = \partial P/\partial y$ throughout Ω, then
$$\oint_{C_1} P\,dx + Q\,dy = \oint_{C_2} P\,dx + Q\,dy.$$

Example 6 Let C_1 be a Jordan curve that does not pass through the origin $(0, 0)$. Show that

$$\oint_{C_1} -\frac{y}{x^2+y^2}\,dx + \frac{x}{x^2+y^2}\,dy = \begin{cases} 0 & \text{if } C_1 \text{ does not enclose the origin} \\ 2\pi & \text{if } C_1 \text{ does enclose the origin}. \end{cases}$$

SOLUTION In this case

$$\frac{\partial P}{\partial y} = \frac{\partial}{\partial y}\left(-\frac{y}{x^2+y^2}\right) = -\left[\frac{(x^2+y^2)1 - 2y^2}{(x^2+y^2)^2}\right] = \frac{y^2-x^2}{(x^2+y^2)^2},$$

$$\frac{\partial Q}{\partial x} = \frac{\partial}{\partial x}\left(\frac{x}{x^2+y^2}\right) = \frac{(x^2+y^2)1 - 2x^2}{(x^2+y^2)^2} = \frac{y^2-x^2}{(x^2+y^2)^2}.$$

Thus
$$\frac{\partial Q}{\partial x} = \frac{\partial P}{\partial y} \quad \text{except at the origin.}$$

If C_1 does not enclose the origin, then $\partial Q/\partial x - \partial P/\partial y = 0$ throughout the region enclosed by C_1, and, by Green's theorem, the line integral is 0.

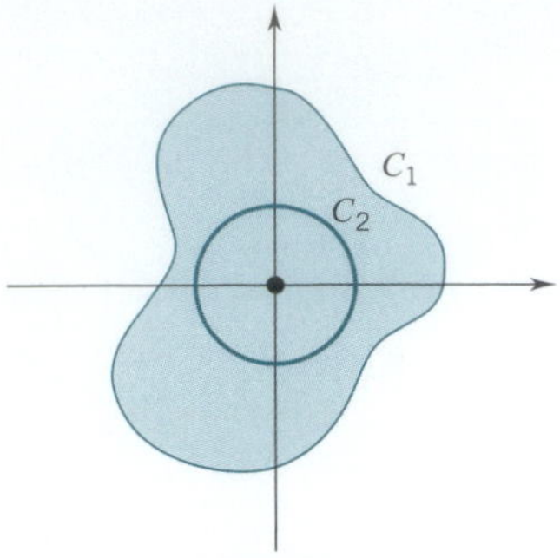

If C_1 does enclose the origin, we draw within the inner region of C_1 a small circle centered at the origin

$$C_2: \quad x^2 + y^2 = a^2. \qquad \text{(Figure 17.5.12)}$$

Figure 17.5.12

Since $\partial Q/\partial x - \partial P/\partial y = 0$ on the annular region bounded by C_1 and C_2, we know from (17.5.4) that the line integral over C_1 equals the line integral over C_2. All we have to show now is that the line integral over C_2 is 2π. This is straightforward. Parametrizing the circle by

$$\mathbf{r}(u) = a\cos u\,\mathbf{i} + a\sin u\,\mathbf{j} \qquad \text{with} \qquad u \in [0, 2\pi],$$

we have

$$\oint_{C_2} -\frac{y}{x^2+y^2}\,dx + \frac{x}{x^2+y^2}\,dy = \int_0^{2\pi} (\sin^2 u + \cos^2 u)\,du = \int_0^{2\pi} du = 2\pi. \quad \square$$

check this ⟶ ↑

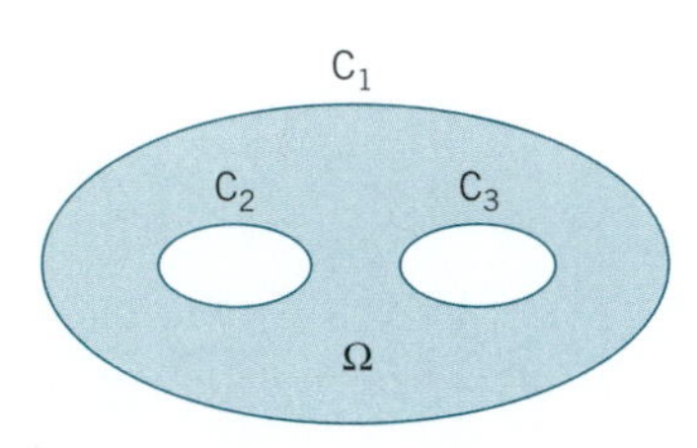

Figure 17.5.13

Figure 17.5.13 shows a region bounded by three Jordan curves: C_2 and C_3, each exterior to the other, both within C_1. For such a region Green's theorem gives

$$\iint_{\Omega} \left(\frac{\partial Q}{\partial x} - \frac{\partial P}{\partial y}\right) dx\,dy =$$

$$\oint_{C_1} P\,dx + Q\,dy + \oint_{C_2} P\,dx + Q\,dy + \oint_{C_3} P\,dx + Q\,dy.$$

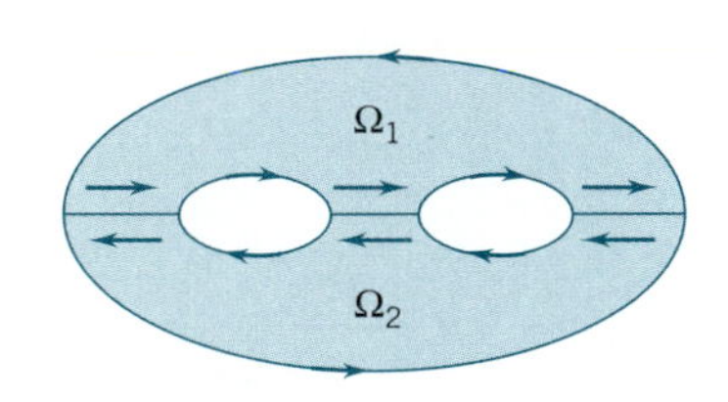

Figure 17.5.14

To see this, break up Ω into two regions by making the crosscuts shown in Figure 17.5.14.

The general formula for configurations of this type reads

$$\iint_{\Omega} \left(\frac{\partial Q}{\partial x} - \frac{\partial P}{\partial y}\right) dx\,dy = \oint_{C_1} P\,dx + Q\,dy + \sum_{i=2}^{n} \oint_{C_i} P\,dx + Q\,dy.$$

EXERCISES 17.5

Evaluate the line integral (a) directly; and (b) by applying Green's theorem.

1. $\oint_C xy\,dx + x^2\,dy$; where C is the triangle with vertices $(0,0), (0,1), (1,1)$.

2. $\oint_C x^2y\,dx + 2y^2\,dy$; where C is the square with vertices $(0,0), (1,0), (1,1), (0,1)$.

3. $\oint_C (3x^2 + y)\,dx + (2x + y^3)\,dy; \quad C: 9x^2 + 4y^2 = 36.$

4. $\oint_C y^2\,dx + x^2\,dy$; where C is the boundary of the region that lies between the curves $y = x$ and $y = x^2$.

Evaluate by Green's theorem.

5. $\oint_C 3y\,dx + 5x\,dy; \quad C: x^2 + y^2 = 1.$

6. $\oint_C 5x\,dx + 3y\,dy; \quad C: (x-1)^2 + (y+1)^2 = 1.$

7. $\oint_C x^2 dy$; where C is the rectangle with vertices $(0,0), (a,0), (a,b), (0,b)$.

8. $\oint_C y^2\,dx$; where C is the rectangle of Exercise 7.

9. $\oint_C (3xy + y^2)\,dx + (2xy + 5x^2)dy$;
$C: (x-1)^2 + (y+2)^2 = 1.$

10. $\oint_C (xy + 3y^2)\,dx + (5xy + 2x^2)\,dy;$
$C: (x-1)^2 + (y+2)^2 = 1.$

11. $\oint_C (2x^2 + xy - y^2)\,dx + (3x^2 - xy + 2y^2)\,dy;$
$C: (x-a)^2 + y^2 = r^2.$

12. $\oint_C (x^2 - 2xy + 3y^2)\,dx + (5x + 1)\,dy;$
$C: x^2 + (y-b)^2 = r^2.$

13. $\oint_C e^x \sin y\,dx + e^x \cos y\,dy;$
$C: (x-a)^2 + (y-b)^2 = r^2.$

14. $\oint_C e^x \cos y\,dx + e^x \sin y\,dy$ where C is the rectangle with vertices $(0,0), (1,0), (1,\pi), (0,\pi)$.

15. $\oint_C 2xy\,dx + x^2 dy$ where C is the cardioid $r = 1 - \cos\theta, \theta \in [0, 2\pi]$.

16. $\oint_C y^2\,dx + 2xy\,dy$ where C is the first quadrant loop of the petal curve $r = 2\sin 2\theta$.

In Exercises 17 and 18, find the area enclosed by the curve by integrating over the curve.

17. The circle $x^2 + y^2 = a^2$.

18. The astroid $x^{2/3} + y^{2/3} = a^{2/3}$.

19. Sketch the region Ω bounded by the curves $xy = 4$ and $x + y = 5$. Then use Green's theorem to find the area of Ω.

20. Sketch the region Ω bounded by the curves $y^2 - x^2 = 5$ and $y = 3$. Then use Green's theorem to find the area of Ω.

21. Let C be a piecewise-smooth Jordan curve. Calculate

$$\oint_C (ay + b)\,dx + (cx + d)\,dy$$

given that C encloses a region of area A.

22. Calculate

$$\oint_C \mathbf{F}(\mathbf{r}) \cdot \mathbf{dr}$$

given that $\mathbf{F}(x,y) = 2y\,\mathbf{i} - 3x\,\mathbf{j}$ and C is the astroid $x^{2/3} + y^{2/3} = a^{2/3}$.

23. Use Green's theorem to find the area under one arch of the cycloid

$$x(\theta) = R(\theta - \sin\theta), \quad y(\theta) = R(1 - \cos\theta).$$

24. Find the Jordan curve C that maximizes the line integral

$$\oint_C y^3\,dx + (3x - x^3)\,dy.$$

25. Complete the proof of Green's theorem for the elementary region of Figure 17.5.2 by showing that

$$\oint_C Q(x,y)dy = \iint_\Omega \frac{\partial Q}{\partial x}(x,y)\,dxdy.$$

26. Suppose that f and g have continuous first-partial derivatives in a simply connected open region Ω. Show that if C is any piecewise-smooth simple closed curve in Ω, then

$$\oint_C [f(\mathbf{r})\nabla g(\mathbf{r}) + g(\mathbf{r})\nabla f(\mathbf{r})] \cdot \mathbf{dr} = 0.$$

27. Let $(\bar{x}, \bar{y})$ be the centroid of a Jordan region with piecewise-smooth boundary C and area A. Show that

$$\bar{x}A = \tfrac{1}{2}\oint_C x^2 dy \qquad \text{and} \qquad \bar{y}A = -\tfrac{1}{2}\oint_C y^2\,dx.$$

28. Let Ω be a plate of constant mass density λ in the form of a Jordan region with a piecewise-smooth boundary C. Show that the moments of inertia of the plate about the coordinate axes are given by the formulas

(17.5.5) $$I_x = -\frac{\lambda}{3}\oint_C y^3\,dx, \quad I_y = \frac{\lambda}{3}\oint_C x^3 dy.$$

29. Let P and Q be continuously differentiable functions on the region Ω of Figure 17.5.13. Given that $\partial P/\partial y = \partial Q/\partial x$ on Ω, find a relation between the line integrals

$$\oint_{C_1} P\,dx + Q\,dy, \quad \oint_{C_2} P\,dx + Q\,dy,$$

$$\oint_{C_3} P\,dx + Q\,dy.$$

30. Show that, if $f = f(x)$ and $g = g(y)$ are everywhere continuously differentiable, then

$$\int_C f(x)\,dx + g(y)\,dy = 0$$

for all piecewise-smooth Jordan curves C.

31. Let C be a piecewise-smooth Jordan curve that does not pass through the origin. Evaluate

$$\oint_C \frac{x}{x^2 + y^2}\,dx + \frac{y}{x^2 + y^2}\,dy$$

(a) if C does not enclose the origin.
(b) if C does enclose the origin.

32. Let C be a piecewise-smooth Jordan curve that does not pass through the origin. Evaluate

$$\oint_C -\frac{y^3}{(x^2 + y^2)^2}\,dx + \frac{xy^2}{(x^2 + y^2)^2}\,dy$$

(a) if C does not enclose the origin.
(b) if C does enclose the origin.

33. Let **v** be a vector field continuously differentiable on the entire plane. Use Green's theorem to verify that if **v** is a gradient field [$\mathbf{v} = \nabla\phi$], then

$$\oint_C \mathbf{v} \cdot d\mathbf{r} = 0$$

for every piecewise-smooth Jordan curve C.

34. Let C be the line segment from the point (x_1, y_1) to the point (x_2, y_2). Show that

$$\int_C -y\,dx + x\,dy = x_1y_2 - x_2y_1.$$

35. Let $(x_1, y_1), (x_2, y_2), \ldots, (x_n, y_n)$ be the vertices of a polygon in counterclockwise order. Show that the area of the polygon is

$$A = \tfrac{1}{2}[(x_1y_2 - x_2y_1) + (x_2y_3 - x_3y_2) + \cdots$$
$$+ (x_{n-1}y_n - x_ny_{n-1}) + (x_ny_1 - x_1y_n)].$$

36. Use the formula of Exercise 35 to find the area of:

(a) The triangle with vertices $(0,0), (2,1), (1,4)$.

(b) The pentagon with vertices $(0,0), (3,1), (2,4), (0,6), (-1,2)$.

■ PROJECT 17.5 The Folium of Descartes

The word *folium* means "leaf" in Latin. The *folium of Descartes*, shown in the figure, was introduced by Descartes in 1638. It is the graph of the equation:

$$x^3 + y^3 = 3axy, \ a > 0.$$

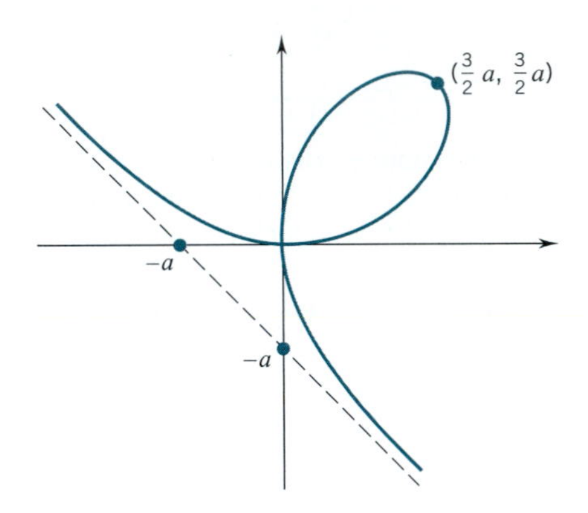

Problem 1. Show that if the parameter t is defined by setting $t = y/x$, then

$$x = \frac{3at}{1+t^3}, \quad y = \frac{3at^2}{1+t^3}, \ t \neq -1,$$

are parametric equations for the folium.

Problem 2. Use this parametric representation of the folium to show that the line $x + y + a = 0$ is an asymptote of the curve. HINT: Show that $x + y \to -a$ as $t \to -1$.

Problem 3. Show that the folium is symmetric with respect to the line $y = x$. Use this fact and the results obtained in Problems 1 and 2 to describe the orientation of the curve as t varies from $-\infty$ to ∞ $(t \neq -1)$.

Problem 4. Express the area of the loop as an integral. Use the third equation in (17.5.2) to find the area of the loop.
HINT: You may want to double the area of the bottom half of the loop.

Problem 5. Show that the area of the loop equals the area of the region between the curve and its asymptote.

The curve defined by

$$x^{2n+1} + y^{2n+1} = (2n+1)ax^ny^n, \quad a > 0,$$

where n is a positive integer greater than 1, is called the *generalized folium of Descartes*. The following is a parametric representation of the generalized folium:

$$x = \frac{(2n+1)at^n}{1+t^{2n+1}}, \quad y = \frac{(2n+1)at^{n+1}}{1+t^{2n+1}}, \ t \neq -1.$$

Problem 6.

a. Use a graphing utility to graph several of these curves.
b. Calculate the area of the loop of the generalized folium.

*SUPPLEMENT TO SECTION 17.5

A JUSTIFICATION OF THE JACOBIAN AREA FORMULA

We based the change of variables for double integrals on the Jacobian area formula [Formula (16.10.1)]. Green's theorem enables us to derive this formula under the conditions spelled out in the following:

Let Γ be a Jordan region in the uv-plane with a piecewise-smooth boundary C_Γ. A vector function $\mathbf{r}(u, v) = x(u, v)\,\mathbf{i} + y(u, v)\,\mathbf{j}$ with continuous second partials maps Γ onto a region Ω of the xy-plane. If $\mathbf{r}$ is one-to-one on the interior of Γ and the Jacobian J of the components of $\mathbf{r}$ is different from zero on the interior of Γ, then

$$\text{area of } \Omega = \iint_\Gamma |J(u, v)|\, du dv.$$

PROOF Suppose that C_Γ is parametrized by $u = u(t), v = v(t)$ with $t \in [a, b]$. Then the boundary of Ω is a piecewise-smooth curve C given by

$$\mathbf{r}[u(t), v(t)] = x[u(t), v(t)]\,\mathbf{i} + y[u(t), v(t)]\,\mathbf{j}, \quad t \in [a, b].$$

By Green's theorem, the area of Ω is

$$\begin{aligned}
\oint_C x\, dy &= \left| \int_a^b x[u(t), v(t)]\, \frac{d}{dt}\left(y[u(t), v(t)]\right) dt \right| \\
&= \left| \int_a^b x[u(t), v(t)] \left(\frac{\partial y}{\partial u}[u(t), v(t)]u'(t) + \frac{\partial y}{\partial v}[u(t), v(t)]v'(t) \right) dt \right| \\
&= \left| \int_a^b \left(x[u(t), v(t)]\frac{\partial y}{\partial u}[u(t), v(t)]u'(t) + x[u(t), v(t)]\frac{\partial y}{\partial v}[u(t), v(t)]v'(t) \right) dt \right| \\
&= \left| \int_{C_\Gamma} x\frac{\partial y}{\partial u}\, du + x\frac{\partial y}{\partial v} dv \right| \\
&\quad \text{(again by Green's theorem)} \\
&= \left| \iint_\Gamma \left[\frac{\partial}{\partial u}\left(x\frac{\partial y}{\partial v} \right) - \frac{\partial}{\partial v}\left(x\frac{\partial y}{\partial u} \right) \right] du dv \right|.
\end{aligned}$$

Now

$$\begin{aligned}
\frac{\partial}{\partial u}\left(x\frac{\partial y}{\partial v} \right) - \frac{\partial}{\partial v}\left(x\frac{\partial y}{\partial u} \right) &= \frac{\partial x}{\partial u}\frac{\partial y}{\partial v} + x\frac{\partial^2 y}{\partial u \partial v} - \frac{\partial x}{\partial v}\frac{\partial y}{\partial u} - x\frac{\partial^2 y}{\partial v \partial u} \\
&= \frac{\partial x}{\partial u}\frac{\partial y}{\partial v} - \frac{\partial x}{\partial v}\frac{\partial y}{\partial u} = J(u, v).
\end{aligned}$$

Therefore

$$\text{area of } \Omega = \left| \iint_\Gamma J(u, v)\, du dv \right| = \iint_\Gamma |J(u, v)|\, du dv,$$

the final equality holding because $J(u, v)$ cannot change sign on Γ. □

■ 17.6 PARAMETRIZED SURFACES; SURFACE AREA

You have seen that a space curve can be parametrized by a vector function $\mathbf{r} = \mathbf{r}(u)$ where u ranges over some interval I of the u-axis (Figure 17.6.1). In an analogous manner, we can parametrize a surface S in space by a vector function $\mathbf{r} = \mathbf{r}(u, v)$ where (u, v) ranges over some region Ω of the uv-plane (Figure 17.6.2).

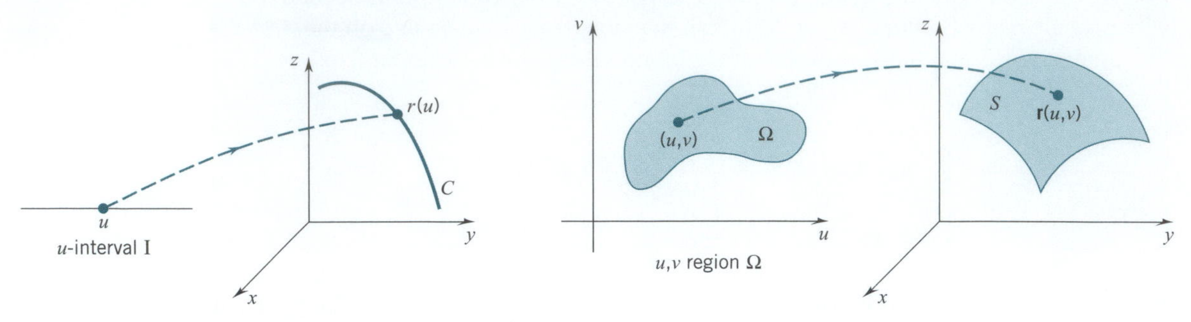

Figure 17.6.1

Figure 17.6.2

Example 1 (*The graph of a function*) Just as the graph of a function

$$y = f(x), \qquad x \in [a, b]$$

can be parametrized by setting

$$\mathbf{r}(u) = u\,\mathbf{i} + f(u)\,\mathbf{j}, \qquad u \in [a, b],$$

the graph of a function

$$z = f(x, y), \qquad (x, y) \in \Omega$$

can be parametrized by setting

$$\mathbf{r}(u, v) = u\,\mathbf{i} + v\,\mathbf{j} + f(u, v)\,\mathbf{k}, \qquad (u, v) \in \Omega.$$

As (u, v) ranges over Ω, the tip of $\mathbf{r}(u, v)$ traces out the surface, which is the graph of f. ❑

Example 2 (*A plane*) If two vectors $\mathbf{a}$ and $\mathbf{b}$ are not parallel, then the set of all linear combinations $u\,\mathbf{a} + v\,\mathbf{b}$ generate a plane p_0 that passes through the origin. We can parametrize this plane by setting

$$\mathbf{r}(u, v) = u\,\mathbf{a} + v\,\mathbf{b}, \quad u, v \text{ real numbers.}$$

The plane p that is parallel to p_0 and passes through the tip of $\mathbf{c}$ can be parametrized by setting

$$\mathbf{r}(u, v) = u\,\mathbf{a} + v\,\mathbf{b} + \mathbf{c}, \quad u, v \text{ real numbers.}$$

Note that the plane contains the lines

$$l_1\colon \mathbf{r}(u, 0) = u\,\mathbf{a} + \mathbf{c} \qquad \text{and} \qquad l_2\colon \mathbf{r}(0, v) = v\,\mathbf{b} + \mathbf{c}. \quad ❑$$

Example 3 (*A sphere*) The sphere of radius a centered at the origin can be parametrized by

$$\mathbf{r}(u, v) = a\cos u\cos v\,\mathbf{i} + a\sin u\cos v\,\mathbf{j} + a\sin v\,\mathbf{k}$$

with (u, v) ranging over the rectangle $R\colon 0 \le u \le 2\pi, -\frac{1}{2}\pi \le v \le \frac{1}{2}\pi$.

To derive this parametrization, we refer to Figure 17.6.3. The points of latitude v (see the figure) form a circle of radius $a\cos v$ on the horizontal plane $z = a\sin v$.

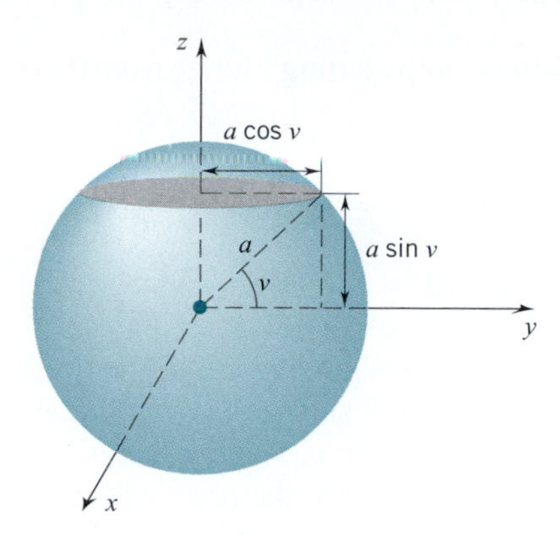

Figure 17.6.3

This circle can be parametrized by

$$\mathbf{R}(u) = a\cos v\,(\cos u\,\mathbf{i} + \sin u\,\mathbf{j}) + a\sin v\,\mathbf{k}, \quad u \in [0, 2\pi].$$

This expands to give

$$\mathbf{R}(u) = a\cos u\cos v\,\mathbf{i} + a\sin u\cos v\,\mathbf{j} + a\sin v\,\mathbf{k}, \quad u \in [0, 2\pi].$$

Letting v range from $-\frac{1}{2}\pi$ to $\frac{1}{2}\pi$, we obtain the entire sphere.

The xyz-equation for this same sphere is $x^2 + y^2 + z^2 = a^2$. It is easy to verify that the parametrization satisfies this equation:

$$\begin{aligned} x^2 + y^2 + z^2 &= a^2\cos^2 u\cos^2 v + a^2\sin^2 u\cos^2 v + a^2\sin^2 v \\ &= a^2(\cos^2 u + \sin^2 u)\cos^2 v + a^2\sin^2 v \\ &= a^2(\cos^2 v + \sin^2 v) = a^2. \quad \square \end{aligned}$$

Example 4 (*A cone*) Figure 17.6.4 shows a right circular cone with vertex semiangle α and slant height s. The points of slant height v (see the figure) form a circle of radius $v\sin\alpha$ on the horizontal plane $z = v\cos\alpha$. This circle can be parametrized by

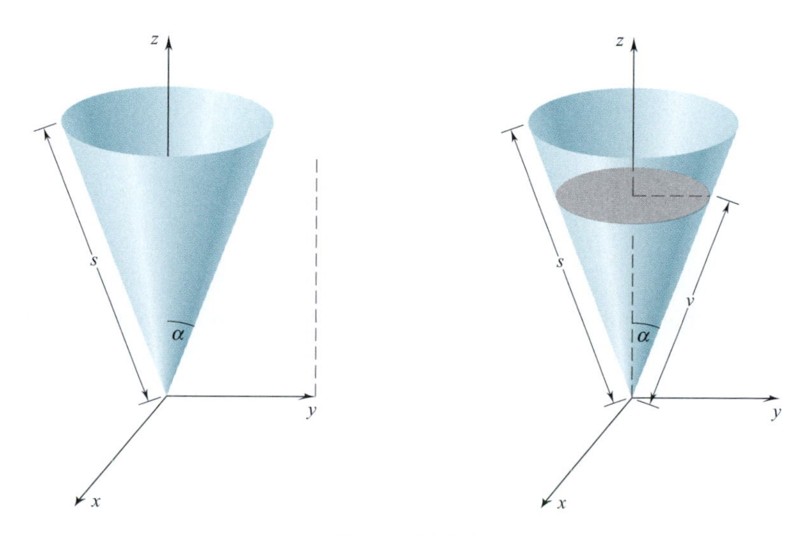

Figure 17.6.4

$$\begin{aligned}\mathbf{R}(u) &= v\sin\alpha(\cos u\,\mathbf{i} + \sin u\,\mathbf{j}) + v\cos\alpha\,\mathbf{k}\\ &= v\cos u\sin\alpha\,\mathbf{i} + v\sin u\sin\alpha\,\mathbf{j} + v\cos\alpha\,\mathbf{k}, \qquad u \in [0, 2\pi].\end{aligned}$$

Since we can obtain the entire cone by letting v range from 0 to s, the cone is parametrized by setting

$$\mathbf{r}(u, v) = v\cos u\sin\alpha\,\mathbf{i} + v\sin u\sin\alpha\,\mathbf{j} + v\cos\alpha\,\mathbf{k},$$

with $0 \leq u \leq 2\pi, 0 \leq v \leq s$. □

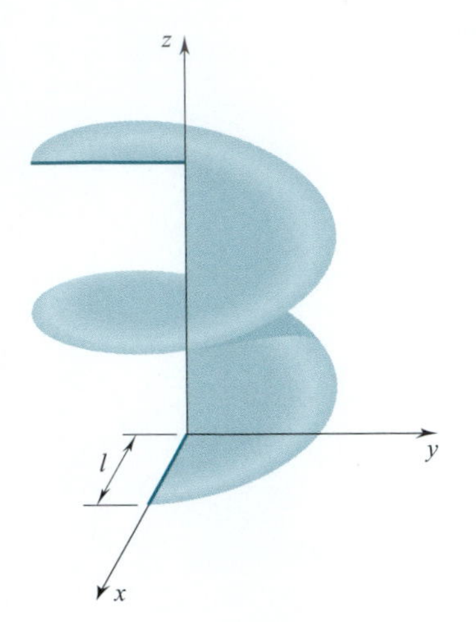

Figure 17.6.5

Example 5 (*A spiral ramp*) A rod of length l initially resting on the x-axis and attached at one end to the z-axis sweeps out a surface by rotating about the z-axis at constant rate ω while climbing at a constant rate b. The surface is pictured in Figure 17.6.5.

To parametrize this surface, we mark the point of the rod at a distance u from the z-axis $(0 \leq u \leq l)$ and ask for the position of this point at time v. At time v the rod will have climbed a distance bv and rotated through an angle ωv. Thus the point will be found at the tip of the vector

$$u(\cos\omega v\,\mathbf{i} + \sin\omega v\,\mathbf{j}) + bv\,\mathbf{k} = u\cos\omega v\,\mathbf{i} + u\sin\omega v\,\mathbf{j} + bv\,\mathbf{k}.$$

The entire surface can be parametrized by setting

$$\mathbf{r}(u, v) = u\cos\omega v\,\mathbf{i} + u\sin\omega v\,\mathbf{j} + bv\,\mathbf{k} \quad \text{with } 0 \leq u \leq l,\ v \geq 0. \quad □$$

The Fundamental Vector Product

Let S be a surface parametrized by a differentiable vector function

$$\mathbf{r} = \mathbf{r}(u, v) = x(u, v)\,\mathbf{i} + y(u, v)\,\mathbf{j} + z(u, v)\,\mathbf{k}.$$

For simplicity, let us suppose that (u, v) varies over the open rectangle $R: a < u < b$, $c < v < d$. Since $\mathbf{r}$ is a function of u and v, we can form the partial with respect to u,

$$\mathbf{r}'_u = \frac{\partial x}{\partial u}\mathbf{i} + \frac{\partial y}{\partial u}\mathbf{j} + \frac{\partial z}{\partial u}\mathbf{k},$$

and we can form the partial with respect to v,

$$\mathbf{r}'_v = \frac{\partial x}{\partial v}\mathbf{i} + \frac{\partial y}{\partial v}\mathbf{j} + \frac{\partial z}{\partial v}\mathbf{k}$$

Now let (u_0, v_0) be a point of R for which

$$\mathbf{r}'_u(u_0, v_0) \times \mathbf{r}'_v(u_0, v_0) \neq \mathbf{0}.$$

(The reason for this condition will be apparent as we go on.) The vector function

$$\mathbf{r}_1(u) = \mathbf{r}(u, v_0), \quad u \in (a, b)$$

(here we are keeping v fixed at v_0) traces out a differentiable curve C_1 that lies on S (Figure 17.6.6). The vector function

$$\mathbf{r}_2(v) = \mathbf{r}(u_0, v), \quad v \in (c, d)$$

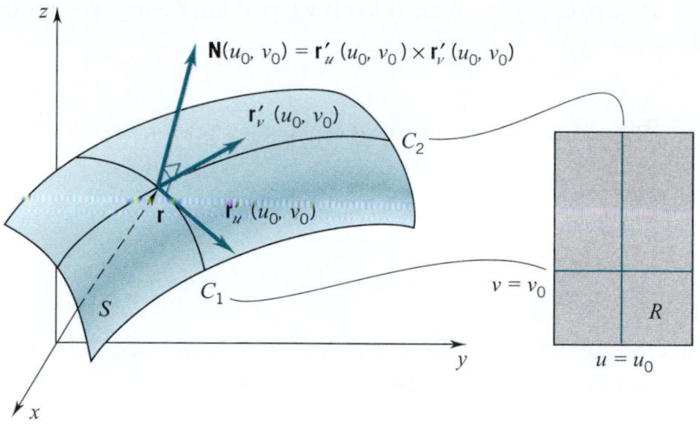

Figure 17.6.6

(this time we are keeping u fixed at u_0) traces out a differentiable curve C_2 that also lies on S. Both curves pass through the tip of $\mathbf{r}(u_0, v_0)$:

$$C_1 \text{ with tangent vector } \mathbf{r}_1'(u_0) = \mathbf{r}_u'(u_0, v_0),$$
$$C_2 \text{ with tangent vector } \mathbf{r}_2'(v_0) = \mathbf{r}_v'(u_0, v_0).$$

The cross product $\mathbf{N}(u_0, v_0) = \mathbf{r}_u'(u_0, v_0) \times \mathbf{r}_v'(u_0, v_0)$, which we have assumed to be different from zero, is perpendicular to both curves at the tip of $\mathbf{r}(u_0, v_0)$ and can be taken as a normal to the surface at that point. We record the result as follows:

(17.6.1) If S is the surface given by a differentiable function $\mathbf{r} = \mathbf{r}(u, v)$, then the vector $\mathbf{N}(u, v) = \mathbf{r}_u'(u, v) \times \mathbf{r}_v'(u, v)$ is perpendicular to the surface at the tip of $\mathbf{r}(u, v)$ and, if different from zero, can be taken as a normal to the surface at this point.

The cross product

$$\mathbf{N} = \mathbf{r}_u' \times \mathbf{r}_v' = \begin{vmatrix} \mathbf{i} & \mathbf{j} & \mathbf{k} \\ \dfrac{\partial x}{\partial u} & \dfrac{\partial y}{\partial u} & \dfrac{\partial z}{\partial u} \\ \dfrac{\partial x}{\partial v} & \dfrac{\partial y}{\partial v} & \dfrac{\partial z}{\partial v} \end{vmatrix}$$

is called the *fundamental vector product* of the surface.

Example 6 For the plane $\mathbf{r}(u, v) = u\,\mathbf{a} + v\,\mathbf{b} + \mathbf{c}$ we have

$$\mathbf{r}_u'(u, v) = \mathbf{a}, \quad \mathbf{r}_v'(u, v) = \mathbf{b} \quad \text{and therefore} \quad \mathbf{N}(u, v) = \mathbf{a} \times \mathbf{b}.$$

The vector $\mathbf{a} \times \mathbf{b}$ is normal to the plane. □

Example 7 We parametrized the sphere $x^2 + y^2 + z^2 = a^2$ by setting

$$\mathbf{r}(u, v) = a \cos u \cos v\,\mathbf{i} + a \sin u \cos v\,\mathbf{j} + a \sin v\,\mathbf{k}$$

with $0 \le u \le 2\pi, -\frac{1}{2}\pi \le v \le \frac{1}{2}\pi$. In this case

$$\mathbf{r}_u'(u, v) = -a \sin u \cos v\,\mathbf{i} + a \cos u \cos v\,\mathbf{j}$$

and $\qquad \mathbf{r}'_v(u, v) = -a\cos u \sin v\,\mathbf{i} - a\sin u \sin v\,\mathbf{j} + a\cos v\,\mathbf{k}.$

Thus

$$\mathbf{N}(u, v) = \begin{vmatrix} \mathbf{i} & \mathbf{j} & \mathbf{k} \\ -a\sin u \cos v & a\cos u \cos v & 0 \\ -a\cos u \sin v & -a\sin u \sin v & a\cos v \end{vmatrix}$$

check this

$$= a\cos v\,(a\cos u \cos v\,\mathbf{i} + a\sin u \cos v\,\mathbf{j} + a\sin v\,\mathbf{k})$$

$$= a\cos v\,\mathbf{r}(u, v).$$

As was to be expected, the fundamental vector product of a sphere, being perpendicular to the sphere, is parallel to the radius vector $\mathbf{r}(u, v)$. ❑

The Area of a Parametrized Surface

A linear function

$$\mathbf{r}(u, v) = u\,\mathbf{a} + v\,\mathbf{b} + \mathbf{c} \qquad (\mathbf{a} \text{ and } \mathbf{b} \text{ not parallel})$$

parametrizes a plane p. Horizontal lines from the uv-plane, lines with equations of the form $v = v_0$, are mapped onto lines parallel to $\mathbf{a}$, and vertical lines, $u = u_0$, are mapped onto lines parallel to $\mathbf{b}$:

$$\mathbf{r}(u, v_0) = \underset{\text{direction vector}}{u\,\mathbf{a}} + \underbrace{v_0\,\mathbf{b} + \mathbf{c}}_{\text{constant}}, \qquad \mathbf{r}(u_0, v) = \underset{\text{direction vector}}{v\,\mathbf{b}} + \underbrace{u_0\,\mathbf{a} + \mathbf{c}}_{\text{constant}}.$$

Thus a rectangle R in the uv-plane with sides parallel to the u and v axes,

$$R: \quad u_1 \le u \le u_2, \quad v_1 \le v \le v_2,$$

is mapped onto a parallelogram on p with sides parallel to $\mathbf{a}$ and $\mathbf{b}$. What is important to us here is that

$$\text{the area of the parallelogram} = \|\mathbf{a} \times \mathbf{b}\| \cdot (\text{the area of } R).$$

To show this, we refer Figure 17.6.7. The parallelogram is generated by the vectors

$$\mathbf{r}(u_2, v_1) - \mathbf{r}(u_1, v_1) = (u_2\,\mathbf{a} + v_1\,\mathbf{b} + \mathbf{c}) - (u_1\,\mathbf{a} + v_1\,\mathbf{b} + \mathbf{c}) = (u_2 - u_1)\,\mathbf{a},$$

$$\mathbf{r}(u_1, v_2) - \mathbf{r}(u_1, v_1) = (u_1\,\mathbf{a} + v_2\,\mathbf{b} + \mathbf{c}) - (u_1\,\mathbf{a} + v_1\,\mathbf{b} + \mathbf{c}) = (v_2 - v_1)\,\mathbf{b}.$$

The area of the parallelogram is thus

$$\|(u_2 - u_1)\,\mathbf{a} \times (v_2 - v_1)\,\mathbf{b}\| = \|\mathbf{a} \times \mathbf{b}\|(u_2 - u_1)(v_2 - v_1)$$

$$= \|\mathbf{a} \times \mathbf{b}\| \cdot (\text{area of } R).$$

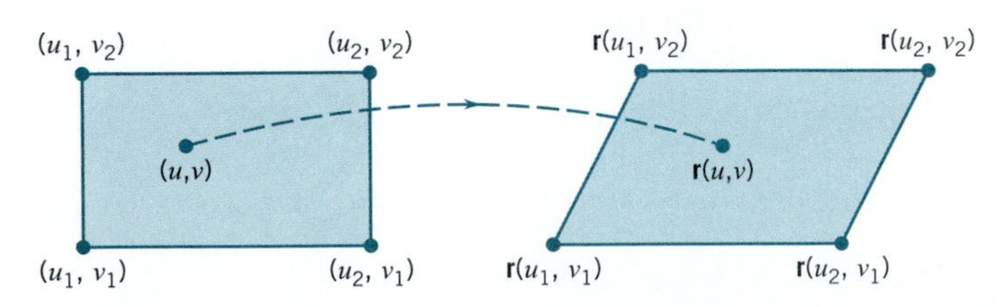

Figure 17.6.7

We can summarize as follows:

(17.6.2)

Let R be a rectangle in the uv-plane with sides parallel to the coordinate axes. If $\mathbf{a}$ and $\mathbf{b}$ are not parallel, the linear function

$$\mathbf{r}(u, v) = u\,\mathbf{a} + v\,\mathbf{b} + \mathbf{c}, \qquad (u, v) \in R$$

parametrizes a parallelogram with sides parallel to $\mathbf{a}$ and $\mathbf{b}$, and

$$\text{the area of the parallelogram} = ||\mathbf{a} \times \mathbf{b}|| \cdot (\text{the area of } R.)$$

More generally, let's suppose that we have a surface S parametrized by a continuously differentiable function

$$\mathbf{r} = \mathbf{r}(u, v), \qquad (u, v) \in \Omega.$$

We assume that Ω is a basic region in the uv-plane and that $\mathbf{r}$ is one-to-one on the interior of Ω. (We don't want $\mathbf{r}$ to cover parts of S more than once.) Also we assume that the fundamental vector product $\mathbf{N} = \mathbf{r}'_u \times \mathbf{r}'_v$ is never zero on the interior of Ω. (We can then use it as a normal.) Under these conditions we call S a *smooth surface* and define

(17.6.3)

$$\text{area of } S = \iint_{\Omega} ||\mathbf{N}(u, v)||\, du dv.$$

We show the reasoning behind this definition in the case where Ω is a rectangle R with sides parallel to the coordinate axes. We begin by breaking up R into N little rectangles $R_1, \ldots, R_N$. This induces a decomposition of S into little pieces $S_1, \ldots, S_N$. Taking (u_i^*, v_i^*) as the center of R_i, we have the tip of $\mathbf{r}(u_i^*, v_i^*)$ in S_i. Since the vector $\mathbf{r}'_u(u_i^*, v_i^*) \times \mathbf{r}'_v(u_i^*, v_i^*)$ is normal to the surface at the tip of $\mathbf{r}(u_i^*, v_i^*)$, we can parametrize the tangent plane at this point by the linear function

$$\mathbf{f}(u, v) = u\,\mathbf{r}'_u(u_i^*, v_i^*) + v\,\mathbf{r}'_v(u_i^*, v_i^*) + [\mathbf{r}(u_i^*, v_i^*) - u_i^*\,\mathbf{r}'_u(u_i^*, v_i^*) - v_i^*\,\mathbf{r}'_v(u_i^*, v_i^*)].$$

(Check that this linear function gives the right plane.) S_i is the portion of S that corresponds to R_i. The portion of the tangent plane that corresponds to this same R_i is a parallelogram with area

$$||\mathbf{r}'_u(u_i^*, v_i^*) \times \mathbf{r}'_v(u_i^*, v_i^*)|| \cdot (\text{area of } R_i) = ||\mathbf{N}(u_i^*, v_i^*)|| \cdot (\text{area of } R_i). \qquad \text{[by (17.6.2)]}$$

Taking this as our estimate for the area of S_i, we have

$$\text{area of } S = \sum_{i=1}^{N} \text{area of } S_i \cong \sum_{i=1}^{N} ||\mathbf{N}(u_i^*, v_i^*)|| \cdot (\text{area of } R_i).$$

This is a Riemann sum for

$$\iint_R ||\mathbf{N}(u, v)||\, du dv.$$

and tends to this integral as the maximal diameter of the R_i tends to zero.

To make sure that Formula (17.6.3) does not violate our previously established notion of area, we must verify that it gives the expected result both for plane regions and for surfaces of revolution. This is done in Examples 9 and 10. By way of introduction we begin with the sphere.

Example 8 (*The surface area of a sphere*) The function

$$\mathbf{r}(u, v) = a\cos u\cos v\,\mathbf{i} + a\sin u\cos v\,\mathbf{j} + a\sin v\,\mathbf{k},$$

with (u, v) ranging over the set $\Omega: 0 \le u \le 2\pi, -\frac{1}{2}\pi \le v \le \frac{1}{2}\pi$, parametrizes a sphere of radius a. For this parametrization

$$\underset{\text{Example 7}}{\mathbf{N}(u, v) = a\cos v\,\mathbf{r}(u, v)} \qquad \text{and} \qquad \underset{-\frac{1}{2}\pi \le v \le \frac{1}{2}\pi}{\|\mathbf{N}(u, v)\| = a^2|\cos v| = a^2\cos v}.$$

According to the new formula,

$$\begin{aligned}\text{area of the sphere} &= \iint_\Omega a^2\cos v\,dudv\\ &= \int_0^{2\pi}\left(\int_{-\frac{1}{2}\pi}^{\frac{1}{2}\pi} a^2\cos v\,dv\right)du = 2\pi a^2\int_{-\frac{1}{2}\pi}^{\frac{1}{2}\pi}\cos v\,dv = 4\pi a^2\end{aligned}$$

which, as you know, is correct. ❑

Example 9 (*The area of a plane region*) If S is a plane region Ω, then S can be parametrized by setting

$$\mathbf{r}(u, v) = u\,\mathbf{i} + v\,\mathbf{j}, \quad (u, v) \in \Omega.$$

Here $\mathbf{N}(u, v) = \mathbf{r}'_u(u, v) \times \mathbf{r}'_v(u, v) = \mathbf{i} \times \mathbf{j} = \mathbf{k}$ and $\|\mathbf{N}(u, v)\| = 1$. In this case (17.6.3) reduces to the familiar formulas

$$A = \iint_\Omega dudv. \quad ❑$$

Example 10 (*The area of a surface of revolution*) Let S be the surface generated by revolving the graph of a function

$$y = f(x), \quad x \in [a, b], \qquad \text{(Figure 17.6.8)}$$

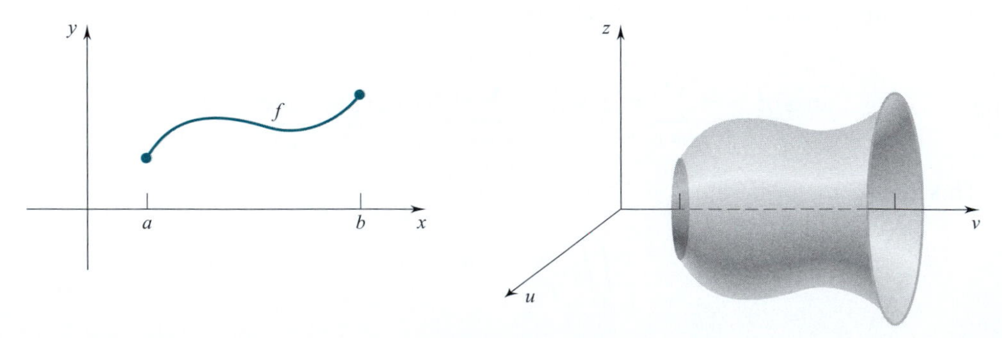

Figure 17.6.8

about the x-axis. We assume that f is positive and continuously differentiable. We can parametrize S by setting.

$$\mathbf{r}(u, v) = v\,\mathbf{i} + f(v)\cos u\,\mathbf{j} + f(v)\sin u\,\mathbf{k}$$

with (u, v) ranging over the set $\Omega: 0 \le u \le 2\pi, a \le v \le b$. (We leave it to you to verify that this is right.) In this case

$$\mathbf{N}(u, v) = \mathbf{r}'_u(u, v) \times \mathbf{r}'_v(u, v) = \begin{vmatrix} \mathbf{i} & \mathbf{j} & \mathbf{k} \\ 0 & -f(v)\sin u & f(v)\cos u \\ 1 & f'(v)\cos u & f'(v)\sin u \end{vmatrix}$$

$$= -f(v)f'(v)\,\mathbf{i} + f(v)\cos u\,\mathbf{j} + f(v)\sin u\,\mathbf{k}.$$

Therefore $||\mathbf{N}(u, v)|| = f(v)\sqrt{[f'(v)]^2 + 1}$ and

$$\text{area of } S = \iint_{\Omega} f(v)\sqrt{[f'(v)]^2 + 1}\,du dv$$

$$= \int_0^{2\pi}\left(\int_a^b f(v)\sqrt{[f'(v)]^2 + 1}\;dv\right)du = \int_a^b 2\pi f(v)\sqrt{[f'(v)]^2 + 1}\,dv.$$

This is in agreement with Formula (9.9.3). ❑

Example 11 (*Spiral ramp*) One turn of the spiral ramp of Example 5 is the surface

$$S: \quad \mathbf{r}(u, v) = u\cos\omega v\,\mathbf{i} + u\sin\omega v\,\mathbf{j} + bv\,\mathbf{k}$$

with (u, v) ranging over the set $\Omega: 0 \le u \le l, 0 \le v \le 2\pi/\omega$. In this case

$$\mathbf{r}'_u(u, v) = \cos\omega v\,\mathbf{i} + \sin\omega v\,\mathbf{j}, \quad \mathbf{r}'_v(u, v) = -\omega u\sin\omega v\,\mathbf{i} + \omega u\cos\omega v\,\mathbf{j} + b\,\mathbf{k}.$$

Therefore

$$\mathbf{N}(u, v) = \begin{vmatrix} \mathbf{i} & \mathbf{j} & \mathbf{k} \\ \cos\omega v & \sin\omega v & 0 \\ -\omega u\sin\omega v & \omega u\cos\omega v & b \end{vmatrix} = b\sin\omega v\,\mathbf{i} - b\cos\omega v\,\mathbf{j} + \omega u\,\mathbf{k},$$

and

$$||\mathbf{N}(u, v)|| = \sqrt{b^2 + \omega^2 u^2}.$$

Thus

$$\text{area of } S = \iint_{\Omega} \sqrt{b^2 + \omega^2 u^2}\,du dv$$

$$= \int_0^{2\pi/\omega}\left(\int_0^l \sqrt{b^2 + \omega^2 u^2}\,du\right)dv = \frac{2\pi}{\omega}\int_0^l \sqrt{b^2 + \omega^2 u^2}\,du.$$

The integral can be evaluated by setting $u = (b/\omega)\tan x$. ❑

The Area of a Surface $z = f(x, y)$

Figure 17.6.9 shows a surface that projects onto a basic region Ω of the xy-plane. Above each point (x, y) of Ω there is one and only one point of S. The surface S is then the graph of a function

$$z = f(x, y), \quad (x, y) \in \Omega.$$

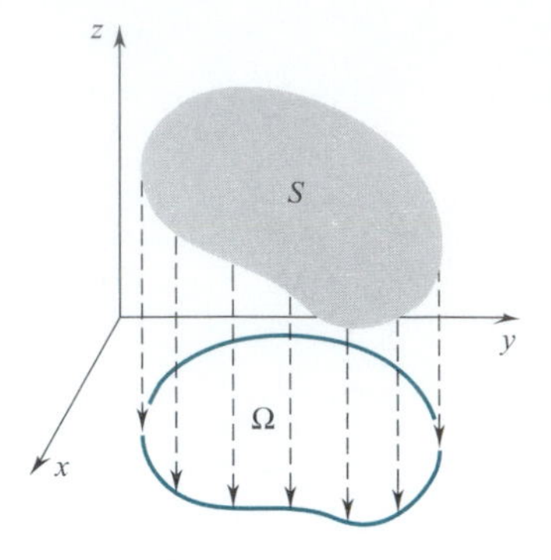

Figure 17.6.9

As we show, if f is continuously differentiable, then

(17.6.4)
$$\text{area of } S = \iint_{\Omega} \sqrt{[f_x(x,y)]^2 + [f_y(x,y)]^2 + 1}\, dxdy.$$

DERIVATION OF FORMULA (17.6.4) We can parametrize S by setting

$$\mathbf{r}(u, v) = u\,\mathbf{i} + v\,\mathbf{j} + f(u, v)\,\mathbf{k}, \quad (u, v) \in \Omega.$$

We may just as well use x and y and write

$$\mathbf{r}(x, y) = x\,\mathbf{i} + y\,\mathbf{j} + f(x, y)\,\mathbf{k}, \quad (x, y) \in \Omega.$$

Clearly $\quad \mathbf{r}'_x(x,y) = \mathbf{i} + f_x(x,y)\,\mathbf{k} \quad \text{and} \quad \mathbf{r}'_y(x,y) = \mathbf{j} + f_y(x,y)\,\mathbf{k}.$

Thus
$$\mathbf{N}(x,y) = \begin{vmatrix} \mathbf{i} & \mathbf{j} & \mathbf{k} \\ 1 & 0 & f_x(x,y) \\ 0 & 1 & f_y(x,y) \end{vmatrix} = -f_x(x,y)\,\mathbf{i} - f_y(x,y)\,\mathbf{j} + \mathbf{k}.$$

Therefore $||\mathbf{N}(x,y)|| = \sqrt{[f_x(x,y)]^2 + [f_y(x,y)]^2 + 1}$ and the formula is verified. ❑

Example 12 Find the surface area of that part of the parabolic cylinder $z = y^2$ that lies over the triangle with vertices $(0, 0), (0, 1), (1, 1)$ in the xy-plane.

SOLUTION Here $f(x, y) = y^2$ so that

$$f_x(x,y) = 0, \quad f_y(x,y) = 2y.$$

The base triangle can be expressed by writing

$$\Omega\colon\ 0 \le y \le 1, \quad 0 \le x \le y.$$

The surface has area

$$\begin{aligned} A &= \iint_{\Omega} \sqrt{[f_x(x,y)]^2 + [f_y(x,y)]^2 + 1}\, dxdy \\ &= \int_0^1 \int_0^y \sqrt{4y^2 + 1}\, dx\, dy \\ &= \int_0^1 y\sqrt{4y^2+1}\, dy = \left[\tfrac{1}{12}(4y^2+1)^{3/2}\right]_0^1 = \tfrac{1}{12}(5\sqrt{5} - 1). \quad ❑ \end{aligned}$$

Example 13 Find the surface area of that part of the hyperbolic paraboloid $z = xy$ that lies inside the cylinder $x^2 + y^2 = a^2$. See Figure 17.6.10.

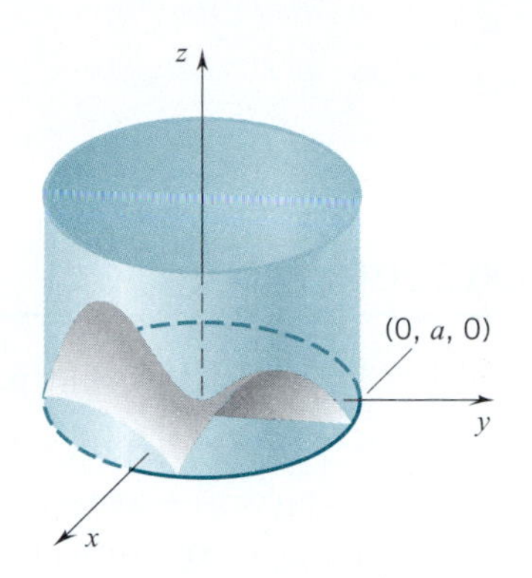

Figure 17.6.10

SOLUTION Here $f(x,y) = xy$, so that $f_x(x,y) = y, \quad f_y(x,y) = x.$
The formula gives

$$A = \iint_{\Omega} \sqrt{y^2 + x^2 + 1}\, dxdy.$$

In polar coordinates the base region takes the form

$$\Gamma\colon\ 0 \le r \le a, \quad 0 \le \theta \le 2\pi.$$

Thus we have

$$A = \iint_{\Gamma} \sqrt{r^2 + 1}\, r\, drd\theta = \int_0^{2\pi}\int_0^a \sqrt{r^2 + 1}\, r\, dr\, d\theta = \tfrac{2}{3}\pi[(a^2 + 1)^{3/2} - 1]. \quad \square$$

There is an elegant version of this last area formula [Formula (17.6.4)] that is geometrically vivid. We know that the vector

$$\mathbf{r}'_x(x,y) \times \mathbf{r}'_y(x,y) = -f_x(x,y)\,\mathbf{i} - f_y(x,y)\,\mathbf{j} + \mathbf{k}$$

is normal to the surface at the point $(x, y, f(x,y))$. The unit vector in that direction, the vector

$$\mathbf{n}(x,y) = \frac{-f_x(x,y)\,\mathbf{i} - f_y(x,y)\,\mathbf{j} + \mathbf{k}}{\sqrt{[f_x(x,y)]^2 + [f_y(x,y)]^2 + 1}},$$

is called the *upper unit normal*. (It is the unit normal with a nonnegative $\mathbf{k}$-component.)

Now let $\gamma(x,y)$ be the angle between $\mathbf{n}(x,y)$ and $\mathbf{k}$ (Figure 17.6.11). Since $\mathbf{n}(x,y)$ and $\mathbf{k}$ are both unit vectors,

$$\cos[\gamma(x,y)] = \mathbf{n}(x,y) \cdot \mathbf{k} = \frac{1}{\sqrt{[f_x(x,y)]^2 + [f_y(x,y)]^2 + 1}}.$$

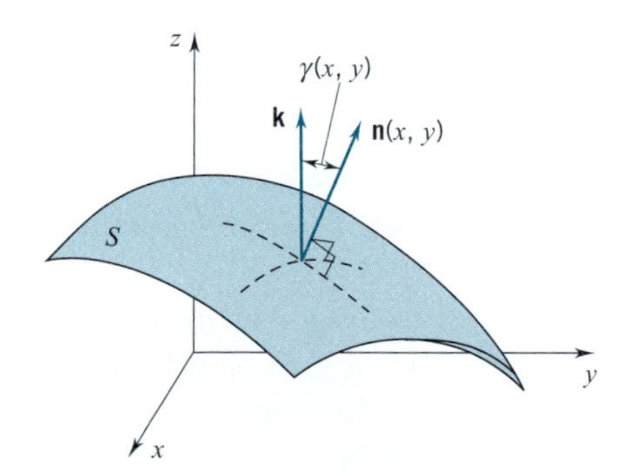

Figure 17.6.11

Taking reciprocals we have

$$\sec[\gamma(x,y)] = \sqrt{[f_x(x,y)]^2 + [f_y(x,y)]^2 + 1}.$$

The area formula can therefore be written

(17.6.5)

$$A = \iint_{\Omega} \sec[\gamma(x,y)]\,dxdy.$$

EXERCISES 17.6

Calculate the fundamental vector product.

1. $\mathbf{r}(u,v) = (u^2 - v^2)\,\mathbf{i} + (u^2 + v^2)\,\mathbf{j} + 2uv\,\mathbf{k}$.

2. $\mathbf{r}(u,v) = u\cos v\,\mathbf{i} + u\sin v\,\mathbf{j} + \mathbf{k}$.

3. $\mathbf{r}(u,v) = (u+v)\,\mathbf{i} + (u+v)\,\mathbf{j} + (u-v)\,\mathbf{k}$.

4. $\mathbf{r}(u,v) = \cos u \sin v\,\mathbf{i} + \sin u \cos v\,\mathbf{j} + u\,\mathbf{k}$.

Find a parametric representation for the surface.

5. The upper half of the ellipsoid $4x^2 + 9y^2 + z^2 = 36$.

6. The part of the cylinder $x^2 + y^2 = 4$ that lies between the planes $z = 1$ and $z = 4$.

7. The part of the sphere $x^2 + y^2 + z^2 = 4$ that lies above the plane $z = \sqrt{2}$.

8. The part of the plane $z = x + 2$ that lies inside the cylinder $x^2 + y^2 = 1$.

9. $y = g(x,z), (x,z) \in \Omega$. **10.** $x = h(y,z), (y,z) \in \Gamma$.

Find an equation in x, y, z for the given surface and identify the surface.

11. $\mathbf{r}(u,v) = a\cos u \cos v\,\mathbf{i} + b\sin u \cos v\,\mathbf{j} + c\sin v\,\mathbf{k}$; $0 \le u \le 2\pi, -\frac{1}{2}\pi \le v \le \frac{1}{2}\pi$.

12. $\mathbf{r}(u,v) = au\cos v\,\mathbf{i} + bu\sin v\,\mathbf{j} + u^2\,\mathbf{k}$; $0 \le u, 0 \le v \le 2\pi$.

13. $\mathbf{r}(u,v) = au\cosh v\,\mathbf{i} + bu\sinh v\,\mathbf{j} + u^2\,\mathbf{k}$; u real, v real.

C 14. Use a graphing utility to draw the surface.

(a) Exercise 11 with $a = 4, b = 3, c = 2$; experiment with other values of a, b, c.

(b) Exercise 12 with $a = 3, b = 2$; experiment with other values of a and b.

(c) Exercise 13 with $a = 3, b = 2$; experiment with viewpoints to obtain a good view of the surface and try other values of a and b.

15. The graph of a continuously differentiable function $y = f(x), x \in [a,b]$ is revolved about the y-axis. Parametrize the surface given that $a \ge 0$.

16. Show that the area of the surface of Exercise 15 is given by the formula

$$A = \int_a^b 2\pi x\sqrt{1 + [f'(x)]^2}\,dx.$$

17. A plane p intersects the xy-plane at an angle γ. (Draw a figure.) Find the area of the region Γ on p given that the projection of Γ onto the xy-plane is a region Ω of area A_Ω.

18. Determine the area of the portion of the plane $x+y+z = a$ that lies within the cylinder $x^2 + y^2 = b^2$.

19. Find the area of the part of the plane $bcx + acy + abz = abc$ that lies within the first octant.

20. Find the area of the surface $z^2 = x^2 + y^2$ from $z = 0$ to $z = 1$.

21. Find the area of the surface $z = x^2 + y^2$ from $z = 0$ to $z = 4$.

Calculate the area of the surface

22. $z^2 = 2xy$ with $0 \le x \le a,\ 0 \le y \le b,\ z \ge 0$.

23. $z = a^2 - (x^2 + y^2)$ with $\frac{1}{4}a^2 \le x^2 + y^2 \le a^2$.

24. $3z^2 = (x+y)^3$ with $x + y \le 2,\ x \ge 0,\ y \ge 0$.

25. $3z = x^{3/2} + y^{3/2}$ with $0 \le x \le 1,\ 0 \le y \le x$.

26. $z = y^2$ with $0 \le x \le 1,\ 0 \le y \le 1$.

27. $x^2 + y^2 + z^2 - 4z = 0$ with $0 \le 3(x^2 + y^2) \le z^2,\ z \ge 2$.

28. $x^2 + y^2 + z^2 - 2az = 0$ with $0 \le x^2 + y^2 \le bz$. Assume $a > b > 0$.

29. (a) Find a formula for the area of a surface that is projectable onto a region Ω of the yz- plane; say,

$$S\colon\ x = g(y,z), \quad (y,z) \in \Omega.$$

Assume that g is continuously differentiable.

(b) Find a formula for the area of a surface that is projectable onto a region Ω of the xz-plane; say,

$$S\colon\ y = h(x,z), \quad (x,z) \in \Omega.$$

Assume that h is continuously differentiable.

30. (a) Determine the fundamental vector product for the cylindrical surface

$$\mathbf{r}(u,v) = a\cos u\,\mathbf{i} + a\sin u\,\mathbf{j} + v\,\mathbf{k};\quad 0 \le u \le 2\pi, 0 \le v \le l.$$

(b) Use your answer to part (a) to find the area of the surface.

31. (a) Determine the fundamental vector product for the cone of Example 4:

$$\mathbf{r}(u,v) = v\cos u \sin\alpha\,\mathbf{i} + v\sin u\sin\alpha\,\mathbf{j} + v\cos\alpha\,\mathbf{k};$$
$$0 \le u \le 2\pi,\ 0 \le v \le s.$$

(b) Use your answer to part (a) to calculate the area of the cone.

C 32. (a) Show that $\mathbf{r}(u) = a\cos u \sin v\mathbf{i} + a\sin u \sin v\mathbf{j} + b\cos v\mathbf{k}$, $0 \le u \le 2\pi, 0 \le v \le \pi$, parametrizes the ellipsoid of revolution

$$\frac{x^2}{a^2} + \frac{y^2}{a^2} + \frac{z^2}{b^2} = 1.$$

(b) Use a graphing utility to draw the surface with $a = 3, b = 4$; experiment with other values of a and b.

(c) Show that the surface area of the ellipsoid is given by the formula

$$A = 2\pi a \int_0^{\pi} \sin v\sqrt{b^2 \sin^2 v + a^2 \cos^2 v}\, dv.$$

33. (a) Show that $\mathbf{r}(u) = a\cos u \cosh v\,\mathbf{i} + b\sin u\cosh v\,\mathbf{j} + c\sinh v\,\mathbf{k}, 0 \le u \le 2\pi, v$ real, parametrizes the hyperboloid of one sheet

$$\frac{x^2}{a^2} + \frac{y^2}{b^2} - \frac{z^2}{c^2} = 1.$$

(b) Use a graphing utility to draw the surface with $a = 3, b = 2, c = 4$; experiment with other values of a, b, c.

(c) Set up a double integral for the surface area of the part of the hyperboloid in part (b) that lies between the planes $z = -3$ and $z = 3$.

34. (a) Show that $\mathbf{r}(u) = a\cos u \sinh v\,\mathbf{i} + b\sin u\sinh v\,\mathbf{j} + c\cosh u\,\mathbf{k}, \quad 0 \le u \le 2\pi, v$ real, parametrizes the hyperboloid of two sheets

$$\frac{x^2}{a^2} + \frac{y^2}{a^2} - \frac{z^2}{b^2} = -1.$$

(b) Use a graphing utility to draw the surface with $a = 3$, $b = 2, c = 4$; experiment with other values of a, b, c.

(c) Explain why the your graph shows only the upper surface of the hyperboloid. Change the parametrization so that your graph displays the lower half of the surface.

35. Let Ω be a plane region in space and let A_1, A_2, A_3 be the areas of the projections of Ω onto the three coordinate planes. Express the area of Ω in terms of A_1, A_2, A_3.

36. Let S be a surface given in cylindrical coordinates by an equation of the form $z = f(r, \theta), (r, \theta) \in \Omega$. Show that if f is continuously differentiable, then

$$\text{area of } S = \iint_{\Omega} \sqrt{r^2[f_r(r,\theta)]^2 + [(f_\theta(r,\theta)]^2 + r^2}\, dr d\theta$$

provided the integrand is never zero on the interior of Ω.

37. The following surfaces are given in cylindrical coordinates. Find the surface area.

(a) $z = r + \theta; \quad 0 \le r \le 1, \quad 0 \le \theta \le \pi.$

(b) $z = r e^{\theta}; \quad 0 \le r \le a, \quad 0 \le \theta \le 2\pi.$

38. Show that, for a flat surface S that is part of the xy-plane, (17.6.3) gives

$$\text{area of } S = \iint_{\Omega} |J(u, v)|\, du\, dv,$$

where J is the Jacobian of the components of a vector function, which is defined on some region Ω and parametrizes S. Except for notation this is Formula (16.10.1).

■ 17.7 SURFACE INTEGRALS

The Mass of a Material Surface

Imagine a thin distribution of matter spread out over a surface S. We call this a *material surface*.

If the mass density (the mass per unit area) is a constant λ throughout, then the total mass of the material surface is the density λ times the area of S:

$$M = \lambda\,(\text{area of } S).$$

If, however, the mass density varies continuously from point to point, $\lambda = \lambda(x, y, z)$, then the total mass must be calculated by integration.

To develop the appropriate integral, we suppose that

$$S\colon\ \mathbf{r} = \mathbf{r}(u, v) = x(u, v)\,\mathbf{i} + y(u, v)\,\mathbf{j} + z(u, v)\,\mathbf{k}, \qquad (u, v) \in \Omega,$$

is a smooth surface, a surface that meets the conditions for area formula (17.6.3).† Our first step is to break up Ω into N little basic regions $\Omega_1, \ldots, \Omega_N$. This decomposes the surface into N little pieces $S_1, \ldots, S_N$. The area of S_i is given by the integral

$$\iint_{\Omega_i} ||\mathbf{N}(u, v)||\, du dv. \qquad \text{[Formula (17.6.3)]}$$

† We repeat the conditions here: $\mathbf{r}$ is continuously differentiable; Ω is a basic region in the uv-plane; $\mathbf{r}$ is one-to-one on the interior of Ω; $\mathbf{N} = \mathbf{r}'_u \times \mathbf{r}'_v$ is never zero on the interior of Ω.

By the mean-value theorem for double integrals, there exists a point (u_i^*, v_i^*) in Ω_i for which

$$\iint_{\Omega_i} ||\mathbf{N}(u, v)||\, dudv = ||\mathbf{N}(u_i^*, v_i^*)||(\text{area of } \Omega_i).$$

It follows that

$$\text{area of } S_i = ||\mathbf{N}(u_i^*, v_i^*)||(\text{area of } \Omega_i).$$

Since the point (u_i^*, v_i^*) is in Ω_i, the tip of $\mathbf{r}(u_i^*, v_i^*)$ is on S_i. The mass density at this point is $\lambda[\mathbf{r}(u_i^*, v_i^*)]$. If S_i is small (which we can guarantee by choosing Ω_i small), then the mass density on S_i is approximately the same throughout. Thus we can estimate M_i, the mass contribution of S_i, by writing

$$M_i \cong \lambda[\mathbf{r}(u_i^*, v_i^*)](\text{area of } S_i) = \lambda[\mathbf{r}(u_i^*, v_i^*)]\ ||\mathbf{N}(u_i^*, v_i^*)||(\text{area of } \Omega_i).$$

Adding up these estimates, we have an estimate for the total mass of the surface:

$$\begin{aligned} M &\cong \sum_{i=1}^{N} \lambda[\mathbf{r}(u_i^*, v_i^*)]\ ||\mathbf{N}(u_i^*, v_i^*)||(\text{area of } \Omega_i) \\ &= \sum_{i=1}^{N} \lambda[x(u_i^*, v_i^*), y(u_i^*, v_i^*), z(u_i^*, v_i^*)]\ ||\mathbf{N}(u_i^*, v_i^*)||(\text{area of } \Omega_i). \end{aligned}$$

This last expression is a Riemann sum for

$$\iint_{\Omega} \lambda[x(u, v), y(u, v), z(u, v)]\ ||\mathbf{N}(u, v)||\, dudv$$

and tends to this integral as the maximal diameter of the Ω_i tends to zero. We can therefore conclude that

(17.7.1)
$$M = \iint_{\Omega} \lambda[x(u, v), y(u, v), z(u, v)]\ ||\mathbf{N}(u, v)||\, dudv.$$

Surface Integrals

The double integral in (17.7.1) can be calculated not only for a mass density function λ but for any scalar field H continuous over S. We call this integral *the surface integral of H over S* and write

(17.7.2)
$$\iint_{S} H(x, y, z)\, d\sigma = \iint_{\Omega} H[x(u, v), y(u, v) z(u, v)]\ ||\mathbf{N}(u, v)||\, dudv$$

Note that, if $H(x, y, z)$ is identically 1, then the right-hand side of (17.7.2) gives the area of S. Thus

(17.7.3)
$$\iint_{S} d\sigma = \text{area of } S.$$

Example 1 Let $\mathbf{a} = a_1\,\mathbf{i}+a_2\,\mathbf{j}+a_3\,\mathbf{k}$ and $\mathbf{b} = b_1\,\mathbf{i}+b_2\,\mathbf{j}+b_3\,\mathbf{k}$ be nonzero vectors. Calculate

$$\iint_S xy\, d\sigma \quad \text{where} \quad S\text{:} \quad \mathbf{r}(u,v) = u\,\mathbf{a} + v\,\mathbf{b};\ \ 0 \le u \le 1,\ \ 0 \le v \le 1.$$

SOLUTION Call the parameter set Ω. Then

$$\iint_S xy\, d\sigma = \iint_\Omega x(u,v)y(u,v)||\mathbf{N}(u,v)||\, dudv.$$

A simple calculation shows that $||\mathbf{N}(u,v)|| = ||\mathbf{a}\times\mathbf{b}||$. Thus

$$\iint_S xy\, d\sigma = ||\mathbf{a}\times\mathbf{b}|| \iint_\Omega x(u,v)y(u,v),\, dudv.$$

To find $x(u,v)$ and $y(u,v)$, we need the **i** and **j** components of $\mathbf{r}(u,v)$. We can get these as follows:

$$\begin{aligned}\mathbf{r}(u,v) = u\,\mathbf{a} + v\,\mathbf{b} &= u(a_1\,\mathbf{i} + a_2\,\mathbf{j} + a_3\,\mathbf{k}) + v(b_1\,\mathbf{i} + b_2\,\mathbf{j} + b_3\,\mathbf{k})\\ &= (a_1u + b_1v)\,\mathbf{i} + (a_2u + b_2v)\,\mathbf{j} + (a_3u + b_3v)\,\mathbf{k}.\end{aligned}$$

Therefore $x(u,v) = a_1u + b_1v$ and $y(u,v) = a_2u + b_2v$. We can now write

$$\begin{aligned}\iint_S xy\, d\sigma &= ||\mathbf{a}\times\mathbf{b}|| \iint_\Omega (a_1u + b_1v)(a_2u + b_2v)\, du\, dv\\ &= ||\mathbf{a}\times\mathbf{b}|| \int_0^1 \left(\int_0^1 \left[a_1a_2u^2 + (a_1b_2 + b_1a_2)uv + b_1b_2v^2\right] du\right) dv\\ &= ||\mathbf{a}\times\mathbf{b}||\ \left[\tfrac{1}{3}a_1a_2 + \tfrac{1}{4}(a_1b_2 + b_1a_2) + \tfrac{1}{3}b_1b_2\right].\quad \square\end{aligned}$$

check this ⟶

Example 2 Calculate

$$\iint_S \sqrt{x^2 + y^2}\, d\sigma$$

where S is the spiral ramp of Example 11, Section 17.6:

$$S\text{:} \quad \mathbf{r}(u,v) = u\cos\omega v\,\mathbf{i} + u\sin\omega v\,\mathbf{j} + bv\,\mathbf{k};\quad 0 \le u \le l,\quad 0 \le v \le 2\pi/\omega.$$

SOLUTION Call the parameter set Ω. As you saw in Example 11, Section 17.6,

$$||\mathbf{N}(u,v)|| = \sqrt{b^2 + \omega^2u^2}.$$

Therefore

$$\iint_S \sqrt{x^2+y^2}\,d\sigma = \iint_\Omega \sqrt{[x(u,v)]^2+[y(u,v)]^2}\,\|\mathbf{N}(u,v)\|\,dudv$$

$$= \iint_\Omega \sqrt{u^2\cos^2\omega v + u^2\sin^2\omega v}\,\sqrt{b^2+\omega^2u^2}\,dudv$$

$$= \iint_\Omega u\sqrt{b^2+\omega^2u^2}\,dudv$$

$u \geq 0$ on Ω

$$= \int_0^{2\pi/\omega}\left(\int_0^l u\sqrt{b^2+\omega^2u^2}\,du\right)dv$$

$$= \frac{2\pi}{\omega}\int_0^l u\sqrt{b^2+\omega^2u^2}\,du = \frac{2\pi}{3\omega^3}\left[(b^2+\omega^2l^2)^{3/2}-b^3\right]. \quad \square$$

Like the other integrals you have studied, the surface integral satisfies a mean-value condition; namely, if the scalar field H is continuous, then there is a point (x_0, y_0, z_0) on S for which

$$\iint_S H(x,y,z)\,d\sigma = H(x_0,y_0,z_0)(\text{ area of } S).$$

we call $H(x_0, y_0, z_0)$ *the average value of H on S*. Thus we can write

(17.7.4)
$$\iint_S H(x,y,z)\,d\sigma = \begin{pmatrix}\text{average value}\\ \text{of } H \text{ on } S\end{pmatrix}\cdot(\text{area of } S).$$

We can also take weighted averages: if H and G are continuous on S and G is nonnegative on S, then there is a point (x_0, y_0, z_0) on S for which

(17.7.5)
$$\iint_S H(x,y,z)G(x,y,z)\,d\sigma = H(x_0,y_0,z_0)\iint_S G(x,y,z)\,d\sigma.$$

As you would expect, we call $H(x_0, y_0, z_0)$ *the G-weighted average of H on S*.

The coordinates of the centroid $(\bar{x}, \bar{y}, \bar{z})$ of a surface are simply averages taken over the surface: for a surface S of area A,

$$\bar{x}A = \iint_S x\,d\sigma,\quad \bar{y}A = \iint_S y\,d\sigma,\quad \bar{z}A = \iint_S z\,d\sigma.$$

In the case of a material surface of mass density $\lambda = \lambda(x, y, z)$, the coordinates of the center of mass (x_M, y_M, z_M) are density-weighted averages: for a surface S of total mass M,

$$x_M M = \iint_S x\lambda(x,y,z)\,d\sigma,\quad y_M M = \iint_S y\lambda(x,y,z)\,d\sigma,\quad z_M M = \iint_S z\lambda(x,y,z)\,d\sigma.$$

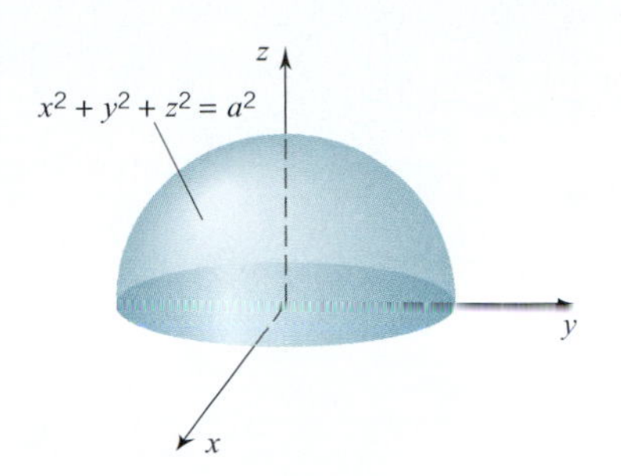

Figure 17.7.1

Example 3 Locate the center of mass of a material surface in the form of a hemispherical shell $x^2 + y^2 + z^2 = a^2$ with $z \geq 0$ given that the mass density is directly proportional to the distance from the xy-plane. See Figure 17.7.1.

SOLUTION The surface S can be parametrized by the function

$$\mathbf{r}(u, v) = a \cos u \cos v\, \mathbf{i} + a \sin u \cos v\, \mathbf{j} + a \sin v\, \mathbf{k}; \quad 0 \leq u \leq 2\pi, 0 \leq v \leq \tfrac{1}{2}\pi.$$

Call the parameter set Ω and recall that $||\mathbf{N}(u, v)|| = a^2 \cos v$ (Example 8, Section 17.6). The density function can be written $\lambda(x, y, z) = kz$, where k is the constant of proportionality. We can calculate the mass as follows:

$$\begin{aligned} M = \iint_S \lambda(x, y, z)\, d\sigma = k \iint_S z\, d\sigma &= k \iint_\Omega z(u, v)||\mathbf{N}(u, v)||\, du dv \\ &= k \int_0^{2\pi} \left(\int_0^{\pi/2} (a \sin v)(a^2 \cos v)\, dv \right) du \\ &= 2\pi k a^3 \int_0^{\pi/2} \sin v \cos v\, dv = \pi k a^3. \end{aligned}$$

By symmetry $x_M = 0$ and $y_M = 0$. To find z_M we write

$$\begin{aligned} z_M M = \iint_S z\lambda(x, y, z)\, d\sigma &= k \iint_S z^2 d\sigma \\ &= k \iint_\Omega [z(u, v)]^2 ||\mathbf{N}(u, v)||\, du\, dv \\ &= k \int_0^{2\pi} \left(\int_0^{\pi/2} (a^2 \sin^2 v)(a^2 \cos v)\, dv \right) du \\ &= 2\pi k a^4 \int_0^{\pi/2} \sin^2 v \cos v\, dv = \tfrac{2}{3}\pi k a^4. \end{aligned}$$

Since $M = \pi k a^3$, we see that $z_M = \frac{2}{3}\pi k a^4 / M = \frac{2}{3}a$. The center of mass is the point $(0, 0, \frac{2}{3}a)$. ❑

Suppose that a material surface S rotates about an axis. The moment of inertia of the surface about that axis is given by the formula

(17.7.6)
$$I = \iint_S \lambda(x, y, z)[R(x, y, z)]^2\, d\sigma$$

where $\lambda = \lambda(x, y, z)$ is the mass density function and $R(x, y, z)$ is the distance from the axis to the point (x, y, z). (As usual, the moments of inertia about the x, y, z axes are denoted by I_x, I_y, I_z.)

Example 4 A homogeneous material surface with mass density 1 is the shape of a spherical shell

$$S: \quad x^2 + y^2 + z^2 = a^2. \qquad \text{(Figure 17.7.2)}$$

Calculate the moment of inertia about the z-axis.

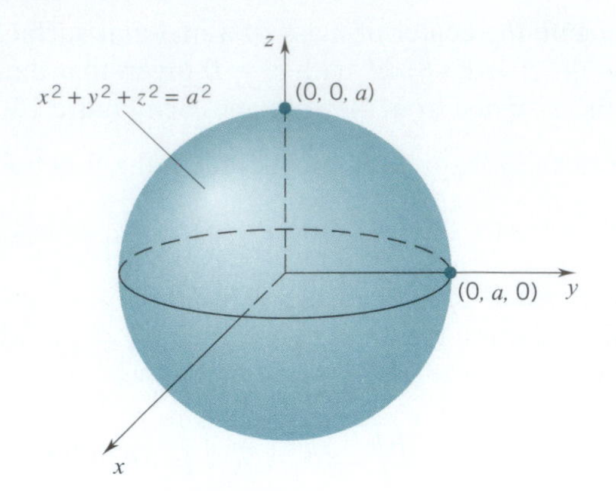

Figure 17.7.2

SOLUTION We parametrize S by setting

$$\mathbf{r}(u, v) = a\cos u \cos v\,\mathbf{i} + a\sin u\cos v\,\mathbf{j} + a\sin v\,\mathbf{k}; \quad 0 \le u \le 2\pi, -\tfrac{1}{2}\pi \le v \le \tfrac{1}{2}\pi.$$

Call the parameter set Ω and recall that $||\mathbf{N}(u, v)|| = a^2\cos v$. We can calculate the moment of inertia as follows:

$$\begin{aligned} I_z = \iint_S (1)(x^2+y^2)\,d\sigma &= \iint_\Omega \left([x(u,v)]^2 + [y(u,v)]^2\right)||\mathbf{N}(u,v)||d\sigma \\ &= \iint_\Omega (a^2\cos^2 v)(a^2\cos v)\,dudv \\ &= a^4\int_0^{2\pi}\left(\int_{-\pi/2}^{\pi/2}\cos^3 v\,dv\right)du \\ &= 2\pi a^4 \int_{-\pi/2}^{\pi/2}\cos^3 v\,dv = \tfrac{8}{3}\pi a^4. \end{aligned}$$

↑ check this

Since the surface has mass $M = A = 4\pi a^2$, it follows that $I_z = \frac{2}{3}Ma^2$. ❑

A surface

$$S:\ z = f(x,y),\ (x,y) \in \Omega$$

can be parametrized by the function

$$\mathbf{r}(x,y) = x\,\mathbf{i} + y\,\mathbf{j} + f(x,y)\,\mathbf{k}, \quad (x,y) \in \Omega.$$

As you saw in Section 17.6, $||\mathbf{N}(x,y)|| = \sec[\gamma(x,y)]$ where $\gamma(x,y)$ is the angle between $\mathbf{k}$ and the upper unit normal. Therefore, for any continuous scalar field H on S,

(17.7.7)
$$\iint_S H(x,y,z)\,d\sigma = \iint_\Omega H(x,y,z)\sec[\gamma(x,y)]\,dxdy.$$

In evaluating this last integral we use the fact that

$$\sec[\gamma(x,y)] = \sqrt{[f_x(x,y)]^2 + [f_y(x,y)]^2 + 1}. \qquad \text{(Section 17.6)}$$

Example 5 Calculate

$$\iint_S \sqrt{x^2+y^2}\, d\sigma \quad \text{with} \quad S: z = xy,\ 0 \le x^2+y^2 \le 1.$$

SOLUTION The base region Ω is the unit disc. The function $z = f(x,y) = xy$ has partial derivatives $f_x(x,y) = y,\ f_y(x,y) = x$. Therefore

$$\sec[\gamma(x,y)] = \sqrt{y^2+x^2+1} = \sqrt{x^2+y^2+1}$$

and
$$\iint_S \sqrt{x^2+y^2}\, d\sigma = \iint_\Omega \sqrt{x^2+y^2}\sqrt{x^2+y^2+1}\, dxdy.$$

We evaluate this last integral by changing to polar coordinates. The region Ω is the set of all (x,y) with polar coordinates $[r,\theta]$ in the set

$$\Gamma: \quad 0 \le \theta \le 2\pi, \quad 0 \le r \le 1.$$

Therefore
$$\begin{aligned}
\iint_S \sqrt{x^2+y^2}\, d\sigma = \iint_\Gamma r\sqrt{r^2+1}\, r\, drd\theta &= \int_0^{2\pi}\left(\int_0^1 r^2\sqrt{r^2+1}\, dr\right) d\theta \\
&= 2\pi \int_0^1 r^2\sqrt{r^2+1}\, dr \\
&= 2\pi \int_0^{\pi/4} \tan^2\phi \sec^3\phi\, d\phi \quad (r = \tan\phi) \\
&= 2\pi \int_0^{\pi/4} [\sec^5\phi - \sec^3\phi]\, d\phi \\
&= \tfrac{1}{4}\pi[3\sqrt{2} - \ln(\sqrt{2}+1)]. \quad \square
\end{aligned}$$

The Flux of a Vector Field

Suppose that
$$S:\ \mathbf{r} = \mathbf{r}(u,v), \quad (u,v) \in \Omega$$

is a smooth surface with a unit normal $\mathbf{n} = \mathbf{n}(x,y,z)$ that is continuous on all of S. Such a surface is called an *oriented surface*.† Note that an oriented surface has two sides: the side with normal $\mathbf{n}$ and the side with normal $-\mathbf{n}$.† If $\mathbf{v} = \mathbf{v}(x,y,z)$ is a vector field continuous on S, then we can form the surface integral

(17.7.8)
$$\iint_S (\mathbf{v} \cdot \mathbf{n})\, d\sigma = \iint_S [\mathbf{v}(x,y,z) \cdot \mathbf{n}(x,y,z)]\, d\sigma.$$

This surface integral is called *the flux of* **v** *across S in the direction of* **n**.

† Not all surfaces have two sides. In Exercise 41 we exhibit a surface (the Möbius band) which has only one side

Note that the flux across a surface depends on the choice of unit normal. If $-\mathbf{n}$ is chosen instead of $\mathbf{n}$, the sign of the flux is reversed:

$$\iint_S (\mathbf{v}\cdot[-\mathbf{n}])\,d\sigma = \iint_S -(\mathbf{v}\cdot\mathbf{n})\,d\sigma = -\iint_S (\mathbf{v}\cdot\mathbf{n})\,d\sigma.$$

Example 6 Calculate the flux of the vector field $\mathbf{v}(x,y,z) = x\,\mathbf{i} + y\,\mathbf{j}$ across the sphere $S: x^2+y^2+z^2=a^2$ in the outward direction (Figure 17.7.3).

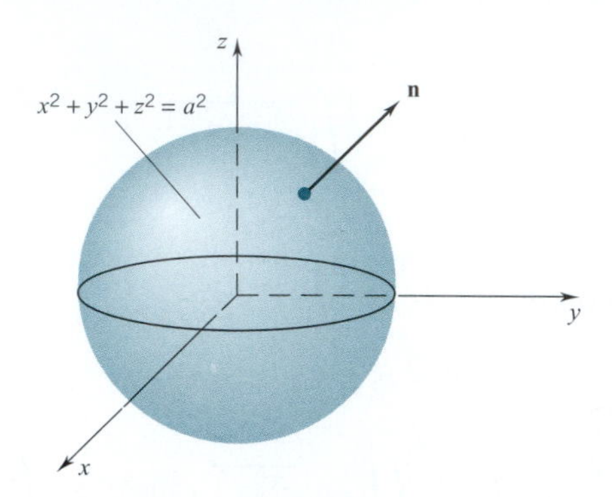

Figure 17.7.3

SOLUTION The outward unit normal is the vector

$$\mathbf{n}(x,y,z) = \frac{1}{a}(x\,\mathbf{i} + y\,\mathbf{j} + z\,\mathbf{k}).$$

Here $$\mathbf{v}\cdot\mathbf{n} = (x\,\mathbf{i}+y\,\mathbf{j})\cdot\frac{1}{a}(x\,\mathbf{i}+y\,\mathbf{j}+z\,\mathbf{k}) = \frac{1}{a}(x^2+y^2).$$

Therefore $$\text{flux out of } S = \frac{1}{a}\iint_S (x^2+y^2)\,d\sigma.$$

To evaluate this integral we use the usual parametrization

$$\mathbf{r}(u,v) = a\cos u\cos v\,\mathbf{i} + a\sin u\cos v\,\mathbf{j} + a\sin v\,\mathbf{k};\quad 0\le u\le 2\pi, -\tfrac{1}{2}\pi \le v\le \tfrac{1}{2}\pi.$$

Recall that $||\mathbf{N}(u,v)|| = a^2\cos v$. Thus

$$\begin{aligned}\text{flux out of } S &= \frac{1}{a}\iint_\Omega ([x(u,v)]^2 + [y(u,v)]^2||\mathbf{N}(u,v)||\,du dv\\ &= \frac{1}{a}\iint_\Omega (a^2\cos^2 u\cos^2 v + a^2\sin^2 u\cos^2 v)(a^2\cos v)\,du dv\\ &= a^3\iint_\Omega \cos^3 v\,du dv\\ &= a^3\int_0^{2\pi}\left(\int_{-\pi/2}^{\pi/2}\cos^3 v\,dv\right)du = 2\pi a^3\int_{-\pi/2}^{\pi/2}\cos^3 v\,dv = \tfrac{8}{3}\pi a^3.\quad\square\end{aligned}$$

If S is the graph of a function $z = f(x,y)$, $(x,y) \in \Omega$ and $\mathbf{n}$ is the upper unit normal, then the flux of the vector field $\mathbf{v} = v_1\mathbf{i} + v_2\mathbf{j} + v_3\mathbf{k}$ across S in the direction of $\mathbf{n}$ is

(17.7.9)
$$\iint_S (\mathbf{v}\cdot\mathbf{n})\,d\sigma = \iint_\Omega (-v_1 f_x - v_2 f_y + v_3)\,dxdy$$

PROOF From Section 17.6 we know that

$$\mathbf{n} = \frac{-f_x\,\mathbf{i} - f_y\,\mathbf{j} + \mathbf{k}}{\sqrt{(f_x)^2 + (f_y)^2 + 1}} = (-f_x\,\mathbf{i} - f_y\,\mathbf{j} + \mathbf{k})\cos\gamma$$

where γ is the angle between $\mathbf{n}$ and $\mathbf{k}$. Thus $\mathbf{v}\cdot\mathbf{n} = (-v_1 f_x - v_2 f_y + v_3)\cos\gamma$ and

$$\iint_S (\mathbf{v}\cdot\mathbf{n})\,d\sigma = \iint_\Omega (\mathbf{v}\cdot\mathbf{n})\sec\gamma\,dx\,dy = \iint_\Omega (-v_1 f_x - v_2 f_y + v_3)\,dxdy. \quad \square$$

Example 7 Let S be the part of the paraboloid $z = 1 - (x^2 + y^2)$ that lies above the unit disc Ω. Calculate the flux of $\mathbf{v} = x\,\mathbf{i} + y\,\mathbf{j} + z\,\mathbf{k}$ across this surface in the direction of the upper unit normal. (See Figure 17.7.4.)

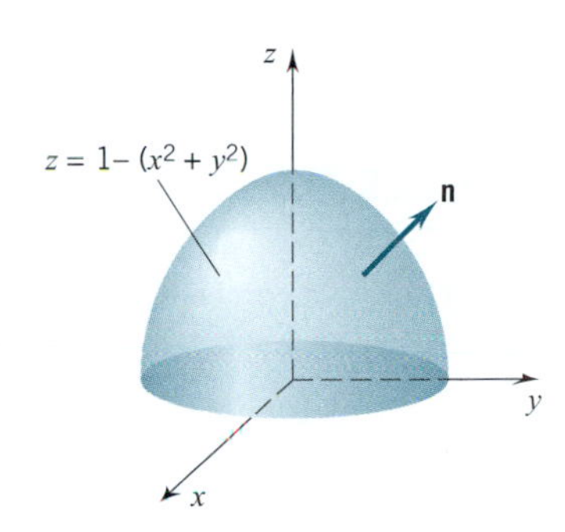

Figure 17.7.4

SOLUTION

$$f(x,y) = 1 - (x^2 + y^2); \qquad f_x = -2x, \qquad f_y = -2y.$$

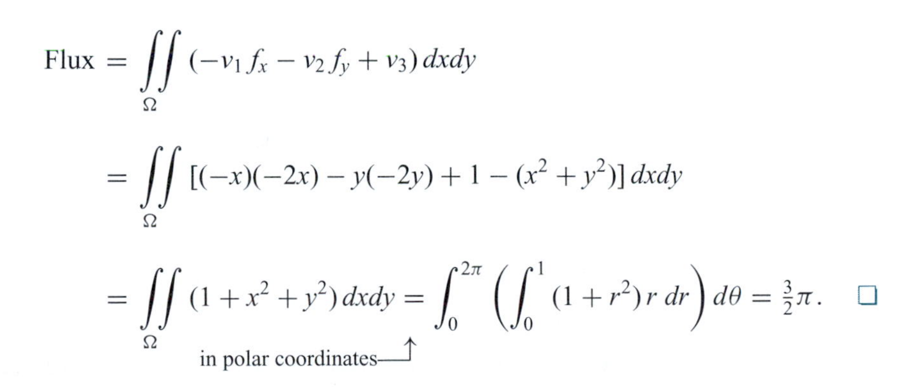

$$\text{Flux} = \iint_\Omega (-v_1 f_x - v_2 f_y + v_3)\,dxdy$$

$$= \iint_\Omega [(-x)(-2x) - y(-2y) + 1 - (x^2 + y^2)]\,dxdy$$

$$= \iint_\Omega (1 + x^2 + y^2)\,dxdy = \int_0^{2\pi}\left(\int_0^1 (1 + r^2)\,r\,dr\right)d\theta = \tfrac{3}{2}\pi. \quad \square$$

in polar coordinates ↑

The flux through a closed *piecewise-smooth oriented surface* (a closed surface that consists of a finite number of smooth oriented pieces joined together at the boundaries) can be evaluated by integrating over each smooth piece and adding up the results.

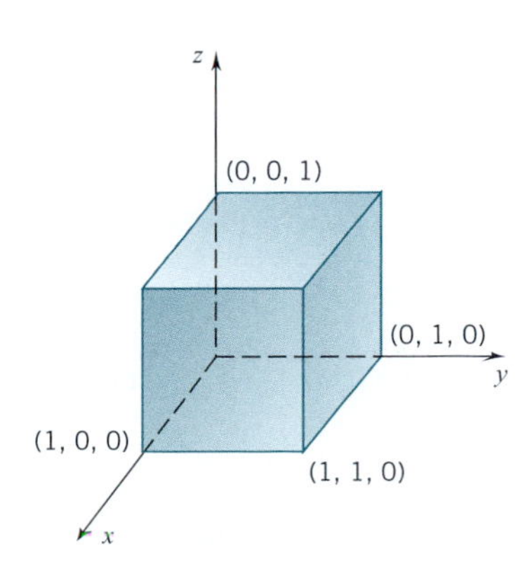

Figure 17.7.5

Example 8 Calculate the total flux of $\mathbf{v}(x,y,z) = xy\,\mathbf{i} + 4yz^2\,\mathbf{j} + yz\,\mathbf{k}$ out of the unit cube: $0 \le x \le 1,\ 0 \le y \le 1,\ 0 \le z \le 1$ (Figure 17.7.5).

SOLUTION The total flux is the sum of the fluxes across the faces of the cube:

Face	$\mathbf{n}$	$\mathbf{v} \cdot \mathbf{n}$	Flux
$x = 0$	$-\mathbf{i}$	$-xy = 0$	0
$x = 1$	$\mathbf{i}$	$xy = y$	$\frac{1}{2}$
$y = 0$	$-\mathbf{j}$	$-4yz^2 = 0$	0
$y = 1$	$\mathbf{j}$	$4yz^2 = 4z^2$	$\frac{4}{3}$
$z = 0$	$-\mathbf{k}$	$-yz = 0$	0
$z = 1$	$\mathbf{k}$	$yz = y$	$\frac{1}{2}$

The total flux is $\frac{1}{2} + \frac{4}{3} + \frac{1}{2} = \frac{7}{3}$. ☐

Flux plays an important role in the study of fluid motion. Imagine a surface S within a fluid and choose a unit normal $\mathbf{n}$. Take $\mathbf{v}$ as the velocity of the fluid that passes through S. You can expect $\mathbf{v}$ to vary from point to point, but, for simplicity, let's assume that $\mathbf{v}$ does not change with time. (Such a time-independent flow is said to be *steady state*.) The flux of $\mathbf{v}$ across S is the average component of $\mathbf{v}$ in the direction of $\mathbf{n}$ times the area of the surface S. This is just the volume of fluid that passes through S in unit time, from the $-\mathbf{n}$ side of S to the $\mathbf{n}$ side of S. If S is a closed surface (such as a cube, a sphere, or an ellipsoid) and $\mathbf{n}$ is chosen as the *outer unit normal*, then the flux across S gives the volume of fluid that flows *out* through S in unit time; if $\mathbf{n}$ is chosen as the *inner unit normal*, then the flux gives the volume of liquid that flows *in* through S in unit time.

EXERCISES 17.7

Evaluate these integrals over the surface S: $z = \frac{1}{2}y^2$; $0 \le x \le 1$, $0 \le y \le 1$.

1. $\iint_S d\sigma$.

2. $\iint_S x^2 d\sigma$.

3. $\iint_S 3y\, d\sigma$.

4. $\iint_S (x - y)\, d\sigma$.

5. $\iint_S \sqrt{2z}\, d\sigma$.

6. $\iint_S \sqrt{1 + y^2}\, d\sigma$.

Evaluate the surface integral.

7. $\iint_S xy\, d\sigma$; S the first octant part of the plane $x + 2y + 3z = 6$.

8. $\iint_S xyz\, d\sigma$; S the first octant part of the plane $x + y + z = 1$.

9. $\iint_S x^2 z\, d\sigma$; S that part of the cylinder $x^2 + z^2 = 1$ which lies between the planes $y = 0$ and $y = 2$, and is above the xy-plane.

10. $\iint_S (x^2 + y^2 + z^2)\, d\sigma$; S that part of the plane $z = x + 2$ which lies inside the cylinder $x^2 + y^2 = 1$.

11. $\iint_S (x^2 + y^2)\, d\sigma$; S the hemisphere $z = \sqrt{1 - (x^2 + y^2)}$.

12. $\iint_S (x^2 + y^2)\, d\sigma$; S that part of the paraboloid $z = 1 - x^2 - y^2$ which lies above the xy-plane.

In Exercises 13–15, find the mass of a material surface in the shape of a triangle $(a, 0, 0), (0, a, 0), (0, 0, a)$ given that the mass density varies as indicated. Take $a > 0$.

13. $\lambda(x, y, z) = k$.

14. $\lambda(x, y, z) = k(x + y)$.

15. $\lambda(x, y, z) = kx^2$.

16. Locate the centroid of the triangle $(a, 0, 0), (0, a, 0), (0, 0, a)$. Take $a > 0$.

17. Locate the centroid of the hemisphere $x^2 + y^2 + z^2 = a^2$, $z \ge 0$.

In Exercises 18 and 19, let S be the parallelogram given by $\mathbf{r}(u, v) = (u + v)\,\mathbf{i} + (u - v)\,\mathbf{j} + 2u\,\mathbf{k}$; $0 \le u \le 1, 0 \le v \le 1$.

18. Find the area of S.

19. Determine the flux of $\mathbf{v} = x\,\mathbf{i} - y\,\mathbf{j}$ across S in the direction of the fundamental vector product.

20. Find the mass of the material surface $S: z = 1 - \frac{1}{2}(x^2 + y^2)$ with $0 \le x \le 1$, $0 \le y \le 1$ given that the density at each point (x, y, z) is proportional to xy.

In Exercises 21–23, calculate the flux out of the sphere $x^2 + y^2 + z^2 = a^2$.

21. $\mathbf{v} = z\,\mathbf{k}$. **22.** $\mathbf{v} = x\,\mathbf{i} + y\,\mathbf{j} + z\,\mathbf{k}$. **23.** $\mathbf{v} = y\,\mathbf{i} - x\,\mathbf{j}$.

24. A homogeneous plate of mass density 1 is in the form of the parallelogram of Exercises 18 and 19. Determine the moments of inertia about the coordinate axes: (a) I_x. (b) I_y. (c) I_z.

In Exercises 25–27, determine the flux across the triangle $(a, 0, 0), (0, a, 0), (0, 0, a), a > 0$, in the direction of the upper unit normal.

25. $\mathbf{v} = x\,\mathbf{i} + y\,\mathbf{j} + z\,\mathbf{k}$.

26. $\mathbf{v} = (x + z)\,\mathbf{k}$.

27. $\mathbf{v} = x^2\,\mathbf{i} - y^2\,\mathbf{j}$.

In Exercises 28–30, determine the flux across $S: z = xy$ with $0 \le x \le 1, 0 \le y \le 2$ in the direction of the upper unit normal.

28. $\mathbf{v} = -xy^2\,\mathbf{i} + z\,\mathbf{j}$.

29. $\mathbf{v} = xz\,\mathbf{j} - xy\,\mathbf{k}$.

30. $\mathbf{v} = x^2y\,\mathbf{i} + z^2\,\mathbf{k}$.

31. Calculate the flux of $\mathbf{v} = x\,\mathbf{i} + y\,\mathbf{j} + z\,\mathbf{k}$ out of the cylinderical surface

$$S: \mathbf{r}(u, v) = a\cos u\,\mathbf{i} + a\sin u\,\mathbf{j} + v\,\mathbf{k};$$
$$0 \le u \le 2\pi, 0 \le v \le l.$$

32. Calculate the flux of the gravitational field.

$$\mathbf{F}(\mathbf{r}) = G\frac{mM}{r^3}\mathbf{r} \quad \text{out of the sphere} \quad x^2 + y^2 + z^2 = a^2.$$

In Exercises 33-36, find the flux across $S: z = \frac{2}{3}(x^{3/2} + y^{3/2})$ with $0 \le x \le 1, 0 \le y \le 1 - x$ in the direction of the upper unit normal.

33. $\mathbf{v} = x\,\mathbf{i} - y\,\mathbf{j} + \frac{3}{2}z\,\mathbf{k}$.

34. $\mathbf{v} = x^2\,\mathbf{i}$.

35. $\mathbf{v} = y^2\,\mathbf{j}$.

36. $\mathbf{v} = y\,\mathbf{i} - \sqrt{xy}\,\mathbf{j}$.

37. The cone

$$\mathbf{r}(u, v) = v\cos u\sin\alpha\,\mathbf{i} + v\sin u\sin\alpha\,\mathbf{j} + v\cos\alpha\,\mathbf{k},$$
$$0 \le u \le 2\pi, \quad 0 \le v \le s,$$

has area $A = \pi s^2 \sin\alpha$. Locate the centroid.

In Exercises 38–40, the mass density of a material cone $z = \sqrt{x^2 + y^2}$ with $0 \le z \le 1$ varies directly as the distance from the z-axis.

38. Find the mass of the cone.

39. Locate the center of mass.

40. Determine the moments of inertia about the coordinate axes: (a) I_x. (b) I_y. (c) I_z.

41. You have seen that, if S is a smooth oriented surface immersed in a fluid of velocity $\mathbf{v}$, then the flux

$$\iint_S (\mathbf{v} \cdot \mathbf{n})\,d\sigma$$

is the volume of fluid that passes through S in unit time from the $-\mathbf{n}$ side of S to the $\mathbf{n}$ side of S. This requires that S be a two-sided surface. There are, however, one-sided surfaces; for example, the Möbius band. To construct a material Möbius band, start with a piece of paper in the form of the rectangle in the figure. Now give the piece of paper a single twist and join the two far edges together so that C coincides with A and D coincides with B. (a) Convince yourself that this surface is one-sided and therefore the notion of flux cannot be applied to it. (b) The surface is not smooth because it is impossible to erect a unit normal $\mathbf{n}$ that varies continuously over the entire surface. Convince yourself of this as follows: erect a unit normal $\mathbf{n}$ at some point P and make a circuit of the surface with $\mathbf{n}$. Note that, as $\mathbf{n}$ returns to P, the direction of $\mathbf{n}$ has been reversed.

In Exercises 42–44, assume that the parallelogram of Exercises 18 and 19 is a material surface with a mass density that varies directly as the square of the distance from the x-axis.

42. Determine the mass.

43. Find the x-coordinate of the center of mass.

44. Find the moment of inertia about the z-axis.

45. Calculate the total flux of $\mathbf{v}(x, y, z) = y\,\mathbf{i} - x\,\mathbf{j}$ out of the solid bounded on the sides by the cylinder $x^2 + y^2 = 1$ and above and below by the planes $z = 1$ and $z = 0$. HINT: Draw a figure.

46. Calculate the total flux of $\mathbf{v}(x, y, z) = y\,\mathbf{i} - x\,\mathbf{j}$ out of the solid bounded above by $z = 4$ and below by $z = x^2 + y^2$.

47. Calculate the total flux of $\mathbf{v}(x, y, z) = x\,\mathbf{i} + y\,\mathbf{j} + z\,\mathbf{k}$ out of the solid bounded above by $z = \sqrt{2 - (x^2 + y^2)}$ and below by $z = x^2 + y^2$.

48. Calculate the total flux of $\mathbf{v}(x, y, z) = xz\,\mathbf{i} + 4xyz^2\,\mathbf{j} + 2z\,\mathbf{k}$ out of the unit cube: $0 \le x \le 1, 0 \le y \le 1, 0 \le z \le 1$.

■ 17.8 THE VECTOR DIFFERENTIAL OPERATOR ∇

Divergence $\nabla \cdot \mathbf{v}$, Curl $\nabla \times \mathbf{v}$

The *vector differential operator* ∇ is defined formally by setting

(17.8.1)
$$\nabla = \frac{\partial}{\partial x}\mathbf{i} + \frac{\partial}{\partial y}\mathbf{j} + \frac{\partial}{\partial z}\mathbf{k}.$$

By "formally," we mean that this is not an ordinary vector. Its "components" are differentiation symbols. As the term "operator" suggests, ∇ is to be thought of as something that "operates" on things. What sorts of things? Scalar fields and vector fields.

Suppose that f is a differentiable scalar field. Then ∇ operates on f as follows:

$$\nabla f = \left(\frac{\partial}{\partial x}\mathbf{i} + \frac{\partial}{\partial y}\mathbf{j} + \frac{\partial}{\partial z}\mathbf{k}\right) f = \frac{\partial f}{\partial x}\mathbf{i} + \frac{\partial f}{\partial y}\mathbf{j} + \frac{\partial f}{\partial z}\mathbf{k}.$$

This is just the *gradient* of f, with which you are already familiar.

How does ∇ operate on vector fields? In two ways. If $\mathbf{v} = v_1\,\mathbf{i} + v_2\,\mathbf{j} + v_3\,\mathbf{k}$ is a differentiable vector field, then, by definition,

(17.8.2)
$$\nabla \cdot \mathbf{v} = \frac{\partial v_1}{\partial x} + \frac{\partial v_2}{\partial y} + \frac{\partial v_3}{\partial z}$$

and

(17.8.3)
$$\nabla \times \mathbf{v} = \begin{vmatrix} \mathbf{i} & \mathbf{j} & \mathbf{k} \\ \dfrac{\partial}{\partial x} & \dfrac{\partial}{\partial y} & \dfrac{\partial}{\partial z} \\ v_1 & v_2 & v_3 \end{vmatrix} = \left(\frac{\partial v_3}{\partial y} - \frac{\partial v_2}{\partial z}\right)\mathbf{i} + \left(\frac{\partial v_1}{\partial z} - \frac{\partial v_3}{\partial x}\right)\mathbf{j} + \left(\frac{\partial v_2}{\partial x} - \frac{\partial v_1}{\partial y}\right)\mathbf{k}.$$

The first "product," $\nabla \cdot \mathbf{v}$, defined in imitation of the ordinary dot product, is called the *divergence* of $\mathbf{v}$:

$$\nabla \cdot \mathbf{v} = \operatorname{div} \mathbf{v}.$$

The second "product," $\nabla \times \mathbf{v}$, defined in imitation of the ordinary cross product, is called the *curl* of $\mathbf{v}$:

$$\nabla \times \mathbf{v} = \operatorname{curl} \mathbf{v}.$$

Interpretation of Divergence and Curl

Suppose we know the divergence of a field and also the curl. What does that tell us? For definitive answers we must wait for the divergence theorem and Stokes's theorem, but, in a preliminary way, we can give you some rough answers right now.

View $\mathbf{v}$ as the velocity field of some fluid. The divergence of $\mathbf{v}$ at a point P gives us an indication of whether the fluid tends to accumulate near P (negative divergence) or tends to move away from P (positive divergence). In the first case, P is sometimes called a *sink*, and in the second case, it is called a *source*. The curl at P measures the rotational tendency of the fluid.

Example 1 Set

$$\mathbf{v}(x, y, z) = \alpha\,\mathbf{r} = \alpha x\,\mathbf{i} + \alpha y\,\mathbf{j} + \alpha z\,\mathbf{k}. \qquad (\alpha \text{ a constant})$$

The divergence is

$$\nabla \cdot \mathbf{v} = \alpha\frac{\partial x}{\partial x} + \alpha\frac{\partial y}{\partial y} + \alpha\frac{\partial z}{\partial z} = 3\alpha.$$

The curl is

$$\nabla \times \mathbf{v} = \begin{vmatrix} \mathbf{i} & \mathbf{j} & \mathbf{k} \\ \dfrac{\partial}{\partial x} & \dfrac{\partial}{\partial y} & \dfrac{\partial}{\partial z} \\ \alpha x & \alpha y & \alpha z \end{vmatrix} = \alpha \begin{vmatrix} \mathbf{i} & \mathbf{j} & \mathbf{k} \\ \dfrac{\partial}{\partial x} & \dfrac{\partial}{\partial y} & \dfrac{\partial}{\partial z} \\ x & y & z \end{vmatrix} = \mathbf{0}$$

because the partial derivatives that appear in the expanded determinant

$$\frac{\partial y}{\partial x}, \qquad \frac{\partial x}{\partial y}, \qquad \text{etc.},$$

are all zero. ❑

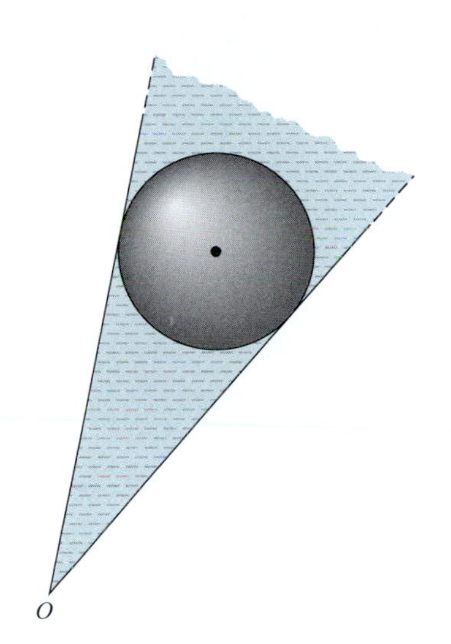

Figure 17.8.1

The field of Example 1

$$\mathbf{v}(x, y, z) = \alpha(x\,\mathbf{i} + y\,\mathbf{j} + z\,\mathbf{k}) = \alpha\,\mathbf{r}$$

can be viewed as the velocity field of a fluid in radial motion—toward the origin if $\alpha < 0$, away from the origin if $\alpha > 0$. Figure 17.8.1 shows a point (x, y, z), a spherical neighborhood of that point, and a cone emanating from the origin that is tangent to the boundary of the neighborhood.

Note two things: all the fluid in the cone stays in the cone, and the speed of the fluid is proportional to its distance from the origin. Therefore, if the divergence 3α is negative, then α is negative, the motion is toward the origin, and the neighborhood *gains fluid* because the fluid coming in is moving more quickly than the fluid going out. (Also, the entry area is larger than the exit area.) If, however, the divergence 3α is positive, then α is positive, the motion is away from the origin, and the neighborhood *loses fluid* because the fluid coming in is moving more slowly than the fluid going out. (Also, the entry area is smaller than the exit area.)

Since the motion is radial, the fluid has no rotational tendency whatsoever, and we would expect the curl to be identically zero. It is.

Example 2 Set

$$\mathbf{v}(x, y, z) = -\omega y\,\mathbf{i} + \omega x\,\mathbf{j}. \qquad (\omega \text{ a positive constant})$$

The divergence is

$$\nabla \cdot \mathbf{v} = -\omega\frac{\partial y}{\partial x} + \omega\frac{\partial x}{\partial y} = 0 + 0 = 0.$$

The curl is

$$\nabla \times \mathbf{v} = \begin{vmatrix} \mathbf{i} & \mathbf{j} & \mathbf{k} \\ \dfrac{\partial}{\partial x} & \dfrac{\partial}{\partial y} & \dfrac{\partial}{\partial z} \\ -\omega y & \omega x & 0 \end{vmatrix} = \left(\omega \frac{\partial x}{\partial x} + \omega \frac{\partial y}{\partial y}\right) \mathbf{k} = 2\omega\,\mathbf{k}. \quad \square$$

The field of Example 2,

$$\mathbf{v}(x, y, z) = -\omega y\,\mathbf{i} + \omega x\,\mathbf{j},$$

is the velocity field of uniform counterclockwise rotation about the z-axis with angular speed ω. You can see this by noting that $\mathbf{v}$ is perpendicular to $\mathbf{r} = x\,\mathbf{i} + y\,\mathbf{j} + z\,\mathbf{k}$:

$$\mathbf{v} \cdot \mathbf{r} = (-\omega y\,\mathbf{i} + x\,\mathbf{j}) \cdot (x\,\mathbf{i} + y\,\mathbf{j} + z\,\mathbf{k}) = -\omega yx + \omega xy = 0$$

and the speed at each point is ωR where R is the radius of rotation:

$$v = \sqrt{\omega^2 y^2 + \omega^2 x^2} = \omega\sqrt{x^2 + y^2} = \omega R.$$

How is the curl, $2\omega\,\mathbf{k}$, related to the rotation? The angular velocity vector (see Exercise 14, Section 13.6) is the vector $\boldsymbol{\omega} = \omega\,\mathbf{k}$. In this case, then, the curl of $\mathbf{v}$ is twice the angular velocity vector.

With this rotation no neighborhood gains any fluid and no neighborhood loses any fluid. As we saw, the divergence is identically zero.

Basic Identities

For vectors we have $\mathbf{a} \times \mathbf{a} = \mathbf{0}$. Is it true that $\nabla \times \nabla = \mathbf{0}$? Define $(\nabla \times \nabla)f$ by

$$(\nabla \times \nabla)f = \nabla \times (\nabla f).$$

THEOREM 17.8.4 THE CURL OF A GRADIENT IS ZERO

If f is a scalar field with continuous second partials, then

$$\nabla \times (\nabla f) = \mathbf{0}.$$

PROOF

$$\nabla \times (\nabla f) = \begin{vmatrix} \mathbf{i} & \mathbf{j} & \mathbf{k} \\ \dfrac{\partial}{\partial x} & \dfrac{\partial}{\partial y} & \dfrac{\partial}{\partial z} \\ \dfrac{\partial f}{\partial x} & \dfrac{\partial f}{\partial y} & \dfrac{\partial f}{\partial z} \end{vmatrix} =$$

$$\left(\frac{\partial^2 f}{\partial y \partial z} - \frac{\partial^2 f}{\partial z \partial y}\right)\mathbf{i} + \left(\frac{\partial^2 f}{\partial z \partial x} - \frac{\partial^2 f}{\partial x \partial z}\right)\mathbf{j} + \left(\frac{\partial^2 f}{\partial x \partial y} - \frac{\partial^2 f}{\partial y \partial x}\right)\mathbf{k} = \mathbf{0}. \quad \square$$

For vectors we have $\mathbf{a} \cdot (\mathbf{a} \times \mathbf{c}) = 0$. The analogous operator formula, $\nabla \cdot (\nabla \times \mathbf{v}) = 0$, is also valid.

THEOREM 17.8.5 THE DIVERGENCE OF A CURL IS ZERO

If the components of the vector field $\mathbf{v} = v_1\mathbf{i} + v_2\mathbf{j} + v_3\mathbf{k}$ have continuous second partials, then

$$\nabla \cdot (\nabla \times \mathbf{v}) = 0.$$

PROOF Again the key is the equality of the mixed partials:

$$\nabla \cdot (\nabla \times \mathbf{v}) = \frac{\partial}{\partial x}\left(\frac{\partial v_3}{\partial y} - \frac{\partial v_2}{\partial z}\right) + \frac{\partial}{\partial y}\left(\frac{\partial v_1}{\partial z} - \frac{\partial v_3}{\partial x}\right) + \frac{\partial}{\partial z}\left(\frac{\partial v_2}{\partial x} - \frac{\partial v_1}{\partial y}\right) = 0,$$

since for each component v_i the mixed partials cancel. Try it for v_1. ❑

The next two identities are product rules. Here f is a scalar field and $\mathbf{v}$ is a vector field.

(17.8.6) $$\nabla \cdot (f\mathbf{v}) = (\nabla f) \cdot \mathbf{v} + f(\nabla \cdot \mathbf{v}). \quad [\text{div}\,(f\mathbf{v}) = (\text{grad}\, f) \cdot \mathbf{v} + f(\text{div}\,\mathbf{v})]$$

(17.8.7) $$\nabla \times (f\mathbf{v}) = (\nabla f) \times \mathbf{v} + f(\nabla \times \mathbf{v}). \quad [\text{curl}\,(f\mathbf{v}) = (\text{grad}\, f) \times \mathbf{v} + f(\text{curl}\,\mathbf{v})]$$

The verification of these identities is left to you in the Exercises.

We know from Example 1 that $\nabla \cdot \mathbf{r} = 3$ and $\nabla \times \mathbf{r} = \mathbf{0}$ at all points of space. Now we can show that if n is an integer, then, for all $\mathbf{r} \neq \mathbf{0}$,

(17.8.8) $$\nabla \cdot (r^n\mathbf{r}) = (n+3)r^n \quad \text{and} \quad \nabla \times (r^n\mathbf{r}) = \mathbf{0}.$$ †

PROOF Recall that $\nabla r^n = nr^{n-2}\mathbf{r}$. [(15.1.5.)] Using (17.8.6), we have

$$\begin{aligned}\nabla \cdot (r^n\mathbf{r}) &= (\nabla r^n) \cdot \mathbf{r} + r^n(\nabla \cdot \mathbf{r})\\ &= (nr^{n-2}\mathbf{r}) \cdot \mathbf{r} + r^n(3)\\ &= nr^{n-2}(\mathbf{r} \cdot \mathbf{r}) + 3r^n = nr^n + 3r^n = (n+3)r^n.\end{aligned}$$

From (15.1.5) you can see that $r^n\mathbf{r}$ is a gradient. Its curl is therefore $\mathbf{0}$ (17.8.4) ❑

The Laplacian

From the operator ∇ we can construct other operators, the most important of which is the Laplacian $\nabla^2 = \nabla \cdot \nabla$. The Laplacian (named after the French mathematician Pierre-Simon Laplace) operates on scalar fields according to the following rule:

(17.8.9) $$\nabla^2 f = \nabla \cdot (\nabla f) = \frac{\partial^2 f}{\partial x^2} + \frac{\partial^2 f}{\partial y^2} + \frac{\partial^2 f}{\partial z^2}.$$ ††

† If n is positive and even, these formulas also hold at $\mathbf{r} = \mathbf{0}$.

†† In some texts, you will see the Laplacian of f written Δf. Unfortunately this can be misread as the increment of f.

Example 3 If $f(x,y,z) = x^2 + y^2 + z^2$, then

$$\nabla^2 f = \frac{\partial^2}{\partial x^2}(x^2+y^2+z^2) + \frac{\partial^2}{\partial y^2}(x^2+y^2+z^2) + \frac{\partial^2}{\partial z^2}(x^2+y^2+z^2)$$

$$= 2+2+2 = 6. \quad \square$$

Example 4 If $f(x,y,z) = e^{xyz}$, then

$$\nabla^2 f = \frac{\partial}{\partial x^2}(e^{xyz}) + \frac{\partial^2}{\partial y^2}(e^{xyz}) + \frac{\partial^2}{\partial z^2}(e^{xyz})$$

$$= \frac{\partial}{\partial x}(yz\,e^{xyz}) + \frac{\partial}{\partial y}(xz\,e^{xyz}) + \frac{\partial}{\partial z}(xy\,e^{xyz})$$

$$= y^2z^2\,e^{xyz} + x^2z^2\,e^{xyz} + x^2y^2\,e^{xyz}$$

$$= (y^2z^2 + x^2z^2 + x^2y^2)\,e^{xyz}. \quad \square$$

Example 5 To calculate $\nabla^2(\sin r) = \nabla^2(\sin\sqrt{x^2+y^2+z^2})$, we could write

$$\frac{\partial^2}{\partial x^2}(\sin\sqrt{x^2+y^2+z^2}) + \frac{\partial^2}{\partial y^2}(\sin\sqrt{x^2+y^2+z^2}) + \frac{\partial^2}{\partial z^2}(\sin\sqrt{x^2+y^2+z^2})$$

and proceed from there. The calculations are straightforward but lengthy. We will proceed in a different way.

Recall that

$$\nabla^2 f = \nabla\cdot\nabla f \quad (17.8.9) \qquad \nabla\cdot(f\mathbf{v}) = (\nabla f)\cdot\mathbf{v} + f(\nabla\cdot\mathbf{v}) \quad (17.8.6)$$

$$\nabla f(r) = f'(r)r^{-1}\mathbf{r} \quad (15.3.11) \qquad \nabla\cdot(r^n\mathbf{r}) = (n+3)r^n \quad (17.8.8)$$

Using these relations, we have

$$\begin{aligned}\nabla^2(\sin r) = \nabla\cdot(\nabla\sin r) &= \nabla\cdot[(\cos r)\,r^{-1}\mathbf{r}]\\ &= [(\nabla\cos r)\cdot r^{-1}\mathbf{r}] + \cos r\,(\nabla\cdot r^{-1}\mathbf{r})\\ &= \{(-\sin r)\,r^{-1}\mathbf{r}]\cdot r^{-1}\mathbf{r}\} + (\cos r)(2r^{-1})\\ &= -\sin r + 2r^{-1}\cos r.\end{aligned}$$

We leave it to you to verify each step. $\square$

EXERCISES 17.8

Calculate $\nabla\cdot\mathbf{v}$ and $\nabla\times\mathbf{v}$.

1. $\mathbf{v}(x,y) = x\,\mathbf{i} + y\,\mathbf{j}$.
2. $\mathbf{v}(x,y) = y\,\mathbf{i} + x\,\mathbf{j}$.
3. $\mathbf{v}(x,y) = \dfrac{x}{x^2+y^2}\,\mathbf{i} + \dfrac{y}{x^2+y^2}\,\mathbf{j}$.
4. $\mathbf{v}(x,y) = \dfrac{y}{x^2+y^2}\,\mathbf{i} + \dfrac{x}{x^2+y^2}\,\mathbf{j}$.
5. $\mathbf{v}(x,y,z) = x\,\mathbf{i} + 2y\,\mathbf{j} + 3z\,\mathbf{k}$.
6. $\mathbf{v}(x,y,z) = yz\,\mathbf{i} + xz\,\mathbf{j} + xy\,\mathbf{k}$.
7. $\mathbf{v}(x,y,z) = xyz\,\mathbf{i} + xz\,\mathbf{j} + z\,\mathbf{k}$.
8. $\mathbf{v}(x,y,z) = x^2y\,\mathbf{i} + y^2z\,\mathbf{j} + xy^2\,\mathbf{k}$.
9. $\mathbf{v}(\mathbf{r}) = r^{-2}\,\mathbf{r}$
10. $\mathbf{v}(\mathbf{r}) = e^x\,\mathbf{r}$.
11. $\mathbf{v}(\mathbf{r}) = e^{r^2}(\mathbf{i}+\mathbf{j}+\mathbf{k})$.
12. $\mathbf{v}(\mathbf{r}) = e^{y^2}\,\mathbf{i} + e^{z^2}\,\mathbf{j} + e^{x^2}\mathbf{k}$.
13. Suppose that f is a differentiable function of one variable and $\mathbf{v}(x,y,z) = f(x)\,\mathbf{i}$. Determine $\nabla\cdot\mathbf{v}$ and $\nabla\times\mathbf{v}$.
14. Show that, if $\mathbf{v}$ is a differentiable vector field of the form $\mathbf{v}(\mathbf{r}) = f(x)\,\mathbf{i} + g(y)\,\mathbf{j} + h(z)\,\mathbf{k}$, then $\nabla\times\mathbf{v} = \mathbf{0}$.

15. Show that divergence and curl are *linear* operators:

$$\nabla \cdot (\alpha \mathbf{u} + \beta \mathbf{v}) = \alpha(\nabla \cdot \mathbf{u}) + \beta(\nabla \cdot \mathbf{v}) \quad \text{and}$$

$$\nabla \times (\alpha \mathbf{u} + \beta \mathbf{v}) = \alpha(\nabla \times \mathbf{u}) + \beta(\nabla \times \mathbf{v}).$$

16. (*Important*) Show that the gravitational field

$$\mathbf{F}(\mathbf{r}) = -\frac{GmM}{r^3}\mathbf{r}$$

has zero divergence and zero curl at each $\mathbf{r} \neq \mathbf{0}$.

A vector field $\mathbf{v}$ with the property that $\nabla \cdot \mathbf{v} = 0$ is said to be *solenoidal*, from the Greek word for "tubular." Exercise 16 shows that the gravitational field is solenoidal. If $\mathbf{v}$ is the velocity field of some fluid and $\nabla \cdot \mathbf{v} = 0$ in a solid T in three-dimensional space, then $\mathbf{v}$ has no sources or sinks within T.

17. Show that the vector field $\mathbf{v}(x, y, z) = (2x + y + 2z)\mathbf{i} + (x + 4y - 3z)\mathbf{j} + (2x - 3y - 6z)\mathbf{k}$ is solenoidal.

18. Show that $\mathbf{v}(x, y, z) = 3x^2\mathbf{i} - y^2\mathbf{j} + (2yz - 6xz)\mathbf{k}$ is solenoidal.

A vector field $\mathbf{v}$ with the property that $\nabla \times \mathbf{v} = \mathbf{0}$ is said to be *irrotational*. Exercise 16 shows that the gravitational field is irrotational. If $\mathbf{v}$ is the velocity field of some fluid, then $\nabla \times \mathbf{v}$ measures the tendency of the fluid to "curl" or rotate about an axis. Thus $\nabla \times \mathbf{v} = \mathbf{0}$ in some solid T in three-dimensional space can be interpreted to mean that the fluid is tending to move in a straight line.

19. Show that the vector field $\mathbf{v}(x, y, z) = x\,\mathbf{i} + y\,\mathbf{j} - 2z\,\mathbf{k}$ is irrotational.

20. Show that the vector field of Exercise 17 is irrotational.

In Exercises 21–26, calculate the Laplacian $\nabla^2 f$.

21. $f(x, y, z) = x^4 + y^4 + z^4$.

22. $f(x, y, z) = xyz$.

23. $f(x, y, z) = x^2y^3z^4$.

24. $f(\mathbf{r}) = \cos r$.

25. $f(\mathbf{r}) = e^r$.

26. $f(\mathbf{r}) = \ln r$.

27. Given a vector field $\mathbf{u}$, the operator $\mathbf{u} \cdot \nabla$ is defined by setting

$$(\mathbf{u} \cdot \nabla) f = \mathbf{u} \cdot \nabla f = u_1 \frac{\partial f}{\partial x} + u_2 \frac{\partial f}{\partial y} + u_3 \frac{\partial f}{\partial z}.$$

Calculate $(\mathbf{r} \cdot \nabla) f$: (a) for $f(\mathbf{r}) = r^2$; (b) for $f(\mathbf{r}) = 1/r$.

28. (Based on Exercise 27) We can also apply $\mathbf{u} \cdot \nabla$ to a vector field $\mathbf{v}$ by applying it to each component. By definition

$$(\mathbf{u} \cdot \nabla)\mathbf{v} = (\mathbf{u} \cdot \nabla v_1)\mathbf{i} + (\mathbf{u} \cdot \nabla v_2)\mathbf{j} + (\mathbf{u} \cdot \nabla v_3)\mathbf{k}.$$

(a) Calculate $(\mathbf{u} \cdot \nabla)\mathbf{r}$ for an arbitrary vector field $\mathbf{u}$.

(b) Calculate $(\mathbf{r} \cdot \nabla)\mathbf{u}$ given that $\mathbf{u} = yz\,\mathbf{i} + xz\,\mathbf{j} + xy\,\mathbf{k}$.

29. Show that, if $f(\mathbf{r}) = g(r)$ and g is twice differentiable, then

$$\nabla^2 f = g''(r) + 2r^{-1}g'(r).$$

30. Verify the following identities.

(a) $\nabla \cdot (f\mathbf{v}) = (\nabla f) \cdot \mathbf{v} + f(\nabla \cdot \mathbf{v})$.

(b) $\nabla \times (f\mathbf{v}) = (\nabla f) \times \mathbf{v} + f(\nabla \times \mathbf{v})$.

(c) $\nabla \times (\nabla \times \mathbf{v}) = \nabla(\nabla \cdot \mathbf{v}) - \nabla^2\mathbf{v}$, where $\nabla^2\mathbf{v} = (\nabla^2 v_1)\mathbf{i} + (\nabla^2 v_2)\mathbf{j} + (\nabla^2 v_3)\mathbf{k}$.

HINT: Begin part (c) by writing out the ith component of each side.

As you saw in Project 14.6, the equation

$$\nabla^2 f = \frac{\partial^2 f}{\partial x^2} + \frac{\partial^2 f}{\partial y^2} + \frac{\partial^2 f}{\partial z^2} = 0$$

is called *Laplace's equation in three dimensions*. A scalar field $f = f(x, y, z)$ with continuous second partial derivatives is said to be *harmonic* on a solid T if it is a solution of Laplace's equation.

31. Show that the scalar field

$$f(x, y, z) = x^2 + 2y^2 - 3z^2 + xy + 2xz - 3yz$$

is harmonic.

32. Show that the scalar field

$$f(x, y, z) = \frac{1}{\sqrt{x^2 + y^2 + z^2}}$$

is harmonic on every solid T that excludes the origin. Except for a constant factor, f is a potential function for the gravitational field. (see Example 1, Section 17.3.)

33. For what nonzero integers n is $f(\mathbf{r}) = r^n$ harmonic on every solid T that excludes the origin?

34. Show that if $f = f(x, y, z)$ satisfies Laplace's equation, then its gradient field is both solenoidal and irrotational.

■ 17.9 THE DIVERGENCE THEOREM

Let Ω be a Jordan region with a piecewise-smooth boundary C, and let P and Q be continuously differentiable scalar fields on an open set containing Ω. Green's theorem allows us to express a double integral over Ω as a line integral over C:

$$\iint_{\Omega} \left(\frac{\partial Q}{\partial x} - \frac{\partial P}{\partial y}\right) dxdy = \oint_C P\,dx + Q\,dy.$$

In vector terms Green's theorem can be written:

(17.9.1)
$$\iint_{\Omega} (\nabla \cdot \mathbf{v})\, dxdy = \oint_C (\mathbf{v} \cdot \mathbf{n})\, ds.$$

Here $\mathbf{n}$ is the outer unit normal and the integral on the right is taken with respect to arc length (Section 17.4).

PROOF Set $\mathbf{v} = Q\,\mathbf{i} - P\,\mathbf{j}$. Then

$$\iint_{\Omega} (\nabla \cdot \mathbf{v})\, dxdy = \iint_{\Omega} \left(\frac{\partial Q}{\partial x} - \frac{\partial P}{\partial y}\right) dxdy.$$

All we have to show then is that

$$\oint_C (\mathbf{v} \cdot \mathbf{n})\, ds = \oint_C P\, dx + Q\, dy.$$

For C traversed counterclockwise, $\mathbf{n} = \mathbf{T} \times \mathbf{k}$ where $\mathbf{T}$ is the unit tangent vector. (Draw a figure.) Thus

$$\mathbf{v} \cdot \mathbf{n} = \mathbf{v} \cdot (\mathbf{T} \times \mathbf{k}) = (-\mathbf{v}) \cdot (\mathbf{k} \times \mathbf{T}) = (-\mathbf{v} \times \mathbf{k}) \cdot \mathbf{T}.$$

$\uparrow$ $\mathbf{a} \cdot (\mathbf{b} \times \mathbf{c}) = (\mathbf{a} \times \mathbf{b}) \cdot \mathbf{c}$

Since $-\mathbf{v} \times \mathbf{k} = (P\,\mathbf{j} - Q\,\mathbf{i}) \times \mathbf{k} = P\,\mathbf{i} + Q\,\mathbf{j}$, we have $\mathbf{v} \cdot \mathbf{n} = (P\,\mathbf{i} + Q\,\mathbf{j}) \cdot \mathbf{T}$.

Therefore

$$\oint_C (\mathbf{v} \cdot \mathbf{n})\, ds = \oint_C [(P\,\mathbf{i} + Q\,\mathbf{j}) \cdot \mathbf{T}]\, ds \underset{(17.4.6)}{=} \oint_C (P\,\mathbf{i} + Q\,\mathbf{j}) \cdot d\mathbf{r} = \oint_C P\, dx + Q\, dy. \quad \square$$

Green's theorem expressed as (17.9.1) has a higher dimensional analog that is known as the divergence theorem.†

THEOREM 17.9.2 THE DIVERGENCE THEOREM

Let T be a solid bounded by a closed oriented surface S which, if not smooth, is piecewise smooth. If the vector field $\mathbf{v} = \mathbf{v}(x, y, z)$ is continuously differentiable throughout T, then

$$\iiint_T (\nabla \cdot \mathbf{v})\, dxdydz = \iint_S (\mathbf{v} \cdot \mathbf{n})\, d\sigma$$

where $\mathbf{n}$ is the outer unit normal.

PROOF We will carry out the proof under the assumption that S is smooth and that any line parallel to a coordinate axis intersects S at most twice.

† This is also called Gauss's theorem after the German mathematician Carl Friedrich Gauss (1777–1855). Often referred to as "The Prince of Mathematicians," Gauss is regarded by many as one of the greatest geniuses of all time.

Our first step is to express the outer unit normal **n** in terms of its direction cosines:

$$\mathbf{n} = \cos\alpha_1\,\mathbf{i} + \cos\alpha_2\,\mathbf{j} + \cos\alpha_3\,\mathbf{k}.$$

Then, for $\mathbf{v} = v_1\,\mathbf{i} + v_2\,\mathbf{j} + v_3\,\mathbf{k}$,

$$\mathbf{v}\cdot\mathbf{n} = v_1\cos\alpha_1 + v_2\cos\alpha_2 + v_3\cos\alpha_3.$$

The idea of the proof is to show that

$$\text{(1)}\qquad \iint_S v_1\cos\alpha_1\,d\sigma = \iiint_T \frac{\partial v_1}{\partial x}\,dxdydz,$$

$$\text{(2)}\qquad \iint_S v_2\cos\alpha_2\,d\sigma = \iiint_T \frac{\partial v_2}{\partial y}\,dxdydz,$$

$$\text{(3)}\qquad \iint_S v_3\cos\alpha_3\,d\sigma = \iiint_T \frac{\partial v_3}{\partial z}\,dxdydz.$$

All three equations can be verified in much the same manner. We will carry out the details only for the third equation.

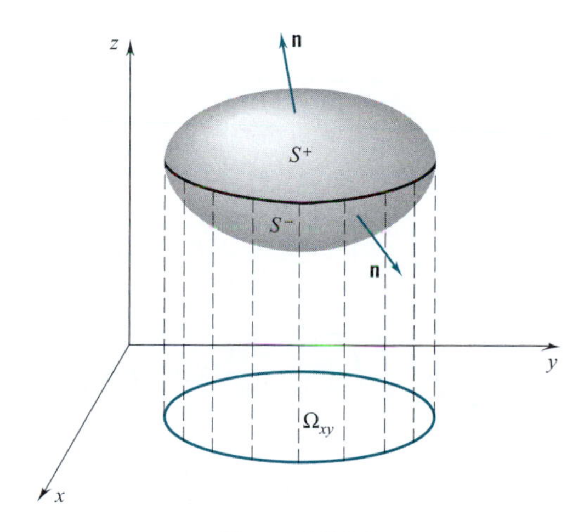

Figure 17.9.1

Let Ω_{xy} be the projection of T onto the xy-plane. (See Figure 17.9.1.) If $(x, y) \in \Omega_{xy}$, then, by assumption, the vertical line through (x, y) intersects S in at most two points, an upper point P^+ and a lower point P^-. (If the vertical line intersects S at only one point P, we set $P = P^+ = P^-$.) As (x, y) ranges over Ω_{xy}, the upper point P^+ describes a surface

$$S^+: \quad z = f^+(x, y), \quad (x, y) \in \Omega_{xy} \qquad \text{(see the figure)}$$

and the lower point describes a surface

$$S^-: \quad z = f^-(x, y), \quad (x, y) \in \Omega_{xy}.$$

By our assumptions, f^+ and f^- are continuously differentiable, $S = S^+ \cup S^-$, and the solid T is the set of all points (x, y, z) with

$$f^-(x,y) \le z \le f^+(x,y), \quad (x,y) \in \Omega_{xy}.$$

Now let λ be the angle between the positive z-axis and the upper unit normal. On S^+ the outer unit normal $\mathbf{n}$ is the upper unit normal. Thus on S^+,

$$\lambda = \alpha_3 \qquad \text{and} \qquad \cos\alpha_3 \sec\gamma = 1.$$

On S^- the outer unit normal $\mathbf{n}$ is the lower unit normal. In this case,

$$\lambda = \pi - \alpha_3 \qquad \text{and} \qquad \cos\alpha_3 \sec\gamma = -1.$$

Thus,

$$\iint_{S^+} v_3 \cos\alpha_3 \, d\sigma \underset{(17.7.7)}{=} \iint_{\Omega_{xy}} v_3 \cos\alpha_3 \sec\gamma \, dxdy = \iint_{\Omega_{xy}} v_3[x, y, f^+(x,y)] \, dxdy$$

and

$$\iint_{S^-} v_3 \cos\alpha_3 \, d\sigma = \iint_{\Omega_{xy}} v_3 \cos\alpha_3 \sec\gamma \, dxdy = -\iint_{\Omega_{xy}} v_3[x, y, f^-(x,y)] \, dxdy.$$

It follows that

$$\begin{aligned}
\iint_S v_3 \cos\alpha_3 \, d\sigma &= \iint_{S^+} v_3 \cos\alpha_3 \, d\sigma + \iint_{S^-} v_3 \cos\alpha_3 \, d\sigma \\
&= \iint_{\Omega xy} (v_3[x, y, f^+(x,y)] - v_3[x, y, f^-(x,y)]) \, dxdy \\
&= \iint_{\Omega xy} \left(\int_{f^-(x,y)}^{f^+(x,y)} \frac{\partial v_3}{\partial z}(x, y, z) \, dz \right) dxdy \\
&= \iiint_T \frac{\partial v_3}{\partial z}(x, y, z) \, dxdydz.
\end{aligned}$$

This confirms Equation (3). Equation (2) can be confirmed by projection onto the xz-plane; Equation (1) can be confirmed by projection onto the yz-plane. □

Divergence as Outward Flux per Unit Volume

Choose a point P and surround it by a closed ball N_ϵ of radius ϵ. According to the divergence theorem,

$$\iiint_{N_\epsilon} (\nabla \cdot \mathbf{v}) \, dxdydz = \text{flux of } \mathbf{v} \text{ out of } N_\epsilon.$$

Thus $\quad$ (average divergence of $\mathbf{v}$ on N_ϵ) (volume of N_ϵ) = flux of $\mathbf{v}$ out of N_ϵ

and $$\text{average divergence of } \mathbf{v} \text{ on } N_\epsilon = \frac{\text{flux of } \mathbf{v} \text{ out of } N_\epsilon}{\text{volume of } N_\epsilon}.$$

Taking the limit of both sides as ϵ shrinks to 0, we have

$$\text{divergence of } \mathbf{v} \text{ at } P = \lim_{\epsilon \to 0^+} \frac{\text{flux of } \mathbf{v} \text{ out of } N_\epsilon}{\text{volume of } N_\epsilon}.$$

In this sense *divergence is outward flux per unit volume.*

Think of $\mathbf{v}$ as the velocity of a fluid. As suggested in Section 17.8, negative divergence at P signals an accumulation of fluid near P:

$$\nabla \cdot \mathbf{v} < 0 \text{ at } P \Longrightarrow \text{flux out of } N_\epsilon < 0 \Longrightarrow \text{net flow into } N_\epsilon.$$

Positive divergence at P signals a flow of liquid away from P:

$$\nabla \cdot \mathbf{v} > 0 \text{ at } P \Longrightarrow \text{flux out of } N_\epsilon > 0 \Longrightarrow \text{net flow out of } N_\epsilon.$$

Points at which the divergence is negative are called *sinks;* points at which the divergence is positive are called *sources*. If the divergence of $\mathbf{v}$ is 0 throughout, then the flow has no sinks and no sources and $\mathbf{v}$ is called *solenoidal* (see Exercises 17.8).

Solids Bounded by Two or More Closed Surfaces

The divergence theorem, stated for solids bounded by a single closed oriented surface, can be extended to solids bounded by several closed surfaces. Suppose, for example, that we start with a solid bounded by a closed oriented surface S_1 and extract from the interior of that solid a solid bounded by a closed oriented surface S_2. The remaining solid, call it T and see Figure 17.9.2, has a boundary S that consists of two pieces: an outer piece S_1 and an inner piece S_2. The key here is to note that the *outer* normal for T points *out* of S_1 but *into* S_2. The divergence theorem can be proved for T by slicing T into two pieces T_1 and T_2 as in Figure 17.9.3 and applying the divergence theorem to each piece:

$$\iiint_{T_1} (\nabla \cdot \mathbf{v})\, dxdydz = \iint_{\text{bdry of } T_1} (\mathbf{v} \cdot \mathbf{n})\, d\sigma,$$

$$\iiint_{T_2} (\nabla \cdot \mathbf{v})\, dxdydz = \iint_{\text{bdry of } T_2} (\mathbf{v} \cdot \mathbf{n})\, d\sigma,$$

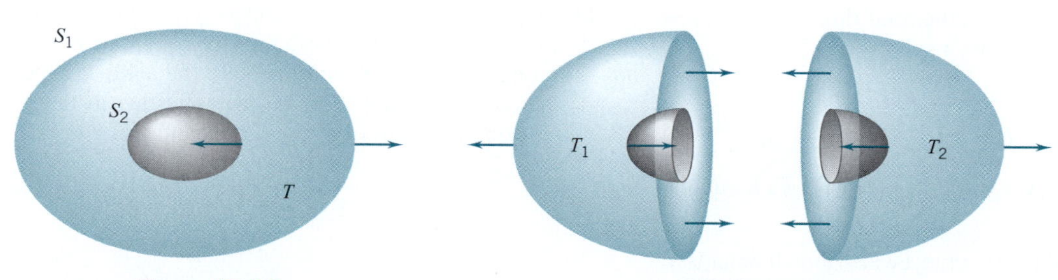

Figure 17.9.2

Figure 17.9.3

The triple integrals over T_1 and T_2 add up to the triple integral over T. When the surface integrals are added together, the integrals along the common cut cancel (because the normals are in opposite directions), and therefore only the integrals over S_1 and S_2 remain. Thus the surface integrals add up to the surface integral over $S = S_1 \cup S_2$ and the divergence theorem still holds:

$$\iiint_T (\nabla \cdot \mathbf{v})\, dxdydz = \iint_S (\mathbf{v} \cdot \mathbf{n})\, d\sigma.$$

EXERCISES 17.9

Calculate the flux $\mathbf{v}$ out of the unit ball $x^2 + y^2 + z^2 \le 1$ by applying the divergence theorem.

1. $\mathbf{v}(x,y,z) = x\,\mathbf{i} + y\,\mathbf{j} + z\,\mathbf{k}$.

2. $\mathbf{v}(x,y,z) = (1-x)\,\mathbf{i} + (2-y)\,\mathbf{j} + (3-z)\,\mathbf{k}$.

3. $\mathbf{v}(x,y,z) = x^2\,\mathbf{i} + y^2\,\mathbf{j} + z^2\,\mathbf{k}$.

4. $\mathbf{v}(x,y,z) = (1-x^2)\,\mathbf{i} + y^2\,\mathbf{j} + z\,\mathbf{k}$.

Verify the divergence theorem on the unit cube $0 \le x \le 1$, $0 \le y \le 1, 0 \le z \le 1$ for the following vector fields.

5. $\mathbf{v}(x,y,z) = x\,\mathbf{i} + y\,\mathbf{j} + z\,\mathbf{k}$.

6. $\mathbf{v}(x,y,z) = xy\,\mathbf{i} + yz\,\mathbf{j} + xz\,\mathbf{k}$.

7. $\mathbf{v}(x,y,z) = x^2\,\mathbf{i} - xz\,\mathbf{j} + z^2\,\mathbf{k}$.

8. $\mathbf{v}(x,y,z) = x\,\mathbf{i} + xy\,\mathbf{j} + xyz\,\mathbf{k}$.

Use the divergence theorem to find the total flux out of the given solid.

9. $\mathbf{v}(x,y,z) = x\,\mathbf{i} + 2y^2\,\mathbf{j} + 3z^2\,\mathbf{k};\quad x^2+y^2 \le 9,\quad 0 \le z \le 1.$

10. $\mathbf{v}(x,y,z) = xy\,\mathbf{i} + yz\,\mathbf{j} + xz\,\mathbf{k};\quad 0 \le x \le 1,$
$0 \le y \le 1-x,\quad 0 \le z \le 1-x-y.$

11. $\mathbf{v}(x,y,z) = x^2\,\mathbf{i} + xy\,\mathbf{j} - 2xz\,\mathbf{k};\quad 0 \le x \le 1,$
$0 \le y \le 1-x,\quad 0 \le z \le 1-x-y.$

12. $\mathbf{v}(x,y,z) = (2xy+2z)\,\mathbf{i} + (y^2+1)\,\mathbf{j} - (x+y)\,\mathbf{k};$
$0 \le x \le 4,\quad 0 \le y \le 4-x,\quad 0 \le z \le 4-x-y.$

13. $\mathbf{v}(x,y,z) = x^2\,\mathbf{i} + y^2\,\mathbf{j} + z^2\,\mathbf{k}$; the cylinder
$x^2+y^2 \le 4,\quad 0 \le z \le 4$, including the top and base.

14. $\mathbf{v}(x,y,z) = 2x\,\mathbf{i} + xy\,\mathbf{j} + xz\,\mathbf{k}$; the ball $x^2+y^2+z^2 \le 4$.

In Exercises 15 and 16, calculate the total flux of $\mathbf{v}(x,y,z) = 2xy\,\mathbf{i} + y^2\,\mathbf{j} + 3yz\,\mathbf{k}$ out of the given solid.

15. The ball: $x^2+y^2+z^2 \le a^2$.

16. The cube: $0 \le x \le a,\quad 0 \le y \le a,\quad 0 \le z \le a$.

17. What is the flux of $\mathbf{v}(x,y,z) = Ax\,\mathbf{i} + By\,\mathbf{j} + Cz\,\mathbf{k}$ out of a solid of volume V?

18. Let T be a basic solid with a piecewise-smooth boundary. Show that if f is harmonic on T (defined in Exercises 17.8), then the flux of ∇f out of T is zero.

19. Let S be a closed smooth surface with continuous unit normal $\mathbf{n} = \mathbf{n}(x,y,z)$. Show that

$$\iint_S \mathbf{n}\, d\sigma = \left(\iint_S n_1\, d\sigma\right)\mathbf{i} + \left(\iint_S n_2\, d\sigma\right)\mathbf{j} + \left(\iint_S n_3\, d\sigma\right)\mathbf{k} = \mathbf{0}.$$

20. Let T be a solid with a piecewise-smooth boundary S and let $\mathbf{n}$ be the outer unit normal.

(a) Verify the identity $\nabla \cdot (f\nabla f) = ||\nabla f||^2 + f(\nabla^2 f)$ and show that, if f is harmonic on T, then

$$\iint_S (f f'_{\mathbf{n}})\, d\sigma = \iiint_T ||\nabla f||^2\, dxdydz$$

where $f'_{\mathbf{n}}$ is the directional derivative $\nabla f \cdot \mathbf{n}$.

(b) Show that, if g is continuously differentiable on T, then

$$\iint_S (g f'_{\mathbf{n}})\, d\sigma = \iiint_T [(\nabla g \cdot \nabla f) + g(\nabla^2 f)]\, dxdydz.$$

21. Let T be a solid with a piecewise-smooth boundary. Show that if f and g have continuous second partials, then the flux of $\nabla f \times \nabla g$ out of T is zero.

22. Let T be a solid with a piecewise-smooth boundary S. Express the volume of T as a surface integral over S.

23. Suppose that a solid T (boundary S, outer unit normal $\mathbf{n}$) is immersed in a fluid. The fluid exerts a pressure $p = p(x,y,z)$ at each point of S, and therefore the solid T experiences a force. The total force on the solid due to the pressure distribution is given by the surface integral

$$\mathbf{F} = -\iint_S p\,\mathbf{n}\, d\sigma.$$

(The formula says that the force on the solid is the average pressure against S times the area of S.) Now choose a coordinate system with the z-axis vertical and assume that the fluid fills a region of space to the level $z = c$. The depth of a point (x, y, z) is then $c - z$ and we have $p(x, y, z) = \rho(c - z)$, where ρ is the weight density of the fluid (the weight per unit volume). Apply the divergence theorem to each component of $\mathbf{F}$ to show that $\mathbf{F} = W\mathbf{k}$, where W is the weight of the fluid displaced by the solid. We call this the *buoyant force* on the solid. (This shows that the object is not pushed from side to side by the pressure and verifies the *principle of Archimedes*: that the buoyant force on an object at rest in a fluid equals the weight of the fluid displaced.)

24. If $\mathbf{F}$ is a force applied at the tip of a radius vector $\mathbf{r}$, then the *torque*, or twisting strength, of $\mathbf{F}$ about the origin is given by the cross product $\boldsymbol{\tau} = \mathbf{r} \times \mathbf{F}$. From physics we learn that the total torque on the solid T of Exercise 23 is given by the formula

$$\boldsymbol{\tau}_{\text{Tot}} = -\iint_S [\mathbf{r} \times \rho(c - z)\,\mathbf{n}]\,d\sigma.$$

Use the divergence theorem to find the components of $\boldsymbol{\tau}_{\text{Tot}}$ and show that $\boldsymbol{\tau}_{\text{Tot}} = \bar{\mathbf{r}} \times \mathbf{F}$ where $\mathbf{F}$ is the buoyant force $W\mathbf{k}$ and $\bar{\mathbf{r}}$ is the centroid of T. (This indicates that for calculating the twisting effect of the buoyant force we can view this force as being applied at the centroid. This is very important in ship design. Imagine, for example, a totally submerged submarine. While the buoyant force acts upward through the centroid of the submarine, gravity acts downward through the center of mass. Suppose the submarine should tilt a bit to one side as depicted in the figure. If the centroid lies above the center of mass, then the buoyant force acts to restore the submarine to an upright position. If, however, the centroid lies below the center of mass, then disaster. Once the submarine has tilted a bit, the buoyant force will make it tilt further. This kind of

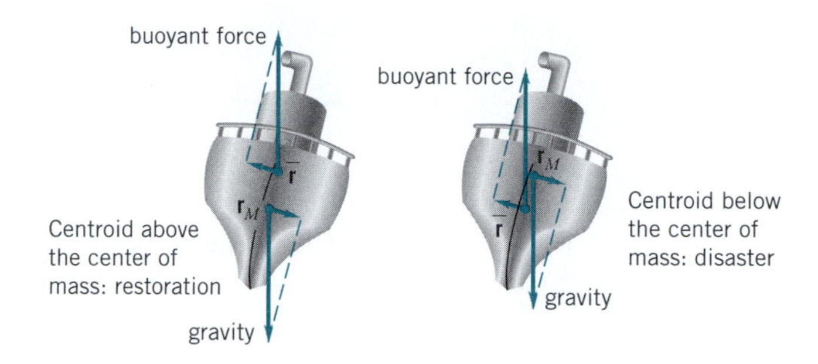

analysis also applies to surface ships, but in this case the buoyant force acts upward through *the centroid of the portion of the ship that is submerged*. This point is called the *center of flotation*. One must design and load a ship to keep the center of flotation above the center of mass. Putting a lot of heavy cargo on the deck, for instance, tends to raise the center of mass and destabilize the ship.)

■ PROJECT 17.9 Static Charges

Consider a point charge q somewhere in space. This charge creates around itself an *electric field* $\mathbf{E}$, which in turn exerts an electric force on every other nearby charge. If we center our coordinate system at q, then the electric field at the point $\mathbf{r}$ can be written

$$\mathbf{E}(\mathbf{r}) = q\frac{\mathbf{r}}{r^3}.$$

This result is found experimentally. Note that this field has the same form as a gravitational field.

Problem 1. Show that $\nabla \cdot \mathbf{E} = 0$ for all $\mathbf{r} \neq \mathbf{0}$.]

We are interested in the flux of $\mathbf{E}$ out of a closed surface S. We assume that S does not pass through q.

Problem 2. Suppose that the charge q is outside S. Then $\mathbf{E}$ is continuously differentiable on the region T bounded by S. Use the divergence theorem to show that

$$\text{flux of } \mathbf{E} \text{ out of } S = 0.$$

If q is inside S, then the divergence theorem does not apply to T directly because $\mathbf{E}$ is not differentiable on all of T. We can circumvent this difficulty by surrounding q by a small sphere S_a of radius a and applying the divergence theorem to the region T' bounded on the outside by S and on the inside by S_a. See the figure.

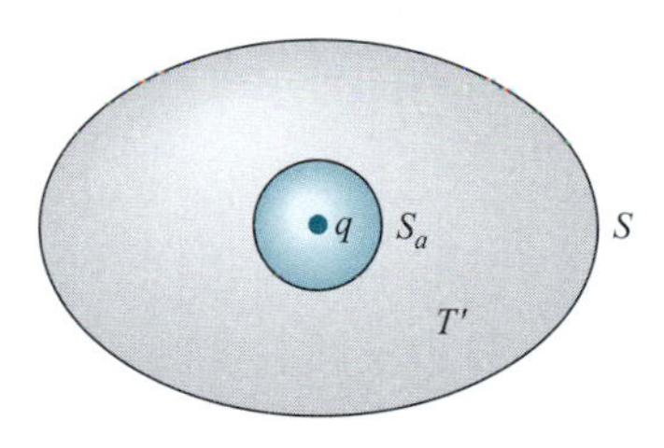

Since $\mathbf{E}$ is continuously differentiable on T',

$$\iiint_T (\nabla \cdot \mathbf{E})\,dxdydz = \text{flux of } \mathbf{E} \text{ out of } S + \text{flux of } \mathbf{E} \text{ into } S_a$$
$$= \text{flux of } \mathbf{E} \text{ out of } S - \text{flux of } \mathbf{E} \text{ out of } S_a.$$

Since $\nabla \cdot \mathbf{E} = 0$ on T', it follows that

$$\text{flux of } \mathbf{E} \text{ out of } S = \text{flux of } \mathbf{E} \text{ out of } S_a.$$

Problem 3. Show that: flux out of $S_a = 4\pi q$.

Thus, if $\mathbf{E}$ is the electric field of a point charge q and S is a closed surface that does not pass through q, then

$$\text{flux of } \mathbf{E} \text{ out of } S = \begin{cases} 0 & \text{if } q \text{ is outside } S \\ 4\pi q & \text{if } q \text{ is inside } S. \end{cases}$$

■ 17.10 STOKES'S THEOREM

We return to Green's theorem

$$\iint_{\Omega} \left(\frac{\partial Q}{\partial x} - \frac{\partial P}{\partial y}\right) dxdy = \oint P\,dx + Q\,dy.$$

Setting $\mathbf{v} = P\,\mathbf{i} + Q\,\mathbf{j} + R\,\mathbf{k}$, we have

$$(\nabla \times \mathbf{v}) \cdot \mathbf{k} = \begin{vmatrix} \mathbf{i} & \mathbf{j} & \mathbf{k} \\ \dfrac{\partial}{\partial x} & \dfrac{\partial}{\partial y} & \dfrac{\partial}{\partial z} \\ P & Q & R \end{vmatrix} \cdot \mathbf{k} = \frac{\partial Q}{\partial x} - \frac{\partial P}{\partial y}.$$

Thus in terms of $\mathbf{v}$, Green's theorem can be written

$$\iint_{\Omega} [(\nabla \times \mathbf{v}) \cdot \mathbf{k}]\,dxdy = \oint_C \mathbf{v}(\mathbf{r}) \cdot \mathbf{dr}.$$

Since any plane can be coordinatized as the xy-plane, this result can be phrased as follows: Let S be a flat surface in space bounded by a Jordan curve C. If $\mathbf{v}$ is continuously differentiable on S, then

$$\iint_{S} [(\nabla \times \mathbf{v}) \cdot \mathbf{n}]\,d\sigma = \oint_C \mathbf{v}(\mathbf{r}) \cdot \mathbf{dr},$$

where $\mathbf{n}$ is a unit normal for S and the line integral is taken in the *positive sense*, meaning in the direction of the unit tangent $\mathbf{T}$ for which $\mathbf{T} \times \mathbf{n}$ points away from the surface. See Figure 17.10.1. (An observer marching along C with the same orientation as $\mathbf{n}$ keeps the surface to his left.) The symbol $\oint_C$ denotes this line integral.

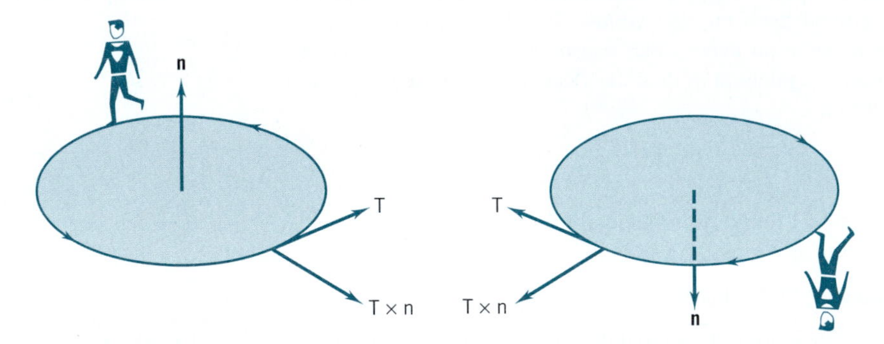

Figure 17.10.1

Figure 17.10.2 shows a *polyhedral surface* S bounded by a closed polygonal path C. The surface S consists of a finite number of flat faces $S_1, \ldots, S_N$ with polygonal boundaries $C_1, \ldots, C_N$ and unit normals $\mathbf{n}_1, \ldots, \mathbf{n}_N$. We choose these unit normals in a consistent manner; that is, they emanate from the same side of the surface. Now let $\mathbf{n} = \mathbf{n}(x, y, z)$ be a vector function of norm 1 which is $\mathbf{n}_1$ on S_1, $\mathbf{n}_2$ on S_2, $\mathbf{n}_3$ on S_3, etc. It is immaterial how $\mathbf{n}$ is defined on the line segments that join the different faces. Suppose now that $\mathbf{v} = \mathbf{v}(x, y, z)$ is a vector function continuously differentiable on an open set that contains S. Then

$$\iint_{S} [(\nabla \times \mathbf{v}) \cdot \mathbf{n}]\,d\sigma = \sum_{i=1}^{N} \iint_{S_i} [(\nabla \times \mathbf{v}) \cdot \mathbf{n}_i]\,d\sigma = \sum_{i=1}^{N} \oint_{C_i} \mathbf{v}(\mathbf{r}) \cdot \mathbf{dr},$$

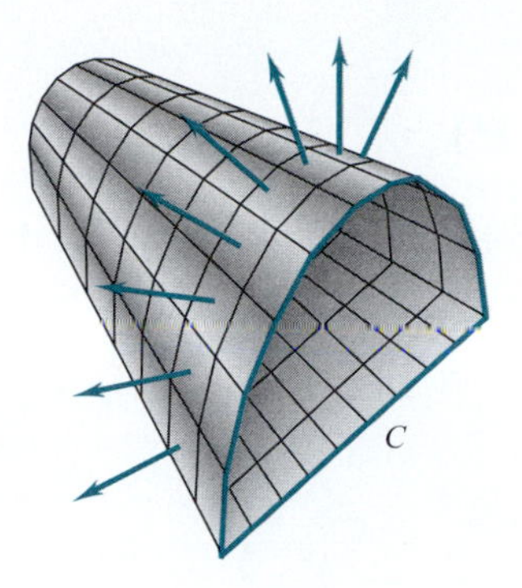

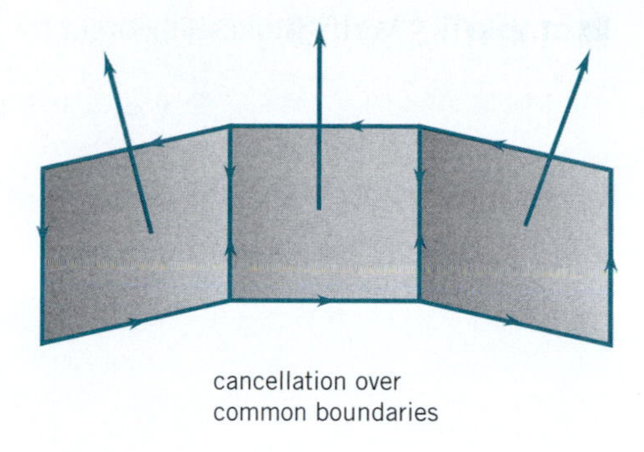

Figure 17.10.2

the integral over C_i being taken in the positive sense with respect to $\mathbf{n}_i$. Now when we add these line integrals, we find that all the line segments that make up the C_i but are not part of C are traversed twice and in opposite directions. (See the figure.) Thus these line segments contribute nothing to the sum of the line integrals and we are left with the integral around C. It follows that for a polyhedral surface S with boundary C

$$\iint_S [(\nabla \times \mathbf{v}) \cdot \mathbf{n}]\, d\sigma = \oint_C \mathbf{v}(\mathbf{r}) \cdot \mathbf{dr}.$$

This result can be extended to smooth oriented surfaces with smooth bounding curves (see Figure 17.10.3) by approximating these configurations with polyhedral configurations of the type considered and using a limit process. In an admittedly informal way we have arrived at Stokes's theorem.

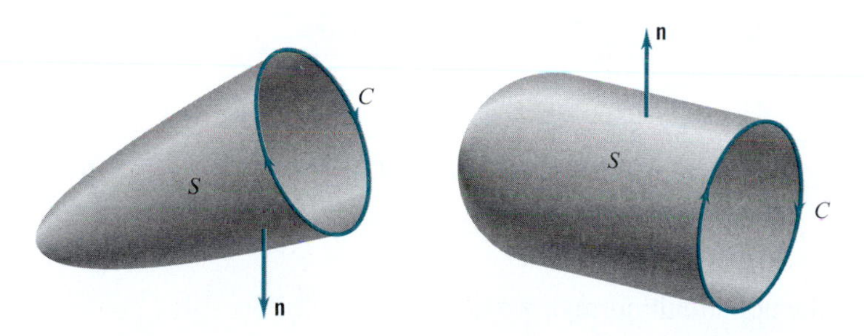

Figure 17.10.3

THEOREM 17.10.1 STOKES'S THEOREM †

Let S be a smooth oriented surface with a smooth bounding curve C. If $\mathbf{v} = \mathbf{v}(x, y, z)$ is a continuously differentiable vector field on an open set that contains S, then

$$\iint_S [(\nabla \times \mathbf{v}) \cdot \mathbf{n}]\, d\sigma = \oint_C \mathbf{v}(\mathbf{r}) \cdot \mathbf{dr},$$

where $\mathbf{n} = \mathbf{n}(x, y, z)$ is a unit normal that varies continuously on S and the line integral is taken in the positive sense with respect to $\mathbf{n}$.

† The result was announced publicly for the first time by George Gabriel Stokes (1819–1903), an Irish mathematician and physicist who, like Green, was a Cambridge professor.

Example 1 Verify Stokes's theorem for

$$\mathbf{v} = -3y\,\mathbf{i} + 3x\,\mathbf{j} + z^4\,\mathbf{k},$$

taking S as the portion of the ellipsoid $2x^2 + 2y^2 + z^2 = 1$ that lies above the plane $z = 1/\sqrt{2}$ (Figure 17.10.4).

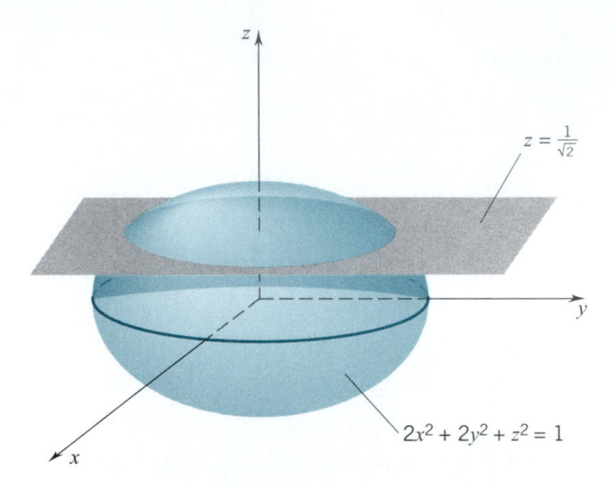

Figure 17.10.4

SOLUTION A little algebra shows that S is the graph of

$$f(x,y) = \sqrt{1 - 2(x^2 + y^2)}$$

with (x, y) restricted to the disc $\Omega: x^2 + y^2 \leq \frac{1}{4}$. Now

$$\nabla \times \mathbf{v} = \begin{vmatrix} \mathbf{i} & \mathbf{j} & \mathbf{k} \\ \dfrac{\partial}{\partial x} & \dfrac{\partial}{\partial y} & \dfrac{\partial}{\partial z} \\ -3y & 3x & z^4 \end{vmatrix} = \left[\frac{\partial}{\partial x}(3x) + \frac{\partial}{\partial y}(3y)\right]\mathbf{k} = 6\,\mathbf{k}.$$

Taking $\mathbf{n}$ as the upper unit normal, we have

$$\begin{aligned} \iint_S [(\nabla \times \mathbf{v}) \cdot \mathbf{n}]\, d\sigma &= \iint_S (6\mathbf{k} \cdot \mathbf{n})\, d\sigma \\ &= \iint_\Omega (-(0)f_x - (0)f_y + 6)\, dxdy \quad \text{(17.7.9)} \\ &= \iint_\Omega 6\, dxdy = 6\,(\text{area of } \Omega) = 6(\tfrac{1}{4}\pi) = \tfrac{3}{2}\pi. \end{aligned}$$

The bounding curve C is the set of all (x, y, z) with $x^2 + y^2 = \frac{1}{4}$ and $z = 1/\sqrt{2}$. We can parametrize C by setting

$$\mathbf{r}(u) = \tfrac{1}{2}\cos u\,\mathbf{i} + \tfrac{1}{2}\sin u\,\mathbf{j} + \frac{1}{\sqrt{2}}\mathbf{k}, \quad u \in [0, 2\pi].$$

Since $\mathbf{n}$ is the upper unit normal, this parametrization gives C in the positive sense.

Thus

$$\begin{aligned}\oint_C \mathbf{v}(\mathbf{r})\cdot d\mathbf{r} &= \int_0^{2\pi} (-\tfrac{3}{2}\sin u\,\mathbf{i} + \tfrac{3}{2}\cos u\,\mathbf{j} + \tfrac{1}{4}\mathbf{k})\cdot(-\tfrac{1}{2}\sin u\,\mathbf{i} + \tfrac{1}{2}\cos u\,\mathbf{j})\,du \\ &= \int_0^{2\pi} (\tfrac{3}{4}\sin^2 u + \tfrac{3}{4}\cos^2 u)\,du = \int_0^{2\pi} \tfrac{3}{4}\,du = \tfrac{3}{2}\pi.\end{aligned}$$

This is the value we obtained for the surface integral. □

Example 2 Verify Stokes's theorem for

$$\mathbf{v} = z^2\,\mathbf{i} - 2x\,\mathbf{j} + y^3\,\mathbf{k},$$

taking S as the upper half of the unit sphere $x^2 + y^2 + z^2 = 1$.

SOLUTION We use the upper unit normal $\mathbf{n} = x\,\mathbf{i} + y\,\mathbf{j} + z\,\mathbf{k}$. Now

$$\nabla\times\mathbf{v} = \begin{vmatrix} \mathbf{i} & \mathbf{j} & \mathbf{k} \\ \dfrac{\partial}{\partial x} & \dfrac{\partial}{\partial y} & \dfrac{\partial}{\partial z} \\ z^2 & -2x & y^3 \end{vmatrix} = 3y^2\,\mathbf{i} + 2z\,\mathbf{j} - 2\,\mathbf{k}.$$

Therefore

$$\begin{aligned}\iint_S [(\nabla\times\mathbf{v})\cdot\mathbf{n}]\,d\sigma &= \iint_S [(3y^2\,\mathbf{i} + 2z\,\mathbf{j} - 2\,\mathbf{k})\cdot(x\,\mathbf{i} + y\,\mathbf{j} + z\,\mathbf{k})]\,d\sigma \\ &= \iint_S (3xy^2 + 2yz - 2z)\,d\sigma \\ &= \iint_S 3xy^2\,d\sigma + \iint_S 2yz\,d\sigma - \iint_S 2z\,d\sigma.\end{aligned}$$

The first integral is zero because S is symmetric about the yz-plane and the integrand is odd with respect to x. The second integral is zero because S is symmetric about the xz-plane and the integrand is odd with respect to y. Thus

$$\iint_S [(\nabla\times\mathbf{v})\cdot\mathbf{n}]\,d\sigma = -\iint_S 2z\,d\sigma = -2\bar{z}(\text{area of } S) = -2(\tfrac{1}{2})2\pi = -2\pi.$$

Exercise 17, Section 17.7 ↑

This is also the value of the integral along the bounding base circle taken in the positive sense: $\mathbf{r}(u) = \cos u\,\mathbf{i} + \sin u\,\mathbf{j}$, $u \in [0, 2\pi]$, and

$$\begin{aligned}\oint_C \mathbf{v}(\mathbf{r})\cdot d\mathbf{r} &= \oint_C z^2\,dx - 2x\,dy = -2\oint_C x\,dy \\ &= -2\int_0^{2\pi} \cos^2 u\,du = -2\pi.\end{aligned}$$ □

Earlier you saw that the curl of a gradient is zero. Using Stokes's theorem we can prove a partial converse.

(17.10.2) If a vector field $\mathbf{v} = \mathbf{v}(x, y, z)$ is continuously differentiable on an open convex† set U and $\nabla\times\mathbf{v} = \mathbf{0}$ on all of U, then $\mathbf{v}$ is the gradient of some scalar field ϕ defined on U.

† *A* set U is said to be *convex* 0 if for each pair of points $p, q \in U$, the line segment $\overline{pq}$ lies entirely in U.

PROOF Choose a point $\mathbf{a}$ in U, and for each point $\mathbf{x}$ in U, define

$$\phi(\mathbf{x}) = \int_{\mathbf{a}}^{\mathbf{x}} \mathbf{v}(\mathbf{r}) \cdot \mathbf{dr}.$$

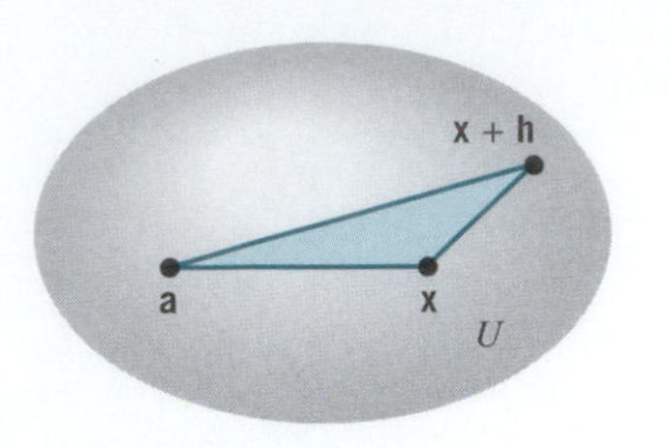

Figure 17.10.5

(This is the line integral from $\mathbf{a}$ to $\mathbf{x}$ taken along the line segment that joins these two points. We know that this line segment lies in U because U is convex.)

Since U is open, $\mathbf{x} + \mathbf{h}$ is in U for all $\mathbf{h}$ sufficiently small. Assume then that $\mathbf{h}$ is sufficiently small for $\mathbf{x} + \mathbf{h}$ to be in U. Since U is convex, the triangular region with vertices at $\mathbf{a}, \mathbf{x}, \mathbf{x} + \mathbf{h}$ lies in U. (See Figure 17.10.5.) Since $\nabla \times \mathbf{v} = \mathbf{0}$ on U, we can conclude from Stokes' theorem that

$$\int_{\mathbf{a}}^{\mathbf{x}} \mathbf{v}(\mathbf{r}) \cdot \mathbf{dr} + \int_{\mathbf{x}}^{\mathbf{x}+\mathbf{h}} \mathbf{v}(\mathbf{r}) \cdot \mathbf{dr} + \int_{\mathbf{x}+\mathbf{h}}^{\mathbf{a}} \mathbf{v}(\mathbf{r}) \cdot \mathbf{dr} = 0.$$

Therefore

$$\begin{aligned} \int_{\mathbf{x}}^{\mathbf{x}+\mathbf{h}} \mathbf{v}(\mathbf{r}) \cdot \mathbf{dr} &= -\int_{\mathbf{x}+\mathbf{h}}^{\mathbf{a}} \mathbf{v}(\mathbf{r}) \cdot \mathbf{dr} - \int_{\mathbf{a}}^{\mathbf{x}} \mathbf{v}(\mathbf{r}) \cdot \mathbf{dr} \\ &= \int_{\mathbf{a}}^{\mathbf{x}+\mathbf{h}} \mathbf{v}(\mathbf{r}) \cdot \mathbf{dr} - \int_{\mathbf{a}}^{\mathbf{x}} \mathbf{v}(\mathbf{r}) \cdot \mathbf{dr}. \end{aligned}$$

By our definition of ϕ, we have

$$\phi(\mathbf{x} + \mathbf{h}) - \phi(\mathbf{x}) = \int_{\mathbf{x}}^{\mathbf{x}+\mathbf{h}} \mathbf{v}(\mathbf{r}) \cdot \mathbf{dr}.$$

We can parametrize the line segment from $\mathbf{x}$ to $\mathbf{x} + \mathbf{h}$ by $\mathbf{r}(u) = \mathbf{x} + u\mathbf{h}$ with $u \in [0, 1]$. Therefore

$$\begin{aligned} \phi(\mathbf{x} + \mathbf{h}) - \phi(\mathbf{x}) &= \int_0^1 [\mathbf{v}(\mathbf{r}(u)) \cdot \mathbf{r}'(u)]\, du \\ &= \int_0^1 [\mathbf{v}(\mathbf{r}(u)) \cdot \mathbf{h}]\, du \\ \text{Theorem 5.8.1} \longrightarrow \quad &= \mathbf{v}(\mathbf{r}(u_0)) \cdot \mathbf{h} \qquad \text{for some } u_0 \text{ in } [0, 1] \\ = \mathbf{v}(\mathbf{x} + u_0\mathbf{h}) \cdot \mathbf{h} &= \mathbf{v}(\mathbf{x}) \cdot \mathbf{h} + [\mathbf{v}(\mathbf{x} + u_0\mathbf{h}) - \mathbf{v}(\mathbf{x})] \cdot \mathbf{h}. \end{aligned}$$

The fact that $\mathbf{v} = \nabla\phi$ follows from observing that $[\mathbf{v}(\mathbf{x} + u_o\mathbf{h}) - \mathbf{v}(\mathbf{x})] \cdot \mathbf{h}$ is $o(\mathbf{h})$:

$$\begin{aligned} \frac{|[\mathbf{v}(\mathbf{x} + u_0\,\mathbf{h}) - \mathbf{v}(\mathbf{x})] \cdot \mathbf{h}|}{||\mathbf{h}||} &\le \frac{||\mathbf{v}(\mathbf{x} + u_0\mathbf{h}) - \mathbf{v}\,(\mathbf{x})||\;||\mathbf{h}||}{||\mathbf{h}||} \\ &= ||\mathbf{v}(\mathbf{x} + u_0\mathbf{h}) - \mathbf{v}(\mathbf{x})|| \to 0 \end{aligned}$$

as $\mathbf{h} \to \mathbf{0}$. ❑

The Normal Component of $\nabla \times$ v as Circulation per Unit Area; Irrotational Flow

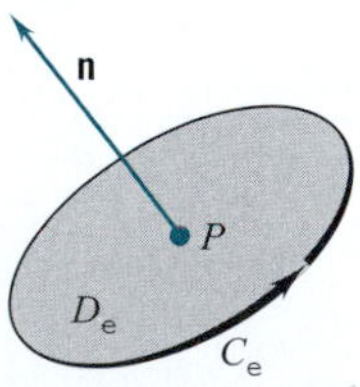

Figure 17.10.6

Interpret $\mathbf{v} = \mathbf{v}(x, y, z)$ as the velocity of a fluid flow. In Section 17.8 we stated that $\nabla \times \mathbf{v}$ measures the rotational tendency of the fluid. Now we can be more precise.

Take a point P within the flow and choose a unit vector $\mathbf{n}$. Let D_ϵ be the ϵ-disc that is centered at P and is perpendicular to $\mathbf{n}$. Let C_ϵ be the circular boundary of D_ϵ directed in the positive sense with respect to $\mathbf{n}$. (See Figure 17.10.6.) By Stokes's theorem,

$$\iint_{D_\epsilon} [(\nabla \times \mathbf{v}) \cdot \mathbf{n}]\, d\sigma = \oint_{C_\epsilon} \mathbf{v}(\mathbf{r}) \cdot \mathbf{dr}.$$

The line integral on the right is called *the circulation* of $\mathbf{v}$ around C_ϵ. Thus we can say that

$$\begin{pmatrix} \text{the average } \mathbf{n}\text{-component of} \\ \nabla \times \mathbf{v} \text{ on } D_\epsilon \end{pmatrix} (\text{ the area of } D_\epsilon) = \text{ the circulation of } \mathbf{v} \text{ around } C_\epsilon.$$

It follows that

$$\text{the average } \mathbf{n}\text{-component of } \nabla \times \mathbf{v} \text{ on } D_\epsilon = \frac{\text{the circulation of } \mathbf{v} \text{ around } C_\epsilon}{\text{the area of } D_\epsilon}.$$

Taking the limit as ϵ shrinks to 0, you can see that

$$\text{the } \mathbf{n}\text{-component of } \nabla \times \mathbf{v} \text{ at } P = \lim_{\epsilon \to 0^-} \frac{\text{the circulation of } \mathbf{v} \text{ around } C_\epsilon}{\text{the area of } D_\epsilon}.$$

At each point P the component of $\nabla \times \mathbf{v}$ in any direction $\mathbf{n}$ is the circulation of $\mathbf{v}$ per unit area in the plane normal to $\mathbf{n}$. If $\nabla \times \mathbf{v} = \mathbf{0}$ identically, the fluid has no rotational tendency, and the flow is called irrotational.

Remark Flux and circulation apply to vector fields where no material substance is flowing. Electromagnetic phenomena result from the action and interaction of two vector fields: the electric field E and the magnetic field B. The four fundamental laws of electromagnetism can be stated as equations that give the flux and circulation of these two fields.

EXERCISES 17.10

In Exercises 1–4, let S be the upper half of the unit sphere $x^2 + y^2 + z^2 = 1$ and take $\mathbf{n}$ as the upper unit normal. Find

$$\iint_S [(\nabla \times \mathbf{v}) \cdot \mathbf{n}]\, d\sigma$$

(a) By direct calculation. (b) By Stokes's theorem.

1. $\mathbf{v}(x, y, z) = x\,\mathbf{i} + y\,\mathbf{j} + z\,\mathbf{k}$.

2. $\mathbf{v}(x, y, z) = y\,\mathbf{i} - x\,\mathbf{j} + z\,\mathbf{k}$.

3. $\mathbf{v}(x, y, z) = z^2\,\mathbf{i} + 2x\,\mathbf{j} - y^3\,\mathbf{k}$.

4. $\mathbf{v}(x, y, z) = 6xz\,\mathbf{i} - x^2\,\mathbf{j} - 3y^2\,\mathbf{k}$.

In Exercises 5–7, let S be the triangular surface with vertices $(2, 0, 0), (0, 2, 0), (0, 0, 2)$ and take $\mathbf{n}$ as the upper unit normal. Find

$$\iint_S [(\nabla \times \mathbf{v}) \cdot \mathbf{n}]\, d\sigma$$

(a) By direct calculation. (b) By Stokes's theorem.

5. $\mathbf{v}(x, y, z) = 2z\,\mathbf{i} - y\,\mathbf{j} + x\,\mathbf{k}$.

6. $\mathbf{v}(x, y, z) = (x^2 + y^2)\,\mathbf{i} + y^2\,\mathbf{j} + (x^2 + z^2)\,\mathbf{k}$.

7. $\mathbf{v}(x, y, z) = x^4\,\mathbf{i} + xy\,\mathbf{j} + z^4\,\mathbf{k}$.

8. Show that if $\mathbf{v} = \mathbf{v}(x, y, z)$ is continuously differentiable everywhere and its curl is identically zero, then

$$\int_C \mathbf{v}(\mathbf{r}) \cdot \mathbf{dr} = 0 \quad \text{for every smooth closed curve } C.$$

9. Let $\mathbf{v} = y\,\mathbf{i} + z\,\mathbf{j} + x^2y^2\,\mathbf{k}$ and let S be the surface $z = x^2 + y^2$ from $z = 0$ to $z = 4$. Calculate the flux of $\nabla \times \mathbf{v}$ in the direction of the lower unit normal $\mathbf{n}$.

10. Let $\mathbf{v} = \frac{1}{2}y\,\mathbf{i} + 2xz\,\mathbf{j} - 3x\,\mathbf{k}$ and let S be the surface $y = 1 - (x^2 + z^2)$ from $y = -8$ to $y = 1$. Calculate the flux of $\nabla \times \mathbf{v}$ in the direction of the unit normal $\mathbf{n}$ with positive $\mathbf{j}$-component.

11. Let $\mathbf{v} = 2x\,\mathbf{i} + 2y\,\mathbf{j} + x^2y^2z^2\,\mathbf{k}$ and let S be the lower half of the ellipsoid.

$$\frac{x^2}{4} + \frac{y^2}{9} + \frac{z^2}{27} = 1.$$

Calculate the flux of $\nabla \times \mathbf{v}$ in the direction of the upper unit normal $\mathbf{n}$.

12. Let S be a smooth closed surface and let $\mathbf{v} = \mathbf{v}(x, y, z)$ be a vector field with second partials continuous on an open convex set that contains S. Show that

$$\iint_S [(\nabla \times \mathbf{v}) \cdot \mathbf{n}]\, d\sigma = 0$$

where $\mathbf{n} = \mathbf{n}(x, y, z)$ is the outer unit normal.

13. The upper half of the ellipsoid $\frac{1}{2}x^2 + \frac{1}{2}y^2 + z^2 = 1$ intersects the cylinder $x^2 + y^2 - y = 0$ in a curve C. Calculate the circulation of $\mathbf{v} = y^3\,\mathbf{i} + (xy + 3xy^2)\,\mathbf{j} + z^4\,\mathbf{k}$ around C by using Stokes's theorem.

14. The sphere $x^2 + y^2 + z^2 = a^2$ intersects the plane $x + 2y + z = 0$ in a curve C. Calculate the circulation of $\mathbf{v} = 2y\,\mathbf{i} - z\,\mathbf{j} + 2x\,\mathbf{k}$ about C by using Stokes's theorem.

15. The paraboloid $z = x^2 + y^2$ intersects the plane $z = y$ in a curve C. Calculate the circulation of $\mathbf{v} = 2z\,\mathbf{i} + x\,\mathbf{j} + y\,\mathbf{k}$ about C using Stokes's theorem.

16. The cylinder $x^2 + y^2 = b^2$ intersects the plane $y + z = a^2$ in a curve C. Assume $a^2 \geq b > 0$. Calculate the circulation of $\mathbf{v} = xy\,\mathbf{i} + yz\,\mathbf{j} + xz\,\mathbf{k}$ about C using Stokes's theorem.

17. Let S be a smooth oriented surface with a smooth bounding curve C and let $\mathbf{a}$ be a fixed vector. Show that

$$\iint_S (2\mathbf{a} \cdot \mathbf{n})\, d\sigma = \oint_C (\mathbf{a} \times \mathbf{r}) \cdot d\mathbf{r}$$

where $\mathbf{n} = \mathbf{n}(x, y, z)$ is a unit normal vector that varies continuously over S and the line integral is taken in the positive sense with respect to $\mathbf{n}$.

18. Let S be a smooth oriented surface with smooth bounding curve C. Show that, if ϕ and ψ are sufficiently differentiable scalar fields, then

$$\iint_S [(\nabla\phi \times \nabla\psi) \cdot \mathbf{n}]\, d\sigma = \oint_C (\phi\nabla\psi) \cdot d\mathbf{r}$$

where $\mathbf{n} = \mathbf{n}(x, y, z)$ is a unit normal that varies continuously on S and the line integral is taken in the positive sense with respect to $\mathbf{n}$.

19. Let S be a smooth oriented surface with a smooth plane bounding curve C and let $\mathbf{v} = \mathbf{v}(x, y, z)$ be a vector field with second partials continuous on an open convex set that contains S. If S does not cross the plane of C, then Stokes's theorem for S follows readily from the divergence theorem and Green's theorem. Carry out the argument.

20. Our derivation of Stokes's theorem was admittedly nonrigorous. The following version of Stokes's theorem lends itself more readily to rigorous proof.

Give a detailed proof of the theorem. HINT: Set $\mathbf{v} = v_1\mathbf{i} + v_2\mathbf{j} + v_3\mathbf{k}$. Then

$$\iint_S [(\nabla \times \mathbf{v}) \cdot \mathbf{n}]\, d\sigma = \iint_S [(\nabla \times v_1\mathbf{i}) \cdot \mathbf{n}]\, d\sigma + \iint_S [(\nabla \times v_2\mathbf{j}) \cdot \mathbf{n}]\, d\sigma + \iint_S [(\nabla \times v_3\mathbf{k}) \cdot \mathbf{n}]\, d\sigma$$

and $$\oint_C \mathbf{v}(\mathbf{r}) \cdot d\mathbf{r} = \int_C v_1\, dx + \int_C v_2\, dy + \int_C v_3\, dz.$$

Show that $$\iint_S [(\nabla \times v_1\mathbf{i}) \cdot \mathbf{n}]\, d\sigma = \int_C v_1\, dx$$

by showing that both integrals can be written

$$\iint_\Gamma \left[\frac{\partial v_1}{\partial u}\frac{\partial x}{\partial v} - \frac{\partial v_1}{\partial v}\frac{\partial x}{\partial u}\right] du\, dv.$$

A similar argument (no need to carry it out) equates the integrals for v_2 and v_3 and proves the theorem.

THEOREM 17.10.3

Let Γ be a Jordan region in the uv-plane with a piecewise-smooth boundary C_Γ given in a counterclockwise sense by a pair of functions $u = u(t)$, $v = v(t)$ with $t \in [a, b]$. Let $\mathbf{R}\,(u, v) = x(u, v)\,\mathbf{i} + y(u, v)\,\mathbf{j} + z(u, v)\,\mathbf{k}$ be a vector function with continuous second partials on Γ. Assume that $\mathbf{R}$ is one-to-one on Γ and that the fundamental vector product $\mathbf{N} = \mathbf{R}'_u \times \mathbf{R}'_v$ is never zero. The surface $S\colon \mathbf{R} = \mathbf{R}\,(u, v), (u, v) \in \Gamma$ is a smooth oriented surface bounded by the oriented space curve $C : \mathbf{r}\,(t) = \mathbf{R}\,(u(t), v(t))$, $t \in [a, b]$. If $\mathbf{v} = \mathbf{v}\,(x, y, z)$ is a vector field continuously differentiable on S, then

$$\iint_S [(\nabla \times \mathbf{v}) \cdot \mathbf{n}]\, d\sigma = \oint_C \mathbf{v}(\mathbf{r}) \cdot d\mathbf{r}$$

where $\mathbf{n}$ is the unit normal in the direction of the fundamental vector product.

■ CHAPTER HIGHLIGHTS

17.1 Line Integrals

smooth curve (p. 1020) work along a curve (p. 1021)
line integral (p. 1021)
invariance under changes in parameter (p. 1022)
piecewise-smooth curve (p. 1023)
work as a line integral (p. 1025)
circulation of a vector field (p. 1027)

17.2 The Fundamental Theorem for Line Integrals

gradient field (p. 1028)
the fundamental theorem (p. 1028)

17.3 Work-Energy Formula; Conservation of Mechanical Energy

work-energy formula (p. 1032)
conservative force fields (p. 1033)
potential energy functions (p. 1033)
conservation of mechanical energy (p. 1033)
differences in potential energy (p. 1034)
escape velocity (p. 1035)
equipotential surfaces (p. 1035)

17.4 Another Notation for Line Integrals; Line Integrals with Respect to Arc Length

$$\int_C P\,dx + Q\,dy + R\,dz \qquad \int_C f(\mathbf{r})\,ds$$

mass, center of mass, moment of inertia of a wire (p. 1038)

17.5 Green's Theorem

Jordan region (p. 1041)
Green's theorem for a Jordan region (p. 1041)
area enclosed by a plane curve (p. 1045)
Green's theorem for regions bounded by several Jordan curves (p. 1047)
a justification of the Jacobian area formula (supplement) (p. 1050)

17.6 Parametrized Surfaces; Surface Area

parametrized surface: $\mathbf{r} = \mathbf{r}(u, v),\ (u, v) \in \Omega$.
fundamental vector product: $\mathbf{N} = \mathbf{r}'_u \times \mathbf{r}'_v$
$\mathbf{N}$ as a normal (p. 1055)
area of a parametrized surface (p. 1056)
area of a surface $z = f(x, y,)$ (p. 1060)
upper unit normal (p. 1061)
secant area formula (p. 1062)

17.7 Surface Integrals

mass of a material surface (p. 1064)
surface integral (p. 1064)
average value on a surface, weighted average (p. 1066)
centroid (p. 1066)
center of mass of a material surface (p. 1066)
moment of inertia (p. 1067)
oriented surface (p. 1069)
flux of a vector field (p. 1069)
closed, piecewise-smooth surface (p. 1071)
one-sided surface, Möbius band (p. 1073)

17.8 The Vector Differential Operator ∇

the operator del: $\nabla = \dfrac{\partial}{\partial x}\mathbf{i} + \dfrac{\partial}{\partial y}\mathbf{j} + \dfrac{\partial}{\partial z}\mathbf{k}$

gradient of f : ∇f
divergence of $\mathbf{v}$: $\nabla \cdot \mathbf{v}$ (p. 1074)
curl of $\mathbf{v}$: $\nabla \times \mathbf{v}$ (p. 1074)
curl of a gradient is zero (p. 1076)
divergence of a curl is zero (p. 1077)

$$\nabla \cdot (f\,\mathbf{v}) = (\nabla f) \cdot \mathbf{v} + f\,(\nabla \cdot \mathbf{v})$$
$$\nabla \times (f\mathbf{v}) = (\nabla f) \times \mathbf{v} + f(\nabla \times \mathbf{v})$$
$$\nabla \cdot (r^n\,\mathbf{r}) = (n + 3)\,r^n$$
$$\nabla \times (r^n\mathbf{r}) = 0$$

Laplacian: $\nabla^2 f = \nabla \cdot \nabla f = \dfrac{\partial^2 f}{\partial x^2} + \dfrac{\partial^2 f}{\partial y^2} + \dfrac{\partial^2 f}{\partial z^2}$

17.9 The Divergence Theorem

divergence theorem for a solid bounded by a single closed surface (p. 1080)
divergence as outward flux per unit volume (p. 1082)
sinks and sources (p. 1083)
divergence theorem for solids bounded by two or more closed surfaces (p. 1083)
buoyant force, principle of Archimedes (p. 1085)
an application to static charges (p. 1085)

17.10 Stokes's Theorem

positive sense along a curve bounding an open surface (p. 1086)
Stokes's theorem (p. 1087)
conditions under which $\nabla \times \mathbf{v} = \mathbf{0}$ implies $\mathbf{v} = \nabla f$
normal component of $\nabla \times \mathbf{v}$ as circulation per unit area
irrotational flow (p. 1090)

Evaluated the repeated integral.

1. $\int_0^1 \int_y^{\sqrt{y}} xy^2\, dx\, dy$

2. $\int_0^1 \int_{-y}^{y} e^{x+y}\, dx\, dy$

3. $\int_0^1 \int_x^{3x} 2ye^{x^3}\, dy\, dx$

4. $\int_{-1}^{2} \int_0^4 \int_0^1 xyz\, dx\, dy\, dz$

5. $\int_0^2 \int_0^{2-3x} \int_0^{x+y} x\, dz\, dy\, dx$

6. $\int_0^{\pi/2} \int_z^{\pi/2} \int_0^{\sin z} 3x^2 \sin y\, dx\, dy\, dz$

7. $\int_{-\pi/2}^{0} \int_0^{2\sin\theta} \int_0^{r^2} r^2 \cos\theta\, dz\, dr\, d\theta$

8. $\int_{-\pi/6}^{\pi/2} \int_0^{\pi/2} \int_0^1 \rho^3 \sin\varphi \cos\varphi\, d\rho\, d\theta\, d\varphi$

Sketch the region Ω corresponding to the repeated integral. Then change the order of integration and evaluate.

9. $\int_0^1 \int_0^{\sqrt{1-x^2}} \frac{1}{\sqrt{1-y^2}}\, dy\, dx$

10. $\int_0^2 \int_{\frac{1}{2}x}^{1} \cos(y^2)\, dy\, dx$

Evaluate the integral.

11. $\iint_\Omega xy\, dxdy;\ \Omega: 0 \le x^2 + y^2 \le 1, x, y > 0$

12. $\iint_\Omega (x-y) dxdy;$ Ω is the region between the curves $y^2 = 3x$ and $y^2 = 4 - x$

13. $\iint_\Omega (x^2 - xy) dxdy;$ Ω is the region between the curves $y = x$ and $y = 3x - x^2$

14. $\iiint_T z\, dxdydz;$ T is the region bounded by the planes $x = 0,\ y = 0,\ z = 0,\ y + z = 1,\ x + z = 1$

15. $\iiint_T xy\, dxdydz;$ T is the region in the first octant bounded by the coordinate planes and the hemisphere $z = \sqrt{4 - x^2 - y^2}$

16. $\iiint_T (x^2 + 2z)\, dxdydz;$ T is the region bounded by the planes $z = 0$ and $y + z = 4$, and the cylinder $y = x^2$

17. Find the volume of the solid in the first octant bounded by $z = x^2 + y^2$, $x + y = 1$, and the coordinate planes.

18. Find the volume of the solid in the first octant bounded by the cylinder $x^2 + y^2 = 9$ and the planes $z = y$ and $z = 0$.

Use polar coordinates to evaluate the integral.

19. $\int_0^2 \int_0^{\sqrt{4-y^2}} e^{\sqrt{x^2+y^2}}\, dx\, dy$

20. $\int_{-1}^{1} \int_0^{\sqrt{1-x^2}} \tan^{-1}(y/x)\, dy\, dx$

Find the mass and center of mass of the plate that occupies the region Ω and has the density function λ.

21. Ω is the region between the curves $y = x$ and $y = \sqrt{x}$; $\lambda(x, y) = 2x$.

22. Ω is the upper half of the cardioid $r = 2(1 + \cos\theta)$; λ is the distance to the pole.

23. A homogeneous plate is in the shape of an isosceles triangle with base b and height h.

(a) Find the centroid of the plate.

(b) Find the moment of inertia about the base.

(c) Find the moment of inertia about the axis of symmetry of the triangle.

Use a triple integral to find the volume of the solid. Use either rectangular, cylindrical or spherical coordinates, whichever is appropriate.

24. The solid bounded above by the paraboloid $z = 4(x^2 + y^2)$, below by the plane $z = -1$, and on the sides by the cylinders $y = x^2$ and $y = x$.

25. The solid bounded above by the elliptic paraboloid $z = 12 - x^2 - 2y^2$ and below by the elliptic paraboloid $z = 2x^2 + y^2$. HINT: Find the curve of intersection of the two surfaces.

26. The solid in the first octant bounded by the plane $2x + y + z = 2$ and inside the cylinder $y^2 + z^2 = 1$.

27. The solid bounded above by the sphere $x^2 + y^2 + z^2 = 4$ and below by the plane $z = 1$.

28. The solid bounded above by the sphere $x^2 + y^2 + z^2 = 4$, on the sides by the cylinder $x^2 + y^2 = 1$, and below by the x, y-plane.

29. A homogeneous solid is in the shape of the region in the first octant bounded by the cylinders $x^2 + z^2 = 1$ and $y^2 + z^2 = 1$.

(a) Find the centroid of the solid. (b) Find I_z.

30. Find the Jacobian of the transformation.

(a) $x = u^2 - v^2,\ y = 2uv$

(b) $x = u^2 + 2vw,\ y = v^2 + 2uw,\ z = uvw$

Evaluate the integral using the suggested transformation.

31. $\iint_\Omega \sin\left(\frac{y - x}{y + x}\right) dxdy;$ Ω is the region in the first quadrant bounded by the line $x + y = 1$ and $x + y = 2$. Let $x = \frac{1}{2}(v - u), u = \frac{1}{2}(v + u)$.

32. $\iiint_T dxdydz;$ T is the solid that lies between the paraboloids $z = x^2 + y^2$ and $z = 4x^2 + 4y^2$, and between the planes $z = 1$ and $z = 4$. Let $x = (r/u)\cos\theta$, $y = (r/u)\sin\theta, z = r^2$.

33. Integrate $\mathbf{h}(x, y) = x^2 y\,\mathbf{i} - xy\,\mathbf{j}$ over the path:

(a) The straight-line segment from $(0, 0)$ to $(1, 1)$.

(b) $\mathbf{r}(u) = u^2\,\mathbf{i} + u^3\,\mathbf{j}, 0 \le u \le 1$.

34. Integrate $\mathbf{h}(x,y) = (2xy^2 + x)\,\mathbf{i} + (2x^2y - 1)\,\mathbf{j}$ over the path:

(a) The straight-line segment from $(-1,2)$ to $(2,4)$.

(b) The polygonal path from $(-1,2)$ to $(0,0)$ to $(2,4)$.

(c) The straight-line segment from $(-1,2)$ to $(0,0)$, then the parabolic path $y = x^2$ from $(0,0)$ to $(2,4)$.

35. Integrate $\mathbf{h}(x,y,z) = \sin y\,\mathbf{i} + xe^{xy}\,\mathbf{j} + \sin z\,\mathbf{k}$ over the curve $\mathbf{r}(u) = u^2\,\mathbf{i} + u\,\mathbf{j} + u^3\,\mathbf{k}, 0 \le u \le 3$.

36. Integrate $\mathbf{h}(x,y,z) = x^2\,\mathbf{i} + xy\,\mathbf{j} + z^2\,\mathbf{k}$ over the curve $\mathbf{r}(u) = \cos u\,\mathbf{i} + \sin u\,\mathbf{j} + u^2\,\mathbf{k}, 0 \le u \le \pi/2$.

37. The force exerted by a charged particle at the origin on a charged particle at a point $x, y, z)$ is

$$\mathbf{F}(x,y,z) = \frac{C(x\,\mathbf{i} + y\,\mathbf{j} + z\,\mathbf{k})}{\sqrt{x^2+y^2+z^2}}, C \text{ constant}$$

Find the work done by $\mathbf{F}$ applied to a particle that moves in a straight line from $(1,0,0)$ to $(3,0,4)$.

38. An object is moving through a force field $\mathbf{F}$ in such a way that its velocity vector at each point (x,y,z) is orthogonal to $\mathbf{F}(x,y,z)$. Show that the work done by $\mathbf{F}$ on the object is 0.

Verify that the vector field $\mathbf{h}$ is a gradient field. Then find the line integral over C in two ways: (a) calculate $\int_C \mathbf{h}(\mathbf{r}) \cdot \mathbf{dr}$ directly; (b) the fundamental theorem of line integrals.

39. $\mathbf{h}(x,y) = (ye^{xy} + 2x)\,\mathbf{i} + (xe^{xy} - 2y)\,\mathbf{j}$; C: $\mathbf{r}(u) = u\,\mathbf{i} + u^2\,\mathbf{j}$; $0 \le u \le 2$.

40. $\mathbf{h}(x,y,z = 4x^3y^3z^2\,\mathbf{i} + 3x^4y^2z^2\,\mathbf{j} + 2x^4y^3z\,\mathbf{k}$. C: $\mathbf{r}(u) = u\,\mathbf{i} + u^2\,\mathbf{j} + u^3\,\mathbf{k}, 0 \le u \le 1$.

41. Evaluate $\int_C y^2\,dx + (x^2 - xy)\,dy$ along the path from $(0,0)$ to $(2,8)$.

(a) C : the straight-line segment.

(b) C : the polygonal path $(0,0)$ to $(2,0)$ to $(2,8)$.

(c) C : the cubic $y = x^3$.

42. Evaluate $\int_C z\,dx + x\,dy + y\,dz$ along the circular helix C : $\mathbf{r}(u) = a\cos u\,\mathbf{i} + a\sin u\,\mathbf{j} + u\,\mathbf{k}, 0 \le u \le 2\pi$.

43. Evaluate $\int_C ye^{xy}\,dx + \cos x\,dy + (xy/z)\,dz$ along the twisted cubic C : $\mathbf{r}(u) = u\,\mathbf{i} + u^2\,\mathbf{j} + u^3\,\mathbf{k}, 0 \le u \le 2$.

44. A wire winds around the z-axis in the shape of the circular helix C : $\mathbf{r}(u) = \sin u\,\mathbf{i} - \cos u\,\mathbf{j} + 4u\,\mathbf{k}, \pi \le u \le 2\pi$. The mass density at the point $P(x,y,z)$ on the wire is equal to the square of the distance from P to the x-axis. Find the mass of the wire.

Verify Green's theorem (a) by calculating the line integral over the simple closed curve C directly, and (b) by calculating the corresponding double integral of the region enclosed by C.

45. $\oint_C xy^2dx - x^2ydy$; C: the closed curve in the first quadrant determined by the parabolas $y = x^2$ and $y^2 = x$.

46. $\oint_C (x^2+y^2)dx + (x^2 - y^2)dy$; C : the unit circle $x^2 + y^2 = 1$.

Evaluate the line integral using Green's theorem.

47. $\oint_C (x - 2y^2)\,dx + 2xy\,dy$; C : the rectangle with vertices $(0,0), (2,0), (2,1), (0,1)$.

48. $\oint_C xy\,dx + (\frac{1}{2}x^2 + xy)dy$; C : the upper semi-ellipse $x^2 + 4y^2 = 1$ together with the interval $[-1, 1]$

49. $\oint_C \ln(x^2+y^2)dx + \ln(x^2+y^2)dy$; C : the boundary of the semi-circular ring determined by $x^2 + y^2 = 1$ and $x^2 + y^2 = 4, y \ge 0$.

50. $\oint_C y^2dx$; C: the cardioid $r = 1 + \sin\theta$.

Find the area of the surface.

51. The part of the sphere $x^2 + y^2 + z^2 = 4$ that is inside the cylinder $x^2 + y^2 = 2x$.

52. The part of the plane $x + y + 2z = 4$ that lies inside the cylinder $x^2 + y^2 = 4$.

53. The part of the cone $z = \sqrt{x^2+y^2}$ that lies between the planes $z = 0$ and $z = 3$.

Evaluate the surface integral.

54. $\iint_S xz\,d\sigma$; S is the first octant part of the plane $x + y + z = 1$

55. $\iint_S (x^2 + y^2 + z^2)d\sigma$; S is the cylinder $y^2 + z^2 = 4$, $0 \le x \le 2$, together with the circular discs at each end.

56. Verify the divergence theorem on the surface $S : y^2 + z^2 = 1, 0 \le x \le 4$, for the vector field $\mathbf{v}(x,y,z) = (x+z)\,\mathbf{i} + (y+z)\,\mathbf{j} + (x+z)\,\mathbf{k}$.

57. Calculate the total flux of $\mathbf{v}(x,y,z) = 2x\,\mathbf{i} + xz\,\mathbf{j} + z^2\,\mathbf{k}$ out of the solid bounded by the paraboloid $z = 9 - x^2 - y^2$ and the xy-plane.

58. Calculate the total flux of $\mathbf{v}(x,y,z) = x^2\,\mathbf{i} - xz\,\mathbf{j} + z^2\,\mathbf{k}$ out of the cube $0 \le x \le a, 0 \le y \le a, 0 \le z \le a, a > 0$.

59. Let S be the hemisphere $z = \sqrt{4 - x^2 - y^2}$ and take $\mathbf{n}$ as the upper unit normal. Let $\mathbf{v}(x,y,z) = z\,\mathbf{i} + x\,\mathbf{j} + y\,\mathbf{k}$ and find

$$\iint_S [(\nabla \times \mathbf{v}) \cdot \mathbf{n}]\,d\sigma$$

(a) by direct calculation; and (b) by Stokes's theorem.

60. Let S be that part of the paraboloid $z = 9 - x^2 - y^2$ for which $z \ge 0$ and take $\mathbf{n}$ as the upper unit normal. Let $\mathbf{v}(x,y,z) = z^3\,\mathbf{i} + x\,\mathbf{j} + y^2\,\mathbf{k}$ and find

$$\iint_S [(\nabla \times \mathbf{v}) \cdot \mathbf{n}]\,d\sigma$$

(a) by direct calculation; and (b) by Stokes's theorem.

CHAPTER

18

ELEMENTARY DIFFERENTIAL EQUATIONS

18.1 INTRODUCTION; REVIEW OF EQUATIONS ALREADY CONSIDERED

Suppose that y is an unknown function of x. If we know y', then we can recover y, up to an additive constant, by integrating y' :

$$y(x) = \int y'(x)\,dx + C = F(x) + C. \qquad \text{(where } F \text{ is an antiderivative for } y'\text{)}$$

If we don't know y' but we know y'', then we can still recover y, but now we have to integrate twice and there will be two arbitrary constants:

$$y'(x) = \int y''(x)\,dx + C_1 = G(x) + C_1 \qquad \text{(where } G \text{ is an antiderivative for } y''\text{)}$$

$$y(x) = \int [G(x) + C_1]\,dx + C_2$$

$$= \int G(x)\,dx + \int C_1\,dx + C_2$$

$$= F(x) + C_1 x + C_2. \qquad \text{(where } F \text{ is an antiderivative for } G\text{)}$$

It often happens in mathematics and in applications to other fields that we don't know y', we don't know y'', we don't know any of the derivatives explicitly, but we do have an equation that relates y to one or more of its derivatives. An equation that relates an unknown function to one or more of its derivatives is called an *ordinary differential equation*.† The recovery of a function y from a differential equation is called *solving* the differential equation. From some differential equations we can recover y completely and describe its action explicitly as a function of x (up to one or more arbitrary constants).

† In contrast to partial differential equations, which arise in the study of functions of several variables.

More frequently, we cannot recover y completely, but we can obtain an equation in x and y which is satisfied by y and involves none of the derivatives of y. Such an equation, carrying one or more arbitrary constants, represents a family of curves called *integral curves* (*solution curves*) of the differential equation.

Finally, there are differential equations from which we can extract no explicit solutions and no integral curves. Such equations have to be approached by other methods.

The *order* of a differential equation is the order of the highest derivative that appears in the equation. The equations used by scientists to model the processes of nature are almost all equations of order one or order two. We can be thankful for that. It would be rather burdensome to have to solve differential equations of order, say, 100.

This is not your first encounter with differential equations. They have appeared off and on throughout the text. We review here the equations with which you are expected to be familiar before you begin the study of this chapter.

First-Order Linear Differential Equations

An equation of the form

$$y' + p(x)\,y = q(x) \tag{1}$$

where p and q are known functions of x defined and continuous on some interval I is called a *first-order linear differential equation with continuous coefficients*. Such an equation is intimately related to the exponential function.

To solve the equation, we multiply it by $e^{H(x)}$ where $H(x)$ is an antiderivative for p. That makes the left side a derivative with respect to x and, as shown in Section *8. 8, eventually leads to the conclusion that

$$y(x) = e^{-H(x)}\left\{\int e^{H(x)} q(x)\,dx + C\right\} \tag{2}$$

where C is an arbitrary constant.

There is no point memorizing this expression. We are led to it naturally once we have multiplied by $e^{H(x)}$. What's important here is that (2) gives us explicitly *every* function that satisfies the differential equation. Because of this, we call (2) the *general solution* of the differential equation (1). By assigning particular values to the constant C, we obtain what are called *particular solutions* of the differential equation.

Separable First-Order Equations

In Section *8. 9 we discussed *separable equations*. These are equations that can be written in the form

$$p(x) + q(y)\,y' = 0 \tag{3}$$

with p and q continuous where defined.

Since y is assumed to be a function of x, we can write

$$p(x) + q(y(x))\,y'(x) = 0.$$

Integration with respect to x gives

$$\int p(x)\,dx + \int q(y(x))\,y'(x)\,dx = C,$$

which leads to

$$\int p(x)\,dx + \int q(y)\,dy = C.$$

We have separated the variables. If P is an antiderivative for p and Q is an antiderivative for q, we have

$$P(x) + Q(y) = C,$$

an equation that we can write as

$$F(x, y) = C. \tag{4}$$

If y is related to y' by equation (3), then y is related to x by equation (4). The curves in the xy-plane generated by (4) are *integral curves* (*solution curves*) of the differential equation. Different values of C give different integral curves.

Each integral curve is a solution of the differential equation in the sense that along the curve the numbers x, y, y' are related as prescribed by the differential equation.

In some cases equation (4) can be solved for y in terms of x, giving us functions that satisfy the differential equation. Usually equation (4) cannot be solved for y in terms of x, and we have to be satisfied with equation (4) and the integral curves that it generates. We can, however, assert the following: any function $y = y(x)$ whose graph lies on one of the integral curves satisfies the differential equation. In Section *8. 9 we showed that the integral curves of the differential equation

$$x + yy' = 0$$

are of the form

$$x^2 + y^2 = C, \qquad C \geq 0.$$

The graphs of the functions

$$y = \sqrt{1 - x^2} \qquad \text{and} \qquad y = -\sqrt{1 - x^2}$$

both lie on the integral curve

$$x^2 + y^2 = 1.$$

As you can readily verify, both functions satisfy the differential equation.

The Equation of Harmonic Motion

In the Exercises for Section 3.6 and Section 4.4 we introduced the *second-order linear differential equation*

$$\frac{d^2y}{dt^2} + \omega^2 y = 0. \qquad (\omega \text{ a constant}) \tag{5}$$

This is the equation of *simple harmonic motion*, motion during which the acceleration remains a constant negative multiple of the displacement from a point of equilibrium. This equation models the motion of a simple pendulum and the up-and-down oscillations of a mass suspended from a spring.

We cited three (completely equivalent) ways to express all the solutions of (5). Here we remind you of one of them:

$$y(t) = A\cos\omega t + B\sin\omega t$$

where A and B are arbitrary constants. Particular solutions of the differential equation are obtained by assigning numerical values to these constants.

EXERCISES 18.1

Determine whether the differential equation has the indicated functions as solutions.

1. $2y' - y = 0$; $y_1(x) = e^{x/2}$, $y_2(x) = x^2 + 2e^{x/2}$.

2. $y' + xy = x$; $y_1(x) = e^{-x^2/2}$, $y_2(x) = 1 + Ce^{-x^2/2}$, C any constant

3. $y' + y = y^2$; $y_1(x) = \dfrac{1}{e^x + 1}$, $y_2(x) = \dfrac{1}{Ce^x + 1}$, C any constant.

4. $y'' + 4y = 0$; $y_1(x) = 2\sin 2x$, $y_2(x) = 2\cos x$.

5. $y'' - 4y = 0$; $y_1(x) = e^{2x}$, $y_2(x) = C\sinh 2x$, C any constant.

6. $y'' - 2y' - 3y = 7e^{3x}$; $y_1(x) = e^{-x} + 2e^{3x}$, $y_2(x) = \frac{7}{4}xe^{3x}$.

Show that the members of the given family of functions are solutions of the differential equation. Then find a member of the family that satisfies the initial condition(s).

7. $y = Ce^{5x}$; $y' = 5y$, $y(0) = 2$.

8. $y = \dfrac{x^2}{3} + \dfrac{C}{x}$; $xy' + y = x^2$, $y(3) = 2$.

9. $y = \dfrac{1}{Ce^x + 1}$; $y' + y = y^2$, $y(1) = -1$.

10. $y = x\ln\dfrac{C}{x}$; $y' = \dfrac{y - x}{x}$, $y(2) = 4$.

11. $y = C_1x + C_2x^{1/2}$; $2x^2y'' - xy' + y = 0$, $y(4) = 1$, $y'(4) = -2$.

12. $y = C_1\sin 3x + C_2\cos 3x$; $y'' + 9y = 0$, $y(\pi/2) = y'(\pi/2) = 1$.

13. $y = C_1x^2 + C_2x^2\ln x$; $x^2y'' - 3xy' + 4y = 0$, $y(1) = 0$, $y'(1) = 1$.

14. $y = C_1 + C_2e^x + C_3e^{2x} + \frac{1}{4}x^2 + \frac{3}{4}x - xe^x$; $y''' - 3y'' + 2y' = x + e^x$, $y(0) = 1$, $y'(0) = -\frac{1}{4}$, $y''(0) = -\frac{3}{2}$.

Identify the differential equation as linear or separable and then solve the equation.

15. $y' + y = 2e^{-2x}$.

16. $y^2 + 1 = yy'\sec^2 x$.

17. $y' = \dfrac{x^2y - y}{y + 1}$.

18. $\dfrac{dy}{dx} = \dfrac{x^3 - 2y}{x}$.

19. $xy' + 2y = \dfrac{\cos x}{x}$.

20. $y' = \dfrac{\ln x}{xy + xy^3}$.

21. $yy' = 4x\sqrt{y^2 + 1}$.

22. $xy' = x^2 + 2y$.

Determine the values of r, if any, such that $y = e^{rx}$ is a solution of the given differential equation.

23. $y' + 3y = 0$.

24. $y'' - 5y' + 6y = 0$.

25. $y'' + 6y' + 9y = 0$.

26. $y''' - 3y' + 2y = 0$.

Determine values of r, if any, such that $y = x^r$ is a solution of the given differential equation.

27. $xy'' + y' = 0$.

28. $x^2y'' + xy' - y = 0$.

29. $4x^2y'' - 4xy' + 3y = 0$.

30. $x^3y''' - 2x^2y'' - 2xy' + 8y = 0$.

An nth-order differential equation together with conditions that are specified at two, or more, points on an interval I is called a *boundary-value problem*. In particular, if conditions are specified at two points, the problem is called a *two-point boundary-value problem*.

31. Each member of the family of functions $y = C_1\sin 4x + C_2\cos 4x$ is a solution of the differential equation $y'' + 16y = 0$.

(a) Find all the members of this family that satisfy the boundary conditions:

$$y(0) = 0, \quad y(\pi/2) = 0.$$

(b) Find all the members of this family that satisfy the boundary conditions:

$$y(0) = 0, \quad y(\pi/8) = 0.$$

32. For each real number r, each member of the family $y = C_1\sin rx + C_2\cos rx$ is a solution of the differential equation $y'' + r^2y = 0$.

(a) Determine the numbers r such that the two-point boundary-value problem

$$y'' + r^2y = 0, \quad y(0) = 0, \quad y(\pi) = 0$$

has a nonzero solution.

(b) Determine the numbers r such that the two-point boundary-value problem

$$y'' + r^2y = 0, \quad y(0) = 0, \quad y(\pi/2) = 0$$

has a nonzero solution.

■ 18.2 BERNOULLI EQUATIONS; HOMOGENEOUS EQUATIONS; NUMERICAL METHODS

Bernoulli Equations

A first-order equation of the form

(18.2.1) $$y' + p(x)y = q(x)y^r$$

where p and q are functions defined and continuous on some interval I, and r is a real number different from 0 and 1, is called a *Bernoulli equation*.† We exclude 0 and 1 because in each of these cases the equation is simply a linear equation.

METHOD OF SOLUTION To solve (18.2.1), we multiply the equation by y^{-r} and obtain

$$y^{-r}y' + p(x)\,y^{1-r} = q(x).$$

We can transform this equation into a linear equation by setting $v = y^{1-r}$. For then

$$v' = (1-r)\,y^{-r}y'$$

and our differential equation becomes

$$\frac{1}{1-r}v' + p(x)\,v = q(x),$$

which we can write as

$$v' + (1-r)p(x)\,v = (1-r)\,q(x).$$

This equation is linear in v, and we can solve for v by the method introduced in Section *8. 8 and reviewed in the introduction to this chapter. Replacing v by y^{1-r}, we have the integral curves in terms of x and y. ❑

Example 1 Find the integral curves of the equation $y' + 4y = 3e^{2x}y^2$.

SOLUTION The equation is a Bernoulli equation with $p(x) = 4,\ q(x) = 3e^{2x},\ r = 2$. We will solve the equation by the method just outlined. Our first step is to multiply the equation by y^{-2} (thereby excluding, at least for the moment, $y = 0$):

$$y^{-2}y' + 4y^{-1} = 3e^{2x}.$$

We set $v = y^{-1}$. Differentiation gives $v' = -y^{-2}y'$ and transforms the equation into

$$-v' + 4v = 3e^{2x},$$

which we write as

$$v' - 4v = -3e^{2x}.$$

We solve this last equation by setting $H(x) = \int(-4)\,dx = -4x$ and multiplying through by $e^{H(x)} = e^{-4x}$. This gives us

$$e^{-4x}v' - 4e^{-4x}v = -3e^{-2x},$$

which we recognize as stating that

$$\frac{d}{dx}(e^{-4x}v) = -3e^{-2x}.$$

Integration gives

$$e^{-4x}\,v = \tfrac{3}{2}e^{-2x} + C,$$

which we write as

$$v = \tfrac{3}{2}e^{2x} + Ce^{4x}.$$

† These equations were introduced by Jacob Bernoulli (1654—1705), who, along with his brother Johann, made many contributions to the development of calculus and its applications.

Replacing v by y^{-1}, we have

$$\frac{1}{y} = \frac{3}{2}e^{2x} + Ce^{4x}.$$

These are the integral curves of our Bernoulli equation. As you can verify, this family of solutions can also be written

$$y = \frac{2}{3e^{2x} + Ke^{4x}} \qquad (K = 2C). \qquad \square$$

Remark In applying our method of solution, we had to exclude $y = 0$. Note that the constant function $y = 0$ is a solution of the differential equation, and this function is not a member of the family of integral curves (there is no value that can be assigned to C to produce $y = 0$). A solution of a differential equation that is not included in the family of integral curves is called a *singular solution*. $\square$

Homogeneous Equations

A first-order differential equation

(18.2.2) $$y' = f(x, y) \qquad (f \text{ continuous})$$

is said to be *homogeneous* if

$$f(tx, ty) = f(x, y) \qquad \text{for all} \quad t \neq 0.$$

Remark The term "homogeneous" requires some explanation. As you may have noted in your study of algebra, a function $H = H(x, y)$ is said to be *homogeneous to degree n* if

$$H(tx, ty) = t^n H(x, y) \qquad \text{for all } t \neq 0.$$

Thus the $f = f(x, y)$ of the homogeneous differential equation (18.2.2) can be viewed as a function *homogeneous to degree* 0:

$$f(tx, ty) = t^0 f(x, y) = f(x, y) \qquad \text{for all } t \neq 0.$$

To construct such a function, simply divide two functions homogeneous to the same degree:

$$\frac{M(tx, ty)}{N(tx, ty)} = \frac{t^n M(x, y)}{t^n N(x, y)} = \frac{M(x, y)}{N(x, y)} \qquad \text{for all } t \neq 0. \qquad \square$$

Now back to our differential equation (18.2.2).

METHOD OF SOLUTION The first step is to write the equation in the form

(1) $$y' = g\left(\frac{y}{x}\right)$$

where g is a continuous function of only one variable. That this can be done is shown as follows: observe that for $x \neq 0$

$$f(x, y) = f\left(x, x\left(\frac{y}{x}\right)\right) = f\left(1, \frac{y}{x}\right)$$

and define

$$g\left(\frac{y}{x}\right) = f\left(1, \frac{y}{x}\right).$$

We now set $y/x = v$ and transform (1) into an equation in v and x:

$$y = vx \qquad \text{gives} \qquad y' = v + v'x$$

and

$$g\left(\frac{y}{x}\right) \qquad \text{becomes} \qquad g(v).$$

Thus equation (1) can be written

$$v + v'x = g(v),$$

which can be rearranged to give

$$\frac{1}{x} + \left[\frac{1}{v - g(v)}\right] v' = 0.$$

This equation is separable. We can solve it by the method introduced in Section *8.9 (reviewed in the introduction to this chapter) and write the resulting integral curves in terms of x and y by substituting y/x for v. ☐

Example 2 Show that the differential equation $y' = \dfrac{3x^2 + y^2}{xy}$ is homogeneous and find the integral curves.

SOLUTION The equation is homogeneous: for $t \neq 0$

$$f(tx, ty) = \frac{3(tx)^2 + (ty)^2}{(tx)(ty)} = \frac{t^2(3x^2 + y^2)}{t^2(xy)} = \frac{3x^2 + y^2}{xy} = f(x, y).$$

Setting $y/x = v$, we have

$$\frac{3x^2 + y^2}{xy} = \frac{3 + (y/x)^2}{y/x} = \frac{3 + v^2}{v},$$

and, since $y = vx$,

$$y' = v + v'x.$$

Our differential equation now reads

$$v + v'x = \frac{3 + v^2}{v}.$$

Multiplying both sides by v and simplifying, we get

$$v'xv = 3,$$

which can be rearranged to give

$$-\frac{1}{x} + \frac{1}{3} v v' = 0.$$

This is a separable equation which we can readily solve:

$$\begin{aligned}
\int -\frac{1}{x}\, dx + \frac{1}{3}\int v v' dx &= 0 \\
-\int \frac{1}{x}\, dx + \frac{1}{3}\int v\, dv &= 0 \\
-\ln|x| + \tfrac{1}{6}v^2 &= C.
\end{aligned}$$

This gives

$$v^2 = 6(\ln|x| + C).$$

Substituting y/x back in for v, we have

$$\frac{y^2}{x^2} = 6(\ln|x| + C).$$

The integral curves take the form

$$y^2 = 6x^2(\ln|x| + C). \quad \square$$

Numerical Methods

As indicated earlier, the differential equations that occur most frequently in applications are of order one or order two. We have presented techniques for solving some first-order equations; methods for solving certain second-order equations are presented in Sections 18.4 and 18.5. However, many of the differential equations that arise in the study of physical phenomena cannot be solved exactly. Therefore, we need other methods to describe solutions or to approximate them. So called *qualitative methods* which are used to determine the behavior of solutions are studied in more advanced courses on differential equations. Here we present two numerical methods for approximating solutions. We will apply these methods to first-order initial value problems: problems in which we are given a first-order differential equation and some point on the solution curve from which we can start the process.

So that you can keep track of the accuracy (or inaccuracy) of the approximations, we will work with an initial-value problem for which an exact solution is available. As you can check, the initial-value problem

(18.2.3) $$y' = x + 2y, \qquad y(0) = 1$$

has the exact solution

(18.2.4) $$y = \tfrac{1}{4}(5\,e^{2x} - 2x - 1).$$

We will use our numerical methods to estimate $y(1)$; the actual value is

$$y(1) = \tfrac{1}{4}(5e^2 - 3) \cong 8.4863.$$

The first method we consider is called the *Euler method*. Figure 18.2.1 shows the graph of a differentiable function f. Given $f(x_0)$, we can estimate $f(x_0 + h)$ by proceeding along the tangent line, the line with slope $f'(x_0)$:

$$f(x_0 + h) \cong f(x_0) + hf'(x_0).$$

Using this estimate for $f(x_0 + h)$, we can go on to estimate $f(x_0 + 2h)$ by proceeding along the line with slope $f'(x_0 + h)$. We can go on to estimate $f(x_0 + 3h)$, $f(x_0 + 4h)$, etc. We illustrate the process in Figure 18.2.2. Presumably, if h is taken small enough, the path of line segments stays fairly close to the graph of f. This way of approximating a curve by line segments is the basis of the Euler method for solving initial-value problems.

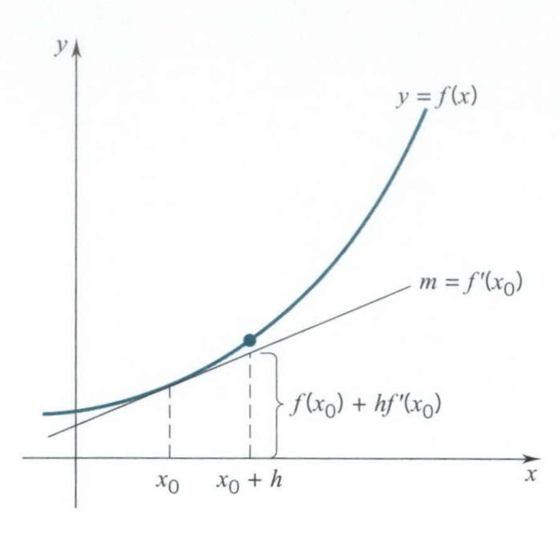

Figure 18.2.1

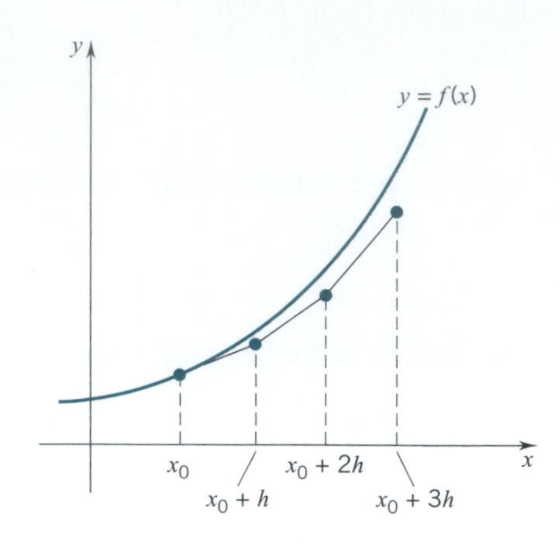

Figure 18.2.2

Consider the initial-value problem

(18.2.5)
$$y' = f(x, y), \qquad y(x_0) = y_0.$$

The point (x_0, y_0) lies on the solution curve. The point on the curve with x-coordinate $x_1 = x_0 + h$ has approximate y-coordinate

$$y_1 = y_0 + hy'(x_0) = y_0 + hf(x_0, y_0).$$

The point on the curve with x-coordinate $x_2 = x_0 + 2h$ has approximate y-coordinate

$$y_2 = y_1 + hy'(x_1) = y_1 + hf(x_1, y_1).$$

Successive steps of h units (to the right if $h > 0$ and to the left if $h < 0$) produce a succession of points near the graph of the solution curve:

(18.2.6)
$$(x_n, y_n): \quad x_n = x_0 + nh, \quad y_n = y_{n-1} + hf(x_{n-1}, y_{n-1}), \qquad (n \geq 1).$$

Example 3 Now let's return to the initial-value problem (18.2.3):

$$y' = x + 2y, \quad y(0) = 1.$$

In the notation of (18.2.5)

$$f(x, y) = x + 2y, \quad x_0 = 0, \quad y_0 = 1.$$

We want to estimate $y(1)$. Setting $h = 0.1$, we will need 10 iterations.

By (18.2.6)

$$x_n = (0.1)n \qquad \text{and} \qquad y_n = y_{n-1} + 0.1(x_{n-1} + 2y_{n-1}), \qquad 1 \leq n \leq 10.$$

The computations are given in the chart that follows. The numbers in the last column represent the actual values (rounded off to four decimal places) as calculated from the exact solution (18.2.4).

n	x_n	y_n	$y(x_n)$
0	0	1	1.0000
1	0.1	1.2000	1.2268
2	0.2	1.4500	1.5148
3	0.3	1.7600	1 .8776
4	0.4	2.1420	2.3319
5	0.5	2.6104	2.8979
6	0.6	3.1825	3.6001
7	0.7	3.8790	4.4690
8	0.8	4.7248	5.5413
9	0.9	5.7497	6.8621
10	1.0	6.9897	8.4863

The error in this estimate for $y(1)$ is about 17.6%. In this case the Euler method has not given us a very accurate estimate. Presumably we could improve on the accuracy of the estimate by using a smaller h, but as we explain below, that's not necessarily the case. □

Computational errors arise in several ways. With each iteration of the procedure we introduce the error inherent in the approximation method, and this error is compounded by the errors introduced in the previous steps. To reduce this contribution to the error, we can use smaller values of h. However, this improvement is made at the expense of increased round-off error. Each calculation produces an error in the last decimal place carried. Increasing the number of times the procedure is iterated by using smaller h increases the number of times this round-off error is made and thus tends to increase the total error.

The second numerical method that we consider is called the *Runge-Kutta method*. This is one of the oldest methods for generating numerical solutions of differential equations. Yet it remains one of the most accurate methods known. In the Euler method we estimated $y(x_n)$ from our estimate for $y(x_{n-1})$ by using the slope of y at x_{n-1}. The idea behind the Runge-Kutta method is to select a slope more representative of the derivative of y on the interval $[x_{n-1}, x_n]$. This is done by selecting intermediate points in the interval and then forming a weighted average of the slopes at these points. Details of the development of the formulas used below can be found in any text on numerical analysis or differential equations.

In the Runge-Kutta method successive points of approximation for the initial-value problem

$$(1) \qquad y' = f(x, y), \qquad y(x_0) = y_0$$

are chosen as follows:

$$(2) \qquad (x_n, y_n) \quad \text{with } x_n = x_0 + nh, \quad y_n = y_{n-1} + hK$$

where

$$\begin{aligned} K_1 &= f(x_{n-1}, y_{n-1}), \\ K_2 &= f(x_{n-1} + \tfrac{1}{2}h, y_{n-1} + \tfrac{1}{2}hK_1), \\ K_3 &= f(x_{n-1} + \tfrac{1}{2}h, y_{n-1} + \tfrac{1}{2}hK_2), \\ K_4 &= f(x_{n-1} + h, y_{n-1} + hK_3), \end{aligned}$$

and

$$K = \tfrac{1}{6}(K_1 + 2K_2 + 2K_3 + K_4).$$

Note that as a consequence of (1), the numbers K_1, K_2, K_3, K_4 give slopes, and K is a weighted average of these slopes. As you can see from (2), K gives the slope of the line segment that joins the approximation (x_{n-1}, y_{n-1}) to the approximation (x_n, y_n).

Example 4 We reconsider the initial-value problem

$$y' = x + 2y, \qquad y(0) = 1,$$

but this time we estimate $y(1)$ by the Runge-Kutta method. Again we have $f(x,y) = x + 2y$, and take $h = 0.1$.

The results of the computations (these can be verified rather quickly on a computer or programmable calculator) are tabulated below. Again, (18.2.4) was used to calculate the actual values of $y(x_n)$ (to four decimal places.)

n	x_n	y_n	$y(x_n)$
0	0	1	1.0000
1	0.1	1.2267	1.2268
2	0.2	1.5148	1.5148
3	0.3	1.8776	1.8776
4	0.4	2.3319	2.3319
5	0.5	2.8978	2.8979
6	0.6	3.6001	3.6001
7	0.7	4.4689	4.4690
8	0.8	5.5412	5.5413
9	0.9	6.8618	6.8621
10	1.0	8.4861	8.4863

The error in this approximation of $y(1)$ is about 0.002%. Runge-Kutta has given us a much better approximation than the Euler method. ❑

EXERCISES 18.2

Find the integral curves.

1. $y' + xy = xy^3$.

2. $y' + y^2(x^2 + x + 1) = y$.

3. $y' = 4y + 2e^x\sqrt{y}$.

4. $2xy\, y' = 1 + y^2$.

5. $(x-2)\, y' + y = 5(x-2)^2 y^{1/2}$.

6. $y\, y' - xy^2 + x = 0$.

Find a solution to the initial value problem.

7. $y' + xy - y^3 e^{x^2} = 0; \quad y(0) = \frac{1}{2}$.

8. $xy' + y - y^2 \ln x = 0; \quad y(1) = 1$.

9. $2x^3 y' = y(y^2 + 3x^2); \quad y(1) = 1$.

10. $y' + y \tan x - y^2 \sec^3 x = 0; \quad y(0) = 3$.

11. Show that the change of variable $\mu = \ln y$ transforms the equation

$$y' - \left(\frac{y}{x}\right) \ln y = xy$$

into a linear equation. Find the integral curves.

12. (a) Show that the change of variable indicated in Exercise 11 transforms

$$y' + yf(x) \ln y = g(x)y$$

into a first-order linear equation.

(b) Find a change of variable which transforms

$$y' \cos y + g(x) \sin y = f(x)$$

into a linear equation:

Verify that the equation is homogeneous, and find the integral curves.

13. $y' = \dfrac{x^2 + y^2}{2xy}$.

14. $y' = \dfrac{y^2}{xy + x^2}$.

15. $y' = \dfrac{x - y}{x + y}$.

16. $y' = \dfrac{x + y}{x - y}$.

17. $y' = \dfrac{x^2(e^y)^{1/x} + y^2}{xy}$.

18. $y' = \dfrac{x^2 + 3y^2}{4xy}$.

19. $y' = \dfrac{y}{x} + \sin\left(\dfrac{y}{x}\right)$.

20. $x\,dy = y\left[1 + \ln\left(\frac{y}{x}\right)\right]dx.$

Find the integral curve that-satisfies the initial condition.

21. $y' = \dfrac{y^3 - x^3}{xy^2}, \qquad y(1) = 2.$

22. $x \sin\left(\frac{y}{x}\right) dy = \left[x + y \sin\left(\frac{y}{x}\right)\right] dx, \quad y(1) = 0.$

In Exercises 23–32, solve the initial value problem (a) by the Euler method, (b) by the Runge-Kutta method. Then solve the initial-value problem and estimate the accuracy of your numerical estimate using

$$\text{relative percentage error} = \frac{y_{\text{actual}} - y_{\text{approx}}}{y_{\text{actual}}} \times 100\%.$$

23. Estimate $y(1)$ if $y' = y$ and $y(0) = 1$, setting $h = 0.2$.

24. Estimate $y(1)$ if $y' = x + y$ and $y(0) = 2$, setting $h = 0.2$.

25. Exercise 23 setting $h = 0.1$.

26. Exercise 24 setting $h = 0.1$.

27. Estimate $y(1)$ if $y' = 2x$ and $y(2) = 5$, setting $h = 0.1$.

28. Estimate $y(0)$ if $y' = 3x^2$ and $y(1) = 2$, setting $h = 0.1$.

29. Estimate $y(2)$ if $y' = 1/(2y)$ and $y(1) = 1$, setting $h = 0.1$.

30. Estimate $y(2)$ if $y' = 1/(3y^2)$ and $y(1) = 1$, setting $h = 0.1$.

31. Exercise 23 setting $h = 0.05$.

32. Exercise 24 setting $h = 0.05$.

■ PROJECT 18.2 Direction Fields

Here we introduce a geometric approach to differential equations of the form $y' = f(x, y)$ that enables us to produce sketches of solution curves without requiring us to calculate any integrals. The approach does not produce equations in x and y; it produces pictures, pictures from which we can gather useful information. Do the curves slant up or do they slant down? Are there any maxima or minima? What is the concavity of the curves? We usually do not get precise answers to such questions, but we can get a good qualitative sense of what the curves look like.

The geometric approach to which we have been alluding is based on the construction of what is called a *direction field* (a *slope field*) for the differential equation. What this means is described below.

If a solution curve for the equation $y' = f(x, y)$ passes through the point (x, y), then it does so with slope $f(x, y)$. We construct a direction field for the differential equation by selecting a grid of points (x_i, y_i), $i = 1, 2 \cdots, n$, and drawing at each point a short line segment with slope $f(x_i, y_i)$. We can then use these little line segments to sketch the solution curve for the initial-value problem

$$y' = f(x, y), \quad y(a) = b$$

by starting at the point (a, b) and following the line segments in both directions. Figure A shows a direction field for the differential equation

$$y' = x - y$$

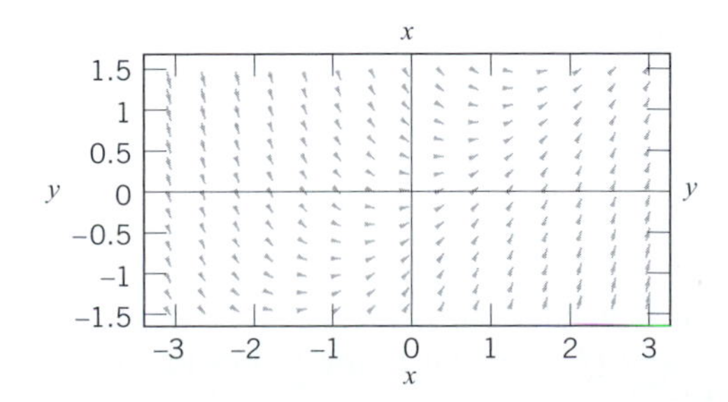

Figure A

drawn within the rectangle $R: -3 \leq x \leq 3, -1.5 \leq y \leq 1.5$ A sketch of the solution curve that satisfies the initial condition $y(0) = 0$ is shown in Figure B.

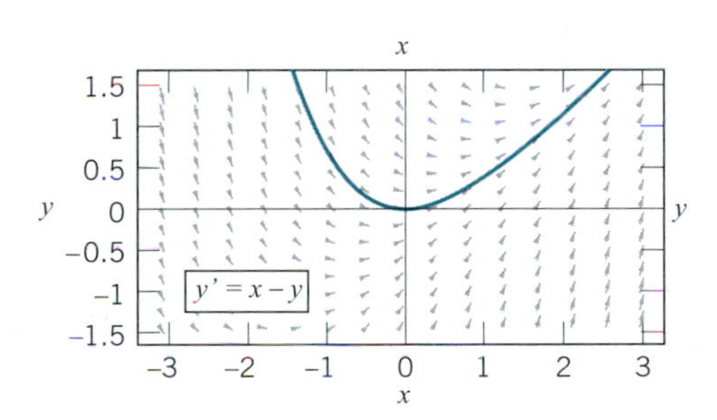

Figure B

REMARK Computer algebra systems usually include a feature for sketching direction fields.

Problem 1. In this problem we consider the initial-value problem: $y' = y$, $y(0) = 1$.

(a) Use a CAS to draw a direction field in the rectangle $R: -3 \leq x \leq 1.5, -1 \leq y \leq 3$.

(b) Use this direction field to sketch the solution curve that satisfies the initial condition. Experiment with other rectangles to obtain additional views of the solution curve.

(c) Solve the initial-value problem by other means and then compare the graph of your solution to the curve you obtained in part (b).

Problem 2. Repeat Problem 1 with $y' = x + 2y$, $y(0) = 1$ and $R: -1 \leq x \leq 2, -1 \leq y \leq 9$.

Problem 3. Repeat Problem 1 with $y' = 2xy$, $y(0) = 1$ and $R: -1.5 \leq x \leq 3, -1 \leq y \leq 8$.

Problem 4. In this problem we consider the initial-value problem: $y' = -4x/y$, $y(1) = 1$.

(a) Use a CAS to draw a direction field in the rectangle $R: -2 \leq x \leq 2, -3 \leq y \leq 3$.

(b) Use this direction field to sketch the solution curve that satisfies the initial condition. Experiment with other rectangles to obtain additional views of the solution curve.

(c) Solve the initial value problem by other means and then compare the graph of your solution to the curve you obtained in part (b).

18.3 EXACT DIFFERENTIAL EQUATIONS; INTEGRATING FACTORS

We begin with two functions $P = P(x, y)$ and $Q = Q(x, y)$, each continuously differentiable on a simply connected region Ω. The differential equation

(18.3.1) $$P(x, y) + Q(x, y)\, y' = 0$$

is said to be *exact* on Ω if

$$\frac{\partial P}{\partial y}(x, y) = \frac{\partial Q}{\partial x}(x, y) \qquad \text{for all } (x, y) \in \Omega.$$

The reason for this terminology is as follows. If the equation

$$P(x, y) + Q(x, y)\, y' = 0$$

is exact, then (by Theorem 15.9.2) the vector-valued function

$$P(x, y)\,\mathbf{i} + Q(x, y)\,\mathbf{j}$$

is "exactly" in the form of a gradient and there is a function F defined on Ω such that

$$\frac{\partial F}{\partial x} = P \qquad \text{and} \qquad \frac{\partial F}{\partial y} = Q.$$

Therefore, we can write (18.3.1) as

(1) $$\frac{\partial F}{\partial x}(x, y) + \frac{\partial F}{\partial y}(x, y)\, y' = 0.$$

Since, by the chain rule [(15.3.6)],

$$\frac{d}{dx}[F(x, y)] = \frac{\partial F}{\partial x}(x, y) + \frac{\partial F}{\partial y}(x, y)\, y',$$

equation (1) can be written

$$\frac{d}{dx}[F(x, y)] = 0.$$

Integrating with respect to x, we have

$$F(x, y) = C.$$

The integral curves of (18.3.1) are the level curves of F.

How F can be obtained from P and Q is shown in the following example. The process was explained in Section 15.9.

Example 1 The differential equation $(xy^2 - x^3) + (x^2y - y)y' = 0$ is everywhere exact: the coefficients

$$P(x,y) = xy^2 - x^3 \qquad \text{and} \qquad Q(x,y) = x^2y - y$$

are everywhere continuously differentiable, and at all points

$$\frac{\partial P}{\partial y} = 2xy = \frac{\partial Q}{\partial x}.$$

To find the integral curves,

$$F(x,y) = C,$$

We set $\qquad \dfrac{\partial F}{\partial x}(x,y) = xy^2 - x^3 \qquad$ and $\qquad \dfrac{\partial F}{\partial y}(x,y) = x^2y - y.$

Integrating $\partial F/\partial x$ with respect to x, we have

$$F(x,y) = \tfrac{1}{2}x^2y^2 - \tfrac{1}{4}x^4 + \phi(y)$$

where $\phi(y)$ is independent of x but may depend on y. Differentiation with respect to y gives

$$\frac{\partial F}{\partial y} = x^2y - \phi'(y).$$

The two equations for $\partial F/\partial y$ can be reconciled by having

$$\phi'(y) = -y$$

and setting $\qquad \phi(y) = -\tfrac{1}{2}y^2.$

The integral curves of the differential equation can be written

$$\tfrac{1}{2}x^2y^2 - \tfrac{1}{4}x^4 - \tfrac{1}{2}y^2 = C.$$

Checking: Differentiation with respect to x gives

$$\tfrac{1}{2}x^2(2yy') + xy^2 - x^3 - yy' = 0$$
$$(xy^2 - x^3) + (x^2y - y)y' = 0.$$

This is the original equation. ❑

If the equation

$$P(x,y) + Q(x,y)y' = 0, \qquad (x,y) \in \Omega,$$

is not exact on Ω, it may be possible to find a function $\mu = \mu(x,y)$ not identically zero such that the equation

$$\mu(x,y)P(x,y) + \mu(x,y)Q(x,y)y' = 0$$

is exact. If μ is never zero on Ω, then any solution of this second equation gives a solution of the first equation. We call $\mu(x,y)$ an *integrating factor.*

Example 2 Consider the differential equation

$$(*) \qquad \left(2y^2 + 3x + \frac{2}{x^2}\right) + \left(2xy - \frac{y}{x}\right) y' = 0$$

on the right half-plane $\Omega = \{(x, y) : x > 0\}$.

The coefficients are continuously differentiable on Ω, but the equation is not exact there:

$$\frac{\partial}{\partial y}\left(2y^2 + 3x + \frac{2}{x^2}\right) = 4y \qquad \text{but} \qquad \frac{\partial}{\partial x}\left(2xy - \frac{y}{x}\right) = 2y + \frac{y}{x^2}.$$

However, multiplication by x gives

$$(**) \qquad (2xy^2 + 3x^2 + 2x^{-1}) + (2x^2y - y)\, y' = 0,$$

and this equation is exact:

$$\frac{\partial}{\partial y}(2xy^2 + 3x + 2x^{-1}) = 4xy = \frac{\partial}{\partial x}(2x^2y - y).$$

Thus we can solve $(*)$ by solving $(**)$. As you can check, the integral curves on Ω (where x remains positive) are of the form

$$x^2y^2 + x^3 + 2 \ln x - \tfrac{1}{2}y^2 = C. \quad \square$$

Let's return to the general situation and write our equation in the form

$$(2) \qquad P + Qy' = 0.$$

If this equation is not exact, how can we find an integrating factor?

Observe first of all that the equation

$$\mu P + \mu Q y' = 0$$

is exact iff

$$\frac{\partial}{\partial y}(\mu P) = \frac{\partial}{\partial x}(\mu Q),$$

and this occurs iff

$$(3) \qquad \mu \frac{\partial P}{\partial y} + P \frac{\partial \mu}{\partial y} = \mu \frac{\partial}{\partial x} Q + Q \frac{\partial \mu}{\partial x}.$$

Thus μ is an integrating factor for (2) iff it satisfies equation (3).

In theory all we have to do to find an integrating factor for (2) is solve equation (3) for μ. Unfortunately equation (3) is a partial differential equation that is usually more difficult to solve than equation (2). To get anywhere, we will have to make assumptions on the nature of μ.

The assumption that μ depends not on both x and y but only on one of these variables simplifies matters considerably. We will *assume* that μ is independent of y. Then equation (3) reduces to

$$\mu \frac{\partial P}{\partial y} = \mu \frac{\partial Q}{\partial x} + \frac{d\mu}{dx} Q$$

and gives

$$\text{(4)} \qquad \frac{1}{\mu}\frac{d\mu}{dx} = \frac{1}{Q}\left(\frac{\partial P}{\partial y} - \frac{\partial Q}{\partial x}\right).$$

Since the left side of (4) is independent of y, the right side is independent of y. As you can check, this equation is satisfied by setting

$$\mu = e^{\int r(x)dx} \qquad \text{where} \qquad r(x) = \frac{1}{Q}\left(\frac{\partial P}{\partial y} - \frac{\partial Q}{\partial x}\right).$$

What does all this mean? It means that

(18.3.2)

> if
> $$r = \frac{1}{Q}\left(\frac{\partial P}{\partial y} - \frac{\partial Q}{\partial x}\right)$$
> is independent of y, then the function
> $$\mu = e^{\int r(x)\,dx}$$
> is an integrating factor for the equation
> $$P + Qy' = 0.$$

PROOF We assume that $r = \dfrac{1}{Q}\left(\dfrac{\partial P}{\partial y} - \dfrac{\partial Q}{\partial x}\right)$ is independent of y and write

$$P\,e^{\int r(x)\,dx} + Qe^{\int r(x)\,dx}y' = 0.$$

All we have to show is that

$$\frac{\partial}{\partial y}\left(P\,e^{\int r(x)\,dx}\right) = \frac{\partial}{\partial x}\left(Q\,e^{\int r(x)\,dx}\right).$$

This can be seen as follows:

$$\begin{aligned}
\frac{\partial}{\partial x}\left(Q\,e^{\int r(x)\,dx}\right) &= Q\,r(x)e^{\int r(x)\,dx} + \frac{\partial Q}{\partial x}e^{\int r(x)\,dx} \\
&= \left[Q\,r(x) + \frac{\partial Q}{\partial x}\right]e^{\int r(x)\,dx} \\
&= \left[\left(\frac{\partial P}{\partial y} - \frac{\partial Q}{\partial x}\right) + \frac{\partial Q}{\partial x}\right]e^{\int r(x)\,dx} \\
&= \frac{\partial P}{\partial y}\,e^{\int r(x)\,dx} = \frac{\partial}{\partial y}\left(P\,e^{\int r(x)\,dx}\right). \quad \square
\end{aligned}$$

Example 3 Earlier we considered the equation

$$\left(2y^2 + 3x + \frac{2}{x^2}\right) + \left(2xy - \frac{y}{x}\right)y' = 0$$

on the right half-plane and found that the equation was not exact there. We made it exact by multiplying through by x. We can obtain this integrating factor by using (18.3.2). In this case

$$r(x) = \frac{1}{Q}\left(\frac{\partial P}{\partial y} - \frac{\partial Q}{\partial x}\right) = \frac{1}{2xy - (y/x)}\left[4y - \left(2y + \frac{y}{x^2}\right)\right] = \frac{2y - (y/x^2)}{2xy - (y/x)} = \frac{1}{x}$$

so that $$e^{\int r(x)\,dx} = e^{\int (1/x)\,dx} = e^{\ln x} = x. \quad \square$$

In the Exercises you will be asked to show that

(18.3.3) if

$$R = -\frac{1}{P}\left(\frac{\partial P}{\partial y} - \frac{\partial Q}{\partial x}\right)$$

is independent of x, then the function

$$\mu = e^{\int R(y)\,dy}$$

is an integrating factor for the equation

$$P + Q\,y' = 0.$$

One final remark. We have been working with equations

$$P(x,y) + Q(x,y)\,y' = 0.$$

Such equations are often written

$$P(x,y)\,dx + Q(x,y)\,dy = 0.$$

To accustom you to this notation, we will use it in some of the Exercises.

EXERCISES 18.3

Find the maximal simply connected region on which the equation is exact (in each of these cases there is one) and find the integral curves.

1. $(xy^2 - y) + (x^2y - x)\,y' = 0.$
2. $e^x \sin y + (e^x \cos y)\,y' = 0.$
3. $(e^y - y\,e^x) + (x\,e^y - e^x)\,y' = 0.$
4. $\sin y + (x \cos y + 1)\,y' = 0.$
5. $\ln y + 2xy + \left(\dfrac{x}{y} + x^2\right)\,y' = 0.$
6. $2x \tan^{-1} y + \left(\dfrac{x^2}{1+y^2}\right)\,y' = 0.$
7. $\left(\dfrac{y}{x} + 6x\right)dx + (\ln x - 2)\,dy = 0.$
8. $e^x + \ln y + \dfrac{y}{x} + \left(\dfrac{x}{y} + \ln x + \sin y\right)\,y' = 0.$
9. $(y^3 - y^2 \sin x - x)\,dx + (3xy^2 + 2y \cos x + e^{2y})\,dy = 0.$
10. $(e^{2y} - y \cos xy) + (2x\,e^{2y} - x \cos xy + 2y)y' = 0.$
11. Let p and q be functions of one variable everywhere continuously differentiable.
    (a) Is the equation $p(x) + q(y)\,y' = 0$ necessarily exact?
    (b) Show that the equation $p(y) + q(x)y' = 0$ is not necessarily exact. Then find an integrating factor.
12. Prove (18.3.3).

Solve the equation using an integrating factor if necessary.

13. $(e^{y-x} - y) + (x\,e^{y-x} - 1)\,y' = 0.$
14. $(x + e^y) - \frac{1}{2}x^2y' = 0.$
15. $(3x^2y^2 + x + e^y) + (2x^3y + y + x\,e^y)\,y' = 0.$
16. $\sin 2x \cos y - (\sin^2 x \sin y)\,y' = 0.$
17. $(y^3 + x + 1) + (3y^2)\,y' = 0.$
18. $(e^{2x+y} - 2y) + (x\,e^{2x-y} + 1)\,y' = 0.$

Find an integral curve that passes through the point (x_0, y_0). Use an integrating factor if necessary.

19. $(x^2 + y) + (x + e^y)\,y' = 0; \quad (x_0, y_0) = (1, 0).$
20. $(3x^2 - 2xy + y^3) + (3xy^2 - x^2)\,y' = 0; \quad (x_0, y_0) = (1, -1).$
21. $(2y^2 + x^2 + 2) + (2xy)\,y' = 0; \quad (x_0, y_0) = (1, 0).$
22. $(x^2 + y) + (3x^2y^2 - x)\,y' = 0; \quad (x_0, y_0) = (1, 1).$
23. $y^3 + (1 + xy^2)\,y' = 0; \quad (x_0, y_0) = (-2, -1).$
24. $(x + y)^2 + (2xy + x^2 - 1)\,y' = 0; \quad (x_0, y_0) = (1, 1).$
25. $[\cosh(x - y^2) + e^{2x}]\,dx + y[1 - 2\cosh(x - y^2)]\,dy = 0;$ $(x_0, y_0) = (2, \sqrt{2}).$
26. In section 8.8 we solved the linear differential equation

    $$y' + p(x)\,y = q(x)$$

    by using the integrating factor

    $$e^{\int p(x)\,dx}.$$

    Show that this integrating factor is obtainable by the methods of this section.

27. (a) Find a value of k, if possible, such that the differential equation

$$(xy^2 + kx^2y + x^3)\,dx + (x^3 + x^2y + y^2)\,dy = 0$$

is everywhere exact.

(b) Find a value of k, if possible, such that the differential equation

$$y\,e^{2xy} + 2x + (kx\,e^{2xy} - 2y)\,y' = 0$$

is everywhere exact.

28. (a) Find functions f and g, not both identically zero, such that the differential equation

$$g(y)\sin x\,dx + y^2 f(x)\,dy = 0$$

is everywhere exact.

(b) Find all functions g such that the differential equation $g(y)\,e^y + xy\,y' = 0$ is everywhere exact.

In Exercises 29–34, solve the given differential equation by any means at your disposal.

29. $y' = y^2x^3$.

30. $y\,y' = 4x\,e^{2x+y}$.

31. $y' + \dfrac{4y}{x} = x^4$.

32. $y' + 2xy - 2x^3 = 0$.

33. $(y\,e^{xy} - 2x)\,dx + \left(\dfrac{2}{y} + x\,e^{xy}\right)\,dy = 0$.

34. $y\,dx + (2xy - e^{-2y})\,dy = 0$.

■ 18.4 THE EQUATION $y'' + ay' + by = 0$

A differential equation of the form

$$y'' + ay' + by = \phi(x)$$

where a and b are real numbers and ϕ is a continuous function on some interval I is a *second-order linear differential equation with constant coefficients.* By a solution of such an equation we mean a function $y = y(x)$ that satisfies the equation for all x in I. In this section we set $\phi = 0$ and consider the *reduced equation*†

(18.4.1)
$$\boxed{y'' + ay' + by = 0.}$$

As you will see, the solutions of the reduced equation are defined for all real x.

The Characteristic Equation

Earlier you saw that the function $y = e^{-ax}$ satisfies the first-order linear equation

$$y' + ay = 0.$$

This suggests that the differential equation

$$y'' + ay' + by = 0$$

may have a solution of the form $y = e^{rx}$.

If $y = e^{rx}$, then

$$y' = r\,e^{rx} \qquad \text{and} \qquad y'' = r^2\,e^{rx}.$$

Substitution into the differential equation gives

$$r^2\,e^{rx} + ar\,e^{rx} + b\,e^{rx} = e^{rx}(r^2 + ar + b) = 0,$$

and since $e^{rx} \neq 0$,

$$r^2 + ar + b = 0.$$

This shows that the function $y = e^{rx}$ satisfies the differential equation iff

$$r^2 + ar + b = 0.$$

†Sometimes referred to as the *homogeneous equation*.

This quadratic equation in r is called the *characteristic equation*.†

The nature of the solutions of the differential equation

$$y'' + ay' + by = 0$$

depends on the nature of the roots of the characteristic equation. There are three cases to be considered.

Case 1: *The characteristic equation has two distinct real roots r_1 and r_2.* In this case both

$$y_1(x) = e^{r_1 x} \qquad \text{and} \qquad y_2(x) = e^{r_2 x}$$

are solutions of the reduced equation.

Case 2: *The characteristic equation has only one real root $r_1 = r_2 = \alpha$.* In this case the characteristic equation $(r - \alpha)^2 = 0$ can be written

$$r^2 - 2\alpha r + \alpha^2 = 0.$$

As you are asked to show in Exercise 33, the substitution $y = u\,e^{\alpha x}$ gives

$$u'' = 0.$$

This equation is satisfied by

$$\text{the constant function } u_1 = 1 \quad \text{and} \quad \text{the identity function } u_2 = x.$$

Thus the reduced equation is satisfied by the products

$$y_1(x) = e^{\alpha x} \qquad \text{and} \qquad y_2(x) = x\,e^{\alpha x}.$$

Case 3: *The characteristic equation has two complex roots $r_1 = \alpha + i\beta$ and $r_2 = \alpha - i\beta$.* In this case the characteristic equation $[r-(\alpha+i\beta)][r-(\alpha-i\beta)] = 0$ can be written

$$r^2 - 2\alpha r + (\alpha^2 + \beta^2) = 0.$$

As you are asked to show in Exercise 33, the substitution $y = u\,e^{\alpha x}$ eliminates α and gives

$$u'' + \beta^2 u = 0.$$

This equation, the equation of harmonic motion, is satisfied by the functions

$$u_1(x) = \cos \beta x \qquad \text{and} \qquad u_2(x) = \sin \beta x.\ \dagger\dagger$$

Thus, the reduced equation is satisfied by the products

$$y_1(x) = e^{\alpha x} \cos \beta x \qquad \text{and} \qquad y_2(x) = e^{\alpha x} \sin \beta x. \quad \square$$

---

†Sometimes called the *auxiliary equation*.

†† We reviewed this in the introduction to this chapter. In any case, the statements are easy to verify.

Linear Combinations of Solutions; Existence and Uniqueness of Solutions; Wronskians

Observe that *if y_1 and y_2 are both solutions of the reduced equation, then every linear combination*

$$u(x) = C_1 y_1(x) + C_2 y_2(x)$$

is also a solution.

PROOF Set

$$u = C_1 y_1 + C_2 y_2$$

and differentiate twice to get

$$u' = C_1 y_1' + C_2 y_2' \qquad \text{and} \qquad u'' = C_1 y_1'' + C_2 y_2''.$$

Since y_1 and y_2 are solutions of (18.4.1),

$$y_1'' + ay_1' + by_1 = 0 \qquad \text{and} \qquad y_2'' + ay_2' + by_2 = 0.$$

Therefore,

$$\begin{aligned} u'' + au' + bu &= (C_1 y_1'' + C_2 y_2'') + a(C_1 y_1' + C_2 y_2') + b(C_1 y_1 + C_2 y_2) \\ &= C_1(y_1'' + ay_1' + by_1) + C_2(y_2'' + ay_2' + by_2) \\ &= C_1(0) + C_2(0) = 0 \quad \square \end{aligned}$$

You have seen how to obtain solutions of the differential equation

$$y'' + ay' + by = 0$$

from the characteristic equation

$$r^2 + ar + b = 0.$$

We can form more solutions by taking linear combinations of these solutions. Question: Are there still other solutions or do all solutions arise in this manner? Answer: All solutions of the reduced equation are linear combinations of the solutions that we have already found.

To show this, we have to go a little deeper into the theory. Our point of departure is a result that we prove in a supplement to this section.

THEOREM 18.4.2 EXISTENCE AND UNIQUENESS THEOREM

Let x_0, α_0, α_1 be arbitrary real numbers. The reduced equation

$$y'' + ay' + by = 0$$

has a unique solution $y = y(x)$ that satisfies the initial conditions

$$y(x_0) = \alpha_0, \qquad y'(x_0) = \alpha_1.$$

Geometrically, the theorem says that there is one and only one solution, the graph of which passes through a prescribed point (x_0, α_0) with a prescribed slope α_1. We assume the result and go on from there.

DEFINITION 18.4.3

Let y_1 and y_2 be two solutions of

$$y'' + ay' + by = 0.$$

The *Wronskian*† of y_1 and y_2 is the function W defined for all real x by

$$W(x) = y_1(x)\,y_2'(x) - y_2(x)\,y_1'(x).$$

Note that the Wronskian can be written as the 2×2 determinant (see Appendix A.2)

$$W(x) = \begin{vmatrix} y_1(x) & y_2(x) \\ y_1'(x) & y_2'(x) \end{vmatrix}.$$

Wronskians have a very special property.

THEOREM 18.4.4

If both y_1 and y_2 are solutions of

$$y'' + ay' + by = 0,$$

then their Wronskian W is either identically zero or never zero.

PROOF Assume that both y_1 and y_2 are solutions of the homogeneous equations and set

$$W = y_1 y_2' - y_2 y_1'.$$

Differentiation gives

$$W' = y_1 y_2'' + y_1' y_2' - y_1' y_2' - y_1'' y_2 = y_1 y_2'' - y_1'' y_2.$$

Since y_1 and y_2 are solutions, we know that

$$y_1'' + ay_1' + by_1 = 0$$

and

$$y_2'' + ay_2' + by_2 = 0.$$

Multiplying the first equation by $-y_2$ and the second equation by y_1, we have

$$\begin{aligned} -y_1''y_2 - ay_1'y_2 - by_1y_2 &= 0 \\ y_2''y_1 + ay_2'y_1 + by_2y_1 &= 0. \end{aligned}$$

We now add these two equations and obtain

$$(y_1y_2'' - y_2y_1'') + a(y_1y_2' - y_2y_1') = 0.$$

† Named after Count Hoëné Wronski, a Polish mathematician (1776–1853).

In terms of the Wronskian, we have

$$W' + aW = 0.$$

This is a first-order linear differential equation with general solution

$$W(x) = Ce^{-ax}.$$

If $C = 0$, then W is identically 0; if $C \neq 0$, then W is never zero. ❑

Remark How can we get a pair of solutions y_1, y_2 with a nonzero Wronskian? Simply choose four numbers $\alpha, \beta, \gamma, \delta$ such that

$$\begin{vmatrix} \alpha & \gamma \\ \beta & \delta \end{vmatrix} \neq 0$$

and let x_0 be any number. Let y_1 be the solution of the reduced equation that satisfies the initial conditions

$$y_1(x_0) = \alpha, \qquad y_1'(x_0) = \beta,$$

and let y_2 be the solution that satisfies the initial conditions

$$y_2(x_0) = \gamma, \qquad y_2'(x_0) = \delta.$$

Then the Wronskian of y_1 and y_2 at x_0 is given by

$$W(x_0) = \begin{vmatrix} \alpha & \gamma \\ \beta & \delta \end{vmatrix} \neq 0,$$

and so $W(x) \neq 0$ for all x. It is often convenient to prescribe the initial conditions $\alpha = 1,\ \beta = 0,\ \gamma = 0,\ \delta = 1$ so that

$$W(x_0) = \begin{vmatrix} 1 & 0 \\ 0 & 1 \end{vmatrix} = 1. \quad ❑$$

THEOREM 18.4.5

Every solution of the equation

$$y'' + ay' + by = 0$$

can be expressed in a unique manner as the linear combination of any two solutions with a nonzero Wronskian.

PROOF Let u be any solution of the equation and let y_1, y_2 be any two solutions with nonzero Wronskian. Choose a number x_0 and form the equations

$$(1) \qquad \begin{aligned} C_1 y_1(x_0) + C_2 y_2(x_0) &= u(x_0) \\ C_1 y_1'(x_0) + C_2 y_2'(x_0) &= u'(x_0). \end{aligned}$$

The Wronskian of y_1 and y_2 at x_0,

$$W(x_0) = y_1(x_0)\, y_2'(x_0) - y_2(x_0)\, y_1'(x_0),$$

is different from zero. This guarantees that the system of equations (1) has a unique solution given by

$$C_1 = \frac{u(x_0)\, y_2'(x_0) - y_2(x_0) u'(x_0)}{y_1(x_0)\, y_2'(x_0) - y_2(x_0)\, y_1'(x_0)}, \qquad C_2 = \frac{y_1(x_0) u'(x_0) - u(x_0)\, y_1'(x_0)}{y_1(x_0)\, y_2'(x_0) - y_2(x_0)\, y_1'(x_0)}.$$

Our work is finished. The function $C_1y_1 + C_2y_2$ is a solution of the equation which by (1) has the same value as u at x_0 and the same derivative. Thus, by Theorem 18.4.2, $C_1y_1 + C_2y_2$ and u cannot be different functions; that is,

$$u = C_1y_1 + C_2y_2.$$

This proves the theorem. ❑

The General Solution

The arbitrary linear combination

$$y = C_1y_1 + C_2y_2$$

of any two solutions with nonzero Wronskian is the *general solution*. By Theorem 18.4.5, we can obtain any *particular solution* by adjusting C_1 and C_2.

THEOREM 18.4.6

Given the equation

$$y'' + ay' + by = 0,$$

we form the characteristic equation

$$r^2 + ar + b = 0.$$

I. If the characteristic equation has two distinct real roots r_1 and r_2, then the general solution takes the form

$$y = C_1e^{r_1x} + C_2e^{r_2x}.$$

II. If the characteristic equation has only one real root, α, then the general solution takes the form

$$y = C_1e^{\alpha x} + C_2xe^{\alpha x} = (C_1 + C_2x)e^{\alpha x}.$$

III. If the characteristic equation has two complex roots

$$r_1 = \alpha + i\beta \quad \text{and} \quad r_2 = \alpha - i\beta,$$

then the general solution takes the form

$$y = C_1\, e^{\alpha x} \cos \beta x + C_2\, e^{\alpha x} \sin \beta x = e^{\alpha x}(C_1 \cos \beta x + C_2 \sin \beta x).$$

PROOF To prove this theorem, it is enough to show that the three solution pairs

$$e^{r_1x},\ e^{r_2x} \quad e^{\alpha x},\ x\,e^{\alpha x} \quad e^{\alpha x}\cos\beta x,\ e^{\alpha x}\sin\beta x$$

all have nonzero Wronskians. The Wronskian of the first pair is the function

$$\begin{aligned} W(x) &= e^{r_1x}\frac{d}{dx}(e^{r_2x}) - \frac{d}{dx}(e^{r_1x})\,e^{r_2x} \\ &= e^{r_1x}r_2\,e^{r_2x} - r_1\,e^{r_1x}\,e^{r_2x} = (r_2 - r_1)\,e^{(r_1+r_2)x}. \end{aligned}$$

$W(x)$ is different from zero since, by assumption, $r_2 \neq r_1$.

We leave it to you to verify that the other two pairs also have nonzero Wronskians. ❑

It's time to look at specific examples.

Example 1 Find the general solution of the equation $y'' + 2y' - 15y = 0$. Then find the particular solution that satisfies the initial condiions

$$y(0) = 0, \qquad y'(0) = -1.$$

SOLUTION The characteristic equation is the quadratic $r^2 + 2r - 15 = 0$. Factoring the left side, we have

$$(r + 5)(r - 3) = 0.$$

There are two real roots: -5 and 3. The general solution takes the form

$$y = C_1\,e^{-5x} + C_2\,e^{3x}.$$

Differentiating the general solution, we have

$$y' = -5C_1\,e^{-5x} + 3C_2\,e^{3x}.$$

The conditions

$$y(0) = 0, \qquad y'(0) = -1$$

are satisfied iff

$$C_1 + C_2 = 0 \qquad \text{and} \qquad -5C_1 + 3C_2 = -1.$$

Solving these two equations simultaneously, we find that

$$C_1 = \tfrac{1}{8}, \quad C_2 = -\tfrac{1}{8}.$$

The solution that satisfies the prescribed side conditions is the function

$$y = \tfrac{1}{8}\,e^{-5x} - \tfrac{1}{8}\,e^{3x}.$$

You should verify this. ❑

Example 2 Find the general solution of the equation $y'' + 4y' + 4y = 0$.

SOLUTION The characteristic equation takes the form $r^2 + 4r + 4 = 0$, which can be written

$$(r + 2)^2 = 0.$$

The number -2 is the only root and

$$y = C_1e^{-2x} + C_2xe^{-2x}.$$

is the general solution. ❑

Example 3 Find the general solution of the equation $y'' + y' + 3y = 0$.

SOLUTION The characteristic equation is $r^2 + r + 3 = 0$. The quadratic formula shows that there are two complex roots:

$$r_1 = -\tfrac{1}{2} + i\tfrac{1}{2}\sqrt{11}, \quad r_2 = -\tfrac{1}{2} - i\tfrac{1}{2}\sqrt{11}.$$

The general solution takes the form

$$y = e^{-x/2}[C_1 \cos(\tfrac{1}{2}\sqrt{11}\,x) + C_2 \sin(\tfrac{1}{2}\sqrt{11}\,x)]. \quad \square$$

In our final example we revisit the equation of simple harmonic motion.

Example 4 Find the general solution of the equation

$$y'' + \omega^2 y = 0 \qquad (\omega \neq 0).$$

SOLUTION The characteristic equation is $r^2 + \omega^2 = 0$ and the roots are

$$r_1 = \omega i, \quad r_2 = -\omega i.$$

Thus the general solution is

$$y = C_1 \cos \omega x + C_2 \sin \omega x. \quad \square$$

Remark As you probably recall, the equation in Example 4 describes the oscillatory motion of an object suspended by a spring under the assumption that there are no forces acting on the spring-mass system other than the restoring force of the spring. This spring-mass problem and some generalizations of it are studied in Section 18.6. In the Exercises you are asked to show that the general solution that we gave above can be written

$$y = A \sin(\omega x + \phi_0)$$

where A and ϕ_0 are constants with $A > 0$ and $\phi_0 \in [0, 2\pi)$. $\square$

EXERCISES 18.4

Find the general solution.

1. $y'' + 2y' - 8y = 0$.
2. $y'' - 13y' + 42y = 0$.
3. $y'' + 8y' + 16y = 0$.
4. $y'' + 7y' + 3y = 0$.
5. $y'' + 2y' + 5y = 0$.
6. $y'' - 3y' + 8y = 0$.
7. $2y'' + 5y' - 3y = 0$.
8. $y'' - 12y = 0$.
9. $y'' + 12y = 0$.
10. $y'' - 3y' + \frac{9}{4}y = 0$.
11. $5y'' + \frac{11}{4}y' - \frac{3}{4}y = 0$.
12. $2y'' + 3y' = 0$.
13. $y'' + 9y = 0$.
14. $y'' - y' - 30y = 0$.
15. $2y'' + 2y' + y = 0$.
16. $y'' - 4y' + 4y = 0$.
17. $8y'' + 2y' - y = 0$.
18. $5y'' - 2y' + y = 0$.

Solve the initial-value problem.

19. $y'' - 5y' + 6y = 0, \quad y(0) = 1, \quad y'(0) = 1$.
20. $y'' + 2y' + y = 0, \quad y(2) = 1, \quad y'(2) = 2$.
21. $y'' + \frac{1}{4}y = 0, \quad y(\pi) = 1, \quad y'(\pi) = -1$.
22. $y'' - 2y' + 2y = 0, \quad y(0) = -1, \quad y'(0) = -1$.
23. $y'' + 4y' + 4y = 0, \quad y(-1) = 2, \quad y'(-1) = 1$.
24. $y'' - 2y' + 5y = 0, \quad y(\pi/2) = 0, \quad y'(\pi/2) = 2$.
25. Find all solutions of the equation $y'' - y' - 2y = 0$ that satisfy the given initial conditions:
(a) $y(0) = 1$. (b) $y'(0) = 1$.
(c) $y(0) = 1, \quad y'(0) = 1$.

26. Prove that the general solution of the differential equation

$$y'' - \omega^2 y = 0 \qquad (\omega > 0)$$

can be written

$$y = C_1 \cosh \omega x + C_2 \sinh \omega x.$$

27. Suppose that the roots r_1 and r_2 of the characteristic equation of (18.4.1) are real and distinct. Then they can be written as $r_1 = \alpha + \beta$ and $r_2 = \alpha - \beta$, where α and β are real. Show that the general solution of the equation (18.4.1) can be expressed in the form

$$y = e^{\alpha x}(C_1 \cosh \beta x + C_2 \sinh \beta x).$$

28. Show that the general solution of the differential equation

$$y'' + \omega^2 y = 0$$

can be written

$$y = A \sin(\omega x + \phi_0),$$

where A and ϕ_0 are constants with $A > 0$ and $\phi_0 \in [0, 2\pi)$.

29. Complete the proof of Theorem 18.4.6 by showing that the following solutions have nonzero Wronskians.

(a) $y_1 = e^{\alpha x}, \quad y_2 = x\,e^{\alpha x}$. (one root case)

(b) $y_1 = e^{\alpha x}\cos \beta x, \quad y_2 = e^{\alpha x}\sin \beta x$. (complex root case)

30. In the absence of any external electromotive force, the current i in a simple electrical circuit varies with time t according to the formula

$$L\frac{d^2 i}{dt^2} + R\frac{di}{dt} + \frac{1}{C}i = 0. \qquad (L, R, C \text{ constants})\dagger$$

Find the general solution of this equation given that $L = 1, R = 10^3$, and

(a) $C = 5 \times 10^{-6}$.

(b) $C = 4 \times 10^{-6}$.

(c) $C = 2 \times 10^{-6}$.

31. Find a differential equation $y'' + ay' + by = 0$ that is satisfied by both functions.

(a) $y_1 = e^{2x}, \quad y_2 = e^{-4x}$.

(b) $y_1 = 3\,e^{-x}, \quad y_2 = 4\,e^{5x}$.

(c) $y_1 = 2\,e^{3x}, \quad y_2 = x\,e^{3x}$.

32. Find a differential equation $y'' + ay' + by = 0$ that is satisfied by both functions.

(a) $y_1 = 2\cos 2x, \quad y_2 = -\sin 2x$.

(b) $y_1 = e^{-2x}\cos 3x, \quad y_2 = 2\,e^{-2x}\sin 3x$.

33. (a) Show that the substitution $y = e^{\alpha x}u$ transforms

$$y'' - 2\alpha y' + \alpha^2 y = 0 \quad \text{into} \quad u'' = 0.$$

† L is inductance, R is resistance, and C is capacitance. If L is given in henrys, R in ohms, C in farads, and t in seconds, then the current is given in amperes.

(b) Show that the substitution $y = e^{\alpha x}u$ transforms

$$y'' - 2\alpha y' + (\alpha^2 + \beta^2)y = 0 \quad \text{into} \quad u'' + \beta^2 u = 0.$$

Exercises 34 and 35 relate to the differential equation $y'' + ay' + by = 0$ where a and b are nonnegative constants.

34. Prove that if a and b are both positive, then $y(x) \to 0$ as $x \to \infty$ for all solutions y of the equation.

35. (a) Prove that if $a = 0$ and $b > 0$, then all solutions of the equation are bounded.

(b) Suppose that $a > 0,\ b = 0$, and $y = y(x)$ is a solution of the equation. Prove that

$$\lim_{x\to\infty} y(x) = k$$

for some constant k. Determine k for the solution that satisfies the initial conditions: $y(0) = y_0,\ y'(0) = y_1$.

36. Let y_1, y_2 be solutions of the reduced equation defined for all real x. Show that if $y_1(a) = y_2(a) = 0$ at some number a, then y_1 and y_2 are scalar multiples of each other.

37. Let y_1, y_2 be solutions of the reduced equation defined for all real x. Show that the Wronskian of y_1, y_2 is zero iff one of these functions is a scalar multiple of the other.

Euler Equations An equation of the form

$$(*) \qquad x^2 y'' + \alpha x y' + \beta y = 0,$$

where α and β are real numbers, is called an *Euler equation.*

38. Show that the Euler equation $(*)$ can be transformed into an equation of the form

$$\frac{d^2 y}{dz^2} + a\frac{dy}{dz} + by = 0,$$

where a and b are real numbers, by means of the change of variable $z = \ln x$. HINT: If $z = \ln x$, then by the chain rule,

$$\frac{dy}{dx} = \frac{dy}{dz}\frac{dz}{dx} = \frac{dy}{dz}\frac{1}{x}.$$

Now calculate d^2y/dx^2 and substitute the result into the differential equation.

In Exercises 39–42, Use the change of variable indicated in Exercise 38 to transform the given equation into an equation with constant coefficients. Find the general solution of that equation, and then express it in terms of x.

39. $x^2y'' - xy' - 8y = 0$.

40. $x^2y'' - 2xy' + 2y = 0$.

41. $x^2y'' - 3xy' + 4y = 0$.

42. $x^2y'' - xy' + 5y = 0$.

*SUPPLEMENT TO SECTION 18.4

PROOF OF THEOREM 18.4.2

Existence: Take two solutions y_1, y_2 with nonzero Wronskian

$$W(x) = y_1(x)\,y_2'(x) - y_2(x)\,y_1'(x).$$

For any numbers x_0, α_0, α_1, the equations

$$C_1y_1(x_0) + C_2y_2(x_0) = \alpha_0$$
$$C_1y_1'(x_0) + C_2y_2'(x_0) = \alpha_1$$

can be solved for C_1 and C_2. For those values of C_1 and C_2, the function

$$y = C_1y_1 + C_2y_2$$

is a solution of (18.4.1) that satisfies the prescribed initial conditions.

Uniqueness: Let us assume that there are two distinct solutions y_1, y_2 that satisfy the same prescribed initial conditions

$$y_1(x_0) = \alpha_0 = y_2(x_0) \quad \text{and} \quad y_1'(x_0) = \alpha_1 = y_2'(x_0).$$

Then the solution $y = y_1 - y_2$ satisfies the initial conditions

$$y\,(x_0) = 0, \; y'(x_0) = 0.$$

Since y_1 and y_2 are, by assumption, distinct functions, there is at least one number x at which y is not zero. Therefore, by the continuity of y there exists an interval I on which y is does not take on the value zero.

Now let u be *any* solution of (18.4.1). The Wronskian of y and u is zero at x_0 :

$$W(x_0) = y\,(x_0)\,u'(x_0) - u(x_0)\,y'(x_0) = (0)u'(x_0) - u(x_0)(0) = 0.$$

Therefore the Wronskian of y and u is everywhere zero. Since $y\,(x) \neq 0$ for all $x \in I$, the quotient u/y is defined on I, and on that interval

$$\frac{d}{dx}\left(\frac{u}{y}\right) = \frac{yu' - uy'}{y^2} = \frac{W}{y^2} = 0, \quad \frac{u}{y} = C, \quad \text{and} \quad u = Cy.$$

We have shown that on the interval I every solution is some scalar multiple of y.

Now let u_1 and u_2 be any two solutions with a nonzero Wronskian W. From what we have just shown, there are constants C_1 and C_2 such that on I

$$u_1 = C_1y \quad \text{and} \quad u_2 = C_2y.$$

Then on I

$$W = u_1u_2' - u_2u_1' = (C_1y)(C_2y') - (C_2y)(C_1y') = C_1C_2(\,yy' - yy') = 0.$$

This contradicts the statement that $W \neq 0$.

The assumption that there are two distinct solutions that satisfy the same prescribed initial conditions has led to a contradiction. This proves uniqueness.

■ 18.5 THE EQUATION $y'' + ay' + by = \phi(x)$

In Section 18.4 we solved the reduced equation

$$y'' + ay' + by = 0.$$

Here we go on to the *complete equation*

(18.5.1)
$$y'' + ay' + by = \phi(x).\dagger$$

The function $\phi = \phi(x)$ that appears on the right will be called the *forcing function*.†† We assume that ϕ is continuous on some interval I and we will solve the differential equation on that interval.

We begin by proving two simple but important results.

(18.5.2) If both y_1 and y_2 are solutions of the complete equation, then their difference $u = y_1 - y_2$ is a solution of the reduced equation.

PROOF If

$$y_1'' + ay_1' + by_1 = \phi(x) \qquad \text{and} \qquad y_2'' + ay_2' + by_2 = \phi(x),$$

then

$$\begin{aligned} u'' + au' + bu &= (y_1'' - y_2'') + a(y_1' - y_2') + b(y_1 - y_2) \\ &= (y_1'' + ay_1' + by_1) - (y_2'' + ay_2' + by_2) \\ &= \phi(x) - \phi(x) = 0 \quad \square \end{aligned}$$

(18.5.3) If y_p is a particular solution of the complete equation, then every solution of the complete equation can be written as a solution of the reduced equation plus y_p.

PROOF Let y_p be a solution of the complete equation. If y is another solution of the complete equation, then, by (18.5.2), $y - y_p$ is a solution of the reduced equation. Obviously

$$y = (y - y_p) + y_p. \quad \square$$

It follows from (18.5.3) that we can obtain the *general solution* of the complete equation by starting with the general solution of the reduced equation and then adding to it a particular solution of the complete equation. The general solution of the complete equation can thus be written

(18.5.4)
$$y = C_1u_1 + C_2u_2 + y_p$$

where u_1, u_2 are any two solutions of the reduced equation which have a nonzero Wronskian and y_p is any particular solution of the complete equation.

The main task before us is to search for functions which can serve as y_p. In this search the following result can sometimes be used to advantage. It is called the

---

† Sometimes referred to as the *nonhomogeneous equation*.

†† Because this is the role played by ϕ in the study of vibrations.

superposition principle.

(18.5.5)

> If y_1 is a solution of
> $$y'' + ay' + by = \phi_1(x)$$
> and y_2 is a solution of
> $$y'' + ay' + by = \phi_2(x),$$
> then $y_1 + y_2$ is a solution of
> $$y'' + ay' + by = \phi_1(x) + \phi_2(x).$$

Using the superposition principle, we can find solutions to an equation in which the forcing function has several terms by finding solutions to equations in which the forcing function has only one term and then adding up the results. Verification of the superposition principle is left to you as an exercise.

We are now ready to describe two methods by which we can find some solution of the complete equation, some function that can play the role of y_p.

Variation of Parameters

The method that we outline here gives particular solutions to all equations of the form

$$y'' + ay' + by = \phi(x). \tag{1}$$

The general solution of the reduced equation

$$y'' + ay' + by = 0$$

can be written

$$y = C_1u_1 + C_2u_2$$

where u_1, u_2 are any two solutions with a nonzero Wronskian and the coefficients C_1, C_2 are arbitrary constants. In the method called *variation of parameters,* we let the coefficients vary. That is, we replace the constants C_1, C_2 by the functions

$$z_1 = z_1(x), \qquad z_2 = z_2(x)$$

and seek solutions of the form

$$y_p = z_1u_1 + z_2u_2. \tag{2}$$

Differentiating (2), we have

$$y_p = z_1u_1' + z_1'u_1 + z_2u_2' + z_2'u_2 = (z_1u_1' + z_2u_2') + (z_1'u_1 + z_2'u_2).$$

We now impose a restriction on z_1, z_2: we require that

$$z_1'u_1 + z_2'u_2 = 0. \tag{3}$$

Having imposed this restriction, we have

$$y_p' = z_1u_1' + z_2u_2'$$

and, differentiating once more,

$$y_p'' = z_1u_1'' + z_1'u_1' + z_2u_2'' + z_2'u_2'.$$

A straightforward calculation that we leave to you shows that y_p will satisfy equation (1) iff

$$(4) \qquad z_1'u_1' + z_2'u_2' = \phi(x).$$

Equations (3) and (4) can now be solved simultaneously for z_1' and z_2'. As you can verify yourself, the unique solutions are

$$(5) \qquad z_1' = -\frac{u_2\phi}{W} \quad \text{and} \quad z_2' = \frac{u_1\phi}{W}$$

where the denominator $W = u_1u_2' - u_2u_1'$ is the Wronskian of u_1 and u_2. The functions z_1, z_2 are now found by integration:

$$z_1 = -\int \frac{u_2\phi}{W}\,dx, \qquad z_2 = \int \frac{u_1\phi}{W}\,dx.$$

The function

$$\text{(18.5.6)} \qquad \boxed{y_p = \left(-\int \frac{u_2(x)\phi(x)}{W}\,dx\right)u_1(x) + \left(\int \frac{u_1(x)\phi(x)}{W}\right)u_2(x)}$$

is a particular solution of the equation

$$y'' + ay' + by = \phi(x).$$

Example 1 Use variation of parameters to find a solution of the equation

$$y'' + y = \tan x, \qquad -\frac{\pi}{2} < x < \frac{\pi}{2}.$$

Then give the general solution.

SOLUTION The reduced equation $y'' + y = 0$ has solutions

$$u_1 = \cos x, \qquad u_2 = \sin x.$$

The Wronskian of these solutions is identically 1:

$$W = u_1u_2' - u_2u_1' = (\cos x)(\cos x) - (-\sin x)\sin x = \cos^2 x + \sin^2 x = 1.$$

Since $\phi(x) = \tan x$, we can set

$$\begin{aligned} z_1 &= -\int \frac{u_2\phi}{W}\,dx \\ &= -\int \frac{\sin x \tan x}{1}\,dx \\ &= -\int \frac{\sin^2 x}{\cos x}\,dx \\ &= \int \frac{\cos^2 x - 1}{\cos x}\,dx = \int (\cos x - \sec x)\,dx = \sin x - \ln|\sec x + \tan x| \end{aligned}$$

and

$$z_2 = \int \frac{u_1\phi}{W}\,dx = \int \frac{\cos x \tan x}{1}\,dx = \int \sin x\,dx = -\cos x.$$

We didn't include any arbitrary constants here. At this stage we are looking for only one solution of the complete equation, not a family of solutions.

By (18.5.6), the function

$$\begin{aligned} y_p &= (\sin x - \ln|\sec x + \tan x|)\cos x + (-\cos x)\sin x \\ &= -(\ln|\sec x + \tan x|)\cos x \end{aligned}$$

is a solution of the complete equation.

The general solution can be written

$$y = C_1 \cos x + C_2 \sin x - (\ln|\sec x + \tan x|)\cos x. \quad \square$$

Example 2 Find the general solution of $y'' = 5y' + 6y = 4e^{2x}$.

SOLUTION The equation $y'' - 5y' + 6y = 0$ has characteristic equation

$$r^2 - 5r + 6 = (r-2)(r-3) = 0.$$

Thus, $u_1(x) = e^{2x}, u_2(x) = e^{3x}$ are solutions. Their Wronskian W is e^{5x}:

$$W = u_1 u_2' - u_2 u_1' = e^{2x}\, 3e^{3x} - e^{3x}\, 2e^{2x} = e^{5x}.$$

Since $\phi(x) = 4e^{2x}$, we have

$$z_1 = -\int \frac{u_2\phi}{W}\,dx = -\int \frac{e^{3x}\, 4e^{2x}}{e^{5x}}\,dx = -\int 4\,dx = -4x$$

and

$$z_2 = \int \frac{u_1\phi}{W}\,dx = \int \frac{e^{2x}\, 4e^{2x}}{e^{5x}}\,dx = \int 4e^{-x}\,dx = -4e^{-x}.$$

Therefore, by (18.5.6),

$$y_p = -4x\,e^{2x} - 4e^{-x}e^{3x} = -4x\,e^{2x} - 4e^{2x}$$

is a solution of the complete equation.

The general solution can be written

$$\begin{aligned} y &= A_1u_1 + A_2u_2 + y_p = A_1e^{2x} + A_2e^{3x} - 4x\,e^{2x} - 4e^{2x} \\ &= (A_1 - 4)e^{2x} + A_2e^{3x} - 4x\,e^{2x} \\ &= C_1e^{2x} + C_2e^{3x} - 4x\,e^{2x}, \qquad (C_1 = A_1 - 4, C_2 = A_2). \quad \square \end{aligned}$$

Undetermined Coefficients

In equations

$$y'' + ay' + by = \phi(x)$$

that arise in the study of physical phenomena, the forcing function is often a polynomial, a sine, a cosine, an exponential, or a simple combination thereof. Particular solutions of such equations can usually be found by what is formally called the method of *undetermined coefficients,* but is probably just as aptly called the method of "informed guessing."

Instead of trying to lay out formal rules of procedure, we take a practical approach and proceed directly to examples. In the first few examples we won't be looking for the general solution of the equation. We will be looking for any solution that we can get hold of, any function that can act as y_p.

Example 3 Suppose that we are faced with the equation

$$(*) \qquad y'' + 2y' + 5y = 10e^{-2x}.$$

From what we know about exponentials, it seems reasonable to guess a solution of the form $y = Ae^{-2x}$. Proceeding with our guess we have

$$y = Ae^{-2x}, \qquad y' = -2Ae^{-2x}, \qquad y'' = 4Ae^{-2x}.$$

Therefore $\qquad y'' + 2y' + 5y = 4Ae^{-2x} + 2(-2Ae^{-2x}) + 5Ae^{-2x} = 5Ae^{-2x}.$

Our exponential function satisfies $(*)$ provided

$$5Ae^{-2x} = 10e^{-2x}.$$

It follows from this equation that $5A = 10$ and so $A = 2$. The function $y_p = 2e^{-2x}$ is a particular solution. ❑

Example 4 This time we seek a solution to the equation

$$(*) \qquad y'' + 2y' + y = 10\cos 3x.$$

Following the lead of Example 3, we may be tempted to try a solution of the form $y = A\cos 3x$. For this function

$$y'' + 2y' + y = -8A\cos 3x - 6A\sin 3x.$$

Verify this. Let us see what happens with $y = B\sin 3x$. For this function

$$y'' + 2y' + y = 6B\cos 3x - 8B\sin 3x.$$

Verify this. Combining these two calculations, we find that the function

$$y = A\cos 3x + B\sin 3x$$

satisfies the equation

$$y'' + 2y' + y = (-8A + 6B)\cos 3x + (-6A - 8B)\sin 3x.$$

We can satisfy the equation $(*)$ by having

$$-8A + 6B = 10 \qquad \text{and} \qquad -6A - 8B = 0.$$

As you can check, these relations lead to $A = -\frac{4}{5}$, $B = \frac{3}{5}$. Therefore, the function

$$y_p = -\tfrac{4}{5}\cos 3x + \tfrac{3}{5}\sin 3x$$

is a particular solution of $(*)$. ❑

Example 5 For the equation

$$(*) \qquad y'' - 5y' + 6y = 4e^{2x}$$

you may be tempted to try to find a solution of the form $y = Ae^{2x}$. This won't work. The characteristic equation of the reduced equation reads $r^2 - 5r + 6 = 0$, which factors into $(r-2)(r-3) = 0$. This tells us that all functions of the form $y = Ae^{2x}$ are solutions of the reduced equation. For such functions the left side of $(*)$ is zero and cannot be $4e^{2x}$. What can we do? We need an e^{2x} and we don't want any sines, cosines, or any other exponentials around. We try a function of the form $y = Axe^{2x}$. Perhaps the left side of the equation, $y'' - 5y' + 6y$, will eliminate the x and leave us with a constant multiple of e^{2x}, which is what we want. Substituting y and its derivatives

$$y' = Ae^{2x} + 2Axe^{2x}, \qquad y'' = 4Ae^{2x} + 4Axe^{2x}$$

into equation $(*)$, we get

$$(4Ae^{2x} + 4Axe^{2x}) - 5(Ae^{2x} + 2Axe^{2x}) + 6(Axe^{2x}) = 4e^{2x},$$

which simplifies to

$$-Ae^{2x} = 4e^{2x}.$$

Thus, $y = Axe^{2x}$ satisfies $(*)$ provided that $A = -4$. The function

$$y_p = -4xe^{2x}$$

is a particular solution $(*)$. Now go back and look at Example 2. ❑

In the next example we will use the superposition principle: the fact that if y_1 is a solution of

$$y'' + ay' + by = \phi_1(x)$$

and y_2 is a solution of

$$y'' + ay' + by = \phi_2(x),$$

then $y_1 + y_2$ is a solution of

$$y'' + ay' + by = \phi_1(x) + \phi_2(x).$$

Example 6 We consider the equation $y'' - 2y' + y = e^x + e^{-x}\sin x$. This time we want the general solution.

First we calculate the general solution of the reduced equation

$$y'' - 2y' + y = 0.$$

Since the characteristic equation $r^2 - 2r + 1 = 0$ has the factored form $(r-1)^2 = 0$, the general solution takes the form

$$y_g = C_1e^x + C_2xe^x.$$

Now we look for some solution of

$$(*) \qquad y'' - 2y' + y = e^x.$$

Functions of the form $y = Ae^x$ or $y = Axe^x$ won't work because they are solutions of the reduced equation. So we go one step further and try $y = Ax^2e^x$. As you can check, substituting this guess into $(*)$ leads to the conclusion that this y is a solution provided $A = \frac{1}{2}$. The function

$$y_1 = \tfrac{1}{2}x^2e^x$$

is a particular solution of $(*)$.

Next we look for a solution of

$$(**) \qquad y'' - 2y' + y = e^{-x} \sin x.$$

The result in Example 4 suggests that we try to find solution of the form

$$y = Ae^{-x} \sin x + Be^{-x} \cos x.$$

For this function, you can verify that

$$y'' - 2y' + y = (3A + 4B)e^{-x} \sin x + (-4A + 3B)e^{-x} \cos x.$$

We can satisfy equation $(**)$ by having

$$3A + 4B = 1 \qquad \text{and} \qquad -4A + 3B = 0.$$

These relations lead to $A = \frac{3}{25}$, $B = \frac{4}{25}$. The function

$$y_2 = \tfrac{3}{25}e^{-x} \sin x + \tfrac{4}{25}e^{-x} \cos x$$

is a particular solution of $(**)$.

Therefore, the general solution of the equation

$$y'' - 2y' + y = e^x + e^{-x} \sin x$$

can be written

$$y = y_g + y_1 + y_2 = C_1e^x + C_2xe^x + \tfrac{1}{2}x^2e^x + \tfrac{3}{25}e^{-x} \sin x + \tfrac{4}{25}e^{-x} \cos x. \quad \square$$

We have approached these problems as tests of ingenuity. There are detailed recipes that give suggested trial solutions for a multitude of forcing functions ϕ. We won't attempt to give them here.

EXERCISES 18.5

Find a particular solution.

1. $y'' + 5y' + 6y = 3x + 4$.

2. $y'' - 3y' - 10y = 5$.

3. $y'' + 2y' + 5y = x^2 - 1$.

4. $y'' + y' - 2y = x^3 + x$.

5. $y'' + 6y' + 9y = e^{3x}$.

6. $y'' + 6y' + 9y = e^{-3x}$.

7. $y'' + 2y' + 2y = e^x$.

8. $y'' + 4y' + 4y = x\,e^{-x}$.

9. $y'' - y' - 12y = \cos x$.

10. $y'' - y' - 12y = \sin x$.

11. $y'' + 7y' + 6y = 3 \cos 2x$.

12. $y'' + y' + 3y = \sin 3x$.

13. $y'' - 2y' + 5y = e^{-x} \sin 2x$.

14. $y'' + 4y' + 5y = e^{2x} \cos x$.

15. $y'' + 6y' + 8y = 3\,e^{-2x}$.

16. $y'' - 2y' + 5y = e^x \sin x$.

Find the general solution.

17. $y'' + y = e^x$.

18. $y'' - 2y' + y = -25 \sin 2x$.

19. $y'' - 3y' - 10y = -x - 1$.

20. $y'' + 4y = x \cos 2x$.

21. $y'' + 3y' - 4y = e^{-4x}$.

22. $y'' + 2y' = 4 \sin 2x$.

23. $y'' + y' - 2y = 3x\,e^x$.

24. $y'' + 4y'' + 4y = x\,e^{-2x}$.

25. Prove the superposition principle (18.5.5).

26. Use (18.5.5) to find a particular solution.

(a) $y'' + 2y' - 15y = x + e^{2x}$.

(b) $y'' - 7y' - 12y = e^{-x} + \sin\,2x$.

27. Find the general solution of the equation

$$y'' - 4y' + 3y = \cosh x.$$

Use variation of parameters to find a particular solution.

28. $y'' + y = 3\sin x \sin 2x$.

29. $y'' - 2y' + y = x\,e^x \cos x$.

30. $y'' + y = \csc x, \quad 0 < x < \pi$.

31. $y'' - 4y' + 4y = \frac{1}{3}x^{-1}e^{2x}, \quad x > 0$.

32. $y'' + 4y = \sec^2 2x$.

33. $y'' + 4y' + 4y = \dfrac{e^{-2x}}{x^2}$.

34. $y'' + 2y' + y = e^{-x}\ln x$.

35. $y'' - 2y' + 2y = e^x \sec x$.

36. Show that the change of variable $y = ve^{kx}$ transforms the equation

$$y'' + ay' + by = (c_n x^n + \cdots + c_1 x + c_0)e^{kx}$$

into

$$v'' + (2k + a)v' + (k^2 + ak + b)v = c_n x^n + \cdots + c_1 x + c_0.$$

37. In Exercise 30 of Section 18.4 we introduced a differential equation for the electrical current in a simple circuit. In the presence of an external electromotive force $F(t)$, the equation takes the form

$$L\frac{d^2 i}{dt^2} + R\frac{di}{dt} + \frac{1}{C}i = F(t).$$

Find the current i given that $F(t) = F_0, i(0) = 0, i'(0) = F_0/L$ and (a) $CR^2 = 4L$; (b) $CR^2 < 4L$.

38. (a) Show that $y_1 = x$, $y_2 = x\ln x$ are solutions of the Euler equation

$$x^2y'' - xy' + y = 0$$

and that their Wronskian is nonzero on $(0, \infty)$.

(b) Use variation of parameters to find a particular solution of the equation

$$x^2y'' - xy' + y = 4x\ln x.$$

39. (a) Show that $y_1 = \sin(\ln x^2)$ and $y_2 = \cos(\ln x^2)$ are solutions of the Euler equation

$$x^2y'' + xy' + 4y = 0.$$

Verify that their Wronskian is nonzero on $(0, \infty)$.

(b) Use variation of parameters to find a particular solution of the equation.

$$x^2y'' + xy' + 4y = \sin(\ln x).$$

■ 18.6 MECHANICAL VIBRATIONS

Simple Harmonic Motion

An object moves along a straight line. Instead of continuing in one direction, it moves back and forth, oscillating about a central point. Call the central point $x = 0$ and denote by $x(t)$ the displacement of the object at time t. If the acceleration is a constant negative multiple of the displacement,

$$a(t) = -kx(t), \quad k > 0,$$

then the object is said to be in *simple harmonic motion.*

Since, by definition,

$$a(t) = x''(t),$$

we have

$$x''(t) = -kx(t),$$

and therefore

$$x''(t) + kx(t) = 0.$$

To emphasize that k is positive, we set $k = \omega^2$ where $\omega = \sqrt{k} > 0$. The equation of motion then takes the form

(18.6.1)

$$\boxed{x''(t) + \omega^2 x(t) = 0.}$$

This is a second-order, linear differential equation with constant coefficients. The characteristic equation reads

$$r^2 + \omega^2 = 0,$$

and the roots are $\pm\omega i$. Therefore, the general solution of (18.6.1) has the form

$$x(t) = C_1 \cos \omega t + C_2 \sin \omega t.$$

A routine calculation shows that the general solution can be written (Exercise 28, Section 18.4)

(18.6.2)
$$\boxed{x(t) = A \sin (\omega t + \phi_0)}$$

where A and ϕ_0 are constants with $A > 0$ and $\phi_0 \in [0, 2\pi)$.

Now let's analyze the motion measuring t in seconds. By adding $2\pi/\omega$ to t, we increase $\omega t + \phi_0$ by 2π :

$$\omega\left(t + \frac{2\pi}{\omega}\right) + \phi_0 = \omega t + \phi_0 + 2\pi.$$

Therefore the motion is *periodic* with *period*

$$T = \frac{2\pi}{\omega}.$$

A complete oscillation takes $2\pi/\omega$ seconds. The reciprocal of the period gives the number of complete oscillations per second. This is called the *frequency*:

$$f = \frac{\omega}{2\pi}.$$

The number ω is called the *angular frequency*. Since $\sin(\omega t + \phi_0)$ oscillates between -1 and 1,

$$x(t) = A \sin (\omega t + \phi_0)$$

oscillates between $-A$ and A. The number A is called the *amplitude* of the motion.

In Figure 18.6.1 we have plotted x against t. The oscillations along the x-axis are now waves in the tx-plane. The period of the motion, $2\pi/\omega$, is the t distance (the time separation) between consecutive wave crests. The amplitude of the motion, A, is the height of the waves measured in x units from $x = 0$. The number ϕ_0 is known as the *phase constant*, or *phase shift*. The phase constant determines the initial displacement (the height of the wave at time $t = 0$.) If $\phi_0 = 0$, the object starts at the center of the interval of motion (the wave starts at the origin of the tx-plane).

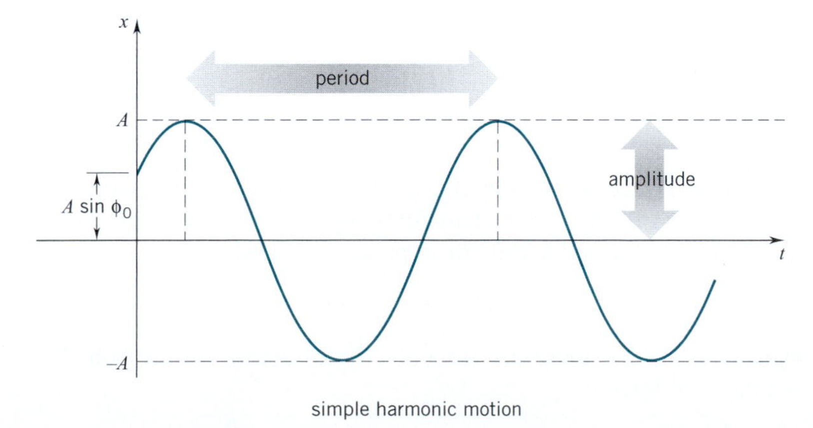

simple harmonic motion

Figure 18.6.1

Example 1 Find an equation for the oscillatory motion of an object, given that the period is $2\pi/3$ and, at time $t = 0$, $x = 1$, $v = x' = 3$.

SOLUTION We begin by setting $x(t) = A\sin(\omega t + \phi_0)$. In general the period is $2\pi/\omega$, so that here

$$\frac{2\pi}{\omega} = \frac{2\pi}{3} \quad \text{and thus} \quad \omega = 3.$$

The equation of motion takes the form

$$x(t) = A\sin(3t + \phi_0).$$

By differentiation

$$v(t) = 3A\cos(3t + \phi_0).$$

The conditions at $t = 0$ give

$$1 = x(0) = A\sin\phi_0, \quad 3 = v(0) = 3A\cos\phi_0$$

and therefore

$$1 = A\sin\phi_0, \quad 1 = A\cos\phi_0.$$

Adding the squares of these equations, we have

$$2 = A^2\sin^2\phi_0 + A^2\cos^2\phi_0 = A^2.$$

Since $A > 0, A = \sqrt{2}$.

To find ϕ_0 we note that

$$1 = \sqrt{2}\sin\phi_0 \quad \text{and} \quad 1 = \sqrt{2}\cos\phi_0.$$

These equations are satisfied by setting $\phi_0 = \frac{1}{4}\pi$. The equation of motion can be written

$$x(t) = \sqrt{2}\sin(3t + \tfrac{1}{4}\pi). \quad \square$$

Undamped Vibrations

A coil spring hangs naturally to a length l_0. When a bob of mass m is attached to it, the spring stretches l_1 inches. The bob is later pulled down an additional x_0 inches and then released. What is the resulting motion? Throughout we refer to Figure 18.6.2, taking the downward direction as positive.

We begin by analyzing the forces acting on the bob at general position x (stage IV). First there is the weight of the bob:

$$F_1 = mg.$$

This is a downward force, and by our choice of coordinate system, positive. Then there is the restoring force of the spring. This force, by Hooke's law, is proportional to the total displacement $l_1 + x$ and acts in the opposite direction:

$$F_2 = -k(l_1 + x) \qquad \text{with } k > 0.$$

If we neglect resistance, these are the only forces acting on the bob. Under these conditions the total force is

$$F = F_1 + F_2 = mg - k(l_1 + x),$$

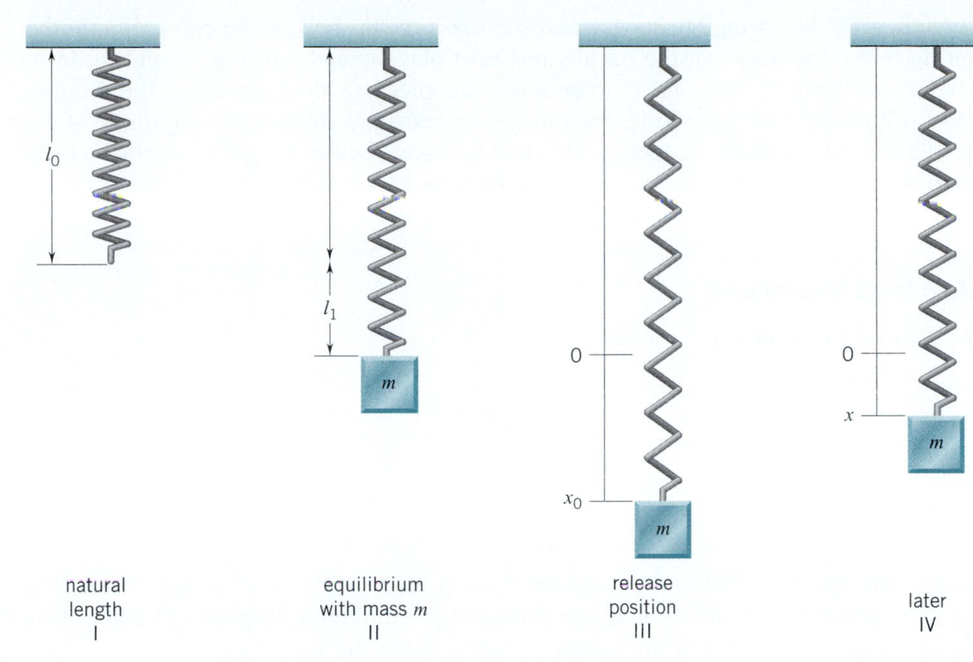

Figure 18.6.2

which we rewrite as

$$(1) \qquad F = (mg - kl_1) - kx.$$

At stage II (Figure 18.6.2) there was equilibrium. The force of gravity, mg, plus the force of the spring, $-kl_1$, must have been 0:

$$mg - kl_1 = 0.$$

Equation (1) can therefore be simplified to read

$$F = -kx.$$

Using Newton's second law,

$$F = ma, \qquad \text{(force = mass × accelration)}$$

we have $\qquad ma = -kx \quad \text{and} \quad \text{thus} \quad a = -\frac{k}{m}x.$

At each time t we have

$$x''(t) = -\frac{k}{m}x(t) \quad \text{and therefore} \quad x''(t) + \frac{k}{m}x(t) = 0.$$

Since $k/m > 0$, we can set $\omega = \sqrt{k/m}$ and write

$$x''(t) + \omega^2 x(t) = 0.$$

The motion of the bob is simple harmonic motion with period $T = 2\pi/\omega$. ❑

There is something remarkable about simple harmonic motion that we have not yet specifically pointed out; namely, that the frequency $f = \omega/2\pi$ is completely independent of the amplitude of the motion. The oscillations of the bob occur with frequency

$$f = \frac{\sqrt{k/m}}{2\pi}. \qquad \text{(here } \omega = \sqrt{k/m}\text{)}$$

By adjusting the spring constant k and the mass of the bob m, we can calibrate the spring-bob system so that the oscillations take place exactly once a second (at least almost exactly). We then have a primitive timepiece (a first cousin of the windup clock.) With the passing of time, friction and air resistance reduce the amplitude of the oscillations but not their frequency. This will be shown below. By giving the bob a little push or pull once in a while (by rewinding our clock), we can restore the amplitude of the oscillations and thus maintain the steady "ticking".

Damped Vibrations

We derived the equation of motion

$$x'' + \frac{k}{m}x = 0$$

from the force equation

$$F = -kx.$$

Unless the spring is frictionless and the motion takes place in a vacuum, there is a resistance to the motion that tends to dampen the vibrations. Experiment shows that the resistance force R is approximately proportional to the velocity x':

$$R = -cx'. \qquad (c > 0)$$

Taking this resistance term into account, the force equation reads

$$F = -k\,x - cx'.$$

Newton's law $F = ma = mx''$ then gives

$$mx'' = -cx' - kx,$$

which we can write as

(18.6.3)
$$\boxed{x'' + \frac{c}{m}x' + \frac{k}{m}x = 0.}$$

This is the equation of motion in the presence of a *damping factor.* To study the motion, we analyze this equation.

The characteristic equation

$$r^2 + \frac{c}{m}r + \frac{k}{m} = 0$$

has roots

$$r = \frac{-c \pm \sqrt{c^2 - 4km}}{2m}.$$

There are three possibilities:

$$c^2 - 4km < 0, \quad c^2 - 4km > 0, \quad c^2 - 4km = 0.$$

Case 1: $c^2 - 4km < 0$. In this case the characteristic equation has two complex roots:

$$r_1 = \frac{c}{2m} + i\omega, \quad r_2 = -\frac{c}{2m} - i\omega \quad \text{where} \quad \omega = \frac{\sqrt{4km - c^2}}{2m}.$$

The general solution of (18.6.3),

$$x = e^{-(c/2m)t}(C_1 \cos \omega t + C_2 \sin \omega t),$$

can also be written

(18.6.4)
$$\boxed{x(t) = A\, e^{(-c/2m)t} \sin\,(\omega t + \phi_0),}$$

where, as before, A and ϕ_0 are constants, $A > 0$, $\phi_0 \in [0, 2\pi)$. This is called the *underdamped case*. The motion is similar to simple harmonic motion, except that the damping term $e^{(-c/2m)t}$ ensures that $x \to 0$ as $t \to \infty$. The vibrations continue indefinitely with constant frequency $2\pi/\omega$ but with diminishing amplitude $A\, e^{(-c/2m)t}$. As $t \to \infty$, the amplitude of the vibrations tends to zero; the vibrations die down. The motion is illustrated in Figure 18.6.3. ❑

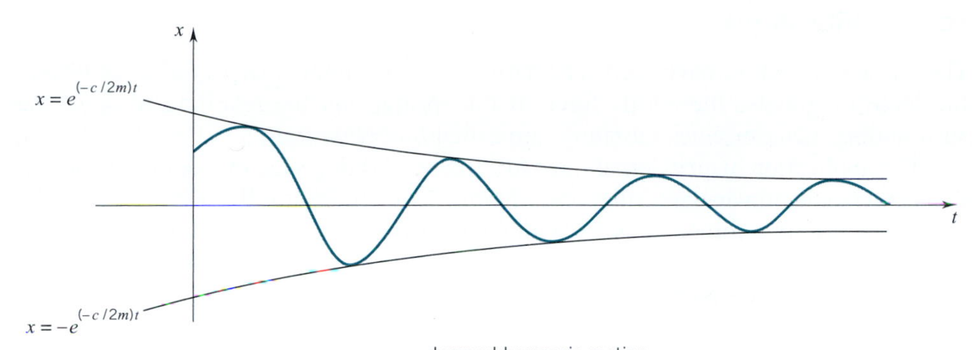

damped harmonic motion

Figure 18.6.3

Case 2: $c^2 - 4km > 0$. In this case the characteristic equation has two distinct real roots:

$$r_1 = \frac{-c + \sqrt{c^2 - 4km}}{2m}, \qquad r_2 = \frac{-c - \sqrt{c^2 - 4km}}{2m}.$$

The general solution takes the form

(18.6.5)
$$\boxed{x = C_1\, e^{r_1 t} + C_2\, e^{r_2 t}.}$$

This is called the *overdamped case*. The motion is nonoscillatory. Since $\sqrt{c^2 - 4km} < \sqrt{c^2} = c$, both r_1 and r_2 are negative. As $t \to \infty$, $x \to 0$. ❑

Case 3: $c^2 - 4km = 0$. In this case the characteristic equation has only one root

$$r_1 = -\frac{c}{2m},$$

and the general solution takes the form

(18.6.6)
$$\boxed{x = C_1\, e^{-(c/2m)t} + C_2\, te^{-(c/2m)t}.}$$

This is called the *critically damped case*. Once again the motion is nonoscillatory. Moreover, as $t \to \infty$, $x \to 0$. ❑

In both the overdamped and critically damped cases, the mass moves slowly back to its equilibrium position ($x \to 0$ as $t \to \infty$). Depending on the initial conditions, the mass may move through the equilibrium once, but only once; there is no oscillatory motion. Two typical examples of the motion are shown in Figure 18.6.4.

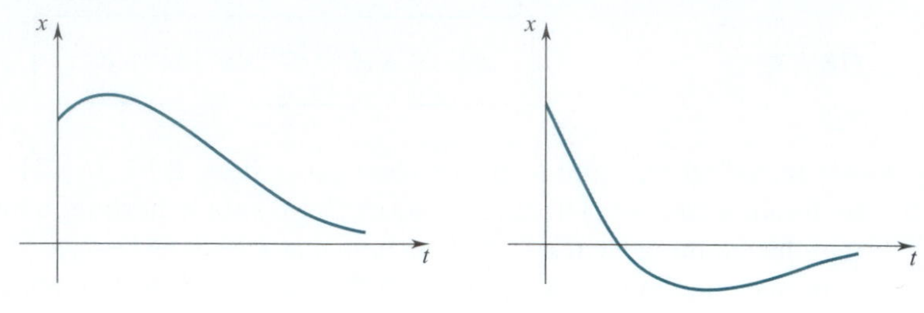

Figure 18.6.4

Forced Vibrations

The vibrations that we have been considering result from the interplay of three forces: the force of gravity, the elastic force of the spring, and the retarding force of the surrounding medium. Such vibrations are called *free vibrations.*

The application of an external force to a freely vibrating system modifies the vibrations and results in what are called *forced vibrations* . In what follows we examine the effect of a pulsating force $F_0 \cos \gamma t$. Without loss of generality we can take both F_0 and γ as positive.

In an undamped system the force equation reads

$$F = -kx + F_0 \cos \gamma t,$$

and the equation of motion takes the form

(18.6.7)
$$x'' + \frac{k}{m}x = \frac{F_0}{m}\cos \gamma t.$$

We set $\omega = \sqrt{k/m}$ and write

(18.6.8)
$$x'' + \omega^2 x = \frac{F_0}{m}\cos \gamma t.$$

As you'll see, the nature of the vibrations depends on the relation between the *applied frequency,* $\gamma/2\pi$, and the *natural frequency* of the system, $\omega/2\pi$.

Case 1: $\gamma \neq \omega$. In this case the method of undetermined coefficients gives the particular solution

$$x_p = \frac{F_0/m}{\omega^2 - \gamma^2}\cos \gamma t.$$

The general equation of motion can thus be written

(18.6.9)
$$x = A \sin(\omega t + \phi_0) + \frac{F_0/m}{\omega^2 - \gamma^2}\cos \gamma t.$$

If ω/γ is rational, the vibrations are periodic. If, on other hand, ω/γ is not rational, then the vibrations are not periodic and the motion, though bounded by

$$|A| + \left|\frac{F_0/m}{\omega^2 - \gamma^2}\right|,$$

can be highly irregular. These motions are illustrated in Figures 18.6.5 and 18.6.6. Each graph is the solution of the initial value problem:

$$x'' + 4x = 3\cos\gamma t,\ x(0) = 0,\ x'(0) = 1. \quad \square$$

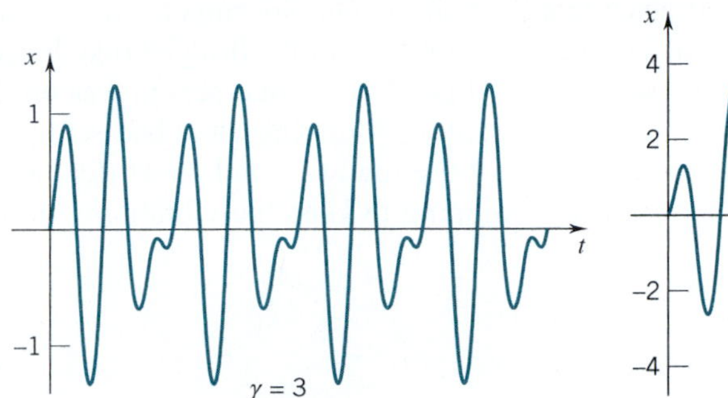

Figure 18.6.5

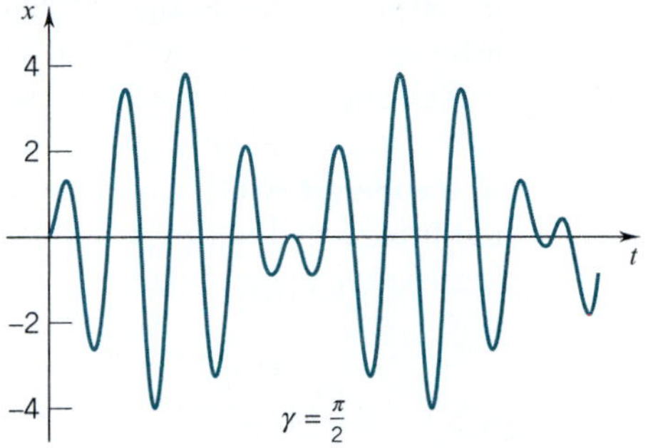

Figure 18.6.6

Case 2: $\gamma = \omega$. In this case the method of undetermined coefficients gives

$$x_p = \frac{F_0}{2\omega m}\, t \sin\omega t$$

and the general solution takes the form

(18.6.10)

$$x = A\sin(\omega t + \phi_0) + \frac{F_0}{2\omega m}\, t\sin\omega t.$$

The undamped system is said to be in *resonance*. The motion is oscillatory but, because of the extra t present in the second-term, it is far from periodic. As $t \to \infty$, the amplitude of vibration increases without bound. The motion is illustrated in Figure 18.6.7. The graph is the solution of the initial value problem

$$x'' + 4x = 3\cos 2t,\ x(0) = 0,\ x'(0) = 1. \quad \square$$

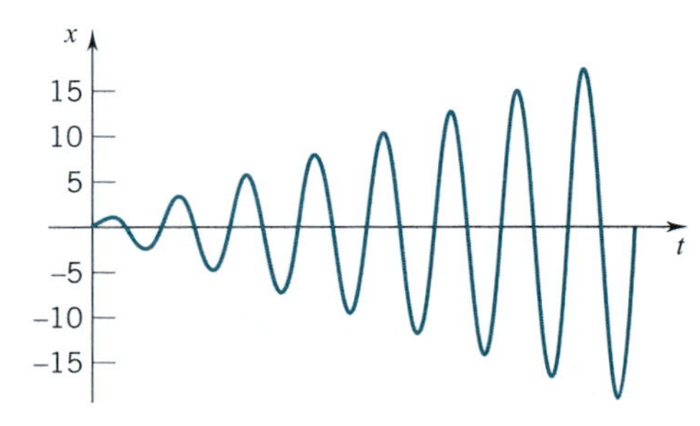

Figure 18.6.7

Undamped systems and unbounded vibrations are mathematical fictions. No real mechanical system is totally undamped, and unbounded vibrations do not occur in nature. Nevertheless, a form of resonance can occur in a real mechanical system. (See Exercises 24–28.) A periodic external force applied to a mechanical system that is insufficiently damped can set up vibrations of very large amplitude. Such vibrations have caused the destruction of some formidable man-made structures. In 1850 the suspension bridge at Angers, France, was destroyed by vibrations set up by the unified step of a column of marching soldiers. More than two hundred French soldiers were killed in that catastrophe. (Soldiers today are told to break ranks before crossing a bridge.) The collapse of the bridge at Tacoma, Washington, is a more recent event. Slender in construction and graceful in design, the Tacoma bridge was opened to traffic on July 1, 1940. The third longest suspension bridge in the world, with a main span of 2800 feet, the bridge attracted many admirers. On November 1 of that same year, after less than five months of service, the main span of the bridge broke loose from its cables and crashed into the water below. (Luckily only one person was on the bridge at the time, and he was able to crawl to safety.) A driving wind had set up vibrations in resonance with the natural vibrations of the roadway, and the stiffening girders of the bridge had not provided sufficient damping to keep the vibrations from reaching destructive magnitude.

EXERCISES 18.6

1. An object is in simple harmonic motion. Find an equation for the motion given that the period is $\frac{1}{4}\pi$ and, at time $t = 0, x = 1$ and $v = 0$. What is the amplitude? What is the frequency?

2. An object is in simple harmonic motion. Find an equation for the motion given that the frequency is $1/\pi$ and, at time $t = 0, x = 0$ and $v = -2$. What is the amplitude? What is the period?

3. An object is in simple harmonic motion with period T and amplitude A. What is the velocity at the central point $x = 0$?

4. An object is in simple harmonic motion with period T. Find the amplitude given that $v = \pm v_0$ at $x = x_0$.

5. An object in simple harmonic motion passes through the central point $x = 0$ at time $t = 0$ and every 3 seconds thereafter. Find the equation of motion given that $v(0) = 5$.

6. Show that simple harmonic motion $x(t) = A \sin(\omega t + \phi_0)$ can just as well be written: (a) $x(t) = A\cos(\omega t + \phi_1)$; (b) $x(t) = B\sin\omega t + C\cos\omega t$.

Exercises 7–12 relate to the motion of the bob depicted in Figure 18.6.2.

7. What is $x(t)$ for the bob of mass m?

8. Find the positions of the bob where it attains: (a) maximum speed; (b) zero speed; (c) maximum acceleration; (d) zero acceleration.

9. Where does the bob take on half of its maximum speed?

10. Find the maximal kinetic energy obtained by the bob. (Remember: $\text{KE} = \frac{1}{2}mv^2$, where m is the mass of the object and v is the speed.)

11. Find the time average of the kinetic energy of the bob during one period T.

12. Express the velocity of the bob in terms of k, m, x_0, and $x(t)$.

13. Given that $x''(t) = 8 - 4x(t)$ with $x(0) = 0$ and $x'(0) = 0$, show that the motion is simple harmonic motion centered at $x = 2$. Find the amplitude and the period.

14. The figure shows a pendulum of mass m swinging on an arm of length L. The angle θ is measured counterclockwise. Neglecting friction and the weight of the arm, we can describe the motion by the equation

$$mL\theta''(t) = -mg\sin\theta(t),$$

which reduces to

$$\theta''(t) = -\frac{g}{L}\sin\theta(t).$$

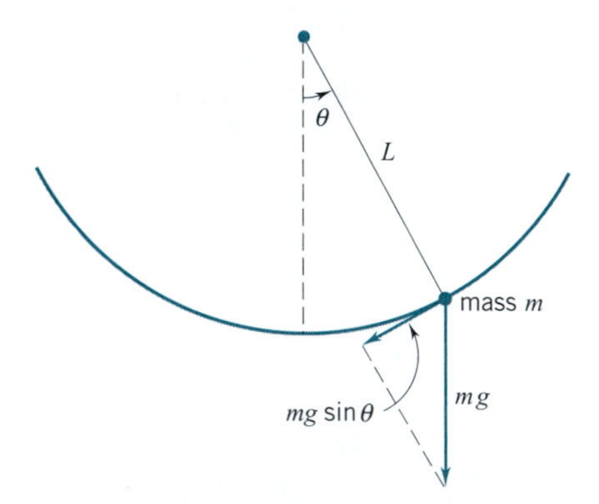

(a) For small angles we replace $\sin\theta$ by θ and write

$$\theta''(t) \cong -\frac{g}{L}\theta(t).$$

Justify this step.

(b) Solve the approximate equation in part (a) part given that the pendulum

(i) is held at an angle $\theta_0 > 0$ and released at time $t = 0$.
(ii) passes through the vertical position at time $t = 0$ and $\theta'(0) = -\sqrt{g/L}\,\theta_1$.

(c) Find L given that the motion repeats itself every 2 seconds.

15. A cylindrical buoy of mass m and radius r centimeters floats with its axis vertical in a liquid of density ρ kilograms per cubic centimeter. Suppose that the buoy is pushed x_0 centimeters down into the liquid. See the figure.

(a) Neglecting friction and given that the buoyancy force is equal to the weight of the liquid displaced, show that the buoy bobs up and down in simple harmonic motion by finding the equation of motion.

(b) Solve the equation obtained in part (a). Specify the amplitude and the period.

16. Explain in detail the connection between uniform circular motion and simple harmonic motion.

17. What is the effect of an increase in the resistance constant c on the amplitude and frequency of the vibrations given by Equation (18.6.4)?

18. Prove that the motion given by (18.6.5) can pass through the equilibrium point at most once. How many times can the motion change directions?

19. Prove that the motion given by (18.6.6) can pass through the equilibrium point at most once. How many times can the motion change directions?

20. Show that, if $\gamma \neq \omega$, then the method of undetermined coefficients applied to (18.6.8) gives

$$x_p = \frac{F_0/m}{\omega^2 - \gamma^2}\cos\gamma t.$$

21. Show that if ω/γ is rational, then the vibrations given by (18.6.9) are periodic.

22. Show that, if $\gamma = \omega$, then the method of undetermined coefficients applied to (18.6.8) gives

$$x_p = \frac{F_0}{2\omega m} t\sin\omega t.$$

Forced Vibrations in a Damped System

Write the equation

$$x'' + \frac{c}{m}x' + \frac{k}{m}x = \frac{F_0}{m}\cos\gamma t$$

as

$$(*) \qquad x'' + 2\alpha x' + \omega^2 x = \frac{F_0}{m}\cos\gamma t.$$

We will assume throughout that $0 < \alpha < \omega$. (For large α the resistance is large and the motion is not as interesting.)

23. Find the general solution of the reduced equation $x'' + 2\alpha x' + \omega^2 x = 0$.

24. Verify that the function

$$x_p = \frac{F_0/m}{(\omega^2-\gamma^2)^2 + 4\alpha^2\gamma^2}[(\omega^2-\gamma^2)\cos\gamma t + 2\alpha\gamma\sin\gamma t]$$

is a particular solution of $(*)$.

25. Determine x_p if $\omega = \gamma$. Show that the amplitude of the vibrations is very large if the resistance constant c is very small.

26. Show that the solution x_p in Exercise 24 can be written

$$x_p = \frac{F_0/m}{\sqrt{(\omega^2-\gamma^2)^2 + 4\alpha^2\gamma^2}}\sin(\gamma t + \phi).$$

27. Show that, if $2\alpha^2 \geq \omega^2$, then the amplitude of vibration of the solution x_p in Exercise 26 decreases as γ increases.

28. Suppose now that $2\alpha^2 \leq \omega^2$.

(a) Find the value of γ that maximizes the amplitude of the solution x_p in Exercise 26.

(b) Determine the frequency that corresponds to this value of γ. (This is called the *resonant frequency* of the damped system).

(c) What is the *resonant amplitude* of the system? (In other words, what is the amplitude of the vibrations if the applied force is at resonant frequency?)

(d) Show that, if c, the constant of resistance, is very small, then the resonant amplitude is very large.

■ CHAPTER HIGHLIGHTS

18.1 Introduction; Review of Equations Already Considered

ordinary differential equation (p. 1096)
order (p. 1097)
first-order linear differential equations (p. 1097)
separable first-order equations (p. 1097)
equations of harmonic motion (p. 1098)

18.2 Bernoulli Equation; Homogeneous Equations; Numerical Methods

Bernoulli equation: $y' + p(x)y = q(x)y^r$ (p. 1099)
method of solution — transformation into a linear equation (p. 1100)
homogeneous function (p. 1101)

homogeneous equation: $y' = f(x, y)$ where f is a homogeneous function to degree 0 (p. 1101)
method of solution — transformation into a separable equation (p. 1101)
Euler method (p. 1103)
Runge-Kutta method (p. 1105)

18.3 Exact Differential Equations; Integrating Factors

exact differential equation: $P(x, y) + Q(x, y)\, y' = 0$ (p. 1108)
integrating factor (p. 1109)
transformation into an exact differential equation (p. 1111, p. 1112)

18.4 The Equation $y'' + ay' + by = 0$

the reduced equation (homogeneous equation) (p. 1113)
the characteristic equation (p. 1113)
linear combinations of solutions (p. 1115)
existence and uniqueness of solutions (p. 1115)
Wronskians (p. 1116)
the general solution of $y'' + ay' + by = 0$ (p. 1118)

18.5 The Equation $y'' + ay' + by = \phi(x)$

the complete equation (nonhomogeneous equation) (p. 1123)
forcing function (p. 1123)
particular solution of $y'' + ay' + by = \phi(x)$ (p. 1123)
the general solution of $y'' + ay' + by = \phi(x)$ (p. 1123)
superposition principle (p. 1124)
variation of parameters (p. 1124)
undetermined coefficients (p. 1125)

18.6 Mechanical Vibrations

simple harmonic motion: $x'' + \omega^2 x = 0$ (p. 1130)
undamped vibrations (p. 1132)
damped vibrations: $x'' + \frac{c}{m}x' + \frac{k}{m}x = 0$ (p. 1132)
damping factor (p. 1134)
underdamped (p. 1135) overdamped (p. 1135)
critically damped (p. 1135)
forced vibrations: $x'' + \omega^2 x = \frac{F_0}{m}\cos \gamma t$ (p. 1136)
applied frequency (p. 1136) natural frequency (p. 1136)

APPENDIX A SOME ADDITIONAL TOPICS

■ A.1 ROTATION OF AXES; EQUATIONS OF SECOND DEGREE

Rotation of Axes

We begin by referring to Figure A.1.1. From the figure,

$$\cos\theta = \frac{x}{r}, \quad \sin\theta = \frac{y}{r}.$$

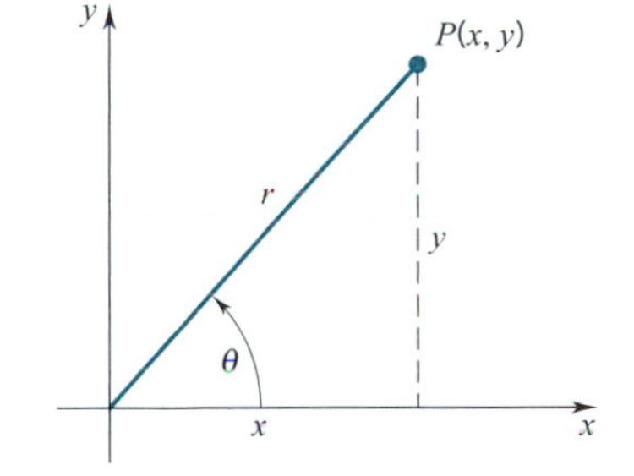

Figure A.1.1

Therefore

(A.1.1)
$$\boxed{x = r\cos\theta, \quad y = r\sin\theta.}$$

These equations come up repeatedly in calculus. In particular, these are the equations that we use to convert polar coordinates to rectangular coordinates.

Consider now a rectangular coordinate system Oxy. By rotating this system about the origin counterclockwise through an angle of α radians, we obtain a new coordinate system OXY. See Figure A.1.2.

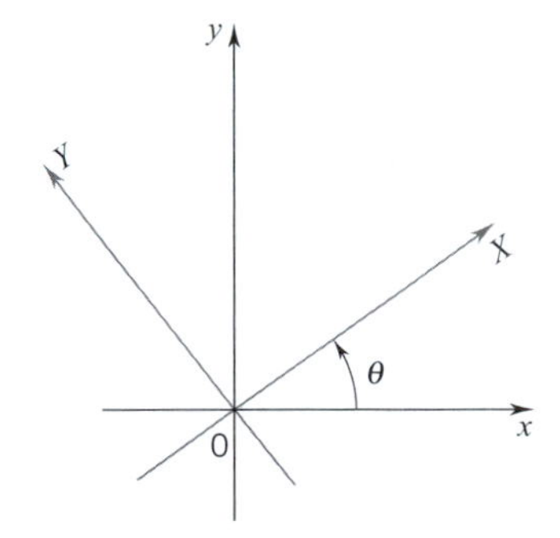

Figure A.1.2

A point P now has two pairs of rectangular coordinates:

(x, y) in the Oxy system and (X, Y) in the OXY system.

Here we investigate the relation between (x, y) and (X, Y). With P as in Figure A.1.3,

$$x = r\cos(\alpha+\beta), \quad y = r\sin(\alpha+\beta)$$

and

$$X = r\cos\beta, \quad Y = r\sin\beta.$$

Since

$$\cos(\alpha+\beta) = \cos\alpha\cos\beta - \sin\alpha\sin\beta,$$
$$\sin(\alpha+\beta) = \sin\alpha\cos\beta + \cos\alpha\sin\beta,$$

we have

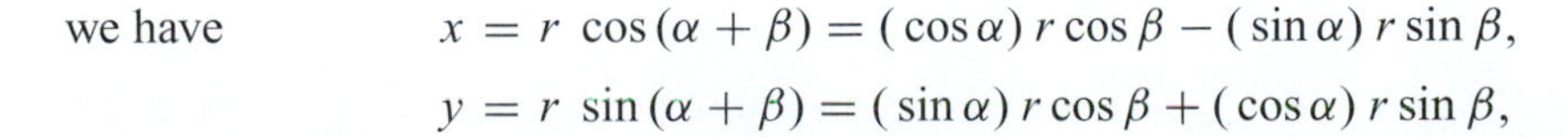

$$x = r\cos(\alpha+\beta) = (\cos\alpha)\,r\cos\beta - (\sin\alpha)\,r\sin\beta,$$
$$y = r\sin(\alpha+\beta) = (\sin\alpha)\,r\cos\beta + (\cos\alpha)\,r\sin\beta,$$

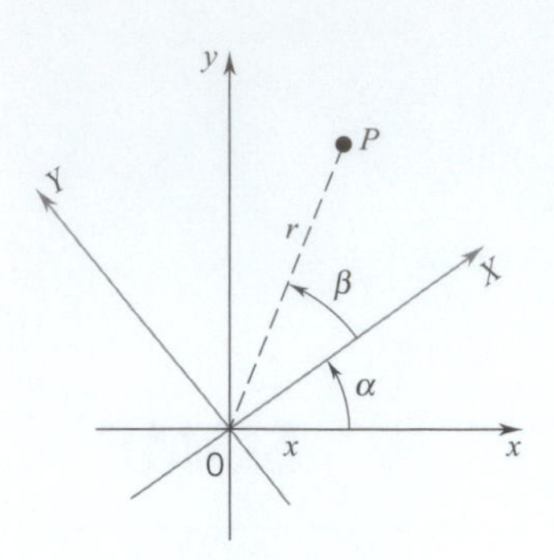

Figure A.1.3

and therefore

(A.1.2)
$$x = (\cos\alpha)X - (\sin\alpha)Y, \quad y = (\sin\alpha)X + (\cos\alpha)Y.$$

These formulas give the algebraic consequences of a counterclockwise rotation of α radians.

Equations of Second Degree

As you know, the graph of an equation of the form

$$ax^2 + cy^2 + dx + ey + f = 0 \qquad (a, c \text{ not both } 0)$$

is a conic section (except for degenerate cases).

The *general equation of second degree in x and y* is an equation of the form

(A.1.3)
$$ax^2 + bxy + cy^2 + dx + ey + f = 0$$

with a, b, c not all 0. The graph of such an equation is still a conic section (again, except for degenerate cases). For example, the graph of

$$xy - 2 = 0$$

is the hyperbola shown in Figure A.1.4.

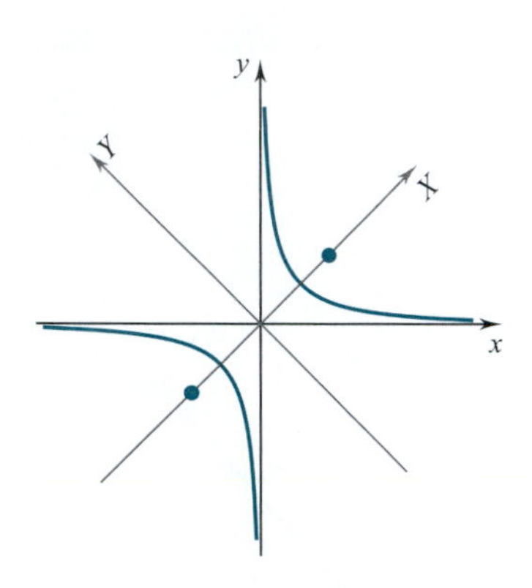

Figure A.1.4

Eliminating the *xy*-Term

Rotations of the coordinate system enable us to simplify equations of the second degree by eliminating the xy-term; that is, if in the Oxy coordinate system, a curve S has an equation of the form

$$(1) \qquad ax^2 + bxy + cy^2 + dx + ey + f = 0 \qquad \text{with} \qquad b \neq 0,$$

then there exists a coordinate system OXY, differing from Oxy by a rotation α, with $0 < \alpha < \pi/2$, such that in the OXY system S has an equation of the form

$$(2) \qquad AX^2 + CY^2 + DX + EY + F = 0,$$

with A and C not both zero. To see this, substitute

$$x = (\cos\alpha)X - (\sin\alpha)Y, \quad y = (\sin\alpha)X + (\cos\alpha)Y$$

in equation (1). This will give you a second-degree equation in X and Y in which the coefficient of XY is

$$-2a\cos\alpha\sin\alpha + b(\cos^2\alpha - \sin^2\alpha) + 2c\cos\alpha\sin\alpha.$$

This can be simplified to

$$(c - a)\sin 2\alpha + b\cos 2\alpha.$$

To eliminate the XY term, we must have this coefficient equal to zero, that is, we must have

$$b\cos 2\alpha = (a - c)\sin 2\alpha,$$

which, with $b \neq 0$, gives

$$\cot 2\alpha = \frac{a-c}{b}.$$

With $0 < \alpha < \frac{\pi}{2}$, we have $0 < 2\alpha < \pi$. Therefore

$$2\alpha = \cot^{-1}\left(\frac{a-c}{b}\right)$$

and

$$\alpha = \frac{1}{2}\cot^{-1}\left(\frac{a-c}{b}\right).$$

We have shown that an equation of the form (1) can be transformed into an equation of the form (2) by rotating the axes through the angle α given by

(A.1.4)

$$\boxed{\alpha = \frac{1}{2}\cot^{-1}\left(\frac{a-c}{b}\right)}$$

We leave it as an exercise to show that the coefficients A and C in (2) are not both zero.

Example 1 In the case of $xy - 2 = 0$, we have

$a = c = 0, b = 1$, and $\alpha = \frac{1}{2}\cot^{-1}(0) = \frac{1}{2}(\frac{\pi}{2}) = \frac{1}{4}\pi$.

Setting

$$x = (\cos\tfrac{1}{4}\pi)X - (\sin\tfrac{1}{4}\pi)Y = \tfrac{1}{2}\sqrt{2}(X - Y),$$

$$y = (\sin\tfrac{1}{4}\pi)X + (\cos\tfrac{1}{4}\pi)Y = \tfrac{1}{2}\sqrt{2}(X + Y),$$

we find that $xy - 2 = 0$ becomes

$$\tfrac{1}{2}(X^2 - Y^2) - 2 = 0,$$

which can be written

$$\frac{X^2}{4} - \frac{Y^2}{4} = 1.$$

This is the equation of a hyperbola in standard position in the OXY system. The hyperbola is shown in Figure A.1.4. ❑

Example 2 In the case of $11x^2 + 4\sqrt{3}xy + 7y^2 - 1 = 0$, we have

$a = 11, b = 4\sqrt{3}$, and $c = 7$.

Thus we choose

$$\alpha = \tfrac{1}{2}\cot^{-1}\left(\frac{11-7}{4\sqrt{3}}\right) = \tfrac{1}{2}\cot^{-1}\left(\frac{1}{\sqrt{3}}\right) = \tfrac{1}{6}\pi.$$

Setting

$$x = (\cos\tfrac{1}{6}\pi)X - (\sin\tfrac{1}{6}\pi)Y = \tfrac{1}{2}(\sqrt{3}X - Y),$$

$$y = (\sin\tfrac{1}{6}\pi)X + (\cos\tfrac{1}{6}\pi)Y = \tfrac{1}{2}(X + \sqrt{3}Y),$$

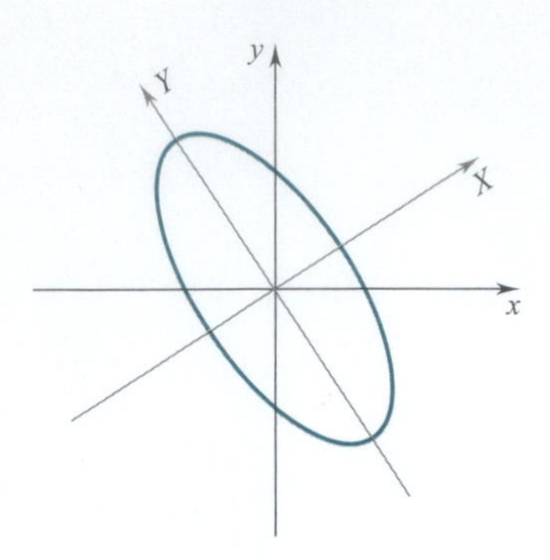

Figure A.1.5

we find that our initial equation simplifies to $13X^2 + 5Y^2 - 1 = 0$, which we can write as

$$\frac{X^2}{(1/\sqrt{13})^2} + \frac{Y^2}{(1/\sqrt{5})^2} = 1.$$

This is the equation of an ellipse. The ellipse is pictured in Figure A.1.5. □

The Discriminant

It is possible to draw general conclusions about the graph of a second-degree equation

$$ax^2 + bxy + cy^2 + dx + ey + f = 0, \quad a, b, c \text{ not all } 0,$$

just from the *discriminant* $\Delta = b^2 - 4ac$. There are three cases:

Case 1. If $\Delta < 0$, the graph is an ellipse, a circle, a point, or empty.

Case 2. If $\Delta > 0$, the graph is a hyperbola or a pair of intersecting lines.

Case 3. If $\Delta = 0$, the graph is a parabola, a line, a pair of lines, or empty.

Below we outline how these assertions can be verified. A useful first step is to rotate the coordinate system so that the equation takes the form

$$AX^2 + CY^2 + DX + EY + F = 0. \tag{3}$$

An elementary but time-consuming computation shows that the discriminant is unchanged by a rotation, so that in this instance we have

$$\Delta = b^2 - 4ac = -4AC.$$

Moreover, A and C cannot both be zero. If $\Delta < 0$, then $AC > 0$, and we can rewrite (3) as

$$\frac{X^2}{C} + \frac{D}{AC}X + \frac{Y^2}{A} + \frac{E}{AC}Y + \frac{F}{AC} = 0.$$

By completing the squares, we obtain an equation of the form

$$\frac{(X-\alpha)^2}{(\sqrt{|C|})^2} + \frac{(Y-\beta)^2}{(\sqrt{|A|})^2} = K.$$

If $K > 0$, we have an ellipse or a circle. If $K = 0$, we have the point (α, β). If $K < 0$, the set is empty.

If $\Delta > 0$, then $AC < 0$. Proceeding as before, we obtain an equation of the form

$$\frac{(X-\alpha)^2}{(\sqrt{|C|})^2} - \frac{(Y-\beta)^2}{(\sqrt{|A|})^2} = K.$$

If $K \neq 0$, we have a hyperbola. If $K = 0$, the equation becomes

$$\left(\frac{X-\alpha}{\sqrt{|C|}} - \frac{Y-\beta}{\sqrt{|A|}}\right)\left(\frac{X-\alpha}{\sqrt{|C|}} + \frac{Y-\beta}{\sqrt{|A|}}\right) = 0,$$

so that we have a pair of lines intersecting at the point (α, β).

If $\Delta = 0$, then $AC = 0$, so that either $A = 0$ or $C = 0$. Since A and C are not both zero, there is no loss in generality in assuming that $A \neq 0$ and $C = 0$. In this case equation (3) reduces to

$$AX^2 + DX + EY + F = 0.$$

Dividing by A and completing the square, we have an equation of the form

$$(X - \alpha)^2 = \beta Y + K.$$

If $\beta \neq 0$, we have a parabola. If $\beta = 0$ and $K = 0$, we have a line. If $\beta = 0$ and $K > 0$, we have a pair of parallel lines. If $\beta = 0$ and $K < 0$, the set is empty.

EXERCISES A.1

In Exercises 1–8, (a) use the discriminant to identify the curve; (b) find a rotation $\alpha \in (0, \pi/2)$ that eliminates the xy-term; (c) rewrite the equation in terms of the new coordinate system; (d) sketch the curve displaying both coordinate systems.

1. $xy = 1$.

2. $xy - y + x = 1$.

3. $11x^2 + 10\sqrt{3}xy + y^2 - 4 = 0$.

4. $52x^2 - 72xy + 73y^2 - 100 = 0$.

5. $x^2 - 2xy + y^2 + x + y = 0$.

6. $3x^2 + 2\sqrt{3}xy + y^2 - 2x + 2\sqrt{3}y = 0$.

7. $x^2 + 2\sqrt{3}xy + 3y^2 + 2\sqrt{3}x - 2y = 0$.

8. $2x^2 + 4\sqrt{3}xy + 6y^2 + (8 - \sqrt{3})x + (8\sqrt{3} + 1)y + 8 = 0$.

In Exercises 9 and 10, find a rotation $\alpha \in (0, \pi/2)$ that eliminates the xy-term. Then find $\cos\alpha$ and $\sin\alpha$.

9. $x^2 + xy + Kx + Ly + M = 0$.

10. $5x^2 + 24xy + 12y^2 + Kx + Ly + M = 0$.

11. Show that after a rotation of axes through an angle α, the coefficients in the equation

$$AX^2 + BXY + CY^2 + DX + EY + F = 0$$

are related to the coefficients in the equation

$$ax^2 + bxy + cy^2 + dx + ey + f = 0, \quad a, b, c \text{ not all } 0,$$

as follows:

$$\begin{aligned}
A &= a\cos^2\alpha + b\cos\alpha\,\sin\alpha + c\sin^2\alpha,\\
B &= 2(c - a)\cos\alpha\sin\alpha + b(\cos^2\alpha - \sin^2\alpha),\\
C &= a\sin^2\alpha - b\cos\alpha\sin\alpha + c\cos^2\alpha,\\
D &= d\cos\alpha + e\sin\alpha,\\
E &= e\cos\alpha - d\sin\alpha,\\
F &= f.
\end{aligned}$$

12. Use the results of Exercise 11 to show the following:

(a) $B^2 - 4AC = b^2 - 4ac$.

(b) If $B = 0$, then A and C cannot both be 0.

■ A.2 DETERMINANTS

By a *matrix* we mean a rectangular arrangement of numbers enclosed in parentheses. For example,

$$\begin{pmatrix} 2 & 4 \\ 3 & 1 \end{pmatrix} \quad \begin{pmatrix} 1 & 6 & 3 \\ 5 & 2 & 2 \end{pmatrix} \quad \begin{pmatrix} 2 & 4 & 0 \\ 4 & 7 & 1 \\ 0 & 1 & 1 \end{pmatrix}$$

are all matrices. The numbers that appear in a matrix are called the *entries*.

Each matrix has a certain number of rows and a certain number of columns. A matrix with m rows and n columns is called an $m \times n$ *matrix*. Thus the first matrix above is a 2×2 matrix, the second a 2×3 matrix, the third a 3×3 matrix. The first and third matrices are called *square*; they have the same number of rows as columns. Here we will be working with square matrices as these are the only ones that have determinants.

We could give a definition of determinant that is applicable to all square matrices, but the definition is complicated and would serve little purpose at this point. Our interest here is in the 2×2 case and in the 3×3 case. We begin with the 2×2 case.

(A.2.1)

> The *determinant* of the matrix
> $$\begin{pmatrix} a_1 & a_2 \\ b_1 & b_2 \end{pmatrix}$$
> is the number $a_1b_2 - a_2b_1$.

We have a special notation for the determinant. We change the parentheses of the matrix to vertical bars:

$$\text{Determinant of} \begin{pmatrix} a_1 & a_2 \\ b_1 & b_2 \end{pmatrix} = \begin{vmatrix} a_1 & a_2 \\ b_1 & b_2 \end{vmatrix} = a_1b_2 - a_2b_1.$$

Thus, for example,

$$\begin{vmatrix} 5 & 8 \\ 4 & 2 \end{vmatrix} = (5 \cdot 2) - (8 \cdot 4) = 10 - 32 = -22$$

$$\text{and} \quad \begin{vmatrix} 4 & 0 \\ 0 & \frac{1}{4} \end{vmatrix} = (4 \cdot \tfrac{1}{4}) - (0 \cdot 0) = 1.$$

We remark on three properties of 2×2 determinants:

1. If the rows or columns of a 2×2 determinant are interchanged, the determinant changes sign:

$$\begin{vmatrix} b_1 & b_2 \\ a_1 & a_2 \end{vmatrix} = - \begin{vmatrix} a_1 & a_2 \\ b_1 & b_2 \end{vmatrix}, \qquad \begin{vmatrix} a_2 & a_1 \\ b_2 & b_1 \end{vmatrix} = - \begin{vmatrix} a_1 & a_2 \\ b_1 & b_2 \end{vmatrix}.$$

PROOF Just note that

$$b_1a_2 - b_2a_1 = -(a_1b_2 - a_2b_1) \quad \text{and} \quad a_2b_1 - a_1b_2 = -(a_1b_2 - a_2b_1). \quad \square$$

2. A common factor can be removed from any row or column and placed as a factor in front of the determinant:

$$\begin{vmatrix} \lambda a_1 & \lambda a_2 \\ b_1 & b_2 \end{vmatrix} = \lambda \begin{vmatrix} a_1 & a_2 \\ b_1 & b_2 \end{vmatrix}, \qquad \begin{vmatrix} \lambda a_1 & a_2 \\ \lambda b_1 & b_2 \end{vmatrix} = \lambda \begin{vmatrix} a_1 & a_2 \\ b_1 & b_2 \end{vmatrix}.$$

PROOF Just note that

$$(\lambda a_1)b_2 - (\lambda a_2)b_1 = \lambda(a_1b_2 - a_2b_1)$$

and

$$(\lambda a_1)b_2 - a_2(\lambda b_1) = \lambda(a_1b_2 - a_2b_1). \quad \square$$

3. If the rows or columns of a 2×2 determinant are the same, the determinant is 0.

PROOF

$$\begin{vmatrix} a_1 & a_2 \\ a_1 & a_2 \end{vmatrix} = a_1a_2 - a_2a_1 = 0, \qquad \begin{vmatrix} a_1 & a_1 \\ b_1 & b_1 \end{vmatrix} = a_1b_1 - a_1b_1 = 0. \quad \square$$

The determinant of a 3×3 matrix is harder to define. One definition is this:

$$\begin{vmatrix} a_1 & a_2 & a_3 \\ b_1 & b_2 & b_3 \\ c_1 & c_2 & c_3 \end{vmatrix} = a_1 b_2 c_3 - a_1 b_3 c_2 + a_2 b_3 c_1 - a_2 b_1 c_3 + a_3 b_1 c_2 - a_3 b_2 c_1.$$

The problem with this definition is that it is hard to remember. What saves us is that the expansion on the right can be conveniently written in terms of 2×2 determinants; namely, the expression on the right can be written

$$a_1(b_2 c_3 - b_3 c_2) - a_2(b_1 c_3 - b_3 c_1) + a_3(b_1 c_2 - b_2 c_1),$$

which turns into

$$a_1 \begin{vmatrix} b_2 & b_3 \\ c_2 & c_3 \end{vmatrix} - a_2 \begin{vmatrix} b_1 & b_3 \\ c_1 & c_3 \end{vmatrix} + a_3 \begin{vmatrix} b_1 & b_2 \\ c_1 & c_2 \end{vmatrix}.$$

We then have

(A.2.2)
$$\begin{vmatrix} a_1 & a_2 & a_3 \\ b_1 & b_2 & b_3 \\ c_1 & c_2 & c_3 \end{vmatrix} = a_1 \begin{vmatrix} b_2 & b_3 \\ c_2 & c_3 \end{vmatrix} - a_2 \begin{vmatrix} b_1 & b_3 \\ c_1 & c_3 \end{vmatrix} + a_3 \begin{vmatrix} b_1 & b_2 \\ c_1 & c_2 \end{vmatrix}.$$

We will take this as our definition. It is called the *expansion of the determinant by elements of the first row.* Note that the coefficients are the entries a_1, a_2, a_3 of the first row, that they occur alternately with + and −signs, and that each is multiplied by a determinant. You can remember which determinant goes with which entry a_i as follows: in the original matrix, mentally cross out the row and column in which the entry a_i is found, and take the determinant of the remaining 2×2 matrix. For example, the determinant that goes with a_3 is

$$\begin{vmatrix} \not{a_1} & \not{a_2} & \not{a_3} \\ b_1 & b_2 & \not{b_3} \\ c_1 & c_2 & \not{c_3} \end{vmatrix} = \begin{vmatrix} b_1 & b_2 \\ c_1 & c_2 \end{vmatrix}.$$

When first starting to work with specific 3×3 determinants, it is a good idea to set up the formula with blank 2×2 determinants:

$$\begin{vmatrix} a_1 & a_2 & a_3 \\ b_1 & b_2 & b_3 \\ c_1 & c_2 & c_3 \end{vmatrix} = a_1 \begin{vmatrix} & \\ & \end{vmatrix} - a_2 \begin{vmatrix} & \\ & \end{vmatrix} + a_3 \begin{vmatrix} & \\ & \end{vmatrix}$$

and then fill in the 2×2 determinants by using the "crossing out" rule explained above.

Example 1

$$\begin{aligned} \begin{vmatrix} 1 & 2 & 1 \\ 0 & 3 & 4 \\ 6 & 2 & 5 \end{vmatrix} &= 1 \begin{vmatrix} 3 & 4 \\ 2 & 5 \end{vmatrix} - 2 \begin{vmatrix} 0 & 4 \\ 6 & 5 \end{vmatrix} + 1 \begin{vmatrix} 0 & 3 \\ 6 & 2 \end{vmatrix} \\ &= 1(15 - 8) - 2(0 - 24) + 1(0 - 18) \\ &= 7 + 48 - 18 = 37. \quad \square \end{aligned}$$

A straightforward (but somewhat laborious) calculation shows that 3×3 determinants have the three properties we proved earlier for 2×2 determinants.

1. If two rows or columns are interchanged, the determinant changes sign.
2. A common factor can be removed from any row or column and placed as a factor in front of the determinant.
3. If two rows or columns are the same, the determinant is 0.

EXERCISES A.2

Evaluate the following determinants.

1. $\begin{vmatrix} 1 & 2 \\ 3 & 4 \end{vmatrix}$.

2. $\begin{vmatrix} 1 & -1 \\ -1 & 1 \end{vmatrix}$.

3. $\begin{vmatrix} 1 & 1 \\ a & a \end{vmatrix}$.

4. $\begin{vmatrix} a & b \\ b & d \end{vmatrix}$.

5. $\begin{vmatrix} 1 & 0 & 3 \\ 2 & 4 & 1 \\ 0 & 1 & 0 \end{vmatrix}$.

6. $\begin{vmatrix} 1 & 0 & 0 \\ 0 & 2 & 0 \\ 0 & 0 & 3 \end{vmatrix}$.

7. $\begin{vmatrix} 0 & 0 & 1 \\ 0 & 2 & 0 \\ 3 & 0 & 0 \end{vmatrix}$.

8. $\begin{vmatrix} a & 0 & 0 \\ b & c & 0 \\ d & e & f \end{vmatrix}$.

9. If A is a matrix, its *transpose* A^T is obtained by interchanging the rows and columns. Thus

$$\begin{pmatrix} a_1 & a_2 \\ b_1 & b_2 \end{pmatrix}^T = \begin{pmatrix} a_1 & b_1 \\ a_2 & b_2 \end{pmatrix}$$

and

$$\begin{pmatrix} a_1 & a_2 & a_3 \\ b_1 & b_2 & b_3 \\ c_1 & c_2 & c_3 \end{pmatrix}^T = \begin{pmatrix} a_1 & b_1 & c_1 \\ a_2 & b_2 & c_2 \\ a_3 & b_3 & c_3 \end{pmatrix}.$$

Show that the determinant of a matrix equals the determinant of its transpose: (a) for the 2×2 case; (b) for the 3×3 case.

Justify the assertions made in Exercises 10–14 by invoking the relevant properties of determinants.

10. $\begin{vmatrix} 1 & 2 & 3 \\ 4 & 5 & 6 \\ 7 & 8 & 9 \end{vmatrix} + \begin{vmatrix} 4 & 5 & 6 \\ 1 & 2 & 3 \\ 7 & 8 & 9 \end{vmatrix} = 0.$

11. $\begin{vmatrix} 1 & 2 & 3 \\ 4 & 5 & 6 \\ 7 & 8 & 9 \end{vmatrix} = \begin{vmatrix} 4 & 5 & 6 \\ 7 & 8 & 9 \\ 1 & 2 & 3 \end{vmatrix}.$

12. $\begin{vmatrix} 1 & 2 & 3 \\ 4 & 5 & 6 \\ 7 & 8 & 9 \end{vmatrix} + \begin{vmatrix} 1 & 2 & 3 \\ 1 & 2 & 3 \\ 7 & 8 & 9 \end{vmatrix} = \begin{vmatrix} 1 & 2 & 3 \\ 4 & 5 & 6 \\ 7 & 8 & 9 \end{vmatrix}.$

13. $\frac{1}{2}\begin{vmatrix} 1 & 0 & 7 \\ 3 & 4 & 5 \\ 2 & 4 & 6 \end{vmatrix} = \begin{vmatrix} 1 & 0 & 7 \\ 3 & 4 & 5 \\ 1 & 2 & 3 \end{vmatrix}.$

14. $\begin{vmatrix} 1 & 2 & 3 \\ x & 2x & 3x \\ 4 & 5 & 6 \end{vmatrix} = 0.$

15. (a) Verify that the equations

$$3x + 4y = 6$$
$$2x - 3y = 7$$

can be solved by the prescription

$$x = \frac{\begin{vmatrix} 6 & 4 \\ 7 & -3 \end{vmatrix}}{\begin{vmatrix} 3 & 4 \\ 2 & -3 \end{vmatrix}}, \quad y = \frac{\begin{vmatrix} 3 & 6 \\ 2 & 7 \end{vmatrix}}{\begin{vmatrix} 3 & 4 \\ 2 & -3 \end{vmatrix}}.$$

(b) More generally, verify that the equations

$$a_1x + a_2y = d$$
$$b_1x + b_2y = e$$

can be solved by the prescription

$$x = \frac{\begin{vmatrix} d & a_2 \\ e & b_2 \end{vmatrix}}{\begin{vmatrix} a_1 & a_2 \\ b_1 & b_2 \end{vmatrix}}, \quad y = \frac{\begin{vmatrix} a_1 & d \\ b_1 & e \end{vmatrix}}{\begin{vmatrix} a_1 & a_2 \\ b_1 & b_2 \end{vmatrix}}$$

provided that the determinant in the denominator is different from 0.

(c) Devise an analogous rule for solving three linear equations in three unknowns.

16. Show that a 3×3 determinant can be " expanded by the elements of the bottom row" as follows:

$$\begin{vmatrix} a_1 & a_2 & a_3 \\ b_1 & b_2 & b_3 \\ c_1 & c_2 & c_3 \end{vmatrix} = c_1\begin{vmatrix} a_2 & a_3 \\ b_2 & b_3 \end{vmatrix} - c_2\begin{vmatrix} a_1 & a_3 \\ b_1 & b_3 \end{vmatrix} + c_3\begin{vmatrix} a_1 & a_2 \\ b_1 & b_2 \end{vmatrix}.$$

HINT: You can check this directly by writing out the values of the determinants on the right, or you can interchange rows twice to bring the bottom row to the top and then expand by the elements of the top row.

APPENDIX B SOME ADDITIONAL PROOFS

In this appendix we present some proofs that many would consider too advanced for the main body of the text. Some details are omitted. These are left to you.

The arguments presented in Sections B.1, B.2, and B.4 require some familiarity with the *least upper bound axiom*. This is discussed in Section 10.1. In addition, Section B.4 requires some understanding of *sequences*, for which we refer you to Sections 10.2 and 10.3.

B.1 THE INTERMEDIATE-VALUE THEOREM

LEMMA B.1.1

Let f be continuous on $[a, b]$. If $f(a) < 0 < f(b)$ or $f(b) < 0 < f(a)$, then there is a number c between a and b for which $f(c) = 0$.

PROOF Suppose that $f(a) < 0 < f(b)$. (The other case can be treated in a similar manner.) Since $f(a) < 0$, we know from the continuity of f that there exists a number ξ such that f is negative on $[a, \xi)$. Let

$$c = \text{lub } \{\xi : f \text{ is negative on } [a, \xi)\}.$$

Clearly, $c \leq b$. We cannot have $f(c) > 0$, for then f would be positive on some interval extending to the left of c, and we know that, to the left of c, f is negative. Incidentally, this argument excludes the possibility that $c = b$ and means that $c < b$. We cannot have $f(c) < 0$, for then there would be an interval $[a, t)$, with $t > c$, on which f is negative, and this would contradict the definition of c. It follows that $f(c) = 0$. ❑

THEOREM B.1.2 THE INTERMEDIATE-VALUE THEOREM

If f is continuous on $[a, b]$ and K is a number between $f(a)$ and $f(b)$, then there is at least one number c between a and b for which $f(c) = K$.

PROOF Suppose, for example, that

$$f(a) < K < f(b).$$

(The other possibility can be handled in a similar manner.) The function

$$g(x) = f(x) - K$$

is continuous on $[a, b]$. Since

$$g(a) = f(a) - K < 0 \qquad \text{and} \qquad g(b) = f(b) - K > 0,$$

we know from the lemma that there is a number c between a and b for which $g(c) = 0$. Obviously, then, $f(c) = K$. ❑

■ B.2 BOUNDEDNESS; EXTREME-VALUE THEOREM

LEMMA B.2.1

If f is continuous on $[a, b]$ then f is bounded on $[a, b]$.

PROOF Consider

$$\{x : x \in [a, b] \text{ and } f \text{ is bounded on } [a, x]\}.$$

It is easy to see that this set is nonempty and bounded above by b. Thus we can set

$$c = \text{lub}\,\{x : f \text{ is bounded on } [a, x]\}.$$

Now we argue that $c = b$. To do so, we suppose that $c < b$. From the continuity of f at c, it is easy to see that f is bounded on $[c - \epsilon, c + \epsilon]$ for some $\epsilon > 0$. Being bounded on $[a, c - \epsilon]$ and on $[c - \epsilon, c + \epsilon]$, it is obviously bounded on $[a, c + \epsilon]$. This contradicts our choice of c. We can therefore conclude that $c = b$. This tells us that f is bounded on $[a, x]$ for all $x < b$. We are now almost through. From the continuity of f, we know that f is bounded on some interval of the form $[b - \epsilon, b]$. Since $b - \epsilon < b$, we know from what we have just proved that f is bounded on $[a, b - \epsilon]$. Being bounded on $[a, b - \epsilon]$ and bounded on $[b - \epsilon, b]$, it is bounded on $[a, b]$. ❑

THEOREM B.2.2 THE EXTREME-VALUE THEOREM

If f is continuous on $[a, b]$, then f takes on both a maximum value M and a minimum value m on $[a, b]$.

PROOF By the lemma, f is bounded on $[a, b]$. Set

$$M = \text{lub}\,\{f(x) : x \in [a, b]\}.$$

We must show that there exists c in $[a, b]$ such that $f(c) = M$. To do this, we set

$$g(x) = \frac{1}{M - f(x)}.$$

If f does not take on the value M, then g is continuous on $[a, b]$ and thus, by the lemma, bounded on $[a, b]$. A look at the definition of g makes it clear that g cannot be bounded on $[a, b]$. The assumption that f does not take on the value M has led to a contradiction. (That f takes a minimum value m can be proved in a similar manner.) ☐

B.3 INVERSES

THEOREM B.3.1 CONTINUITY OF THE INVERSE

Let f be a one-to-one function defined on an interval (a, b). If f is continuous, then its inverse f^{-1} is also continuous.

PROOF If f is continuous, then, being one-to-one, f either increases throughout (a, b) or it decreases throughout (a, b). The proof of this assertion we leave to you.

Suppose now that f increases throughout (a, b). Let's take c in the domain of f^{-1} and show that f^{-1} is continuous at c.

We first observe that $f^{-1}(c)$ lies in (a, b) and choose $\epsilon > 0$ sufficiently small so that $f^{-1}(c) - \epsilon$ and $f^{-1}(c) + \epsilon$ also lie in (a, b). We seek a $\delta > 0$ such that

$$\text{if } c - \delta < x < c + \delta, \qquad \text{then} \qquad f^{-1}(c) - \epsilon < f^{-1}(x) < f^{-1}(c) + \epsilon.$$

This condition can be met by choosing δ to satisfy

$$f(f^{-1}(c) - \epsilon) < c - \delta \qquad \text{and} \qquad c + \delta < f(f^{-1}(c) + \epsilon),$$

for then, if $c - \delta < x < c + \delta$,

$$f(f^{-1}(c) - \epsilon) < x < f(f^{-1}(c) + \epsilon),$$

and, since f^{-1} also increases,

$$f^{-1}(c) - \epsilon < f^{-1}(x) < f^{-1}(c) + \epsilon.$$

The case where f decreases throughout (a, b) can be handled in a similar manner. ☐

THEOREM B.3.2 DIFFERENTIABILITY OF THE INVERSE

Let f be a one-to-one function differentiable on an open interval I. Let a be a point of I and let $f(a) = b$. If $f'(a) \neq 0$, then f^{-1} is differentiable at b and

$$(f^{-1})'(b) = \frac{1}{f'(a)}.$$

PROOF (Here we use the characterization of derivative spelled out in Theorem 3.5.8.) We take $\epsilon > 0$ and show that there exists a $\delta > 0$ such that

$$\text{if} \quad 0 < |t - b| < \delta, \qquad \text{then} \qquad \left|\frac{f^{-1}(t) - f^{-1}(b)}{t - b} - \frac{1}{f'(a)}\right| < \epsilon.$$

Since f is differentiable at a, there exists a $\delta_1 > 0$ such that

$$\text{if} \quad 0 < |x - a| < \delta_1, \qquad \text{then} \qquad \left|\frac{1}{\dfrac{f(x) - f(a)}{x - a}} - \frac{1}{f'(a)}\right| < \epsilon,$$

and therefore $$\left|\frac{x-a}{f(x)-f(a)}-\frac{1}{f'(a)}\right|<\epsilon.$$

By the previous theorem, f^{-1} is continuous at b. Hence there exists a $\delta > 0$ such that

$$\text{if} \quad 0<|t-b|<\delta, \qquad \text{then} \qquad 0<|f^{-1}(t)-f^{-1}(b)|<\delta_1,$$

and therefore $$\left|\frac{f^{-1}(t)-f^{-1}(b)}{t-b}-\frac{1}{f'(a)}\right|<\epsilon. \quad \square$$

■ B.4 THE INTEGRABILITY OF CONTINUOUS FUNCTIONS

The aim here is to prove that, if f is continuous on $[a, b]$, then there is one and only one number I that satisfies the inequality

$$L_f(P) \leq I \leq U_f(P) \qquad \text{for all partitions } P \text{ of } [a, b].$$

DEFINITION B.4.1

A function f is said to be *uniformly continuous* on $[a, b]$, if for each $\epsilon > 0$ there exists $\delta > 0$ such that

$$\text{if } x, y \in [a, b] \text{ and } |x-y|<\delta, \text{ then } |f(x)-f(y)|<\epsilon.$$

For convenience, let's agree to say that *the interval* $[a, b]$ *has the property* P_ϵ if there exist sequences $\{x_n\}$, $\{y_n\}$ satisfying

$$x_n, y_n \in [a, b], \quad |x_n - y_n| < 1/n, \quad |f(x_n)-f(y_n)| \geq \epsilon.$$

LEMMA B.4.2

If f is not uniformly continuous on $[a, b]$, then $[a, b]$ has the property P_ϵ for some $\epsilon > 0$.

PROOF If f is not uniformly continuous on $[a, b]$, then there is at least one $\epsilon > 0$ for which there is no $\delta > 0$ such that

$$\text{if } x, y \in [a, b] \text{ and } |x-y|<\delta, \text{ then } |f(x)-f(y)|<\epsilon.$$

The interval $[a, b]$ has the property P_ϵ for that choice of ϵ. The details of the argument are left to you. $\square$

LEMMA B.4.3

Let f be continuous on $[a, b]$. If $[a, b]$ has the property P_ϵ, then at least one of the subintervals $[a, \frac{1}{2}(a+b)]$, $[\frac{1}{2}(a+b), b]$ has the property P_ϵ.

PROOF Let's suppose that the lemma is false. For convenience, we let $c = \frac{1}{2}(a+b)$, so that the halves become $[a, c]$ and $[c, b]$. Since $[a, c]$ fails to have the property P_ϵ, there exists an integer p such that

$$\text{if } x, y \in [a, c] \quad \text{and} \quad |x - y| < 1/p, \quad \text{then} \quad |f(x) - f(y)| < \epsilon.$$

Since $[c, b]$ fails to have the property P_ϵ, there exists an integer q such that

$$\text{if } x, y \in [c, b] \quad \text{and} \quad |x - y| < 1/q, \quad \text{then} \quad |f(x) - f(y)| < \epsilon.$$

Since f is continuous at c, there exists an integer r such that, if $|x - c| < 1/r$, then $|f(x) - f(c)| < \frac{1}{2}\epsilon$. Set $s = \max\{p, q, r\}$ and suppose that

$$x, y \in [a, b], \quad |x - y| < 1/s.$$

If x, y are both in $[a, c]$ or both in $[c, b]$, then

$$|f(x) - f(y)| < \epsilon.$$

The only other possibility is that $x \in [a, c]$ and $y \in [c, b]$. In this case we have

$$|x - c| < 1/r, \quad |y - c| < 1/r,$$

and thus

$$|f(x) - f(c)| < \tfrac{1}{2}\epsilon, \quad |f(y) - f(c)| < \tfrac{1}{2}\epsilon.$$

By the triangle inequality, we again have

$$|f(x) - f(y)| < \epsilon.$$

In summary, we have obtained the existence of an integer s with the property that

$$x, y \in [a, b], \quad |x - y| < 1/s \quad \text{implies} \quad |f(x) - f(y)| < \epsilon.$$

Hence $[a, b]$ does not have the property P_ϵ. This is a contradiction and proves the lemma. □

THEOREM B.4.4

If f is continuous on $[a, b]$, then f is uniformly continuous on $[a, b]$.

PROOF We suppose that f is not uniformly continuous on $[a, b]$ and base our argument on a mathematical version of the classical maxim "Divide and conquer".

By the first lemma of this section, we know that $[a, b]$ has the property P_ϵ for some $\epsilon > 0$. We bisect $[a, b]$ and note by the second lemma that one of the halves, say $[a_1, b_1]$, has the property P_ϵ. We then bisect $[a_1, b_1]$ and note that one of the halves, say $[a_2, b_2]$, has the property P_ϵ. Continuing in this manner, we obtain a sequence of intervals $[a_n, b_n]$, each with the property P_ϵ. Then for each n, we can choose $x_n, y_n \in [a_n, b_n]$ such that

$$|x_n - y_n| < 1/n \quad \text{and} \quad |f(x_n) - f(y_n)| \geq \epsilon.$$

Since

$$a \leq a_n \leq a_{n+1} < b_{n+1} \leq b_n \leq b,$$

we see that sequences $\{a_n\}$ and $\{b_n\}$ are both bounded and monotonic. Thus they are convergent. Since $b_n - a_n \to 0$, we see that $\{a_n\}$ and $\{b_n\}$ both converge to the same limit, say L. From the inequality

$$a_n \le x_n \le y_n \le b_n,$$

we conclude that

$$x_n \to L \quad \text{and} \quad y_n \to L.$$

This tells us that

$$|f(x_n) - f(y_n)| \to |f(L) - f(L)| = 0,$$

which contradicts the statement that $|f(x_n) - f(y_n)| \ge \epsilon$ for all n. ❑

LEMMA B.4.5

If P and Q are partitions of $[a, b]$, then $L_f(P) \le U_f(Q)$.

PROOF $P \cup Q$ is a partition of $[a, b]$ that contains both P and Q. It is obvious then that

$$L_f(P) \le L_f(P \cup Q) \le U_f(P \cup Q) \le U_f(Q). \quad ❑$$

From this lemma it follows that the set of all lower sums is bounded above and has a least upper bound L. The number L satisfies the inequality

$$L_f(P) \le L \le U_f(P) \quad \text{for all partitions } P$$

and is clearly the least of such numbers. Similarly, we find that the set of all upper sums is bounded below and has a greatest lower bound U. The number U satisfies the inequality

$$L_f(P) \le U \le U_f(P) \quad \text{for all partitions } P$$

and is clearly the largest of such numbers.

We are now ready to prove the basic theorem.

THEOREM B.4.6 THE INTEGRABILITY THEOREM

If f is continuous on $[a, b]$, then there exists one and only one number I that satisfies the inequality

$$L_f(P) \le I \le U_f(P) \quad \text{for all partitions } P \text{ of } [a, b].$$

PROOF We know that

$$L_f(P) \le L \le U \le U_f(P) \quad \text{for all } P,$$

so that existence is no problem. We will have uniqueness if we can prove that

$$L = U.$$

To do this, we take $\epsilon > 0$ and note that f, being continuous on $[a, b]$, is uniformly continuous on $[a, b]$. Thus there exists a $\delta > 0$ such that, if

$$x, y \in [a, b] \quad \text{and} \quad |x - y| < \delta, \quad \text{then} \quad |f(x) - f(y)| < \frac{\epsilon}{b - a}.$$

We now choose a partition $P = \{x_0, x_1, \dots, x_n\}$ for which $\max \Delta x_i < \delta$. For this partition P, we have

$$\begin{aligned} U_f(P) - L_f(P) &= \sum_{i=1}^{n} M_i \Delta x_i - \sum_{i=1}^{n} m_i \Delta x_i \\ &= \sum_{i=1}^{n} (M_i - m_i) \Delta x_i \\ &< \sum_{i=1}^{n} \frac{\epsilon}{b-a} \Delta x_i = \frac{\epsilon}{b-a} \sum_{i=1}^{n} \Delta x_i = \frac{\epsilon}{b-a}(b-a) = \epsilon. \end{aligned}$$

Since $\quad U_f(P) - L_f(P) < \epsilon \quad$ and $\quad 0 \le U - L \le U_f(P) - L_f(P),$

you can see that

$$0 \le U - L < \epsilon.$$

Since ϵ was chosen arbitrarily, we must have $U - L = 0$ and $L = U$. ☐

■ B.5 THE INTEGRAL AS THE LIMIT OF RIEMANN SUMS

For the notation we refer to Section 5.2.

THEOREM B.5.1

If f is continuous on $[a, b]$, then

$$\int_a^b f(x)\,dx = \lim_{||P|| \to 0} S^*(P).$$

PROOF Let $\epsilon > 0$. We must show that there exists a $\delta > 0$ such that

$$\text{if} \quad ||P|| < \delta, \quad \text{then} \quad \left| S^*(P) - \int_a^b f(x)\,dx \right| < \epsilon.$$

From the proof of Theorem B.4.6 we know that there exists a $\delta > 0$ such that

$$\text{if} \quad ||P|| < \delta, \quad \text{then} \quad U_f(P) - L_f(P) < \epsilon.$$

For such P we have

$$U_f(P) - \epsilon < L_f(P) \le S^*(P) \le U_f(P) < L_f(P) + \epsilon.$$

This gives

$$\int_a^b f(x)\,dx - \epsilon < S^*(P) < \int_a^b f(x)\,dx + \epsilon,$$

and therefore

$$\left| S^*(P) - \int_a^b f(x)\,dx \right| < \epsilon. \quad \square$$

CHAPTER 12

SECTION 12.1

1.

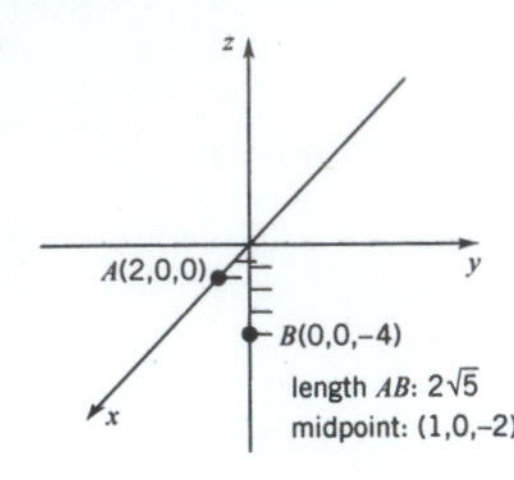

3.

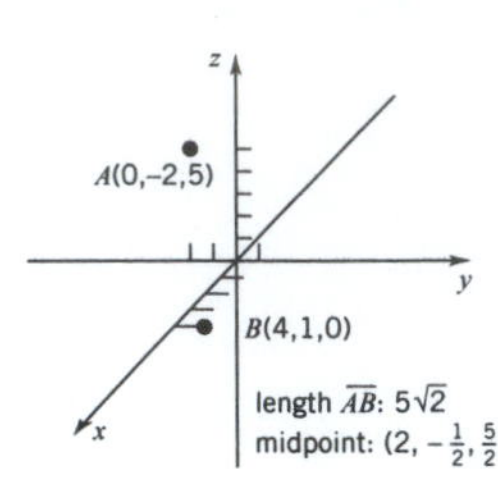

5. $z = -2$ **7.** $y = 1$ **9.** $x = 3$

11. $x^2 + (y-2)^2 + (z+1)^2 = 9$ **13.** $(x-2)^2 + (y-4)^2 + (z+4)^2 = 36$ **15.** $(x-3)^2 + (y-2)^2 + (z-2)^2 = 13$

17. $(x-2)^2 + (y-3)^2 + (z+4)^2 = 25$ **19.** center $(-2, 4, 1)$, radius 4 **21.** center: $(3, -5, 1)$; radius: 6

23. $(2, 3, -5)$ **25.** $(-2, 3, 5)$ **27.** $(-2, 3, -5)$ **29.** $(-2, -3, -5)$ **31.** $(2, -5, 5)$ **33.** $(-2, 1, -3)$

35. $(x-3)^2 + (y-3)^2 + (z-3)^2 = 9, \quad (x-7)^2 + (y-7)^2 + (z-7)^2 = 49$

37. not a sphere; the equation is equivalent to $(x-2)^2 + (y+2)^2 + (z+3)^2 = -3$

39. $d(P,R) = \sqrt{14}, \quad d(Q,R) = \sqrt{45}, \quad d(P,Q) = \sqrt{59}, \quad [d(P,R)]^2 + [d(Q,R)]^2 = [d(P,Q)]^2$

41. The sphere of radius 2 centered at the origin, together with its interior

43. A rectangular box with sides on the coordinate planes and dimensions $1 \times 2 \times 3$, together with its interior

45. A circular cylinder with base the circle $x^2 + y^2 = 4$ and height 4, together with its interior

47. (a) $x = a_1 + t(b_1 - a_1), \quad y = a_2 + t(b_2 - a_2), \quad z = a_3 + t(b_3 - a_3)$ (b) $t = \frac{1}{2}$

49. $(x-1)^2 + (y-1)^2 + (z + \frac{5}{2})^2 = \frac{53}{4}$

SECTION 12.3

1. $(3, 4, -2), \sqrt{29}$ **3.** $(0, -2, -1), \sqrt{5}$ **5.** $(-1, -4, 7)$ **7.** $(5, 2, -8)$ **9.** $3\mathbf{i} - 4\mathbf{j} + 6\mathbf{k}$ **11.** $-3\mathbf{i} - \mathbf{j} + 8\mathbf{k}$

13. 5 **15.** 3 **17.** $\sqrt{6}$ **19.** (a) **a, c, d** (b) **a, c** (c) **a** and **c** both have direction opposite to **d** **21.** $(\frac{3}{5}, -\frac{4}{5}, 0)$

23. $\frac{1}{3}\mathbf{i} - \frac{2}{3}\mathbf{j} + \frac{2}{3}\mathbf{k}$ **25.** $\dfrac{1}{\sqrt{14}}\mathbf{i} - \dfrac{3}{\sqrt{14}}\mathbf{j} - \dfrac{2}{\sqrt{14}}\mathbf{k}$ **27.** (i) $\mathbf{a} + \mathbf{b}$ (ii) $-(\mathbf{a} + \mathbf{b})$ (iii) $\mathbf{a} - \mathbf{b}$ (iv) $\mathbf{b} - \mathbf{a}$

29. (a) $\mathbf{i} - 3\mathbf{j} + 10\mathbf{k}$ (b) $A = -2, B = \frac{3}{2}, C = -\frac{7}{2}$ **31.** $\alpha = \pm 3$ **33.** $\alpha = \pm\frac{1}{3}\sqrt{6}$ **35.** $\pm\frac{2}{13}\sqrt{13}(3\mathbf{j} + 2\mathbf{k})$

37. (a) the parallelogram is a rectangle
(b) simplify $\sqrt{(a_1 - b_1)^2 + (a_2 - b_2)^2 + (a_3 - b_3)^2} = \sqrt{(a_1 + b_1)^2 + (a_2 + b_2)^2 + (a_3 + b_3)^2}$

39. (a)

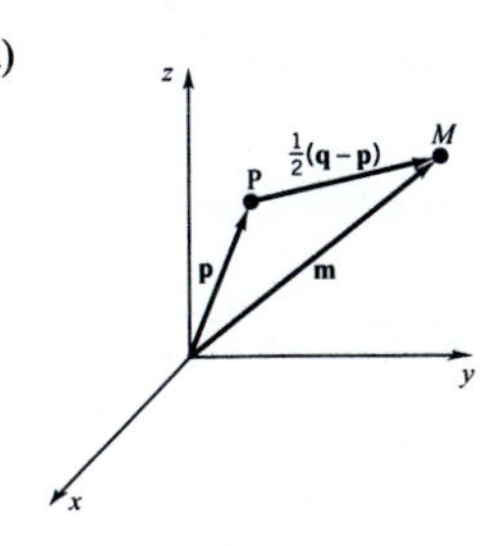

$\mathbf{m} = \mathbf{p} + \frac{1}{2}(\mathbf{q} - \mathbf{p})$

(b) Let $P(x_1, y_1, z_1)$, $Q = (x_2, y_2, z_2)$, and $M = (x_m, y_m, z_m)$. Then

$$(x_m, y_m, z_m) = (x_1, y_1, z_1) + \tfrac{1}{2}(x_2 - x_1, y_2 - y_1, z_2 - z_1)$$
$$= \left(\frac{x_1 + x_2}{2}, \frac{y_1 + y_2}{2}, \frac{z_1 + z_2}{2}\right).$$

41. (a) $\|\mathbf{r}-\mathbf{a}\| = 3$ where $\mathbf{a} = a_1\mathbf{i} + a_2\mathbf{j} + a_3\mathbf{k}$

(b) $\|\mathbf{r}\| \leq 2$

(c) $\|\mathbf{r}-\mathbf{a}\| \leq 1$ where $\mathbf{a} = a_1\mathbf{i} + a_2\mathbf{j} + a_3\mathbf{k}$

SECTION 12.4

1. -1 **3.** 0 **5.** -1 **7.** $\mathbf{a}\cdot\mathbf{b}$ **9.** $\mathbf{a}\cdot(\mathbf{b}+\mathbf{c})$

11. (a) $\mathbf{a}\cdot\mathbf{b} = 5$, $\mathbf{a}\cdot\mathbf{c} = 8$, $\mathbf{b}\cdot\mathbf{c} = 18$ (b) $\cos\sphericalangle(\mathbf{a},\mathbf{b}) = \frac{1}{14}\sqrt{70}$, $\cos\sphericalangle(\mathbf{a},\mathbf{c}) = \frac{8}{25}\sqrt{5}$, $\cos\sphericalangle(\mathbf{b},\mathbf{c}) = \frac{9}{35}\sqrt{14}$

(c) $\text{comp}_{\mathbf{b}}\mathbf{a} = \frac{5}{14}\sqrt{14}$, $\text{comp}_{\mathbf{c}}\mathbf{a} = \frac{8}{5}$ (d) $\mathbf{proj_b a} = \frac{5}{14}(3\mathbf{i}-\mathbf{j}+2\mathbf{k})$, $\mathbf{proj_c a} = \frac{8}{25}(4\mathbf{i}+3\mathbf{k})$ **13.** $\frac{1}{2}\mathbf{i} + \frac{1}{2}\sqrt{2}\mathbf{j} - \frac{1}{2}\mathbf{k}$ **15.** $\frac{1}{3}\pi$

17. $\frac{1}{3}\pi, \frac{2}{3}\pi, \frac{1}{4}\pi$ **19.** 2.2 radians, or 126.3° **21.** 2.5 radians, or 145.3°

23. angles: 38.51°, 95.52°, 45.97°; perimeter: $P \cong 15.924$ **25.** $\cos\alpha = \frac{1}{3}, \cos\beta = \frac{2}{3}, \cos\gamma = \frac{2}{3}$; $\alpha \cong 70.5°, \beta \cong 48.2°, \gamma \cong 48.2°$

27. $\cos\alpha = \frac{3}{13}, \cos\beta = \frac{12}{13}, \cos\gamma = \frac{4}{13}$; $\alpha \cong 76.7°, \beta \cong 22.6°, \gamma \cong 72.1°$ **29.** $x = \pm 4$ **31.** $x = 0, x = 4$

33. (a) The direction angles of a vector always satisfy $\cos^2\alpha + \cos^2\beta + \cos^2\gamma = 1$, and, as you can check, $\cos^2\frac{1}{4}\pi + \cos^2\frac{1}{6}\pi + \cos^2\frac{2}{3}\pi \neq 1$.
(b) The relation $\cos^2\alpha + \cos^2\frac{1}{4}\pi + \cos^2\frac{1}{4}\pi = 1$ gives

$$\cos^2\alpha + \tfrac{1}{2} + \tfrac{1}{2} = 1, \quad \cos\alpha = 0, \quad a_1 = \|\mathbf{a}\| \cos\alpha = 0.$$

35. $\pi-\alpha, \pi-\beta, \pi-\gamma$ **37.** $\mathbf{u} = \pm\frac{1}{165}\sqrt{165}\,(8\mathbf{i}+\mathbf{j}-10\mathbf{k})$ **39.** $\theta = \cos^{-1}(\frac{1}{3}\sqrt{3}) \cong 0.96$ radians

41. (a) $\mathbf{proj_b}\,\alpha\mathbf{a} = (\alpha\mathbf{a}\cdot\mathbf{u_b})\mathbf{u_b} = \alpha(\mathbf{a}\cdot\mathbf{u_b})\mathbf{u_b} = \alpha\mathbf{proj_b\,a}$
(b) $\mathbf{proj_b}(\mathbf{a}+\mathbf{c}) = [(\mathbf{a}+\mathbf{c})\cdot\mathbf{u_b}]\mathbf{u_b}$
$= (\mathbf{a}\cdot\mathbf{u_b} + \mathbf{c}\cdot\mathbf{u_b})\mathbf{u_b}$
$= (\mathbf{a}\cdot\mathbf{u_b})\mathbf{u_b} + (\mathbf{c}\cdot\mathbf{u_b})\mathbf{u_b} = \mathbf{proj_b\,a} + \mathbf{proj_b\,c}$

43. (a) for $\mathbf{a} \neq 0$ the following statements are equivalent:

$$\mathbf{a}\cdot\mathbf{b} = \mathbf{a}\cdot\mathbf{c}, \quad \mathbf{b}\cdot\mathbf{a} = \mathbf{c}\cdot\mathbf{a},$$
$$\mathbf{b}\cdot\frac{\mathbf{a}}{\|\mathbf{a}\|} = \mathbf{c}\cdot\frac{\mathbf{a}}{\|\mathbf{a}\|}, \quad \mathbf{b}\cdot\mathbf{u_a} = \mathbf{c}\cdot\mathbf{u_a},$$

$$(\mathbf{b}\cdot\mathbf{u_a})\mathbf{u_a} = (\mathbf{c}\cdot\mathbf{u_a})\mathbf{u_a}, \quad \mathbf{proj_a b} = \mathbf{proj_a c}. \quad \mathbf{a}\cdot\mathbf{b} = \mathbf{a}\cdot\mathbf{c} \text{ but } \mathbf{b} \neq \mathbf{c}.$$

(b) $\mathbf{b} = (\mathbf{b}\cdot\mathbf{i})\mathbf{i} + (\mathbf{b}\cdot\mathbf{j})\mathbf{j} + (\mathbf{b}\cdot\mathbf{k})\mathbf{k} = (\mathbf{c}\cdot\mathbf{i})\mathbf{i} + (\mathbf{c}\cdot\mathbf{j})\mathbf{j} + (\mathbf{c}\cdot\mathbf{k})\mathbf{k} = \mathbf{c}$.

45. (a) Express the norms as dot products.
(b) The following statements are equivalent:

$$\mathbf{a}\perp\mathbf{b}, \quad \mathbf{a}\cdot\mathbf{b} = 0, \quad \|\mathbf{a}+\mathbf{b}\|^2 - \|\mathbf{a}-\mathbf{b}\|^2 = 0, \quad \|\mathbf{a}+\mathbf{b}\| = \|\mathbf{a}-\mathbf{b}\|.$$

(c) By (b), the relation $\|\mathbf{a}+\mathbf{b}\| = \|\mathbf{a}-\mathbf{b}\|$ gives $\mathbf{a}\perp\mathbf{b}$. The relation $\mathbf{a}+\mathbf{b}\perp\mathbf{a}-\mathbf{b}$ gives

$$0 = (\mathbf{a}+\mathbf{b})\cdot(\mathbf{a}-\mathbf{b}) = \|\mathbf{a}\|^2 - \|\mathbf{b}\|^2 \quad \text{and thus} \quad \|\mathbf{a}\| = \|\mathbf{b}\|.$$

The parallelogram is a square since it has two adjacent sides of equal length that meet at right angles.

47. $\|\mathbf{a}+\mathbf{b}\|^2 = (\mathbf{a}+\mathbf{b})\cdot(\mathbf{a}+\mathbf{b}) = \mathbf{a}\cdot\mathbf{a} + 2\mathbf{a}\cdot\mathbf{b} + \mathbf{b}\cdot\mathbf{b}$
$\|\mathbf{a}-\mathbf{b}\|^2 = (\mathbf{a}-\mathbf{b})\cdot(\mathbf{a}-\mathbf{b}) = \mathbf{a}\cdot\mathbf{a} - 2\mathbf{a}\cdot\mathbf{b} + \mathbf{b}\cdot\mathbf{b}$
and the result follows.

49. $\dfrac{\mathbf{a}\cdot\mathbf{c}}{\|\mathbf{a}\|\,\|\mathbf{c}\|} = \|\mathbf{a}\|\,\|\mathbf{b}\| + \mathbf{a}\cdot\mathbf{b} = \dfrac{\mathbf{b}\cdot\mathbf{c}}{\|\mathbf{b}\|\,\|\mathbf{c}\|}$

51. If $\mathbf{a}\perp\mathbf{b}$ and $\mathbf{a}\perp\mathbf{c}$, then $\mathbf{a}\cdot\mathbf{b} = 0$ and $\mathbf{a}\cdot\mathbf{c} = 0$, so that

$$\mathbf{a},\cdot\ (\alpha\mathbf{b}+\beta\mathbf{c}) = \alpha(\mathbf{a}\cdot\mathbf{b}) + \beta(\mathbf{a}\cdot\mathbf{c}) = 0.$$

Thus $\mathbf{a}\perp(\alpha\mathbf{b}+\beta\mathbf{c})$.

53. Existence of decomposition: $\mathbf{a} = (\mathbf{a} \cdot \mathbf{u_b})\mathbf{u_b} + [\mathbf{a} - (\mathbf{a} \cdot \mathbf{u_b})\mathbf{u_b}]$. Uniqueness of decomposition: suppose that $\mathbf{a} = \mathbf{a}_{\|} + \mathbf{a}_{\perp} = \mathbf{A}_{\|} + \mathbf{A}_{\perp}$. Then the vector $\mathbf{a}_{\|} - \mathbf{A}_{\|} = \mathbf{A}_{\perp} - \mathbf{a}_{\perp}$ is both parallel to **b** and perpendicular to **b**. (Exercises 37 and 38.) Therefore it is zero. Consequently $\mathbf{A}_{\|} = \mathbf{a}_{\|}$ and $\mathbf{A}_{\perp} = \mathbf{a}_{\perp}$.

55. Place center of sphere at the origin.

$$\begin{aligned} \overrightarrow{P_1Q} \cdot \overrightarrow{P_2Q} &= (-\mathbf{a} + \mathbf{b}) \cdot (\mathbf{a} + \mathbf{b}) \\ &= -\|\mathbf{a}\|^2 + \|\mathbf{b}\|^2 \\ &= 0. \end{aligned}$$

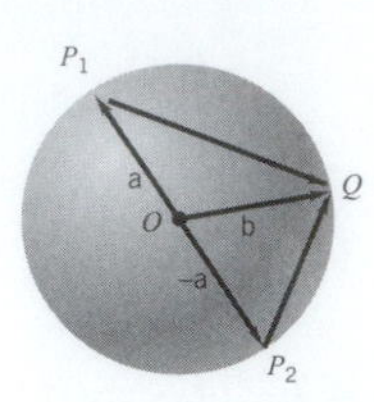

SECTION 12.5

1. $-2\mathbf{k}$ **3.** $\mathbf{i} + \mathbf{j} + \mathbf{k}$ **5.** $-3\mathbf{i} - \mathbf{j} - 2\mathbf{k}$ **7.** -1 **9.** 0 **11.** 1 **13.** $3\mathbf{i} - 2\mathbf{j} - 3\mathbf{k}$ **15.** $\mathbf{i} + \mathbf{j} - 2\mathbf{k}$ **17.** -3

19. $5\mathbf{i} - 4\mathbf{j} - \mathbf{k}$ **21.** $\left(\frac{1}{\sqrt{6}}, \frac{-1}{\sqrt{6}}, \frac{-2}{\sqrt{6}}\right), \left(\frac{-1}{\sqrt{6}}, \frac{1}{\sqrt{6}}, \frac{2}{\sqrt{6}}\right)$ **23.** $\mathbf{N} = 3\mathbf{j}$; area $= \frac{3}{2}$ **25.** $\mathbf{N} = 8\mathbf{i} + 4\mathbf{j} + 4\mathbf{k}$; area $= 2\sqrt{6}$ **27.** 1

29. 2 **31.** $-2(\mathbf{a} \times \mathbf{b})$ **33.** $\mathbf{a} = \mathbf{0}$ **35.** $\begin{vmatrix} \alpha & \beta \\ \gamma & \delta \end{vmatrix} (\mathbf{a} \times \mathbf{b})$ **37.** $\mathbf{a} \cdot (\mathbf{b} \times \mathbf{c}) = (\mathbf{a} \times \mathbf{b}) \cdot \mathbf{c} = (\mathbf{c} \times \mathbf{a}) \cdot \mathbf{b} = (\mathbf{b} \times \mathbf{c}) \cdot \mathbf{a} = (\mathbf{a} \times -\mathbf{c}) \cdot \mathbf{b}$,
$\mathbf{a} \cdot (\mathbf{c} \times \mathbf{b}) = \mathbf{c} \cdot (\mathbf{b} \times \mathbf{a}) = (-\mathbf{a} \times \mathbf{b}) \cdot \mathbf{c}$

39. $\mathbf{a} \times \mathbf{b}$ is perpendicular to the plane determined by **a** and **b**;
c is in this plane iff $\mathbf{a} \times \mathbf{b} \cdot \mathbf{c} = 0$.

41. $\mathbf{a} \cdot \mathbf{b} = \mathbf{a} \cdot \mathbf{c}$ implies $\mathbf{a} \cdot (\mathbf{b} - \mathbf{c}) = 0$; **a** is perpendicular to $\mathbf{b} - \mathbf{c}$.
$\mathbf{a} \times \mathbf{b} = \mathbf{a} \times \mathbf{c}$ implies $\mathbf{a} \times (\mathbf{b} - \mathbf{c}) = 0$; **a** is parallel to $\mathbf{b} - \mathbf{c}$.
Since $\mathbf{a} \neq \mathbf{0}$, it follows that $\mathbf{b} - \mathbf{c} = \mathbf{0}$, or $\mathbf{b} = \mathbf{c}$.

43. $\mathbf{c} \times \mathbf{a} = \|\mathbf{a}\|^2\mathbf{b}$ **45.** either $\mathbf{a} = \mathbf{0}$ or $\mathbf{b} = \mathbf{0}$

47. The result follows from Exercise 46.

SECTION 12.6

1. P and Q **3.** $\mathbf{r}(t) = (3\mathbf{i} + \mathbf{j}) + t\mathbf{k}$ **5.** $\mathbf{r}(t) = t(x_1\mathbf{i} + y_1\mathbf{j} + z_1\mathbf{k})$ **7.** $x(t) = 1 + t, \quad y(t) = -t, \quad z(t) = 3 + t$

9. $x(t) = 2, \quad y(t) = t, \quad z(t) = 3$ **11.** $\mathbf{r}(t) = (-\mathbf{i} + 2\mathbf{j} - 3\mathbf{k}) + t(2\mathbf{i} + \mathbf{j} + 4\mathbf{k})$ **13.** intersect at $(1, 3, 1)$

15. skew **17.** parallel **19.** skew **21.** $P(1, 2, 0)$, $\frac{1}{4}\pi$ rad **23.** $(x_0 - [d_1/d_3]z_0, y_0 - [d_2/d_3]z_0, 0)$

25. The lines are parallel. **27.** $\mathbf{r}(t) = (2\mathbf{i} + 7\mathbf{j} - \mathbf{k}) + t(2\mathbf{i} - 5\mathbf{j} + 4\mathbf{k}), \quad 0 \le t \le 1$ **29.** $\mathbf{u} = -\frac{2}{3}\mathbf{i} + \frac{1}{3}\mathbf{j} + \frac{2}{3}\mathbf{k}, \quad 9 \le t \le 15$

31. triples of the form $X(u) = 3 + au$, $Y(u) = -1 + bu$, $Z(u) = 8 + cu$ with $2a - 4b + 6c = 0$ **33.** 1

35. $\sqrt{69/14} \cong 2.22$ **37.** $\sqrt{3} \cong 1.73$ **39.** (a) 1 (b) $\sqrt{3}$ **41.** $\mathbf{r}(t) = \frac{1}{11}(7\mathbf{i} + 4\mathbf{j} - \mathbf{k}) \pm t[\frac{1}{11}\sqrt{11}(\mathbf{i} - \mathbf{j} + 3\mathbf{k})]$

43. $0 < t < s$ **45.** $d(l_1, l_2) = \dfrac{10}{\sqrt{285}}$

SECTION 12.7

1. Q **3.** $x - 4y + 3z - 2 = 0$ **5.** $3x - 2y + 5z - 9 = 0$ **7.** $y - z - 2 = 0$ **9.** $x_0(x - x_0) + y_0(y - y_0) + z_0(z - z_0) = 0$

11. $\frac{1}{\sqrt{30}}(2\mathbf{i} - \mathbf{j} + 5\mathbf{k}), \quad -\frac{1}{\sqrt{30}}(2\mathbf{i} - \mathbf{j} + 5\mathbf{k})$ **13.** $\frac{1}{15}x + \frac{1}{12}y - \frac{1}{10}z = 1$ **15.** $\frac{1}{2}\pi$ **17.** $\cos\theta = \frac{2}{21}\sqrt{42} \cong 0.617, \quad \theta \cong 0.91$ rad

19. coplanar **21.** not coplanar **23.** $\dfrac{2}{\sqrt{21}}$ **25.** $\frac{22}{5}$ **27.** $x + z = 2$ **29.** $3x - 4z - 5 = 0$

31. $\dfrac{x - x_0}{A} = \dfrac{y - y_0}{B} = \dfrac{z - z_0}{C}$ **33.** $(x - x_0)/d_1 = (y - y_0)/d_2, \quad (y - y_0)/d_2 = (z - z_0)/d_3$ **35.** $x(t) = t, \quad y(t) = t, \quad z(t) = -t$

37. $P(-\frac{19}{14}, \frac{15}{7}, \frac{17}{14})$ **39.** $10x - 7y + z = 0$ **41.** circle centered at P with radius $\|\mathbf{N}\|$

43. If $\alpha > 0$, then P_1 lies on the same side of the plane as the tip of $\mathbf{N}$;

if $\alpha < 0$, then P_1 and the tip of $\mathbf{N}$ lie on opposite sides of the plane

45. $\mathbf{a}\cdot\mathbf{b}\times\mathbf{c} = 0$

47. (a) $(4, 0, 0), (0, 5, 0), (0, 0, 2)$
(b) $5x + 4y = 20, x + 2z = 4, 2y + 5z = 10$
(c) $\pm\frac{1}{\sqrt{141}}(5\mathbf{i} + 4\mathbf{j} + 10\mathbf{k})$
(d)

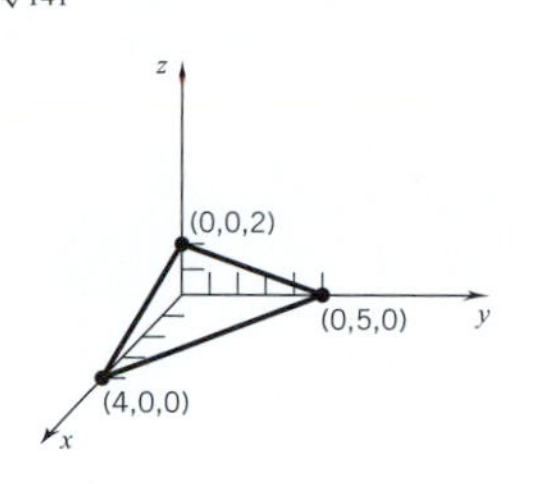

49. (a) (4,0,0), (0,0,6), no y-intercept
(b) $x = 4,\ 3x + 2z = 12,\ z = 6$
(c) $\pm\frac{1}{\sqrt{13}}(3\mathbf{i} + 2\mathbf{k})$
(d)

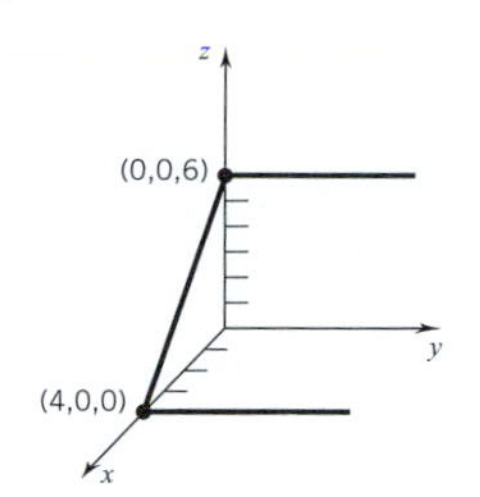

51. $10x + 4y + 5z = 20$ **53.** $5x + 3y = 15$

CHAPTER 13

SECTION 13.1

1. $\mathbf{f}'(t) = 2\mathbf{i} - \mathbf{j} + 3\mathbf{k}$ **3.** $\mathbf{f}'(t) = -\dfrac{1}{2\sqrt{1-t}}\mathbf{i} + \dfrac{1}{2\sqrt{1+t}}\mathbf{j} + \dfrac{1}{(1-t)^2}\mathbf{k}$ **5.** $\mathbf{f}'(t) = \cos t\,\mathbf{i} - \sin t\,\mathbf{j} + \sec^2 t\mathbf{k}$.

7. $-\dfrac{1}{1-t}\mathbf{i} - \sin t\,\mathbf{j} + 2t\,\mathbf{k}$ **9.** $12t\,\mathbf{j} + 2\mathbf{k}$ **11.** $-4\cos 2t\,\mathbf{i} - 4\sin 2t\,\mathbf{j}$

13. (a) $\mathbf{i}$
(b) $\mathbf{i} - \mathbf{j} + \frac{5}{\sqrt{2}}\mathbf{k}$

15. $\mathbf{i} + 3\mathbf{j}$

17. $(e-1)\mathbf{i} + (1 - 1/e)\mathbf{k}$

19. $\dfrac{\pi}{4}\mathbf{i} + \tan(1)\,\mathbf{j}$

21. $\frac{1}{2}\mathbf{i} + \mathbf{j}$ **23.** $0\mathbf{i} + 0\mathbf{j} + 0\mathbf{k} = \mathbf{0}$

25. (a) $\mathbf{i} + \frac{e-1}{2}\,\mathbf{j}$
(b) $\left[5 + \ln\left(\frac{4}{9}\right)\right]\mathbf{i} + \left[\frac{-5}{36} + \ln\left(\frac{9}{4}\right)\right]\mathbf{j} + \frac{295}{2592}\mathbf{k}$

27.

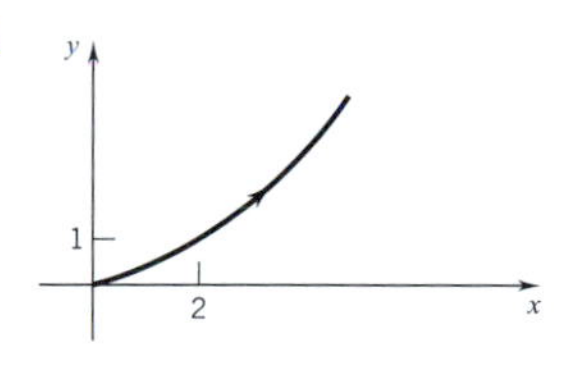

29.

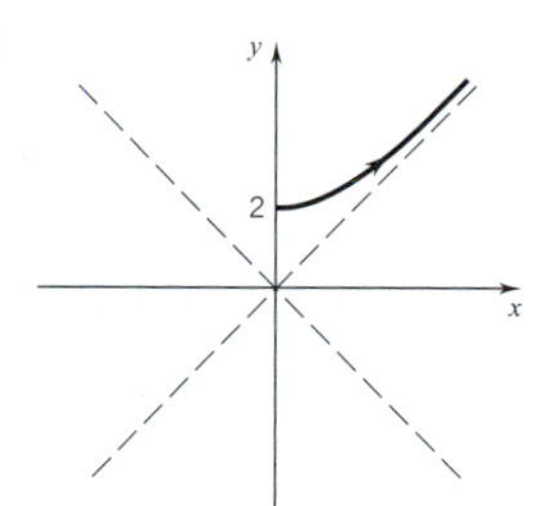

31.

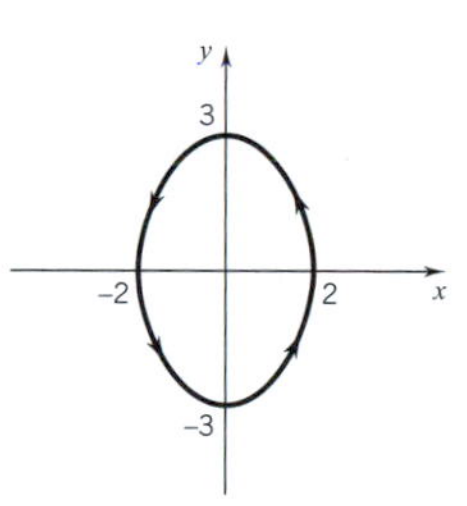

33.

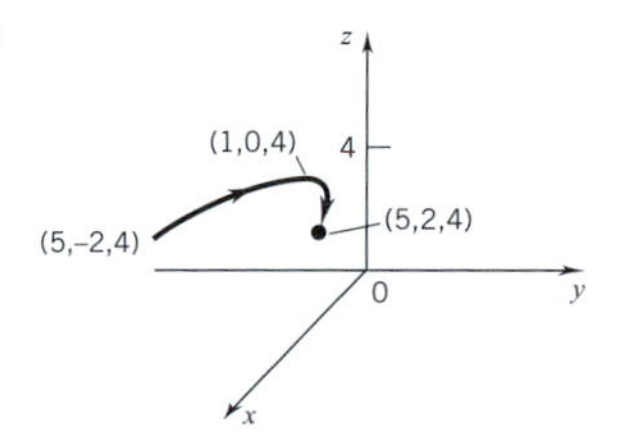

35.

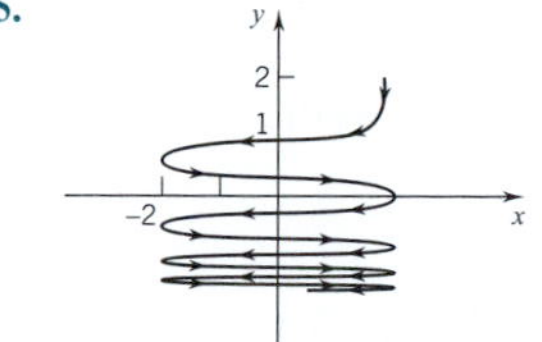

37.

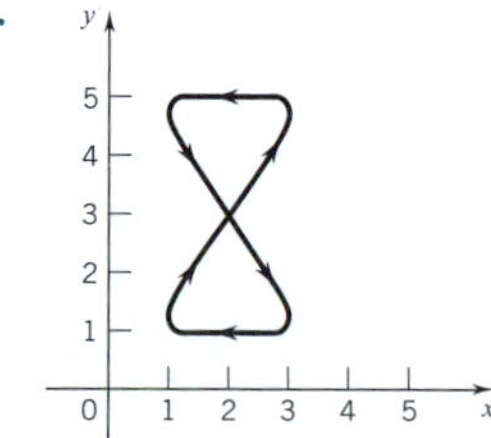

39. (a) $\mathbf{f}(t) = 3\cos t\,\mathbf{i} + 2\sin t\,\mathbf{j}$ (b) $\mathbf{f}(t) = 3\cos t\,\mathbf{i} - 2\sin t\,\mathbf{j}$ **41.** (a) $\mathbf{f}(t) = t\,\mathbf{i} + t^2\mathbf{j}$ (b) $\mathbf{f}(t) = -t\,\mathbf{i} + t^2\,\mathbf{j}$

43. $\mathbf{f}(t) = (1+2t)\mathbf{i} + (4+5t)\mathbf{j} + (-2+8t)\mathbf{k}, 0 \le t \le 1$

45. $\mathbf{f}'(t_0) = \mathbf{i} + m\mathbf{j}$ $\int_a^b \mathbf{f}(t)\,dt = \frac{1}{2}(b^2 - a^2)\mathbf{i} + A\mathbf{j}$, $\int_a^b \mathbf{f}'(t)\,dt = (b-a)\mathbf{i} + (d-c)\mathbf{j}$ 47. $\mathbf{f}(t) = t\,\mathbf{i} + (\frac{1}{3}t^3 + 1)\mathbf{j} - \mathbf{k}$ 49. $\mathbf{f}(t) = e^{\alpha t}\mathbf{c}$

51. (a) if $\mathbf{f}'(t_0) = \mathbf{0}$ on an interval, then the derivative of each component is 0 on that interval, each component is constant on that interval, and therefore $\mathbf{f}$ itself is constant on that interval

(b) set $\mathbf{h}(t) = \mathbf{f}(t) - \mathbf{g}(t)$ and apply part (a)

53. set $\mathbf{f}(t) = f_1(t)\mathbf{i} + f_2(t)\mathbf{j} + f_3(t)\mathbf{k}$, and apply (3. 1. 4) to f_1, f_2, f_3 55. no; as a counterexample set $\mathbf{f}(t) = \mathbf{i} = \mathbf{g}(t)$

57. $||\mathbf{f}(t)||^2 = \mathbf{f}(t) \cdot \mathbf{f}(t)$

$$2||\mathbf{f}(t)||\frac{d||\mathbf{f}(t)||}{dt} = 2\mathbf{f}(t) \cdot \mathbf{f}'(t)$$

$$\frac{d||\mathbf{f}(t)||}{dt} = \frac{\mathbf{f}(t) \cdot \mathbf{f}'(t)}{||\mathbf{f}(t)||}$$

59. (c) Increasing a causes the object to move faster around the cylinder; increasing b increases the rate at which the object rises.

61. (d) Increasing b increases the number of "peaks;" if $A = B$, overall shape is circular; if $A \neq B$, overall shape is elliptical.

SECTION 13.2

1. $\mathbf{f}'(t) = \mathbf{b}$, $\mathbf{f}''(t) = \mathbf{0}$ 3. $\mathbf{f}'(t) = 2\,e^{2t}\,\mathbf{i} - \cos\,t\,\mathbf{j}$, $\mathbf{f}''(t) = 4\,e^{2t}\,\mathbf{i} + \sin t\,\mathbf{j}$ 5. $\mathbf{f}'(t) = (3t^2 - 8t^3)\mathbf{j}, \mathbf{f}''(t) = (6t - 24t^2)\mathbf{j}$

7. $\mathbf{f}'(t) = -2t\,\mathbf{i} + e^t(t+1)\mathbf{k}$, $\mathbf{f}''(t) = -2\,\mathbf{i} + e^t(t+2)\mathbf{k}$

9. $\mathbf{f}'(t) = (\sin t + t\,\cos t + 2\,\sin 2t)\mathbf{i} + (2\,\cos 2t - \cos t + t\sin t)\mathbf{j} - 3\,\sin 3t\,\mathbf{k}$

$\mathbf{f}''(t) = (2\,\cos\,t - t\,\sin\,t + 4\,\cos\,2t)\mathbf{i} + (-4\,\sin 2t + 2\,\sin t + t\,\cos t)\mathbf{j} - 9\,\cos 3t\,\mathbf{k}$

11. $\mathbf{f}'(t) = \frac{1}{2}\sqrt{t}\,\mathbf{g}'(\sqrt{t}) + \mathbf{g}(\sqrt{t})$, $\mathbf{f}''(t) = \frac{1}{4}\mathbf{g}''(\sqrt{t}) + \frac{3}{4}(1/\sqrt{t})\mathbf{g}'(\sqrt{t})$ 13. $-\sin t\,e^{\cos t}\,\mathbf{i} + \cos t\,e^{\sin t}\,\mathbf{j}$ 15. $4\,e^{2t} - 4\,e^{-2t}$

17. $(\mathbf{a} \times \mathbf{d}) + (\mathbf{b} \times \mathbf{c}) + 2t(\mathbf{b} \times \mathbf{d})$ 19. $(\mathbf{a} \cdot \mathbf{d}) + (\mathbf{b} \cdot \mathbf{c}) + 2t(\mathbf{b} \cdot \mathbf{d})$ 21. $\mathbf{r}(t) = \mathbf{a} + t\mathbf{b}$ 23. $\mathbf{r}(t) = \frac{1}{2}t^2\mathbf{a} + \frac{1}{6}t^3\mathbf{b} + t\mathbf{c} + \mathbf{d}$

25. $\mathbf{r}''(t) = -\sin t\,\mathbf{i} - \cos t\,\mathbf{j} = -\mathbf{r}(t)$; no. 27. $\mathbf{r}(t) \cdot \mathbf{r}'(t) = 0$, $\mathbf{r}(t) \times \mathbf{r}'(t) = \mathbf{k}$

29. $\frac{d}{dt}[\mathbf{f}(t) \times \mathbf{f}'(t)] = [\mathbf{f}(t) \times \mathbf{f}''(t)] + \underbrace{[\mathbf{f}'(t) \times \mathbf{f}'(t)]}_{0} = \mathbf{f}(t) \times \mathbf{f}''(t)$

31. $[\mathbf{f} \cdot \mathbf{g} \times \mathbf{h}]' = \mathbf{f}' \cdot (\mathbf{g} \times \mathbf{h}) + \mathbf{f} \cdot (\mathbf{g} \times \mathbf{h})' = \mathbf{f}' \cdot (\mathbf{g} \times \mathbf{h}) + \mathbf{f} \cdot [\mathbf{g}' \times \mathbf{h} + \mathbf{g} \times \mathbf{h}']$ and the result follows.

33. The following four statements are equivalent: $||\mathbf{r}(t)|| = \sqrt{\mathbf{r}(t) \cdot \mathbf{r}(t)}$ is constant, $\mathbf{r}(t) \cdot \mathbf{r}(t)$ is constant, $d/dt\,[\mathbf{r}(t) \cdot \mathbf{r}(t)] = 2[\mathbf{r}(t) \cdot \mathbf{r}'(t)] = 0$ identically, $\mathbf{r}(t) \cdot \mathbf{r}'(t) = 0$ identically.

35. $$\frac{[\mathbf{f}(t+h) \times \mathbf{g}(t+h)] - [\mathbf{f}(t) \times \mathbf{g}(t)]}{h} = \left(\mathbf{f}(t+h) \times \left[\frac{\mathbf{g}(t+h) - \mathbf{g}(t)}{h}\right]\right) + \left(\left[\frac{\mathbf{f}(t+h) - \mathbf{f}(t)}{h}\right] \times \mathbf{g}(t)\right).$$

Now take the limit as $h \to 0$.

(Appeal to Theorem 13.1.3.)

SECTION 13.3

1. $\pi\,\mathbf{j} + \mathbf{k}$, $R(u) = (\mathbf{i} + 2\mathbf{k}) + u(\pi\mathbf{j} + \mathbf{k})$ 3. $\mathbf{b} - 2\mathbf{c}$, $\mathbf{R}(u) = (\mathbf{a} - \mathbf{b} + \mathbf{c}) + u(\mathbf{b} - 2\mathbf{c})$

5. $4\mathbf{i} - \mathbf{j} + 4\mathbf{k}$, $\mathbf{R}(u) = (2\mathbf{i} + 5\mathbf{k}) + u(4\mathbf{i} - \mathbf{j} + 4\mathbf{k})$

7. $-\sqrt{2}\mathbf{i} + \frac{3\sqrt{2}}{2}\mathbf{j} + \mathbf{k}$, $\mathbf{R}(u) = \left(\sqrt{2}\mathbf{i} + \frac{3\sqrt{2}}{2}\mathbf{j} + \frac{\pi}{4}\mathbf{k}\right) + u\left(-\sqrt{2}\mathbf{i} + \frac{3\sqrt{2}}{2}\mathbf{j} + \mathbf{k}\right)$

9. The scalar components $x(t) = at$ and $y(t) = bt^2$ satisfy the equation $a^2y(t) = a^2(bt^2) = b(at)^2 = b[x(t)]^2$ and generate the parabola $a^2y = bx^2$.

11. (a) $P(0, 1)$ (b) $P(1, 2)$ (c) $P(-1, 2)$

13. The tangent line at $t = t_0$ has the form $\mathbf{R}(u) = \mathbf{r}(t_0) + u\mathbf{r}'(t_0)$. If $\mathbf{r}'(t_0) = \alpha\mathbf{r}(t_0)$, then

$$\mathbf{R}(u) = \mathbf{r}(t_0) + u\,\alpha\,\mathbf{r}(t_0) = (1 + u\alpha)\mathbf{r}(t_0).$$

The tangent line passes through the origin at $u = -1/\alpha$.

15. $\pi/2 \cong 1.57$, or $90°$

17. $P(1, 2, -2)$; $\cos^{-1}\left(\frac{1}{5}\sqrt{5}\right) \cong 1.11\text{rad}$

19. (a) $\mathbf{r}(t) = a\cos t\,\mathbf{i} + b\sin t\,\mathbf{j}$ (b) $\mathbf{r}(t) = a\cos t\,\mathbf{i} - b\sin t\,\mathbf{j}$

(c) $\mathbf{r}(t) = a\cos 2t\,\mathbf{i} + b\sin 2t\,\mathbf{j}$ (d) $\mathbf{r}(t) = a\cos 3t\,\mathbf{i} - b\sin 3t\,\mathbf{j}$

21. $\mathbf{r}'(t) = t^3\,\mathbf{i} + 2t\,\mathbf{j}$

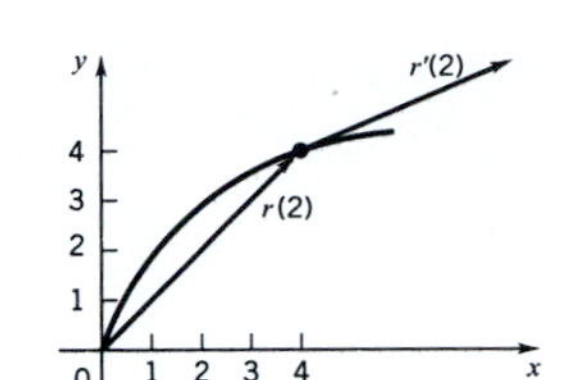

23. $\mathbf{r}'(t) = 2e^{2t}\,\mathbf{i} - 4e^{-4t}\,\mathbf{j}$

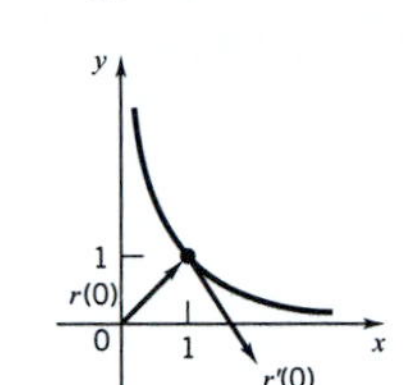

25. $\mathbf{r}'(t) = -2\sin t\,\mathbf{i} + 3\cos t\,\mathbf{j}$

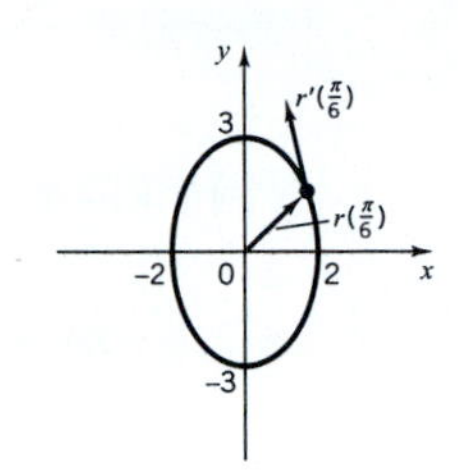

27. $\mathbf{r}(t) = (t^2 + 1)\,\mathbf{i} + t\,\mathbf{j}$, $t \geq 1$; or, $\mathbf{r}(t) = \sec^2 t\,\mathbf{i} + \tan t\,\mathbf{j}$, $t \in [\frac{1}{4}\pi, \frac{1}{2}\pi)$

29. $\mathbf{r}(t) = \cos t \sin 3t\,\mathbf{i} + \sin t\,\sin 3t\,\mathbf{j}$, $t \in [0, \pi]$

31.

$y^3 = x^2$

There is no tangent vector at the origin.

33. $(2, 4, 8)$; $\cos^{-1}\left(\dfrac{24}{\sqrt{21}\sqrt{161}}\right) \cong 1.15$ rad

35. $\mathbf{T}(1) = \dfrac{1}{\sqrt{2}}\mathbf{j} + \dfrac{1}{\sqrt{2}}\mathbf{k}$, $+\mathbf{N}(1) = \dfrac{-1}{\sqrt{2}}\mathbf{j} + \dfrac{1}{\sqrt{2}}\mathbf{k}$, $x - 1 = 0$

37. $\frac{1}{5}\sqrt{5}(-2\mathbf{i} + \mathbf{k})$, $-\mathbf{j}$, $x + 2z = \frac{1}{2}\pi$

39. $\frac{1}{14}\sqrt{14}(\mathbf{i} + 2\mathbf{j} + 3\mathbf{k})$, $\frac{1}{266}\sqrt{266}(-11\,\mathbf{i} - 8\,\mathbf{j} + 9\mathbf{k})$, $3x - 3y + z = 1$

41. $\mathbf{T}(0) = \dfrac{1}{\sqrt{3}}\mathbf{i} + \dfrac{1}{\sqrt{3}}\mathbf{j} + \dfrac{1}{\sqrt{3}}\mathbf{k}$, $\mathbf{N}(0) = \dfrac{1}{\sqrt{2}}\mathbf{i} - \dfrac{1}{\sqrt{2}}\mathbf{j}$, $x + y - 2z + 1 = 0$

43. $\mathbf{T}_1 = \dfrac{\mathbf{R}'(u)}{||\mathbf{R}'(u)||} = -\dfrac{\mathbf{r}'(a + b - u)}{||\mathbf{r}'(a + b - u)||} = -\mathbf{T}$. Then $\mathbf{T}_1'(u) = \mathbf{T}'(a + b - u)$ and thus $\mathbf{N}_1 = \mathbf{N}$.

45. Let $\mathbf{T}$ be the unit tangent at the tip of $\mathbf{R}(u) = \mathbf{r}(\phi(u))$ as calculated from the parametrization $\mathbf{r}$, and let $\mathbf{T}_1$ be the unit tangent at the same point as calculated from the parametrization $\mathbf{R}$. Then

$$\mathbf{T}_1 = \frac{\mathbf{R}'(u)}{||\mathbf{R}'(u)||} = \frac{\mathbf{r}'(\phi(u))\phi'(u)}{||\mathbf{r}'(\phi(u))\phi'(u)||} = \frac{\mathbf{r}'(\phi(u))}{||\mathbf{r}'(\phi(u))||} = \mathbf{T}.$$

$\phi'(u) > 0$ ⟶ ↑

This shows the invariance of the unit tangent. The invariance of the principal normal and the osculating plane follows directly from the invariance of the unit tangent.

47. (a) Let $t = \Psi(v) = 2\pi - v^2$. When t increases from 0 to 2π, v decreases from $\sqrt{2\pi}$ to 0.

(b) $\mathbf{T}_\mathbf{r}\left(\dfrac{\pi}{4}\right) = -\dfrac{1}{\sqrt{10}}\mathbf{i} + \dfrac{1}{\sqrt{10}}\mathbf{j} + \dfrac{2}{\sqrt{5}}\mathbf{k}$, $\mathbf{T}_\mathrm{R}\left(\dfrac{\sqrt{7\pi}}{2}\right) = \dfrac{1}{\sqrt{10}}\mathbf{i} - \dfrac{1}{\sqrt{10}}\mathbf{j} - \dfrac{2}{\sqrt{5}}\mathbf{k}$

$\mathbf{N}_r\left(\dfrac{\pi}{4}\right) = -\dfrac{\sqrt{2}}{2}\mathbf{i} - \dfrac{\sqrt{2}}{2}\mathbf{j}$; $\mathbf{N}_\mathrm{R}\left(\dfrac{\sqrt{7\pi}}{2}\right) = -\dfrac{\sqrt{2}}{2}\mathbf{i} - \dfrac{\sqrt{2}}{2}\mathbf{j}$

49. (a) $x = 1 - t, y = 1 + t, z = -\dfrac{\sqrt{2}}{2} - \dfrac{5\sqrt{2}}{2}t$

(c) the tangent line is parallel to the xy-plane at the points where $t = \dfrac{(2n + 1)\pi}{10}$, $n = 0, 1, 2, \ldots, 9$.

SECTION 13.4

1. $\frac{52}{3}$ **3.** $2\pi\sqrt{a^2+b^2}$ **5.** $\ln(1+\sqrt{2})$ **7.** $\frac{1}{27}(13\sqrt{13}-8)$ **9.** $\sqrt{2}(e^{\pi}-1)$ **11.** $6+\frac{1}{2}\sqrt{2}\ln(2\sqrt{2}+3)$ **13.** e^2 **15.** $4\pi^2$

17. differentiate $s(t)=\int_a^t \sqrt{[x'(u)]^2+[y'(u)]^2+[z'(u)]^2}\,du$ **19.** see Exercise 16, differentiate $s(x)=\int_a^x \sqrt{1+[f'(t)]^2}\,dt$

21. Let L be the length as computed from **r** and L^* the length as computed from **R**. Then

$$L^* = \int_c^d \|\mathbf{R}'(u)\|du = \int_c^d \|\mathbf{r}'(\phi(u))\|\phi'(u)du) = \int_a^b \|\mathbf{r}'(t)\|dt = L.$$

$$t=\phi(u) \uparrow$$

23. (a) $s=5t$ (b) $\mathbf{R}(s)=3\cos\left(\frac{s}{5}\right)\mathbf{i}+3\sin\left(\frac{s}{5}\right)\mathbf{j}+\frac{4s}{5}\mathbf{k}$ (c) $Q(-3,0,4\pi)$

(d) $\mathbf{R}'(s)=\frac{-3}{5}\sin\left(\frac{s}{5}\right)\mathbf{i}+\frac{3}{5}\cos\left(\frac{s}{5}\right)\mathbf{j}+\frac{4}{5}\mathbf{k},\ \|\mathbf{R}'(s)\|=1$

25. 0.5077 **27.** 22.0939 **29.** $L \cong 17.6286$

SECTION 13.5

1. $v/r,\ v^2/r$ **3.** $\|\mathbf{r}''(t)\| = a^2|b\sin at| = a^2|y(t)|$ **5.** $y=\cos\pi x,\quad 0\le x\le 2$

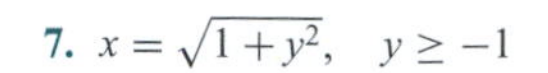

7. $x=\sqrt{1+y^2},\quad y\ge -1$

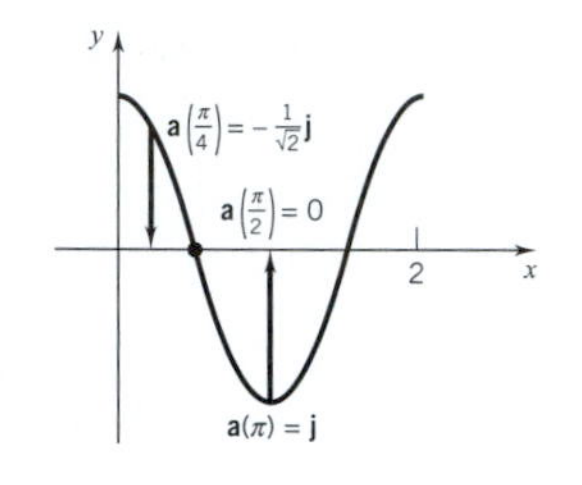

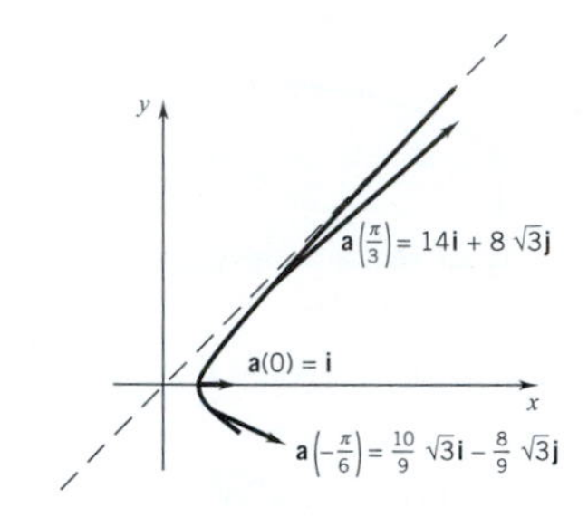

9. (a) (x_0,y_0,z_0) (b) $\alpha\cos\theta\,\mathbf{j}+\alpha\sin\theta\,\mathbf{k}$ (c) $|\alpha|$ (d) $-32\mathbf{k}$

(e) arc from parabola $z=z_0+(\tan\theta)(y-y_0)-16(y-y_0)^2/(\alpha^2\cos^2\theta)$ in the plane $x=x_0$

11. $\|\mathbf{r}(t)\|$ is constant iff $\|\mathbf{r}(t)\|^2=\mathbf{r}(t)\cdot\mathbf{r}(t)$ is constant;

$\mathbf{r}(t)\cdot\mathbf{r}(t)$ is constant iff $\frac{d}{dt}[\mathbf{r}(t)\cdot\mathbf{r}(t)]=2\mathbf{r}(t)\cdot\mathbf{r}'(t)=0.$

13. $\dfrac{e^{-x}}{(1+e^{-2x})^{3/2}}$ **15.** $\dfrac{2}{(1+4x)^{3/2}}$ **17.** $|\cos x|$ **19.** $\dfrac{|\sin x|}{(1+\cos^2 x)^{3/2}}$ **21.** $\frac{5}{2}\sqrt{5}$ **23.** $5\sqrt{5}$ **25.** $\dfrac{10\sqrt{10}}{3}$

27. $(\frac{1}{2}\sqrt{2},\frac{1}{2}\ln\frac{1}{2})$ **29.** $\dfrac{1}{(1+t^2)^{3/2}}$ **31.** $\dfrac{12|t|}{(4+9t^4)^{3/2}}$ **33.** $\frac{1}{2}\sqrt{2}\,e^{-t}$ **35.** $\dfrac{2+t^2}{(1+t^2)^{3/2}}$ **37.** $\sqrt{2}$ **39.** $\dfrac{a^4b^4}{(b^4x^2+a^4y^2)^{3/2}}$

41. $\kappa=\frac{1}{3}\sqrt{2}e^{-t},\quad a_{\mathbf{T}}=\sqrt{3}e^t,\quad a_{\mathbf{N}}=\sqrt{2}e^t$ **43.** $\kappa=1,\quad a_{\mathbf{T}}=0,\quad a_{\mathbf{N}}=4$ **45.** $\kappa=\dfrac{2}{2+t^2}\quad a_{\mathbf{T}}=2t,\quad a_{\mathbf{N}}=4+2t^2$

47. $\kappa=\dfrac{1}{8}\sqrt{\dfrac{2}{1-t^2}};a_{\mathbf{T}}=0;\ a_{\mathbf{N}}=\dfrac{1}{2}\sqrt{\dfrac{2}{1-t^2}}$ **49.** $a_{\mathbf{T}}=\dfrac{6t+12t^3}{\sqrt{1+t^2+t^4}};\quad a_{\mathbf{N}}=6\sqrt{\dfrac{1+4t^2+t^4}{1+t^2+t^4}}$ **51.** $\dfrac{e^{-a\theta}}{\sqrt{1+a^2}}$

53. $\dfrac{3}{2\sqrt{2a^2(1-\cos\theta)}}=\dfrac{3}{2\sqrt{2ar}}$ **55.** (a) $s(\theta)=4R|\cos\frac{1}{2}\theta|$ (b) $\rho(\theta)=4R\sin\frac{1}{2}\theta$ (c) $\rho^2+s^2=16R^2$ **57.** $9\rho^2+s^2=16a^2$

SECTION 13.6

1. (a) $\mathbf{r}'(0)=b\omega\,\mathbf{j}$ (b) $\mathbf{r}''(t)=\omega^2\mathbf{r}(t)$ (c) The torque is **0** and the angular momentum is constant.

3. (a) $\mathbf{v}(t)=2\mathbf{j}+(\alpha/m)t\mathbf{k}$ (b) $v(t)=(1/m\sqrt{4m^2+\alpha^2t^2}$ (c) $\mathbf{p}(t)=2m\,\mathbf{j}+\alpha t\mathbf{k}$

(d) $\mathbf{r}(t_1)=[2t+y_0]\mathbf{j}+[(\alpha/2m)t^2+z_0]\mathbf{k},\quad t\ge 0,\quad z=(\alpha/8m)(y-y_0)^2+z_0,\quad y\ge y_0,\quad x=0$

5. $\mathbf{F}(t) = 2m\mathbf{k}$ **7.** (a) $\pi b\,\mathbf{j} + \mathbf{k}$ (b) $\sqrt{\pi^2 b^2 + 1}$ (c) $-\pi^2 a\,\mathbf{i}$ (d) $m(\pi b\mathbf{j} + \mathbf{k})$
(e) $m[b(1-\pi)\,\mathbf{i} - 2a\,\mathbf{j} + 2\pi ab\mathbf{k}]$ (f) $-m\pi^2 a[\mathbf{j} - b\mathbf{k}]$

9. We have $m\mathbf{v} = m\mathbf{v}_1 + m\mathbf{v}_2$ and $\frac{1}{2}mv^2 = \frac{1}{2}mv_1^2 + \frac{1}{2}mv_2^2$. Therefore $\mathbf{v} = \mathbf{v}_1 + \mathbf{v}_2$ and $v^2 = v_1^2 + v_2^2$. Since

$$v^2 = \mathbf{v} \bullet \mathbf{v} = (\mathbf{v}_1 + \mathbf{v}_2) \bullet (\mathbf{v}_1 + \mathbf{v}_2) = v_1^2 + v_2^2 + 2(\mathbf{v}_1 \bullet \mathbf{v}_2),$$

we have $\mathbf{v}_1 \bullet \mathbf{v}_2 = 0$ and $\mathbf{v}_1 \perp \mathbf{v}_2$.

11. Here $\mathbf{r}''(t) = \mathbf{a}, \mathbf{r}'(t) = \mathbf{v}(0) + t\mathbf{a}, \mathbf{r}(0) + t\mathbf{v}(0) + \frac{1}{2}t^2\mathbf{a}$. If neither $\mathbf{v}(0)$ nor $\mathbf{a}$ is zero, the displacement $\mathbf{r}(t) - \mathbf{r}(0)$ is a linear combination of $\mathbf{v}(0)$ and $\mathbf{a}$ and thus remains on the plane determined by these vectors. The equation of this plane can be written

$$[\mathbf{a} \times \mathbf{v}(0)] \bullet [\mathbf{r} - \mathbf{r}(0)] = 0.$$

[If either $\mathbf{v}(0)$ or $\mathbf{a}$ is zero, the motion is restricted to a straight line; if both of these vectors are zero, the particle remains at its initial position $\mathbf{r}(0)$.]

13. $\mathbf{r}(t) = \mathbf{i} + t\,\mathbf{i} + (qE_0/2m)t^2\mathbf{k}$ **15.** $\mathbf{r}(t) = (1 + t^3/6m)\,\mathbf{i} + (t^4/12m)\,\mathbf{j} + t\,\mathbf{k}$.

17. $\frac{d}{dt}(\frac{1}{2}mv^2) = mv\frac{dv}{dt} = m\left(\mathbf{v} \bullet \frac{d\mathbf{v}}{dt}\right) = m\frac{d\mathbf{v}}{dt} \bullet \mathbf{v} = \mathbf{F} \bullet \frac{d\mathbf{r}}{dt} = 4r^2\left(\mathbf{r} \bullet \frac{d\mathbf{r}}{dt}\right) = 4r^2\left(r\frac{dr}{dt}\right) = 4r^3\frac{dr}{dt} = \frac{d}{dt}(r^4)$. Therefore $d/dt(\frac{1}{2}mv^2 - r^4) = 0$ and $\frac{1}{2}mv^2 - r^4$ is a constant E. Evaluating E from $t = 0$, we find that $E = 2m$. Thus $\frac{1}{2}mv^2 - r^4 = 2m$ and $v = \sqrt{4 + (2/m)r^4}$.

SECTION 13.7

1. about 61. 1% of an earth year **3.** set $x = r\cos\theta, \quad y = r\sin\theta$

5. Substitute

$$r = \frac{a}{1 + e\cos\theta}, \quad \left(\frac{dr}{d\theta}\right)^2 = \frac{(a\,e\sin\theta)^2}{(1 + e\cos\theta)^4}$$

into the right side of the equation and you will see that, with a and e^2 as given, the expression reduces to E.

CHAPTER 14

SECTION 14.1

1. dom (f) = the first and third quadrants, including the axes; range $(f) = [0, \infty)$

3. dom (f) = the set of all points (x, y) not on the line $y = -x$; range $(f) = (-\infty, 0) \cup (0, \infty)$ **5.** dom (f) = the entire plane; range $(f) = (-1, 1)$

7. dom (f) = the first and third quadrants, excluding the axes; range $(f) = (-\infty, \infty)$

9. dom (f) = the set of all points (x, y) with $x^2 < y$; in other words, the set of all points of the plane above the parabola $y = x^2$; range $(f) = (0, \infty)$

11. dom (f) = the set of all points (x, y) with $-3 \le x \le 3, -2 \le y \le 2$ (a rectangle); range $(f) = [-2, 3]$

13. dom (f) = the set of all points (x, y, z) not on the plane $x + y + z = 0$; range $(f) = \{-1, 1\}$

15. dom (f) = the set of all points (x, y, z) with $|y| < |x|$; range $(f) = (-\infty, 0]$

17. dom (f) = the set of all points (x, y) such that $x^2 + y^2 < 9$; in other words, the set of all points of the plane inside the circle $x^2 + y^2 = 9$; range $(f) = \left[\frac{2}{3}, \infty\right)$

19. dom (f) = the set of all points (x, y, z) with $x + 2y + 3z > 0$; in other words, the set of all points in space that lie on the same side of the plane $x + 2y + 3z = 0$ as the point $(1, 1, 1)$; range $(f) = (-\infty, \infty)$

21. dom (f) = all of space; range $(f) = (0, \infty)$

23. dom $(f) = \{x : x \ge 0\}$; range $(f) = [0, \infty)$
dom $(g) = \{(x, y) : x \ge 0, y \text{ real}\}$; range $(g) = [0, \infty)$
dom $(h) = \{x, y, z) : x \ge 0, y, z \text{ real}\}$; range $(h) = [0, \infty)$

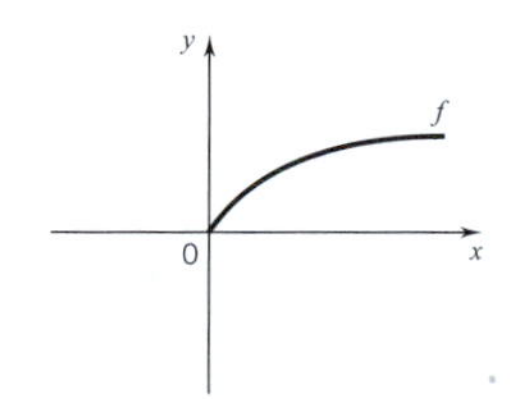

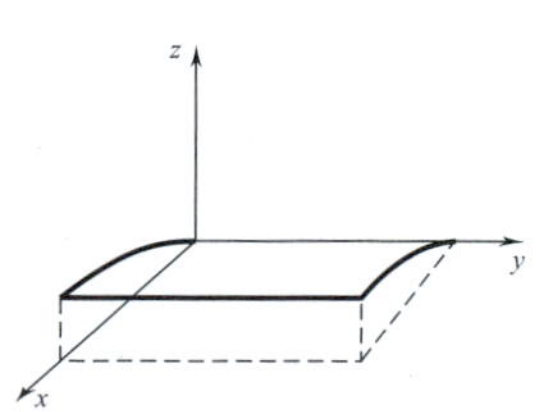

25. dom $(f) = \{(x, y) : 1 - 4xy \ge 0\}$ **27.** $\lim_{h \to 0} \frac{f(x + h, y) - f(x, y)}{h} = 4x; \quad \lim_{h \to 0} \frac{f(x, y + h) - f(x, y)}{h} = -1$

29. $\lim_{h\to 0} \dfrac{f(x+h,y)-f(x,y)}{h} = 3-y;\quad \lim_{h\to 0} \dfrac{f(x,y+h)-f(x,y)}{h} = -x+4y$

31. $\lim_{h\to 0} \dfrac{f(x+h,y)-f(x,y)}{h} = -y\sin(xy);\quad \lim_{h\to 0} \dfrac{f(x,y+h)-f(x,y)}{h} = -x\sin(xy)$

33. (a) $\dfrac{3y}{(x+y)(x+h+y)}$
(b) $-\dfrac{3x}{(x+y)(x+y+h)}$
(c) $\dfrac{3y}{(x+y)^2};\quad -\dfrac{3x}{(x+y)^2}$

35. (a) $f(x,y) = x^2y$ (b) $f(x,y) = \pi x^2 y$ (c) $f(x,y) = 2|y|$ **37.** $V = \dfrac{lh(10-lh)}{l+h}$ **39.** $V = \pi r^2 h + \frac{4}{3}\pi r^3$

SECTION 14.2

1. an elliptic cone **3.** a parabolic cylinder **5.** a hyperboloid of one sheet **7.** sphere of radius 2 centered at the origin

9. an elliptic paraboloid **11.** a hyperbolic paraboloid

13.

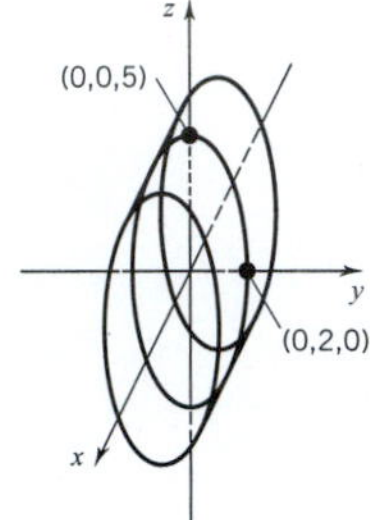

15.

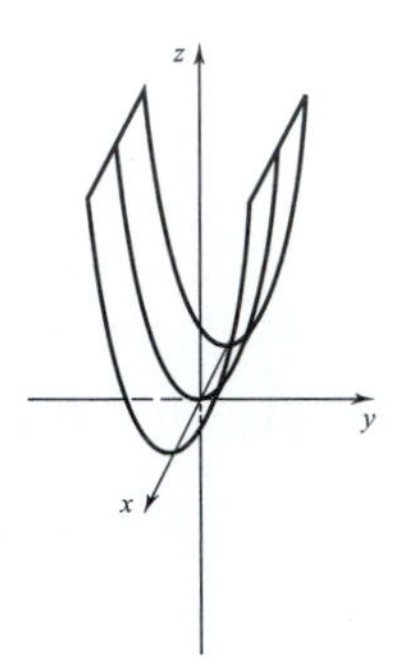

17.

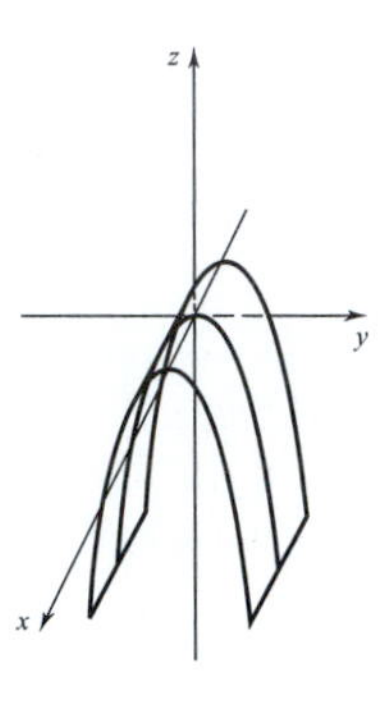

19.

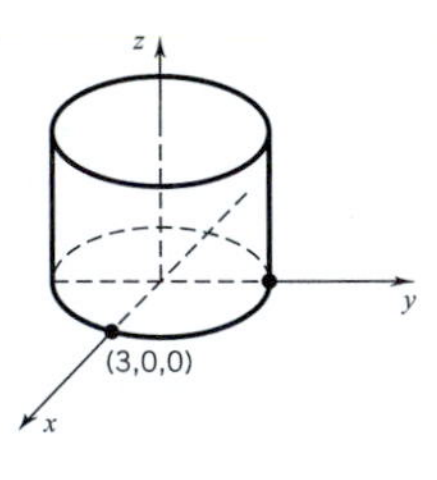

21.

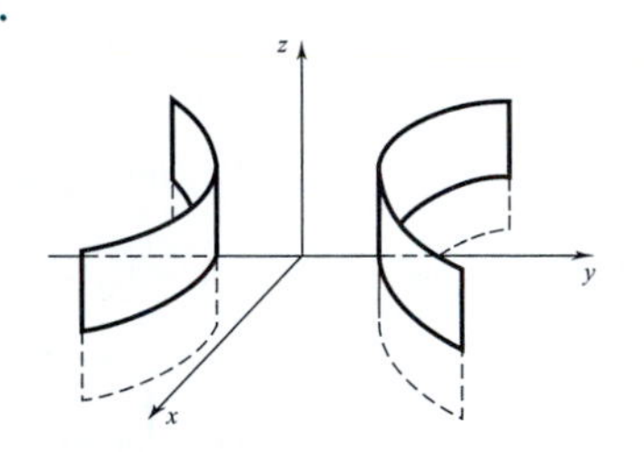

23.

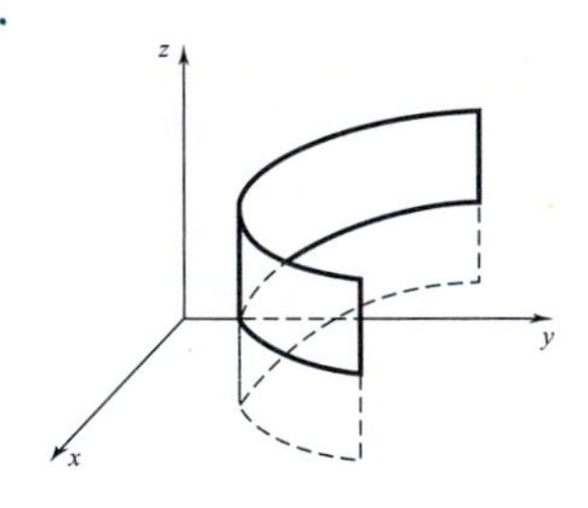

25. elliptic paraboloid, xy-trace: the origin, xz-trace: the parabola $x^2 = 4z$, yz-trace: the parabola $y^2 = 9z$, surface has the form of Figure 14.2.5

27. an elliptic cone, xy-trace: the origin, xz-trace: the lines $x = \pm 2z$, yz-trace: the lines $y = \pm 3z$, surface has the form of Figure 14.2.4

29. a hyperboloid of two sheets, xy-trace: none, xz-trace: the hyperbola $4z^2 - x^2 = 4$, yz-trace: the hyperbola $9z^2 - y^2 = 9$, surface has the form of Figure 14.2.3

31. hyperboloid of two sheets, xy-trace: the hyperbola $\dfrac{x^2}{4} - \dfrac{y^2}{9} = 1$, xz-trace: the hyperbola $\dfrac{x^2}{4} - z^2 = 1$, yz-trace: none, see Figure 14.2.3 for an example

33. elliptic paraboloid, xz-trace: the origin, yz-trace: the parabola $z^2 = 4y$, xy-trace: the parabola $x^2 = 9y$, surface has the form of Figure 14.2.5

35. hyperboloid of two sheets, xy-trace: the hyperbola $\dfrac{y^2}{4} - \dfrac{x^2}{9} = 1$, xz–trace: none, yz-trace: the hyperbola $\dfrac{y^2}{4} - z^2 = 1$, see Figure 14.2.3 for an example

37. paraboloid of revolution, xy-trace: the origin, xz-trace: the parabola $x^2 = 4z$, yz-trace: the parabola $y^2 = 4z$, surface has the form of figure 14.2.5.

39. (a) an elliptic paraboloid (opening up if A and B are both positive, opening down if A and B are both negative)
(b) a hyperbolic paraboloid (c) the xy-plane if A and B are both zero; otherwise, a parabolic cylinder

41. $x^2 + y^2 - 4z = 0$ (paraboloid of revolution) **43.** (a) a circle (b) (i) $\sqrt{x^2 + y^2} = -3z$ (ii) $\sqrt{x^2 + z^2} = \frac{1}{3}y$

45. the line $5x + 7y = 30$ **47.** the circle $x^2 + y^2 = \frac{5}{4}$ **49.** the ellipse $x^2 + 2y^2 = 2$ **51.** the parabola $x^2 = -4(y - 1)$

53. set $\frac{x}{a} = \cos u \cos v$, $\frac{y}{b} = \cos u \sin v$, $\frac{z}{c} = \sin u$. **55.** set $\frac{x}{a} = v \cos u$, $\frac{y}{b} = v \sin u$, $\frac{z}{c} = v$

SECTION 14.3

1. lines of slope 1: $y = x - c$

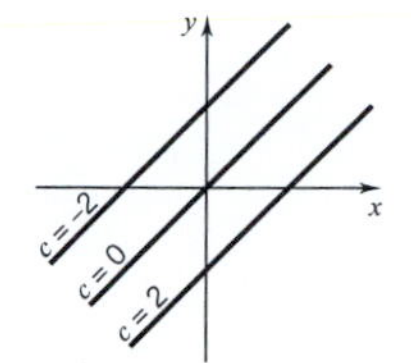

3. parabolas, $y = x^2 - c$

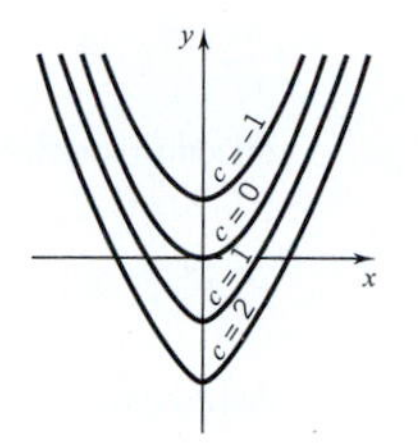

5. the y-axis and the lines $y = \left(\frac{1-c}{c}\right)x$, the origin omitted throughout

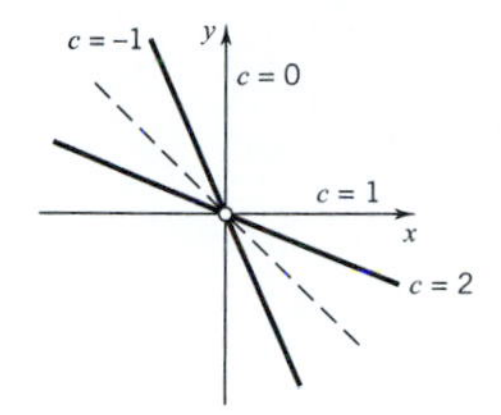

7. the cubics $y = x^3 - c$

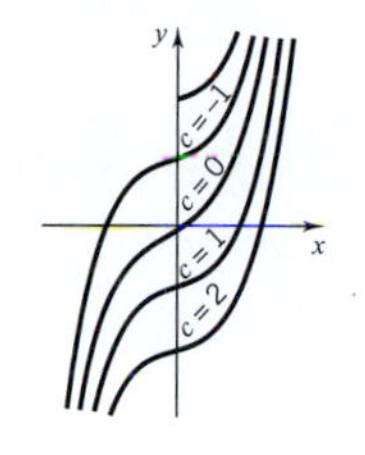

9. the lines $y = \pm x$ and the hyperbolas $x^2 - y^2 = c$

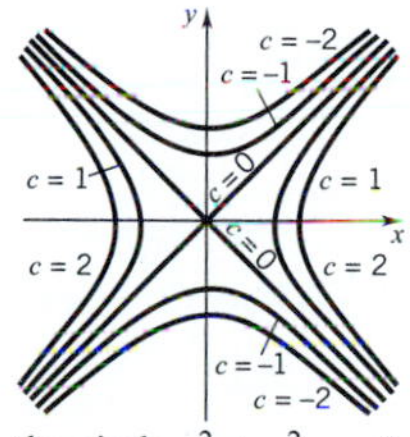

11. pairs of horizontal lines $y = \pm\sqrt{c}$ and the x-axis

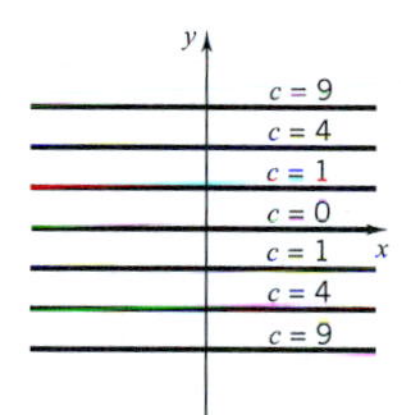

13. the circle $x^2 + y^2 = e^c$ with c real

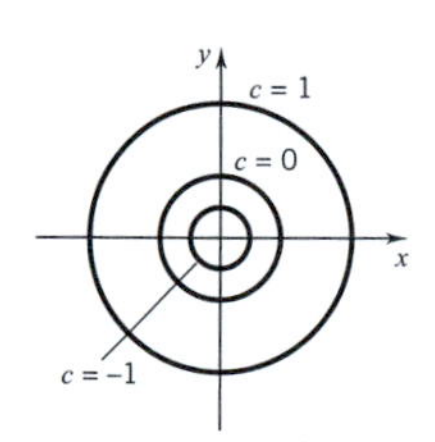

15. the curves $y = e^{cx^2}$ with the point (0, 1) omitted

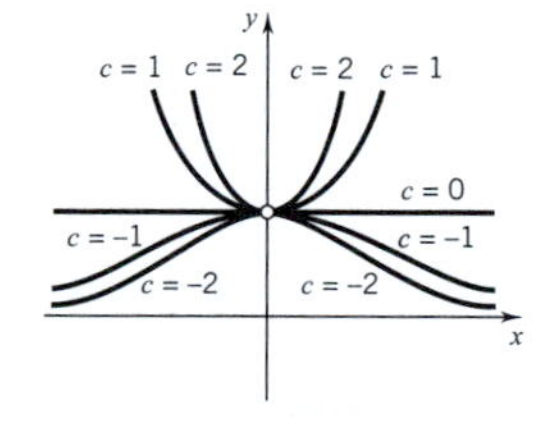

17. the coordinate axes and pairs of lines $y = \pm(\sqrt{1-c}/\sqrt{c})x$, the origin omitted throughout

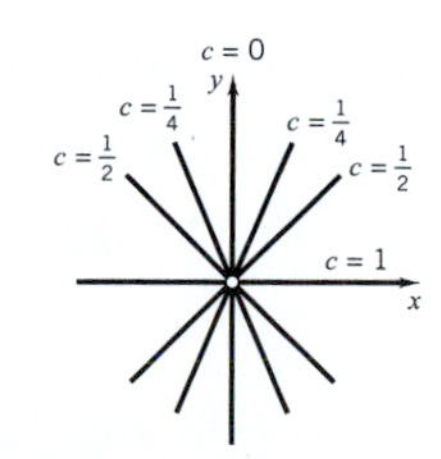

19. $x + 2y + 3z = 0$, plane through the origin **21.** $z = \sqrt{x^2 + y^2}$, the upper nappe of the circular cone $z^2 = x^2 + y^2$; Figure 14.2.4

23. the elliptic paraboloid $\dfrac{x^2}{(3\sqrt{2})^2} + \dfrac{y^2}{(2\sqrt{2})^2} = z$; Figure 14.2.5

25. (i) hyperboloid of two sheets; Figure 14.2.3 (ii) circular cone; Figure 14.2.4 (iii) hyperboloid of one sheet; Figure 14.2.2

27. $4x^2 + y^2 = 1$ **29.** $y^2 \tan^{-1} x = \pi$ **31.** $x^2 + 2y^2 - 2xyz = 13$

35. (a) $\dfrac{3x + 2y + 1}{4x^2 + 9} = \dfrac{3}{5}$ (b) $x^2 + 2y^2 - z^2 = 21$ **37.** $x^2 + y^2 + z^2 = \dfrac{GmM}{c}$. The surfaces of constant gravitational force are concentric spheres.

39. (a) $T(x, y) = \dfrac{k}{x^2 + y^2}$ (b) $x^2 + y^2 = \dfrac{k}{c}$, concentric circles (c) $10°$ **41.** F **43.** A **45.** E

SECTION 14.4

1. $\dfrac{\partial f}{\partial x} = 6x - y, \quad \dfrac{\partial f}{\partial y} = 1 - x$ **3.** $\dfrac{\partial \rho}{\partial \phi} = \cos\phi \cos\theta, \quad \dfrac{\partial \rho}{\partial \theta} = -\sin\phi \sin\theta$ **5.** $\dfrac{\partial f}{\partial x} = e^{x-y} + e^{y-x}, \quad \dfrac{\partial f}{\partial y} = -e^{x-y} - e^{y-x}$

7. $\dfrac{\partial g}{\partial x} = \dfrac{(AD - BC)y}{(Cx + Dy)^2}, \quad \dfrac{\partial g}{\partial y} = \dfrac{(BC - AD)x}{(Cx + Dy)^2}$ **9.** $\dfrac{\partial u}{\partial x} = y + z, \quad \dfrac{\partial u}{\partial y} = x + z, \quad \dfrac{\partial u}{\partial z} = x + y$

11. $\dfrac{\partial f}{\partial x} = z\cos(x - y), \quad \dfrac{\partial f}{\partial y} = -z\cos(x - y), \quad \dfrac{\partial f}{\partial z} = \sin(x - y)$

13. $\dfrac{\partial \rho}{\partial \theta} = e^{\theta+\phi}[\cos(\theta - \phi) - \sin(\theta - \phi)], \quad \dfrac{\partial \rho}{\partial \phi} = e^{\theta+\phi}[\cos(\theta - \phi) + \sin(\theta - \phi)]$

15. $\dfrac{\partial f}{\partial x} = 2xy\sec(xy) + x^2y^2\sec(xy)\tan(xy), \quad \dfrac{\partial f}{\partial y} = x^2\sec(xy) + x^3y\sec(xy)\tan(xy)$ **17.** $\dfrac{\partial h}{\partial x} = \dfrac{y^2 - x^2}{(x^2 + y^2)^2}, \quad \dfrac{\partial h}{\partial y} = -\dfrac{2xy}{(x^2 + y^2)^2}$

19. $\dfrac{\partial f}{\partial x} = \dfrac{\sin y\,(\cos x + x\sin x)}{y\cos^2 x}, \quad \dfrac{\partial f}{\partial y} = \dfrac{x\,(y\cos y - \sin y)}{y^2\cos x}$ **21.** $\dfrac{\partial h}{\partial x} = 2f(x)f'(x)g(y), \quad \dfrac{\partial h}{\partial y} = [f(x)]^2 g'(y)$

23. $\dfrac{\partial f}{\partial x} = (y^2 \ln z)z^{xy^2}, \quad \dfrac{\partial f}{\partial y} = (2xy \ln z)z^{xy^2}, \quad \dfrac{\partial f}{\partial z} = xy^2z^{xy^2 - 1}$

25. $\dfrac{\partial h}{\partial r} = 2r\,e^{2t}\cos(\theta - t), \quad \dfrac{\partial h}{\partial \theta} = -r^2 e^{2t}\sin(\theta - t), \quad \dfrac{\partial h}{\partial t} = r^2 e^{2t}[2\cos(\theta - t) + \sin(\theta - t)]$

27. $\dfrac{\partial f}{\partial x} = -\dfrac{yz}{x^2 + y^2}, \quad \dfrac{\partial f}{\partial y} = \dfrac{xz}{x^2 + y^2}, \quad \dfrac{\partial f}{\partial z} = \tan^{-1}(y/x)$ **29.** $f_x(0, e) = 1, \quad f_y(0, e) = e^{-1}$ **31.** $f_x(1, 2) = \frac{2}{9}, \quad f_y(1, 2) = -\frac{1}{9}$

33. $f_x(x, y) = 2xy, \quad f_y(x, y) = x^2$ **35.** $f_x(x, y) = \dfrac{2}{x}, \quad f_y(x, y) = \dfrac{1}{y}$

37. $f_x(x, y) = -\dfrac{1}{(x - y)^2}, \quad f_y(x, y) = \dfrac{1}{(x - y)^2}$ **39.** $f_x(x, y, z) = y^2z, \quad f_y(x, y, z) = 2xyz, \quad f_z(x, y, z) = xy^2$

41. (b) $x = x_0, \quad z - z_0 = f_y(x_0, y_0)(y - y_0)$ **43.** $x = 2, \quad z - 5 = 2(y - 1)$ **45.** $y = 2, \quad z - 9 = 6(x - 3)$

47. (a) $m_x = -6$; tangent line: $y = 2, \quad z = -6x + 13$
(b) $m_y = 18$; tangent line: $x = 1, z = 18y - 29$

49. $u_x = v_y = 2x, \quad u_y = -v_x = -2y$ **51.** $u_x = v_y = \dfrac{x}{x^2 + y^2}, \quad u_y = -v_x = \dfrac{y}{x^2 + y^2}$

53. (a) f depends only on y (b) f depends only on x

55. (a) $50\sqrt{3}$ in.^{2} (b) $50\sqrt{3}$ (c) 50 (d) $\frac{5}{18}\pi$ in.^{2} (e) -2

57. (a) y_0-section: $\mathbf{r}(x) = x\mathbf{i} + y_0\mathbf{j} + f(x_0, y_0)\mathbf{k}$

tangent line: $\mathbf{R}(t) = [x_0\mathbf{i} + y_0\mathbf{j} + f(x_0, y_0)\mathbf{k}] + t\left[\mathbf{i} + \dfrac{\partial f}{\partial x}(x_0, y_0)\mathbf{k}\right]$

(b) x_0-section: $\mathbf{r}(y) = x_0\mathbf{i} + y\mathbf{j} + f(x_0, y)\mathbf{k}$

tangent line: $\mathbf{R}(t) = [x_0\mathbf{i} + y_0\mathbf{j} + f(x_0, y_0)\mathbf{k}] + t\left[\mathbf{j} + \dfrac{\partial f}{\partial y}(x_0, y_0)\mathbf{k}\right]$

(c) For (x, y, z) in the plane

$$[(x-x_0)\mathbf{i}+(y-y_0)\mathbf{j}+(z-f(x_0,y_0))\mathbf{k}]\bullet\left[\left(\mathbf{i}+\frac{\partial f}{\partial x}(x_0,y_0)\mathbf{k}\right)\times\left(\mathbf{j}+\frac{\partial f}{\partial y}(x_0,y_0)\mathbf{k}\right)\right]=0.$$

From this it follows that

$$z-f(x_0,y_0)=(x-x_0)\frac{\partial f}{\partial x}(x_0,y_0)+(y-y_0)\frac{\partial f}{\partial y}(x_0,y_0).$$

59. (a) Set $u = ax + by$. Then $\dfrac{\partial w}{\partial x} = ag'(u)$ and $\dfrac{\partial w}{\partial y} = bg'(u)$. (b) Set $u = x^m y^n$. Then $\dfrac{\partial w}{\partial x} = mx^{m-1}y^n g'(u)$ and $\dfrac{\partial w}{\partial y} = nx^m y^{n-1} g'(u)$.

61. $V\dfrac{\partial P}{\partial V} = V\left(-\dfrac{kT}{V^2}\right) = -k\dfrac{T}{V} = -P;\ V\dfrac{\partial P}{\partial V} + T\dfrac{\partial P}{\partial T} = -k\dfrac{T}{V} + T\left(\dfrac{k}{V}\right) = 0$

SECTION 14.5

1. interior $= \{(x,y) : 2 < x < 4, 1 < y < 3\}$. (the inside of the rectangle)
boundary = the union of the four line segments that bound the rectangle
set is closed

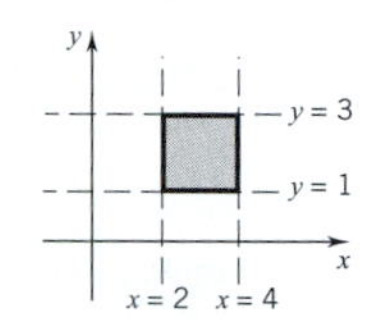

3. interior = the entire set (region between two concentric circles)
boundary $= \{(x,y) : x^2 + y^2 = 1 \text{ or } x^2 + y^2 = 4\}$ (the two circles)
set is open)

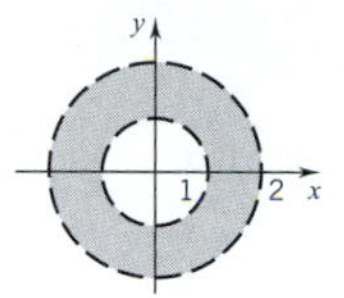

5. interior $= \{(x,y) : 1 < x^2 < 4\} = \{(x,y) : -2 < x < 1\} \cup \{(x,y) : 1 < x < 2\}$
(two vertical strips without the boundary lines)
boundary $= \{(x,y) : x = -2, x = -1, x = 1, \text{ or } x = 2\}$ (four vertical lines)
set is neither open nor closed

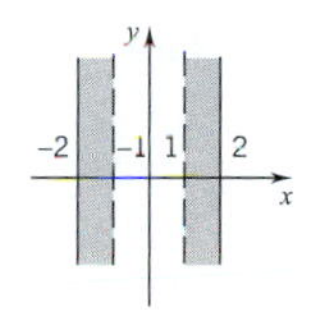

7. interior $= \{(x,y) : y < x^2\}$ (region below the parabola)
boundary $= \{(x,y) : y = x^2\}$ (the parabola)
set is closed

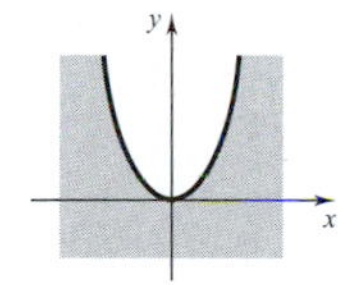

9. interior $= \{(x,y,z) : x^2 + y^2 < 1, 0 < z < 4\}$
(the inside of a cylinder)
boundary = the total surface of the cylinder
(the curved part, the top, the bottom)
set is closed

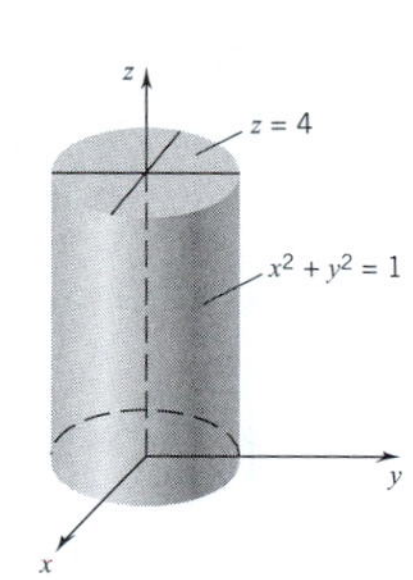

11. (a) ϕ (b) S (c) closed

13. interior $= \{x : 1 < x < 3\}$, boundary $= \{1, 3\}$; set is closed

15. interior = the entire set, boundary $= \{1\}$; set is open

17. interior $= \{x : |x| > 1\}$, boundary $= \{1, -1\}$; set is neither open nor closed

19. interior $= \phi$, boundary $= \{\text{the entire set}\} \cup \{0\}$; the set is neither open nor closed

SECTION 14.6

1. $\dfrac{\partial^2 f}{\partial x^2} = 2A, \quad \dfrac{\partial^2 f}{\partial y^2} = 2C, \quad \dfrac{\partial^2 f}{\partial y \partial x} = \dfrac{\partial^2 f}{\partial x \partial y} = 2B$

3. $\dfrac{\partial^2 f}{\partial x^2} = Cy^2 e^{xy}, \quad \dfrac{\partial^2 f}{\partial y^2} = Cx^2 e^{xy}, \quad \dfrac{\partial^2 f}{\partial y \partial x} = \dfrac{\partial^2 f}{\partial x \partial y} = Ce^{xy}(xy+1)$

5. $\dfrac{\partial^2 f}{\partial x^2} = 2, \quad \dfrac{\partial^2 f}{\partial y^2} = 4(x + 3y^2 + z^3), \quad \dfrac{\partial^2 f}{\partial z^2} = 6z(2x + 2y^2 + 5z^3)$

$\dfrac{\partial^2 f}{\partial x \partial y} = \dfrac{\partial^2 f}{\partial y \partial x} = 4y, \quad \dfrac{\partial^2 f}{\partial z \partial x} = \dfrac{\partial^2 f}{\partial x \partial z} = 6z^2, \quad \dfrac{\partial^2 f}{\partial z \partial y} = \dfrac{\partial^2 f}{\partial y \partial z} = 12yz^2$

7. $\dfrac{\partial^2 f}{\partial x^2} = \dfrac{1}{(x+y)^2} - \dfrac{1}{x^2}, \quad \dfrac{\partial^2 f}{\partial y^2} = \dfrac{1}{(x+y)^2}, \quad \dfrac{\partial^2 f}{\partial y \partial x} = \dfrac{\partial^2 f}{\partial x \partial y} = \dfrac{1}{(x+y)^2}$

9. $\dfrac{\partial^2 f}{\partial x^2} = 2(y+z), \quad \dfrac{\partial^2 f}{\partial y^2} = 2(x+z), \quad \dfrac{\partial^2 f}{\partial z^2} = 2(x+y)$; the second mixed partials are all $2(x+y+z)$

11. $\dfrac{\partial^2 f}{\partial x^2} = y(y-1)x^{y-2}, \quad \dfrac{\partial^2 f}{\partial y^2} = (\ln x)^2 x^y, \quad \dfrac{\partial^2 f}{\partial y \partial x} = \dfrac{\partial^2 f}{\partial x \partial y} = x^{y-1}(1 + y \ln x)$

13. $\dfrac{\partial^2 f}{\partial x^2} = y e^x, \quad \dfrac{\partial^2 f}{\partial y^2} = x e^y, \quad \dfrac{\partial^2 f}{\partial y \partial x} = e^y + e^x = \dfrac{\partial^2 f}{\partial x \partial y}$

15. $\dfrac{\partial^2 f}{\partial x^2} = \dfrac{y^2 - x^2}{(x^2+y^2)^2}, \quad \dfrac{\partial^2 f}{\partial y^2} = \dfrac{x^2 - y^2}{(x^2+y^2)^2}, \quad \dfrac{\partial^2 f}{\partial y \partial x} = -\dfrac{2xy}{(x^2+y^2)^2} = \dfrac{\partial^2 f}{\partial x \partial y}$

17. $\dfrac{\partial^2 f}{\partial x^2} = -2y^2 \cos 2xy, \quad \dfrac{\partial^2 f}{\partial y^2} = -2x^2 \cos 2xy, \quad \dfrac{\partial^2 f}{\partial y \partial x} = -[\sin 2xy + 2xy \cos 2xy] = \dfrac{\partial^2 f}{\partial x \partial y}$

19. $\dfrac{\partial^2 f}{\partial x^2} = 0, \quad \dfrac{\partial^2 f}{\partial y^2} = xz \sin y, \quad \dfrac{\partial^2 f}{\partial z^2} = -xy \sin z$

$\dfrac{\partial^2 f}{\partial y \partial x} = \sin z - z \cos y = \dfrac{\partial^2 f}{\partial x \partial y}$

$\dfrac{\partial^2 f}{\partial z \partial x} = y \cos z - \sin y = \dfrac{\partial^2 f}{\partial x \partial z}$

$\dfrac{\partial^2 f}{\partial z \partial y} = x \cos z - x \cos y = \dfrac{\partial^2 f}{\partial y \partial z}$

21. $x^2 \dfrac{\partial^2 u}{\partial x^2} + 2xy \dfrac{\partial^2 u}{\partial x \partial y} + y^2 \dfrac{\partial^2 u}{\partial y^2} = x^2\left(\dfrac{-2y^2}{(x+y)^3}\right) + 2xy\left(\dfrac{2xy}{(x+y)^3}\right) + y^2\left(\dfrac{-2x^2}{(x+y)^3}\right) = 0$

23. (a) no, since $\dfrac{\partial^2 f}{\partial x \partial y} \neq \dfrac{\partial^2 f}{\partial y \partial x}$ (b) no, since $\dfrac{\partial^2 f}{\partial x \partial y} \neq \dfrac{\partial^2 f}{\partial y \partial x}$ for $x \neq y$

25.

$$\frac{\partial^3 f}{\partial x^2 \partial y} \underset{\text{by defnition}}{=} \frac{\partial}{\partial x}\left(\frac{\partial^2 f}{\partial x \partial y}\right) \underset{(14.6.5)}{=} \frac{\partial}{\partial x}\left(\frac{\partial^2 f}{\partial y \partial x}\right) \underset{\text{by definition}}{=} \frac{\partial^2}{\partial x \partial y}\left(\frac{\partial f}{\partial x}\right) \underset{(14.6.5)}{=} \frac{\partial^2}{\partial y \partial x}\left(\frac{\partial f}{\partial x}\right) \underset{\text{by definition}}{=} \frac{\partial}{\partial y}\left(\frac{\partial^2 f}{\partial x^2}\right) \underset{\text{by definition}}{=} \frac{\partial^3 f}{\partial y \partial x^2}$$

27. (a) 0 (b) 0 (c) $\dfrac{m}{1+m^2}$ (d) 0 (e) $\dfrac{f'(0)}{1+[f'(0)]^2}$ (f) $\frac{1}{4}\sqrt{3}$ (g) does not exist

29. (a) $\dfrac{\partial g}{\partial x}(0,0) = \lim\limits_{h \to 0} \dfrac{g(h,0) - g(0,0)}{h} = \lim\limits_{h \to 0} 0 = 0, \quad \dfrac{\partial g}{\partial y}(0,0) = \lim\limits_{h \to 0} \dfrac{g(0,h) - g(0,0)}{h} - \lim\limits_{h \to 0} = 0$

(b) as (x, y) tends to $(0, 0)$ along the x-axis, $g(x, y) = g(x, 0) = 0$ tends to 0;
as (x, y) tends to $(0, 0)$ along the line $y = x$, $g(x, y) = g(x, x) = \frac{1}{2}$ tends to $\frac{1}{2}$

31. For $y \neq 0$, $\dfrac{\partial f}{\partial x}(0,y) = \lim_{h\to 0}\dfrac{f(h,y)-f(0,y)}{h} = \lim_{h\to 0}\dfrac{y(y^2-h^2)}{h^2+y^2} = y$. Since $\dfrac{\partial f}{\partial x}(0,0) = \lim_{h\to 0}\dfrac{f(h,0)-f(0,0)}{h} = \lim_{h\to 0} 0 = 0$,

we have $\dfrac{\partial f}{\partial x}(0,y) = y$ for all y. For $x \neq 0$, $\dfrac{\partial f}{\partial y}(x,0) = \lim_{h\to 0}\dfrac{f(x,h)-f(x,0)}{h} = \lim_{h\to 0}\dfrac{x(h^2-x^2)}{x^2+h^2} = -x$.

Since $\dfrac{\partial f}{\partial y}(0,0) = \lim_{h\to 0}\dfrac{f(0,h)-f(0,0)}{h} = \lim_{h\to 0} 0 = 0$, we have $\dfrac{\partial f}{\partial y}(x,0) = -x$ for all x.

Therefore $\dfrac{\partial^2 f}{\partial y\partial x}(0,y) = 1$ for all y and $\dfrac{\partial^2 f}{\partial x\partial y}(x,0) = -1$ for all x. In particular, $\dfrac{\partial^2 f}{\partial y\partial x}(0,0) = 1$, while $\dfrac{\partial^2 f}{\partial x\partial y}(0,0) = -1$

33. f must have the form: $f(x,y) = g(x) + h(y)$

CHAPTER 15

SECTION 15.1

1. $(6x-y)\,\mathbf{i} + (1-x)\,\mathbf{j}$ **3.** $e^{xy}[(xy+1)\,\mathbf{i} + x^2\,\mathbf{j}]$ **5.** $[2y^2\sin(x^2+1) + 4x^2y^2\cos(x^2+1)]\,\mathbf{i} + 4xy\sin(x^2+1)\,\mathbf{j}$

7. $(e^{x-y}+e^{y-x})(\mathbf{i} - \mathbf{j})$ **9.** $(z^2+2xy)\,\mathbf{i} + (x^2+2yz)\,\mathbf{j} + (y^2+2xz)\,\mathbf{k}$ **11.** $e^{-z}(2xy\,\mathbf{i} + x^2\,\mathbf{j} - x^2y\,\mathbf{k})$

13. $e^{x+2y}\cos(z^2+1)\,\mathbf{i} + 2e^{x+2y}\cos(z^2+1)\,\mathbf{j} - 2ze^{x+2y}\sin(z^2+1)\,\mathbf{k}$ **15.** $\left[2y\cos(2xy) + \dfrac{2}{x}\right]\mathbf{i} + 2x\cos(2xy)\,\mathbf{j} + \dfrac{1}{z}\mathbf{k}$

17. $\nabla f = -\,\mathbf{i} + 18\,\mathbf{j}$ **19.** $\frac{4}{5}\,\mathbf{i} + \frac{2}{5}\,\mathbf{j}$ **21.** $\mathbf{i}$ **23.** $\nabla f = -\frac{1}{2}\sqrt{2}(\mathbf{i} + 2\,\mathbf{j} + \mathbf{k})$ **25.** $\mathbf{i} + \frac{3}{5}\,\mathbf{j} - \frac{4}{5}\mathbf{k}$

27. (a) $\nabla f(0,2) = 4\,\mathbf{i}$

(b) $\nabla f(\pi/4, \pi/6) = \left(-1 - \dfrac{-1+\sqrt{3}}{2\sqrt{2}}\right)\mathbf{i} + \left(-\dfrac{1}{2} + \dfrac{-1+\sqrt{3}}{\sqrt{2}}\right)\mathbf{j}$

(c) $\nabla f(1,e) = (1-2e)\,\mathbf{i} - 2\,\mathbf{j}$

29. $(6x-y)\,\mathbf{i} + (1-x)\,\mathbf{j}$ **31.** $(2xy+z^2)\,\mathbf{i} + (2yz+x^2)\,\mathbf{j} + (2xz+y^2)\,\mathbf{k}$

33. $f(x,y) = x^2y + y$ **35.** $f(x,y) = \dfrac{x^2}{2} + x\sin y - y^2$ **37.** (a) $(1/r^2)\,\mathbf{r}$

(b) $[(\cos r)/r]\mathbf{r}$

(c) $(e^r/r)\mathbf{r}$

39. (a) $(0,0)$ (b)

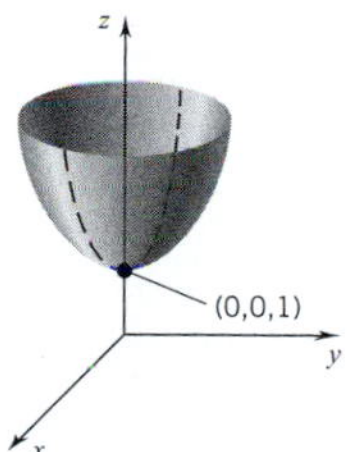

(c) f has an absolute minimum at (0,0)

41. (a) Let $\mathbf{c} = c_1\mathbf{i} + c_2\mathbf{j} + c_3\mathbf{k}$. First, we take $\mathbf{h} = h\mathbf{i}$. Since $\mathbf{c}\bullet\mathbf{h}$ is $o(\mathbf{h})$.

$$0 = \lim_{h\to 0}\frac{\mathbf{c}\bullet\mathbf{h}}{||\mathbf{h}||} = \lim_{h\to 0}\frac{c_1 h}{h} = c_1.$$

Similarly, $c_2 = 0$ and $c_3 = 0$.

(b) $(\mathbf{y}-\mathbf{z})\bullet\mathbf{h} = [f(\mathbf{x}+\mathbf{h}) - f(\mathbf{x}) - \mathbf{z}\bullet\mathbf{h}] + [\mathbf{y}\bullet\mathbf{h} - f(\mathbf{x}+\mathbf{h}) + f(\mathbf{x})] = o(\mathbf{h}) + o(\mathbf{h}) = o(\mathbf{h})$, so that, by part (a), $\mathbf{y} - \mathbf{z} = 0$.

43. (a) In Section 14.6 we showed that f was not continuous at (0, 0). It is therefore not differentiable at (0, 0).

(b) For $(x,y) \neq (0,0)$, $\dfrac{\partial f}{\partial x} = \dfrac{2y(y^2-x^2)}{(x^2+y^2)^2}$. As (x,y) tends to $(0,0)$ along the y-axis, $\partial f/\partial x = 2/y$ tends to ∞.

SECTION 15.2

1. $-2\sqrt{2}$ **3.** $\frac{1}{5}(7-4e)$ **5.** $\frac{1}{4}\sqrt{2}(a-b)$ **7.** $\dfrac{2}{\sqrt{65}}$ **9.** $\frac{2}{3}\sqrt{6}$ **11.** $-3\sqrt{2}$ **13.** $\dfrac{\sqrt{3}\pi}{12}$ **15.** $-(x^2+y^2)^{-1/2}$

17. (a) $\sqrt{2}[a(B-A) + b(C-B)]$ (b) $\sqrt{2}[a(A-B) + b(B-C)]$ **19.** $-\frac{7}{5}\sqrt{5}$ **21.** $\dfrac{18}{\sqrt{14}}$ or $\dfrac{-18}{\sqrt{14}}$

23. increases most rapidly in the direction of $\frac{1}{\sqrt{2}}\mathbf{i}+\frac{1}{\sqrt{2}}\mathbf{j}$, rate of change $2\sqrt{2}$; decreases most rapidly in the direction of $-\frac{1}{\sqrt{2}}\mathbf{i}-\frac{1}{\sqrt{2}}\mathbf{j}$, rate of change $-2\sqrt{2}$

25. increases most rapidly in the direction of $\frac{1}{\sqrt{6}}\mathbf{i}-\frac{2}{\sqrt{6}}\mathbf{j}+\frac{1}{\sqrt{6}}\mathbf{k}$, rate of change 1; decreases most rapidly in the direction of $-\frac{1}{\sqrt{6}}\mathbf{i}+\frac{2}{\sqrt{6}}\mathbf{j}-\frac{1}{\sqrt{6}}\mathbf{k}$, rate of change -1

27. $\nabla f = f'(x_0)\mathbf{i}$. If $f'(x_0) \neq 0$, the gradient points in the direction in which f increases: to the right if $f'(x_0) > 0$, to the left if $f'(x_0) < 0$.

29. (a) $\lim_{h\to 0}\frac{f(h,0)-f(0,0)}{h} = \lim_{h\to 0}\frac{\sqrt{h^2}}{h} = \lim_{h\to 0}\frac{|h|}{h}$ does not exist

(b) no; by Theorem 15.2.5 f cannot be differentiable at $(0,0)$

31. (a) $-\frac{2}{3}\sqrt{97}$ (b) $-\frac{8}{3}$ (c) $-\frac{26}{3}\sqrt{2}$

33. (a) its projection onto the xy-plane is the curve $y = x^3$ from $(1,1)$ to $(0,0)$

(b) its projection onto the xy-plane is the curve $y = -2x^3$ from $(1,-2)$ to $(0,0)$

35. its projection onto the xy-plane is the curve $(b^2)^{a^2}x^{b^2} = (a^2)^{b^2}y^{a^2}$ from (a^2, b^2) to $(0,0)$

37. the curve $y = \ln|\sqrt{2}\sin x|$ in the direction of decreasing x

39. (a) 16 (b) 4 (c) $\frac{16}{17}\sqrt{17}$

(d) The limits computed in (a) and (b) are not directional derivatives. In (a) and (b) we have, in essence, computed $\nabla f(2,4)\cdot\mathbf{r}_0$ taking $\mathbf{r}_0 = \mathbf{i}+4\mathbf{j}$ in (a) and $\mathbf{r}_0 = \frac{1}{4}\mathbf{i}+\mathbf{j}$ in (b). In neither case is $\mathbf{r}_0$ a unit vector.

41. (b) $\frac{2\sqrt{3}-3}{2}$

43. $\nabla(fg) = \left(f\frac{\partial g}{\partial x}+g\frac{\partial f}{\partial x}\right)\mathbf{i}+\left(f\frac{\partial g}{\partial y}+g\frac{\partial f}{\partial y}\right)\mathbf{j} = f\nabla g + g\nabla f$

45. $\nabla f^n = \frac{\partial f^n}{\partial x}\mathbf{i}+\frac{\partial f^n}{\partial y}\mathbf{j} = nf^{n-1}\frac{\partial f}{\partial x}\mathbf{i}+nf^{n-1}\frac{\partial f}{\partial y}\mathbf{j} = nf^{n-1}\nabla f$

SECTION 15.3

1. $C = (\frac{1}{3}, \frac{5}{3})$

3. (a) $f(x,y,z) = a_1x + a_2y + a_3z + C$

(b) $f(x,y,z) = g(x,y,z) + a_1x + a_2y + a_3z + C$

5. (a) U is not connected (b) (i) $g(\mathbf{x}) = f(\mathbf{x}) - 1$ (ii) $g(\mathbf{x}) = -f(\mathbf{x})$

7. e^t

9. $\frac{-2\sin 2t}{1+\cos^2 2t}$

11. $t^t\left[\frac{1}{t}+\ln t+(\ln t)^2\right]+\frac{1}{t}$

13. $3t^2 - 5t^4$

15. $2\omega(b^2-a^2)\sin\omega t\cos\omega t + b\omega$

17. $\sin 2t - 3\cos 2t$

19. $e^{t/2}(\frac{1}{2}\sin 2t + 2\cos 2t) + e^{2t}(2\sin\frac{1}{2}t + \frac{1}{2}\cos\frac{1}{2}t)$

21. $e^{t^2}\left[2t\sin\pi t + \pi\cos\pi t\right]$

23. $1 - 4t + 6t^2 - 4t^3$

25. increasing $\frac{1288}{3}\pi$ in.^{3}/ sec

27. 41.34 sq in. / sec

29. $\frac{\partial u}{\partial s} = 2s\cos^2 t - t\sin s\cos t - st\cos s\cos t$

$\frac{\partial u}{\partial t} = -2s^2\sin t\cos t + st\sin s\sin t - s\sin s\cos t$

31. $\frac{\partial u}{\partial s} = 4s^3t^2\tan(s+t^2) + s^4t^2\sec^2(s+t^2)$;

$\frac{\partial u}{\partial t} = 2s^4t\tan(s+t^2) + 2s^4t^3\sec^2(s+t^2)$

33. $\frac{\partial u}{\partial s} = 2s\cos^2 t - \sin(t-s)\cos t + s\cos t\cos(t-s) + 2t^2\sin s\cos s$

$\frac{\partial u}{\partial t} = -2s^2\sin t\cos t + s\sin(t-s)\sin t - s\cos t\cos(t-s) + 2t\sin^2 s$

35. $\frac{d}{dt}[f(\mathbf{r}(t))] = \left[\nabla f(\mathbf{r}(t))\cdot\frac{\mathbf{r}'(t)}{||\mathbf{r}'(t)||}\right]||\mathbf{r}'(t)|| = f'_{\mathbf{u}(t)}(\mathbf{r}(t))\,||\mathbf{r}'(t)||$ where $\mathbf{u}(t) = \frac{\mathbf{r}'(t)}{||\mathbf{r}'(t)||}$

37. (a) $(\cos r)\frac{\mathbf{r}}{r}$ (b) $(r\cos r + \sin r)\frac{\mathbf{r}}{r}$

39. (a) $(r\cos r - \sin r)\frac{\mathbf{r}}{r^3}$ (b) $\left(\frac{\sin r - r\cos r}{\sin^2 r}\right)\frac{\mathbf{r}}{r}$

41. (a) See the figure

(b) $\dfrac{\partial u}{\partial r} = \dfrac{\partial u}{\partial x}\left(\dfrac{\partial x}{\partial w}\dfrac{\partial w}{\partial r} + \dfrac{\partial x}{\partial t}\dfrac{\partial t}{\partial r}\right) + \dfrac{\partial u}{\partial y}\left(\dfrac{\partial y}{\partial w}\dfrac{\partial w}{\partial r} + \dfrac{\partial y}{\partial t}\dfrac{\partial t}{\partial r}\right) + \dfrac{\partial u}{\partial z}\left(\dfrac{\partial z}{\partial w}\dfrac{\partial w}{\partial r} + \dfrac{\partial z}{\partial t}\dfrac{\partial t}{\partial r}\right)$

To obtain $\dfrac{\partial u}{\partial s}$, replace each r by s.

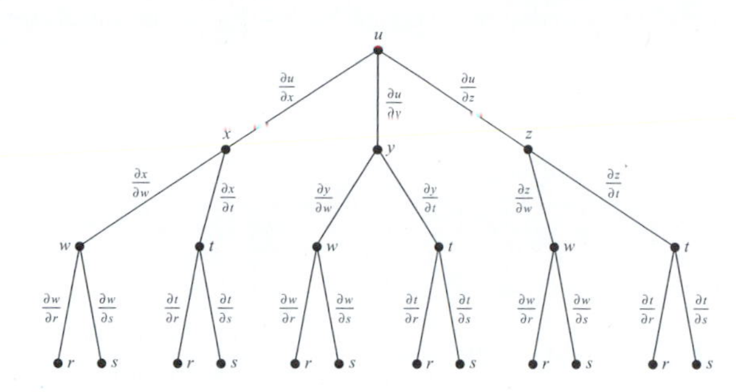

43. $\dfrac{du}{dt} = \dfrac{\partial u}{\partial x}\dfrac{dx}{dt} + \dfrac{\partial u}{\partial y}\dfrac{dy}{dt}$

$\dfrac{d^2u}{dt^2} = \dfrac{\partial u}{\partial x}\dfrac{d^2x}{dt^2} + \dfrac{dx}{dt}\left[\dfrac{\partial^2 u}{\partial x^2}\dfrac{dx}{dt} + \dfrac{\partial^2 u}{\partial y \partial x}\dfrac{dy}{dt}\right] + \dfrac{\partial u}{\partial y}\dfrac{d^2y}{dt^2} + \dfrac{dy}{dt}\left[\dfrac{\partial^2 u}{\partial x \partial y}\dfrac{dx}{dt} + \dfrac{\partial^2 u}{\partial y^2}\dfrac{dy}{dt}\right]$ and the result follows.

45. (a) $\dfrac{\partial u}{\partial r} = \dfrac{\partial u}{\partial x}\dfrac{\partial x}{\partial r} + \dfrac{\partial u}{\partial y}\dfrac{\partial y}{\partial r} = \dfrac{\partial u}{\partial x}\cos\theta + \dfrac{\partial u}{\partial y}\sin\theta, \quad \dfrac{\partial u}{\partial \theta} = \dfrac{\partial u}{\partial x}\dfrac{\partial x}{\partial \theta} + \dfrac{\partial u}{\partial y}\dfrac{\partial y}{\partial \theta} = \dfrac{\partial u}{\partial x}(-r\sin\theta) + \dfrac{\partial u}{\partial y}(r\cos\theta)$

(b) $\left(\dfrac{\partial u}{\partial r}\right)^2 = \left(\dfrac{\partial u}{\partial x}\right)^2\cos^2\theta + 2\dfrac{\partial u}{\partial x}\dfrac{\partial u}{\partial y}\cos\theta\sin\theta + \left(\dfrac{\partial u}{\partial y}\right)^2\sin^2\theta, \; \dfrac{1}{r^2}\left(\dfrac{\partial u}{\partial \theta}\right)^2 = \left(\dfrac{\partial u}{\partial x}\right)^2\sin^2\theta - 2\dfrac{\partial u}{\partial x}\dfrac{\partial u}{\partial y}\cos\theta\sin\theta + \left(\dfrac{\partial u}{\partial y}\right)^2\cos^2\theta,$

$\left(\dfrac{\partial u}{\partial r}\right)^2 + \dfrac{1}{r^2}\left(\dfrac{\partial u}{\partial \theta}\right)^2 = \left(\dfrac{\partial u}{\partial x}\right)^2(\cos^2\theta + \sin^2\theta) + \left(\dfrac{\partial u}{\partial y}\right)^2(\sin^2\theta + \cos^2\theta) = \left(\dfrac{\partial u}{\partial x}\right)^2 + \left(\dfrac{\partial u}{\partial y}\right)^2$

47. Solve the equations in Exercise 45 (a) for $\dfrac{\partial u}{\partial x}, \dfrac{\partial u}{\partial y}$:

$\dfrac{\partial u}{\partial x} = \dfrac{\partial u}{\partial r}\cos\theta - \dfrac{1}{r}\dfrac{\partial u}{\partial \theta}\sin\theta; \quad \dfrac{\partial u}{\partial y} = \dfrac{\partial u}{\partial r}\sin\theta + \dfrac{1}{r}\dfrac{\partial u}{\partial \theta}\cos\theta$

Then $\nabla u = \dfrac{\partial u}{\partial x}\mathbf{i} + \dfrac{\partial u}{\partial y}\mathbf{j} = \dfrac{\partial u}{\partial r}(\cos\theta\,\mathbf{i} + \sin\theta\,\mathbf{j}) + \dfrac{1}{r}\dfrac{\partial u}{\partial \theta}(-\sin\theta\,\mathbf{i} + \cos\theta\,\mathbf{j})$

49. $\nabla u = r(2 - \sin 2\theta)\,\mathbf{e}_r - r\cos 2\theta\,\mathbf{e}_\theta$

51. From Exercise 45(a);

$\dfrac{\partial^2 u}{\partial r^2} = \dfrac{\partial^2 u}{\partial x^2}\cos^2\theta + 2\dfrac{\partial^2 u}{\partial y \partial x}\sin\theta\cos\theta + \dfrac{\partial^2 u}{\partial y^2}\sin^2\theta$

$\dfrac{\partial^2 u}{\partial \theta^2} = \dfrac{\partial^2 u}{\partial x^2}r^2\sin^2\theta - 2\dfrac{\partial^2 u}{\partial y \partial x}r^2\sin\theta\cos\theta + \dfrac{\partial^2 u}{\partial y^2}r^2\cos^2\theta - r\left(\dfrac{\partial u}{\partial x}\cos\theta + \dfrac{\partial u}{\partial y}\sin\theta\right).$

The term in parentheses is just $\dfrac{\partial u}{\partial r}$, and the result follows.

53. $\dfrac{dy}{dx} = -\dfrac{e^y + y\,e^x - 4xy}{x\,e^y + e^x - 2x^2}$

55. $\dfrac{dy}{dx} = \dfrac{\cos xy - xy\sin xy - y\sin x}{x^2\sin xy - \cos x}$

57. $\dfrac{\partial z}{\partial x} = -\dfrac{2x - yz(x^2+y^2+z^2)\sin xyz}{2z - xy(x^2+y^2+z^2)\sin xyz}; \quad \dfrac{\partial z}{\partial y} = -\dfrac{2y - xz(x^2+y^2+z^2)\sin xyz}{2z - xy(x^2+y^2+z^2)\sin xyz}$

59. $\dfrac{\partial \mathbf{u}}{\partial s} = \dfrac{\partial \mathbf{u}}{\partial x}\dfrac{\partial x}{\partial s} + \dfrac{\partial \mathbf{u}}{\partial y}\dfrac{\partial y}{\partial s}, \quad \dfrac{\partial \mathbf{u}}{\partial t} = \dfrac{\partial \mathbf{u}}{\partial x}\dfrac{\partial x}{\partial t} + \dfrac{\partial \mathbf{u}}{\partial y}\dfrac{\partial y}{\partial t}$

where

$\dfrac{\partial \mathbf{u}}{\partial x} = \dfrac{\partial u_1}{\partial x}\mathbf{i} + \dfrac{\partial u_2}{\partial x}\mathbf{j}, \quad \dfrac{\partial \mathbf{u}}{\partial y} = \dfrac{\partial u_1}{\partial y}\mathbf{i} + \dfrac{\partial u_2}{\partial y}\mathbf{j}$

SECTION 15.4

1. normal vector $\mathbf{i}+\mathbf{j}$; tangent vector $\mathbf{i}-\mathbf{j}$
tangent line $x+y+2=0$; normal line $x-y=0$

3. normal vector $\sqrt{2}\,\mathbf{i}-5\,\mathbf{j}$; tangent vector $5\,\mathbf{i}+\sqrt{2}\,\mathbf{j}$
tangent line $\sqrt{2}x-5y+3=0$; normal line $5x+\sqrt{2}y-6\sqrt{2}=0$

5. normal vector $7\,\mathbf{i}-17\,\mathbf{j}$; tangent vector $17\,\mathbf{i}+7\,\mathbf{j}$
tangent line $7x-17y+6=0$; normal line $17x+7y-82=0$

7. normal vector $\mathbf{i}-\mathbf{j}$; tangent vector $\mathbf{i}+\mathbf{j}$
tangent line $x-y-3=0$; normal line $x+y+1=0$

9 0.

11. $4x-5y+4z=0;\quad x=1+4t,\ y=2-5t,\quad z=\frac{3}{2}+4t$

13. $x+ay-z-1=0;\quad x=1+t,\ y=\frac{1}{a}+at,\quad z=1-t$

15. $2x+2y-z=0;\quad x=2t,\ y=2t,\quad z=-t$

17. $b^2c^2x_0x-a^2c^2y_0y-a^2b^2z_0z-a^2-b^2-c^2=0;$
$x=x_0+2b^2c^2x_0t,\ \ y=y_0-2a^2c^2y_0t,\quad z=z_0-2a^2bc^2z_0t$

19. $(a^2/b, b^2/a, 3ab)$ **21.** $(0,0,0)$ **23.** $(\frac{1}{3},\frac{11}{6},-\frac{1}{12})$ **25.** $\dfrac{x-x_0}{\partial f/\partial x(x_0,y_0,z_0)}=\dfrac{y-y_0}{\partial f/\partial y(x_0,y_0,z_0)}=\dfrac{z-z_0}{\partial f/\partial z(x_0,y_0,z_0)}$

27. the tangent planes meet at right angles and therefore the normals ∇F and ∇G must meet at right angles:

$$\frac{\partial F}{\partial x}\frac{\partial G}{\partial x}+\frac{\partial F}{\partial y}\frac{\partial G}{\partial y}+\frac{\partial F}{\partial z}\frac{\partial G}{\partial z}=0$$

29. $\frac{9}{2}a^3\ (V=\frac{1}{3}Bh)$ **31.** approx. 0.528 rad **33.** $3x+4y+6z=22,\quad 6x+y-z=11$

35. (1, 1, 2) lies on both surfaces and the normals at this point are perpendicular.

37. (a) $3x+4y+6=0$ (b) $\mathbf{r}(t)=(4t-2)\mathbf{i}-3t\mathbf{j}+(43t^2-16t+6)\mathbf{k}$ (c) $\mathbf{R}(s)=(2\mathbf{i}-3\mathbf{j}+33\mathbf{k})+s(4\mathbf{i}-3\mathbf{j}+70\mathbf{k})$
(d) $4x-18y-z=29$ (e) $\mathbf{r}(t)=t\mathbf{i}-(\frac{3}{4}t+\frac{3}{2})\mathbf{j}+(\frac{35}{2}t-2)\,\mathbf{k};\ l=l'$

39. (a) $2\mathbf{i}+2\mathbf{j}+4\mathbf{k};\quad x=1+2t,\quad y=2+2t,\quad z=2+4t$
(b) $x+y+2z-7=0$

41. (c) $\nabla f(x,y)=\mathbf{0}$ at $(0,0),(\pm 1,0),(0,\pm 1),(1,\pm 1),(-1,\pm 1)$

SECTION 15.5

1. $(1,0)$ gives a local max of 1 **3.** $(-2,1)$ gives a local min of -2 **5.** $(4,-2)$ gives a local min of -10

7. $(0,0)$ is a saddle point; $(2,2)$ gives a local min of -8 **9.** $(1,\frac{3}{2})$ is a saddle point; $(5,\frac{27}{2})$ gives a local min of $-\frac{117}{4}$

11. $(0,n\pi)$ for integral n are saddle points; no local extreme values **13.** $(1,-1)$ and $(-1,1)$ are saddle points; no local extreme values

15. $(\frac{1}{2},4)$ gives a local min of 6 **17.** $(1,1)$ gives a local min of 3 **19.** $(1,0)$ gives a local min of -1; $(-1,0)$ gives a local max of 1

21. $(0,0)$ is a saddle point; $(1,0)$ and $(-1,0)$ give a local min of -3

23. (π,π) is a saddle point; $\left(\frac{\pi}{2},\frac{\pi}{2}\right)$ and $\left(\frac{3\pi}{2},\frac{3\pi}{2}\right)$ give a local maximum of 1;
$\left(\frac{\pi}{2},\frac{3\pi}{2}\right)$ and $\left(\frac{3\pi}{2},\frac{\pi}{2}\right)$ give a local minimum of -1.

25. (a) $f_x=2x+ky, f_y=2y+kx;\quad f_x(0,0)=f_y(0,0)=0$ independent of k (b) $|k|>2$ (c) $|k|<2$ (d) $|k|=2$

27. $(\frac{32}{9},-\frac{16}{9},\frac{32}{9});\ \frac{16}{3}$ **29.** $\dfrac{\sqrt{114}}{6}$

33. $(0,0)$ is a saddle point; $(1,1)$ gives a local maximum of 3.

35. $(-1,0)$ gives a local maximum of 1; $(1,0)$ gives a local minimum of -1.

SECTION 15.6

1. $(1,1)$ gives absolute min of -1; $(2,4)$ gives absolute max of 10

3. $(4,-2)$ gives absolute min of -13 ; $(0,-3)$ gives absolute max of 8

5. $(\sqrt{2}, -\sqrt{2})$ and $(-\sqrt{2}, \sqrt{2})$ give absolute min of 0; $(\sqrt{2}, \sqrt{2})$ and $(-\sqrt{2}, -\sqrt{2})$ give absolute max of 12

7. $(1, 1)$ gives absolute min of 0; $\left(-\sqrt{2}, -\sqrt{2}\right)$ gives absolute max of $6 + 4\sqrt{2}$

9. $(1, 0)$ gives absolute min of -1; $(-1, 0)$ gives absolute max of 1

11. absolute min of 0 along the lines $x = 0$ and $x = 2$; $(1, 0)$ gives absolute max of 2

13. $(\sqrt{2}, 2)$ gives absolute min of $-8 - 4\sqrt{2}$; $(-1, 1)$ gives absolute max of 1

15. $(0, 1)$ gives absolute min of -1; $(0, -1)$ gives absolute max of 1

17. absolute min of 0 along the line $y = x, 0 \le x \le 4$; $(0, 12)$ gives absolute max of 144

19. $x = 6, y = 6, z = 6$; maximum $= 216$ **21.** $\frac{1}{27}$

23. (a) $(0, 0)$ (b) no local extremes as $(0, 0)$ is a saddle point
(c) $(1, 0)$ and $(-1, 0)$ give absolute max of $\frac{1}{4}$; $(0, 1)$ and $(0, -1)$ give absolute min of $-\frac{1}{9}$

25. (a) saddle point, $f(x, y) = 0$ along the plane curve $y = x^{2/3}$ (see figure)
(b) $(0, 0)$ gives a local max of 3

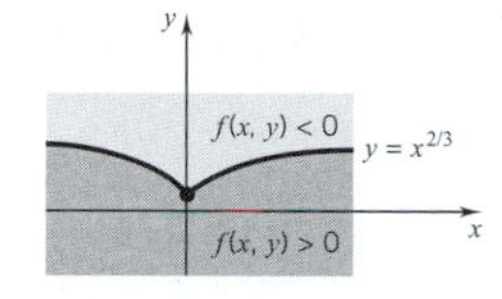

27. $\left(\dfrac{x_1 + x_2 + x_3}{3}, \dfrac{y_1 + y_2 + y_3}{3}\right)$ **29.** $\theta = \frac{1}{6}\pi, x = (2 - \sqrt{3})P, y = \frac{1}{6}(3 - \sqrt{3})P$ **31.** $\frac{2}{3}\sqrt{6}$ **33.** $4 \times 4 \times 6$ m

35. $V = \dfrac{xy(S - 2xy)}{2(x + y)}$ has a maximum value when $x = y = z = \sqrt{\dfrac{S}{6}}$. **37.** (a) $y = x - \frac{2}{3}$ (b) $y = \frac{14}{13}x^2 - \frac{19}{13}$

39. (a) cross section 18×18 inches; length 36 inches
(b) radius of cross section $36/\pi$ inches; length 36 inches

41. $x = 4$ in., $\theta = \dfrac{\pi}{3}$

SECTION 15.7

1. 2 **3.** $-\frac{1}{2}ab$ **5.** $\frac{2}{9}\sqrt{3}ab^2$ **7.** 1 **9.** $\frac{1}{9}\sqrt{3}abc$ **11.** $19\sqrt{2}$ **13.** $\frac{1}{27}abc$ **15.** 1

17. closest point $(\frac{2}{3}, \frac{1}{3}, \frac{2}{3})$; furthest point $(-\frac{2}{3}, -\frac{1}{3}, -\frac{2}{3})$ **19.** $f(3, -2, 1) = 14$ **21.** $|D|(A^2 + B^2 + C^2)^{-1/2}$

23. $4A^2(a^2 + b^2 + c^2)^{-1}$, where A is the area of the triangle and a, b, c, are the sides **25.** $(2^{-1/3}, -2^{-1/3})$ **27.** hint is given

29. (a) $f(\frac{k}{2}, \frac{k}{2}) = \frac{k}{2}$ is the maximum value (b) $(xy)^{1/2} = f(x, y) \le f(\frac{k}{2}, \frac{k}{2}) = \frac{k}{2} = \dfrac{x + y}{2}$

31. Same argument as Exercises 29 and 30: $f\left(\dfrac{k}{n}, \dfrac{k}{n}, \ldots, \dfrac{k}{n}\right) = \dfrac{k}{n}$ is the maximum value of $f(x_1, x_2, \ldots, x_n) = (x_1 x_2 \cdots x_n)^{1/n}$

33. radius $\sqrt[3]{\dfrac{V}{2\pi}}$; height $2\sqrt[3]{\dfrac{V}{2\pi}}$ **35.** $\dfrac{abc}{27}$ **37.** $4 \times 4 \times 6$ **39.** $\sqrt{\dfrac{S}{3}} \times \sqrt{\dfrac{S}{3}} \times \dfrac{1}{2}\sqrt{\dfrac{S}{3}}$

41. (a) cross section 18×18 inches; length 36 inches (b) radius of cross section $36/\pi$ inches; length 36 inches

43. $Q_1 = 10{,}000, Q_2 = 20{,}000, Q_3 = 30{,}000$

SECTION 15.8

1. $df = (3x^2y - 2xy^2)\Delta x + (x^3 - 2x^2y)\Delta y$ **3.** $df = (\cos y + y \sin x)\Delta x - (x \sin y + \cos x)\Delta y$

5. $df = \Delta x - (\tan z)\Delta y - (y \sec^2 z)\Delta z$

7. $df = \dfrac{y(y^2 + z^2 - x^2)}{(x^2 + y^2 + z^2)^2}\Delta x + \dfrac{x(x^2 + z^2 - y^2)}{(x^2 + y^2 + z^2)^2}\Delta y - \dfrac{2xyz}{(x^2 + y^2 + z^2)^2}\Delta z$

9. $df = [\cos(x + y) + \cos(x - y)]\Delta x + [\cos(x + y) - \cos(x - y)]\Delta y$

11. $df = (y^2 z\, e^{xz} + \ln z)\Delta x + 2y\, e^{xz}\Delta y + \left(xy^2\, e^{xz} + \frac{x}{z}\right)\Delta z$ **13.** $\Delta u = -7.15, du = -7.50$ **15.** $\Delta u = 2.896; du = 2.5$

17. $22\frac{249}{352}$ taking $u = x^{1/2}y^{1/4}$, $x = 121$, $y = 16$, $\Delta x = 4$, $\Delta y = 1$

19. $\frac{1}{14}\sqrt{2\pi}$ taking $u = \sin x \cos y$, $x = \pi$, $y = \frac{1}{4}\pi$, $\Delta x = -\frac{1}{7}\pi$, $\Delta y = -\frac{1}{20}\pi$ **21.** $f(2.9, 0.01) \cong 8.67$

23. $f(2.94, 1.1, 0.92) \cong 2.3391$ **25.** $dz = -\frac{1}{90}$, $\Delta z = -\frac{1}{93}$ **27.** decreases about 13.6π in.^{2} **29.** $S \cong 246.8$

31. (a) $dv = 0.24$ (b) $\Delta V = 0.22077$ **33.** $dT = 2.9$ **35.** $dA = 3\,\Delta x + 12.5\Delta\theta$; more sensitive to a change in θ

37. (a) $\Delta h = -\dfrac{(2r+\Delta r)h}{(r+\Delta r)^2}\Delta r$, $\Delta h \cong -\left(\dfrac{2h}{r}\right)\Delta r$ (b) $\Delta h = -\dfrac{(2r+h+\Delta r)}{r+\Delta r}\Delta r$, $\Delta h \cong -\left(\dfrac{2r+h}{r}\right)\Delta r$

39. (a) 1.962cm (b) 12.75cm^2 **41.** $2.23 \le s \pm |\Delta s| \le 2.27$

SECTION 15.9

1. $f(x,y) = \frac{1}{2}x^2y^2 + C$ **3.** $f(x,y) = xy + C$ **5.** not a gradient **7.** $f(x,y) = \sin x + y\cos x + C$

9. $f(x,y) = e^x \cos y^2 + C$ **11.** $f(x,y) = xy\, e^x + e^{-y} + C$ **13.** not a gradient

15. $f(x,y) = x + xy^2 + \frac{1}{2}x^2y^2 + \frac{1}{2}y^2 + y + C$ **17.** $f(x,y) = \sqrt{x^2+y^2} + C$ **19.** $f(x,y) = \frac{1}{3}x^3 \sin^{-1} y + y - y\ln y + C$

21. (a) yes (b) yes (c) no **23.** $f(x,y) = Ce^{x+y}$ **25.** (a), (b), (c) routine; (d) $f(x,y,z) = x^2 + yz + C$

27. $f(x,y,z) = x^2 + y^2 - z^2 + xy + yz + C$ **29.** $f(x,y,z) = xy^2z^3 + x + \frac{1}{2}y^2 + z + C$ **31.** $\mathbf{F}(\mathbf{r}) = \nabla\left(G\frac{mM}{r}\mathbf{r}\right)$

CHAPTER 16

SECTION 16.1

1. 819 **3.** 0 **5.** $a_2 - a_1$ **7.** $(a_2 - a_1)(b_2 - b_1)$ **9.** $a_2^2 - a_1^2$ **11.** $(a_2^2 - a_1^2)(b_2 - b_1)$

13. $2n(a_2 - a_1) - 3m(b_2 - b_1)$ **15.** $(a_2 - a_1)(b_2 - b_1)(c_2 - c_1)$ **17.** $a_{111} + a_{222} + \cdots + a_{nnn} = \sum_{p=1}^{n} a_{ppp}$

SECTION 16.2

1. $L_f(P) = 2\frac{1}{4}$, $U_f(P) = 5\frac{3}{4}$ **3.** (a) $L_f(P) = \sum_{i=1}^{m}\sum_{j=1}^{n}(x_{i-1} + 2y_{i-1})\,\Delta x_i\,\Delta y_j$, $U_f(P) = \sum_{i=1}^{m}\sum_{j=1}^{n}(x_i + 2y_j)\,\Delta x_i\,\Delta y_j$

(b) $I = 4$; the volume of the prism bounded above by the plane $z = x + 2y$ and below by R

5. $L_f(P) = -\frac{7}{24}$, $U_f(P) = \frac{7}{24}$ **7.** (a) $L_f(P) = \sum_{i=1}^{m}\sum_{j=1}^{n} 4x_{i-1}y_{j-1}\Delta x_i\,\Delta y_j$, $U_f(P) = \sum_{i=1}^{m}\sum_{j=1}^{n} 4x_i y_j \Delta x_i\,\Delta y_j$ (b) $I = b^2d^2$

9. (a) $L_f(P) = \sum_{i=1}^{m}\sum_{j=1}^{n} 3(x_{i-1}^2 - y_j^2)\,\Delta x_i\,\Delta y_j$, $U_f P = \sum_{i=1}^{m}\sum_{j=1}^{n} 3(x_i^2 - y_{j-1}^2)\,\Delta x_i\,\Delta y_j$ **11.** $\iint_\Omega dxdy = \int_a^b \phi(x)\,dx$

(b) $I = bd(b^2 - d^2)$

13. Suppose $f(x_0, y_0) \ne 0$. Assume $f(x_0, y_0) > 0$. Since f is continuous, there exists a disc Ω_ϵ with radius ϵ centered at (x_0, y_0) such that $f(x,y) > 0$ on Ω_ϵ. Let R be a rectangle contained in Ω_ϵ. Then $\iint_R f(x,y)\,dx\,dy > 0$, which contradicts the hypothesis. **15.** 6

17. By Theorem 16.2.10, there exists a point (x_1, y_1) in D_r such that

$$\iint_{D_r} f(x,y)\,dxdy = f(x_1,y_1)\iint_{D_r} dxdy = f(x_1,y_1)\pi r^2$$

Thus $f(x_1,y_1) = \dfrac{1}{\pi r^2}\iint_{D_r} f(x,y)\,dxdy$. As $r \to 0$, $(x_1,y_1) \to (x_0,y_0)$ and $f(x_1,y_1) \to f(x_0,y_0)$ since f is continuous. The result follows.

19. $\frac{1}{8}$ of a sphere of radius 2; $\frac{4}{3}\pi$

21. tetrahedron bounded by the coordinate planes and the plane $\dfrac{x}{3}+\dfrac{y}{2}+\dfrac{z}{6}=1; 6$

SECTION 16.3

1. 1 **3.** $\frac{9}{2}$ **5.** $\frac{1}{24}$ **7.** 2 **9.** $\frac{1}{4}\pi^2+\frac{1}{64}\pi^4$ **11.** $\frac{2}{27}$ **13.** $\frac{512}{15}$ **15.** 0 **17.** $\frac{1}{4}(e^4-1)$

19. $\displaystyle\int_0^1\int_{y^{1/2}}^{y^{1/4}} f(x,y)\,dx\,dy$

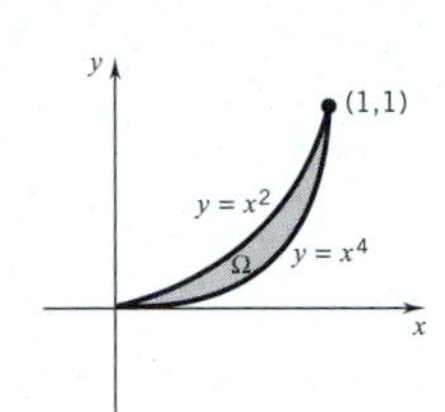

21. $\displaystyle\int_{-1}^{0}\int_{-x}^{1} f(x,y)dy\,dx+\int_0^1\int_x^1 f(x,y)\,dy\,dx$

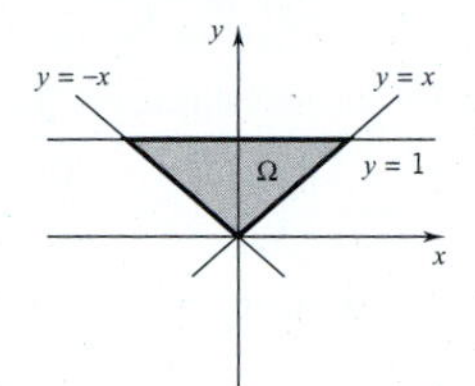

23. $\displaystyle\int_1^2\int_1^y f(x,y)\,dx\,dy+\int_2^4\int_{y/2}^{y} f(x,y)\,dx\,dy+\int_4^8\int_{y/2}^{4} f(x,y)\,dx\,dy$ **25.** 9 **27.** $\frac{1}{160}$

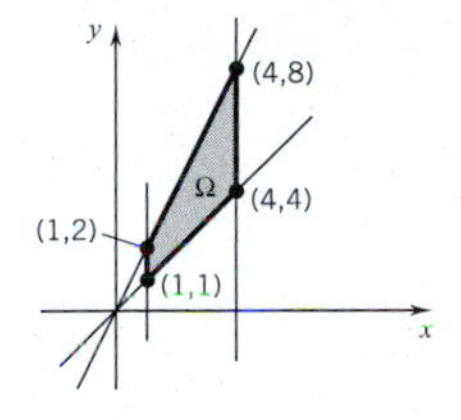

29. $\frac{2}{3}(\cos\frac{1}{2}-\cos 1)$ **31.** $1-\ln 2$ **33.** $\ln 4-\frac{1}{2}$ **35.** 4 **37.** $\frac{2}{15}$

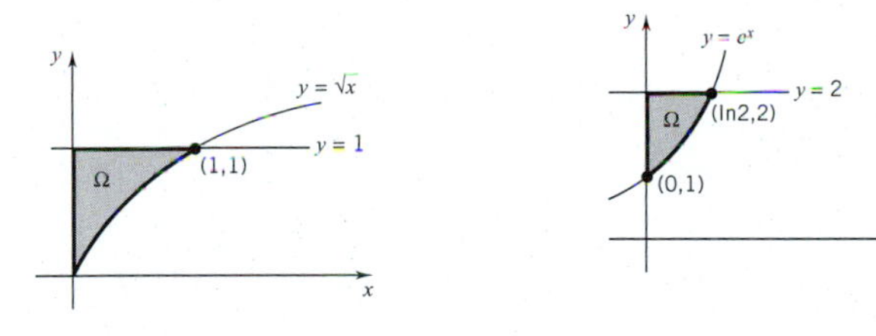

39. 3π **41.** $\frac{1}{6}$ **43.** $\frac{11}{70}$ **45.** $\frac{2}{3}a^3$ **47.** $\frac{1}{2}(e-1)$ **49.** $\frac{1}{12}(e-1)$ **51.** $\frac{2}{3}$ **53.** 1

55. $$\iint_R f(x)g(y)\,dx\,dy=\int_c^d\int_a^b f(x)g(y)\,dx\,dy=\int_c^d\left(\int_a^b f(x)g(y)\,dx\right)dy=\int_c^d g(y)\left(\int_a^b f(x)\,dx\right)dy$$
$$=\left(\int_a^b f(x)\,dx\right)\left(\int_c^d g(y)\,dy\right)$$

57. We have $R: -a\le x\le a, c\le y\le d$. Set $f(x,y)=g_y(x)$. For each fixed $y\in[c,d]$, g_y is an odd function. Thus

$$\int_{-a}^{a} g_y(x)\,dx=0.$$

Therefore

$$\iint_R f(x,y)\,dx\,dy=\int_c^d\int_{-a}^{a} f(x,y)\,dx\,dy=\int_c^d\int_{-a}^{a} g_y(x)\,dx\,dy=\int_c^d 0\,dy=0.$$

59. Note that $\Omega = \{(x, y) : 0 \le x \le y, \quad 0 \le y \le 1\}$. Set $\Omega' = \{(x, y) : 0 \le y \le x, \quad 0 \le x \le 1\}$.

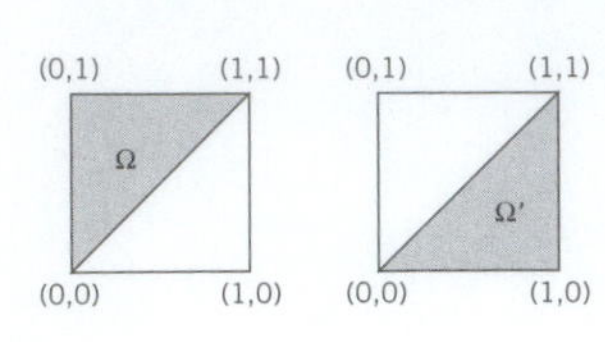

$$\iint_{\Omega} f(x)f(y) = dx\,dy = \int_0^1 \int_0^y f(x)f(y)\,dx\,dy$$

$$= \int_0^1 \int_0^x f(y)f(x)\,dy\,dx$$

x and y are dummy variables ⟶ ↑

$$= \iint_{\Omega'} f(x)f(y)\,dx\,dy.$$

Note that Ω and Ω' overlap and their union is the unit square $R : \{(x, y) : 0 \le x \le 1, \quad 0 \le y \le 1\}$. If $\int_0^1 f(x)\,dx = 0$, then

$$0 = \left(\int_0^1 f(x)\,dx\right)\left(\int_0^1 f(y)\,dy\right) = \iint_R f(x)f(y)\,dx\,dy$$

by Exercise 55 ⟶ ↑

$$= \iint_{\Omega} f(x)f(y)\,dx\,dy + \iint_{\Omega} f(x)f(y)\,dx\,dy$$

$$= 2\iint_{\Omega} f(x)f(y)\,dx\,dy$$

and therefore $\iint_{\Omega} f(x)f(y)\,dx\,dy = 0.$

61. Let M be the maximum value of $|f(x, y)|$ on Ω.

$$\int_{\phi_1(x+h)}^{\phi_2(x+h)} = \int_{\phi_1(x+h)}^{\phi_1(x)} + \int_{\phi_1(x)}^{\phi_2(x)} + \int_{\phi_2(x)}^{\phi_2(x+h)}$$

$$|F(x+h) - F(x)| = \left|\int_{\phi_1(x+h)}^{\phi_2(x+h)} f(x, y)\,dy - \int_{\phi_1(x)}^{\phi_2(x)} f(x, y)\,dy\right|$$

$$= \left|\int_{\phi_1(x+h)}^{\phi_1(x)} f(x, y)\,dy + \int_{\phi_2(x)}^{\phi_2(x+h)} f(x, y)\,dy\right| \le \left|\int_{\phi_1(x+h)}^{\phi_1(x)} f(x, y)\,dy\right| + \left|\int_{\phi_2(x)}^{\phi_2(x+h)} f(x, y)\,dy\right|$$

$$\le |\phi_1(x) - \phi_1(x+h)|M + |\phi_2(x+h) - \phi_2(x)|M.$$

The expression on the right tends to 0 as h tends to 0 since ϕ_1 and ϕ_2 are continuous.

63. (a) $\int_1^2 \int_{x^2-2x+2}^{1+\sqrt{x-1}} 1\,dy\,dx = \frac{1}{3}$

(b) $\int_1^2 \int_{y^2-2y+2}^{1+\sqrt{y-1}} 1\,dx\,dy = \frac{1}{3}$

SECTION 16.4

1. $\frac{1}{6}$ **3.** 6 **5.** (a) $\pi \sin 1$ (b) $\pi(\sin 4 - \sin 1)$ **7.** (a) $\frac{2}{3}$ (b) $\frac{14}{3}$ **9.** $\frac{1}{3}\pi$ **11.** $\frac{\pi}{6} - \frac{\sqrt{3}}{8}$ **13.** $\frac{\pi}{2}(\sin 1 - \cos 1)$

15. $\frac{\pi}{2}$ **17.** $\frac{3\pi}{4}$ **19.** $\frac{4\pi}{3} + 2\sqrt{3}$ **21.** 4 **23.** $b^3\pi$ **25.** $\frac{16}{3}\sqrt{3}\pi$ **27.** $\frac{2}{3}(8 - 3\sqrt{3})\pi$ **29.** 2π **31.** $\frac{1}{3}\pi a^2 b$ **33.** $\frac{\pi}{2}$

SECTION 16.5

1. $M = \frac{2}{3}, \quad x_M = 0, \quad y_M = \frac{1}{2}$ **3.** $M = \frac{1}{6}, \quad x_M = \frac{4}{7}, \quad y_M = \frac{3}{4}$ **5.** $M = \frac{32}{3}, \quad x_M = \frac{16}{3}, \quad y_M = \frac{9}{7}$

7. $M = \frac{5}{8}, \quad x_M = \frac{4}{5}, \quad y_m = \frac{152}{75}$ **9.** $M = \dfrac{5\pi}{3}, \quad x_M = \dfrac{21}{20}, \quad y_M = 0$

11. $I_x = \frac{1}{12}MW^2, \quad I_y = \frac{1}{12}ML^2, \quad I_z = \frac{1}{12}M(L^2 + W^2); \quad K_x = \frac{1}{6}\sqrt{3}W, \quad K_y = \frac{1}{6}\sqrt{3}L, \quad K_z = \frac{1}{6}\sqrt{3}\sqrt{L^2 + W^2}$

13. $x_M = \frac{1}{6}L, \quad y_M = 0$ **15.** $I_x = I_y = \frac{1}{4}MR^2, \quad I_z = \frac{1}{2}MR^2; \quad K_x = K_y = \frac{1}{2}R, \ K_z = \frac{1}{2}\sqrt{2}R$

17. center the disc at a distance $\sqrt{I_0 - \frac{1}{2}MR^2} / \sqrt{M}$ from the origin **19.** $I_x = \frac{1}{4}Mb^2, \ I_y = \frac{1}{4}Ma^2, \ I_z = \frac{1}{4}M(a^2 + b^2)$

21. $I_x = \frac{1}{10}, \quad I_y = \frac{1}{16}, \quad I_z = \frac{13}{80}$ **23.** $I_x = \dfrac{33\pi}{40}, \quad I_y = \dfrac{93\pi}{40}, \quad I_z = \dfrac{63\pi}{20}$

25. (a) $\frac{1}{4}M(r_2^2 + r_1^2)$ (b) $\frac{1}{4}M(r_2^2 + 5r_1^2)$ (c) $\frac{1}{4}M(5r_2^2 + r_1^2)$ **27.** $\frac{1}{2}M(r_2^2 + r_1^2)$ **29.** $x_M = 0, \quad y_M = R/\pi$

31. on the diameter through P at a distance $\frac{6}{5}R$ from P

33. Suppose Ω, a basic region of area A, is broken up into n basic regions $\Omega_1, \cdots, \Omega_n$ with areas $A_1, \cdots, A_n$. Then

$$\bar{x}A = \iint_{\Omega} x\,dx\,dy = \sum_{i=1}^{n}\left(\iint_{\Omega_i} x\,dx\,dy\right) = \sum_{i=1}^{n}\bar{x}_i A_i = \bar{x}_1 A_1 + \cdots + \bar{x}_n A_n.$$

The second formula follows in the same manner.

SECTION 16.6

1. they are equal **3.** $\iiint_{\Pi} \alpha\,dx\,dy\,dz = \alpha \iiint_{\Pi} dx\,dy\,dz = \alpha(\text{volume of }\Pi) = \alpha(a_2 - a_1)(b_2 - b_1)(c_2 - c_1)$ **5.** $\frac{1}{4}a^2b^2c$

7. $\bar{x} = \dfrac{A^2BC - a^2bc}{ABC - abc}, \ \bar{y} = \dfrac{AB^2C - ab^2c}{ABC - abc}, \ \bar{z} = \dfrac{ABC^2 - abc^2}{ABC - abc}$

9. $M = \frac{1}{2}Ka^4$ where K is the constant of proportionality for the density function **11.** $I_z = \frac{2}{3}Ma^2$

SECTION 16.7

1. abc **3.** $\frac{2}{3}$ **5.** 16 **7.** $\frac{1}{3}$ **9.** $\frac{47}{24}$

11.
$$\iiint_{\Pi} f(x)g(y)h(z)\,dx\,dy\,dz = \int_{c1}^{c2}\left[\int_{b1}^{b2}\left(\int_{a1}^{a2} f(x)g(y)h(z)\,dx\right)dy\right]dz$$
$$= \int_{c_1}^{c_2}\left[\int_{b_1}^{b_2} g(y)h(z)\left(\int_{a_1}^{a_2} f(x)\,dx\right)dy\right]dz$$
$$= \int_{c_1}^{c_2}\left[h(z)\left(\int_{a_1}^{a_2} f(x)\,dx\right)\left(\int_{b_1}^{b_2} g(y)\,dy\right)dz\right]$$
$$= \left(\int_{a_1}^{a_2} f(x)\,dx\right)\left(\int_{b_1}^{b_2} g(y)\,dy\right)\left(\int_{c_1}^{c_2} h(z)\,dz\right)$$

13. 8 **15.** $(\frac{2}{3}a, \frac{2}{3}b, \frac{2}{3}c)$ **17.**

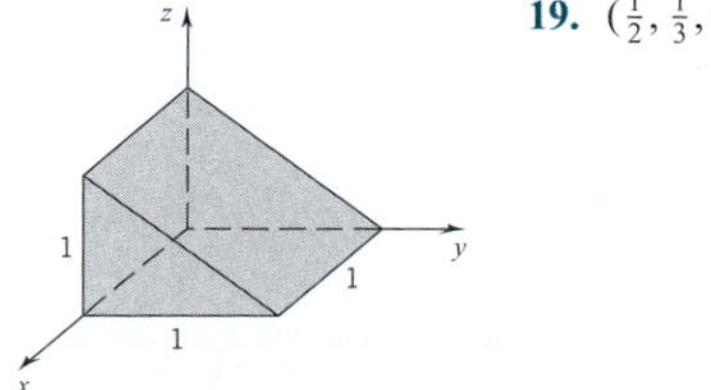

19. $(\frac{1}{2}, \frac{1}{3}, \frac{1}{3})$

21. $\int_{-r}^{r}\int_{-\sqrt{r^2-x^2}}^{\sqrt{r^2-x^2}}\int_{-\sqrt{r^2-(x^2+y^2)}}^{\sqrt{r^2-(x^2+y^2)}} k(r-\sqrt{x^2+y^2+z^2})\,dz\,dy\,dx$ **23.** $\int_{0}^{1}\int_{-\sqrt{x-x^2}}^{\sqrt{x-x^2}}\int_{-2x-3y-10}^{1-y^2} dz\,dy\,dx$

25. $\int_{-1}^{1}\int_{-2\sqrt{2-2x^2}}^{2\sqrt{2-2x^2}}\int_{3x^2+y^2/4}^{4-x^2-y^2/4} k(z-3x^2-\frac{1}{4}y^2)\,dz\,dy\,dx$ **27.** $\frac{28}{3}$ **29.** $\frac{1}{270}$ **31.** $\frac{12}{5}$ **33.** $V=\frac{8}{3},\ (\frac{11}{10},\frac{9}{4},\frac{11}{20})$

35. $V=\frac{27}{2},\ (\frac{1}{2},\frac{3}{2},\frac{12}{5})$ **37.** $V=\frac{1}{6}abc,\ (\frac{1}{4}a,\frac{1}{4}b,\frac{1}{4}c)$ **39.** (a) $\frac{1}{3}M(a^2+b^2)$ (b) $\frac{1}{12}M(a^2+b^2)$ (c) $\frac{1}{3}Ma^2+\frac{1}{12}Mb^2$

41. $M=\frac{1}{2}k,\quad (\frac{7}{12},\frac{34}{45},\frac{37}{90})$ **43.** (a) 0 by symmetry (b) $\frac{4}{3}\pi a_4$ **45.** $8\int_0^a\int_0^{\sqrt{a^2-x^2}}\int_0^{\sqrt{a^2-x^2-y^2}} dz\,dy\,dz=\frac{4}{3}\pi a^3$

47. $M=\frac{128}{15}k$

49. $M=\frac{135}{4}k,\ (\frac{1}{2},\frac{9}{5},\frac{12}{5})$ **51.** (a) $V=\int_0^6\int_{z/2}^{3}\int_x^{6-x} dy\,dx\,dz$

(b) $V=\int_0^3\int_0^{2x}\int_x^{6-x} dy\,dz\,dx$

(c) $V=\int_0^6\int_{z/2}^{3}\int_{z/2}^{y} dx\,dy\,dz+\int_0^6\int_3^{(12-z)/2}\int_{z/2}^{6-y} dx\,dy\,dz$

53. (a) $V=\iint_{\Omega_{yz}} 2y\,dy\,dz$ (b) $V=\iint_{\Omega_{yz}}\left(\int_{-y}^{y}dx\right)dy\,dz$ (c) $V=\int_0^4\int_{-\sqrt{4-y}}^{\sqrt{4-y}}\int_{-y}^{y} dx\,dz\,dy$ (d) $V=\int_{-2}^{2}\int_0^{4-z^2}\int_{-y}^{y} dx\,dy\,dz$

55. (a) 6.80703 (b) $\frac{16\sqrt{3}}{3}(4\sqrt{2}-2)\cong 33.7801$

SECTION 16.8

1. $r^2+z^2=9$ **3.** $z=2r$ **5.** $4r^2=z^2$ **7.** 2π **9.** 8π

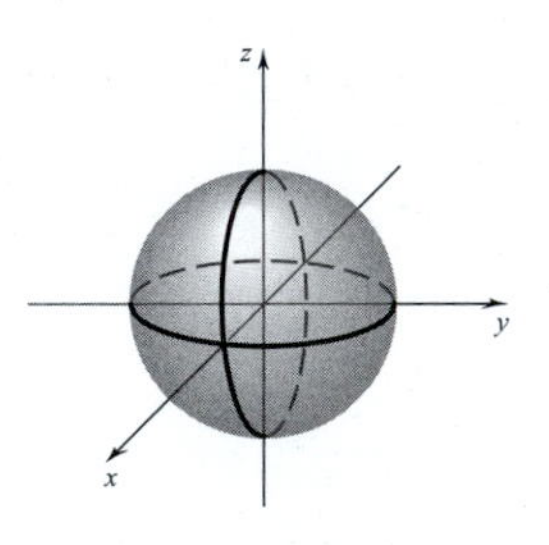

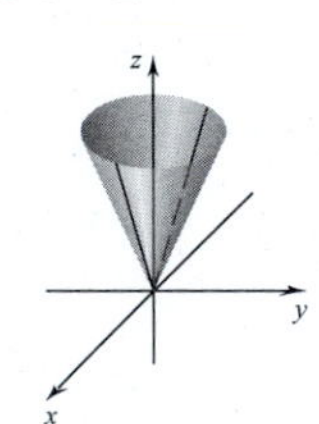

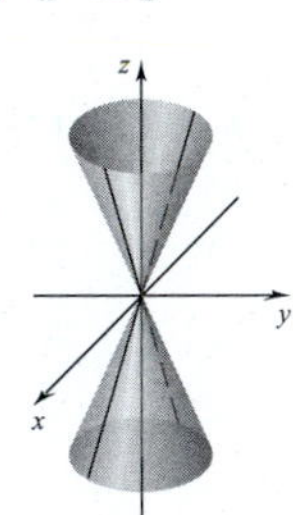

11. $\frac{1}{6}(8-3\sqrt{3})\pi$ **13.** $\frac{9\pi^2}{8}$ **15.** $\frac{\pi}{2}(1-\cos 1)\cong 0.7221$ **17.** $\frac{32}{9}a^3$ **19.** $\frac{1}{36}a^3(9\pi-16)$ **21.** $\frac{1}{32}\pi$ **23.** $\frac{1}{3}\pi(2-\sqrt{3})$

25. $\frac{32}{3}\pi\sqrt{2}$ **27.** $M=\frac{1}{2}k\pi R^2h^2$ **29.** $\frac{1}{2}MR^2$ **31.** Inverting the cone and placing the vertex at the origin, we have

$$V=\int_0^h\int_0^{2\pi}\int_0^{(R/h)z} r\,dr\,d\theta\,dz=\frac{1}{3}\pi R^2h.$$

33. $\frac{3}{10}MR^2$ **35.** $\frac{1}{2}\pi$ **37.** $\frac{1}{4}k\pi$

SECTION 16.9

1. $(\sqrt{3},\frac{1}{4}\pi,\cos^{-1}[\frac{1}{3}\sqrt{3}])$ **3.** $\left(\frac{3}{4},\frac{3}{4}\sqrt{3},\frac{3}{2}\sqrt{3}\right)$ **5.** $(\rho,\theta,\phi)=\left(\frac{4\sqrt{6}}{3},\frac{\pi}{4},\frac{\pi}{3}\right)$ 7. $(x,y,z)=(0,0,3)$

9. the circular cylinder $x^2+y^2=1$; the radius of the cylinder is 1 and the axis is the z-axis **11.** the lower nappe of the cone $z^2=x^2+y^2$

13. horizontal plane one unit above the xy-plane **15.** T: sphere centered at the origin, radius 2; $\frac{32\pi}{3}$

17. T: the portion of the sphere $x^2+y^2+z^2=9$ that lies between the planes $z=0$ and $z=\frac{3}{2}\sqrt{3}$; $\frac{9}{4}\pi\sqrt{3}$ **19.** $\frac{\pi}{3}(\sqrt{2}-1)$

21. $\frac{243\pi}{20}$ **23.** $V=\frac{4}{3}\pi R^3$ **25.** $V=\frac{2}{3}\alpha R^3$

27. $M=\frac{1}{6}k\pi h[(r^2+h^2)^{3/2}-h^3]$ **29.** (a) $\frac{2}{5}MR^2$ (b) $\frac{7}{5}MR^2$ **31.** (a) $\frac{2}{5}M\left(\frac{R_2^5-R_1^5}{R_2^3-R_1^3}\right)$ (b) $\frac{2}{3}MR^2$ (c) $\frac{5}{3}MR^2$ **33.** $V=\frac{2}{3}\pi(1-\cos\alpha)a^3$

35. (a) $\rho=2R\cos\phi$ (b) $0\le\theta\le2\pi,\quad 0\le\phi\le\frac{1}{4}\pi,\quad R\sec\theta\le\rho\le2R\cos\phi$ **37.** $V=\frac{1}{3}(16-6\sqrt{2})\pi$

39. Encase T in a spherical wedge W. W has spherical coordinates in a box Π that contains S. Define f to be zero outside of T. Then $F(\rho,\theta,\phi)=f(\rho\sin\phi\cos\theta,\rho\sin\phi\sin\theta,\rho\cos\phi)$ is zero outside of S and

$$\iiint_T f(x,y,z)\,dx\,dy\,dz = \iiint_W f(x,y,z)\,dx\,dy\,dz$$
$$= \iiint_\Pi F(\rho,\theta,\phi)\rho^2\sin\phi\,d\rho\,d\theta\,d\phi = \iiint_S F(\rho,\theta,\phi)\rho^2\sin\phi\,d\rho\,d\theta\,d\phi.$$

41. $\mathbf{F}=\frac{GmM}{R^2}(\sqrt{2}-1)\mathbf{k}$

SECTION 16.10

1. $ad-bc$ **3.** $2(v^2-u^2)$ **5.** $-3u^2v^2$ **7.** abc **9.** r **11.** $w(1+w\cos v)$ **13.** $\frac{1}{2}$ **15.** 0 **17.** $\frac{2}{3}$

19. (a) $A=3\ln 2$ (b) $\bar{x}=\frac{7}{9\ln 2},\ \bar{y}=\frac{14}{9\ln 2}$ **21.** $I_x=\frac{4}{75}M,\ I_y=\frac{14}{75}M,\ I_z=\frac{18}{75}M$ **23.** $A=\frac{32}{15}$ **25.** $A=\pi/\sqrt{65}$

27. $V=\frac{4}{3}\pi abc$ **29.** $I_x=\frac{1}{5}M(b^2+c^2),\ I_y=\frac{1}{5}M(a^2+c^2),\ I_z=\frac{1}{5}M(a^2+b^2)$

CHAPTER 17

SECTION 17.1

1. (a) 1 (b) −2 **3.** 0 **5.** (a) $-\frac{17}{6}$ (b) $\frac{17}{6}$ **7.** −8 **9.** $-\pi$ **11.** (a) 1 (b) $\frac{23}{21}$

13. (a) $2+\sin 2-\cos 3$ (b) $\frac{4}{5}+\sin 1-\cos 1$ **15.** 26 **17.** $\frac{1}{3}$ **19.** $\frac{8\pi^3}{3}$

21. $\int_C \mathbf{q}\cdot d\mathbf{r}=\int_a^b[\mathbf{q}\cdot\mathbf{r}'(u)]\,du+\int_a^b\frac{d}{du}[\mathbf{q}\cdot\mathbf{r}(u)]\,du=[\mathbf{q}\cdot\mathbf{r}(b)]-[\mathbf{q}\cdot\mathbf{r}(a)]=\mathbf{q}\cdot[\mathbf{r}(b)-\mathbf{r}(a)]$

$\int_C \mathbf{r}\cdot d\mathbf{r}=\int_a^b[\mathbf{r}(u)\cdot\mathbf{r}'(u)]\,du=\frac{1}{2}\int_a^b\frac{d}{du}[\mathbf{r}(u)\cdot\mathbf{r}(u)]\,du=\frac{1}{2}\int_a^b\frac{d}{du}(\|\mathbf{r}(u)\|^2)\,du=\frac{1}{2}(\|\mathbf{r}(b)\|^2-\|\mathbf{r}(a)\|^2)$

23. $\int_C \mathbf{f}(\mathbf{r})\cdot d\mathbf{r}=\int_a^b[\,\mathbf{f}(\mathbf{r}(u))\cdot\mathbf{r}'(u)]\,du=\int_a^b[f(u)\mathbf{i}\cdot\mathbf{i}\,]\,du=\int_a^b f(u)\,du$ **25.** $|W|$ = area of ellipse

27. force at time $t=m\,\mathbf{r}''(t)=m(2\beta\,\mathbf{j}+6\gamma t\mathbf{k})$; $W=(2\beta^2+\frac{9}{2}\gamma^2)m$ **29.** 0 **31.** (a) $\left(\frac{1}{\sqrt{5}}-\frac{1}{\sqrt{14}}\right)c$ (b) $\frac{4}{5}c$ **33.** $\alpha=\frac{15}{6}$

SECTION 17.2

1. 0 **3.** −1 **5.** 0 **7.** 0 **9.** 0 **11.** 4 **13.** e^3-2e^2+3 **15.** e^5-2e^2+1 **17.** 2π **19.** 14 **21.** 0

23. Set $f(x,y,z)=g(x)$ and $C:\mathbf{r}(u)=u\,\mathbf{i},\ u\in[a,b]$. In this case, $\nabla f(\mathbf{r}(u))=g'(x(u))\mathbf{i}=g'(u)\mathbf{i}$ and $\mathbf{r}'(u)=\mathbf{i}$, so that

$$\int_C \nabla f(\mathbf{r})\cdot d\mathbf{r}=\int_a^b[\nabla f(\mathbf{r}(u))\cdot\mathbf{r}'(u)]\,du=\int_a^b g'(u)\,du \quad\text{and}\quad f(\mathbf{r}(b))-f(\mathbf{r}(a))=g(b)-g(a).$$

The statement $\int_C \nabla f(\mathbf{r})\cdot d\mathbf{r}=f(\mathbf{r}(b))-f(\mathbf{r}(a))$ reduces to $\int_a^b g'(u)\,du=g(b)-g(a)$.

25. (a) $\mathbf{F}(\mathbf{r}) = cx\sqrt{x^2+y^2-z^2}\ \mathbf{i} + cy\sqrt{x^2+y^2+z^2}\ \mathbf{j} + cz\sqrt{x^2+y^2+z^2}\ \mathbf{k}$; $||\mathbf{F}(\mathbf{r})|| = cr^2$
(b) $f(x,y,z) = \frac{c}{3}(x^2+y^2+z^2)^{3/2}$

27. $W = mG\left(\dfrac{1}{r_2} - \dfrac{1}{r_1}\right)$ **29.** $f(x,y,z) = \dfrac{mGr_0^2}{r_0+z}$

SECTION 17.3

1. If f is continuous, then $-f$ is continuous and has antiderivatives u. The scalar fields $U(x,y,z) = u(x)$ are potential functions for $\mathbf{F}$:

$$\nabla U = \frac{\partial U}{\partial x}\,\mathbf{i} + \frac{\partial U}{\partial y}\,\mathbf{j} + \frac{\partial U}{\partial z}\,\mathbf{k} = \frac{du}{dx}\,\mathbf{i} = -f\,\mathbf{i} = -\mathbf{F}.$$

3. The scalar field $U(x,y,z) = cz + d$ is a potential energy function for $\mathbf{F}$. We know that the total mechanical energy remains constant. Thus, for any times t_1 and t_2,

$$\tfrac{1}{2}m[v(t_1)]^2 + U(\mathbf{r}(t_1)) = \tfrac{1}{2}m[v(t_2)]^2 + U(\mathbf{r}(t_2)).$$

This gives

$$\tfrac{1}{2}m[v(t_1)]^2 + cz(t_1) + d = \tfrac{1}{2}m[v(t_2)]^2 + cz(t_2) + d.$$

Solve this equation for $v(t_2)$ and you have the desired formula.

5. (a) We know that $-\nabla U$ points in the direction of maximum decrease of U. Thus $\mathbf{F} = -\nabla U$ attempts to drive objects toward a region where U has lower values. (b) At a point where U has a minimum, $\nabla U = \mathbf{0}$ and therefore $\mathbf{F} = \mathbf{0}$.

7. (a) By conservation of energy $\frac{1}{2}mv^2 + U = E$. Since E is constant and U is constant, v is constant.
(b) ∇U is perpendicular to any surface where U is constant. Obviously so is $\mathbf{F} = -\nabla U$.

9. $f(x,y,z) = -\dfrac{k}{\sqrt{x^2+y^2+z^2}}$ is a potential function for $\mathbf{F}$. The work done by $\mathbf{F}$ moving an object along C is $W = \displaystyle\int_C \mathbf{F}(\mathbf{r})\cdot d\mathbf{r} = \int_a^b \nabla f\cdot d\mathbf{r} =$ $f(\mathbf{r}(b)) - f(\mathbf{r}(a))$. Since $\mathbf{r}(a) = (x_0,y_0,z_0)$ and $\mathbf{r}(b) = (x_1,y_1,z_1)$ are points on the unit sphere, $f(\mathbf{r}(b)) = f(\mathbf{r}(a)) = -k$, and so $W = 0$.

SECTION 17.4

1. $\frac{1}{2}$ **3.** $\frac{9}{2}$ **5.** $\frac{11}{6}$ **7.** 2 **9.** 16 **11.** $\frac{104}{5}$ **13.** 4 **15.** 4 **17.** 2 **19.** 3 **21.** $\frac{176}{3}$ **23.** 56 **25.** $\frac{1177}{30}$

27. (a) $\dfrac{\partial P}{\partial y} = 6x - 4y = \dfrac{\partial Q}{\partial x}$ (b) 7 (c) $\frac{-37}{3}$ **29.** (a) $M = 2ka^2, x_M = y_M = \frac{1}{8}a(\pi+2)$ (b) $I_x = ka^4 = \frac{1}{2}Ma^2$

31. (a) $I_z = 2ka^4 = Ma^2$ (b) $I = \frac{1}{3}ka^4 = \frac{1}{6}Ma^2$

33. (a) $L = 2\pi\sqrt{a^2+b^2}$ (b) $x_M = y_M = 0, z_M = \pi b$ (c) $I_x = I_y = \frac{1}{6}M(3a^2 + 8b^2\pi^2), I_z = Ma^2$ **35.** $M = \frac{2}{3}\pi k\sqrt{a^2+b^2}(3a^2 + 4\pi^2 b^2)$

SECTION 17.5

1. $\frac{1}{6}$ **3.** 6π **5.** 2π **7.** a^2b **9.** 7π **11.** $5a\pi r^2$ **13.** 0 **15.** 0 **17.** πa^2

19. $\frac{15}{2} - 4\ln 4$ **21.** $(c-a)A$ **23.** $3\pi R^2$

25. Taking Ω to be of type II, we have

$$\iint_\Omega \frac{\partial Q}{\partial x}(x,y)\,dx\,dy = \int_c^d \int_{\psi_1(y)}^{\psi_2(y)} \frac{\partial Q}{\partial x}(x,y)\,dx\,dy = \int_c^d \{Q[\psi_2(y),y] - Q[\psi_1(y),y]\}\,dy$$

$$\stackrel{*}{=} \int_c^d Q[\psi_2(y),y]\,dy - \int_c^d Q[\psi_1(y),y]\,dy.$$

Set $C_3 : \mathbf{r}_3(u) = \psi_1(u)\,\mathbf{i} + u\,\mathbf{j},\ u\in[c,d]$ and $C_4 : \mathbf{r}_4(u) = \psi_2(u)\,\mathbf{i} + u\,\mathbf{j},\ u\in[c,d]$. Then

$$\oint_C Q(x,y)\,dy = \int_{C_4} Q(x,y)\,dy - \int_{C_3} Q(x,y)\,dy = \int_c^d Q[\psi_2(u),u]\,du - \int_c^d Q[\psi_1(u),u]\,du.$$

Comparison with $*$ proves the result.

27. by (16.5.3), $\bar{x}A = \iint_\Omega x\,dxdy$ and $\bar{y}A = \iint_\Omega y\,dxdy$;

by Green's theorem, $\iint_\Omega x\,dxdy = \oint_C \frac{1}{2}x^2 dy$ and $\iint_\Omega y\,dxdy = -\oint_C \frac{1}{2}y^2 dx$

29. $\oint_{C_1} = \oint_{C_2} + \oint_{C_3}$

31. (a) 0 (b) 0

33. If Ω is the region bounded by C, then

$$\oint_C \mathbf{v} \cdot \mathbf{dr} = \oint_C \frac{\partial \phi}{\partial x} dx + \frac{\partial \psi}{\partial y} dy = \iint_\Omega \left\{ \frac{\partial}{\partial x}\left(\frac{\partial \phi}{\partial y}\right) - \frac{\partial}{\partial y}\left(\frac{\partial \phi}{\partial x}\right) \right\} dx\,dy$$

is zero by equality of mixed partials.

35. $A = \frac{1}{2}\oint_C (-y\,dx + x\,dy)$

$= \frac{1}{2}\left[\int_{C_1} + \int_{C_2} + \cdots + \int_{C_n}\right]$

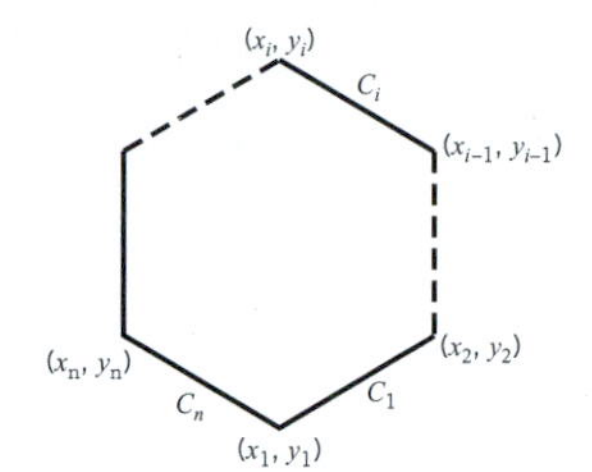

Now

$$\int_{C_i} (-y\,dx + x\,dy) = \int_0^1 [-(y_i + u(y_{i+1} - y_i))(x_{i+1} - x_i) + (x_i + u(x_{i+1} - x_i))(y_{i+1} - y_i)]\,du$$

$$= x_i y_{i+1} - x_{i+1} y_i, \quad i = 1, 2, \ldots, n\ (x_{n+1} = x_1, y_{n+1} = y_1).$$

Thus $A = \frac{1}{2}[(x_1y_2 - x_2y_1) + (x_2y_3 - x_3y_2) + \cdots + (x_ny_1 - x_1y_n)]$.

SECTION 17.6

1. $4[(u^2 - v^2)\mathbf{i} - (u^2 + v^2)\mathbf{j} + 2uv\mathbf{k}]$ **3.** $2(\mathbf{j} - \mathbf{i})$ **5.** $\mathbf{r}(u, v) = 3\cos u \cos v\,\mathbf{i} + 2\sin u \cos v\,\mathbf{j} + 6\sin v\,\mathbf{k}$, $u \in [0, 2\pi]$, $v \in [0, \pi/2]$

7. $\mathbf{r}(u, v) = 2\cos u \cos v\,\mathbf{i} + 2\sin u \cos v\,\mathbf{j} + 2\sin v\,\mathbf{k}$, $u \in [0, 2\pi]$, $v \in (\pi/4, \pi/2]$ **9.** $\mathbf{r}(u, v) = u\,\mathbf{i} + g(u, v)\,\mathbf{j} + v\,\mathbf{k}$, $(u, v) \in \Omega$

11. $x^2/a^2 + y^2/b^2 + z^2/c^2 = 1$; ellipsoid **13.** $x^2/a^2 - y^2/b^2 = z$; hyperbolic paraboloid

15. $\mathbf{r}(u, v) = v\cos u\,\mathbf{i} + v\sin u\,\mathbf{j} + f(v)\mathbf{k}; 0 \le u \le 2\pi, a \le v \le b$ **17.** area of $\Gamma = A_\Omega \sec\gamma$ **19.** $\frac{1}{2}\sqrt{a^2b^2 + a^2c^2 + b^2c^2}$

21. $\frac{1}{6}\pi(17\sqrt{17} - 1)$ **23.** $\frac{1}{6}\pi\left[(4a^2 + 1)^{3/2} - (a^2 + 1)^{3/2}\right]$ **25.** $\frac{1}{15}(36\sqrt{6} - 50\sqrt{5} + 32)$ **27.** 4π

29. (a) $\iint_\Omega \sqrt{\left[\frac{\partial g}{\partial y}(y, z)\right]^2 + \left[\frac{\partial g}{\partial z}(y, z)\right]^2 + 1}\,dydz = \iint_\Omega \sec[\alpha(y, z)]\,dydz$

where α is the angle between the unit normal with positive $\mathbf{i}$ component and the x-axis

(b) $\iint_\Omega \sqrt{\left[\frac{\partial h}{\partial x}(x, z)\right]^2 + \left[\frac{\partial h}{\partial z}(x, z)\right]^2 + 1}\,dxdz = \iint_\Omega \sec[\beta(x, z)]\,dxdz$

where β is the angle between the unit normal with positive $\mathbf{j}$ component and the y-axis

31. (a) $\mathbf{N}(u, v) = v\cos u \sin\alpha \cos\alpha\,\mathbf{i} + v\sin u \sin\alpha \cos\alpha\,\mathbf{j} - v\sin^2\alpha\,\mathbf{k}$ (b) $A = \pi s^2 \sin\alpha$

33. (c) $A = \int_0^{2\pi}\int_{-\ln 2}^{\ln 2} ||\mathbf{N}(u, v)||\,dudv$

$$= \int_0^{2\pi}\int_{-\ln 2}^{\ln 2} \sqrt{64\cos^2 u \cosh^2 v + 144\sin^2 u \cosh^2 v + 36\cosh^2 v \sinh^2 v}\,dudv$$

35. $A = \sqrt{A_1^2 + A_2^2 + A_3^2}$

37. (a) $\frac{1}{4}\sqrt{2\pi}[\sqrt{6} + \ln(\sqrt{2} + \sqrt{3})]$

(b) $\frac{1}{2}a^2[\sqrt{2e^{4\pi} + 1} - \sqrt{3} + 2\pi + \ln(1 + \sqrt{3}) - \ln(1 + \sqrt{2e^{4\pi} + 1})]$

SECTION 17.7

1. $\frac{1}{2}[\sqrt{2}+\ln(1+\sqrt{2})]$ **3.** $2\sqrt{2}-1$ **5.** $\frac{1}{3}[2\sqrt{2}-1]$ **7.** $\dfrac{9\sqrt{14}}{2}$ **9.** $\frac{4}{3}$ **11.** $\dfrac{4\pi}{3}$ **13.** $\frac{1}{2}\sqrt{3}a^2k$

15. $\frac{1}{12}\sqrt{3}a^4k$ **17.** $(0,0,\frac{1}{2}a)$ **19.** 2 **21.** $\frac{4}{3}\pi a^3$ **23.** 0 **25.** $\dfrac{1}{2}a^3$ **27.** 0 **29.** $-\frac{3}{2}$ **31.** $2\pi/a^2$

33. $\frac{8}{35}$ **35.** $-\frac{4}{63}$ **37.** $\bar{x}=\bar{y}=0,\quad \bar{z}=\frac{2}{3}s\cos\alpha$ **39.** $x_M=y_M=0,\quad z_M=\frac{3}{4}$ **41.** no answer required **43.** $x_M=\frac{11}{9}$

45. Total flux out of the solid is 0. It is clear from a diagram that the outer unit normal to the cylindrical side of the solid is given by $\mathbf{n}=x\,\mathbf{i}+y\,\mathbf{j}$, in which case $\mathbf{v}\cdot\mathbf{n}=0$. The outer unit normals to the top and bottom of the solid are $\mathbf{k}$ and $-\mathbf{k}$, respectively. So, here as well, $\mathbf{v}\cdot\mathbf{n}=0$ and the total flux is 0.

47. $(4\sqrt{2}-\frac{7}{2})\pi$

SECTION 17.8

1. $\nabla\cdot\mathbf{v}=2,\ \nabla\times\mathbf{v}=\mathbf{0}$ **3.** $\nabla\cdot\mathbf{v}=0,\ \nabla\times\mathbf{v}=\mathbf{0}$ **5.** $\nabla\cdot\mathbf{v}=6,\quad \nabla\times\mathbf{v}=\mathbf{0}$

7. $\nabla\cdot\mathbf{v}=yz+1,\ \nabla\times\mathbf{v}=-x\,\mathbf{i}+xy\,\mathbf{j}+(1-x)z\mathbf{k}$ **9.** $\nabla\cdot\mathbf{v}=1/r^2,\ \nabla\times\mathbf{v}=\mathbf{0}$

11. $\nabla\cdot\mathbf{v}=2(x+y+z)e^{r^2},\ \nabla\times\mathbf{v}=2e^{r^2}[(y-z)\,\mathbf{i}-(x-z)\,\mathbf{j}+(x-y)\mathbf{k}]$ **13.** $\nabla\cdot\mathbf{v}=f'(x),\quad \nabla\times\mathbf{v}=\mathbf{0}$ **15.** use components

17. $\nabla\cdot\mathbf{F}=\dfrac{\partial P}{\partial x}+\dfrac{\partial Q}{\partial y}+\dfrac{\partial R}{\partial z}=2+4-6=0$ **19.** $\nabla\times\mathbf{F}=\begin{vmatrix}\mathbf{i} & \mathbf{j} & \mathbf{k}\\ \dfrac{\partial}{\partial x} & \dfrac{\partial}{\partial y} & \dfrac{\partial}{\partial z}\\ x & y & -2z\end{vmatrix}=\mathbf{0}$ **21.** $\nabla^2 f=12\,(x^2+y^2+z^2)$

23. $\nabla^2 f=2y^3z^4+6x^2yz^4+12x^2y^3z^2$ **25.** $\nabla^2 f=e^r(1+2r^{-1})$ **27.** (a) $2r^2$ (b) $-1/r$

29.
$$\begin{aligned}\nabla^2 f&=\nabla^2 g(r)=\nabla\cdot(\nabla g(r))=\nabla\cdot(g'(r)r^{-1}\mathbf{r})\\&=[(\nabla g'(r))\cdot r^{-1}\mathbf{r}]+g'(r)(\nabla\cdot r^{-1}\mathbf{r})\\&=\{[g''(r)r^{-1}\mathbf{r}]\cdot r^{-1}\mathbf{r}\}+g'(r)(2r^{-1})=g''(r)+2r^{-1}g'(r)\end{aligned}$$

31. no answer required **33.** $n=-1$

SECTION 17.9

1. $\displaystyle\iint_S(\mathbf{v}\cdot\mathbf{n})\,d\sigma=\iiint_T(\nabla\cdot\mathbf{v})\,dxdydz=\iiint_T 3\,dxdydz=3V=4\pi$

3. $\displaystyle\iint_S(\mathbf{v}\cdot\mathbf{n})\,d\sigma=\iiint_T(\nabla\cdot\mathbf{v})\,dxdydz=\iiint_T 2(x+y+z)\,dxdydz.$

The flux is zero since the function $f(x,y,z)=2(x+y+z)$ satisfies the relation $f(-x,-y,-z)=-f(x,y,z)$ and T is symmetric about the origin.

5.

Face	n	v • n	Flux	
$x=0$	$-\mathbf{i}$	0	0	
$x=1$	$\mathbf{i}$	1	1	
$y=0$	$-\mathbf{j}$	0	0	total flux = 3
$y=1$	$\mathbf{j}$	1	1	
$z=0$	$-\mathbf{k}$	0	0	
$z=1$	$\mathbf{k}$	1	1	

$\displaystyle\iiint_T(\nabla\cdot\mathbf{v})\,dxdydz=\iiint_T 3\,dxdydz=3V=3$

7.

Face	n	v • n	Flux	
$x=0$	$-\mathbf{i}$	0	0	
$x=1$	$\mathbf{i}$	1	1	
$y=0$	$-\mathbf{j}$	xz	fluxes added up to 0	total flux = 2
$y=1$	$\mathbf{j}$	$-xz$		
$z=0$	$-\mathbf{k}$	0	0	
$z=1$	$\mathbf{k}$	1	1	

$\displaystyle\iiint_T(\nabla\cdot\mathbf{v})\,dxdydz=\iiint_T 2(x+z)\,dxdydz=2(\bar{x}+\bar{z})V=2\left(\tfrac{1}{2}+\tfrac{1}{2}\right)1=2$

9. flux $\displaystyle=\iiint_T(1+4y+6z)\,dxdydz=(1+4\bar{y}+6\bar{z})V=(1+0+3)9\pi=36\pi$ **11.** $\frac{1}{24}$ **13.** $64\,\pi$ **15.** 0 **17.** $(A+B+C)V$

19. Let T be the solid enclosed by S and set $n = n_1\,\mathbf{i} + n_2\,\mathbf{j} + n_3\,\mathbf{k}$.

$$\iint_S n_1\,d\sigma = \iint_S (\mathbf{i}\bullet\mathbf{n})\,d\sigma = \iiint_T (\nabla\bullet\mathbf{i})\,dxdydz = \iiint_T 0\,dxdydz = 0.$$

Similarly $\iint_S n_2\,d\sigma = 0$ and $\iint_S n_3\,d\sigma = 0$.

21. A routine computation shows that $\nabla\bullet(\nabla f\times\nabla g) = 0$. Therefore

$$\iint_S [(\nabla f\times\nabla g)\bullet\mathbf{n}]d\sigma = \iiint_T [\nabla\bullet(\nabla f\times\nabla g)]\,dxdydz = 0.$$

23. Set $\mathbf{F} = F_1\,\mathbf{i} + F_2\,\mathbf{j} + F_3\mathbf{k}$.

$$F_1 = \iint_S [\rho(z-c)\mathbf{i}\bullet\mathbf{n}]\,d\sigma = \iiint_T [\nabla\bullet\rho(z-c)\mathbf{i}]\,dxdydz = \iiint_T \underbrace{\frac{\partial}{\partial x}[\rho(z-c)]}_{0}\,dxdydz = 0.$$

Similarly, $F_2 = 0$.

$$F_3 = \iint_S [\rho(z-c)\mathbf{k}\bullet\mathbf{n}]d\sigma = \iiint_T [\nabla\bullet\rho(z-c)\mathbf{k}]\,dxdydz = \iiint_T \frac{\partial}{\partial z}[(\rho(z-c)]\,dxdydz = \iiint_T \rho\,dxdydz = W.$$

SECTION 17.10

For Exercises 1 and 3 : $\mathbf{n} = x\,\mathbf{i} + y\,\mathbf{j} + z\,\mathbf{k}$ and $C : \mathbf{r}(u) = \cos u\,\mathbf{i} + \sin u\,\mathbf{j},\ u\in[0, 2\pi]$.

1. (a) $\iint_S [(\nabla\times\mathbf{v})\bullet\mathbf{n}]d\sigma = \iint_S (\mathbf{0}\bullet\mathbf{n})\,d\sigma = 0.$

(b) S is bounded by the unit circle $C : \mathbf{r}(u) = \cos u\,\mathbf{i} + \sin u\,\mathbf{j},\ u\in[0, 2\pi]$.

$\oint_c \mathbf{v}(\mathbf{r})\bullet d\mathbf{r} = 0$ since $\mathbf{v}$ is a gradient.

3. (a) $\iint_S [(\nabla\times\mathbf{v})\bullet\mathbf{n}]\,d\sigma = \iint_S [(-3y^2\,\mathbf{i} + 2z\,\mathbf{j} + 2\,\mathbf{k})\bullet(x\,\mathbf{i} - y\,\mathbf{j} + z\,\mathbf{k})]\,d\sigma$

$$= \iint_S (-3xy^2 + 2yz + 2z)\,d\sigma$$

$$= \underbrace{\iint_S (-3xy^2)\,d\sigma}_{0\text{ by symmetry}} + \underbrace{\iint_S 2yz\,d\sigma}_{0\text{ by symmetry}} + \underbrace{\iint_S 2z\,d\sigma}_{2\bar{z}A = 2\pi} = 2\pi$$

by Ex.17, Section 17.7

(b) $\oint_C \mathbf{v}(\mathbf{r})\bullet d\mathbf{r} = \oint_C z^2\,dx + 2x\,dy = \oint_C 2x\,dy = \int_0^{2\pi} 2\cos^2 u\,du = 2\pi$

For Exercises 5 and 7 take $S : z = 2 - x - y$ with $0\le x\le 2, 0\le y\le 2-x$ and C as the triangle $(2,0,0), (0,2,0), (0,0,2)$. Then $C = C_1\cup C_2\cup C_3$ with

$$\begin{aligned} C_2 : \mathbf{r}_1(u) &= 2(1-u)\mathbf{i} + 2u\,\mathbf{j}, u\in[0,1],\\ C_2 : \mathbf{r}_2(u) &= 2(1-u)\mathbf{j} + 2u\,\mathbf{k}, u\in[0,1],\\ C_3 : \mathbf{r}_3(u) &= 2(1-u)\mathbf{k} + 2u\,\mathbf{i}, u\in[0,1] \end{aligned}$$

$\mathbf{n} = \frac{1}{3}\sqrt{3}(\mathbf{i} + \mathbf{j} + \mathbf{k})$.

5. (a) $\iint_S [(\nabla\times\mathbf{v})\bullet\mathbf{n}]\,d\sigma = \iint_S \frac{1}{3}\sqrt{3}\,d\sigma = \frac{1}{3}\sqrt{3}A = 2$

(b) $\oint_C \mathbf{v}(r) \cdot d\mathbf{r} = \left(\int_{C_1} + \int_{C_2} + \int_{C_3}\right) \mathbf{v}(\mathbf{r}) \cdot dr = -2 + 2 + 2 = 2$

7. (a) $\iint_S [(\nabla \times \mathbf{v}) \cdot \mathbf{n}]\, d\sigma \iint_S [y\mathbf{k} \cdot \frac{1}{3}\sqrt{3}(\mathbf{i} + \mathbf{j} + \mathbf{k})]\, d\sigma = \frac{1}{3}\sqrt{3} \iint_S y\, d\sigma = \frac{1}{3}\sqrt{3}\bar{y}A = \frac{4}{3}$ **9.** 4π **11.** 0 **13.** $\pm\frac{1}{8}\pi$ **15.** $\pm\frac{1}{4}\pi$

(b) $\oint_C \mathbf{v}(\mathbf{r}) \cdot d\mathbf{r} = \left(\int_{C_1} + \int_{C_2} + \int_{C_3}\right) \mathbf{v}(\mathbf{r}) \cdot d\mathbf{r} = (\frac{4}{3} - \frac{32}{5}) + \frac{32}{5} + 0 = \frac{4}{3}$

17. Straightforward calculation shows that

$$\nabla \times (\mathbf{a} \times \mathbf{r}) = \nabla \times [(a_2z - a_3y)\,\mathbf{i} + (a_3x - a_1z)\,\mathbf{j} + (a_1y - a_2x)\,\mathbf{k}] = 2\mathbf{a}.$$

19. In the plane of C, the curve C bounds some Jordan region that we call Ω. The surface $S \cup \Omega$ is a piecewise-smooth surface that bounds a solid T. Note that $\nabla \times \mathbf{v}$ is continuously differentiable on T. Thus, by the divergence theorem,

$$\iiint_T [\nabla \cdot (\nabla \times \mathbf{v})]\, dxdydz = \iint_{S\cup\Omega} [(\nabla \times \mathbf{v}) \cdot \mathbf{n}]\, d\sigma$$

where $\mathbf{n}$ is the outer unit normal. Since the divergence of a curl is identically zero, we have

$$\iint_{S\cup\Omega} [(\nabla \times \mathbf{v}) \cdot \mathbf{n}] d\sigma = 0.$$

Now $\mathbf{n}$ is $\mathbf{n}_1$ on S and $\mathbf{n}_2$ on Ω. Thus

$$\iint_S [(\nabla \times \mathbf{v}) \cdot \mathbf{n}_1]\, d\sigma + \iint_\Omega [(\nabla \times \mathbf{v}) \cdot \mathbf{n}_2]\, d\sigma = 0.$$

This gives

$$\iint_S [(\nabla \times \mathbf{v}) \cdot \mathbf{n}_1]\, d\sigma = \iint_\Omega [(\nabla \times \mathbf{v}) \cdot (-\mathbf{n}_2)]\, d\sigma = \oint_c \mathbf{v}(\mathbf{r}) \cdot d\mathbf{r}$$

where C is traversed in a positive sense with respect to $-\mathbf{n}_2$ and therefore in a positive sense with respect to $\mathbf{n}_1(-\mathbf{n}_2$ points toward $S)$.

CHAPTER 18

SECTION 18.1

1. y_1 is; y_2 is not **3.** both y_1 and y_2 are solutions **5.** both y_1 and y_2 are solutions

7. $y = 2\,e^{5x}$ **9.** $y = \dfrac{1}{-2\,e^{x-1} + 1}$ **11.** $y = -\dfrac{17}{4}x + 9x^{1/2}$ **13.** $y = x^2 \ln x$ **15.** linear; $y = Ce^{-x} - 2e^{-2x}$

17. separable; $y + \ln|y| = \frac{1}{3}x^3 - x + C$ **19.** linear; $y = \dfrac{\sin x}{x^2} + \dfrac{C}{x^2}$ **21.** separable; $y^2 + 1 = (2x^2 + C)^2$

23. $r = -3$ **25.** $r = -3$ **27.** $r = 0$ **29.** $r = \frac{1}{2}$ or $r = \frac{3}{2}$ **31.** (a) $y = C_1 \sin 4x$ (b) $y(0) = 0$

SECTION 18.2

1. $y^2 = \dfrac{1}{1 + C\,e^{x^2}}$ **3.** $y = (C\,e^{2x} - e^x)^2$ **5.** $y = \left[(x-2)^2 + \dfrac{C}{\sqrt{x-2}}\right]^2$ **7.** $y^{-2} = 4e^{x^2} - 2x\,e^{x^2}$ **9.** $y^2 = \dfrac{x^3}{2-x}$

11. $\ln y = x^2 + Cx$ **13.** $y^2 - x^2 = Cx$ **15.** $x^2 - 2xy - y^2 = C$ **17.** $y + x = x\,e^{y/x}[C - \ln x]$ **19.** $1 - \cos[y/x] = c\,x \sin[y/x]$

21. $y^3 + 3x^3 \ln|x| = 8x^3$ **23.** (a) 2.48832, rel error 8.46% (b) 2.71825, rel error 0.001%

25. (a) 2.59374, rel error 4.58% (b) 2.71828, rel error 0% **27.** (a) 1.9, rel error 5.0% (b) 2.0 rel error 0%

29. (a) 1.42052, rel error −0.45% (b) 1.41421, rel error 0% **31.** (a) 2.65330, rel error 2.39% (b) 2.71828, rel error 0%

SECTION 18.3

1. the whole plane; $\dfrac{x^2y^2}{2} - xy = C$

3. the whole plane; $x\,e^y - y\,e^x = C$

5. the upper half-plane; $x \ln y + x^2 y = C$

7. the right half-plane; $y \ln x + 3x^2 - 2y = C$

9. the whole plane; $xy^3 + y^2 \cos x - \frac{1}{2}x^2 + \frac{1}{2}e^{2y} = C$

11. (a) yes (b) $\dfrac{1}{p(y)q(x)}$ $(p(y)q(x) \neq 0)$ **13.** $x\,e^y - y\,e^x = C$ **15.** $x^3y^2 + \frac{1}{2}x^2 + x\,e^y + \frac{1}{2}y^2 = C$ **17.** $y^3\,e^x + x\,e^x = C$

19. $x^3 + 3xy + 3\,e^y = 4$ **21.** $4x^2y^2 + x^4 + 4x^2 = 5$ **23.** $xy - \dfrac{1}{y} = 3$ **25.** $\sinh(x - y^2) + \frac{1}{2}\,e^{2x} + \frac{1}{2}y^2 = \frac{1}{2}e^4 + 1$

27. (a) $k = 3$ (b) $k = 1$ **29.** $y = \dfrac{-4}{x^4 + C}$ **31.** $y = \frac{1}{9}x^5 + Cx^{-4}$ **33.** $e^{xy} - x^2 + 2\ln|y| = C$

SECTION 18.4

1. $y = C_1\,e^{-4x} + C_2\,e^{2x}$ **3.** $y = C_1\,e^{-4x} + C_2\,x\,e^{-4x}$ **5.** $y = e^{-x}(C_1 \cos 2x + C_2 \sin 2x)$ **7.** $y = C_1\,e^{(1/2)x} + C_2\,e^{-3x}$

9. $y = C_1 \cos 2\sqrt{3}x + C_2 \sin 2\sqrt{3}x$ **11.** $y = C_1\,e^{(1/5)x} + C_2\,e^{-(3/4)x}$ **13.** $y = C_1 \cos 3x + C_2 \sin 3x$

15. $y = e^{-(1/2)x}(C_1 \cos \frac{1}{2}x + C_2 \sin \frac{1}{2}x)$ **17.** $y = C_1\,e^{(1/4)x} + C_2\,e^{-(1/2)x}$ **19.** $y = 2\,e^{2x} - e^{3x}$ **21.** $y = 2\cos\dfrac{x}{2} + \sin\dfrac{x}{2}$

23. $y = 7\,e^{-2(x+1)} + 5x\,e^{-2(x-1)}$ **25.** (a) $y = C\,e^{2x} + (1 - C)\,e^{-x}$ (b) $y = C\,e^{2x} + (2C - 1)\,e^{-x}$ (c) $y = \frac{2}{3}\,e^{2x} + \frac{1}{3}\,e^{-x}$

27. $\alpha = \dfrac{r_1 + r_2}{2}; \beta = \dfrac{r_1 - r_2}{2}$

$y = k_1\,e^{r_1 x} + k_2\,e^{r_2 x} = e^{\alpha x}(C_1 \cosh \beta x + C_2 \sinh \beta x)$, where $k_1 = \dfrac{C_1 + C_2}{2}$, $k_2 = \dfrac{C_1 - C_2}{2}$.

29. (a) The Wronskian of $y_1 = e^{\alpha x}$, $y_2 = x\,e^{\alpha x}$ is:

$W(x) = e^{\alpha x}\,[e^{\alpha x} + \alpha\,x\,e^{\alpha x}] - x\,e^{\alpha x}\,[\alpha\,e^{\alpha x}] = e^{2\alpha x} \neq 0.$

(b) The Wronskian of $y_1 = e^{\alpha x}\cos \beta x$, $y_2 = e^{\alpha x}\sin \beta x$, $\beta \neq 0$, is

$W(x) = e^{\alpha x}\cos \beta x[\alpha\,e^{\alpha x}\sin \beta x + \beta\,e^{\alpha x}\cos \beta x] - e^{\alpha x}\sin \beta x[\alpha\,e^{\alpha x}\cos \beta x - \beta\,e^{\alpha x}\sin \beta x] = \beta\,e^{2\alpha x} \neq 0.$

31. (a) $y'' + 2y' - 8y = 0$ (b) $y'' - 4y' - 5y = 0$ (c) $y'' - 6y' + 9y = 0$

33. Set $y = e^{\alpha x}u$. Then $y' = \alpha e^{\alpha x}u + e^{\alpha x}u'$ and $y'' = \alpha^2 e^{\alpha x}u + 2\alpha e^{\alpha x}u' + e^{\alpha x}u''$.

(a) Substituting into the differential equation yields $e^{\alpha x}u'' = 0$, which implies $u'' = 0$.

(b) Substituting into the differential equation yields $e^{\alpha x}(u'' + \beta^2 u) = 0$, which implies $u'' + \beta^2 u = 0$.

35. (a) If $a = 0$, $b > 0$, then the general solution is $y = C_1 \cos \sqrt{b}\,x + C_2 \sin \sqrt{b}\,x = A\cos(\sqrt{b}\,x + \phi)$, where A and ϕ are constants. Clearly $|y(x)| \leq |A|$ for all x.

(b) If $a > 0$, $b = 0$, then the general solution is $y = C_1 + C_2\,e^{-ax}$ and $\lim_{x\to\infty} y(x) = C_1$. The solution which satisfies the conditions $y(0) = y_0, y'(0) = y_1$ is $y = y_0 + \dfrac{y_1}{a} - \dfrac{y_1}{a}e^{-ax}$; $k = y_0 + \dfrac{y_1}{a}$.

37. If $y_2 = k\,y_1$, then $W(y_1, y_2) = \begin{vmatrix} y_1 & ky_1 \\ y_1' & ky_1' \end{vmatrix} = 0$. Suppose that $W(y_1, y_2) = 0$. Let I be an interval on which y_1 is nonzero. Then

$$\left(\frac{y_2}{y_1}\right)' = \frac{y_1y_2' - y_2y_1'}{y_1^2} = \frac{W(y_1, y_2)}{y_1^2} = 0,$$

which implies $\dfrac{y_2}{y_1} = k$ constant.

39. $y = C_1\,x^4 + C_2\,x^{-2}$ **41.** $y = c_1\,x^2 + C_2\,x^2 \ln x$

SECTION 18.5

1. $y = \frac{1}{2}x + \frac{1}{4}$ **3.** $y = \frac{1}{5}x^2 - \frac{4}{25}x - \frac{27}{125}$ **5.** $y = \frac{1}{36}e^{3x}$ **7.** $y = \frac{1}{5}e^x$ **9.** $y = -\frac{13}{170}\cos x - \frac{1}{170}\sin x$

11. $y = \frac{3}{100}\cos 2x + \frac{21}{100}\sin 2x$ **13.** $y = \frac{1}{20}e^{-x}\sin 2x + \frac{1}{10}e^{-x}\cos 2x$ **15.** $y = \frac{3}{2}x\,e^{-2x}$ **17.** $y = C_1\cos x + C_2\sin x + \frac{1}{2}e^x$

19. $y = C_1e^{5x} + C_2e^{-2x} + \frac{1}{10}x + \frac{7}{100}$ **21.** $y = C_1e^x + C_2e^{-4x} - \frac{1}{5}x\,e^{-4x}$ **23.** $y = C_1e^{-2x} + C_2e^x + \frac{1}{2}x^2e^x - \frac{1}{3}x\,e^x$

25. Let $z = y_1 + y_2$. Then

$$z'' + az' + bz = (y_1'' + y_2'') + a(y_1' + y_2') + b(y_1 + y_2)$$

$$= (y_1'' + ay_1' + by_1) + (y_2'' + ay_2' + by_2) = \phi_1 + \phi_2$$

27. $y = C_1e^{-3x} + C_2e^{-x} + \frac{1}{4}x\,e^{-x} + \frac{1}{16}e^x$

29. $y = 2e^x\sin x - x\,e^x\cos x$ **31.** $y = \frac{1}{3}x\ln|x|\,e^{2x}$ **33.** $y = -\ln|x|e^{-2x}$

35. $y = e^x(x\sin x + \cos x\ln|\cos x|)$

37. (a) $i(t) = -CF_0\,e^{-(R/2L)t} + \dfrac{F_0}{2L}(2 - RC)\,t\,e^{-(R/2L)t} + CF_0$

(b) $i(t) = e^{-(R/2L)t}\left[\dfrac{F_0(2 - RC)}{2L\beta}\sin\beta t - CF_0\cos\beta t\right] + CF_0,$ where $\beta\sqrt{\dfrac{4L - CR^2}{4L^2C}}$

39. (a) $y_1y_2' - y_2y_1' = -\dfrac{2}{x} \neq 0$ (b) $y = \frac{1}{3}\sin(\ln x)$

SECTION 18.6

1. $x(t) = \sin(8t + \frac{1}{2}\pi);\quad A = 1,\quad f = 4/\pi$ **3.** $\pm 2\pi A/T$ **5.** $x(t) = (15/\pi)\sin\frac{1}{3}\pi t$ **7.** $x(t) = x_0\sin(t\sqrt{k/m} + \frac{1}{2}\pi)$

9. at $x = \pm\frac{1}{2}\sqrt{3}\,x_0$ **11.** $\frac{1}{4}k\,x_0^2$

13. Set $y(t) = x(t) - 2$. Equation $x''(t) = 8 - 4x(t)$ can be written $y''(t) + 4y(t) = 0$. This is simple harmonic motion centered at $y = 0$, which is $x = 2$.

$$y(t) = A\sin(2t + \phi_0).$$

The condition $x(0) = 0$ gives $y(0) = -2$ and thus

$$A\sin\phi_0 = -2. \qquad (*)$$

Since $y'(t) = x'(t)$ and $y'(t) = 2A\cos(2t + \phi_0)$, the condition $x'(0) = 0$ gives $y'(0) = 0$, and thus

$$2A\cos\phi_0 = 0. \qquad (**)$$

Equations $(*)$ and $(**)$ are satisfied by $A = 2$, $\phi_0 = \frac{3}{2}\pi$. The equation of motion can therefore be written

$$y(t) = 2\sin(2t + \tfrac{3}{2}\pi).$$

The amplitude is 2 and the period is π.

15. (a) $x''(t) + \omega^2x(t) = 0$ with $\omega = r\sqrt{\pi\rho/m}$ (b) $x(t) = x_0\sin(r\sqrt{\pi p/m}t + \frac{1}{2}\pi)$, taking downward as positive; $A = x_0$, $T = (2/r)\sqrt{m\pi/\rho}$

17. amplitude and frequency both decrease

19. at most once; at most once

21. if $\omega/\gamma = m/n$, then $m/\omega = n/\gamma$ is a period

23. $x = e^{-\alpha t}[c_1\cos\sqrt{\alpha^2 - \omega^2}\,t) + c_2\sin(\sqrt{\alpha^2 - \omega^2}\,t)]$ or equivalently $x = A\,e^{-\alpha t}[\sin(\sqrt{\alpha^2 - \omega^2}\,t) + \phi_0]$

25. $x_p = \dfrac{F_0}{2\alpha\gamma m}\sin\gamma t;$ as $c = 2\alpha m \to 0^+$, the amplitude $\left|\dfrac{F_0}{2\alpha\gamma m}\right| \to \infty$

27. $(\omega^2 - \gamma^2)^2 + 4\alpha^2\gamma^2 = \omega^4 + \gamma^4 + 2\gamma^2(2\alpha^2 - \omega^2)$ increases as γ increases

SKILL MASTERY REVIEW

ANSWERS, SKILL MASTERY REVIEW FIVE (p. 940)

1. $-9\,\mathbf{i}-5\,\mathbf{j}-2\,\mathbf{k}$ **3.** $\sqrt{90}$ **5.** $\cos\theta=\frac{2}{\sqrt{714}}$ **7.** $\pm\frac{1}{\sqrt{293}}(6\,\mathbf{i}-\mathbf{j}+16\,\mathbf{k})$

9. (a) $x=5+4t,\ y=6-7t,\ z=-3+5t$
(b) $x=5-10t,\ y=6-5t,\ z=-3+t$
(c) $x-3y-5z-2=0$

11. (a) no (b) yes **13.** $2x+3y-4z-19=0$ **15.** $2x-2y-z-6=0$ **17.** $\mathbf{f}'(t)=2e^{2t}\,\mathbf{i}+\dfrac{2t}{t^2+1}\,\mathbf{j};\ \mathbf{f}''(t)=4e^{2t}\,\mathbf{i}+\dfrac{2-2t^2}{(t^2+1)^2}\,\mathbf{j}$

19. $\mathbf{f}'(t)=2\cosh 2t\,\mathbf{i}+(te^{-t}-e^{-t})\,\mathbf{j}+\sinh t\,\mathbf{k};$
$\mathbf{f}''(t)=4\sinh 2t\,\mathbf{i}+(2e^{-t}-te^{-t})\,\mathbf{j}+\cosh t\,\mathbf{k}$

21. $\mathbf{v}(t)=2\,\mathbf{i}+\dfrac{1}{t}\,\mathbf{j}-2t\,\mathbf{k};\ \|\mathbf{v}(t)\|=2t+\dfrac{1}{t};$
$\mathbf{a}(t)=-\dfrac{1}{t^2}\,\mathbf{j}-2\,\mathbf{k}$

23. $\mathbf{f}(t)=\left(\frac{1}{3}t^3+1\right)\mathbf{i}+\left(\frac{1}{2}e^{2t}+t-\frac{7}{2}\right)\mathbf{j}+\left(\frac{1}{3}[2t+1]^{3/2}+\frac{8}{3}\right)\mathbf{k}$

25.

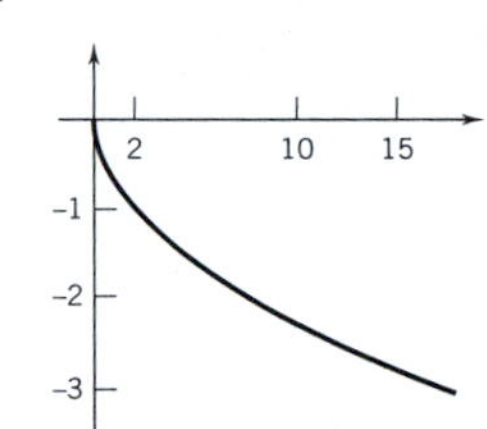

27.

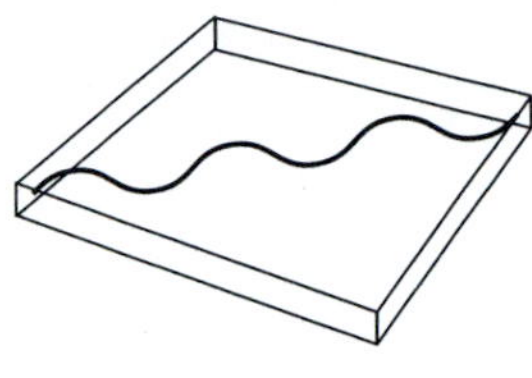

29. $\mathbf{r}'\left(\frac{\pi}{3}\right)=-\sqrt{3}\,\mathbf{i}-\mathbf{j}+\mathbf{k};\ x=-\frac{1}{2}-\sqrt{3}t,\ y=\frac{\sqrt{3}}{2}-t,\ z=\frac{\pi}{3}+t$

31. $\mathbf{T}(t)=-\frac{1}{\sqrt{2}}\sin t\,\mathbf{i}-\frac{1}{\sqrt{2}}\sin t\,\mathbf{j}-\cos t\,\mathbf{k}$
$\mathbf{N}(t)=-\frac{1}{\sqrt{2}}\cos t\,\mathbf{i}-\frac{1}{\sqrt{2}}\cos t\,\mathbf{j}+\sin t\,\mathbf{k}$

33. $\frac{38}{3}$

35. $\sqrt{2}\sinh 1$ **37.** (a) $\dfrac{1}{2(1+e^{-2t})^{3/2}}$ (b) $\dfrac{1}{|t|(1+t^2)^{3/2}}$ **39.** $\kappa=1,\ a_T=0,\ a_N=1$

41. (a) $\text{dom}(f)=\{(x,y):x^2-y^2>1\};\ \text{range}(f)=(-\infty,\infty)$
(b) $\text{dom}(f)=\{(x,y,z):x^2+y^2<z\};\ \text{range}(f)=[0,\infty)$

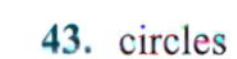

43. circles

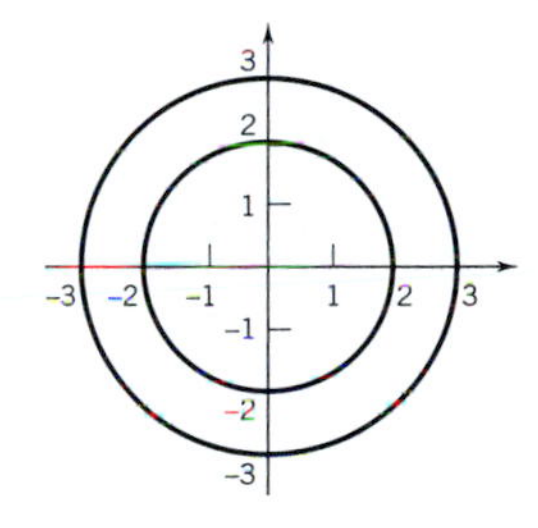

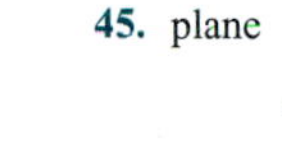

45. plane

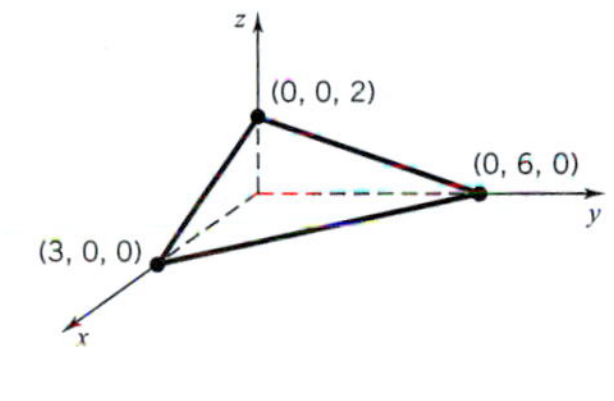

47. $\dfrac{\partial z}{\partial x}=x^2y^2\cos(xy^2)+2x\sin(xy^2);\ \dfrac{\partial z}{\partial y}=2x^3y\cos(xy^2)$

49. $g_x=\dfrac{x}{x^2+y^2+z^2};\ g_y=\dfrac{y}{x^2+y^2+z^2};\ g_z=\dfrac{z}{x^2+y^2+z^2}$

51. $f_{xx}=4yz^3+y^2z^2e^{xyz};\ f_{zx}=12xyz^2+ye^{xyz}+xy^2ze^{xyz}$
$f_{yz}=6x^2z^2+xe^{xyz}+x^2yze^{xyz}$

53. (a) $\nabla f=(4x-4y)\,\mathbf{i}+(-4x+3y^2)\,\mathbf{j}$
(b) $\nabla f=\dfrac{y^3-x^2y}{(x^2+y^2)^2}\,\mathbf{i}+\dfrac{x^3-xy^2}{(x^2+y^2)^2}\,\mathbf{j}$

55. $\frac{2}{\sqrt{5}}$ **57.** $e\sqrt{5}$ **59.** tangent plane: $-\frac{1}{\sqrt{2}}(x-1)+\frac{1}{\sqrt{2}}(y+1)-(z-\sqrt{2})=0;$
normal line: $x=1-\frac{1}{\sqrt{2}}t,\ y=-1+\frac{1}{\sqrt{2}}t,\ z=\sqrt{2}-t$

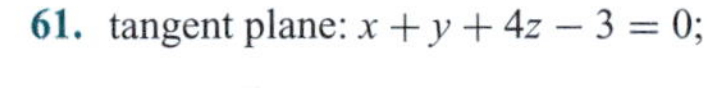

61. tangent plane: $x+y+4z-3=0;$
normal line: $x=t,\ y=-1+t,\ z=1+4t$

63. saddle points: $(5,0),\ (-3,0)$, local min: $(1,4)$ **65.** saddle point: $\left(3,\frac{1}{3}\right)$

67. absolute max: $f(3,3)=6$; absolute min: $f(1,2)=-3$ **69.** (a) not a gradient (b) $f(x,y)=-\dfrac{y}{x}+x^4-x-y\cos 3x+y^3+2y+C$

71. maximum: $f(4,2,4)=f(-4,2,4)=20;$
minimum: $f(-4,-2,4)=f(4,-2,-4)=-20$

73. length = width = $\dfrac{10}{\sqrt[3]{75}}\simeq 2.37$, height = $\dfrac{12}{\sqrt[3]{75}}\cong 2.85$

75. hottest: $T\left(\pm\frac{\sqrt{3}}{2},-\frac{1}{2}\right)=\frac{9}{4}$; coldest: $T\left(0,\frac{1}{2}\right)=-\frac{1}{4}$

ANSWERS, SKILL MASTERY REVIEW SIX (p. 1094)

1. $\frac{1}{40}$ **3.** $\frac{8}{3}(e-1)$ **5.** $-\frac{2}{3}$ **7.** $-\frac{16}{15}$ **9.** 1 **11.** $\frac{1}{8}$ **13.** $-\frac{8}{15}$ **15.** $\frac{32}{15}$ **17.** $\frac{1}{6}$ **19.** $\frac{1}{2}(1+e^2)\pi$

21. $M = \frac{2}{15}; \bar{x} = \frac{15}{28}, \bar{y} = \frac{15}{24}$

23. Introduce a coordinate system with the y-axis the axis of symmetry and the base of the triangle on the x-axis.

(a) $\bar{x} = 0, \bar{y} = \frac{1}{3}h$ (b) $\frac{1}{6}Mh^2$ (c) $\frac{1}{24}Mb^2$

25. 24π **27.** $\frac{5}{3}\pi$ **29.** (a) $\bar{x} = \bar{y} = \frac{9}{64}\pi, \bar{z} = \frac{3}{8}$ (b) $\frac{8}{15}M$ **31.** 0 **33.** (a) $-\frac{1}{12}$ (b) $-\frac{11}{72}$ **35.** $\frac{2}{3} - 6\cos 3 + 2\sin 3 + \frac{1}{3}e^{27} - \cos 27$

37. 4 **39.** $e^8 - 13$ **41.** (a) $\frac{32}{3}$ (b) -32 (c) $-\frac{608}{35}$ **43.** $\frac{1}{3}e^8 + 4\sin 2 + 2\cos 2 + \frac{17}{3}$ **45.** $-\frac{1}{3}$ **47.** 6 **49.** -4

51. $4\pi - 8$ **53.** $6\pi\sqrt{2}$ **55.** $\frac{224}{3}\pi$ **57.** 324π **59.** 4π

INDEX

W

X

(*continued from the front*)

INVERSE TRIGONOMETRIC FUNCTIONS

64. $\int \sin^{-1} u\, du = \sin^{-1} u + \sqrt{1-u^2} + C$

65. $\int \cos^{-1} u\, du = u\cos^{-1} u - \sqrt{1-u^2} + C$

66. $\int \tan^{-1} u\, du = u\tan^{-1} u - \frac{1}{2}\ln(1+u^2) + C$

67. $\int \cot^{-1} u\, du = u\cot^{-1} u + \frac{1}{2}\ln(1+u^2) + C$

68. $\int \sec^{-1} u\, du = u\sec^{-1} u - \ln|u+\sqrt{u^2-1}| + C$

69. $\int \csc^{-1} u\, du = u\csc^{-1} u + \ln|u+\sqrt{u^2-1}| + C$

70. $\int u\sin^{-1} u\, du = \frac{1}{4}(2u^2-1)\sin^{-1} u + u\sqrt{1-u^2} + C$

71. $\int u\tan^{-1} u\, du = \frac{1}{2}(u^2+1)\tan^{-1} u - \frac{1}{2}u + C$

72. $\int u\cos^{-1} u\, du = \frac{1}{4}(2u^2-1)\cos^{-1} u - u\sqrt{1-u^2} + C$

73. $\int u^n \sin^{-1} u\, du = \frac{1}{n+1}\left[u^{n+1}\sin^{-1} u - \int \frac{u^{n+1}du}{\sqrt{1-u^2}}\right], n \neq -1$

74. $\int u^n \cos n^{-1} u\, du = \frac{1}{n+1}\left[u^{n+1}\cos^{-1} u - \int \frac{u^{n+1}\,du}{\sqrt{1-u^2}}\right], n \neq -1$

75. $\int u^n \tan^{-1} u\, du = \frac{1}{n+1}\left[u^{n+1}\tan^{-1} u - \int \frac{u^{n+1}du}{\sqrt{1-u^2}}\right], n \neq -1$

$\sqrt{a^2+u^2}$, $a > 0$

76. $\int \frac{du}{a^2+u^2} = \frac{1}{a}\tan^{-1}\frac{u}{a} + C$

77. $\int \frac{du}{\sqrt{a^2+u^2}} = \ln|u+\sqrt{a^2+u^2}| + C$

78. $\int \sqrt{a^2+u^2}\, du = \frac{u}{2}\sqrt{a^2+u^2} + \frac{a^2}{2}\ln|u+\sqrt{a^2+u^2}| + C$

79. $\int u^2\sqrt{a^2+u^2}\, du = \frac{u}{8}(a^2+2u^2)\sqrt{a^2+u^2} - \frac{a^4}{8}\ln|u+\sqrt{a^2+u^2}| + C$

80. $\int \frac{\sqrt{a^2+u^2}}{u}\, du = \sqrt{a^2+u^2} - a\ln\left|\frac{a+\sqrt{a^2+u^2}}{u}\right| + C$

81. $\int \frac{\sqrt{a^2+u^2}}{u^2}\, du = -\frac{\sqrt{a^2+u^2}}{u} + \ln|u+\sqrt{a^2+u^2}| + C$

82. $\int \frac{u^2\, du}{\sqrt{a^2+u^2}} = \frac{u}{2}\sqrt{a^2+u^2} - \frac{a^2}{2}\ln|u+\sqrt{a^2+u^2}| + C$

83. $\int \frac{du}{u\sqrt{a^2+u^2}} = -\frac{1}{a}\ln\left|\frac{a+\sqrt{a^2+u^2}}{u}\right| + C$

84. $\int \frac{du}{u^2\sqrt{a^2+u^2}} = -\frac{\sqrt{a^2+u^2}}{a^2u} + C$

85. $\int \frac{du}{(a^2+u^2)^{3/2}} = \frac{u}{a^2\sqrt{a^2+u^2}} + C$

$\sqrt{a^2-u^2}$, $a > 0$

86. $\int \frac{du}{\sqrt{a^2-u^2}} = \sin^{-1}\frac{u}{a} + C$

87. $\int \sqrt{a^2-u^2}\, du = \frac{u}{2}\sqrt{a^2-u^2} + \frac{a^2}{2}\sin^{-1}\frac{u}{a} + C$

88. $\int u^2\sqrt{a^2-u^2}\, du = \frac{u}{8}(2u^2-a^2)\sqrt{a^2-u^2} + \frac{a^4}{8}\sin^{-1}\frac{u}{a} + C$